STUDENT SOLUTIONS MANUAL

FOR BOTH

CALCULUS
SINGLE VARIABLE

AND

CALCULUS
MULTIVARIABLE

BY
GIOVANNI VIGLINO

BY
MARION BERGER

June 2017

C O N T E N T S

CALCULUS
SINGLE VARIABLE

CALCULUS
MULTIVARIABLE

CHAPTER 14 VECTOR CALCULUS

PREFACE

This manual contains the solutions to all of the odd-numbered exercises in the two textbooks:

CALCULUS
SINGLE VARIABLE

and

CALCULUS
MULTIVARIABLE

by
Giovanni Viglino

CHAPTER 1
Preliminaries

§1.1 Sets and Functions

1. $f(x) = 3x^2 + x$ is a polynomial, which is defined everywhere: $D_f = (-\infty, \infty)$.

3. $k(x) = \dfrac{x}{(x-1)(x+100)}$ is defined where the denominator is not zero. The denominator is zero at $x = 1$ and $x = -100$: $D_k = (-\infty, -100) \cup (-100, 1) \cup (1, \infty)$.

5. $h(x) = \sqrt{x+7}$ is defined where $x + 7 \geq 0 \Rightarrow x \geq -7$: $D_h = [-7, \infty)$.

7. $f(x) = 2x + 3$, and $g(x) = -x + 1 \Rightarrow f(2) = 2(2) + 3 = 7$ and $g(2) = -2 + 1 = -1$:

    $(f + g)(2) = f(2) + g(2) = 7 + (-1) = 6.$

    $(f - g)(2) = f(2) - g(2) = 7 - (-1) = 7 + 1 = 8.$

    $(fg)(2) = f(2)g(2) = 7(-1) = -7.$

    $\left(\dfrac{f}{g}\right)(2) = \dfrac{f(2)}{g(2)} = \dfrac{7}{-1} = -7.$

    $(2f)(2) = 2f(2) = 2(7) = 14.$

    $(f \circ g)(2) = f(g(2)) = f(-1) = 2(-1) + 3 = -2 + 3 = 1.$

    $(g \circ f)(2) = g(f(2)) = g(7) = -7 + 1 = -6.$

9. $f(x) = x^2 + x$, and $g(x) = -x + 2 \Rightarrow f(2) = 2^2 + 2 = 6$ and $g(2) = -2 + 2 = 0$:

    $(f + g)(2) = f(2) + g(2) = 6 + 0 = 6.$

    $(f - g)(2) = f(2) - g(2) = 6 - 0 = 6.$

    $(fg)(2) = f(2)g(2) = 6(0) = 0.$

    $\left(\dfrac{f}{g}\right)(2) = \dfrac{f(2)}{g(2)} = \dfrac{6}{0}$ is undefined.

    $(2f)(2) = 2f(2) = 2(6) = 12.$

    $(f \circ g)(2) = f(g(2)) = f(0) = 0^2 + 0 = 0.$

    $(g \circ f)(2) = g(f(2)) = g(6) = -6 + 2 = -4.$

11. $f(x) = \dfrac{1}{x+5}$, and $g(x) = x + 3 \Rightarrow f(2) = \left(\dfrac{1}{2+5}\right) = \dfrac{1}{7}$ and $g(2) = 2 + 3 = 5$:

$(f+g)(2) = f(2) + g(2) = \dfrac{1}{7} + 5 = \dfrac{36}{7}$.

$(f-g)(2) = f(2) - g(2) = \dfrac{1}{7} - 5 = -\dfrac{34}{7}$.

$(fg)(2) = f(2)g(2) = \dfrac{1}{7} \cdot 5 = \dfrac{5}{7}$.

$\left(\dfrac{f}{g}\right)(2) = \dfrac{f(2)}{g(2)} = \dfrac{\frac{1}{7}}{5} = \dfrac{1}{35}$.

$(2f)(2) = 2f(2) = 2\left(\dfrac{1}{7}\right) = \dfrac{2}{7}$.

$(f \circ g)(2) = f(g(2)) = f(5) = \dfrac{1}{5+5} = \dfrac{1}{10}$.

$(g \circ f)(2) = g(f(2)) = g\left(\dfrac{1}{7}\right) = \dfrac{1}{7} + 3 = \dfrac{22}{7}$.

13. $f(x) = x + 3$, and $g(x) = -2x + 1$:

$(f+g)(x) = f(x) + g(x) = x + 3 + (-2x + 1) = x + 3 - 2x + 1 = -x + 4$.

$(f-g)(x) = f(x) - g(x) = x + 3 - (-2x + 1) = x + 3 + 2x - 1 = 3x + 2$.

$(fg)(x) = f(x)g(x) = (x + 3)(-2x + 1) = -2x^2 - 6x + x + 3 = -2x^2 - 5x + 3$.

$\left(\dfrac{f}{g}\right)(x) = \dfrac{f(x)}{g(x)} = \dfrac{x+3}{-2x+1}$.

$(2f)(x) = 2f(x) = 2(x + 3) = 2x + 6$.

$(f \circ g)(x) = f(g(x)) = f(-2x + 1) = (-2x + 1) + 3 = -2x + 4$.

$(g \circ f)(x) = g(f(x)) = g(x + 3) = -2(x + 3) + 1 = -2x - 6 + 1 = -2x - 5$.

15. $f(x) = x^2 + x - 1$, and $g(x) = x + 3$:

$(f+g)(x) = f(x) + g(x) = x^2 + x - 1 + x + 3 = x^2 + 2x + 2$.

$(f-g)(x) = f(x) - g(x) = x^2 + x - 1 - (x + 3) = x^2 + x - 1 - x - 3 = x^2 - 4$.

$(fg)(x) = f(x)g(x) = (x^2 + x - 1)(x + 3) = x^3 + 3x^2 + x^2 + 3x - x - 3 = x^3 + 4x^2 + 2x - 3$.

$\left(\dfrac{f}{g}\right)(x) = \dfrac{f(x)}{g(x)} = \dfrac{x^2 + x - 1}{x + 3}$.

$(2f)(x) = 2f(x) = 2(x^2 + x - 1) = 2x^2 + 2x - 2$.

$$(f \circ g)(x) = f(g(x)) = f(x+3) = (x+3)^2 + (x+3) - 1 = x^2 + 6x + 9 + x + 3 - 1$$
$$= x^2 + 7x + 11.$$

$$(g \circ f)(x) = g(f(x)) = g(x^2 + x - 1) = x^2 + x - 1 + 3 = x^2 + x + 2.$$

17. $f(x) = \dfrac{1}{-x+2}$, and $g(x) = x^2 + 3$:

$$(f + g)(x) = f(x) + g(x) = \frac{1}{-x+2} + x^2 + 3 = \frac{1 + (-x+2)(x^2+3)}{-x+2}$$
$$= \frac{1 - x^3 + 2x^2 - 3x + 6}{-x+2} = \frac{-x^3 + 2x^2 - 3x + 7}{-x+2}.$$

$$(f - g)(x) = f(x) - g(x) = \frac{1}{-x+2} - (x^2 + 3) = \frac{1 - (-x+2)(x^2+3)}{-x+2}$$
$$= \frac{1 - (-x^3 + 2x^2 - 3x + 6)}{-x+2} = \frac{x^3 - 2x^2 + 3x - 5}{-x+2}.$$

$$(fg)(x) = f(x)g(x) = \left(\frac{1}{-x+2}\right)(x^2 + 3) = \frac{x^2+3}{-x+2}.$$

$$\left(\frac{f}{g}\right)(x) = \frac{f(x)}{g(x)} = \frac{\frac{1}{-x+2}}{x^2+3} = \frac{1}{(-x+2)(x^2+3)} = \frac{1}{-x^3+2x^2-3x+6}.$$

$$(2f)(x) = 2f(x) = 2\left(\frac{1}{-x+2}\right) = \frac{2}{-x+2}.$$

$$(f \circ g)(x) = f(g(x)) = f(x^2+3) = \frac{1}{-(x^2+3)+2} = \frac{1}{-x^2-1}.$$

$$(g \circ f)(x) = g(f(x)) = g\left(\frac{1}{-x+2}\right) = \left(\frac{1}{-x+2}\right)^2 + 3 = \frac{1}{(-x+2)^2} + 3$$
$$= \frac{1 + 3(-x+2)^2}{(-x+2)^2} = \frac{1 + 3(x^2 - 4x + 4)}{(-x+2)^2} = \frac{1 + 3x^2 - 12x + 12}{(-x+2)^2}$$
$$= \frac{3x^2 - 12x + 13}{(-x+2)^2}.$$

19. (a) $f(x) = 2x + 5$ and $g(x) = -3x - 1 \Rightarrow$

$$(f \circ g)(x) = f(g(x)) = f(-3x - 1) = 2(-3x - 1) + 5 = -6x - 2 + 5 = -6x + 3.$$

(b) From (a): $(f \circ g)(2) = -6(2) + 3 = -12 + 3 = -9.$

Without (a): $(f \circ g)(2) = f(g(2)) = f(-3 \cdot 2 - 1) = f(-7) = 2(-7) + 5 = -9.$

21. (a) $f(x) = x^2 + x$ and $g(x) = x + 1$ $\Rightarrow$

$(f \circ g)(x) = f(g(x)) = f(x+1) = (x+1)^2 + (x+1) = x^2 + 2x + 1 + x + 1 = x^2 + 3x + 2.$

(b) From (a): $(f \circ g)(2) = 2^2 + 3(2) + 2 = 12.$

Without (a): $(f \circ g)(2) = f(g(2)) = f(2+1) = f(3) = 3^2 + 3 = 12.$

23. $f(x) = 3x + 9$ $\Rightarrow$

$f(2) = 3(2) + 9 = 15.$

$f(2+h) = 3(2+h) + 9 = 6 + 3h + 9 = 15 + 3h.$

$f(x+h) = 3(x+h) + 9 = 3x + 3h + 9.$

25. $f(x) = -x^2 + x + 1$ $\Rightarrow$

$f(2) = -2^2 + 2 + 1 = -4 + 3 = -1.$

$f(2+h) = -(2+h)^2 + (2+h) + 1 = -(4 + 4h + h^2) + 3 + h = -4 - 4h - h^2 + 3 + h$
$\qquad = -h^2 - 3h - 1.$

$f(x+h) = -(x+h)^2 + (x+h) + 1 = -(x^2 + 2xh + h^2) + x + h + 1$
$\qquad = -x^2 - 2xh - h^2 + x + h + 1.$

27. $f(x) = \dfrac{x^2}{2x + 3}$ $\Rightarrow$

$f(2) = \dfrac{2^2}{2(2) + 3} = \dfrac{4}{7}.$

$f(2+h) = \dfrac{(2+h)^2}{2(2+h) + 3} = \dfrac{4 + 4h + h^2}{4 + 2h + 3} = \dfrac{h^2 + 4h + 4}{7 + 2h}.$

$f(x+h) = \dfrac{(x+h)^2}{2(x+h) + 3} = \dfrac{x^2 + 2xh + h^2}{2x + 2h + 3}.$

29. $f(x) = 3x + 9$ $\Rightarrow$

$f(x+h) = 3(x+h) + 9 = 3x + 3h + 9$

$\dfrac{f(x+h) - f(x)}{h} = \dfrac{(3x + 3h + 9) - (3x + 9)}{h} = \dfrac{3\cancel{x} + 3h + \cancel{9} - 3\cancel{x} - \cancel{9}}{h} = \dfrac{3\cancel{h}}{\cancel{h}} = 3$, as $h \neq 0$.

31. $f(x) = -x^2 + x + 1$ $\Rightarrow$

$f(x+h) = -(x+h)^2 + x + h + 1 = -(x^2 + 2xh + h^2) + x + h + 1$
$\qquad = -x^2 - 2xh - h^2 + x + h + 1$

$\dfrac{f(x+h) - f(x)}{h} = \dfrac{(-\cancel{x^2} - 2xh - h^2 + \cancel{x} + h + \cancel{1}) - (-\cancel{x^2} + \cancel{x} + \cancel{1})}{h}$

$\qquad = \dfrac{\cancel{h}(-2x - h + 1)}{\cancel{h}} = -2x - h + 1$, as $h \neq 0$.

33. $f(x) = \dfrac{x^2}{2x+3}$ $\Rightarrow$

$$f(x+h) = \frac{(x+h)^2}{2(x+h)+3} = \frac{x^2+2xh+h^2}{2x+2h+3}$$

$$\frac{f(x+h)-f(x)}{h} = \frac{\dfrac{x^2+2xh+h^2}{2x+2h+3} - \dfrac{x^2}{2x+3}}{h} = \frac{(x^2+2xh+h^2)(2x+3) - x^2(2x+2h+3)}{h(2x+2h+3)(2x+3)}$$

$$= \frac{2x^3 + 4x^2h + 2xh^2 + 3x^2 + 6xh + 3h^2 - 2x^3 - 2x^2h - 3x^2}{h(2x+2h+3)(2x+3)}$$

$$= \frac{h(2x^2 + 2xh + 6x + 3h)}{h(2x+2h+3)(2x+3)} = \frac{2x^2 + 2xh + 6x + 3h}{(2x+2h+3)(2x+3)}, \text{ as } h \neq 0.$$

35. $f(x) = \begin{cases} x^2 & \text{if } x < 1 \\ x+1 & \text{if } x \geq 1 \end{cases}$

Since $0 < 1$, $f(0) = 0^2 = 0$.

Since $1 \geq 1$, $f(1) = 1 + 1 = 2$.

37. $f(x) = \begin{cases} 3x-5 & \text{if } x < 0 \\ x^2 & \text{if } 0 \leq x < 5 \\ -2x & \text{if } 5 \leq x < 10 \end{cases}$

Since $-1 < 0$, $f(-1) = 3(-1) - 5 = -3 - 5 = -8$.

Since $0 \leq 1 < 5$, $f(1) = 1^2 = 1$.

Since $5 \leq 7 < 10$, $f(7) = -2(7) = -14$.

Since 10 is not in the domain, $f(10)$ is undefined.

39. $|ab| = |a||b|$ because the only possible difference between ab and $|a||b|$ is that ab might be the negative of $|a||b|$, and taking the absolute value of ab erases any such discrepancy. Likewise, if $b \neq 0$, then $\left|\dfrac{a}{b}\right| = \dfrac{|a|}{|b|}$.

41. $|a| = |a - b + b| \leq |a-b| + |b|$ by the Triangle Inequality, Theorem 1.1(b). Then $|a| \leq |a-b| + |b| \Rightarrow |a| - |b| \leq |a-b|$. Turning that around yields $|a-b| \geq |a| - |b|$.

43. $f(x) = 3x^3 \Rightarrow f(-x) = 3(-x)^3 = -3x^3 = -f(x) \Rightarrow$ f is an odd function.

45. $f(x) = 3x^2 + 1 \Rightarrow f(-x) = 3(-x)^2 + 1 = 3x^2 + 1 = f(x) \Rightarrow$ f is an even function.

47. $f(x) = x^3 + x + 1 \Rightarrow f(-x) = (-x)^3 + (-x) + 1 = -x^3 - x + 1$, which is not f, and not $-f$. Conclusion: f is neither an even nor an odd function.

§1.2 One-to-One Functions and their Inverses

1. We assume that $f(a) = f(b)$ and show that this can only be true if $a = b$:

$$-5a - 1 = -5b - 1$$
$$-5a - 1 + 1 = -5b - 1 + 1$$
$$-5a = -5b$$
$$-\tfrac{1}{5}(-5a) = -\tfrac{1}{5}(-5b)$$
$$a = b$$

3. We assume that $f(a) = f(b)$ and show that this can only be true if $a = b$:

$$a^3 + 1 = b^3 + 1$$
$$a^3 + 1 - 1 = b^3 + 1 - 1$$
$$a^3 = b^3$$
$$(a^3)^{\frac{1}{3}} = (b^3)^{\frac{1}{3}}$$
$$a = b$$

5. We assume that $f(a) = f(b)$ and show that this can only be true if $a = b$:

$$\frac{4}{2a - 3} = \frac{4}{2b - 3}$$
$$4(2b - 3) = 4(2a - 3)$$
$$2b - 3 = 2a - 3$$
$$2b - 3 + 3 = 2a - 3 + 3$$
$$2a = 2b$$
$$a = b$$

7. We assume that $f(a) = f(b)$ and show that this can only be true if $a = b$:

$$\sqrt{a + 1} + 2 = \sqrt{b + 1} + 2$$
$$\sqrt{a + 1} + 2 - 2 = \sqrt{b + 1} + 2 - 2$$
$$\sqrt{a + 1} = \sqrt{b + 1}$$
$$(\sqrt{a + 1})^2 = (\sqrt{b + 1})^2$$
$$a + 1 = b + 1$$
$$a + 1 - 1 = b + 1 - 1$$
$$a = b$$

9. We assume that $f(a) = f(b)$ and show that this can only be true if $a = b$:

$$\sqrt{\frac{3a}{2a + 1}} = \sqrt{\frac{3b}{2b + 1}}$$
$$\left(\sqrt{\frac{3a}{2a + 1}}\right)^2 = \left(\sqrt{\frac{3b}{2b + 1}}\right)^2$$
$$\frac{3a}{2a + 1} = \frac{3b}{2b + 1}$$
$$3a(2b + 1) = 3b(2a + 1)$$
$$6ab + 3a = 6ab + 3b$$
$$6ab + 3a - 6ab = 6ab + 3b - 6ab$$
$$3a = 3b$$
$$a = b$$

11. Solving $f(x) = x^2 - x - 6 = 0$:
$(x - 3)(x + 2) = 0 \Rightarrow x = 3, -2$, so we choose $a = 3$, and $b = -2$. Then

$$f(3) = f(-2) = 0 \quad \text{and} \quad 3 \neq -2.$$

13. Determining suitable values of a and b with $a \neq b$:

$$\frac{a}{a^2 + 1} = \frac{b}{b^2 + 1}$$
$$a(b^2 + 1) = b(a^2 + 1)$$
$$ab^2 + a = a^2b + b$$
$$ab^2 - a^2b = b - a$$
$$ab(b - a) = b - a$$
$$ab = 1$$

Any values of a and b for which $ab = 1$ and $a \neq b$ will work: For example, $a = 2$ and $b = \frac{1}{2}$. [Checking: $f(2) = \frac{2}{2^2 + 1} = \frac{2}{5}$ and $f(\frac{1}{2}) = \frac{\frac{1}{2}}{\frac{1}{4} + 1} = \frac{4}{5} \cdot \frac{1}{2} = \frac{2}{5}$].

15. $f(x) = -x + 1$

$\qquad y = -x + 1$

$\qquad x = 1 - y$

$\qquad y = 1 - x = f^{-1}(x)$

$(f \circ f^{-1})(x) = f(f^{-1}(x)) = f(1 - x)$

$\qquad\qquad\qquad\qquad = -(1 - x) + 1$

$\qquad\qquad\qquad\qquad = -1 + x + 1$

$\qquad\qquad\qquad\qquad = x$

$(f^{-1} \circ f)(x) = f^{-1}(f(x)) = f^{-1}(-x + 1)$

$\qquad\qquad\qquad\qquad = 1 - (-x + 1)$

$\qquad\qquad\qquad\qquad = 1 + x - 1$

$\qquad\qquad\qquad\qquad = x$

17. $f(x) = \dfrac{3}{2x - 5}$

$\qquad y = \dfrac{3}{2x - 5}$

$\qquad 2x - 5 = \dfrac{3}{y}$

$\qquad 2x = \dfrac{3}{y} + 5 = \dfrac{5y + 3}{y}$

$\qquad x = \dfrac{5y + 3}{2y}$

$\qquad y = \dfrac{5x + 3}{2x} = f^{-1}(x)$

$(f \circ f^{-1})(x) = f(f^{-1}(x)) = f\left(\dfrac{5x + 3}{2x}\right)$

$\qquad = \dfrac{3}{2\!\!\!/ \cdot \dfrac{5x + 3}{2\!\!\!/x} - 5} \cdot \dfrac{x}{x}$

$\qquad = \dfrac{3x}{5\!\!\!/x + 3 - 5\!\!\!/x} = \dfrac{3x}{3} = x$

$(f^{-1} \circ f)(x) = f^{-1}(f(x)) = f^{-1}\left(\dfrac{3}{2x - 5}\right)$

$\qquad = \dfrac{5 \cdot \dfrac{3}{2x - 5} + 3}{2 \cdot \dfrac{3}{2x - 5}} \cdot \dfrac{2x - 5}{2x - 5}$

$\qquad = \dfrac{15 + 3(2x - 5)}{6} = \dfrac{1\!\!\!/5 + 6x - 1\!\!\!/5)}{6} = x$

19. $f(x) = \dfrac{5x + 2}{x - 3}$

$\qquad y = \dfrac{5x + 2}{x - 3}$

$\qquad y(x - 3) = 5x + 2$

$\qquad xy - 3y = 5x + 2$

$\qquad xy - 5x = 3y + 2$

$\qquad x(y - 5) = 3y + 2$

$\qquad x = \dfrac{3y + 2}{y - 5}$

$\qquad y = \dfrac{3x + 2}{x - 5} = f^{-1}(x)$

$(f \circ f^{-1})(x) = f(f^{-1}(x)) = f\left(\dfrac{3x + 2}{x - 5}\right)$

$\qquad = \dfrac{5 \cdot \dfrac{3x + 2}{x - 5} + 2}{\dfrac{3x + 2}{x - 5} - 3} \cdot \dfrac{x - 5}{x - 5}$

$\qquad = \dfrac{5(3x + 2) + 2(x - 5)}{3x + 2 - 3(x - 5)}$

$\qquad = \dfrac{15x + 1\!\!\!/0 + 2x - 1\!\!\!/0}{3\!\!\!/x + 2 - 3\!\!\!/x + 15}$

$\qquad = \dfrac{17x}{17} = x$

(continued)

$$(f^{-1} \circ f)(x) = f^{-1}(f(x)) = f^{-1}\left(\frac{5x+2}{x-3}\right)$$

$$= \frac{3 \cdot \dfrac{5x+2}{x-3} + 2}{\dfrac{5x+2}{x-3} - 5} \cdot \frac{x-3}{x-3}$$

$$= \frac{3(5x+2) + 2(x-3)}{5x+2 - 5(x-3)}$$

$$= \frac{15x + \cancel{6} + 2x - \cancel{6}}{\cancel{5}x + 2 - \cancel{5}x + 15}$$

$$= \frac{17x}{17} = x$$

21.

$$f(x) = 2\sqrt{x+3}$$
$$y = 2\sqrt{x+3}$$
$$\frac{y}{2} = \sqrt{x+3}$$
$$\left(\frac{y}{2}\right)^2 = (\sqrt{x+3})^2$$
$$\tfrac{1}{4}y^2 = x+3$$
$$x = \tfrac{1}{4}y^2 - 3$$
$$y = \tfrac{1}{4}x^2 - 3 = f^{-1}(x)$$

$$(f \circ f^{-1})(x) = f(f^{-1}(x)) = f\left(\tfrac{1}{4}x^2 - 3\right)$$
$$= 2\sqrt{(\tfrac{1}{4}x^2 - 3) + 3}$$
$$= 2\sqrt{(\tfrac{1}{4}x^2)}$$
$$= 2 \cdot \frac{x}{2}, \qquad \text{as } x \geq 0$$
$$= x$$

$$(f^{-1} \circ f)(x) = f^{-1}(f(x)) = f^{-1}(2\sqrt{x+3})$$
$$= \tfrac{1}{4}(2\sqrt{x+3})^2 - 3$$
$$= \tfrac{1}{4} \cdot 4(x+3) - 3$$
$$= (x+3) - 3$$
$$= x$$

23.

25.

27.

§1.3 Equations and Inequalities

1(a). $3x - 5 = 2x + 7$
$$3x - 2x = 7 + 5$$
$$x = 12$$

1(b). $3x - 5 < 2x + 7$
$$3x - 2x < 7 + 5$$
$$x < 12$$
$$(-\infty, 12)$$

1(c). $3x - 5 \geq 2x + 7$
$$3x - 2x \geq 7 + 5$$
$$x \geq 12$$
$$[12, \infty)$$

3(a). $\dfrac{1}{3} - \dfrac{2x-1}{6} = \dfrac{1}{2} - \dfrac{2(3x+2)}{3}$

$$6\left[\dfrac{1}{3} - \dfrac{2x-1}{6}\right] = 6\left[\dfrac{1}{2} - \dfrac{2(3x+2)}{3}\right]$$

$$2 - (2x-1) = 3 - 4(3x+2)$$

$$2 - 2x + 1 = 3 - 12x - 8$$

$$12x - 2x = 3 - 8 - 2 - 1$$

$$10x = -8$$

$$x = -\dfrac{8}{10} = -\dfrac{4}{5}$$

3(b). $\dfrac{1}{3} - \dfrac{2x-1}{6} \le \dfrac{1}{2} - \dfrac{2(3x+2)}{3}$

$$6\left[\dfrac{1}{3} - \dfrac{2x-1}{6}\right] \le 6\left[\dfrac{1}{2} - \dfrac{2(3x+2)}{3}\right]$$

$$2 - (2x-1) \le 3 - 4(3x+2)$$

$$2 - 2x + 1 \le 3 - 12x - 8$$

$$12x - 2x \le 3 - 8 - 2 - 1$$

$$10x \le -8$$

$$x \le -\dfrac{8}{10} \; \left(= -\dfrac{4}{5}\right)$$

$$\left(-\infty, -\dfrac{4}{5}\right]$$

5(a). $x^4 - x^3 - 6x^2 = 0$

$$x^2(x^2 - x - 6) = 0$$

$$x^2(x-3)(x+2) = 0$$

$$x = 0, 3, -2$$

5(b).   From (a) and the sign chart,

$x^2(x-3)(x+2) \le 0$ in $[-2, 3]$.

5(c).   From the sign chart in (b), $x^2(x-3)(x+2) > 0$ in $(-\infty, -2) \cup (3, \infty)$.

7(a).   As -1 is a root:

$$x^3 - 2x - 1 = 0$$

$$(x+1)(x^2 - x - 1) = 0$$

$$x = -1, \text{ or}$$

$$x = \dfrac{1 \pm \sqrt{1 - 4(1)(-1)}}{2}$$

$$= \dfrac{1 \pm \sqrt{5}}{2}$$

7(b).   From (a) and the sign chart,

$(x+1)(x^2 - x - 1) \le 0$ in

$$\left(-\infty, -1\right] \cup \left[\dfrac{1-\sqrt{5}}{2}, \dfrac{1+\sqrt{5}}{2}\right].$$

7(c).   From the sign chart in (b), $(x+1)(x^2 - x - 1) > 0$ in $\left(-1, \dfrac{1-\sqrt{5}}{2}\right) \cup \left(\dfrac{1+\sqrt{5}}{2}, \infty\right)$.

9(a). $(1-x)(2x+3)(x+2) = 0$

$$x = 1, -\dfrac{3}{2}, -2$$

9(b).   From the sign chart,

$(1-x)(2x+3)(x+2) < 0$ in

$$\left(-2, -\dfrac{3}{2}\right) \cup (1, \infty).$$

11(a). $(1-x)^{21}(2x+3)^{30}(x+2)^2 = 0$

$$x = 1, -\dfrac{3}{2}, -2$$

11(b).   From the sign chart,

$(1-x)^{21}(2x+3)^{30}(x+2)^2 \le 0$ in

$$[1, \infty) \cup \left\{-2, -\dfrac{3}{2}\right\}.$$

13(a). $-x^4(2x+3)^3(-x+1)^5 = 0$
$$x = 0, -\tfrac{3}{2}, 1$$

13(b). From the sign chart,

$$-x^4(2x+3)^3(-x+1)^5 \le 0$$

in $\left[-\tfrac{3}{2}, 1\right]$.

15(a).   To solve $3(x^2-1)^2 = 10(x^2-1)+8$, let $Y = x^2-1$. The equation then becomes $3Y^2 - 10Y - 8 = 0 \Rightarrow (3Y+2)(Y-4) = 0 \Rightarrow Y = -\tfrac{2}{3}$ or $Y = 4$. Substituting back:

$$x^2 - 1 = -\tfrac{2}{3}$$
$$x^2 = \tfrac{1}{3}$$
$$x = \pm\sqrt{\tfrac{1}{3}} = \pm\tfrac{1}{\sqrt{3}} = \pm\tfrac{\sqrt{3}}{3}$$

$$x^2 - 1 = 4$$
$$x^2 = 5$$
$$x = \pm\sqrt{5}$$

The solutions are $x = \pm\tfrac{\sqrt{3}}{3}, \pm\sqrt{5}$.

15(b). To solve $3(x^2-1)^2 \le 10(x^2-1)+8 \Rightarrow 3(x^2-1)^2 - 10(x^2-1) - 8 \le 0$, use part (a) to rewrite the inequality as $\left(x - \tfrac{\sqrt{3}}{3}\right)\left(x + \tfrac{\sqrt{3}}{3}\right)\left(x - \sqrt{5}\right)\left(x + \sqrt{5}\right) \le 0$. From the sign chart,

, the solution is $\left[-\sqrt{5}, -\tfrac{\sqrt{3}}{3}\right] \cup \left[\tfrac{\sqrt{3}}{3}, \sqrt{5}\right]$.

17(a).

$$\frac{1}{x+1} = \frac{1}{2x}$$
$$2x = x + 1$$
$$x = 1$$

17(b).

$$\frac{1}{x+1} > \frac{1}{2x}$$
$$\frac{1}{x+1} - \frac{1}{2x} > 0$$
$$\frac{2x - x - 1}{2x(x+1)} > 0$$
$$\frac{x - 1}{2x(x+1)} > 0$$

From the sign chart,

$$\frac{1}{x+1} > \frac{1}{2x}$$

in

$(-1, 0) \cup (1, \infty)$.

17(c).   From the sign chart in (b),

$$\frac{1}{x+1} \le \frac{1}{2x}$$ in

$(-\infty, -1) \cup (0, 1]$.

19(a).

$$x = \frac{2}{x} + 1$$
$$x^2 = 2 + x$$
$$x^2 - x - 2 = 0$$
$$(x-2)(x+1) = 0$$
$$x = 2, -1$$

19(b).

$$x \ge \frac{2}{x} + 1$$
$$x - \frac{2}{x} - 1 \ge 0$$
$$\frac{x^2 - 2 - x}{x} \ge 0$$
$$\frac{(x-2)(x+1)}{x} \ge 0$$

From the sign chart,

$$x \ge \frac{2}{x} + 1$$

in

$[-1, 0) \cup [2, \infty)$.

19(c).   From the sign chart in (b),

$$x \le \frac{2}{x} + 1$$ in

$(-\infty, -1] \cup (0, 2]$.

21. Since 1 is a solution of the polynomial equation, and therefore a root of the polynomial, $x - 1$ is a factor of the polynomial. Likewise, $x - 2$ and $x - 3$ are factors, and there are no other nonconstant factors (or there would be additional solutions of the equation). Therefore, any nonzero, constant multiple of the polynomial $(x - 1)(x - 2)(x - 3)$ will satisfy the requirements. So one choice is: $(x - 1)(x - 2)(x - 3) = 0$.

23. As 1, 2, and 5 appear in the solution, we begin with the polynomial $(x-1)(x-2)(x-5)$. To determine whether it should be > 0 or < 0, we chart the sign of the polynomial, and decide that < 0 works: $(x - 1)(x - 2)(x - 5) < 0$ [as would $-(x - 1)(x - 2)(x - 5) > 0$ or any positive constant multiple of either inequality].

25.   As 1, 2, 5, and 9 appear in the solution, we begin with the polynomial $(x - 1)(x - 2)(x - 5)(x - 9)$. To determine whether it should be > 0 or < 0, we chart the sign of the polynomial, and decide that > 0 will work: $(x - 1)(x - 2)(x - 5)(x - 9) > 0$ [as would $-(x-1)(x-2)(x-5)(x-9) < 0$ or any positive constant multiple of either inequality].

27.   As 1, 2, and 5 appear in the solution, we begin with the factors $(x - 1)$, $(x - 2)$, and $(x - 5)$. As 2 is not included in the solution set, but 1 and 5 are, we begin with the rational function $\dfrac{(x - 1)(x - 5)}{x - 2}$. To determine whether it should be ≥ 0 or ≤ 0, we chart the sign of the function, and determine that $\leq$ works: $\dfrac{(x - 1)(x - 5)}{x - 2} \leq 0$ [as would $-\dfrac{(x - 1)(x - 5)}{x - 2} \geq 0$ or any positive constant multiple of either inequality].

§1.4 Trigonometry

1. $\theta = 30° = \overset{1}{\cancel{30}°} \cdot \dfrac{\pi}{\underset{6}{\cancel{180}°}} = \dfrac{\pi}{6}$

3. $\theta = 60° = \overset{}{\cancel{60}°} \cdot \dfrac{\pi}{\underset{3}{\cancel{180}°}} = \dfrac{\pi}{3}$

5. $\theta = 120° = \overset{2}{\cancel{120}°} \cdot \dfrac{\pi}{\underset{3}{\cancel{180}°}} = \dfrac{2\pi}{3}$

7. $\theta = -150° = \overset{-5}{\cancel{-150}°} \cdot \dfrac{\pi}{\underset{6}{\cancel{180}°}} = -\dfrac{5\pi}{6}$

9. $\theta = \dfrac{\pi}{4} = \dfrac{\cancel{\pi}}{\cancel{4}} \cdot \dfrac{\overset{45}{\cancel{180}°}}{\cancel{\pi}} = 45°$

11. $\theta = -\dfrac{\pi}{6} = \dfrac{-\cancel{\pi}}{\cancel{6}} \cdot \dfrac{\overset{30}{\cancel{180}°}}{\cancel{\pi}} = -30°$

13. $\theta = \dfrac{\pi}{2} = \dfrac{\cancel{\pi}}{\cancel{2}} \cdot \dfrac{\overset{90}{\cancel{180}°}}{\cancel{\pi}} = 90°$

15. $\theta = \dfrac{7\pi}{6} = \dfrac{7\cancel{\pi}}{\cancel{6}} \cdot \dfrac{\overset{30}{\cancel{180}°}}{\cancel{\pi}} = 210°$

17. The angle $810°$ is coterminal with $810° - 2(360)° = 810° - 720° = 90°$. As $\sin 810° = \sin 90°$, is the y-coordinate of the point $(0,1)$ on the unit circle, $\sin 810°$ is 1.

19. Since $\frac{17}{2} = 8 + \frac{1}{2}$, the angle $\frac{17\pi}{2} = 8\pi + \frac{\pi}{2}$ is coterminal with $\frac{\pi}{2}$. Then

$$\cot \frac{17\pi}{2} = \cot \frac{\pi}{2} = \frac{\cos \frac{\pi}{2}}{\sin \frac{\pi}{2}} = \frac{0}{1} = 0.$$

21. The angle 11π is coterminal with $11\pi - 5(2\pi) = \pi$. Thus $\sec 11\pi = \sec \pi = \frac{1}{\cos \pi} = \frac{1}{-1} = -1$, as $\cos \pi$ is the x-coordinate of the leftmost point, $(-1, 0)$, on the unit circle.

23. $\tan(-360°) = \tan 0° = 0$, as $0°$ is the rightmost point, $(1,0)$, on the unit circle, and $\tan 0° = \frac{\sin 0°}{\cos 0°} = \frac{0}{1} = 0.$

25.

The angle $135°$ lies in the second quadrant (see figure), with a reference angle of $180° - 135° = 45°$. In the second quadrant, the sine is positive. Thus

$$\sin 135° = \sin 45° = \frac{\text{opp}}{\text{hyp}} = \frac{1}{\sqrt{2}}.$$

27.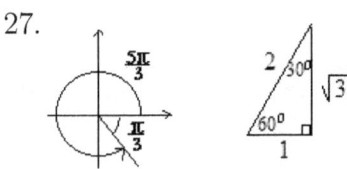

Since $\frac{17\pi}{3} = \left(4 + \frac{5}{3}\right)\pi = 2(2\pi) + \frac{5\pi}{3}$, it is coterminal with $\frac{5\pi}{3}$ (300°) which lies in QIV. There, the tangent is negative, so that

$$\tan \frac{17\pi}{3} = \tan \frac{5\pi}{3} = -\tan \frac{\pi}{3} = -\frac{\text{opp}}{\text{adj}} = -\frac{\sqrt{3}}{1} = -\sqrt{3}.$$

29.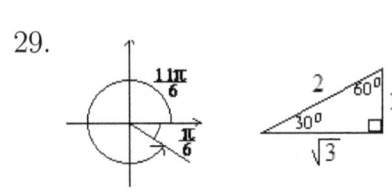

From the leftmost figure, we know that $\frac{11\pi}{6}$ lies in the fourth quadrant with a reference angle of $\frac{\pi}{6}$ (30°). The cosine (and therefore the secant) is positive in QIV. Thus,

$$\sec\left(\frac{11\pi}{6}\right) = \sec \frac{\pi}{6} = \frac{1}{\cos \frac{\pi}{6}} = \frac{1}{\frac{\text{adj}}{\text{hyp}}} = \frac{1}{\frac{\sqrt{3}}{2}} = \frac{2}{\sqrt{3}}.$$

31.

From the leftmost figure, we know that $-510°$ lies in the third quadrant with a reference angle of $30°$. The tangent is positive in QIII. Thus,

$$\tan(-510°) = \tan 30° = \frac{\text{opp}}{\text{adj}} = \frac{1}{\sqrt{3}}.$$

33. $\dfrac{2\sin^2 x}{\sin 2x} = \dfrac{2\sin x \sin x}{2\sin x \cos x} = \dfrac{\sin x}{\cos x} = \tan x.$

35. $\dfrac{(\sin x + \cos x)^2 - 1}{\sin 2x} = \dfrac{\sin^2 x + 2\sin x \cos x + \cos^2 x - 1}{2\sin x \cos x} = \dfrac{2\sin x \cos x}{2\sin x \cos x} = 1.$

37. $\dfrac{\sin^2 \frac{x}{2} \cos^2 \frac{x}{2}}{1 + \cos 2x} = \dfrac{(\sin \frac{x}{2} \cos \frac{x}{2})^2}{1 + (2\cos^2 x - 1)} = \dfrac{(\frac{1}{2}\sin x)^2}{2\cos^2 x} = \dfrac{\frac{1}{4}\sin^2 x}{2\cos^2 x} = \dfrac{1}{8}\left(\dfrac{\sin x}{\cos x}\right)^2 = \dfrac{1}{8}\tan^2 x.$

39. $\dfrac{\tan x + \cot x}{\csc 2x} = \dfrac{\frac{\sin x}{\cos x} + \frac{\cos x}{\sin x}}{\frac{1}{\sin 2x}} = \left(\dfrac{\sin x}{\cos x} + \dfrac{\cos x}{\sin x}\right) \cdot 2\sin x \cos x$

$$= \dfrac{\sin^2 x \cancel{+}^{1} \cos^2 x}{\cancel{\sin} x \cancel{\cos} x} \cdot 2\cancel{\sin} x \cancel{\cos} x = 1 \cdot 2 = 2.$$

41. $(\sec x + \tan x)^2(\sec x - \tan x) = (\sec x + \tan x)[(\sec x + \tan x)(\sec x - \tan x)]$

$$= (\sec x + \tan x)(\sec^2 x \cancel{-}^{1} \tan^2 x) = \sec x + \tan x.$$

CHAPTER 2
Limits and Continuity

§2.1 The Limit: An Intuitive Introduction

1. $\lim\limits_{x \to 3} \dfrac{x^2 - 5}{x + 3} = \dfrac{3^2 - 5}{3 + 3} = \dfrac{4}{6} = \dfrac{2}{3}.$

3. $\lim\limits_{x \to 5} \dfrac{x^2 - 5}{x + 5} = \dfrac{5^2 - 5}{5 + 5} = \dfrac{20}{10} = 2.$

5. $\lim\limits_{x \to 5} \dfrac{x^2 - 25}{x^2 - 3x - 10} = \lim\limits_{x \to 5} \dfrac{(x - 5)(x + 5)}{(x - 5)(x + 2)} = \lim\limits_{x \to 5} \dfrac{x + 5}{x + 2} = \dfrac{5 + 5}{5 + 2} = \dfrac{10}{7}.$

7. $\lim\limits_{x \to 2} \dfrac{x^2 + 3x - 10}{x^2 - 4x + 4} = \lim\limits_{x \to 2} \dfrac{(x + 5)(x - 2)}{(x - 2)^2} = \lim\limits_{x \to 2} \dfrac{x + 5}{x - 2}$ which does not exist, as the denominator tends to 0 while the numerator tends to 7.

9. $\lim\limits_{x \to -1} \dfrac{2x^3 + 5x^2 + 3x}{x^2 - 3x - 4} = \lim\limits_{x \to -1} \dfrac{x(2x + 3)(x + 1)}{(x - 4)(x + 1)} = \lim\limits_{x \to -1} \dfrac{x(2x + 3)}{(x - 4)} = \dfrac{(-1)[2(-1) + 3]}{-1 - 4} = \dfrac{1}{5}.$

11. $\lim\limits_{x \to 1} \dfrac{x^2 - 2x + 1}{x^2 - 1} = \lim\limits_{x \to 1} \dfrac{(x - 1)^2}{(x - 1)(x + 1)} = \lim\limits_{x \to 1} \dfrac{x - 1}{x + 1} = \dfrac{1 - 1}{1 + 1} = \dfrac{0}{2} = 0.$

13. $\lim\limits_{x \to 1} \dfrac{x^2 - 1}{x^3 - x^2 + 2x - 2} = \lim\limits_{x \to 1} \dfrac{(x - 1)(x + 1)}{x^2(x - 1) + 2(x - 1)} = \lim\limits_{x \to 1} \dfrac{(x + 1)(x - 1)}{(x^2 + 2)(x - 1)}$

$\qquad = \lim\limits_{x \to 1} \dfrac{x + 1}{x^2 + 2} = \dfrac{1 + 1}{1 + 2} = \dfrac{2}{3}.$

15. $\lim\limits_{x \to 1} \dfrac{x^2 - 1}{x^3 - 2x^2 + 1} = \lim\limits_{x \to 1} \dfrac{(x - 1)(x + 1)}{(x - 1)(x^2 - x - 1)} = \lim\limits_{x \to 1} \dfrac{x + 1}{x^2 - x - 1} = \dfrac{1 + 1}{1 - 1 - 1} = \dfrac{2}{-1} = -2.$

17. $\lim\limits_{x \to 3} \dfrac{\sqrt{1 + x} - 2}{x - 3} = \lim\limits_{x \to 3} \left(\dfrac{\sqrt{1 + x} - 2}{x - 3} \cdot \dfrac{\sqrt{1 + x} + 2}{\sqrt{1 + x} + 2} \right) = \lim\limits_{x \to 3} \dfrac{1 + x - 4}{(x - 3)(\sqrt{1 + x} + 2)}$

$\qquad = \lim\limits_{x \to 3} \dfrac{x - 3}{(x - 3)(\sqrt{1 + x} + 2)} = \lim\limits_{x \to 3} \dfrac{1}{\sqrt{1 + x} + 2} = \dfrac{1}{\sqrt{1 + 3} + 2} = \dfrac{1}{4}.$

19. $\lim\limits_{x \to 0} \dfrac{x^2 - 1}{\sqrt{1 + x} - 1}$ does not exist, as the denominator tends to 0 while the numerator tends to -1.

21. $\displaystyle\lim_{x\to 0}\left(\frac{1}{x}-\frac{1}{x\sqrt{x+1}}\right)=\lim_{x\to 0}\left(\frac{\sqrt{x+1}-1}{x\sqrt{x+1}}\right)=\lim_{x\to 0}\left(\frac{\sqrt{x+1}-1}{x\sqrt{x+1}}\cdot\frac{\sqrt{x+1}+1}{\sqrt{x+1}+1}\right)$

$\displaystyle=\lim_{x\to 0}\frac{x+1-1}{x\sqrt{x+1}(\sqrt{x+1}+1)}=\lim_{x\to 0}\frac{\not{x}}{\not{x}\sqrt{x+1}(\sqrt{x+1}+1)}$

$\displaystyle=\lim_{x\to 0}\frac{1}{\sqrt{x+1}(\sqrt{x+1}+1)}=\frac{1}{\sqrt{1}(\sqrt{1}+1)}=\frac{1}{2}.$

23. $\displaystyle\lim_{x\to -1}\frac{\sqrt{x^2+8}-3}{x+1}=\lim_{x\to -1}\left(\frac{\sqrt{x^2+8}-3}{x+1}\cdot\frac{\sqrt{x^2+8}+3}{\sqrt{x^2+8}+3}\right)=\lim_{x\to -1}\frac{x^2+8-9}{(x+1)(\sqrt{x^2+8}+3)}$

$\displaystyle=\lim_{x\to -1}\frac{x^2-1}{(x+1)(\sqrt{x^2+8}+3)}=\lim_{x\to -1}\frac{(x-1)(x\not{+}1)}{(x\not{+}1)(\sqrt{x^2+8}+3)}$

$\displaystyle=\lim_{x\to -1}\frac{x-1}{\sqrt{x^2+8}+3}=\frac{-1-1}{\sqrt{1+8}+3}=\frac{-2}{6}=-\frac{1}{3}.$

25. $\displaystyle\lim_{h\to 0}\frac{(x+h)^3-x^3}{h}=\lim_{h\to 0}\frac{\not{x^3}+3x^2h+3xh^2+h^3-\not{x^3}}{h}=\lim_{h\to 0}\frac{\not{h}(3x^2+3xh+h^2)}{\not{h}}$

$\displaystyle=\lim_{h\to 0}3x^2+3xh+h^2=3x^2.$

27. $\displaystyle\lim_{x\to 0}\frac{\sin^2 x+\sin x}{\cot x}=\lim_{x\to 0}\frac{\sin^2 x+\sin x}{\frac{\cos x}{\sin x}}=\lim_{x\to 0}(\sin^2 x+\sin x)\left(\frac{\sin x}{\cos x}\right)=(0+0)(\tfrac{0}{1})=0.$

29. $\displaystyle\lim_{x\to \frac{\pi}{2}}\frac{1-\sin x}{\cos x}=\lim_{x\to \frac{\pi}{2}}\left(\frac{1-\sin x}{\cos x}\cdot\frac{1+\sin x}{1+\sin x}\right)=\lim_{x\to \frac{\pi}{2}}\frac{1-\sin^2 x}{(\cos x)(1+\sin x)}$

$\displaystyle=\lim_{x\to \frac{\pi}{2}}\frac{\cos^{\not{2}} x}{(\not{\cos} x)(1+\sin x)}=\lim_{x\to \frac{\pi}{2}}\frac{\cos x}{1+\sin x}=\frac{0}{1+1}=\frac{0}{2}=0.$

31. The function is not continuous at $x=2$, as it is not defined there. Since $\displaystyle\lim_{x\to 2} f(x)=4$, as $\displaystyle\lim_{x\to 2^-} f(x)=\lim_{x\to 2^+} f(x)=4$, defining $f(2)=4$ makes the function continuous at $x=2$: The discontinuity is therefore removable.

33. The function is not continuous at $x=2$, since it is not defined at $x=2$. The discontinuity is a jump discontinuity, since $\displaystyle\lim_{x\to 2^-} f(x)=\lim_{x\to 2^-} x+1=2+1=3$, while $\displaystyle\lim_{x\to 2^+} f(x)=\lim_{x\to 2^+} x^2=2^2=4\neq 3$.

35. The limit exists and is 5, as $\displaystyle\lim_{x\to 3^-} f(x)=\lim_{x\to 3^+} f(x)=5$. The function is not continuous at $x=3$, because $f(3)=2\neq 5$. The discontinuity is removable, by defining $f(3)=5$.

37. $\lim\limits_{x \to 3^-} f(x) = 2$ and $\lim\limits_{x \to 3^+} f(x) = 5 \neq 2$, means that the limit does not exist and the function is therefore discontinuous at $x = 3$. The discontinuity is a jump discontinuity as the left-hand and right-hand limits exist but are unequal.

39. For example:

41. For example:

§2.2 The Definition of a Limit

1. $\lim\limits_{x \to 1} 2x = 2 \cdot 1 = 2$

$$0 < |x - 1| < \delta \Rightarrow |2x - 2| < 3$$
$$0 < |x - 1| < \delta \Rightarrow 2|x - 1| < 3$$
$$0 < |x - 1| < \delta \Rightarrow |x - 1| < \frac{3}{2}$$

The largest $\delta > 0$ is $\delta = \frac{3}{2}$.

3. $\lim\limits_{x \to 2} (x - 5) = 2 - 5 = -3$

$$0 < |x - 2| < \delta \Rightarrow |x - 5 - (-3)| < 1$$
$$0 < |x - 2| < \delta \Rightarrow |x - 5 + 3| < 1$$
$$0 < |x - 2| < \delta \Rightarrow |x - 2| < 1$$

The largest $\delta > 0$ is $\delta = 1$.

5. $\lim\limits_{x \to 2} (x^2 + 1) = 2^2 + 1 = 5$

$$0 < |x - 2| < \delta \Rightarrow |x^2 + 1 - 5| < 1$$
$$0 < |x - 2| < \delta \Rightarrow |x^2 - 4| < 1$$

$|x^2 - 4| < 1 \Rightarrow -1 < x^2 - 4 < 1 \Rightarrow 4 - 1 < x^2 < 4 + 1 \Rightarrow 3 < x^2 < 5$. For x near 2, $\sqrt{3} < x < \sqrt{5} \Rightarrow \sqrt{3} - 2 < x - 2 < \sqrt{5} - 2$. As $\sqrt{3} - 2 \approx -0.27$ and $\sqrt{5} - 2 \approx 0.24$ then $|x - 2| < \sqrt{5} - 2$, so $\delta = \sqrt{5} - 2$.

7. Given $\epsilon > 0$, we seek $\delta > 0$ so that

$$0 < |x - 1| < \delta \Rightarrow |f(x) - 2| < \epsilon$$
$$0 < |x - 1| < \delta \Rightarrow |5x - 3 - 2| < \epsilon$$
$$0 < |x - 1| < \delta \Rightarrow |5x - 5| < \epsilon$$
$$0 < |x - 1| < \delta \Rightarrow 5|x - 1| < \epsilon$$
$$0 < |x - 1| < \delta \Rightarrow |x - 1| < \frac{\epsilon}{5}$$

Take $\delta = \frac{\epsilon}{5}$.

9. Given $\epsilon > 0$, we seek $\delta > 0$ so that

$$0 < |x + 2| < \delta \Rightarrow |f(x) - 1| < \epsilon$$
$$0 < |x + 2| < \delta \Rightarrow |-x - 1 - 1| < \epsilon$$
$$0 < |x + 2| < \delta \Rightarrow |-x - 2| < \epsilon$$
$$0 < |x + 2| < \delta \Rightarrow |x + 2| < \epsilon$$

Take $\delta = \epsilon$.

11. Given $\epsilon > 0$, we seek $\delta > 0$ so that

$$0 < \left|x - \frac{1}{2}\right| < \delta \Rightarrow \left|f(x) - \frac{5}{4}\right| < \epsilon$$
$$0 < \left|x - \frac{1}{2}\right| < \delta \Rightarrow \left|\frac{1}{2}x + 1 - \frac{5}{4}\right| < \epsilon$$
$$0 < \left|x - \frac{1}{2}\right| < \delta \Rightarrow \left|\frac{1}{2}x - \frac{1}{4}\right| < \epsilon$$
$$0 < \left|x - \frac{1}{2}\right| < \delta \Rightarrow \left|\frac{1}{2}\left(x - \frac{1}{2}\right)\right| < \epsilon$$
$$0 < \left|x - \frac{1}{2}\right| < \delta \Rightarrow \frac{1}{2}\left|x - \frac{1}{2}\right| < \epsilon$$
$$0 < \left|x - \frac{1}{2}\right| < \delta \Rightarrow \left|x - \frac{1}{2}\right| < 2\epsilon$$

Take $\delta = 2\epsilon$.

13. Given $\epsilon > 0$, we seek $\delta > 0$ so that
$$0 < |x - 2| < \delta \Rightarrow |f(x) - 4| < \epsilon$$
$$0 < |x - 2| < \delta \Rightarrow |x^2 - 4| < \epsilon$$
$$0 < |x - 2| < \delta \Rightarrow |(x - 2)(x + 2)| < \epsilon$$
$$0 < |x - 2| < \delta \Rightarrow |x - 2||x + 2| < \epsilon$$
$$0 < |x - 2| < \delta \Rightarrow |x - 2| < \frac{\epsilon}{|x + 2|}$$

One possibility: For $1 < x < 3$, $1 + 2 < x + 2 < 3 + 2 \Rightarrow 3 < x + 2 < 5$, so that $|x - 2| < \frac{\epsilon}{5}$. As $1 < x < 3 \Rightarrow 1 - 2 < x - 2 < 3 - 2 \Rightarrow -1 < x - 2 < 1 \Rightarrow |x - 2| < 1$. Take $\delta = \min(1, \frac{\epsilon}{5})$.

15. Given $\epsilon > 0$, we seek $\delta > 0$ so that
$$0 < |x - 3| < \delta \Rightarrow |f(x) - 8| < \epsilon$$
$$0 < |x - 3| < \delta \Rightarrow |x^2 - 1 - 8| < \epsilon$$
$$0 < |x - 3| < \delta \Rightarrow |x^2 - 9| < \epsilon$$
$$0 < |x - 3| < \delta \Rightarrow |(x - 3)(x + 3)| < \epsilon$$
$$0 < |x - 3| < \delta \Rightarrow |x - 3||x + 3| < \epsilon$$
$$0 < |x - 3| < \delta \Rightarrow |x - 3| < \frac{\epsilon}{|x + 3|}$$

One possibility: For $2 < x < 4$, $2 + 3 < x + 3 < 4 + 3 \Rightarrow 5 < x + 3 < 7$, so that $|x - 3| < \frac{\epsilon}{7}$. As $2 < x < 4 \Rightarrow 2 - 3 < x - 3 < 4 - 3 \Rightarrow -1 < x - 3 < 1 \Rightarrow |x - 3| < 1$. Take $\delta = \min(1, \frac{\epsilon}{7})$.

17. Given $\epsilon > 0$, we seek $\delta > 0$ so that
$$0 < |x - c| < \delta \Rightarrow |f(x) - d| < \epsilon$$
$$0 < |x - c| < \delta \Rightarrow |d - d| < \epsilon$$
$$0 < |x - c| < \delta \Rightarrow 0 < \epsilon$$

which is true for any $\delta > 0$. For example, take $\delta = 1$.

19.
$$\lim_{x \to 2} 3x^2 = 3 \lim_{x \to 2} x^2$$
$$= 3 (\lim_{x \to 2} x)(\lim_{x \to 2} x)$$
$$= 3 \cdot 2 \cdot 2$$
$$= 12.$$

21. $\lim\limits_{x \to -2} (2x + 1)^3 = \left[\lim\limits_{x \to -2} (2x + 1) \right]^3$
$$= \left[\lim_{x \to -2} 2x + \lim_{x \to -2} 1 \right]^3$$
$$= \left[2 \lim_{x \to -2} x + 1 \right]^3$$
$$= [2(-2) + 1]^3$$
$$= (-3)^3 = -27.$$

23. $\lim\limits_{x \to 3} (x^3 - 25)^3 = \left[\lim\limits_{x \to -2} (x^3 - 25) \right]^3$
$$= \left[\lim_{x \to 3} x^3 - \lim_{x \to 3} 25 \right]^3$$
$$= \left[\left(\lim_{x \to 3} x \right)^3 - 25 \right]^3$$
$$= (3^3 - 25)^3$$
$$= 2^3 = 8.$$

25. For $f(x)$ to be continuous at $x = 1$, $\lim\limits_{x \to 1} f(x)$ must equal $f(1)$: $f(1) = b + a + 3$ and $\lim\limits_{x \to 1^-} f(x) = \lim\limits_{x \to 1^-} (ax^2 - b) = a - b$, while $\lim\limits_{x \to 1^+} f(x) = \lim\limits_{x \to 1^+} (bx^3 + ax + 3) = b + a + 3$. Thus we need to have $a - b = b + a + 3 \Rightarrow 2b = -3 \Rightarrow b = -\frac{3}{2}$. Any value of a works.

27. For example: Let $g(x) = \begin{cases} x, & x \neq 1 \\ 2, & x = 1 \end{cases}$ and let $f(x) = \begin{cases} x, & x \neq 1 \\ \frac{1}{2}, & x = 1 \end{cases}$ Then $(gf)(x) = \begin{cases} x^2, & x \neq 1 \\ 1, & x = 1 \end{cases}$ is continuous at $x = 1$, as $\lim\limits_{x \to 1} (gf)(x) = \lim\limits_{x \to 1} x^2 = 1 = (gf)(1)$, but neither f nor g is continuous at $x = 1$.

29.   For example: Let $f(x) = \begin{cases} 1, & x \neq 0 \\ 0, & x = 0 \end{cases}$ and let $g(x) = \begin{cases} x, & x \neq 0 \\ 1, & x = 0 \end{cases}$ Then

$(g \circ f)(x) = \begin{cases} g(1), & x \neq 0 \\ g(0), & x = 0 \end{cases}$ that is, $(g \circ f)(x) = \begin{cases} 1, & x \neq 0 \\ 1, & x = 0 \end{cases}$ so that $(g \circ f)(x) = 1$ is continuous at $x = 0$, but neither f nor g is continuous at $x = 0$.

31.   Suppose that $L \neq M$. Let $\epsilon = \frac{|L-M|}{4}$ which is positive. As $\lim\limits_{x \to c} f(x) = L$ there exists a δ_1 such that if $0 < |x - c| < \delta_1$ then $|f(x) - L| < \epsilon$. Likewise, as $\lim\limits_{x \to c} f(x) = M$ there exists a δ_2 such that if $0 < |x - c| < \delta_2$ then $|f(x) - M| < \epsilon$. Let $\delta = \min(\delta_1, \delta_2)$. Then, if $0 < |x - c| < \delta$:

$$|L - M| = |L - M + f(x) - f(x)| = |[f(x) - M] + [L - f(x)]|$$
$$\leq |f(x) - M| + |L - f(x)| \quad \text{(by the Triangle Inequality)}$$
$$\leq |f(x) - M| + |f(x) - L|$$
$$< \epsilon + \epsilon = 2 \cdot \frac{|L - M|}{4} = \frac{|L - M|}{2}$$

a contradiction!

33.   It suffices to show that $\lim\limits_{x \to c} \dfrac{1}{g(x)} = \dfrac{1}{M}$, because applying Theorem 2.3(c) then yields

$$\lim_{x \to c} \frac{f(x)}{g(x)} = \left[\lim_{x \to c} f(x)\right] \cdot \left[\lim_{x \to c} \frac{1}{g(x)}\right] = L \cdot \frac{1}{M} = \frac{L}{M}$$

We need to show that for any $\epsilon > 0$ there is a $\delta > 0$ such that if $0 < |x - c| < \delta$ then $\left|\dfrac{1}{g(x)} - \dfrac{1}{M}\right| < \epsilon$, that is, $\left|\dfrac{M - g(x)}{Mg(x)}\right| < \epsilon$. Since $\lim\limits_{x \to c} g(x) = M$, there exists a $\delta_1 > 0$ such that if $0 < |x - c| < \delta_1$ then $|g(x) - M| < \dfrac{|M|}{2} \Rightarrow$

$$|M| = |M - g(x) + g(x)| \leq |M - g(x)| + |g(x)| < \frac{|M|}{2} + |g(x)|$$

which means that for $0 < |x - c| < \delta_1$ then $|g(x)| > \dfrac{|M|}{2} \Rightarrow \dfrac{1}{|g(x)|} < \dfrac{2}{|M|}$ so that

$\dfrac{1}{|Mg(x)|} = \dfrac{1}{|M||g(x)|} < \dfrac{1}{|M|} \cdot \dfrac{2}{|M|} = \dfrac{2}{M^2}$. Since $\lim\limits_{x \to c} g(x) = M$, there is a $\delta_2 > 0$ such that

if $0 < |x - c| < \delta_2$ then $|g(x) - M| < \dfrac{M^2}{2}\epsilon$. Let $\delta = \min(\delta_1, \delta_2)$. Then if $0 < |x - c| < \delta$,

$$\left|\frac{1}{g(x)} - \frac{1}{M}\right| = \left|\frac{M - g(x)}{Mg(x)}\right| < \frac{2}{M^2} \cdot \frac{M^2}{2}\epsilon = \epsilon.$$

35.   Given $\lim\limits_{x \to c} f(x) = f(c)$ and $\lim\limits_{x \to c} g(x) = g(c) \neq 0$. By Theorem 2.3(d) $\lim\limits_{x \to c} \dfrac{f(x)}{g(x)} = \dfrac{\lim\limits_{x \to c} f(x)}{\lim\limits_{x \to c} g(x)} = \dfrac{f(c)}{g(c)} = \left(\dfrac{f}{g}\right)(c).$

37.   Suppose $\lim\limits_{x \to c} f(x) = 0$. Given any $\epsilon > 0$ there is a $\delta > 0$ such that if $0 < |x - c| < \delta$ then $|f(x) - 0| < \epsilon$. But $|f(x) - 0| = |f(x)| = ||f(x)| - 0|$. Thus $||f(x)| - 0| < \epsilon$ for such x, which proves that. $\lim\limits_{x \to c} f(x) = 0 \Rightarrow \lim\limits_{x \to c} |f(x)| = 0$.

   Conversely, suppose $\lim\limits_{x \to c} |f(x)| = 0$. Given any $\epsilon > 0$ there is a $\delta > 0$ such that if $0 < |x - c| < \delta$ then $||f(x)| - 0| < \epsilon$. But $||f(x)| - 0| = |f(x)| = |f(x) - 0|$. Thus $|f(x) - 0| < \epsilon$ for such x, which proves that. $\lim\limits_{x \to c} |f(x)| = 0 \Rightarrow \lim\limits_{x \to c} f(x) = 0$.

39.   A rational function is a quotient of polynomials, and polynomials are continuous functions. From Theorem 2.4(d), a quotient of continuous functions is continuous where its denominator is nonzero, i.e. the rational function is continuous where it is defined (i.e. in its domain), and that is precisely what it means to say a rational function is continuous.

CHAPTER 3
The Derivative

§3.1 Tangent Lines and the Derivative

1. $f'(2) = \lim\limits_{h \to 0} \dfrac{f(2+h) - f(2)}{h} = \lim\limits_{h \to 0} \dfrac{4(2+h)^2 - 4(2)^2}{h} = \lim\limits_{h \to 0} \dfrac{4(4 + 4h + h^2) - 4(4)}{h}$

$= \lim\limits_{h \to 0} \dfrac{\cancel{16} + 16h + 4h^2 - \cancel{16}}{h} = \lim\limits_{h \to 0} \dfrac{\cancel{h}(16 + 4h)}{\cancel{h}} = \lim\limits_{h \to 0} 16 + 4h = 16.$

3. $f'(2) = \lim\limits_{h \to 0} \dfrac{f(2+h) - f(2)}{h} = \lim\limits_{h \to 0} \dfrac{-(2+h)^2 + 3(2+h) - 1 - [-2^2 + 3(2) - 1]}{h}$

$= \lim\limits_{h \to 0} \dfrac{-(4 + 4h + h^2) + 6 + 3h - 1 - [-4 + 6 - 1]}{h}$

$= \lim\limits_{h \to 0} \dfrac{-\cancel{4} - 4h - h^2 + \cancel{6} + 3h - \cancel{1} + \cancel{4} - \cancel{6} + \cancel{1}]}{h} = \lim\limits_{h \to 0} \dfrac{-h - h^2}{h}$

$= \lim\limits_{h \to 0} \dfrac{\cancel{h}(-1 - h)}{\cancel{h}} = \lim\limits_{h \to 0} -1 - h = -1.$

5. $f'(2) = \lim\limits_{h \to 0} \dfrac{f(2+h) - f(2)}{h}$

$= \lim\limits_{h \to 0} \dfrac{55 - 55}{h}$

$= \lim\limits_{h \to 0} \dfrac{0}{h}$

$= \lim\limits_{h \to 0} 0 = 0.$

7. Near $x = 2$, the function $f(x)$ is identical to the simpler function $x + 1$:

$f'(2) = \lim\limits_{h \to 0} \dfrac{f(2+h) - f(2)}{h}$

$= \lim\limits_{h \to 0} \dfrac{2 + h + 1 - (2 + 1)}{h}$

$= \lim\limits_{h \to 0} \dfrac{\cancel{3} + h - \cancel{3}}{h}$

$= \lim\limits_{h \to 0} \dfrac{\cancel{h}}{\cancel{h}} = \lim\limits_{h \to 0} 1 = 1.$

9. $f'(2) = \lim\limits_{h \to 0} \dfrac{f(2+h) - f(2)}{h}$

$= \lim\limits_{h \to 0} \dfrac{\sqrt{3(2+h) + 1} - \sqrt{3(2) + 1}}{h} = \lim\limits_{h \to 0} \dfrac{\sqrt{7 + 3h} - \sqrt{7}}{h}$

$= \lim\limits_{h \to 0} \dfrac{\sqrt{7 + 3h} - \sqrt{7}}{h} \cdot \dfrac{\sqrt{7 + 3h} + \sqrt{7}}{\sqrt{7 + 3h} + \sqrt{7}} = \lim\limits_{h \to 0} \dfrac{7 + 3h - 7}{h(\sqrt{7 + 3h} + \sqrt{7})}$

$= \lim\limits_{h \to 0} \dfrac{3\cancel{h}}{\cancel{h}(\sqrt{7 + 3h} + \sqrt{7})} = \lim\limits_{h \to 0} \dfrac{3}{\sqrt{7 + 3h} + \sqrt{7}} = \dfrac{3}{2\sqrt{7}}.$

11. $\left.\dfrac{dy}{dx}\right|_{x=2} = \lim\limits_{\Delta x \to 0} \dfrac{f(2 + \Delta x) - f(2)}{\Delta x}$

$= \lim\limits_{\Delta x \to 0} \dfrac{\dfrac{3(2 + \Delta x)}{2 + \Delta x + 1} - \dfrac{3(2)}{2 + 1}}{\Delta x}$

$= \lim\limits_{\Delta x \to 0} \dfrac{\dfrac{6 + 3\Delta x}{3 + \Delta x} - 2}{\Delta x}$

$= \lim\limits_{\Delta x \to 0} \dfrac{6 + 3\Delta x - 2(3 + \Delta x)}{\Delta x(3 + \Delta x)}$

$= \lim\limits_{\Delta x \to 0} \dfrac{\cancel{6} + 3\Delta x - \cancel{6} - 2\Delta x}{\Delta x(3 + \Delta x)}$

$= \lim\limits_{\Delta x \to 0} \dfrac{\cancel{\Delta x}}{\cancel{\Delta x}(3 + \Delta x)}$

$= \lim\limits_{\Delta x \to 0} \dfrac{1}{3 + \Delta x} = \dfrac{1}{3}.$

13. $f'(x) = \lim\limits_{h \to 0} \dfrac{f(x + h) - f(x)}{h}$

$= \lim\limits_{h \to 0} \dfrac{(x + h) - x}{h}$

$= \lim\limits_{h \to 0} \dfrac{h}{h}$

$= \lim\limits_{h \to 0} 1 = 1.$

15. $f'(x) = \lim\limits_{h \to 0} \dfrac{f(x + h) - f(x)}{h}$

$= \lim\limits_{h \to 0} \dfrac{3(x + h)^2 - 3x^2}{h}$

$= \lim\limits_{h \to 0} \dfrac{3(x^2 + 2xh + h^2) - 3x^2}{h}$

$= \lim\limits_{h \to 0} \dfrac{3\cancel{x^2} + 6xh + 3h^2 - 3\cancel{x^2}}{h}$

$= \lim\limits_{h \to 0} \dfrac{\cancel{h}(6x + 3h)}{\cancel{h}}$

$= \lim\limits_{h \to 0} 6x + 3h = 6x.$

17. $f'(x) = \lim\limits_{h \to 0} \dfrac{f(x + h) - f(x)}{h}$

$= \lim\limits_{h \to 0} \dfrac{[-2(x + h)^2 + (x + h) - 2] - (-2x^2 + x - 2)}{h}$

$= \lim\limits_{h \to 0} \dfrac{-2(x^2 + 2xh + h^2) + x + h - 2 + 2x^2 - x + 2}{h}$

$= \lim\limits_{h \to 0} \dfrac{-2\cancel{x^2} - 4xh - 2h^2 + \cancel{x} + h - \cancel{2} + 2\cancel{x^2} - \cancel{x} + \cancel{2}}{h}$

$= \lim\limits_{h \to 0} \dfrac{-4xh - 2h^2 + h}{h}$

$= \lim\limits_{h \to 0} \dfrac{\cancel{h}(-4x - 2h + 1)}{\cancel{h}}$

$= \lim\limits_{h \to 0} -4x - 2h + 1 = -4x + 1.$

19.

$$f'(x) = \lim_{h \to 0} \frac{f(x+h) - f(x)}{h}$$

$$= \lim_{h \to 0} \frac{\dfrac{2(x+h)+3}{x+h+1} - \dfrac{2x+3}{x+1}}{h}$$

$$= \lim_{h \to 0} \frac{(2x+2h+3)(x+1) - (2x+3)(x+h+1)}{(x+1)(x+h+1)h}$$

$$= \lim_{h \to 0} \frac{2x^2 + 2x + 2xh + 2h + 3x + 3 - (2x^2 + 2xh + 2x + 3x + 3h + 3)}{(x+1)(x+h+1)h}$$

$$= \lim_{h \to 0} \frac{-h}{(x+1)(x+h+1)\,h} = \lim_{h \to 0} -\frac{1}{(x+1)(x+h+1)} = \lim_{h \to 0} -\frac{1}{(x+1)^2}.$$

21.

$$f'(x) = \lim_{h \to 0} \frac{f(x+h) - f(x)}{h}$$

$$= \lim_{h \to 0} \frac{\sqrt{(x+h)+3} - \sqrt{x+3}}{h}$$

$$= \lim_{h \to 0} \frac{\sqrt{x+h+3} - \sqrt{x+3}}{h} \cdot \frac{\sqrt{x+h+3} + \sqrt{x+3}}{\sqrt{x+h+3} + \sqrt{x+3}}$$

$$= \lim_{h \to 0} \frac{x + h + 3 - (x + 3)}{h(\sqrt{x+h+3} + \sqrt{x+3})} = \lim_{h \to 0} \frac{h}{h(\sqrt{x+h+3} + \sqrt{x+3})}$$

$$= \lim_{h \to 0} \frac{1}{\sqrt{x+h+3} + \sqrt{x+3}} = \frac{1}{2\sqrt{x+3}}.$$

23.

$$f'(x) = \lim_{h \to 0} \frac{f(x+h) - f(x)}{h}$$

$$= \lim_{h \to 0} \frac{\dfrac{x+h}{(x+h)^2+1} - \dfrac{x}{x^2+1}}{h}$$

$$= \lim_{h \to 0} \frac{(x+h)(x^2+1) - x[(x+h)^2+1]}{[(x+h)^2+1](x^2+1)h}$$

$$= \lim_{h \to 0} \frac{x^3 + x + x^2 h + h - x[x^2 + 2xh + h^2 + 1]}{[(x+h)^2+1](x^2+1)h}$$

$$= \lim_{h \to 0} \frac{x^3 + x + x^2 h + h - x^3 - 2x^2 h - xh^2 - x}{[(x+h)^2+1](x^2+1)h}$$

$$= \lim_{h \to 0} \frac{-x^2 h + h - xh^2}{[(x+h)^2+1](x^2+1)h} = \lim_{h \to 0} \frac{h(-x^2+1-xh)}{[(x+h)^2+1](x^2+1)\,h}$$

$$= \lim_{h \to 0} \frac{-x^2+1-xh}{[(x+h)^2+1](x^2+1)} = \frac{1-x^2}{(x^2+1)^2}.$$

25. $\dfrac{dy}{dx} = \lim\limits_{\Delta x \to 0} \dfrac{f(x + \Delta x) - f(x)}{\Delta x}$

$= \lim\limits_{\Delta x \to 0} \dfrac{[-(x + \Delta x)^2 - (x + \Delta x)] - (-x^2 - x)}{\Delta x}$

$= \lim\limits_{\Delta x \to 0} \dfrac{-(x^2 + 2x\Delta x + \Delta x^2) - \cancel{x} - \Delta x + x^2 + \cancel{x}}{\Delta x}$

$= \lim\limits_{\Delta x \to 0} \dfrac{-\cancel{x^2} - 2x\Delta x - \Delta x^2 - \Delta x + \cancel{x^2}}{\Delta x}$

$= \lim\limits_{\Delta x \to 0} \dfrac{\cancel{\Delta x}(-2x - \Delta x - 1)}{\cancel{\Delta x}}$

$= \lim\limits_{\Delta x \to 0} \; -2x - \Delta x - 1 = -2x - 1.$

27. $\dfrac{dy}{dx} = \lim\limits_{\Delta x \to 0} \dfrac{f(x + \Delta x) - f(x)}{\Delta x}$

$= \lim\limits_{\Delta x \to 0} \dfrac{\dfrac{1}{\sqrt{2(x + \Delta x) + 3}} - \dfrac{1}{\sqrt{2x + 3}}}{\Delta x}$

$= \lim\limits_{\Delta x \to 0} \dfrac{\dfrac{1}{\sqrt{2x + 2\Delta x + 3}} - \dfrac{1}{\sqrt{2x + 3}}}{\Delta x}$

$= \lim\limits_{\Delta x \to 0} \dfrac{\sqrt{2x + 3} - \sqrt{2x + 2\Delta x + 3}}{\Delta x \sqrt{2x + 3}\,\sqrt{2x + 2\Delta x + 3}} \cdot \dfrac{\sqrt{2x + 3} + \sqrt{2x + 2\Delta x + 3}}{\sqrt{2x + 3} + \sqrt{2x + 2\Delta x + 3}}$

$= \lim\limits_{\Delta x \to 0} \dfrac{(\sqrt{2x + 3} - \sqrt{2x + 2\Delta x + 3})(\sqrt{2x + 3} + \sqrt{2x + 2\Delta x + 3})}{\Delta x \sqrt{2x + 3}\,\sqrt{2x + 2\Delta x + 3}\,(\sqrt{2x + 3} + \sqrt{2x + 2\Delta x + 3})}$

$= \lim\limits_{\Delta x \to 0} \dfrac{\cancel{2x} + \cancel{3} - (\cancel{2x} + 2\Delta x + \cancel{3})}{\Delta x \sqrt{2x + 3}\,\sqrt{2x + 2\Delta x + 3}\,(\sqrt{2x + 3} + \sqrt{2x + 2\Delta x + 3})}$

$= \lim\limits_{\Delta x \to 0} \; -\dfrac{2\cancel{\Delta x}}{\cancel{\Delta x}\sqrt{2x + 3}\,\sqrt{2x + 2\Delta x + 3}\,(\sqrt{2x + 3} + \sqrt{2x + 2\Delta x + 3})}$

$= \lim\limits_{\Delta x \to 0} \; -\dfrac{2}{\sqrt{2x + 3}\,\sqrt{2x + 2\Delta x + 3}\,(\sqrt{2x + 3} + \sqrt{2x + 2\Delta x + 3})}$

$= -\dfrac{\cancel{2}}{(2x + 3)(\cancel{2}\sqrt{2x + 3})} = -\dfrac{1}{(2x + 3)^{\frac{3}{2}}}.$

29. $f'(1) = \lim\limits_{h \to 0} \dfrac{f(1+h) - f(1)}{h}$

$= \lim\limits_{h \to 0} \dfrac{[(1+h)^2 + 2(1+h)] - (1^2 + 2 \cdot 1)}{h}$

$= \lim\limits_{h \to 0} \dfrac{\cancel{1} + 2h + h^2 + \cancel{2} + 2h - \cancel{3}}{h}$

$= \lim\limits_{h \to 0} \dfrac{h^2 + 4h}{h} = \lim\limits_{h \to 0} \dfrac{\cancel{h}(h+4)}{\cancel{h}}$

$= \lim\limits_{h \to 0} h + 4 = 4.$

As the slope of the tangent line at $x = 1$ was found to be 4, we know the equation of the tangent line is of the form $y = 4x + b$. When $x = 1, y = f(1) = 3 \Rightarrow 3 = 4(1) + b \Rightarrow b = -1 \Rightarrow y = 4x - 1$ is the equation of the tangent line.

31. The tangent line at $x = 4$ or $x = 7$ appears to be horizontal so its slope is 0. Since $f'(x)$ is the slope of the tangent line at x, both $f'(4)$ and $f'(7)$ are zero. The tangent line at $x = 2$ appears to rise about 1 unit as one moves 1 unit to the right, so its slope is about 1, i.e. $f'(2) \approx 1$.

33. (a) As the right-hand limit of f at $x = 3$ is different than the left-hand limit there, the limit does not exist at $x = 3$. The same is true at $x = 4$. (The limit exists at $x = 2$ since the right-hand and left-hand limits exist and are equal.)

(b) Since the limit does not exist at $x = 3$ or at $x = 4$, f is not continuous there. It is also discontinuous at $x = 2$, since the limit exists but is not the same as $f(2)$. The function is not defined at $x = 1$, so f is also discontinuous there. Thus f fails to be continuous at $x = 1, 2, 3$, and 4.

(c) The function can only be differentiable where it is continuous. Since the graph has no abrupt changes of direction in the intervals in which f is continuous, f is differentiable there as well. Thus f fails to be differentiable at $x = 1, 2, 3$, and 4.

35. The slope of the graph of f appears to be 1 from $x = 0$ to $x = 2$. Then the graph flattens out and becomes horizontal from $x = 5$ on, so the derivative decreases from 1 to 0 from $x = 5$ on.

37. There are many such functions. Here is one: To have no limit at 0, our function will have a jump discontinuity there. To have a limit at 1 but not be continuous there, our function's definition at 1 will be different than its limit as x approaches 1. To be continuous at 2 but not differentiable there, our function will have a sharp point at 2.

39. There are many such functions. Here is one: To have a positive derivative between 0 and 2 and a negative derivative between 2 and 4, where f is differentiable, the graph of f must be rising from 0 to 2, and falling from 2 to 4. As f is to be discontinuous at 1 (and hence not differentiable there), we will put a jump discontinuity at 1. For our function to be continuous at 2 but not differentiable there, we will put a sharp point at 2 where the slope is changing sign from positive to negative.

41. As f is continuous at $x = 2$, to show that f is not differentiable at $x = 2$, we show that the right-hand and left-hand limits of the difference quotient are unequal:

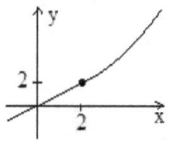

$$\lim_{h \to 0^-} \frac{f(2+h) - f(2)}{h}$$

$$= \lim_{h \to 0^-} \frac{(2+h) - (2)}{h}$$

$$= \lim_{h \to 0^-} \frac{h}{h} = \lim_{h \to 0^-} 1 = 1.$$

$$\lim_{h \to 0^+} \frac{f(2+h) - f(2)}{h}$$

$$= \lim_{h \to 0^+} \frac{[(2+h)^2 - 2] - (2^2 - 2)}{h}$$

$$= \lim_{h \to 0^+} \frac{4 + 4h + h^2 - 2 - 2}{h}$$

$$= \lim_{h \to 0^+} \frac{4h + h^2}{h} = \lim_{h \to 0^+} \frac{h(4+h)}{h}$$

$$= \lim_{h \to 0^+} 4 + h = 4 \neq 1.$$

43. To show that f is differentiable at $x = 1$, we show that f is continuous at 1, and that the right-hand and left-hand limits of the difference quotient are equal:

$\lim_{x \to 1^-} f(x) = \lim_{x \to 1^-} x^2 = 1^2 = 1.$
$\lim_{x \to 1^+} f(x) = \lim_{x \to 1^+} 2x - 1 = 2 - 1 = 1 \Rightarrow \lim_{x \to 1}; f(x) = 1$
and $f(1) = 1 \Rightarrow f$ is continuous at $x = 1$.

$$\lim_{h \to 0^-} \frac{f(1+h) - f(1)}{h}$$

$$= \lim_{h \to 0^-} \frac{(1+h)^2 - (1)^2}{h}$$

$$= \lim_{h \to 0^-} \frac{1 + 2h + h^2 - 1}{h}$$

$$= \lim_{h \to 0^-} \frac{h(2+h)}{h} = \lim_{h \to 0^-} 2 + h = 2.$$

$$\lim_{h \to 0^+} \frac{f(1+h) - f(1)}{h}$$

$$= \lim_{h \to 0^+} \frac{[2(1+h) - 1] - [2(1) - 1]}{h}$$

$$= \lim_{h \to 0^+} \frac{2 + 2h - 1 - 1}{h}$$

$$= \lim_{h \to 0^+} \frac{2h}{h} = \lim_{h \to 0^+} 2 = 2.$$

§3.2 Differentiation Formulas

1. $\begin{aligned} f'(x) &= (3x^5 + 4x^3 - 7)' \\ &= 3(x^5)' + 4(x^3)' - (7)' \\ &= 3(5x^4) + 4(3x^2) - 0 \\ &= 15x^4 + 12x^2. \end{aligned}$

3. $\begin{aligned} g'(x) &= (7x^3 + 5x^2 - 4x + x^{-4} + 1)' \\ &= 7(x^3)' + 5(x^2)' - 4(x)' + (x^{-4})' + (1)' \\ &= 7(3x^2) + 5(2x) - 4 - 4x^{-4-1} + 0 \\ &= 21x^2 + 10x - 4 - 4x^{-5} \\ &= 21x^2 + 10x - 4 - \frac{4}{x^5}. \end{aligned}$

5. $\begin{aligned} g'(x) &= (-x^7 + 2x^2 - x^{-1} - x^{-2} + 101)' \\ &= -(x^7)' + 2(x^2)' - (x^{-1})' - (x^{-2})' + (101)' \\ &= -(7x^6) + 2(2x) - (-1)x^{-1-1} - (-2)x^{-2-1} + 0 \\ &= -7x^6 + 4x + x^{-2} + 2x^{-3} \\ &= -7x^6 + 4x + \frac{1}{x^2} + \frac{2}{x^3}. \end{aligned}$

7. $\begin{aligned} h'(x) &= \left(\frac{x^5 + 3x^4 - 5x^2}{x^2} \right)' \\ &= (x^3 + 3x^2 - 5)' \\ &= (x^3)' + 3(x^2)' - (5)' \\ &= 3x^2 + 3(2x) + 0 \\ &= 3x^2 + 6x. \end{aligned}$

9. $f'(x) = (\sqrt{x} + 2)' = (x^{\frac{1}{2}} + 2)' = (x^{\frac{1}{2}})' + (2)' = \frac{1}{2}x^{\frac{1}{2}-1} + 0 = \frac{1}{2}x^{-\frac{1}{2}} = \frac{1}{2x^{\frac{1}{2}}} = \frac{1}{2\sqrt{x}}.$

11. $\begin{aligned} F'(x) &= \left(\frac{3x^2 + 2x - 5}{x + 4} \right)' = \frac{(x+4)(3x^2 + 2x - 5)' - (3x^2 + 2x - 5)(x+4)'}{(x+4)^2} \\ &= \frac{(x+4)(3 \cdot 2x + 2 + 0) - (3x^2 + 2x - 5)(1+0)}{(x+4)^2} \\ &= \frac{(x+4)(6x+2) - (3x^2 + 2x - 5)}{(x+4)^2} = \frac{6x^2 + 2x + 24x + 8 - 3x^2 - 2x + 5}{(x+4)^2} \\ &= \frac{3x^2 + 24x + 13}{(x+4)^2}. \end{aligned}$

13. $\begin{aligned} f'(x) &= \left(\frac{5}{3x^2 + 1} \right)' = \frac{(3x^2 + 1)(5)' - 5(3x^2 + 1)'}{(3x^2 + 1)^2} = \frac{(3x^2 + 1) \cdot 0 - 5[3(2x) + 0]}{(3x^2 + 1)^2} \\ &= -\frac{30x}{(3x^2 + 1)^2}. \end{aligned}$

15. $\begin{aligned} K'(x) &= [(4x^4 + 2x^3 + x^2)(x^3 + x + 1)]' \\ &= (4x^4 + 2x^3 + x^2)(x^3 + x + 1)' + (4x^4 + 2x^3 + x^2)'(x^3 + x + 1) \\ &= (4x^4 + 2x^3 + x^2)(3x^2 + 1) + (16x^3 + 6x^2 + 2x)(x^3 + x + 1) \\ &= 12x^6 + 4x^4 + 6x^5 + 2x^3 + 3x^4 + x^2 + 16x^6 + 16x^4 + 16x^3 + 6x^5 + 6x^3 \\ &\quad + 6x^2 + 2x^4 + 2x^2 + 2x \\ &= 28x^6 + 12x^5 + 25x^4 + 24x^3 + 9x^2 + 2x. \end{aligned}$

17. $h'(x) = \left(\dfrac{\sqrt{x}+1}{\sqrt{x}-1}\right)'$

$= \dfrac{(\sqrt{x}-1)(\sqrt{x}+1)' - (\sqrt{x}+1)(\sqrt{x}-1)'}{(\sqrt{x}-1)^2}$

$= \dfrac{(\sqrt{x}-1)(x^{\frac{1}{2}}+1)' - (\sqrt{x}+1)(x^{\frac{1}{2}}-1)'}{(\sqrt{x}-1)^2}$

$= \dfrac{(\sqrt{x}-1)(\frac{1}{2}x^{-\frac{1}{2}}) - (\sqrt{x}+1)(\frac{1}{2}x^{-\frac{1}{2}})}{(\sqrt{x}-1)^2}$

$= \dfrac{(\sqrt{x}-1) - (\sqrt{x}+1)}{2\sqrt{x}(\sqrt{x}-1)^2} = \dfrac{-\cancel{2}}{\cancel{2}\sqrt{x}(\sqrt{x}-1)^2}$

$= -\dfrac{1}{\sqrt{x}(\sqrt{x}-1)^2}.$

19. $f'(x) = (x^5 - 3x^2 - x + 1)'$

$= 5x^4 - 3(2x) - 1$

$f''(x) = 5(4x^3) - 6 = 20x^3 - 6$

$f''(2) = 20(2^3) - 6$

$= 160 - 6 = 154.$

21. $\dfrac{dy}{dx} = \dfrac{d}{dx}(2x^4 - x - 1)$

$= 2(4x^3) - 1 = 8x^3 - 1$

$\dfrac{d^2y}{dx^2} = \dfrac{d}{dx}(8x^3 - 1)$

$= 8(3x^2) = 24x^2$

$\dfrac{d^2y}{dx^2}\bigg|_{x=2} = 24(2^2) = 96.$

23. $[f(x) + g(x)]' = f'(x) + g'(x)$ at $x = 1$ is $f'(1) + g'(1) = 6 + 2 = 8.$

25. $\left[\dfrac{f(x)}{g(x)}\right]' = \dfrac{g(x)f'(x) - f(x)g'(x)}{[g(x)]^2}$ at $x = 1$ is $\dfrac{g(1)f'(1) - f(1)g'(1)}{(g(1))^2} = \dfrac{2(6) - 3(2)}{2^2} = \dfrac{3}{2}.$

27. $[f(x) + g(x) + h(x)]' = f'(x) + g'(x) + h'(x)$ at $x = 2$ is $f'(2) + g'(2) + h'(2) = 0 + 2 + 1 = 3.$

29. $[f(x) \cdot g(x) + h(x)]' = f'(x)g(x) + f(x)g'(x) + h'(x)$ at $x = 2$ is

$$f'(2)g(2) + f(2)g'(2) + h'(2) = 0(5) + 6(2) + 1 = 13.$$

31. $\left[\dfrac{f(x) \cdot g(x)}{h(x)}\right]' = \dfrac{h(x)[f(x)g(x)]' - f(x)g(x)h'(x)}{[h(x)]^2}$

$= \dfrac{h(x)[f'(x)g(x) + f(x)g'(x)] - f(x)g(x)h'(x)}{[h(x)]^2}$

at $x = 1$ is

$\dfrac{h(1)[f'(1)g(1) + f(1)g'(1)] - f(1)g(1)h'(1)}{[h(1)]^2} = \dfrac{6[6(2) + 3(2)] - 3(2)(1)}{6^2} = \dfrac{6(18) - 6}{36} = \dfrac{17}{6}.$

33. $\left[\dfrac{g(t)}{h(t)} + g(t)\right]' = \dfrac{h(t)g'(t) - g(t)h'(t)}{[h(t)]^2} + g'(t)$ at $t = g(1) = 2$ is

$\dfrac{h(2)g'(2) - g(2)h'(2)}{[h(2)]^2} + g'(2) = \dfrac{2(2) - 5(1)}{2^2} + 2 = \dfrac{-1}{4} + 2 = \dfrac{7}{4}.$

35. $f(x) = 3x^2 - x - 1 \Rightarrow f'(x) = 6x - 1$, so the slope of the tangent line at $x = 1$ is $f'(1) = 6(1) - 1 = 5$. When $x = 1, y = f(1) = 3 - 1 - 1 = 1$, so the equation of the tangent line is $y - 1 = 5(x - 1) \Rightarrow y = 5x - 4$.

37. $f(x) = \dfrac{x^5 + 2x}{x^4} = x + 2x^{-3} \Rightarrow f'(x) = 1 + 2(-3)x^{-4} = 1 - 6x^{-4}$, so the slope of the tangent line at $x = -1$ is $f'(-1) = 1 - 6(-1)^{-4} = 1 - 6 = -5$. When $x = -1, y = f(-1) = -1 + 2(-1)^{-3} = -1 - 2 = -3$, so the equation of the tangent line is $y - (-3) = -5[x - (-1)] \Rightarrow y = -5x - 8$.

39. $f(x) = \frac{2}{3}x^3 - \frac{1}{2}x^2 - x + 1 \Rightarrow f'(x) = \frac{2}{3}(3x^2) - \frac{1}{2}(2x) - 1 = 2x^2 - x - 1 = 0 \Rightarrow$ $(2x+1)(x-1) = 0 \Rightarrow x = -\frac{1}{2}$ or 1, and there are two points where the tangent is horizontal. When $x = -\frac{1}{2}$: $y = f(-\frac{1}{2}) = \frac{2}{3}(-\frac{1}{2})^3 - \frac{1}{2}(-\frac{1}{2})^2 - (-\frac{1}{2}) + 1 = \frac{2}{3}(-\frac{1}{8}) - \frac{1}{8} + \frac{1}{2} + 1$
$= -\frac{1}{12} - \frac{1}{8} + \frac{1}{2} + 1 = \frac{-2 - 3 + 12 + 24}{24} = \frac{31}{24} \Rightarrow (-\frac{1}{2}, \frac{31}{24})$.
When $x = 1, y = \frac{2}{3} - \frac{1}{2} - 1 + 1 = \frac{2}{3} - \frac{1}{2} = \frac{1}{6} \Rightarrow (1, \frac{1}{6})$.

41. $f(x) = x^3 - \frac{5}{2}x^2 + 1 \Rightarrow f'(x) = 3x^2 - \frac{5}{2}(2x) = 3x^2 - 5x$. Setting $f'(x) = 2 \Rightarrow$ $3x^2 - 5x = 2 \Rightarrow 3x^2 - 5x - 2 = 0 \Rightarrow (3x + 1)(x - 2) = 0 \Rightarrow x = -\frac{1}{3}$ or 2. When $x = -\frac{1}{3}, y = f(-\frac{1}{3}) = (-\frac{1}{3})^3 - \frac{5}{2}(-\frac{1}{3})^2 + 1 = -\frac{1}{27} - \frac{5}{18} + 1 = \frac{-2 - 15 + 54}{54} = \frac{37}{54} \Rightarrow (-\frac{1}{3}, \frac{37}{54})$. When $x = 2, y = 2^3 - \frac{5}{2}(2^2) + 1 = 8 - 10 + 1 = -1 \Rightarrow (2, -1)$.

43. $f(x) = x^3 + x^2 - 100 \Rightarrow f'(x) = 3x^2 + 2x$. Setting $f'(x) = -4 \Rightarrow 3x^2 + 2x = -4 \Rightarrow$ $3x^2 + 2x + 4 = 0$, whose discriminant, $b^2 - 4ac = 2^2 - 4(3)(4) < 0$. As the graph of $y = 3x^2 + 2x + 4$ is a parabola that opens upward, with no x-intercepts, the quadratic equation has no solutions, i.e. $f'(x)$ is never -4.

45. $f(x) = \sqrt{x} + 2 \Rightarrow f'(x) = (x^{\frac{1}{2}} + 2)' = \frac{1}{2}x^{-\frac{1}{2}} = \dfrac{1}{2\sqrt{x}}$. Since the slope of the tangent line at $x = c$ is $\dfrac{1}{2\sqrt{c}}$, a y-intercept $b = -4$ would result in the tangent line equation: $y = \dfrac{1}{2\sqrt{c}} x - 4$. Since the point $(c, \sqrt{c} + 2)$ of tangency must also be on the tangent line: $\sqrt{c} + 2 = \dfrac{1}{2\sqrt{c}} c - 4 \Rightarrow \sqrt{c} + 2 = \frac{1}{2}\sqrt{c} - 4 \Rightarrow \frac{1}{2}\sqrt{c} = -6$ which is impossible, as the value of a square root is always nonnegative.

47. $f(x) = \frac{1}{3}x^3 + x^2 + x \Rightarrow f'(x) = x^2 + 2x + 1$. Since the slope is to be 4: $f'(x) = 4 \Rightarrow x^2 + 2x + 1 = 4 \Rightarrow x^2 + 2x - 3 = 0 \Rightarrow (x + 3)(x - 1) = 0 \Rightarrow x = -3$ or 1. When $x = -3, y = f(-3) = \frac{1}{3}(-3)^3 + (-3)^2 - 3 = -9 + 9 - 3 = -3 \Rightarrow (-3, -3)$. When $x = 1, y = \frac{1}{3} + 1 + 1 = \frac{7}{3} \Rightarrow (1, \frac{7}{3})$. We have to determine which, if any, of these points lies on the tangent line $y = 4x + 9$:
Testing $(-3, -3)$: When $x = -3, 4x + 9 = 4(-3) + 9 = -12 + 9 = -3$ —yes.
Testing $(1, \frac{7}{3})$: When $x = 1, 4x + 9 = 4 + 9 = 13 \neq \frac{7}{3}$ —no! Only one such point: $(-3, -3)$.

49. $p(x) = ax^2 + bx + c \Rightarrow p(1) = a + b + c = -4$, and $p'(x) = 2ax + b \Rightarrow p'(1) = 2a + b = 11$. Lastly, $p''(x) = 2a \Rightarrow p''(1) = 2a = 6 \Rightarrow a = 3$. From $2a + b = 11$ and $a = 3$, then $2(3) + b = 11 \Rightarrow b = 5$. From $a + b + c = -4$ and $a = 3, b = 5$, then $3 + 5 + c = -4 \Rightarrow c = -12 \Rightarrow p(x) = 3x^2 + 5x - 12$.

51. $p(x) = ax^3 + bx^2 + cx + d \Rightarrow p'(x) = 3ax^2 + 2bx + c$, so the slope at $(1,0)$ is $p'(1) = 3a + 2b + c = 3$ (*), the slope of the tangent line $y = 3x - 3$ at that point. As $(1,0)$ lies on the graph of p : $p(1) = 0 \Rightarrow a + b + c + d = 0$ (**). Likewise at $(0,1)$: the slope of the tangent line $y = -2x + 1$ is -2, so $p'(0) = -2 \Rightarrow c = -2$. As $(0,1)$ is on the graph of p : $p(0) = 1 \Rightarrow d = 1$. From equation (*), $3a + 2b - 2 = 3 \Rightarrow 3a + 2b = 5$, and from equation (**), $a + b - 2 + 1 = 0 \Rightarrow a + b = 1 \Rightarrow a = 1 - b$. Substituting into $3a + 2b = 5$, then $3(1 - b) + 2b = 5 \Rightarrow 3 - b = 5 \Rightarrow b = -2$. Then $a = 1 - b = 1 - (-2) = 3$. Thus $p(x) = 3x^3 - 2x^2 - 2x + 1$.

53. $f(x) = \dfrac{2x^3 - x^2}{x} = 2x^2 - x$, for all $x \neq 0$, and we are interested in $x = 2$. The slope of the tangent there is $f'(2)$ with $f'(x) = 4x - 1 \Rightarrow f'(2) = 4(2) - 1 = 7$. Then the normal line there has a slope of $-\frac{1}{7}$. When $x = 2$, $y = f(2) = 2(2^2) - 2 = 6$. The normal line must also contain the point $(2,6)$, so $y = -\frac{1}{7}x + b \Rightarrow 6 = -\frac{1}{7}(2) + b \Rightarrow b = 6 + \frac{2}{7} = \frac{44}{7}$. The equation of the normal line is $y = -\frac{1}{7}x + \frac{44}{7}$.

55. $\left[\dfrac{f(x)}{g(x)}\right]' = \lim\limits_{h \to 0} \dfrac{\dfrac{f(x+h)}{g(x+h)} - \dfrac{f(x)}{g(x)}}{h} = \lim\limits_{h \to 0} \dfrac{g(x)f(x+h) - f(x)g(x+h)}{h[g(x)g(x+h)]}$

$= \lim\limits_{h \to 0} \dfrac{g(x)f(x+h) - g(x)f(x) + g(x)f(x) - f(x)g(x+h)}{h[g(x)g(x+h)]}$

$= \lim\limits_{h \to 0} \dfrac{g(x)[f(x+h) - f(x)] - f(x)[g(x+h) - g(x)]}{h[g(x)g(x+h)]}$

$= \lim\limits_{h \to 0} \dfrac{1}{g(x)g(x+h)} \left[g(x)\lim\limits_{h \to 0} \dfrac{f(x+h) - f(x)}{h} - f(x)\lim\limits_{h \to 0} \dfrac{g(x+h) - g(x)}{h}\right]$

$= \dfrac{1}{[g(x)]^2}[g(x)f'(x) - f(x)g'(x)] = \dfrac{g(x)f'(x) - f(x)g'(x)}{[g(x)]^2}$.

57. $(fgh)'(x) = \{[f(x)g(x)]h(x)\}' = [f(x)g(x)]h'(x) + [f(x)g(x)]'[h(x)]$

$= f(x)g(x)h'(x) + [f(x)g'(x) + f'(x)g(x)][h(x)]$

$= f(x)g(x)h'(x) + f(x)g'(x)h(x) + f'(x)g(x)h(x)$.

59. Let $P(n)$ be the proposition: $1 + 2 + 3 + \cdots + n = \frac{n(n+1)}{2}$.

  I. Case $n = 1$: $1 = \frac{1(1+1)}{2}$ —check!

  II. Assume $P(k)$ is true: i.e. $1 + 2 + 3 + \cdots + k = \frac{k(k+1)}{2}$.

  III. We have only to show that $P(k+1)$ is true: i.e. $1+2+3+\cdots+(k+1) = \frac{(k+1)(k+2)}{2}$.

$$[1 + 2 + 3 + \cdots + k] + (k+1) = \frac{k(k+1)}{2} + (k+1) = (k+1)\left(\frac{k}{2} + 1\right)$$

$$= (k+1)\left(\frac{k+2}{2}\right) = \frac{(k+1)(k+2)}{2}.$$

61. Let $P(n)$ be the proposition that if the functions $f_1, f_2, \ldots, f_n$ are differentiable, then so is their sum, and $[f_1(x) + f_2(x) + \cdots + f_n(x)]' = f_1'(x) + f_2'(x) + \cdots + f_n'(x)$.

  I. Theorem 3.2(d) assures us that the proposition is true for $n = 2$.

  II. Assume $P(k)$ is true: i.e. $[f_1(x) + f_2(x) + \cdots + f_k(x)]' = f_1'(x) + f_2'(x) + \cdots + f_k'(x)$.

  III. We have only to show that $P(k+1)$ is true: i.e.

$$[f_1(x) + f_2(x) + \cdots + f_{k+1}(x)]' = f_1'(x) + f_2'(x) + \cdots + f_{k+1}'(x).$$

  Let $h(x) = f_1(x) + f_2(x) + \cdots + f_k(x)$, then

$$[f_1(x) + f_2(x) + \cdots + f_{k+1}(x)]' = [h(x) + f_{k+1}(x)]' = h'(x) + f_{k+1}'(x) \quad \text{by I.}$$

$$= [f_1(x) + f_2(x) + \cdots + f_k(x)]' + f_{k+1}'(x)$$

$$= [f_1'(x) + f_2'(x) + \cdots + f_k'(x)] + f_{k+1}'(x) \quad \text{by II.}$$

$$= f_1'(x) + f_2'(x) + \cdots + f_{k+1}'(x).$$

63. Let $P(n)$ be the proposition that the nth derivative of x^n is $n!$, for any positive integer n.

  I. $P(1)$ is the proposition that the derivative of x is 1, which follows from Theorem 3.2(b).

  II. Assume $P(k)$ is true: i.e. $\dfrac{d^k}{dx^k}(x^k) = k!$

  III. We have only to show that $P(k+1)$ is true: i.e. $\dfrac{d^{k+1}}{dx^{k+1}}(x^{k+1}) = (k+1)!$

$$\frac{d^{k+1}}{dx^{k+1}}(x^{k+1}) = \frac{d^k}{dx^k}\left[\frac{d}{dx}(x^{k+1})\right] = \frac{d^k}{dx^k}[(k+1)x^k] \quad \text{then, by Theorem 3.2(c),}$$

$$= (k+1)\frac{d^k}{dx^k}(x^k) = (k+1)k! \quad \text{(by II.)} \quad = (k+1)!$$

§3.3 Derivatives of Trigonometric Functions and the Chain Rule

1. $f'(x) = [(x^2 + 3x - 10)^{15}]'$
$$= 15(x^2 + 3x - 10)^{14}(x^2 + 3x - 10)'$$
$$= 15(x^2 + 3x - 10)^{14}(2x + 3)$$
$$= 15(2x + 3)(x^2 + 3x - 10)^{14}.$$

3. $\dfrac{d}{dx}f(x) = \dfrac{d}{dx}\sqrt{x^3 + 2x} = \dfrac{d}{dx}(x^3 + 2x)^{\frac{1}{2}}$
$$= \frac{1}{2}(x^3 + 2x)^{\frac{1}{2}-1}\frac{d}{dx}(x^3 + 2x)$$
$$= \frac{1}{2}(x^3 + 2x)^{-\frac{1}{2}}(3x^2 + 2)$$
$$= \frac{3x^2 + 2}{2\sqrt{x^3 + 2x}}.$$

5. $f'(x) = \left(\dfrac{3x}{\sqrt{x+1}}\right)'$
$$= \frac{\sqrt{x+1}\cdot(3x)' - 3x(\sqrt{x+1})'}{(\sqrt{x+1})^2}$$
$$= \frac{3\sqrt{x+1} - 3x[\frac{1}{2}(x+1)^{\frac{1}{2}-1}](x+1)'}{x+1}$$
$$= \frac{3\sqrt{x+1} - 3x[\frac{1}{2}(x+1)^{-\frac{1}{2}}]\cdot 1}{x+1}$$
$$= \frac{3\sqrt{x+1} - \dfrac{3x}{2\sqrt{x+1}}}{x+1}$$
$$= \frac{6(x+1) - 3x}{2(x+1)^{\frac{1}{2}}(x+1)} = \frac{3x+6}{2(x+1)^{\frac{3}{2}}}.$$

7. $f'(x) = [\sin(2x^2 + 1)]'$
$$= \cos(2x^2 + 1)\cdot(2x^2 + 1)'$$
$$= \cos(2x^2 + 1)\cdot(4x)$$
$$= 4x\cos(2x^2 + 1).$$

9. $f'(x) = [\sin(\cos x)]'$
$$= \cos(\cos x)\cdot(\cos x)'$$
$$= \cos(\cos x)\cdot(-\sin x)$$
$$= -\sin x\cos(\cos x).$$

11. $f'(x) = [x\cos(\sin x)]'$
$$= x[\cos(\sin x)]' + (x)'\cos(\sin x)$$
$$= x[-\sin(\sin x)](\sin x)' + 1\cdot\cos(\sin x)$$
$$= -x\sin(\sin x)\cos x + \cos(\sin x)$$
$$= -x\cos x\sin(\sin x) + \cos(\sin x).$$

13. $f'(x) = \left(\dfrac{\sin^2 x}{\cos x}\right)'$
$$= \frac{\cos x(\sin^2 x)' - \sin^2 x(\cos x)'}{\cos^2 x}$$
$$= \frac{\cos x[2\sin x(\sin x)'] - \sin^2 x(-\sin x)}{\cos^2 x}$$
$$= \frac{\cos x[2\sin x\cos x] + \sin^3 x}{\cos^2 x}$$
$$= \frac{2\sin x\cos^2 x + \sin^3 x}{\cos^2 x}.$$

15. $f'(x) = \left\{\tan[\sin(x^2 + x - 1)]\right\}' = \sec^2[\sin(x^2 + x - 1)] \cdot [\sin(x^2 + x - 1)]'$

$\qquad = \sec^2[\sin(x^2 + x - 1)]\cos(x^2 + x - 1) \cdot (x^2 + x - 1)'$

$\qquad = \sec^2[\sin(x^2 + x - 1)]\cos(x^2 + x - 1) \cdot (2x + 1)$

$\qquad = (2x + 1)\cos(x^2 + x - 1)\sec^2[\sin(x^2 + x - 1)].$

17. $f'(x) = \left[\sqrt{\sec(2x+3)}\right]' = \left\{[\sec(2x+3)]^{\frac{1}{2}}\right\}' = \frac{1}{2}[\sec(2x+3)]^{-\frac{1}{2}}[\sec(2x+3)]'$

$\qquad = \frac{1}{2}[\sec(2x+3)]^{-\frac{1}{2}}\sec(2x+3)\tan(2x+3) \cdot (2x+3)'$

$\qquad = \frac{\not{2}}{\not{2}}[\sec(2x+3)]^{-\frac{1}{2}}\sec(2x+3)\tan(2x+3) \cdot \not{2}$

$\qquad = \sqrt{\sec(2x+3)}\tan(2x+3).$

19. $\dfrac{d}{dx}f(x) = \dfrac{d}{dx}\sin(\cos^2 x) = \cos(\cos^2 x) \cdot \dfrac{d}{dx}(\cos^2 x) = \cos(\cos^2 x)2\cos x \cdot \dfrac{d}{dx}\cos x$

$\qquad = \cos(\cos^2 x)2\cos x(-\sin x) = -2\sin x\cos x\cos(\cos^2 x).$

21. $f'(x) = [\cot^2(\cos x^2)]' = 2\cot(\cos x^2)[\cot(\cos x^2)]'$

$\qquad = 2\cot(\cos x^2)[-\csc^2(\cos x^2)](\cos x^2)' = -2\cot(\cos x^2)\csc^2(\cos x^2)(-\sin x^2)(x^2)'$

$\qquad = 2\cot(\cos x^2)\csc^2(\cos x^2)(\sin x^2)(2x) = 4x\sin x^2\cot(\cos x^2)\csc^2(\cos x^2).$

23. $f'(x) = [\csc(\cos x^2)]^{\frac{2}{3}}$

$\qquad = \frac{2}{3}[\csc(\cos x^2)]^{\frac{2}{3}-1}[\csc(\cos x^2)]'$

$\qquad = \frac{2}{3}[\csc(\cos x^2)]^{-\frac{1}{3}}[-\csc(\cos x^2)\cot(\cos x^2)] \cdot (\cos x^2)'$

$\qquad = -\frac{2}{3}[\csc(\cos x^2)]^{-\frac{1}{3}+1}\cot(\cos x^2)(-\sin x^2) \cdot (x^2)'$

$\qquad = \frac{2}{3}[\csc(\cos x^2)]^{\frac{2}{3}}\cot(\cos x^2)(\sin x^2) \cdot 2x$

$\qquad = \frac{4}{3}x\sin x^2\cot(\cos x^2)[\csc(\cos x^2)]^{\frac{2}{3}}.$

25. We are to find $f'(2)$ ·

$\qquad f'(x) = \left[\dfrac{1}{(3x-5)^3}\right]' = [(3x-5)^{-3}]' = -3(3x-5)^{-3-1} \cdot (3x-5)'$

$\qquad\qquad = -3(3x-5)^{-4}(3) = -\dfrac{9}{(3x-5)^4}$

$\qquad f'(2) = -\dfrac{9}{[3(2)-5]^4} = -\dfrac{9}{1} = -9.$

27. $f'(x) = (3x^2 + x)' = 6x + 1$, and

$$g'(x) = \left(\frac{1}{x+1}\right)' = [(x+1)^{-1}]' = -(x+1)^{-2} \cdot 1 = -\frac{1}{(x+1)^2}.$$

(i) With the chain rule:

$$(g \circ f)'(x) = g'[f(x)]f'(x) = -\frac{1}{[(3x^2+x)+1]^2} \cdot (6x+1) = -\frac{6x+1}{(3x^2+x+1)^2}.$$

(ii) Without the chain rule:

$$(g \circ f) = g(3x^2 + x) = \frac{1}{3x^2+x+1} \Rightarrow$$

$$(g \circ f)'(x) = \frac{(3x^2+x+1)(1)' - 1(3x^2+x+1)'}{(3x^2+x+1)^2} = \frac{0-(6x+1)}{(3x^2+x+1)^2} = -\frac{6x+1}{(3x^2+x+1)^2}.$$

29. $(g \circ f)'(0) = g'[f(0)]f'(0) = g'(1) \cdot 2 = 2(2) = 4.$

31. $(g \circ g)'(0) = g'[g(0)]g'(0) = g'(2) \cdot 1 = 2(1) = 2.$

33. $(f \circ f)'(1) = f'[f(1)]f'(1) = f'(3) \cdot 6 = 3(6) = 18.$

35. $f'(x) = \left[\left(\frac{x}{2x+5}\right)^2\right]' = 2\left(\frac{x}{2x+5}\right) \cdot \left(\frac{x}{2x+5}\right)' = \left(\frac{2x}{2x+5}\right) \cdot \left[\frac{(2x+5)(x)' - x(2x+5)'}{(2x+5)^2}\right]$

$$= \left(\frac{2x}{2x+5}\right) \cdot \left[\frac{2x+5-x(2)}{(2x+5)^2}\right] = \left(\frac{2x}{2x+5}\right) \cdot \left[\frac{5}{(2x+5)^2}\right] = \frac{10x}{(2x+5)^3} \Rightarrow$$

$f'(-2) = \frac{10(-2)}{[2(-2)+5]^3} = \frac{-20}{1} = -20 \Rightarrow$ the equation of the tangent line at $x = -2$ is

$y = -20x + b$. When $x = -2$, $y = f(-2) = \left(\frac{-2}{2(-2)+5}\right)^2 = 4 \Rightarrow 4 = -20(-2) + b \Rightarrow$

$b = -36 \Rightarrow y = -20x - 36$ is the equation of the tangent line at $x = -2$.

37. $f'(x) = (\sin^2 x)' = 2(\sin x)(\sin x)' = 2 \sin x \cos x \Rightarrow$

$f'(\frac{\pi}{2}) = 2 \sin \frac{\pi}{2} \cos \frac{\pi}{2} = 2 \cdot 1 \cdot 0 = 0 \Rightarrow$ the equation of the tangent line at $x = \frac{\pi}{2}$ is $y = b$.

When $x = \frac{\pi}{2}$, $y = \sin^2 \frac{\pi}{2} = 1^2 = 1 \Rightarrow y = 1$ is the equation of the tangent line at $x = \frac{\pi}{2}$.

39. The tangent is horizontal when its slope is 0, so we set $f'(x) = 0$:

$0 = f'(x) = (\sin x + \sqrt{3} \cos x)' = \cos x + \sqrt{3}(-\sin x) \Rightarrow \sqrt{3} \sin x = \cos x \Rightarrow$

$\frac{1}{\sqrt{3}} = \frac{\sin x}{\cos x} = \tan x \Rightarrow x = \frac{\pi}{6}$ or $x = \frac{7\pi}{6}$, as $0 \le x < 2\pi$.

41. $f'(x) = \left[\frac{1}{(2x-3)^2}\right]' = [(2x-3)^{-2}]' = -2(2x-3)^{-3}(2x-3)' = -2(2x-3)^{-3}(2)$

$= -\frac{4}{(2x-3)^3}.$ As the tangent line has a slope of -4, $-\frac{4}{(2x-3)^3} = -4 \Rightarrow$

$(2x-3)^3 = 1 \Rightarrow 2x - 3 = 1 \Rightarrow x = 2 \Rightarrow y = f(2) = 1.$ As the line $y = -4x + 9$

also contains the point $(2,1)$, that line is tangent to the graph of f at $x = 2$. The point of tangency is $(2,1)$.

43. $f'(x) = (x^2 \sin x)' = (x^2)' \sin x + x^2 (\sin x)' = 2x \sin x + x^2 \cos x \implies$

   $f'(\frac{\pi}{2}) = 2(\frac{\pi}{2}) \sin \frac{\pi}{2} + (\frac{\pi}{2})^2 \cos \frac{\pi}{2} = \pi(1) + (\frac{\pi}{2})^2(0) = \pi$, so the slope of the normal line at

   $x = \frac{\pi}{2}$ is $-\frac{1}{\pi}$. Since $f(\frac{\pi}{2}) = (\frac{\pi}{2})^2 \sin \frac{\pi}{2} = (\frac{\pi}{2})^2$, the equation of the normal line is: $y = -\frac{1}{\pi}x + b$,

   with $(\frac{\pi}{2})^2 = -(\frac{1}{\pi})(\frac{\pi}{2}) + b \implies b = (\frac{\pi}{2})^2 + \frac{1}{2} \implies y = -\frac{1}{\pi}x + (\frac{\pi}{2})^2 + \frac{1}{2}$, i.e. $y = -\frac{1}{\pi}x + \frac{\pi^2 + 2}{4}$.

45. $\lim\limits_{x \to 2} [-(x-2)^2] = -(2-2)^2 = 0$, so by the Pinching Theorem, $\lim\limits_{h \to 0} f(x) = 0$.

47. Since $-1 \le \sin t \le 1$, then for $x \ne 0, -1 \le \sin \frac{1}{x} \le 1$. For $x > 0$, $-x \le x \sin \frac{1}{x} \le x$, and for $x < 0$, $-x \ge x \sin \frac{1}{x} \ge x$. As $\lim\limits_{x \to 0} x = \lim\limits_{x \to 0} (-x) = 0$, by the Pinching Theorem, $\lim\limits_{x \to 0^+} x \sin \frac{1}{x} = \lim\limits_{x \to 0^-} x \sin \frac{1}{x} = 0 \implies \lim\limits_{x \to 0} x \sin \frac{1}{x}$.

49. Since $-1 \le \sin t \le 1$, then for $x \ne 1, -1 \le \sin \frac{100}{x-1} \le 1$. Then $-(x-1)^2 \le (x-1)^2 \sin \frac{100}{x-1} \le (x-1)^2$, for $x \ne 1$. As $\lim\limits_{x \to 1} -(x-1)^2 = \lim\limits_{x \to 1} (x-1)^2 = 0$, by the Pinching Theorem, $\lim\limits_{x \to 1} (x-1)^2 \sin \frac{100}{x-1} = 0$.

51. As $\lim\limits_{x \to 0} \dfrac{\sin x}{x} = 1$, $\lim\limits_{x \to 0} \dfrac{3 \sin x}{x} = 3 \lim\limits_{x \to 0} \dfrac{\sin x}{x} = 3 \cdot 1 = 3$.

53. As $\lim\limits_{x \to 0} \dfrac{\sin cx}{cx} = 1$, $1 = \lim\limits_{x \to 0} \dfrac{\sin 3x}{3x} = \lim\limits_{x \to 0} (\frac{1}{3} \cdot \dfrac{\sin 3x}{x}) = \frac{1}{3} \lim\limits_{x \to 0} \dfrac{\sin 3x}{x} \implies$

   $\lim\limits_{x \to 0} \dfrac{\sin 3x}{x} = 3$.

55. As $\lim\limits_{x \to 0} \dfrac{\sin cx}{cx} = 1$, $1 = \lim\limits_{x \to 0} \dfrac{\sin 7x}{7x} = \lim\limits_{x \to 0} (\frac{1}{7} \cdot \dfrac{\sin 7x}{x}) = \frac{1}{7} \lim\limits_{x \to 0} \dfrac{\sin 7x}{x} \implies$

   $\frac{1}{3} = \frac{1}{3} \cdot \frac{1}{7} \lim\limits_{x \to 0} (\dfrac{\sin 7x}{x}) = \frac{1}{7} \lim\limits_{x \to 0} (\frac{1}{3} \cdot \dfrac{\sin 7x}{x}) = \frac{1}{7} \lim\limits_{x \to 0} (\dfrac{\sin 7x}{3x}) \implies \lim\limits_{x \to 0} (\dfrac{\sin 7x}{3x}) - \frac{7}{3}$.

57. $\lim\limits_{x \to 0} \dfrac{\sin^2(2x)}{x^2} = \lim\limits_{x \to 0} (\dfrac{\sin 2x}{x})^2 = \lim\limits_{x \to 0} [\dfrac{\sin 2x}{x} \cdot \dfrac{\sin 2x}{x}] = [\lim\limits_{x \to 0} \dfrac{\sin 2x}{x}] \cdot [\lim\limits_{x \to 0} \dfrac{\sin 2x}{x}]$

   $= [2 \cdot \frac{1}{2} \lim\limits_{x \to 0} \dfrac{\sin 2x}{x}][2 \cdot \frac{1}{2} \lim\limits_{x \to 0} \dfrac{\sin 2x}{x}] = [2 \lim\limits_{x \to 0} \dfrac{\sin 2x}{2x}][2 \lim\limits_{x \to 0} \dfrac{\sin 2x}{2x}]$

   $= [2(1)][2(1)] = 4$.

59. By CheckYourUnderstanding 2.11: If f is continuous at b and if $\lim\limits_{x \to a} g(x) = b$, then $\lim\limits_{x \to a} f(g(x)) = f[\lim\limits_{x \to a} g(x)]$.

$$\text{As } \lim_{x \to 0} \frac{\cos x - 1}{x} = 0, \text{ then } \lim_{x \to 0} \cos\left[\frac{\cos x - 1}{x}\right] = \cos 0 = 1.$$

61. As $\lim\limits_{x \to 0} \dfrac{\sin cx}{cx} = 1$, $1 = \lim\limits_{x \to 0} \dfrac{\sin \pi x}{\pi x} = \lim\limits_{x \to 0}\left(\dfrac{1}{\pi} \cdot \dfrac{\sin \pi x}{x}\right) = \dfrac{1}{\pi}\lim\limits_{x \to 0}\dfrac{\sin \pi x}{x} \Rightarrow$

$\lim\limits_{x \to 0} \dfrac{\sin \pi x}{x} = \pi$. By CheckYourUnderstanding 2.11: If f is continuous at b and if

$\lim\limits_{x \to a} g(x) = b$, then $\lim\limits_{x \to a} f(g(x)) = f[\lim\limits_{x \to a} g(x)]$. Therefore $\lim\limits_{x \to 0} \sin\left(\dfrac{\sin \pi x}{x}\right) = \sin \pi = 0$.

63. $\dfrac{d}{dr}r_e = \dfrac{d}{dr}\left[\left(1 + \dfrac{r}{12}\right)^{12} - 1\right] = 12\left(1 + \dfrac{r}{12}\right)^{11}\dfrac{d}{dr}\left(1 + \dfrac{r}{12}\right) = \cancel{12}\left(1 + \dfrac{r}{12}\right)^{11} \cdot \dfrac{1}{\cancel{12}} = \left(1 + \dfrac{r}{12}\right)^{11}$.

65. $(h \circ g \circ f)'(x) = [(h \circ g) \circ f]'(x) = (h \circ g)'[f(x)] \cdot f'(x) = h'(g[f(x)])g'[f(x)] \cdot f'(x)$
$\qquad = h'[(g \circ f)(x)] \cdot g'[f(x)] \cdot f'(x)$.

§3.4 Implicit Differentiation

1. At $(1,1)$: $y = \sqrt{x} = x^{\frac{1}{2}}$
$y' = \frac{1}{2}x^{-\frac{1}{2}} = \dfrac{1}{2\sqrt{x}}$
When $x = 1$, $y' = \dfrac{1}{2\sqrt{1}} = \dfrac{1}{2}$
At $(1,-1)$:
$\quad y = -\sqrt{x} = -x^{\frac{1}{2}}$
$\quad y' = -\frac{1}{2}x^{-\frac{1}{2}} = -\dfrac{1}{2\sqrt{x}}$
When $x = 1$, $y' = -\dfrac{1}{2\sqrt{1}} = -\dfrac{1}{2}$

Differentiating $x = y^2$:

$1 = 2yy' \Rightarrow y' = \dfrac{1}{2y}$
At $(1,1)$: $y' = \dfrac{1}{2(1)} = \dfrac{1}{2}$
At $(1,-1)$: $y' = \dfrac{1}{2(-1)} = -\dfrac{1}{2}$

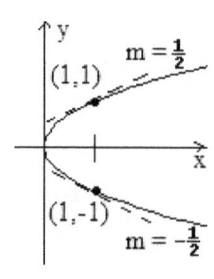

3. At $\left(\frac{1}{2}, \frac{\sqrt{3}}{2}\right)$: $y = \sqrt{1 - x^2} = (1 - x^2)^{\frac{1}{2}}$
$y' = \frac{1}{2}(1 - x^2)^{-\frac{1}{2}}(-2x) = -\dfrac{x}{\sqrt{1 - x^2}}$
When $x = \frac{1}{2}$, $y' = -\dfrac{\frac{1}{2}}{\sqrt{1 - \left(\frac{1}{2}\right)^2}} = -\dfrac{1}{\sqrt{3}}$
At $\left(\frac{1}{2}, -\frac{\sqrt{3}}{2}\right)$: $y = -\sqrt{1 - x^2} = -(1 - x^2)^{\frac{1}{2}}$
$y' = -\frac{1}{2}(1 - x^2)^{-\frac{1}{2}}(-2x) = \dfrac{x}{\sqrt{1 - x^2}}$
When $x = \frac{1}{2}$, $y' = \dfrac{\frac{1}{2}}{\sqrt{1 - \left(\frac{1}{2}\right)^2}} = \dfrac{1}{\sqrt{3}}$

Differentiating $x^2 + y^2 = 1$:

$2x + 2yy' = 0 \Rightarrow y' = -\dfrac{x}{y}$
At $\left(\frac{1}{2}, \frac{\sqrt{3}}{2}\right)$:
$y' = -\dfrac{\frac{1}{2}}{\frac{\sqrt{3}}{2}} = -\dfrac{1}{\sqrt{3}}$
At $\left(\frac{1}{2}, -\frac{\sqrt{3}}{2}\right)$:
$y' = -\dfrac{\frac{1}{2}}{-\frac{\sqrt{3}}{2}} = \dfrac{1}{\sqrt{3}}$

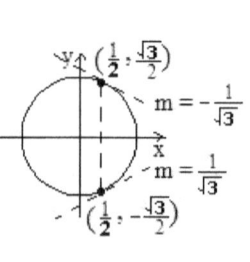

5. At $(1, \frac{\sqrt{3}}{2})$:
$$y = \sqrt{1 - \frac{x^2}{4}} = (1 - \frac{x^2}{4})^{\frac{1}{2}}$$
$$y' = \frac{1}{2}(1 - \frac{x^2}{4})^{-\frac{1}{2}}(-\frac{2x}{4}) = -\frac{x}{4\sqrt{1 - \frac{x^2}{4}}}$$

When $x = 1$, $y' = -\frac{1}{4\sqrt{1 - \frac{1^2}{4}}} = -\frac{1}{2\sqrt{3}}$.

At $(1, -\frac{\sqrt{3}}{2})$:
$$y = -\sqrt{1 - \frac{x^2}{4}} = -(1 - \frac{x^2}{4})^{\frac{1}{2}}$$
$$y' = -\frac{1}{2}(1 - \frac{x^2}{4})^{-\frac{1}{2}}(-\frac{2x}{4}) = \frac{x}{4\sqrt{1 - \frac{x^2}{4}}}$$

When $x = 1$, $y' = \frac{1}{4\sqrt{1 - \frac{1^2}{4}}} = \frac{1}{2\sqrt{3}}$.

Differentiating $\frac{x^2}{4} + y^2 = 1$:
$$\frac{1}{2}x + 2yy' = 0 \Rightarrow y' = -\frac{x}{4y}.$$
At $(1, \frac{\sqrt{3}}{2})$, $y' = -\frac{1}{4(\frac{\sqrt{3}}{2})} = -\frac{1}{2\sqrt{3}}$.
At $(1, -\frac{\sqrt{3}}{2})$, $y' = -\frac{1}{4(\frac{-\sqrt{3}}{2})} = \frac{1}{2\sqrt{3}}$.

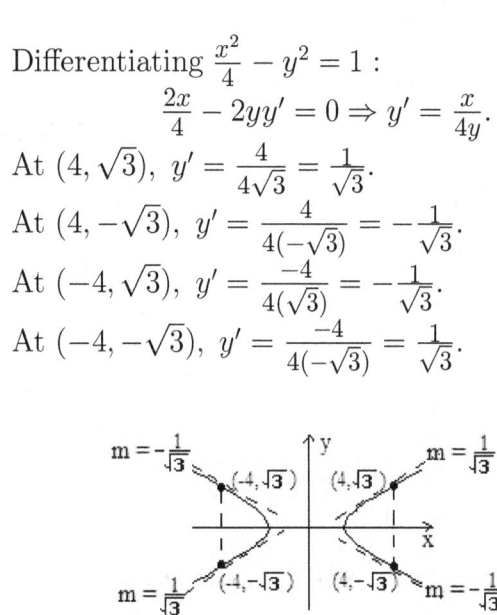

7. At $(\pm 4, \sqrt{3})$:
$$y = \sqrt{\frac{x^2}{4} - 1} = (\frac{x^2}{4} - 1)^{\frac{1}{2}}$$
$$y' = \frac{1}{2}(\frac{x^2}{4} - 1)^{-\frac{1}{2}}(\frac{2x}{4}) = \frac{x}{4\sqrt{\frac{x^2}{4} - 1}}$$

When $x = 4$, $y' = \frac{4}{4\sqrt{\frac{4^2}{4} - 1}} = \frac{1}{\sqrt{3}}$.

When $x = -4$, $y' = \frac{-4}{4\sqrt{\frac{(-4)^2}{4} - 1}} = -\frac{1}{\sqrt{3}}$.

At $(\pm 4, -\sqrt{3})$:
$$y = -\sqrt{\frac{x^2}{4} - 1} = -(\frac{x^2}{4} - 1)^{\frac{1}{2}}$$
$$y' = -\frac{1}{2}(\frac{x^2}{4} - 1)^{-\frac{1}{2}}(\frac{2x}{4}) = -\frac{x}{4\sqrt{\frac{x^2}{4} - 1}}$$

When $x = 4$, $y' = -\frac{4}{4\sqrt{\frac{4^2}{4} - 1)}} = -\frac{1}{\sqrt{3}}$.

When $x = -4$, $y' = -\frac{-4}{4\sqrt{\frac{4^2}{4} - 1)}} = \frac{1}{\sqrt{3}}$.

Differentiating $\frac{x^2}{4} - y^2 = 1$:
$$\frac{2x}{4} - 2yy' = 0 \Rightarrow y' = \frac{x}{4y}.$$
At $(4, \sqrt{3})$, $y' = \frac{4}{4\sqrt{3}} = \frac{1}{\sqrt{3}}$.
At $(4, -\sqrt{3})$, $y' = \frac{4}{4(-\sqrt{3})} = -\frac{1}{\sqrt{3}}$.
At $(-4, \sqrt{3})$, $y' = \frac{-4}{4(\sqrt{3})} = -\frac{1}{\sqrt{3}}$.
At $(-4, -\sqrt{3})$, $y' = \frac{-4}{4(-\sqrt{3})} = \frac{1}{\sqrt{3}}$.

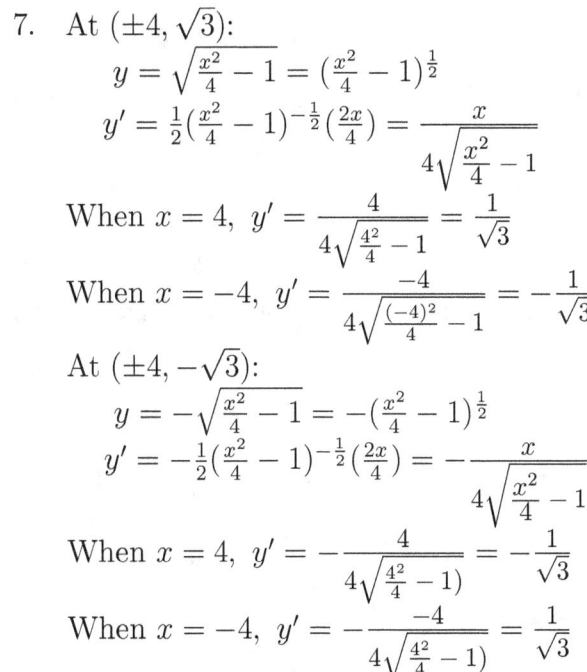

9. Differentiating $x^2y^2 = 4$: $2xy^2 + x^2(2yy') = 0 \Rightarrow y + xy' = 0 \Rightarrow y' = -\frac{y}{x}$. At $(1,2)$, $y' = -\frac{2}{1} = -2$. The equation of the tangent line: $y = -2x + b$ with $2 = -2(1) + b \Rightarrow b = 4 \Rightarrow y = -2x + 4$.

11. Differentiating $x^2 + xy^2 - y = 4$: $2x + x(2yy') + y^2 - y' = 0$ $\Rightarrow$
$(2xy - 1)y' = -2x - y^2$ $\Rightarrow$ $y' = \dfrac{-2x - y^2}{2xy - 1}$. At $(2,0)$, $y' = \dfrac{-4}{-1} = 4$. The equation of the
tangent line: $y = 4x + b$ with $0 = 4(2) + b$ $\Rightarrow$ $b = -8$ $\Rightarrow$ $y = 4x - 8$.

13. Differentiating $x^3 + y^3 = 6xy$: $3x^2 + 3y^2y' = 6[xy' + y]$ $\Rightarrow$ $x^2 + y^2y' = 2xy' + 2y$ $\Rightarrow$
$(2x - y^2)y' = x^2 - 2y$ $\Rightarrow$ $y' = \dfrac{x^2 - 2y}{2x - y^2}$. At $(3,3)$, $y' = \dfrac{9 - 6}{6 - 9} = -1$. The equation of the
tangent line: $y = -x + b$ with $3 = -3 + b$ $\Rightarrow$ $b = 6$ $\Rightarrow$ $y = -x + 6$.

15. Differentiating $y^4 - xy = x^2 - 1$: $4y^3y' - [xy' + y] = 2x$ $\Rightarrow$ $4y^3y' - xy' - y = 2x$ $\Rightarrow$
$(4y^3 - x)y' = 2x + y$ $\Rightarrow$ $y' = \dfrac{2x + y}{4y^3 - x}$. At $(1,0)$, $y' = \dfrac{2}{-1} = -2$. The equation of the tangent
line: $y = -2x + b$ with $0 = -2 + b$ $\Rightarrow$ $b = 2$ $\Rightarrow$ $y = -2x + 2$.

17. Differentiating $x^2 + y^3 = 2y + 3$: $2x + 3y^2y' = 2y'$ $\Rightarrow$ $(2 - 3y^2)y' = 2x$ $\Rightarrow$
$y' = \dfrac{2x}{2 - 3y^2}$. At $(2,1)$, $y' = \dfrac{4}{-1} = -4$. The equation of the tangent line: $y = -4x + b$ with
$1 = -8 + b$ $\Rightarrow$ $b = 9$ $\Rightarrow$ $y = -4x + 9$.

19. Differentiating $x + \cos y = xy$: $1 - (\sin y)y' = xy' + y$ $\Rightarrow$ $(x + \sin y)y' = 1 - y$ $\Rightarrow$
$y' = \dfrac{1 - y}{x + \sin y}$. At $(0, \frac{3\pi}{2})$, $y' = \dfrac{1 - \frac{3\pi}{2}}{-1} = \dfrac{3\pi - 2}{2}$. The equation of the tangent line:
$y = (\frac{3\pi - 2}{2})x + b$ with $\frac{3\pi}{2} = b$ $\Rightarrow$ $y = (\frac{3\pi - 2}{2})x + \frac{3\pi}{2}$.

21. Differentiating $x\cos^2 y = \sin y$: $\cos^2 y + x[2(\cos y)(-\sin y)y'] = (\cos y)y'$ $\Rightarrow$
$\cos y \cos y - 2x(\sin y \cos y)y' = (\cos y)y'$ $\Rightarrow$ $(2x \sin y + 1)y' = \cos y$ $\Rightarrow$ $y' = \dfrac{\cos y}{2x \sin y + 1}$. At
$(0,0)$, $y' = \frac{1}{1} = 1$. The equation of the tangent line: $y = x + b$ with $0 = 0 + b$ $\Rightarrow$ $y = x$.

23. Differentiating $x\sin y - y\cos 2x = 2x$:
$x(\cos y)y' + \sin y - y'\cos 2x - y(-\sin 2x)(2) = 2$ $\Rightarrow$ $y'[x\cos y - \cos 2x] = 2 - \sin y - 2y\sin 2x$ $\Rightarrow$
$y' = \dfrac{2 - \sin y - 2y\sin 2x}{x\cos y - \cos 2x}$. At $(\frac{\pi}{2}, \pi)$, $y' = \dfrac{2 - 0 - 0}{-\frac{\pi}{2} - (-1)} = \dfrac{4}{2 - \pi}$. The equation of the tangent
line: $y = \left(\dfrac{4}{2 - \pi}\right)x + b$ with $\pi = \left(\dfrac{4}{2 - \pi}\right)\left(\frac{\pi}{2}\right) + b$ $\Rightarrow$ $b = \pi - \dfrac{2\pi}{2 - \pi} = \pi[\dfrac{2 - \pi - 2}{2 - \pi}] = \dfrac{\pi^2}{\pi - 2}$ $\Rightarrow$
$y = \left(\dfrac{4}{2 - \pi}\right)x + \dfrac{\pi^2}{\pi - 2}$.

25.   A horizontal tangent line occurs where $y' = 0$. Differentiating $x^2 + y^3 - 3y - 2 = 0$:
$2x + 3y^2 y' - 3y' = 0 \Rightarrow (3 - 3y^2)y' = 2x \Rightarrow y' = \frac{2x}{3 - 3y^2}$. Setting $y' = 0$: $\frac{2x}{3 - 3y^2} = 0 \Rightarrow x = 0$
and $y \neq \pm 1$. To find the corresponding values of y for $x = 0$, we have to solve the equation
$y^3 - 3y - 2 = 0$. Realizing that $y = -1$ is a solution, and then dividing the cubic polynomial
by $y + 1$ yields: $(y + 1)(y^2 - y - 2) = 0 \Rightarrow (y + 1)[(y - 2)(y + 1)] = 0 \Rightarrow y = 2$ (as $y \neq \pm 1$).
There is only one point: $(0, 2)$.

27.   Differentiating $\sqrt{x^2 + y} = y^2 - 7x \Rightarrow (x^2 + y)^{\frac{1}{2}} = y^2 - 7x$:

$$\frac{1}{2}(x^2 + y)^{-\frac{1}{2}}(2x + y') = 2yy' - 7 \Rightarrow \frac{x}{\sqrt{x^2 + y}} + \left[\frac{1}{2\sqrt{x^2 + y}}\right]y' = 2yy' - 7 \Rightarrow$$

$$(2y - \frac{1}{2\sqrt{x^2 + y}})y' = \frac{x}{\sqrt{x^2 + y}} + 7 \Rightarrow y' = \frac{\frac{x}{\sqrt{x^2 + y}} + 7}{2y - \frac{1}{2\sqrt{x^2 + y}}}. \text{ At } (1,3), \ y' = \frac{\frac{1}{2} + 7}{6 - \frac{1}{4}} \cdot \frac{4}{4} = \frac{30}{23}.$$

Then the normal line has a slope of $-\frac{23}{30}$. The equation of the normal line: $y = -\frac{23}{30}x + b$
with $3 = -\frac{23}{30} + b \Rightarrow b = 3 + \frac{23}{30} = \frac{113}{30} \Rightarrow y = -\frac{23}{30}x + \frac{113}{30}$.

29.   Differentiating $xy^2 = yx^2 + 3x - 2y : x\left(2y\frac{dy}{dx}\right) + y^2 = y(2x) + (x^2)\frac{dy}{dx} + 3 - 2\frac{dy}{dx} \Rightarrow$

$$(2xy - x^2 + 2)\frac{dy}{dx} = 2xy + 3 - y^2 \Rightarrow \frac{dy}{dx} = \frac{2xy + 3 - y^2}{2xy - x^2 + 2}.$$

31.   Differentiating $y^2 = \frac{xy}{x + 1}$:

$$2y\frac{dy}{dx} = \frac{(x + 1)\frac{d}{dx}(xy) - xy\frac{d}{dx}(x + 1)}{(x + 1)^2} = \frac{(x + 1)(x\frac{dy}{dx} + y) - (xy)(1)}{(x + 1)^2} \Rightarrow$$

$$2y\frac{dy}{dx}(x + 1)^2 = x(x + 1)\frac{dy}{dx} + xy + y - xy \Rightarrow \left[2y(x + 1)^2 - x(x + 1)\right]\frac{dy}{dx} = y \Rightarrow$$

$$[2y(x^2 + 2x + 1) - x^2 - x]\frac{dy}{dx} = y \Rightarrow \left[2x^2y + 4xy + 2y - x^2 - x\right]\frac{dy}{dx} = y \Rightarrow$$

$$\frac{dy}{dx} = \frac{y}{2x^2y + 4xy + 2y - x^2 - x}.$$

33.   Differentiating $y^2 \sin y = x + y : y^2(\cos y)\frac{dy}{dx} + \left(2y\frac{dy}{dx}\right)\sin y = 1 + \frac{dy}{dx}$

$$(y^2 \cos y + 2y \sin y - 1)\frac{dy}{dx} = 1 \Rightarrow \frac{dy}{dx} = \frac{1}{y^2 \cos y + 2y \sin y - 1}.$$

35. Differentiating $x^2 + y^2 = 25$: $2x + 2yy' = 0 \Rightarrow y' = -\dfrac{x}{y} \Rightarrow y'' = -\dfrac{y \cdot 1 - xy'}{y^2}$

At $(3, -4)$, $y' = -\dfrac{3}{-4} = \dfrac{3}{4}$ so that $y'' = -\dfrac{-4 - 3(\frac{3}{4})}{(-4)^2} = \dfrac{25}{64}.$

37.   Differentiating $\sqrt{x} + \sqrt{y} = 1 \Rightarrow x^{\frac{1}{2}} + y^{\frac{1}{2}} = 1 : \frac{1}{2}x^{-\frac{1}{2}} + \frac{1}{2}y^{-\frac{1}{2}}y' = 0 \Rightarrow$

$y' = -\dfrac{x^{-\frac{1}{2}}}{y^{-\frac{1}{2}}} = -\dfrac{\sqrt{y}}{\sqrt{x}} \Rightarrow y'' = -\dfrac{\sqrt{x}\left(\frac{1}{2}y^{-\frac{1}{2}}y'\right) - \sqrt{y}\left(\frac{1}{2}x^{-\frac{1}{2}}\right)}{x} \cdot \dfrac{2\sqrt{x}\sqrt{y}}{2\sqrt{x}\sqrt{y}} = -\dfrac{xy' - y}{2x^{\frac{3}{2}}\sqrt{y}}.$

At $(\frac{1}{4}, \frac{1}{4})$, $y' = -\dfrac{\sqrt{\frac{1}{4}}}{\sqrt{\frac{1}{4}}} = -\dfrac{\frac{1}{2}}{\frac{1}{2}} = -1$, so that $y'' = -\dfrac{\frac{1}{4} \cdot (-1) - \frac{1}{4}}{2(\frac{1}{4})^{\frac{3}{2}}\sqrt{\frac{1}{4}}} = -\dfrac{-\frac{1}{2}}{\frac{1}{8}} = 4.$

39.   Differentiating $x^3 - y^3 = 1 : 3x^2 - 3y^2y' = 0 \Rightarrow y' = \dfrac{x^2}{y^2} \Rightarrow y'' = \dfrac{y^2(2x) - x^2(2yy')}{y^4}$

$\Rightarrow y'' = \dfrac{2xy^2 - 2x^2y(\frac{x^2}{y^2})}{y^4} \cdot \dfrac{y}{y} = \dfrac{2xy^3 - 2x^4}{y^5}.$

41.   Differentiating $xy^3 = 12 : y^3 + x(3y^2y') = 0 \Rightarrow y' = -\dfrac{y^3}{3xy^2} = -\dfrac{y}{3x} \Rightarrow$

$y'' = -\dfrac{3xy' - y(3)}{9x^2} = -\dfrac{3x(-\frac{y}{3x}) - 3y}{9x^2} = \dfrac{4y}{9x^2}.$

43.   First we find the points of intersection of the two curves, and then show that the slopes of the curves at each of those points are negative reciprocals of each other: $xy = 2 \Rightarrow y = \frac{2}{x}$, substituting this into the equation of the other curve, $x^2 - \left(\frac{2}{x}\right)^2 = 3 \Rightarrow x^4 - 3x^2 - 4 = 0 \Rightarrow (x^2 - 4)(x^2 + 1) = 0 \Rightarrow x^2 - 4 = 0 \Rightarrow x = \pm 2$. When $x = 2, y = 1$ and when $x = -2, y = -1$. There are two points of intersection, $(2, 1)$ and $(-2, -1)$.

Differentiating to find the slope of $xy = 2 : xy' + y = 0 \Rightarrow y' = -\frac{y}{x}$. Differentiating to find the slope of $x^2 - y^2 = 3 : 2x - 2yy' = 0 \Rightarrow y' = \frac{x}{y}$. We see that the slopes are negative reciprocals at every point (x, y) and therefore at their two points of intersection, $(2,1)$ and $(-2, -1)$.

45.   First we find the points of intersection of the two curves, and then show that the slopes of the curves at each of those points are negative reciprocals of each other: $2x + 3y = 0 \Rightarrow y = -\frac{2}{3}x$, substituting this into the equation of the other curve,

$x^2 + \left(-\frac{2}{3}x\right)^2 = 4 \Rightarrow \frac{13}{9}x^2 = 4 \Rightarrow x^2 = \frac{36}{13} \Rightarrow x = \pm\dfrac{6}{\sqrt{13}}.$

When $x = \dfrac{6}{\sqrt{13}}$, $y = -\frac{2}{3} \cdot \dfrac{6}{\sqrt{13}} = -\dfrac{4}{\sqrt{13}}$. When $x = -\dfrac{6}{\sqrt{13}}$, $y = -\frac{2}{3}\left(-\dfrac{6}{\sqrt{13}}\right) = \dfrac{4}{\sqrt{13}}.$

There are two points of intersection, $\left(\frac{6}{\sqrt{13}}, -\frac{4}{\sqrt{13}}\right)$ and $\left(-\frac{6}{\sqrt{13}}, \frac{4}{\sqrt{13}}\right)$.

Differentiating to find the slope of $x^2 + y^2 = 4$: $2x + 2yy' = 0 \Rightarrow y' = -\frac{x}{y}$. The slope of the line $2x + 3y = 0 \Rightarrow y = -\frac{2}{3}x$ is $-\frac{2}{3}$ everywhere.

At $\left(\frac{6}{\sqrt{13}}, -\frac{4}{\sqrt{13}}\right)$, $y' = -\frac{x}{y} = -\frac{\frac{6}{\sqrt{13}}}{-\frac{4}{\sqrt{13}}} = \frac{6}{4} = \frac{3}{2}$ which is the negative reciprocal of $-\frac{2}{3}$.

At $\left(-\frac{6}{\sqrt{13}}, \frac{4}{\sqrt{13}}\right)$, $y' = -\frac{x}{y} = -\frac{\frac{-6}{\sqrt{13}}}{\frac{4}{\sqrt{13}}} = \frac{6}{4} = \frac{3}{2}$ which is the negative reciprocal of $-\frac{2}{3}$.

47. The slope of the radius drawn to the point (x, y) on the circle $(x - x_0)^2 + (y - y_0)^2 = r^2$ is the slope of the line joining (x_0, y_0) and (x, y), namely $m_r = \frac{y - y_0}{x - x_0}$. To find the slope of the tangent to the circle, differentiate the equation $(x - x_0)^2 + (y - y_0)^2 = r^2$:
$$2(x - x_0) + 2(y - y_0)(y - y_0)' = 0 \Rightarrow 2(x - x_0) + 2(y - y_0)y' = 0 \Rightarrow y' = -\frac{x - x_0}{y - y_0}$$
which is the negative reciprocal of m_r.

§3.5 Related Rates

1. Given $\frac{dx}{dt} = 1$ cm/min.

(a) Since $V = x^3$, then $\frac{dV}{dt} = 3x^2\frac{dx}{dt}$. When $x = 50$, $\frac{dV}{dt} = 3(50)^2(1) = 7500$ cm^3/min.

(b) Let S denote the surface area of the cube. Then $S = 6x^2 \Rightarrow \frac{dS}{dt} = 12x\frac{dx}{dt}$.

When $x = 50$, $\frac{dS}{dt} = 12(50)(1) = 600$ cm^2/min.

(c) When $S = 6x^2 = 2400$, $x^2 = 400 \Rightarrow x = 20$. Then
$$\frac{dV}{dt} = 3x^2\frac{dx}{dt} = 3(20)^2(1) = 1200 \text{ cm}^3/\text{min}.$$

3. Given $\frac{dr}{dt} = -1$ cm/min.

(a) Since $V = \frac{4}{3}\pi r^3$, then $\frac{dV}{dt} = \frac{4}{3}\pi(3r^2)\frac{dr}{dt} = 4\pi r^2\frac{dr}{dt}$. When $r = 50$,
$$\frac{dV}{dt} = 4\pi(50)^2(-1) = -10,000\pi \text{ cm}^3/\text{min}.$$

(b) As $S = 4\pi r^2 \Rightarrow \frac{dS}{dt} = 4\pi(2r)\frac{dr}{dt} = 8\pi r\frac{dr}{dt}$.

When $r = 50$, $\frac{dS}{dt} = 8\pi(50)(-1) = -400\pi$ cm^2/min.

(c) When $S = 4\pi r^2 = 1600\pi$, $\frac{dV}{dt} = 4\pi r^2\frac{dr}{dt} = 1600\pi(-1) = -1600\pi$ cm^3/min.

5. Given $\dfrac{dr}{dt} = -4$ ft/min. and $\dfrac{dh}{dt} = 2$ ft/min.

Since $V = \pi r^2 h$, then $\dfrac{dV}{dt} = \pi\left[(2r)\dfrac{dr}{dt}h + r^2\dfrac{dh}{dt}\right]$. When $r = 2$ and $h = 3$,

$$\dfrac{dV}{dt} = \pi[2(2)(-4)(3) + 2^2(2)] = \pi(-48 + 8) = -40\pi \text{ ft}^3/\text{min.}$$

7. (a) Given information about the diameter of the circle, we express the area in terms of the

diameter D : $A = \pi r^2 \Rightarrow A = \pi\left(\dfrac{D}{2}\right)^2 \Rightarrow A = \dfrac{\pi}{4}D^2$. Then $\dfrac{dA}{dt} = \dfrac{\pi}{4}(2D)\dfrac{dD}{dt} = \dfrac{\pi}{2}(D)\dfrac{dD}{dt}$.

When the diameter $D = 6$ and $\dfrac{dD}{dt} = 4$, $\dfrac{dA}{dt} = \dfrac{\pi}{2}(6)(4) = 12\pi$ ft^2/sec.

(b) As $\dfrac{dA}{dt} = \dfrac{\pi}{2}(D)\dfrac{dD}{dt}$, when $D = 3$ and $\dfrac{dA}{dt} = 30$, $30 = \dfrac{\pi}{2}(3)\dfrac{dD}{dt} \Rightarrow \dfrac{dD}{dt} = \dfrac{20}{\pi}$ ft/sec.

9. Given $\dfrac{dl}{dt} = 25$ and $\dfrac{dw}{dt} = -c$, with $c > 0$. To find c such that $\dfrac{dA}{dt} = 250$ when $l = 25$

and $w = 20$. Differentiating $A = lw$: $\dfrac{dA}{dt} = l\dfrac{dw}{dt} + w\dfrac{dl}{dt}$. Then $250 = 25(-c) + 20(25) \Rightarrow$

$10 = -c + 20 \Rightarrow c = 10$.

11. Since the perimeter is constant at 42, $2l + 2w = 42 \Rightarrow l + w = 21 \Rightarrow w = 21 - l$.

Then $A = lw = l(21 - l) = 21l - l^2 \Rightarrow \dfrac{dA}{dt} = 21\dfrac{dl}{dt} - 2l\dfrac{dl}{dt} = (21 - 2l)\dfrac{dl}{dt}$. Given that $\dfrac{dl}{dt} = 2$,

when the rectangle is a square: $4l = 42 \Rightarrow l = \dfrac{21}{2} \Rightarrow \dfrac{dA}{dt} = \left[21 - 2\left(\dfrac{21}{2}\right)\right](2) = 0$ in.^{2}/sec.

13. Let x be the length of a side of the triangle (see figure), then the area is

$A = \frac{1}{2}xh = \left[\frac{1}{2}x\right]\left[\frac{\sqrt{3}}{2}x\right] = \dfrac{\sqrt{3}}{4}x^2$. Given $\dfrac{dx}{dt} = 1$ in./min.

(a) Differentiating $A = \dfrac{\sqrt{3}}{4}x^2$: $\dfrac{dA}{dt} = \dfrac{\sqrt{3}}{4}(2x)\dfrac{dx}{dt} = \dfrac{\sqrt{3}}{2}x\dfrac{dx}{dt}$.

When $x = 1$, $\dfrac{dA}{dt} = \dfrac{\sqrt{3}}{2}(1)(1) = \dfrac{\sqrt{3}}{2}$ in.^{2}/min.

(b) The perimeter $P = 3x \Rightarrow \dfrac{dP}{dt} = 3\dfrac{dx}{dt} = 3(1) = 3$ in./min.

(c) Since the sides are all increasing at the same rate, the angles remain 60°, so the
rate of change is 0 radians/min.

15. (See figure) Given $\dfrac{dx}{dt} = 1$ ft/sec, we are to find $\dfrac{d\theta}{dt}$ when the top of the

ladder is 9 ft from the ground. As $\cos\theta = \dfrac{x}{12}$, differentiating: $-\sin\theta\dfrac{d\theta}{dt} = \dfrac{1}{12}\dfrac{dx}{dt}$.

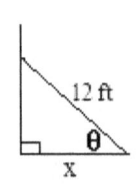

When the top of the ladder is 9 ft from the ground, $\sin\theta = \dfrac{9}{12}$ so that

$-\dfrac{9}{12}\dfrac{d\theta}{dt} = \dfrac{1}{12}(1) \Rightarrow \dfrac{d\theta}{dt} = -\dfrac{1}{9}$ radians/sec.

17. (See figure) Given $\frac{dx}{dt} = 3$ in./min. and $\frac{dh}{dt} = -3$ in./min. The area

$A = \frac{1}{2}xh \;\Rightarrow\; \frac{dA}{dt} = \frac{1}{2}\left[x\frac{dh}{dt} + h\frac{dx}{dt}\right] = \frac{1}{2}(-3x + 3h) = \frac{3}{2}(h - x).$

   (a) When $x = 7$ and $h = 6$, $\quad \frac{dA}{dt} = \frac{3}{2}(6 - 7) = -\frac{3}{2}$ in.^{2}/min.

   (b) When $x = 6$ and $h = 7$, $\quad \frac{dA}{dt} = \frac{3}{2}(7 - 6) = \frac{3}{2}$ in.^{2}/min.

19. (See figure) $\tan\frac{\theta}{2} = \frac{12}{h} \;\Rightarrow\; \sec^2\frac{\theta}{2}\left[\frac{d}{dt}\left(\frac{\theta}{2}\right)\right] = -12h^{-2}\frac{dh}{dt} \;\Rightarrow$

$\sec^2\frac{\theta}{2}(\frac{1}{2})\frac{d\theta}{dt} = -\left(\frac{12}{h^2}\right)\frac{dh}{dt}.$ Given $\frac{dh}{dt} = 1$ in./min. When $h = 12$, the

left-hand right triangle is an isosceles right triangle, so $\frac{\theta}{2} = \frac{\pi}{4} \;\Rightarrow$

$\sec^2\frac{\pi}{4}(\frac{1}{2})\frac{d\theta}{dt} = -\frac{12}{12^2}(1) \;\Rightarrow\; (\sqrt{2})^2(\frac{1}{2})\frac{d\theta}{dt} = -\frac{1}{12} \;\Rightarrow\; \frac{d\theta}{dt} = -\frac{1}{12}$ radians/min.

21. Given $V = \frac{4}{3}\pi r^3$ and $S = 4\pi r^2$, and $\frac{dV}{dt} = -kS$, with $k > 0$. Since $\frac{dV}{dt} = \frac{d}{dt}\left(\frac{4}{3}\pi r^3\right) = $

$\frac{4}{3}\pi(3r^2)\frac{dr}{dt} = 4\pi r^2\frac{dr}{dt},$ we have $4\pi r^2\frac{dr}{dt} = -k(4\pi r^2) \;\Rightarrow\; \frac{dr}{dt} = -k < 0.$

23. (a) As $h = 2r \;\Rightarrow\; r = \frac{h}{2},\; V = \frac{1}{3}\pi r^2 h = \frac{1}{3}\pi\left(\frac{h}{2}\right)^2 h = \frac{\pi}{12}h^3.$

      Then $\frac{dV}{dt} = \frac{\pi}{12}(3h^2)\frac{dh}{dt} = \frac{\pi}{4}(h^2)\frac{dh}{dt}.$ When $h = 2$, as $\frac{dV}{dt} = 2$ ft^3/min.,

          $2 = \frac{\pi}{4}(2^2)\frac{dh}{dt} \;\Rightarrow\; \frac{dh}{dt} = \frac{2}{\pi}$ ft/min.

   (b) From (a), $\frac{dh}{dt} = \frac{2}{\pi}$ ft/min., and since $h = 2r \;\Rightarrow\; \frac{dh}{dt} = 2\frac{dr}{dt},\; \frac{dr}{dt} = \frac{1}{\pi}$ ft/min.

   (c) $V = 35$ ft$^3 \;\Rightarrow\; 35 = \frac{\pi}{12}h^3 \;\Rightarrow\; h^3 = \frac{12(35)}{\pi} \;\Rightarrow\; h = \left(\frac{420}{\pi}\right)^{\frac{1}{3}}.$

      Then from $\frac{dV}{dt} = \frac{\pi}{4}(h^2)\frac{dh}{dt},\; 2 = \frac{\pi}{4}\left(\frac{420}{\pi}\right)^{\frac{2}{3}}\frac{dh}{dt} \;\Rightarrow\; \frac{dh}{dt} = \frac{8}{\pi}\left(\frac{\pi}{420}\right)^{\frac{2}{3}} = \frac{8}{\pi^{\frac{1}{3}}(420)^{\frac{2}{3}}}$ ft/min.

       $\frac{dr}{dt} = \frac{1}{2}\frac{dh}{dt} = \frac{4}{\pi^{\frac{1}{3}}(420)^{\frac{2}{3}}}$ ft/min.

25. As $PV = k$, a constant, and $1000 \cdot 1 = k \;\Rightarrow\; k = 1000$, so $PV = 1000$. When

$V = 800, P = \frac{1000}{800} = \frac{5}{4}.$ Differentiating $PV = 1000:$ $P\frac{dV}{dt} + V\frac{dP}{dt} = 0.$ When $\frac{dV}{dt} = -4,$

then $\frac{5}{4}(-4) + 800\frac{dP}{dt} = 0 \;\Rightarrow\; \frac{dP}{dt} = \frac{5}{800} = \frac{1}{160}$ lb/in.^{2}/min.

27. (See figure) Let $t = 0$ at 1 PM. $D = \sqrt{x^2 + y^2} = (x^2 + y^2)^{\frac{1}{2}} \Rightarrow$

$$\frac{dD}{dt} = \frac{1}{2}(x^2 + y^2)^{-\frac{1}{2}}\left(2x\frac{dx}{dt} + 2y\frac{dy}{dt}\right) = \frac{x\frac{dx}{dt} + y\frac{dy}{dt}}{\sqrt{x^2 + y^2}}.$$

Also, $\dfrac{dx}{dt} = -250$ and $\dfrac{dy}{dt} = 300$ for $0 < x < 1000$.

(a) At 1:08 PM, $t = 8$: $x = 1000 - (8-5)(250) = 1000 - 750 = 250$, $y = 8(300) = 2400 \Rightarrow$

$$\frac{dD}{dt} = \frac{250(-250) + 2400(300)}{\sqrt{250^2 + 2400^2}} = \frac{657,500}{\sqrt{5,822,500}} \approx 272 \text{ ft/min.}$$

(b) (See figure) At 1:10 PM, $t = 10$:
$x = (10 - 5)(250) - 1000 = 1250 - 1000 = 250$,
$y = 10(300) = 3,000$.

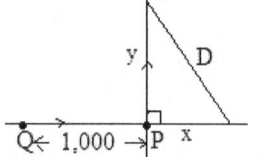

Also, $\dfrac{dx}{dt} = 250$ and $\dfrac{dy}{dt} = 300$.

$$\frac{dD}{dt} = \frac{250(250) + 3000(300)}{\sqrt{250^2 + 3000^2}} = \frac{962,500}{\sqrt{9,062,500}} \approx 320 \text{ ft/min.}$$

29. (See figure) Suppose water is being pumped into the tank at a rate of $k > 0$ in.^{3}/sec, and let V be the volume of water in the tank, when the height of water is h and the radius there is r. As the two right triangles depicted in the figure are similar: $\dfrac{h}{r} = \dfrac{120}{10} = 12 \Rightarrow r = \dfrac{h}{12}$.

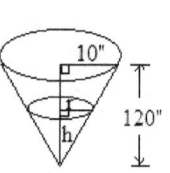

Then $V = \dfrac{1}{3}\pi r^2 h = \dfrac{\pi}{3}\left(\dfrac{h}{12}\right)^2 h = \dfrac{\pi h^3}{3(12)^2} = \dfrac{\pi}{432}h^3$, and $\dfrac{dV}{dt} = \dfrac{\pi}{432}(3h^2)\dfrac{dh}{dt} = \dfrac{\pi}{144}h^2\dfrac{dh}{dt}$

(a) When $h = 40$ and $\dfrac{dh}{dt} = 6$, $\dfrac{dV}{dt} = \dfrac{\pi}{144}(40)^2(6) = \dfrac{200\pi}{3}$. Since water is leaking out at 3 in.^{3}/sec., and flowing into the tank at k in.^{3}/sec., given the water level is rising, $k > 3 \Rightarrow \dfrac{dV}{dt} = k - 3 \Rightarrow k - 3 = \dfrac{200\pi}{3} \Rightarrow k = \left(\dfrac{200\pi}{3} + 3\right)$ in.^{3}/sec.

(b) When $\dfrac{dV}{dt} = -1$, $k - 3 = -1 \Rightarrow k = 2$ in.^{3}/sec.

31. Let V be the volume of water in the pool, h the height of the water, and x as shown in the figure. Given $\dfrac{dV}{dt} = 500$ ft^3/min., and that the pool is 10 ft. wide. As the two right triangles in the figure are similar:

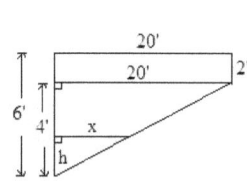

$\dfrac{20}{4} = \dfrac{x}{h} \Rightarrow x = 5h.$

(a) When $h = 2$ ft (which is less than 4 ft), $V = \dfrac{1}{2}xh(10) = 5(5h)h = 25h^2$. Differentiating:

$\dfrac{dV}{dt} = 25(2h)\dfrac{dh}{dt} = 50h\dfrac{dh}{dt}$. Thus, $500 = 50(2)\dfrac{dh}{dt}$ $\Rightarrow$ $\dfrac{dh}{dt} = 5$ ft/min.

(b) (See figure) Given $h = 5$ ft (which is *more* than 4 ft). Then $V = V_1 + V_2$.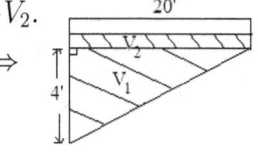

$V_1 = 25h^2 = 25(4^2) = 400$ and $V_2 = (h-4)(10)(20) = 200(h-4)$ $\Rightarrow$

$V = 400 + 200(h-4) = 200h - 400$. Thus $\dfrac{dV}{dt} = 200\dfrac{dh}{dt}$.

But $\dfrac{dV}{dt} = 500$ $\Rightarrow$ $500 = 200\dfrac{dh}{dt}$ $\Rightarrow$ $\dfrac{dh}{dt} = \frac{5}{2}$ ft/min.

CHAPTER 4
The Mean Value Theorem and Applications

§4.1 The Mean Value Theorem

1. The function $f(x) = x^2$, being a polynomial, is continuous on every closed interval and differentiable inside. On $[-1, 1]$, $\dfrac{f(b) - f(a)}{b - a} = \dfrac{f(1) - f(-1)}{1 - (-1)} = \dfrac{1^2 - (-1)^2}{2} = \dfrac{0}{2} = 0$. As $f'(x) = 2x$, then $2c = 0 \Rightarrow c = 0$ which lies inside $(-1, 1)$.

3. The function $f(x) = x^2$, being a polynomial, is continuous on every closed interval and differentiable inside. On $[-3, 2]$, $\dfrac{f(b) - f(a)}{b - a} = \dfrac{f(2) - f(-3)}{2 - (-3)} = \dfrac{2^2 - (-3)^2}{5} = \dfrac{-5}{5} = -1$. As $f'(x) = 2x$, then $2c = -1 \Rightarrow c = -\frac{1}{2}$ which lies inside $(-3, 2)$.

5. The function $f(x) = x^3$, being a polynomial, is continuous on every closed interval and differentiable inside. On $[-1, 0]$, $\dfrac{f(b) - f(a)}{b - a} = \dfrac{f(0) - f(-1)}{0 - (-1)} = \dfrac{0^3 - (-1)^3}{1} = 1$. As $f'(x) = 3x^2$, then $3c^2 = 1 \Rightarrow c^2 = \frac{1}{3} \Rightarrow c = \pm\frac{1}{\sqrt{3}}$. But we need $-1 < c < 0$, so $c = -\frac{1}{\sqrt{3}}$.

7. The function $f(x) = \sqrt{x + 5}$ is continuous where $x + 5 \geq 0 \Rightarrow x \geq -5$, so it is continuous on $[-1, 4]$. f is differentiable for $x > -5$, so on $(-1, 4)$.
$$\frac{f(b) - f(a)}{b - a} = \frac{\sqrt{4 + 5} - \sqrt{-1 + 5}}{4 - (-1)} = \frac{3 - 2}{5} = \frac{1}{5}. \quad \text{As } f'(x) = [(x + 5)^{\frac{1}{2}}]',$$
$f'(x) = \frac{1}{2}(x + 5)^{-\frac{1}{2}}(x + 5)' = \dfrac{1}{2\sqrt{x + 5}}$, then $\dfrac{1}{2\sqrt{c + 5}} = \frac{1}{5} \Rightarrow 2\sqrt{c + 5} = 5 \Rightarrow$
$\sqrt{c + 5} = \frac{5}{2} \Rightarrow c + 5 = \frac{25}{4} \Rightarrow c = \frac{25}{4} - 5 = \frac{5}{4}$, and indeed $\frac{5}{4}$ lies inside $(-1, 4)$.

9. The function $f(x) = \dfrac{x}{x + 1}$ is continuous on any interval not containing -1, so it is continuous on $[0, 1]$. f is differentiable inside any interval not containing -1, so inside $(0, 1)$.
$$\frac{f(b) - f(a)}{b - a} = \frac{\dfrac{1}{1 + 1} - \dfrac{0}{0 + 1}}{1 - 0} = \frac{\frac{1}{2}}{1} = \frac{1}{2}. \quad \text{Then } f'(x) = \frac{(x + 1)(1) - x(1)}{(x + 1)^2} = \frac{1}{(x + 1)^2}.$$
Thus $\frac{1}{2} = \dfrac{1}{(c + 1)^2} \Rightarrow (c + 1)^2 = 2 \Rightarrow c + 1 = \pm\sqrt{2} \Rightarrow c = -1 \pm \sqrt{2}$. But we require $0 < c < 1$, so $c = -1 + \sqrt{2}$.

11.   Differentiating: $f'(x) = (x^{-2})' - 0 = -2x^{-3} = -\dfrac{2}{x^3}$ which can not equal zero, in any interval. Verifying: $f(2) = \dfrac{1}{2^2} - \dfrac{1}{4} = \dfrac{1}{4} - \dfrac{1}{4} = 0$ and $f(-2) = \dfrac{1}{(-2)^2} - \dfrac{1}{4} = \dfrac{1}{4} - \dfrac{1}{4} = 0$. The function $f(x)$ is not defined at $x = 0$ and $-2 \leq 0 \leq 2$, so it is certainly not continuous on $[-2, 2]$ and not differentiable in $(-2, 2)$.

13.   (a)

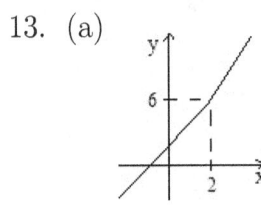

(b) No: The function $f(x)$ is continuous on $[0, 3]$, as seen in (a), but is not differentiable at $x = 2$, since $\lim\limits_{x \to 2^-} f'(x) = 2$, while $\lim\limits_{x \to 2^+} f'(x) = 3$.

(c)   No: $\dfrac{f(b) - f(a)}{b - a} = \dfrac{f(3) - f(0)}{3 - 0} = \dfrac{[3(3)] - [2(0) + 2]}{3} = \dfrac{7}{3}$, and $f'(x)$ is either $(2x + 2)' = 2$ or $(3x)' = 3$ only.

15.   If there exist x_1 and x_2 with $a < x_1 < x_2 < b$ and $f(x_1) = f(x_2) = 0$, then by Rolle's Theorem, there exists a c with $x_1 < c < x_2$ such that $f'(c) = 0$.

17.   Let $f(x) = x^3 + 6x^2 + 15x - 23$. Since f is a polynomial function, it is continuous and differentiable everywhere. As $f(0) = -23 < 0$ and $f(2) = 2^3 + 6(2^2) + 15(2) - 23 > 0$, by the Intermediate Value Theorem, there exists a c in the interval (0,2) with $f(c) = 0$. That is, c is a root of the equation $x^3 + 6x^2 + 15x - 23 = 0$.

   It remains to show that there is no other root. Suppose both a and b, with $a < b$ are solutions of the equation, i.e. $f(a) = f(b) = 0$. Then, f satisfies Rolle's Theorem on $[a, b]$, which means that there is a c in (a, b) with $f'(c) = 0$. But $f'(x) = 3x^2 + 12x + 15$ is never zero, as its discriminant, $b^2 - 4ac = 12^2 - 4(3)15 < 0$.

19.   Let $f(x) = 2x - 1 - \sin x$. Then f is continuous and differentiable everywhere. As $f(0) = -1 < 0$ and $f(\pi) = 2\pi - 1 > 0$, by the Intermediate Value Theorem, there exists a c in the interval $(0, \pi)$ with $f(c) = 0$. That is, c is a root of the equation $2x = 1 + \sin x$.

   It remains to show that there is no other root. Suppose both a and b, with $a < b$ are solutions of the equation, i.e. $f(a) = f(b) = 0$. Then, f satisfies Rolle's Theorem on $[a, b]$, which means that there is a c in (a, b) with $f'(c) = 0$. But $f'(x) = 2 - \cos x \geq 1$ for all x as $-1 \leq \cos x \leq 1$.

21.   Let $f(x) = x^4 + 50x^2 - 300 = 0$. Assume that $f(x) = x^4 + 50x^2 - 300 = 0$ has three solutions, x_1, x_2, x_3, with $x_1 < x_2 < x_3$. Since $f(x_1) = f(x_2) = f(x_3) = 0$, we can apply Rolle's Theorem in $[x_1, x_2]$ and in $[x_2, x_3]$ to conclude that there exist c_1 in (x_1, x_2) and c_2 in (x_2, x_3) such that $f'(c_1) = f'(c_2) = 0$. But this is impossible, as $f'(x) = 4x^3 + 100x = 4x(x^2 + 25) = 0$ has only one solution, $x = 0$.

23. If $a = b$, the inequality holds as both sides are zero. Suppose $a \neq b$, say, $a < b$. Apply the Mean Value Theorem on $[a, b]$ to $f(x) = \sin x$ which is continuous and differentiable everywhere. Then there is a c with $a < c < b$ such that $f'(c) = \dfrac{f(b) - f(a)}{b - a}$, that is, $\cos c = \dfrac{\sin b - \sin a}{b - a}$. As $|\cos x| \leq 1$ for all x, $\left| \dfrac{\sin b - \sin a}{b - a} \right| \leq 1 \Rightarrow |\sin b - \sin a| \leq |b - a|$.

25. Let $s_1(t)$ and $s_2(t)$ be the distances (positions) of runners 1 and 2, respectively, from the starting line, t seconds after the start of the race. Let t_0 denote the time the finish line is reached. Let $f(t) = s_1(t) - s_2(t)$ on $[0, t_0]$. Since $f(0) = f(t_0) = 0$, by Rolle's Theorem, there is a t_c in $(0, t_0)$ such that $f'(t_c) = 0 \Rightarrow s_1'(t_c) = s_2'(t_c)$, i.e. they are running at the same speed at $t = t_c$.

27. The given information implies that f is differentiable in $[0, 2]$, and therefore continuous in $[0, 2]$, so f satisfies the hypotheses of the Mean Value Theorem there. Thus there exists a c in $(0, 2)$ for which $f'(c) = \dfrac{f(2) - f(0)}{2 - 0} = \dfrac{f(2) - 6}{2}$. As $f'(x) \geq 1$ in $[0, 2]$, $f'(c) \geq 1 \Rightarrow \dfrac{f(2) - 6}{2} \geq 1 \Rightarrow f(2) - 6 \geq 2 \Rightarrow f(2) \geq 8$.

29. Let $h(x) = f(x) - g(x)$. The given information implies that h, being the difference of two differentiable functions, is differentiable in $[a, b]$, and therefore continuous in $[a, b]$, and as $h(a) = h(b) = 0$, h satisfies the hypotheses of Rolle's Theorem on $[a, b]$. Hence there is a c with $a < c < b$ such that $h'(c) = 0 \Rightarrow f'(c) = g'(c)$. Since the slope of the tangent line is the derivative, this means that the slopes of the tangent lines to the graphs of f and g at $x = c$ are equal, i.e. the lines are parallel.

§4.2 Graphing Functions

1. $f(x) = 2x^2 + 7x + 4$ is defined for all x. The y-intercept is 4. There are two x-intercepts: $x = \dfrac{-b \pm \sqrt{b^2 - 4ac}}{2a} = \dfrac{-7 \pm \sqrt{7^2 - 4(2)(4)}}{2(2)} = \dfrac{-7 \pm \sqrt{17}}{4}$. As $x \to \pm\infty$, the graph resembles that of $2x^2$.

   $f'(x) = 4x + 7 = 0 \Rightarrow x = -\dfrac{7}{4}$. Sign f':

   There is a minimum at $-\dfrac{7}{4}$:

   $f(-\tfrac{7}{4}) = 2(-\tfrac{7}{4})^2 + 7(-\tfrac{7}{4}) + 4 = 2(\tfrac{49}{16}) - \tfrac{49}{4} + 4 = \dfrac{49 - 98 + 32}{8} = -\dfrac{17}{8}$.

   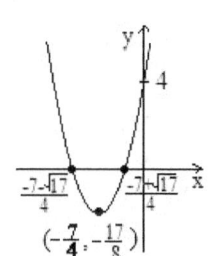

   $f''(x) = 4 > 0$ means that the graph is always concave up, and there are no inflection points.

3. $f(x) = x^3 + 2x^2 = x^2(x+2)$ is defined for all x. The intercepts are $(0,0)$ and $(-2,0)$. As $x \to \pm\infty$, the graph resembles that of x^3.

$f'(x) = 3x^2 + 4x = x(3x+4) = 0 \Rightarrow x = 0, -\frac{4}{3}$. Sign f':

There is a maximum at $x = -\frac{4}{3}$ and a minimum at $x = 0$:

$f\left(-\frac{4}{3}\right) = \left(-\frac{4}{3}\right)^2\left(-\frac{4}{3}+2\right) = \frac{16}{9}\left(\frac{2}{3}\right) = \frac{32}{27}.$ $f(0) = 0.$

$f''(x) = 6x + 4 = 0 \Rightarrow x = -\frac{2}{3}$. Sign f'':

There is an inflection point at $x = -\frac{2}{3}$:

$f\left(-\frac{2}{3}\right) = \left(-\frac{2}{3}\right)^2\left(-\frac{2}{3}+2\right) = \frac{4}{9}\left(\frac{4}{3}\right) = \frac{16}{27}.$

5. $f(x) = -\frac{1}{3}x^3 + 3x^2 - 8x = -\frac{1}{3}x(x^2 - 9x + 24)$ is defined for all x. The only intercept is $(0,0)$, as the quadratic factor has no zeros $(b^2 - 4ac = 81 - 4(24) < 0)$. As $x \to \pm\infty$, $f(x)$ resembles $-\frac{1}{3}x^3$.

$f'(x) = -x^2 + 6x - 8 = -(x-4)(x-2) = 0 \Rightarrow x = 2, 4$. Sign f':

There is a minimum at $x = 2$ and a maximum at $x = 4$:

$f(2) = -\frac{1}{3}(2)^3 + 3(2)^2 - 8(2) = -\frac{8}{3} - 4 = -\frac{20}{3}.$

$f(4) = -\frac{1}{3}(4)^3 + 3(4)^2 - 8(4) = -\frac{64}{3} + 16 = -\frac{16}{3}.$

$f''(x) = -2x + 6 = 0 \Rightarrow x = 3$. Sign f'':

There is an inflection point at $x = 3$:

$f(3) = -\frac{1}{3}(3)^3 + 3(3)^2 - 8(3) = 18 - 24 = -6.$

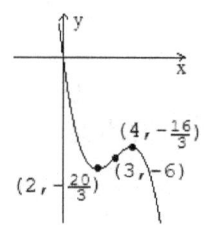

7. $f(x) = \frac{1}{4}x^4 + x^3 = x^3\left(\frac{1}{4}x + 1\right)$ is defined for all x. The intercepts are $(0,0)$ and $(-4,0)$. As $x \to \pm\infty$, $f(x)$ resembles $\frac{1}{4}x^4$.

$f'(x) = x^3 + 3x^2 = x^2(x+3) = 0 \Rightarrow x = 0, -3$. Sign f':

There is a minimum at $x = -3$:

$f(-3) = \frac{1}{4}(-3)^4 + (-3)^3 = \frac{81}{4} - 27 = \frac{81 - 108}{4} = -\frac{27}{4}.$

$f''(x) = 3x^2 + 6x = 3x(x+2) = 0 \Rightarrow x = 0, -2.$

Sign f'':

There are two inflection points: at $x = 0$ and at $x - 2$:

$f(-2) = \frac{1}{4}(-2)^4 + (-2)^3 = 4 - 8 = -4$, and $f(0) = 0.$

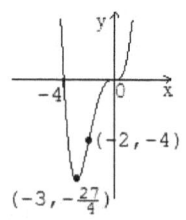

9. $f(x) = \dfrac{x}{2x+1}$ is defined for $x \neq -\frac{1}{2}$. The only intercept is $(0,0)$. The line $x = -\frac{1}{2}$ is a vertical asymptote. Sign f: . From the sign information for f, we conclude that the graph goes to $+\infty$ as x approaches the vertical asymptote from the left, and to $-\infty$ as x approaches from the right. The graph goes from below the x-axis to above the x-axis as you move from left to right across the x-intercept at 0.

$f'(x) = \dfrac{(2x+1) - x(2)}{(2x+1)^2} = \dfrac{1}{(2x+1)^2}$ is never zero. Sign f':

There are no maxima or minima.

$f''(x) = [(2x+1)^{-2}]' = -2(2x+1)^{-3} \cdot 2 = -\dfrac{4}{(2x+1)^3}$:

Sign f'': concave up $\underset{-\frac{1}{2}}{+\ c\ -}$ down . There are no inflection points.

11. $f(x) = \dfrac{2x+1}{x}$ is defined for $x \neq 0$. The only intercept is $-\left(\frac{1}{2},0\right)$. The y-axis is a vertical asymptote. Sign f: $\underset{-\frac{1}{2}\ \ 0}{+\ c\ -\ c\ +}$. From the sign information for f, we conclude that the graph goes to $-\infty$ as x approaches the vertical asymptote from the left, and to $+\infty$ as x approaches from the right. The graph goes from above the x-axis to below the x-axis as you move from left to right across the x-intercept at $-\frac{1}{2}$.

$f'(x) = \dfrac{x(2) - (2x+1)(1)}{x^2} = -\dfrac{1}{x^2}$ is never zero. Sign f': dec. $\underset{0}{-\ n\ -}$ dec.

There are no maxima or minima.

$f''(x) = [-x^{-2}]' = 2x^{-3} = \dfrac{2}{x^3}$: Sign f'': concave down $\underset{0}{-\ c\ +}$ up

There are no inflection points.

13. $f(x) = x + \dfrac{1}{x^2} = \dfrac{x^3+1}{x^2}$ is defined for $x \neq 0$. The x-intercept: $x^3 + 1 = 0 \Rightarrow x = -1$. The y-axis is a vertical asymptote. Sign f: $\underset{-1\ \ 0}{-\ c\ +\ n\ +}$. From the sign information for f, we conclude that the graph goes to $+\infty$ as x approaches the vertical asymptote from either side. The graph goes from below the x-axis to above the x-axis as you move from left to right across the x-intercept at -1. As $x \Rightarrow \pm\infty$, the graph approaches the oblique asymptote $y = x$, because $\dfrac{1}{x^2} \to 0$.

$f'(x) = (x + x^{-2})' = 1 - 2x^{-3} = 1 - \dfrac{2}{x^3} = \dfrac{x^3 - 2}{x^3} = 0 \Rightarrow x = 2^{\frac{1}{3}}$.

Sign f': inc. dec. inc. $\underset{0\ \ 2^{1/3}}{+\ c\ -\ c\ +}$. There is a minimum at $x = 2^{\frac{1}{3}}$.

$f(2^{\frac{1}{3}}) = \dfrac{2+1}{2^{\frac{2}{3}}} = \dfrac{3}{2^{\frac{2}{3}}}$.

$f''(x) = (1 - 2x^{-3})' = 6x^{-4} = \dfrac{6}{x^4}$: Sign f'': concave up up $\underset{0}{+\ n\ +}$

There are no inflection points.

15. $f(x) = \dfrac{x}{x^2 - 1} = \dfrac{x}{(x-1)(x+1)}$ is defined for $x \neq \pm 1$. The only intercept is (0,0).
There are two vertical asymptotes: $x = -1$ and $x = 1$. Sign f: $\underset{-1 \quad 0 \quad 1}{-\overset{c}{\circ}+\overset{c}{\bullet}-\overset{c}{\circ}+}$. From the
sign information for f, we conclude that the graph goes to $-\infty$ as $x \to -1$ or 1 from the
left, and to $+\infty$ as $x \to -1$ or 1 from the right. The graph goes from above the x-axis to
below the x-axis as you move from left to right across the x-intercept at 0. As $x \Rightarrow \pm\infty$,
the graph approaches the horizontal asymptote $y = 0$, because the degree of the numerator
of f is less than that of the denominator.

$$f'(x) = \frac{(x^2-1)(1) - x(2x)}{(x^2-1)^2} = \frac{-x^2 - 1}{(x^2-1)^2} = -\frac{x^2+1}{(x-1)^2(x+1)^2}.$$

Sign f': $\underset{-1 \quad\; 1}{\overset{\text{dec.}\;\;\text{dec.}\;\;\text{dec.}}{-\underset{\circ}{\text{n}}-\underset{\circ}{\text{n}}-}}$. There are no maxima or minima.

$$f''(x) = \frac{(x^2-1)^{\cancel{2}}(-2x) - (-x^2-1)[2(x^2\cancel{+}1)(2x)]}{(x^2-1)^{\cancel{4}\,3}}$$

$$= \frac{-2x(x^2 - 1 - 2x^2 - 2)}{(x^2-1)^3} = \frac{2x(x^2+3)}{(x-1)^3(x+1)^3}$$

Sign f'': $\overset{\text{concave down}\;\;\text{up}\;\;\text{down}\;\;\text{up}}{\underset{-1 \quad\; 0 \quad\; 1}{-\underset{\circ}{c}+\underset{\bullet}{c}-\underset{\circ}{c}+}}$

There is an inflection point at the origin.

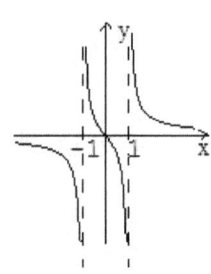

17. $f(x) = \dfrac{x^2}{x^2 - 9} = \dfrac{x^2}{(x-3)(x+3)}$ is defined for $x \neq \pm 3$. The only intercept is (0,0).
There are two vertical asymptotes: $x = -3$ and $x = 3$. Sign f: $\underset{-3 \quad 0 \quad 3}{+\overset{c}{\circ}-\overset{\text{n}}{\bullet}-\overset{c}{\circ}+}$. From the
sign information for f, we conclude that the graph goes to $-\infty$ as $x \to -3$ from the right
and 3 from the left, and to $+\infty$ as $x \to -3$ from the left and 3 from the right. The graph is
tangent to the x-axis at 0 and lies below the x-axis near the origin. As $x \to \pm\infty$, the graph
approaches the horizontal asymptote $y = 1$, because $f(x) = 1 + \dfrac{9}{x^2-9}$ and $\dfrac{9}{x^2-9} \to 0$.

$$f'(x) = \frac{(x^2-9)(2x) - x^2(2x)}{(x^2-9)^2} = -\frac{18x}{(x-3)^2(x+3)^2}.$$ Sign f': $\overset{\text{inc.}\;\;\text{inc.}\;\;\text{dec.}\;\;\text{dec.}}{\underset{-3 \quad\; 0 \quad\; 3}{+\underset{\text{n}}{}+\underset{\circ}{c}-\underset{\text{n}}{}-}}$

There is a maximum at the origin. $f(0) = 0$.

$$f''(x) = -\frac{(x^2-9)^{\cancel{2}}(18) - 18x(2)[(x^2\cancel{+}9)(2x)]}{(x^2-9)^{\cancel{4}\,3}}$$

$$= -\frac{18(x^2 - 9 - 4x^2)}{(x^2-9)^3} = \frac{54(x^2+3)}{(x-3)^3(x+3)^3}$$

Sign f'': $\overset{\text{concave up}\;\;\text{down}\;\;\text{up}}{\underset{-3 \quad\; 3}{+\underset{\circ}{c}-\underset{\circ}{c}+}}$. There are no inflection points.

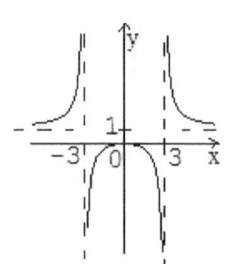

19. $f(x) = \dfrac{x^3 - x}{x^3 - x^2} = \dfrac{x(x-1)(x+1)}{x^2(x-1)} = \dfrac{x+1}{x}$ is defined for $x \neq 0$ or 1. The x-intercept
is -1. There is a vertical asymptote: $x = 0$. Sign f: 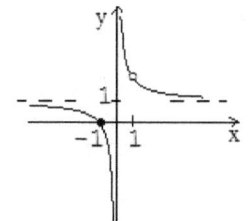. From the sign information for f, we conclude that the graph goes to $-\infty$ as $x \to 0$ from the left, and to $+\infty$
as $x \to 0$ from the right. The graph goes from positive to negative as the graph moves from
left to right through $x = -1$. As $x \to \pm\infty$, the graph approaches the horizontal asymptote
$y = 1$, because $f(x) = 1 + \dfrac{1}{x}$ and $\dfrac{1}{x} \to 0$.

$f'(x) = \dfrac{x(1) - (x+1)(1)}{x^2} = -\dfrac{1}{x^2}$. Sign f': $\overset{\text{dec.}\ \ \text{dec.}\ \text{dec.}}{\underset{0 \quad 1}{\rule{2cm}{0.4pt}}}$

There are no maxima or minima.

$f''(x) = (-x^{-2})' = 2x^{-3} = \dfrac{2}{x^3}$. Sign f'': $\overset{\text{concave down}\quad \text{up}\quad \text{up}}{\underset{0\quad\ \ 1}{\rule{2cm}{0.4pt}}}$

There are no inflection points.

21. $f(x) = \sqrt{x^2 - 1} = \sqrt{(x-1)(x+1)} \geq 0$ is defined for $x^2 - 1 \geq 0 \;\Rightarrow\; (x-1)(x+1) \geq 0$
Domain f: $\underset{-1\quad\ \ 1}{\rule{2cm}{0.4pt}}$ i.e. $D_f = (-\infty, -1) \cup (1, \infty)$. There are two x-intercepts, ± 1, and
no y-intercept. As $x \to \pm\infty$, $f(x)$ resembles the graph of $\sqrt{x^2} = |x|$.

$f'(x) = \left[(x^2 - 1)^{\frac{1}{2}}\right]' = \dfrac{1}{2}(x^2 - 1)^{-\frac{1}{2}}(2x) = \dfrac{x}{\sqrt{x^2 - 1}}$.

Sign f': $\overset{\text{dec.}\qquad\ \text{inc.}}{\underset{-1\quad\ \ 1}{\rule{2cm}{0.4pt}}}$. There are no maxima or minima.

$f''(x) = \dfrac{\sqrt{x^2 - 1}(1) - x\left(\frac{1}{2}\right)(x^2-1)^{-\frac{1}{2}}(2x)}{x^2 - 1} \cdot \dfrac{\sqrt{x^2-1}}{\sqrt{x^2-1}}$

$\quad = \dfrac{x^2 - 1 - x^2}{(x^2-1)^{\frac{3}{2}}} = -\dfrac{1}{(x^2-1)^{\frac{3}{2}}}$. There are no inflection points.

Sign f'': $\overset{\text{concave down}\qquad\text{down}}{\underset{-1\quad\ \ 1}{\rule{2cm}{0.4pt}}}$

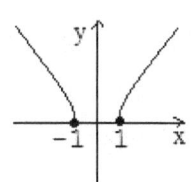

23. $f(x) = x\sqrt{x - 1}$ is defined for $x - 1 \geq 0 \;\Rightarrow\; x \geq 1$. Sign f: $\underset{1}{\rule{2cm}{0.4pt}}^{+}$. The only
intercept is $x = 1$ ($x = 0 < 1$ is not in the domain). As $x \to \pm\infty$, the graph of $f(x)$
resembles that of $x\sqrt{x} = x^{\frac{3}{2}}$.

$f'(x) = \left[x(x-1)^{\frac{1}{2}}\right]' = \left\{x\left[\frac{1}{2}(x-1)^{-\frac{1}{2}}(1)\right] + (x-1)^{\frac{1}{2}}\right\} \cdot \dfrac{2(x-1)^{\frac{1}{2}}}{2(x-1)^{\frac{1}{2}}} = \dfrac{x + 2(x-1)}{2\sqrt{x-1}} = \dfrac{3x - 2}{2\sqrt{x-1}}$

Sign f': $\underset{1}{\rule{2cm}{0.4pt}}^{\ +}$. There are no maxima or minima.

$f''(x) = \dfrac{2\sqrt{x-1}(3) - (3x-2)(2) \cdot \dfrac{1}{2\sqrt{x-1}}}{4(x-1)} \cdot \dfrac{\sqrt{x-1}}{\sqrt{x-1}} = \dfrac{6(x-1) - 3x + 2}{4(x-1)^{\frac{3}{2}}} = \dfrac{3x - 4}{4(x-1)^{\frac{3}{2}}}$

Sign f'': concave down up

There is an inflection point at $x = \frac{4}{3}$: $f\left(\frac{4}{3}\right) = \frac{4}{3}\sqrt{\frac{4}{3} - 1} = \frac{4}{3\sqrt{3}}$.

25. $f(x) = x^{\frac{1}{3}}(x + 4)$ is defined for all x. The intercepts are $x = 0$ and $x = -4$.

Sign f: $\frac{+ \quad \overset{c}{-} \quad \overset{c}{+}}{-4 \quad 0}$. As $x \to \pm\infty$, the graph of $f(x)$ resembles that of $x^{\frac{1}{3}}(x) = x^{\frac{4}{3}}$.

$$f'(x) = x^{\frac{1}{3}}(1) + (x+4)\left(\frac{1}{3}x^{-\frac{2}{3}}\right) = x^{\frac{1}{3}} + \frac{x+4}{3x^{\frac{2}{3}}} = \frac{3x + x + 4}{3x^{\frac{2}{3}}} = \frac{4(x+1)}{3x^{\frac{2}{3}}}$$

Sign f': dec. inc. inc. $\frac{- \quad \overset{c}{-} \quad + \quad \overset{n}{+}}{-1 \quad 0}$. There is a minimum at $x = -1$: $f(-1) = (-1)^{\frac{1}{3}}(-1+4) = -3$.

$$f''(x) = \frac{3x^{\frac{2}{3}}(4) - (4x+4)(\not{3})\left(\frac{2}{\not{3}}\right)x^{-\frac{1}{3}}}{9x^{\frac{4}{3}}} \cdot \frac{x^{\frac{1}{3}}}{x^{\frac{1}{3}}} = \frac{12x - 8x - 8}{9x^{\frac{5}{3}}} = \frac{4(x-2)}{9x^{\frac{5}{3}}}$$

Sign f'': concave up down up $\frac{+ \quad \overset{c}{-} \quad \overset{c}{+}}{0 \quad 2}$

There are inflection points at $x = 0$ and $x = 2$: $f(0) = 0$, $f(2) = 2^{\frac{1}{3}}(2 + 4) = 6 \cdot 2^{\frac{1}{3}}$.

27. $f(x) = x - 6x^{\frac{1}{3}}$ is defined for all x. The intercepts are $x = 0$ and $x = \pm 6^{\frac{3}{2}} = \pm 6\sqrt{6}$.

Sign f: $\frac{- \quad \overset{c}{+} \quad \overset{c}{-} \quad \overset{c}{+}}{-6\sqrt{6} \quad 0 \quad 6\sqrt{6}}$. As $x \to \pm\infty$, the graph of $f(x)$ resembles that of x.

$$f'(x) = 1 - 2x^{-\frac{2}{3}} = 1 - \frac{2}{x^{\frac{2}{3}}} = \frac{x^{\frac{2}{3}} - 2}{x^{\frac{2}{3}}} = 0 \Rightarrow x = \pm 2^{\frac{3}{2}} = \pm 2\sqrt{2}.$$

Sign f': inc. dec. dec. inc. $\frac{+ \quad \overset{c}{-} \quad \overset{n}{-} \quad \overset{c}{+}}{-2\sqrt{2} \quad 0 \quad 2\sqrt{2}}$

There is a maximum at $x = -2\sqrt{2}$:

$$f(-2\sqrt{2}) = -2^{\frac{3}{2}} - 6(-2^{\frac{3}{2}})^{\frac{1}{3}} = -2\sqrt{2} + 6\sqrt{2} = 4\sqrt{2}.$$

There is a minimum at $x = 2\sqrt{2}$:

$$f(2\sqrt{2}) = 2^{\frac{3}{2}} - 6(2^{\frac{3}{2}})^{\frac{1}{3}} = 2\sqrt{2} - 6\sqrt{2} = -4\sqrt{2}.$$

$$f''(x) = -2\left(-\frac{2}{3}x^{-\frac{5}{3}}\right) = \frac{4}{3x^{\frac{5}{3}}}.$$ Sign f'': concave down up $\frac{- \quad \overset{c}{+}}{0}$

The origin is a point of inflection.

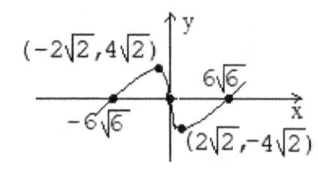

29. $f(x) = x^{\frac{2}{3}}\left(x - \frac{5}{2}\right)$ is defined for all x. The intercepts are $x = 0$ and $x = \frac{5}{2}$.

Sign f: $\frac{- \quad \overset{n}{-} \quad \overset{c}{+}}{0 \quad \frac{5}{2}}$. As $x \to \pm\infty$, the graph of $f(x)$ resembles that of $x^{\frac{5}{3}}$.

$$f'(x) = x^{\frac{2}{3}}(1) + \left(x - \frac{5}{2}\right)\left(\frac{2}{3}x^{-\frac{1}{3}}\right) = x^{\frac{2}{3}} + \frac{2\left(x - \frac{5}{2}\right)}{3x^{\frac{1}{3}}} = x^{\frac{2}{3}} + \frac{2x - 5}{3x^{\frac{1}{3}}} = \frac{3x + 2x - 5}{3x^{\frac{1}{3}}} = \frac{5(x-1)}{3x^{\frac{1}{3}}}$$

Sign f': inc. dec. inc. $\frac{+ \quad \overset{c}{-} \quad \overset{c}{+}}{0 \quad 1}$. There is a maximum at $x = 0$: $f(0) = 0$.

There is a minimum at $x = 1$: $f(1) = 1^{\frac{2}{3}}(1 - \frac{5}{2}) = -\frac{3}{2}$.

$$f''(x) = \frac{3x^{\frac{1}{3}}(5) - (5x - 5)x^{-\frac{2}{3}}}{9x^{\frac{2}{3}}} \cdot \frac{x^{\frac{2}{3}}}{x^{\frac{2}{3}}} = \frac{15x - 5x + 5}{9x^{\frac{4}{3}}} = \frac{10x + 5}{9x^{\frac{4}{3}}} = \frac{5(2x + 1)}{9x^{\frac{4}{3}}}.$$

Sign f'':

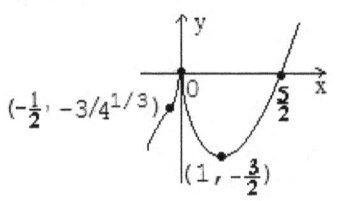

There is an inflection point at $x = -\frac{1}{2}$:

$$f\left(-\frac{1}{2}\right) = \left(-\frac{1}{2}\right)^{\frac{2}{3}}\left(-\frac{1}{2} - \frac{5}{2}\right) = -\frac{3}{4^{\frac{1}{3}}}.$$

31. $f(x) = \dfrac{x^2 - 1}{2x^2 + x} = \dfrac{(x - 1)(x + 1)}{x(2x + 1)}$ is defined for $x \neq 0$ or $-\frac{1}{2}$. The x-intercepts are $x = \pm 1$. There is no y-intercept. Sign f: There are two vertical asymptotes, $x = -\frac{1}{2}$ and the y-axis. From the sign information for f, we conclude that the graph goes to $-\infty$ as $x \to -\frac{1}{2}$ from the left, and as $x \to 0$ from the right. The graph goes to $+\infty$ as $x \to -\frac{1}{2}$ from the right, and as $x \to 0$ from the left. Moving from left to right, the graph goes from positive to negative through $x = -1$, and from negative to positive through $x = 1$. As $x \to \pm\infty, y \approx \left(\dfrac{x^2}{2x^2}\right)$, so the graph approaches the horizontal asymptote: $y = \frac{1}{2}$.

$$f'(x) = \frac{(2x^2 + x)(2x) - (x^2 - 1)(4x + 1)}{(2x^2 + x)^2} = \frac{4x^3 + 2x^2 - (4x^3 + x^2 - 4x - 1)}{x^2(2x + 1)^2} = \frac{x^2 + 4x + 1}{x^2(2x + 1)^2}$$

$x^2 + 4x + 1 = 0 \Rightarrow x = \dfrac{-4 \pm \sqrt{16 - 4}}{2} = \dfrac{-4 \pm 2\sqrt{3}}{2} = -2 \pm \sqrt{3}$. Sign f':

There is a maximum at $x = -2 - \sqrt{3}$:

$$f(-2 - \sqrt{3}) = \frac{(-2 - \sqrt{3})^2 - 1}{2(-2 - \sqrt{3})^2 + (-2 - \sqrt{3})} = \frac{4 + 4\sqrt{3} + 3 - 1}{2(4 + 4\sqrt{3} + 3) - 2 - \sqrt{3}}.$$

$$= \frac{6 + 4\sqrt{3}}{12 + 7\sqrt{3}} \cdot \frac{12 - 7\sqrt{3}}{12 - 7\sqrt{3}} = \frac{72 - 42\sqrt{3} + 48\sqrt{3} - 28(3)}{144 - 49(3)}$$

$$= \frac{-12 + 6\sqrt{3}}{-3} = 4 - 2\sqrt{3}$$

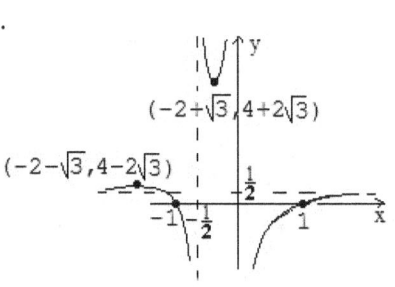

There is a minimum at $x = -2 + \sqrt{3}$:

$$f(-2 + \sqrt{3}) = \frac{(-2 + \sqrt{3})^2 - 1}{2(-2 + \sqrt{3})^2 + (-2 + \sqrt{3})} = \frac{4 - 4\sqrt{3} + 3 - 1}{2(4 - 4\sqrt{3} + 3) - 2 + \sqrt{3}}.$$

$$= \frac{6 - 4\sqrt{3}}{12 - 7\sqrt{3}} \cdot \frac{12 + 7\sqrt{3}}{12 + 7\sqrt{3}} = \frac{72 + 42\sqrt{3} - 48\sqrt{3} - 28(3)}{144 - 49(3)}$$

$$= \frac{-12 - 6\sqrt{3}}{-3} = 4 + 2\sqrt{3}$$

33. $f(x) = x^3 - 3x^2 + 3$ on $[-1, 1]$. $f'(x) = 3x^2 - 6x = 3x(x - 2) = 0 \Rightarrow x = 0$ or $x = 2$, but 2 is not in $[-1, 1]$. So the absolute max and min occur at 0 or an endpoint. Testing: $f(0) = 3$, $f(-1) = (-1)^3 - 3(-1)^2 + 3 = -1 - 3 + 3 = -1$, and $f(1) = 1^3 - 3(1)^2 + 3 = 1 - 3 + 3 = 1$. Thus, the absolute maximum occurs at 0, and the absolute minimum occurs at -1.

35. $f(x) = 3x^5 - 20x^3 + 2$ in $[-3, 3]$. $f'(x) = 15x^4 - 60x^2 = 15x^2(x^2 - 4) = 0 \Rightarrow x = 0$ or ± 2, all in the given interval. The absolute max and min occur at 0, ± 2, or ± 3. Testing:

$$f(0) = 2$$
$$f(-2) = 3(-2)^5 - 20(-2)^3 + 2 = -96 + 160 + 2 = 66$$
$$f(2) = 3(2)^5 - 20(2)^3 + 2 = 3(32) - 160 + 2 = 98 - 160 = -62$$
$$f(-3) = 3(-3)^5 - 20(-3)^3 + 2 = 3(-243) - 20(-27) + 2 = -729 + 540 + 2 = -187$$
$$f(3) = 3(3)^5 - 20(3)^3 + 2 = 3(243) - 20(27) + 2 = 729 - 540 + 2 = 191$$

Thus, the absolute maximum occurs at 3, and the absolute minimum occurs at -3.

37. $f(x) = 3x^5 - 20x^3 + 2$ on $[-1, 3]$. From Exercise #35, in $[-1, 3]$, $f'(x) = 0 \Rightarrow x = 0$ or 2. Therefore, the absolute max and min occur at 0, 2, -1, or 3. From Exercise #35, $f(0) = 2$, $f(2) = -62$, and $f(3) = 191$. Calculating: $f(-1) = 3(-1)^5 - 20(-1)^3 + 2 = -3 + 20 + 2 = 19$. Thus, the absolute maximum occurs at 3, and the absolute minimum occurs at 2.

39. One possible solution:

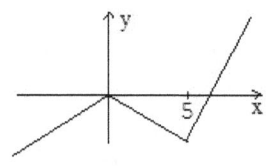

41. First derivative $> 0 \Rightarrow$ increasing. Second derivative $> 0 \Rightarrow$ concave up. One possible solution:

43. First derivative $> 0 \Rightarrow$ increasing. Second derivative $< 0 \Rightarrow$ concave down. One possible solution:

45. One possible solution:

47. One possible solution:

49. One possible solution:

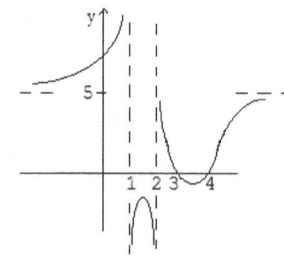

51. $L(t) = 15t^2 - t^3, 0 \le t \le 7$. The rate of change of learning is $L'(t) = 30t - 3t^2$. This function begins to decline when its derivative becomes negative, i.e. when $L''(t) < 0$: $L''(t) = 30 - 6t < 0 \Rightarrow 6t > 30 \Rightarrow t > 5$. Answer: at 5 weeks.

53.   Let $f(x) = \dfrac{x^2 - 1}{x^2 + x - 2} = \dfrac{(x-1)(x+1)}{(x-1)(x+2)} = \dfrac{x+1}{x+2}$, for $x \ne 1$. There is no vertical asymptote at $x = 1$ because both numerator and denominator are zero there. Since the denominator is 0 at $x = -2$, but the numerator is not, a vertical asymptote occurs there.

55.   (a) A horizontal tangent line occurs where the derivative of a function is zero. The derivative of a cubic polynomial is a quadratic polynomial, which cannot have more than two zeros. Thus the graph of a cubic polynomial cannot have more than two horizontal tangent lines.

   (b) (i) One possible function is $p(x) = \frac{1}{3}x^3 + x + 1$, since $p'(x) = x^2 + 1 \ge 1$ for all x (so can never be zero).

   (ii) One possible function is $p(x) = x^3$, since $p'(x) = 3x^2 = 0 \Rightarrow x = 0$. The graph of p only has a horizontal tangent at the origin.

   (iii) One possible function is $p(x) = \frac{1}{3}x^3 + \frac{1}{2}x^2 - 2x + 1 \Rightarrow p'(x) = x^2 + x - 2 = (x+2)(x-1) = 0 \Rightarrow x = -2$ or 1. The graph of p has exactly two horizontal tangent lines, at $x = -2$ and at $x = 1$.

   (c) Suppose $f(x) = ax^3 + bx^2 + cx + d$ assumes a local minimum at both x_1 and x_2, with $x_1 < x_2$, [note that $f'(x_1) = f'(x_2) = 0$]. It follows that the sign of $f'(x)$ must be positive immediately to the right of both points, and negative immediately to their left. In particular, $f'(x) > 0$ just to the right of x_1 and < 0 just to the left of x_2. Since f' is continuous, by the Intermediate Value Theorem, there is an x_3 with $x_1 < x_3 < x_2$ with $f'(x_3) = 0$, but this is impossible, since f' is a quadratic polynomial which can have at most 2 zeros.

   A similar argument shows that there can not be two local maxima.

57.   (a) $f(x) = ax^3 + bx^2 + cx + d \Rightarrow f'(x) = 3ax^2 + 2bx + c \Rightarrow f''(x) = 6ax + 2b = 0 \Rightarrow x = -\dfrac{2b}{6a} = -\dfrac{b}{3a}$, and since the sign of f'' changes about this point, the concavity changes about this point. There are no other such points possible.

   (b) (i) From (a), $x = -\dfrac{b}{3a} = 1 \Rightarrow -b = 3a \Rightarrow b = -3a$. Then $f(x) = ax^3 - 3ax^2 + cx + d$. Since $y = 2$ when $x = 1$, $2 = a - 3a + c + d \Rightarrow d = 2 + 2a - c \Rightarrow f(x) = ax^3 - 3ax^2 + cx + (2 + 2a - c)$. For example, take $a = 1$ and $c = 0$, then $f(x) = x^3 - 3x^2 + 4$.

   (ii) From (i), $f(x) = ax^3 - 3ax^2 + cx + (2 + 2a - c) \Rightarrow f'(x) = 3ax^2 - 6ax + c$. For a maximum at $x = -1$, $f'(-1) = 0 = 3a(-1)^2 - 6a(-1) + c \Rightarrow 9a + c = 0 \Rightarrow c = -9a \Rightarrow f(x) = ax^3 - 3ax^2 - 9ax + (2 + 11a)$ [note $f''(x) = 6ax - 6a < 0$ (a max) for $x = -1$, and $a > 0$]. For example, take $a = 1$: $f(x) = x^3 - 3x^2 - 9x + 13$.

   (iii) From (ii), $f(x) = ax^3 - 3ax^2 - 9ax + (2 + 11a)$ with $a > 0$. For a minimum at $x = 3$, $f'(3) = 0 = 3a(3)^2 - 6a(3) - 9a = 0 \Rightarrow 27a - 27a = 0$ and $f''(3) > 0 \Rightarrow 6a(3) - 6a > 0 \Rightarrow a > 0$. Again, for example, take $a = 1$: $f(x) = x^3 - 3x^2 - 9x + 13$.

§4.3 Optimization

1. $P(x) = -\dfrac{x^3}{30} + 9x^2 + 400x - 75000 \Rightarrow P'(x) = -\dfrac{1}{10}x^2 + 18x + 400 = 0 \Rightarrow$
$x^2 - 180x - 4000 = 0 \Rightarrow (x - 200)(x + 20) = 0 \Rightarrow x = 200$, as $x > 0$. Then
$P'(x) = -\dfrac{1}{10}(x - 200)(x + 20)$.   Sign P': $\underset{0\ \ 200\ \ 500}{\vdash\!\!+\overset{c}{\bullet}-\!\!\dashv}$.   The absolute maximum occurs at
$x = 200$: 200 units should be produced.

3. $C(t) = \dfrac{0.2t}{0.9t^2 + 5t + 3} \cdot \dfrac{10}{10} = \dfrac{2t}{9t^2 + 50t + 30}$

$\quad C'(t) = \dfrac{(9t^2 + 50t + 30)(2) - 2t(18t + 50)}{(9t^2 + 50t + 30)^2} = \dfrac{18t^2 + 100t + 60 - 36t^2 - 100t}{(9t^2 + 50t + 30)^2}$

$\quad\quad = \dfrac{-18t^2 + 60}{(9t^2 + 50t + 30)^2} = 0 \Rightarrow 18t^2 = 60 \Rightarrow t^2 = \dfrac{60}{18} = \dfrac{10}{3} \Rightarrow t = \sqrt{\dfrac{10}{3}}$

Sign C': $\underset{0\ \ \sqrt{\frac{10}{3}}}{\vdash\!\!+\overset{c}{\bullet}-\!\!\dashv}$.   The maximum occurs at $\sqrt{\dfrac{10}{3}} \approx 1.8$ hours after administration, and the
maximum concentration is $C(1.8) = \dfrac{0.2(1.8)}{0.9(1.8)^2 + 5(1.8) + 3} \approx 0.024$ mg/cm^3.

5. $K(t) = 25t^2 - 150t + 700, 0 \le t \le 7$. $K'(t) = 50t - 150 = 0 \Rightarrow t = 3$.
Sign K': $\underset{0\ \ 3\ \ 7}{\vdash\!-\overset{c}{\bullet}+\!\dashv}$

   (a)   The minimum concentration occurs at $t = 3$: $K(3) = 25(9) - 450 + 700 = 475$
bacteria/cm^3.
   (b)   The maximum concentration occurs at an endpoint:
$K(0) = 700$ and $K(7) = 25(49) - 150(7) + 700 = 1225 - 1050 + 700 = 875$, so the maximum
concentration occurs after 7 days, and is 875 bacteria/cm^3.

7. $R(x) = x^2\left(\dfrac{a}{2} - \dfrac{x}{3}\right) = \dfrac{a}{2}x^2 - \dfrac{1}{3}x^3 \Rightarrow R'(x) = ax - x^2$. We want x for which $R'(x)$
is maximum: $(R'(x))' = R''(x) = a - 2x = 0 \Rightarrow x = \dfrac{a}{2}$.   Sign R'': $\underset{0\ \ \frac{a}{2}\ \ a}{\vdash\!\!+\overset{c}{\bullet}-\!\!\dashv}$. The absolute
maximum sensitivity occurs at a dosage of one-half the maximum dosage.

9.   Price per pound is $90 - 3x$. Let $R(x)$ denote the revenue from the sale of $10 + x$ pounds.
Then Revenue is price per pound times the number of pounds: $R(x) = (90 - 3x)(10 + x) =$
$900 + 60x - 3x^2 \Rightarrow R'(x) = 60 - 6x = 0 \Rightarrow x = 10$.   Sign R': $\underset{0\ \ 10}{\overset{\text{inc.\ dec.}}{\vdash\!\!+\overset{c}{\bullet}-}}$
Revenue starts to decrease when $x = 10$.

11.   Let x be the number of \$5 decreases of price. Then $250x$ is the number of additional units sold per month. Revenue = price per unit times the number of units sold:
$R = (950 - 5x)(25000 + 250x) = 23,750,000 - 125,000x + 237,500x - 1250x^2$
$\quad = 23,750,000 + 112,500x - 1250x^2$.
$R' = 112,500 - 2500x = 0 \Rightarrow x = \dfrac{112500}{2500} = 45$. Sign R': $\underset{0\ \ 45}{\vdash\!\!+\overset{c}{\bullet}-}$. The absolute maximum
revenue occurs when the price is $950 - 5(45) = 950 - 225 = \$725$ per unit.

13. Let $x > 0$ be the number and S the sum: $S = x + \frac{1}{x}$. $S' = (x + x^{-1})' = 1 - x^{-2} = 1 - \frac{1}{x^2}$.
Then $S' = 0 \Rightarrow \frac{1}{x^2} = 1 \Rightarrow x = 1$, as $x > 0$. Sign S': $\underset{0 \quad 1}{\vdash\;-\;\overset{c}{\bullet}\;+\;}$. The smallest sum occurs for the number 1.

15.   Let A denote the area of the rectangle: $A = 2x(4 - x^2) = 8x - 2x^3, 0 < x < 2$. $A' = 8 - 6x^2 = 0 \Rightarrow x^2 = \frac{4}{3} \Rightarrow x = \frac{2}{\sqrt{3}}$.
Sign A': $\underset{0 \quad \frac{2}{\sqrt{3}} \quad 2}{\vdash\;+\;\overset{c}{\bullet}\;-\;}$.

Maximum area is $A(\frac{2}{\sqrt{3}}) = 2\left(\frac{2}{\sqrt{3}}\right)\left[4 - \left(\frac{2}{\sqrt{3}}\right)^2\right] = \frac{4}{\sqrt{3}}\left(\frac{8}{3}\right) = \frac{32}{3\sqrt{3}} \cdot \frac{\sqrt{3}}{\sqrt{3}} = \frac{32\sqrt{3}}{9}$ sq. units.

17. Let A denote the area of the triangle:
$$A = \frac{1}{2}x\sqrt{16 - x^2}, \quad 0 < x < 4.$$

$$A' = \frac{1}{2}\left[x \cdot \frac{1}{2\sqrt{16 - x^2}} \cdot -2x + \sqrt{16 - x^2}\right] = \frac{1}{2}\left[-\frac{x^2}{\sqrt{16 - x^2}} + \sqrt{16 - x^2}\right]$$

$$= \frac{1}{2}\left[\frac{-x^2 + 16 - x^2}{\sqrt{16 - x^2}}\right] = \frac{1}{2}\left[\frac{16 - 2x^2}{\sqrt{16 - x^2}}\right] = \frac{8 - x^2}{\sqrt{16 - x^2}} = 0 \Rightarrow x^2 = 8 \Rightarrow x = 2\sqrt{2}, \text{ as}$$

$x > 0$. Sign A': $\underset{0 \quad 2\sqrt{2} \quad 4}{\vdash\;+\;\overset{c}{\bullet}\;-\;}$

The maximum area is $A(2\sqrt{2}) = \frac{1}{2}(2\sqrt{2})\sqrt{16 - (2\sqrt{2})^2} = \sqrt{2} \cdot 2\sqrt{2} = 4$ sq. inches.

19. Let A denote the outer area. Then $A = (x + 6)(y + 8)$, and $xy = 1200$
$\Rightarrow y = \frac{1200}{x} \Rightarrow A = (x + 6)\left(\frac{1200}{x} + 8\right) = 1248 + 8x + 7200x^{-1} \Rightarrow$
$A' = 8 - \frac{7200}{x^2} = 0 \Rightarrow 8x^2 = 7200 \Rightarrow x^2 = 900 \Rightarrow x = 30$, as $x > 0$.
Sign A': $\underset{0 \quad 30}{\vdash\;-\;\overset{c}{\bullet}\;+\;}$. Minimum paper means minimum total area, and
this happens when the outer dimensions are $30 + 6 = 36$ and $\frac{1200}{30} + 8 = 48$, i.e. 36 inches wide by 48 inches high.

21. Let V be the volume of the crate: $V = x(2x)(h) = 2x^2h$.
As $x + 2x + h = 288$, $h = 288 - 3x$. Substituting into V:
$V = 2x^2(288 - 3x) = 576x^2 - 6x^3 \Rightarrow$
$V' = 1152x - 18x^2 = x(1152 - 18x) = 0 \Rightarrow x = \frac{1152}{18} = 64$.
Sign V': $\underset{0 \quad 64}{\vdash\;+\;\overset{c}{\bullet}\;-\;}$. Maximum volume occurs with a width of 64 inches. Then
$h = 288 - 3(64) = 96$. The dimensions are 64 in. wide by 128 in. long by 96 in. high.

23.

Let P denote the length of fencing of the garden:
$P = 2x + 2y + x = 3x + 2y$. From $A = 15000 = xy$,
$y = \frac{15,000}{x} \Rightarrow P = 3x + 2\left(\frac{15000}{x}\right) = 3x + \frac{30,000}{x}$

$P' = (3x + 30,000x^{-1})' = 3 - \frac{30,000}{x^2} = 0 \Rightarrow 3x^2 = 30,000 \Rightarrow x^2 = 10,000 \Rightarrow x = 100$.

Sign P': $\begin{array}{ccc} - & c & + \\ \hline 0 & 100 & \end{array}$. Minimum fencing occurs when $x = 100 \Rightarrow y = \frac{15,000}{100} = 150$, i.e.

when the dimensions of the garden are 100 ft by 150 ft with the inner fence parallel to the 100 ft side.

25.

Let L be the combined cable lengths to A and B: $L = d_1 + d_2$,
where $d_1 = \sqrt{5^2 + (9-x)^2}$ and $d_2 = \sqrt{3^2 + x^2}$, see figure.

$L' = \left\{ [5^2 + (9-x)^2]^{\frac{1}{2}} + (3^2 + x^2)^{\frac{1}{2}} \right\}'$

$= \frac{1}{\cancel{2}} [5^2 + (9-x)^2]^{-\frac{1}{2}} [\cancel{2}(9-x)(-1)] + \frac{1}{\cancel{2}} (3^2 + x^2)^{-\frac{1}{2}} (\cancel{2}x)$

$= \frac{x-9}{\sqrt{25 + (9-x)^2}} + \frac{x}{\sqrt{x^2 + 9}} = 0 \Rightarrow$

$(x-9)\sqrt{x^2 + 9} = -x\sqrt{25 + (9-x)^2} \Rightarrow$ after squaring, $(x-9)^2(x^2+9) = x^2[25 + (9-x)^2]$

$(x^2 - 18x + 81)(x^2 + 9) = x^2(25 + 81 - 18x + x^2) \Rightarrow$

$\cancel{x^4} + 9x^2 - \cancel{18x^3} - 162x + 81x^2 + 729 = 106x^2 - \cancel{18x^3} + \cancel{x^4}$

$16x^2 + 162x - 729 = 0 \Rightarrow (8x - 27)(2x + 27) = 0 \Rightarrow x = \frac{27}{8}$, as $x > 0$.

Sign L': $\begin{array}{ccc} - & c & + \\ \hline 0 & \frac{27}{8} & 9 \end{array}$. The transformer should be located $\frac{27}{8} \approx 3.38$ miles north of a.

27.

Lighthouse

2 mi

A   P   Home

|←x→|←5−x→|

|← 5 mi →|

Suppose he walks from home to the point P (see figure) and then rows directly to the Lighthouse from P. As time $=$ distance/speed, the total time $T = \frac{5-x}{5} + \frac{\sqrt{2^2 + x^2}}{3}$, where $0 \le x \le 5$.

$T' = -\frac{1}{5} + \frac{1}{3} \cdot \frac{1}{\cancel{2}} \cdot \frac{1}{\sqrt{4 + x^2}} (\cancel{2}x) = -\frac{1}{5} + \frac{x}{3\sqrt{4 + x^2}} = 0 \Rightarrow \frac{1}{5} = \frac{x}{3\sqrt{4 + x^2}} \Rightarrow 3\sqrt{4 + x^2} = 5x$,

then squaring: $9(4 + x^2) = 25x^2 \Rightarrow 16x^2 = 36 \Rightarrow x^2 = \frac{36}{16} \Rightarrow x = \frac{6}{4} = \frac{3}{2}$.

Sign T': $\begin{array}{ccc} - & c & + \\ \hline 0 & \frac{3}{2} & 5 \end{array}$. The minimum time occurs when P is 1.5 miles from A. The minimum

time is $T\left(\frac{3}{2}\right) = \frac{5 - \frac{3}{2}}{5} + \frac{\sqrt{2^2 + \left(\frac{3}{2}\right)^2}}{3} = \frac{7}{10} + \frac{5}{6} = \frac{21 + 25}{30} = \frac{46}{30} = \frac{23}{15}$ hrs $= 1 + \frac{8}{15}$ hrs. But

$\frac{8}{15}$ hrs $= \frac{8}{15} \cdot 60 = 32$ min. The minimum time is 1 hr 32 min.

29. 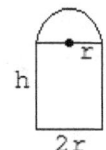 Maximum light for maximum total area: $A = 2rh + \frac{1}{2}\pi r^2$. As the perimeter is 15 feet: $P = 15 = 2r + 2h + \pi r = (2+\pi)r + 2h \Rightarrow$ $h = \frac{15-(2+\pi)r}{2}$. Substituting into A: $A = 2r\left[\frac{15-(2+\pi)r}{2}\right] + \frac{1}{2}\pi r^2$.

Thus $A = 15r - (2+\pi)r^2 + \frac{\pi}{2}r^2 = 15r - \left(2+\frac{\pi}{2}\right)r^2 \Rightarrow$

$A' = 15 - \left(2+\frac{\pi}{2}\right)\cdot 2r = 0 \Rightarrow \left(2+\frac{\pi}{2}\right)\cdot 2r = 15 \Rightarrow (4+\pi)r = 15 \Rightarrow r = \frac{15}{4+\pi}$

Sign A': . The maximum area occurs when $r = \frac{15}{4+\pi}$.

Then $h = \frac{15-(2+\pi)\frac{15}{4+\pi}}{2} = \frac{15[(4+\pi)-(2+\pi)]}{2(4+\pi)} = \frac{15}{4+\pi}$. The dimensions of the base should be $2r = \frac{30}{4+\pi}$ feet wide by $h = \frac{15}{4+\pi}$ feet high.

31. The additional cost is $4x$ so that now $C = 55x + 700x^{-1} + 4x = 59x + 700x^{-1}$. Then $C' = 59 - 700x^{-2} = 0 \Rightarrow x^2 = \frac{700}{59} \Rightarrow x = \sqrt{\frac{700}{59}} \approx 3.44$. Again 4 machines are required.

33. Strength $S = kwd^2$, for some constant k, where d is the depth of the beam and w is its width (see figure). From the right triangle in the figure, $w^2 + d^2 = (2r)^2$, where r is the radius of the log. Then $d^2 = 4r^2 - w^2 \Rightarrow S = kw(4r^2 - w^2) = k(4r^2 w - w^3)$. Differentiating with respect to w (r is a constant):

$S' = k(4r^2 - 3w^2) = 0 \Rightarrow w^2 = \frac{4}{3}r^2 \Rightarrow w = \frac{2}{\sqrt{3}}r$. Sign S': . The strongest beam

has a width of $\frac{2}{\sqrt{3}}r$ units and a depth $d = \sqrt{4r^2 - \frac{4}{3}r^2} = \sqrt{\frac{8}{3}}r = \frac{2\sqrt{2}}{\sqrt{3}}r$ units, i.e. when $d = \sqrt{2}w$.

35. Let A be the given, constant, area. We seek the minimum perimeter $P = 2\ell + 2w$. From $A = \ell w$, $\ell = \frac{A}{w} \Rightarrow P = \frac{2A}{w} + 2w = 2Aw^{-1} + 2w$. Then $P' = -2Aw^{-2} + 2 = 0 \Rightarrow w^2 = A \Rightarrow w = \sqrt{A}$.

Sign P': . Minimum perimeter occurs when $w = \sqrt{A}$, and then $\ell = \frac{A}{\sqrt{A}} = \sqrt{A}$,

i.e. minimum perimeter occurs when $w = \ell$, the rectangle is a square.

37. We want to maximize $A = 4xy$ (note that xy is the area of that part of the rectangle lying in the first quadrant). From $x^2 + y^2 = r^2$ we have: $y = \sqrt{r^2 - x^2}$; so that $A = 4x\sqrt{r^2 - x^2}$. To simplify calculations we find x for which

$$A^2 = 16x^2(r^2 - x^2) = 16(r^2 x^2 - x^4) \text{ is greatest:}$$

$[16(r^2x^2 - x^4)]' = 0 \Rightarrow 16(2r^2x - 4x^3) = 0 \Rightarrow 32x(r^2 - 2x^2) = 0 \Rightarrow x = 0$ or $2x^2 = r^2 \Rightarrow$
$x^2 = \frac{1}{2}r^2 \Rightarrow x = \pm\frac{r}{\sqrt{2}}$. Since x has to be positive ($x = 0$ would yield a minimum area), we
conclude that maximum area occurs when $x = \frac{r}{\sqrt{2}} \Rightarrow y = \sqrt{r^2 - (\frac{r}{\sqrt{2}})^2} = \frac{r}{\sqrt{2}}$, so the figure
is a square.

39.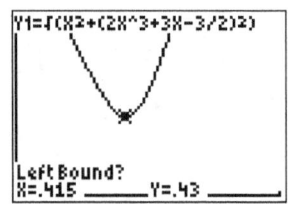

Let D be the distance between the point $\left(0, \frac{1}{2}\right)$ and the curve $y = 2x^3 + 3x - 1$. Then $D = \sqrt{(x - 0)^2 + \left(2x^3 + 3x - 1 - \frac{1}{2}\right)^2}$.
From the calculator glimpse of the graph of D, the minimum distance is $D \approx 0.43$ units.

41.

As the slope of the line segment joining $(x, x+3)$ with $(t, \sqrt{t})$ has a slope of -1, then $-1 = \dfrac{x + 3 - \sqrt{t}}{x - t} \Rightarrow$
$-x + t = x + 3 - \sqrt{t} \Rightarrow$
$2x = t - 3 + \sqrt{t} \Rightarrow x = \frac{1}{2}(t - 3 + \sqrt{t})$

The distance, D, between the two points is $\sqrt{(x + 3 - \sqrt{t})^2 + (x - t)^2}$. Substituting for x in

terms of t: $D = \sqrt{\left(\frac{1}{2}t - \frac{3}{2} + \frac{1}{2}\sqrt{t} + 3 - \sqrt{t}\right)^2 + \left(\frac{1}{2}t - \frac{3}{2} + \frac{1}{2}\sqrt{t} - t\right)^2}$

$$= \sqrt{\left(\frac{1}{2}t + \frac{3}{2} - \frac{1}{2}\sqrt{t}\right)^2 + \left(-\frac{1}{2}t - \frac{3}{2} + \frac{1}{2}\sqrt{t}\right)^2}$$

From the calculator glimpse above, the minimum distance D is obtained when $t = 0.25 \Rightarrow$
$\sqrt{t} = 0.5$. Since the line $y = -x + b$ contains the point $(0.25, 0.5)$: $0.5 = -0.25 + b \Rightarrow b = 0.75$.

43.

From the solution to CYU 4.24, $h \approx 4.1$, $y \approx 2.9$, and $z^2 \approx (4.1)^2 + (x - 2.9)^2$. Now, D is halfway from C to the line joining A and B (see figure), so $s^2 \approx \left(\frac{4.1}{2}\right)^2 + (x - 2.9)^2$.

$$P = K\left\{\frac{1}{x^2 + 10} + \frac{1}{4\left[(12 - x)^2 + 10\right]}\right.$$
$$\left. + \frac{1}{2\left[(4.1)^2 + (x - 2.9)^2 + 10\right]} + \frac{1}{4\left[(\frac{4.1}{2})^2 + (x - 2.9)^2 + 10\right]}\right\}$$

From the calculator glimpse of the graph of P, minimum pollution occurs about 9.3 miles from A.

CHAPTER 5
Integration

§5.1 The Indefinite Integral

1. Since $3x$ is an antiderivative of 3, $\int 3\,dx = 3x + C$.

3. $\int (6x^5 + 5x^4)\,dx = \int 6x^5\,dx + \int 5x^4\,dx$. As x^6 is an antiderivative of $6x^5$, $\int 6x^5\,dx = x^6 + C$. Likewise, as x^5 is an antiderivative of $5x^4$, $\int 5x^4\,dx = x^5 + C$. Then $\int (6x^5 + 5x^4)\,dx = x^6 + x^5 + C$.

5. $\displaystyle \int \left(\frac{x^4}{5} - \frac{3}{x^5}\right) dx = \int \left(\frac{1}{5}x^4 - 3x^{-5}\right) dx = \int \left(\frac{1}{5}x^4\right) dx - \int 3x^{-5}\,dx = \frac{1}{5}\int x^4\,dx - 3\int x^{-5}\,dx$

$$= \frac{1}{5} \cdot \frac{x^{4+1}}{4+1} - 3 \cdot \frac{x^{-5+1}}{-5+1} + C = \frac{1}{5} \cdot \frac{x^5}{5} - 3 \cdot \frac{x^{-4}}{-4} + C = \frac{1}{25}x^5 + \frac{3}{4x^4} + C.$$

7. $\displaystyle \int \left(3x^4 - 4x^{-4} + \frac{2}{x^5}\right) dx = \int 3x^4\,dx - \int 4x^{-4}\,dx + \int 2x^{-5}\,dx$

$$= 3\int x^4\,dx - 4\int x^{-4}\,dx + 2\int x^{-5}\,dx$$

$$= 3 \cdot \frac{x^5}{5} - 4 \cdot \frac{x^{-4+1}}{-4+1} + 2 \cdot \frac{x^{-5+1}}{-5+1} + C$$

$$= 3 \cdot \frac{x^5}{5} - 4 \cdot \frac{x^{-3}}{-3} + 2 \cdot \frac{x^{-4}}{-4} + C = \frac{3}{5}x^5 + \frac{4}{3x^3} - \frac{1}{2x^4} + C.$$

9. $\displaystyle \int x^2(2x - 5)\,dx = \int (2x^3 - 5x^2)\,dx = \int 2x^3\,dx - \int 5x^2\,dx = 2\int x^3\,dx - 5\int x^2\,dx$

$$= 2 \cdot \frac{x^4}{4} - 5 \cdot \frac{x^3}{3} + C = \frac{1}{2}x^4 - \frac{5}{3}x^3 + C.$$

11. $\displaystyle \int x(x-1)(x+1)\,dx = \int x(x^2 - 1)\,dx = \int (x^3 - x)\,dx = \frac{1}{4}x^4 - \frac{1}{2}x^2 + C.$

13. $\displaystyle \int \frac{x^6 + x^2 - x^{-2}}{2x^4}\,dx = \frac{1}{2}\int (x^2 + x^{-2} - x^{-6})\,dx = \frac{1}{2}\left(\frac{x^3}{3} + \frac{x^{-2+1}}{-2+1} + \frac{x^{-6+1}}{-6+1}\right) + C$

$$= \frac{1}{2}\left(\frac{x^3}{3} + \frac{x^{-1}}{1} - \frac{x^{-5}}{5}\right) + C = \frac{1}{6}x^3 - \frac{1}{2x} + \frac{1}{10x^5} + C.$$

15. $\displaystyle \int \frac{(x^4 + x)(x + 1)}{x^4}\,dx = \int \frac{x^5 + x^4 + x^2 + x}{x^4}\,dx = \int (x + 1 + x^{-2} + x^{-3})\,dx$

$$= \frac{x^2}{2} + x + \frac{x^{-1}}{-1} + \frac{x^{-2}}{-2} + C = \frac{1}{2}x^2 + x - \frac{1}{x} - \frac{1}{2x^2} + C.$$

17. $\displaystyle \int x^{-\frac{3}{5}}\,dx = \frac{x^{-\frac{3}{5}+1}}{-\frac{3}{5}+1} + C = \frac{x^{\frac{2}{5}}}{\frac{2}{5}} + C = \frac{5}{2}x^{\frac{2}{5}} + C.$

19. $\int \sqrt{x}(x^2 + x - 3)\, dx = \int (x^{\frac{5}{2}} + x^{\frac{3}{2}} - 3x^{\frac{1}{2}})\, dx = \frac{x^{\frac{5}{2}+1}}{\frac{5}{2}+1} + \frac{x^{\frac{3}{2}+1}}{\frac{3}{2}+1} - 3 \cdot \frac{x^{\frac{1}{2}+1}}{\frac{1}{2}+1} + C$

$\qquad = \frac{x^{\frac{7}{2}}}{\frac{7}{2}} + \frac{x^{\frac{5}{2}}}{\frac{5}{2}} - 3 \cdot \frac{x^{\frac{3}{2}}}{\frac{3}{2}} + C = \frac{2}{7}x^{\frac{7}{2}} + \frac{2}{5}x^{\frac{5}{2}} - 2x^{\frac{3}{2}} + C.$

21. $\int \frac{x(2x^{\frac{1}{3}} + x^3)}{x^{\frac{2}{3}}}\, dx = \int x^{\frac{1}{3}}(2x^{\frac{1}{3}} + x^3)\, dx = \int (2x^{\frac{2}{3}} + x^{\frac{10}{3}})\, dx = 2 \cdot \frac{x^{\frac{2}{3}+1}}{\frac{2}{3}+1} + \frac{x^{\frac{10}{3}+1}}{\frac{10}{3}+1} + C$

$\qquad = 2 \cdot \frac{x^{\frac{5}{3}}}{\frac{5}{3}} + \frac{x^{\frac{13}{3}}}{\frac{13}{3}} + C = \frac{6}{5}x^{\frac{5}{3}} + \frac{3}{13}x^{\frac{13}{3}} + C.$

23. $\int (\sec x \tan x - \sec^2 x)\, dx = \int \sec x \tan x\, dx - \int \sec^2 x\, dx = \sec x - \tan x + C.$

25. Let $I = \int \frac{\sec x - \tan x}{\cot x}\, dx = \int \frac{\sec x}{\cot x}\, dx - \int \frac{\tan x}{\cot x}\, dx = \int \sec x \tan x\, dx - \int \tan^2 x\, dx$

$\qquad \int \sec x \tan x\, dx = \sec x + C$

$\qquad \int \tan^2 x\, dx = \int (\sec^2 x - 1)\, dx = \int \sec^2 x\, dx - \int dx = \tan x - x + C$

$\qquad$ Therefore, $I = \sec x - \tan x + x + C.$

27. $f'(x) = 3x + 5 \Rightarrow f(x) = \int (3x + 5)\, dx = 3 \cdot \frac{x^2}{2} + 5x + C = \frac{3}{2}x^2 + 5x + C.$ As $f(5) = 1,\ 1 = \frac{3}{2}(5^2) + 5(5) + C \Rightarrow 1 = \frac{3}{2}(25) + 25 + C \Rightarrow 2 = 75 + 50 + 2C \Rightarrow 2C = -123 \Rightarrow C = -\frac{123}{2} \Rightarrow f(x) = \frac{3}{2}x^2 + 5x - \frac{123}{2}.$

29. $f'(x) = 3x^2 + 5x \Rightarrow f(x) = \int (3x^2 + 5x)\, dx = 3 \cdot \frac{x^3}{3} + 5 \cdot \frac{x^2}{2} + C = x^3 + \frac{5}{2}x^2 + C.$ As $f(1) = 1,\ 1 = 1^3 + \frac{5}{2}(1^2) + C \Rightarrow C = -\frac{5}{2} \Rightarrow f(x) = x^3 + \frac{5}{2}x^2 - \frac{5}{2}.$

31. $f'(x) = x^3 + 5x - 2 \Rightarrow f(x) = \int (x^3 + 5x - 2)\, dx = \frac{x^4}{4} + 5 \cdot \frac{x^2}{2} - 2x + C = \frac{1}{4}x^4 + \frac{5}{2}x^2 - 2x + C.$ As $f(0) = 1,\ 1 = C \Rightarrow f(x) = \frac{1}{4}x^4 + \frac{5}{2}x^2 - 2x + 1.$

33. $f'(x) = \frac{3x^2 + 5x}{x^4} = 3x^{-2} + 5x^{-3} \Rightarrow$

$\qquad f(x) = \int (3x^{-2} + 5x^{-3})\, dx = 3 \cdot \frac{x^{-1}}{-1} + 5 \cdot \frac{x^{-2}}{-2} + C = -\frac{3}{x} - \frac{5}{2x^2} + C.$

$\qquad$ As $f(1) = 2,\ 2 = -3 - \frac{5}{2} + C \Rightarrow C = \frac{15}{2} \Rightarrow f(x) = -\frac{3}{x} - \frac{5}{2x^2} + \frac{15}{2}.$

35. $f'(x) = (2x + 3)(x - 1) = 2x^2 + x - 3 \Rightarrow$

$\qquad f(x) = \int (2x^2 + x - 3)\, dx = 2 \cdot \frac{x^3}{3} + \frac{x^2}{2} - 3x + C = \frac{2}{3}x^3 + \frac{1}{2}x^2 - 3x + C.$

$\qquad$ As $f(1) = 0,\ 0 = \frac{2}{3} + \frac{1}{2} - 3 + C \Rightarrow C = -\frac{4}{6} - \frac{3}{6} + \frac{18}{6} = \frac{11}{6} \Rightarrow f(x) = \frac{2}{3}x^3 + \frac{1}{2}x^2 - 3x + \frac{11}{6}.$

37. $f''(x) = 3x + 5 \Rightarrow f'(x) = \int (3x + 5)\, dx = 3 \cdot \frac{x^2}{2} + 5x + C = \frac{3}{2}x^2 + 5x + C.$ As $f'(0) = 1,\ 1 = C \Rightarrow f'(x) = \frac{3}{2}x^2 + 5x + 1 \Rightarrow f(x) = \int (\frac{3}{2}x^2 + 5x + 1)\, dx = \frac{3}{2} \cdot \frac{x^3}{3} + 5 \cdot \frac{x^2}{2} + x + C = \frac{1}{2}x^3 + \frac{5}{2}x^2 + x + C.$ As $f(1) = 1,\ 1 = \frac{1}{2} + \frac{5}{2} + 1 + C \Rightarrow C = -3 \Rightarrow f(x) = \frac{1}{2}x^3 + \frac{5}{2}x^2 + x - 3.$

39. To verify that $\int x(x^2 - 1)^4 \, dx = \frac{1}{10}(x^2 - 1)^5 + C$, we only have to differentiate:
$\left[\frac{1}{10}(x^2 - 1)^5 + C\right]' = \frac{1}{10} \cdot 5(x^2 - 1)^4(2x) + 0 = x(x^2 - 1)^4.$

41. To show that $y = \cos x$ is a solution of the equation $(y')^2 + y^2 - 1 = 0$, is to show that the given y and its derivative satisfy the equation: $y = \cos x \Rightarrow y' = -\sin x$
$(y')^2 + y^2 - 1 = (-\sin x)^2 + (\cos x)^2 - 1 = \sin^2 x + \cos^2 x - 1 = 0$, as $\sin^2 x + \cos^2 x = 1$.

43. The slope of the tangent line to the graph of f at $(x, f(x))$ is the derivative, $f'(x)$. Thus $f'(x) = x^2 \Rightarrow f(x) = \int x^2 \, dx = \frac{x^3}{3} + C = \frac{1}{3}x^3 + C$. Since the graph passes through the point $(1,5)$, $f(1) = 5 \Rightarrow 5 = \frac{1}{3}(1)^3 + C \Rightarrow C = 5 - \frac{1}{3} = \frac{14}{3} \Rightarrow f(x) = \frac{1}{3}x^3 + \frac{14}{3}$.

45. The height (in feet) of the stone t seconds after it was released is given, in general, by the formula: $s(t) = -16t^2 + v_0 t + s_0$, where v_0 is the initial velocity of the stone, and s_0 is its initial height.
  From the given information, $v_0 = 0$ (stone was dropped), and $s_0 = 3200 \Rightarrow s(t) = -16t^2 + 3200 \Rightarrow v(t) = s'(t) = -32t$. On impact, $s(t) = 0 \Rightarrow -16t^2 + 3200 = 0 \Rightarrow 16t^2 = 3200 \Rightarrow t^2 = 200 \Rightarrow t = \sqrt{200} = 10\sqrt{2}$. Thus, on impact, the stone's speed is $|v(10\sqrt{2})| = |-32(10\sqrt{2})| = 320\sqrt{2}$ ft/sec.

47. The height (in feet) of the object t seconds after it was released is given, in general, by the formula: $s(t) = -16t^2 + v_0 t + s_0$, where v_0 is the initial velocity of the object, and s_0 is its initial height.
  From the given information, $v_0 = -16$, and $s_0 = 96 \Rightarrow s(t) = -16t^2 - 16t + 96$. When the object hits the ground, $s(t) = 0 \Rightarrow -16t^2 - 16t + 96 = 0 \Rightarrow t^2 + t - 6 = 0 \Rightarrow (t + 3)(t - 2) = 0 \Rightarrow t = 2$. So the impact velocity is $v(2)$, where $v(t) = s'(t) = -32t - 16 \Rightarrow v(2) = -32(2) - 16 = -80 \Rightarrow$ impact speed is 80 ft/sec. Three-quarters of its impact speed is $\frac{3}{4}(80) = 60$ ft/sec.
  For an object propelled upward from the ground at 60 ft/sec, its height at any time t is given by: $s(t) = -16t^2 + 60t$. At maximum height, $v(t) = 0$. But $v(t) = s'(t) = -32t + 60 = 0 \Rightarrow t = \frac{60}{32} = \frac{15}{8}$ seconds. Its height is then $s\left(\frac{15}{8}\right) = -16\left(\frac{15}{8}\right)^2 + 60\left(\frac{15}{8}\right) = \frac{225}{4} \approx 56.3$ ft.

49. (a) $s(t) = t^3 - t = t(t^2 - 1) = t(t - 1)(t + 1)$, $t \geq 0$. Sign $s(t)$: $\frac{-\ \ \ +}{0\ \ \ \ \ 1}$. The sign chart tells us that the particle is located to the left of the origin for the first minute, and then to the right of the origin, thereafter. The direction of motion changes when the particle's velocity is 0: $v(t) = s'(t) = 3t^2 - 1 = 0 \Rightarrow t^2 = \frac{1}{3} \Rightarrow t = \frac{1}{\sqrt{3}}$. At that time, the particle's

position is $s\left(\frac{1}{\sqrt{3}}\right) = \left(\frac{1}{\sqrt{3}}\right)^3 - \left(\frac{1}{\sqrt{3}}\right) = \frac{1}{3\sqrt{3}} - \frac{1}{\sqrt{3}} = \frac{1}{\sqrt{3}}\left(\frac{1}{3} - 1\right) = -\frac{2}{3\sqrt{3}} \cdot \frac{\sqrt{3}}{\sqrt{3}} = -\frac{2\sqrt{3}}{9}$ (see figure):

  (b) From (a), the particle moves to the right when $t > \frac{1}{\sqrt{3}}$, and to the left for $0 < t < \frac{1}{\sqrt{3}}$.

(c) Acceleration $a(t) = v'(t) = (3t^2 - 1)' = 6t > 0$, which means that the velocity is always increasing. As speed is the magnitude of velocity, the particle is speeding up when $v(t) > 0$, i.e. $t > \frac{1}{\sqrt{3}}$, and slowing down when $v(t) < 0$, i.e. $0 < t < \frac{1}{\sqrt{3}}$.

(d) Total distance traveled is the sum of the distance traveled to the left $(0 < t < \frac{1}{\sqrt{3}})$ and the distance traveled to the right $(\frac{1}{\sqrt{3}} < t < 5)$.

To the left: $|s(\frac{1}{\sqrt{3}}) - s(0)| = |-\frac{2\sqrt{3}}{9} - 0| = \frac{2\sqrt{3}}{9}$

To the right: $|s(\frac{1}{\sqrt{3}}) - s(5)| = |-\frac{2\sqrt{3}}{9} - (5^3 - 5)| = \frac{2\sqrt{3}}{9} + 120$

Total distance is $\frac{2\sqrt{3}}{9} + \frac{2\sqrt{3}}{9} + 120 = \frac{4\sqrt{3}}{9} + 120$ meters.

51.   When the brakes are applied, $a(t) = -30 \Rightarrow v(t) = \int a(t)\,dt = -30t + v_0$. As $v_0 = 88$, $v(t) = -30t + 88$. Then $s(t) = \int v(t)\,dt = -30 \cdot \frac{t^2}{2} + 88t + s_0 = -15t^2 + 88t + s_0$. When stopped, $v(t) = 0 \Rightarrow -30t + 88 = 0 \Rightarrow t = \frac{88}{30} = \frac{44}{15}$. The car's position then is $s\left(\frac{44}{15}\right) = -15\left(\frac{44}{15}\right)^2 + 88\left(\frac{44}{15}\right) + s_0$, so the distance traveled is:
$$\left|s\left(\frac{44}{15}\right) - s_0\right| = \left|-15\left(\frac{44}{15}\right)^2 + 88\left(\frac{44}{15}\right)\right| = \frac{1936}{15} \approx 129.1 \text{ ft.}$$

53. Since the object is tossed from ground level, $s(t) = -16t^2 + v_0 t$.
(a) $v(t) = s'(t) = -32t + v_0$. At maximum height, $v(t) = 0 \Rightarrow -32t + v_0 = 0 \Rightarrow t = \frac{v_0}{32}$.
The maximum height attained, $M = s(\frac{v_0}{32}) = -16(\frac{v_0}{32})^2 + v_0(\frac{v_0}{32}) = -\frac{v_0^2}{64} + \frac{v_0^2}{32} = \frac{v_0^2}{64}$ ft.

(b) When the object is at height h,
$$s(t) = h \Rightarrow -16t^2 + v_0 t = h \Rightarrow 16t^2 - v_0 t + h = 0 \Rightarrow t = \frac{v_0 \pm \sqrt{v_0^2 - 64h}}{32}.$$
The velocity at height h, when the object is going up, is
$$v\left(\frac{v_0 - \sqrt{v_0^2 - 64h}}{32}\right) = -32\left(\frac{v_0 - \sqrt{v_0^2 - 64h}}{32}\right) + v_0 = \sqrt{v_0^2 - 64h}$$
The velocity at height h, when the object is coming down, is
$$v\left(\frac{v_0 + \sqrt{v_0^2 - 64h}}{32}\right) = -32\left(\frac{v_0 + \sqrt{v_0^2 - 64h}}{32}\right) + v_0 = -\sqrt{v_0^2 - 64h}$$
As speed is the magnitude of velocity, the speed going up equals the speed coming down.

§5.2 The Definite Integral

1. $\int_0^1 3\,dx = 3x\Big|_0^1 = 3(1 - 0) = 3.$

3. $\int_{-1}^1 (3 + 3x)\,dx = 3\int_{-1}^1 (1 + x)\,dx = 3(x + \frac{x^2}{2})\Big|_{-1}^1 = 3\left[\left(1 + \frac{1^2}{2}\right) - \left(-1 + \frac{(-1)^2}{2}\right)\right]$
$= 3(1 + \frac{1}{2} + 1 - \frac{1}{2}) = 6.$

5. $\int_1^2 (x^2 + 3x - 1)\, dx = \left(\frac{1}{3}x^3 + \frac{3}{2}x^2 - x\right)\Big|_1^2 = \left[\frac{1}{3}(2^3) + \frac{3}{2}(2^2) - 2\right] - \left[\frac{1}{3}(1^3) + \frac{3}{2}(1^2) - 1\right]$

$\qquad = \frac{8}{3} + 6 - 2 - \frac{1}{3} - \frac{3}{2} + 1 = 5 + \frac{7}{3} - \frac{3}{2} = \frac{30 + 14 - 9}{6} = \frac{35}{6}.$

7. $\int_1^2 x(3x - 2)\, dx = \int_1^2 (3x^2 - 2x)\, dx = (x^3 - x^2)\Big|_1^2 = (2^3 - 2^2) - (1^3 - 1^2) = 4 - 0 = 4.$

9. $\int_0^{-1}(3x - 1)(x - 1)\, dx = \int_0^1 (3x^2 - 4x + 1)\, dx = \left(x^3 - 4 \cdot \frac{x^2}{2} + x\right)\Big|_0^{-1} = (x^3 - 2x^2 + x)\Big|_0^{-1}$

$\qquad = [(-1)^3 - 2(-1)^2 - 1] - 0 = -1 - 2 - 1 = -4.$

11. $\int_1^2 \frac{(x^4 + x)(x + 1)}{x^4}\, dx = \int_1^2 \frac{x^5 + x^4 + x^2 + x}{x^4}\, dx = \int_1^2 (x + 1 + x^{-2} + x^{-3})\, dx$

$\qquad = \left(\frac{x^2}{2} + x + \frac{x^{-1}}{-1} + \frac{x^{-2}}{-2}\right)\Big|_1^2 = \left(\frac{x^2}{2} + x - \frac{1}{x} - \frac{1}{2x^2}\right)\Big|_1^2$

$\qquad = \left(2 + 2 - \frac{1}{2} - \frac{1}{8}\right) - \left(\frac{1}{2} + \cancel{1} - \cancel{1} - \frac{1}{2}\right)$

$\qquad = 4 - \frac{1}{2} - \frac{1}{8} = \frac{32 - 4 - 1}{8} = \frac{27}{8}.$

13. $\int_1^2 \sqrt{x}\, dx = \int_1^2 x^{\frac{1}{2}}\, dx = \frac{x^{\frac{3}{2}}}{\frac{3}{2}}\Big|_1^2 = \frac{2}{3}x^{\frac{3}{2}}\Big|_1^2 = \frac{2}{3}\left(2^{\frac{3}{2}} - 1^{\frac{3}{2}}\right) = \frac{2}{3}\left(2\sqrt{2} - 1\right) = \frac{4\sqrt{2} - 2}{3}.$

15. $\int_0^1 \sqrt{x}(x^2 + x - 3)\, dx = \int_0^1 (x^{\frac{5}{2}} + x^{\frac{3}{2}} - 3x^{\frac{1}{2}})\, dx = \left(\frac{x^{\frac{7}{2}}}{\frac{7}{2}} + \frac{x^{\frac{5}{2}}}{\frac{5}{2}} - 3 \cdot \frac{x^{\frac{3}{2}}}{\frac{3}{2}}\right)\Big|_0^1$

$\qquad = \left(\frac{2}{7}x^{\frac{7}{2}} + \frac{2}{5}x^{\frac{5}{2}} - 2x^{\frac{3}{2}}\right)\Big|_0^1 = \left(\frac{2}{7} + \frac{2}{5} - 2\right) - 0$

$\qquad = \frac{10 + 14 - 70}{35} = -\frac{46}{35}.$

17. $\int_{-\frac{\pi}{4}}^{\frac{\pi}{2}} \cos x\, dx = \sin x\Big|_{-\frac{\pi}{4}}^{\frac{\pi}{2}} = \sin \frac{\pi}{2} - \sin\left(-\frac{\pi}{4}\right) = 1 + \frac{1}{\sqrt{2}}.$

19. $\int_{\frac{\pi}{4}}^{\frac{\pi}{3}} \tan x \sec x\, dx = \sec x\Big|_{\frac{\pi}{4}}^{\frac{\pi}{3}} = \sec \frac{\pi}{3} - \sec \frac{\pi}{4} = \frac{1}{\cos \frac{\pi}{3}} - \frac{1}{\cos \frac{\pi}{4}} = \frac{1}{\frac{1}{2}} - \frac{1}{\frac{1}{\sqrt{2}}} = 2 - \sqrt{2}.$

21. 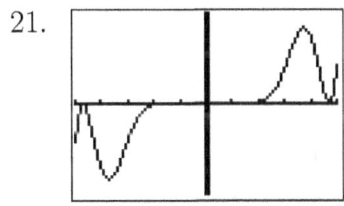 From the calculator glimpse of the graph of the integrand in a window of $[-5, 5] \times [-600, 600]$, we see the graph of an odd function over a symmetric interval. Integrating from -5 to 0 will result in the negative of the value of the integral from 0 to 5. The value of the integral: $\int_{-5}^5 x^5 \cos^2 x\, dx$ is therefore zero.

23.

Area $= \int_0^1 x^3\,dx = \frac{1}{4}x^4\Big|_0^1 = \frac{1}{4}(1^4 - 0) = \frac{1}{4}$.

25.

Area $= \int_1^2 \sqrt{x}\,dx = \int_1^2 x^{\frac{1}{2}}\,dx = \dfrac{x^{\frac{3}{2}}}{\frac{3}{2}}\Big|_1^2 = \frac{2}{3}x^{\frac{3}{2}}\Big|_1^2 = \frac{2}{3}(2^{\frac{3}{2}} - 1^{\frac{3}{2}}) = \frac{4\sqrt{2} - 2}{3}$.

27.

Area $= \int_{-1}^2 |x - 1|\,dx = \int_{-1}^1 -(x - 1)\,dx + \int_1^2 (x - 1)\,dx$

$= \left(-\frac{1}{2}x^2 + x\right)\Big|_{-1}^1 + \left(\frac{1}{2}x^2 - x\right)\Big|_1^2$

$= \left[\left(-\frac{1}{2} + 1\right) - \left(-\frac{1}{2} - 1\right)\right] + \left[\left(\frac{1}{2} \cdot 2^2 - 2\right) - \left(\frac{1}{2} - 1\right)\right]$

$= \frac{1}{2} + \frac{3}{2} + 0 + \frac{1}{2} = \frac{5}{2}$.

29. Let $V(t)$ be the resale value of the car. Then $V'(t) = 1200 + 600t + 4t^3$, for $0 \le t \le 7$
(a) Depreciation in first 3 years:

$V(3) - V(0) = \int_0^3 V'(t)\,dt = \left(1200t + 600 \cdot \dfrac{t^2}{2} + t^4\right)\Big|_0^3$

$= 1200(3) + 300(3^2) + 3^4 - 0 = 3600 + 2700 + 81 = \$6,381.$

(b) Depreciation in 3rd year:

$V(3) - V(2) = \int_2^3 V'(t)\,dt = \left(1200t + 600 \cdot \dfrac{t^2}{2} + t^4\right)\Big|_2^3$

$[\text{from (a)}] = 6381 - [1200(2) + 300(2^2) + 2^4]$

$= 6381 - [2400 + 1200 + 16] = 6381 - 3616 = \$2,765.$

31. $N(t) = \int N'(t)\,dt = \int \dfrac{t}{100}\,dt = \dfrac{1}{100} \cdot \dfrac{t^2}{2} + C = \dfrac{t^2}{200} + C.$ $N(0) = 200 \Rightarrow N(t) = \dfrac{t^2}{200} + 200.$

Find t with $N(t) = 400$: $\dfrac{t^2}{200} + 200 = 400 \Rightarrow \dfrac{t^2}{200} = 200 \Rightarrow t^2 = 200^2 \Rightarrow t = 200$ days.

33. Earnings for 5 years (60 months) with the \$2,000 machine:

$\int_0^{60}(190 + 12t)\,dt = \left(190t + 12 \cdot \dfrac{t^2}{2}\right)\Big|_0^{60} = 190(60) + 6(60)^2 = \$33,000$

Earnings for 5 years (60 months) with the \$3,000 machine:

$\int_0^{60}(250 + 20t)\,dt = \left(250t + 20 \cdot \dfrac{t^2}{2}\right)\Big|_0^{60} = 250(60) + 10(60)^2 = \$51,000$

For an additional \$1,000 initial expense when buying the \$3,000 machine, there is an \$18,000 increase in earnings over 5 years: Buy the \$3,000 machine.

35. $\int_1^4 3g(x)\,dx = 3\int_1^4 g(x)\,dx = 3\cdot 9 = 27.$

37. $\int_1^4 [2g(x) - f(x)]\,dx = \int_1^4 2g(x)\,dx - \int_1^4 f(x)\,dx = 2\int_1^4 g(x)\,dx - \left[\int_1^2 f(x)\,dx + \int_2^4 f(x)\,dx\right]$
$$= 2\cdot 9 - [5+7] = 18 - 12 = 6.$$

39. $\left(\int_1^3 f(x)\,dx\right)\left(\int_3^3 g(x)\,dx\right) = 0$, as $\left(\int_3^3 g(x)\,dx\right) = 0.$

41. Given $T(x) = \int_5^x \sqrt{3t^4 + 1}\,dt$, then $T'(x) = \sqrt{3x^4 + 1}.$

43. Given $T(x) = \displaystyle\int_1^x \frac{\sin t}{t^2 + 1}\,dt$, then $T'(x) = \dfrac{\sin x}{x^2 + 1}.$

45. (a) $T'(x) = \frac{d}{dx}\int_1^x (t^2 + t)\,dt = x^2 + x.$

(b) $T(x) = \int_1^x (t^2 + t)\,dt = \left(\frac{1}{3}t^3 + \frac{1}{2}t^2\right)\Big|_1^x = \frac{1}{3}x^3 + \frac{1}{2}x^2 - \left(\frac{1}{3} + \frac{1}{2}\right) = \frac{1}{3}x^3 + \frac{1}{2}x^2 - \frac{5}{6}.$
$T'(x) = \frac{1}{3}\cdot 3x^2 + \frac{1}{2}\cdot 2x - 0 = x^2 + x$ [same as in (a)].

(c) $T'(x) = \frac{d}{dx}\int_5^x (t^2 + t)\,dt = x^2 + x.$
$T(x) = \int_5^x (t^2 + t)\,dt = \left(\frac{1}{3}t^3 + \frac{1}{2}t^2\right)\Big|_5^x = \frac{1}{3}x^3 + \frac{1}{2}x^2 - \left(\frac{1}{3}\cdot 5^3 + \frac{1}{2}\cdot 5^2\right).$
$T'(x) = \frac{1}{3}\cdot 3x^2 + \frac{1}{2}\cdot 2x - 0 = x^2 + x.$

47. Let $T(x) = \int_a^x f(t)\,dt$, and let $H(x) = T[g(x)] \Rightarrow H'(x) = T'[g(x)]g'(x) = f[g(x)]g'(x).$

49. $H'(x) = \dfrac{d}{dx}\displaystyle\int_5^{2x} \sqrt{3t^4 + 1}\,dt = \sqrt{3(2x)^4 + 1}\cdot\dfrac{d}{dx}(2x) = \sqrt{3(16x^4) + 1}\cdot 2 = 2\sqrt{48x^4 + 1}.$

51. $H'(x) = \dfrac{d}{dx}\displaystyle\int_{x^2}^{\sin x} \dfrac{dt}{x^2 + 1} = \dfrac{d}{dx}\left[\int_{x^2}^1 \dfrac{dt}{x^2 + 1} + \int_1^{\sin x} \dfrac{dt}{x^2 + 1}\right]$
$= \dfrac{d}{dx}\left[-\int_1^{x^2} \dfrac{dt}{x^2 + 1} + \int_1^{\sin x} \dfrac{dt}{x^2 + 1}\right] = -\dfrac{1}{(x^2)^2 + 1}\cdot 2x + \dfrac{1}{\sin^2 x + 1}\cdot\cos x$
$= -\dfrac{2x}{x^4 + 1} + \dfrac{\cos x}{\sin^2 x + 1}.$

53. (a) $H(x) = \int_1^{x^2} t(t - 5)\,dt \Rightarrow H'(x) = x^2(x^2 - 5)\cdot 2x = 2x^3(x^2 - 5) = 2x^5 - 10x^3.$

(b) $H(x) = \int_1^{x^2} (t^2 - 5t)\,dt = \left(\frac{1}{3}t^3 - \frac{5}{2}t^2\right)\Big|_1^{x^2} = \frac{1}{3}x^6 - \frac{5}{2}x^4 - \left(\frac{1}{3} - \frac{5}{2}\right) = \frac{1}{3}x^6 - \frac{5}{2}x^4 + \frac{13}{6}.$
$H'(x) = \frac{1}{3}\cdot 6x^5 - \frac{5}{2}\cdot 4x^3 = 2x^5 - 10x^3$ [same as in (a)].

55. $y = \int_1^x t\sin t\,dt \Rightarrow \dfrac{dy}{dx} = x\sin x \Rightarrow \dfrac{d^2y}{dx^2} = x\left(\dfrac{d}{dx}\sin x\right) + \left(\dfrac{d}{dx}x\right)\sin x = x\cos x + \sin x.$

57. $y = \int_1^{x^2} \tan t \, dt \;\Rightarrow\; \frac{dy}{dx} = (\tan x^2)(2x) = 2x \tan x^2 \;\Rightarrow$

$$\frac{d^2y}{dx^2} = \frac{d}{dx}(2x \tan x^2) = 2 \tan x^2 + 2x(\sec^2 x^2)(2x) = 2 \tan x^2 + 4x^2 \sec^2 x^2.$$

59. $\displaystyle\int_0^2 [f(x) - g(x)] \, dx$

$$= \lim_{n \to \infty} \left[\int_0^{1-\frac{1}{n}} [f(x) - g(x)] \, dx + \int_{1-\frac{1}{n}}^{1+\frac{1}{n}} [f(x) - g(x)] \, dx + \int_{1+\frac{1}{n}}^2 [f(x) - g(x)] \, dx \right]$$

$$= \lim_{n \to \infty} \left[0 + \int_{1-\frac{1}{n}}^{1+\frac{1}{n}} [f(x) - g(x)] \, dx + 0 \right]$$

$$\leq \lim_{n \to \infty} \left\{ \max|f(x) - g(x)| [(1 + \tfrac{1}{n}) - (1 - \tfrac{1}{n})] \right\} < \lim_{n \to \infty} \frac{2}{n} = 0.$$

§5.3 The Substitution Method

1. Let $u = x - 5$, then $du = dx \;\Rightarrow\; \int (x-5)^{15} \, dx = \int u^{15} \, du = \frac{1}{16} u^{16} + C = \frac{1}{16}(x-5)^{16} + C.$

3. Let $u = 2x - 5$, then $du = 2dx \;\Rightarrow\; dx = \frac{1}{2} du \;\Rightarrow$

$$\int \frac{dx}{(2x-5)^{15}} = \frac{1}{2} \int u^{-15} \, du = \frac{1}{2} \left(\frac{u^{-14}}{-14} \right) + C = -\frac{1}{28(2x-5)^{14}} + C.$$

5. Let $u = x^2 + 5$, then $du = 2x dx \;\Rightarrow\; x dx = \frac{1}{2} du \;\Rightarrow$

$$\int x(x^2+5)^{15} \, dx = \frac{1}{2} \int u^{15} \, du = \frac{1}{2} \left(\frac{1}{16} u^{16} \right) + C = \frac{1}{32}(x^2+5)^{16} + C.$$

7. Let $u = 5x^2 - 4$, then $du = 5(2x)dx = 10x dx \;\Rightarrow\; x dx = \frac{1}{10} du \;\Rightarrow$

$$\int \frac{x}{\sqrt{5x^2-4}} \, dx = \int \frac{\frac{1}{10} du}{\sqrt{u}} = \frac{1}{10} \int u^{-\frac{1}{2}} \, du = \frac{1}{10} \cdot \frac{u^{\frac{1}{2}}}{\frac{1}{2}} + C = \frac{1}{5}\sqrt{5x^2-4} + C.$$

9. Let $I = \displaystyle\int \left[3x + \frac{x}{(x^2-3)^2} \right] dx = \int 3x \, dx + \int \frac{x}{(x^2-3)^2} \, dx = 3 \cdot \frac{1}{2}x^2 + \int \frac{x}{(x^2-3)^2} \, dx$

Let $u = x^2 - 3$, then $du = 2x dx \;\Rightarrow\; x dx = \frac{1}{2} du \;\Rightarrow$

$$\int \frac{x}{(x^2-3)^2} \, dx = \frac{1}{2} \int u^{-2} \, du = \frac{1}{2} \left(\frac{u^{-1}}{-1} \right) + C = -\frac{1}{2u} + C = -\frac{1}{2(x^2-3)} + C$$

Therefore, $I = \frac{3}{2}x^2 - \frac{1}{2(x^2-3)} + C.$

11. Let $u = \tan x$, then $du = \sec^2 x dx \;\Rightarrow\; \int \sec^2 x \tan x \, dx = \int u \, du = \frac{1}{2}u^2 + C = \frac{1}{2}\tan^2 x + C.$

[Note: Another choice: $u = \sec x$ leads to an equivalent answer: $\frac{1}{2}\sec^2 x + C$, since

$\frac{1}{2}\sec^2 x + C = \frac{1}{2}(1 + \tan^2 x) + C = \frac{1}{2} + \frac{1}{2}\tan^2 x + C = \frac{1}{2}\tan^2 x + C.$]

13. Let $u = x + 1$, then $du = dx$ and $x = u - 1 \Rightarrow$

$$\int \frac{x^2}{(x+1)^4}\, dx = \int \frac{(u-1)^2}{u^4}\, du = \int \frac{u^2 - 2u + 1}{u^4}\, du = \int \left(u^{-2} - 2u^{-3} + u^{-4}\right) du$$

$$= \frac{u^{-1}}{-1} - 2 \cdot \frac{u^{-2}}{-2} + \frac{u^{-3}}{-3} + C = -\frac{1}{u} + \frac{1}{u^2} - \frac{1}{3u^3} + C$$

$$= -\frac{1}{x+1} + \frac{1}{(x+1)^2} - \frac{1}{3(x+1)^3} + C.$$

15. Let $u = x - 3$, then $du = dx$ and $x = u + 3 \Rightarrow$

$$\int \frac{x^2 + x - 1}{\sqrt{x-3}}\, dx = \int \frac{(u+3)^2 + u + 3 - 1}{u^{\frac{1}{2}}}\, du = \int \frac{u^2 + 7u + 11}{u^{\frac{1}{2}}}\, du$$

$$= \int \left(u^{\frac{3}{2}} + 7u^{\frac{1}{2}} + 11u^{-\frac{1}{2}}\right) du = \frac{u^{\frac{5}{2}}}{\frac{5}{2}} + 7 \cdot \frac{u^{\frac{3}{2}}}{\frac{3}{2}} + 11 \cdot \frac{u^{\frac{1}{2}}}{\frac{1}{2}} + C$$

$$= \frac{2}{5}u^{\frac{5}{2}} + \frac{14}{3}u^{\frac{3}{2}} + 22\sqrt{u} + C = \frac{2}{5}(x-3)^{\frac{5}{2}} + \frac{14}{3}(x-3)^{\frac{3}{2}} + 22\sqrt{x-3} + C.$$

17. Let $u = x^3 + 2$, then $du = 3x^2 dx \Rightarrow x^2 dx = \frac{1}{3}du$. When $x = -1, u = (-1)^3 + 2 = 1$; when $x = 0, u = 2$:

$$\int_{-1}^{0} x^2(x^3 + 2)^5\, dx = \frac{1}{3}\int_{1}^{2} u^5\, du = \frac{1}{3}\left(\frac{u^6}{6}\right)\Big|_{1}^{2} = \frac{1}{18}(2^6 - 1^6) = \frac{63}{18}.$$

19. Let $u = x^2 + 4$, then $du = 2x dx \Rightarrow x dx = \frac{1}{2}du$. When $x = 0, u = 4$; when $x = 2, u = 2^2 + 4 = 8$:

$$\int_{0}^{2} \frac{x}{\sqrt{5(x^2+4)}}\, dx = \frac{1}{2}\int_{4}^{8} \frac{du}{\sqrt{5u}} = \frac{1}{2\sqrt{5}}\int_{4}^{8} u^{-\frac{1}{2}}\, du = \frac{1}{2\sqrt{5}} \cdot \frac{u^{\frac{1}{2}}}{\frac{1}{2}}\Big|_{4}^{8} = \frac{1}{\sqrt{5}}(\sqrt{8} - \sqrt{4})$$

$$= \frac{1}{\sqrt{5}}(2\sqrt{2} - 2) \cdot \frac{\sqrt{5}}{\sqrt{5}} = \frac{2}{5}(\sqrt{10} - \sqrt{5}).$$

21. Let $u = x^2 + 5$, then $du = 2x dx \Rightarrow x dx = \frac{1}{2}du$. When $x = 2, u = 2^2 + 5 = 9$; when $x = -1, u = (-1)^2 + 5 = 6$:

$$\int_{2}^{-1} \frac{x}{(x^2+5)^2}\, dx = \frac{1}{2}\int_{9}^{6} \frac{du}{u^2} = \frac{1}{2}\int_{9}^{6} u^{-2}\, du = \frac{1}{2} \cdot \frac{u^{-1}}{-1}\Big|_{9}^{6} = \frac{1}{2} \cdot \frac{-1}{u}\Big|_{9}^{6} = \frac{1}{2}\left(-\frac{1}{6} + \frac{1}{9}\right)$$

$$= \frac{1}{2}\left(-\frac{1}{18}\right) = -\frac{1}{36}.$$

23. $\displaystyle\int_{-1}^{1} x \cos x^2\, dx = 0$, as it is the integral of an odd function over a symmetric interval. Verification:

Let $u = x^2$, then $du = 2x dx \Rightarrow x dx = \frac{1}{2}du$. When $x = -1, u = 1$; when $x = 1, u = 1$.

$$\int_{-1}^{1} x \cos x^2\, dx = \frac{1}{2}\int_{1}^{1} \cos u\, du = 0, \text{ by Definition 5.3.}$$

25. Let $u = x^2$, then $du = 2xdx \Rightarrow xdx = \frac{1}{2}du$. When $x = 0, u = 0$; when $x = \sqrt{\frac{\pi}{4}}, u = \frac{\pi}{4}$:

$$\int_0^{\sqrt{\frac{\pi}{4}}} x\sec^2 x^2 \, dx = \frac{1}{2}\int_0^{\frac{\pi}{4}} \sec^2 u \, du = \frac{1}{2}\tan u\Big|_0^{\frac{\pi}{4}} = \frac{1}{2}(\tan\frac{\pi}{4} - \tan 0) = \frac{1}{2}(1) = \frac{1}{2}.$$

27. Let $I = \int_0^1 \sqrt{x^3 + x^2} \, dx = \int_0^1 \sqrt{x^2(x+1)} \, dx = \int_0^1 x\sqrt{x+1} \, dx$, as $x \geq 0 \Rightarrow \sqrt{x^2} = x$.
Let $u = x + 1$, then $du = dx$ and $x = u - 1$. When $x = 0, u = 1$; when $x = 1, u = 2$.

$$I = \int_1^2 (u-1)u^{\frac{1}{2}} \, du = \int_1^2 (u^{\frac{3}{2}} - u^{\frac{1}{2}}) \, du = \left(\frac{u^{\frac{5}{2}}}{\frac{5}{2}} - \frac{u^{\frac{3}{2}}}{\frac{3}{2}}\right)\Big|_1^2 = \left[\frac{2}{5}(2^{\frac{5}{2}}) - \frac{2}{3}(2^{\frac{3}{2}})\right] - \left[\frac{2}{5} - \frac{2}{3}\right]$$

$$= \frac{8\sqrt{2}}{5} - \frac{4\sqrt{2}}{3} + \frac{4}{15} = \frac{24\sqrt{2} - 20\sqrt{2} + 4}{15} = \frac{4}{15}(\sqrt{2} + 1).$$

29. The area is $A = \int_{-1}^1 x^2\sqrt{x^3 + 10} \, dx$. Let $u = x^3 + 10$, then $du = 3x^2 dx \Rightarrow x^2 dx = \frac{1}{3}du$.
When $x = -1, u = -1 + 10 = 9$; when $x = 1, u = 1 + 10 = 11$:

$$A = \frac{1}{3}\int_9^{11} \sqrt{u} \, du = \frac{1}{3}\int_9^{11} u^{\frac{1}{2}} \, du = \frac{1}{3}\cdot\frac{u^{\frac{3}{2}}}{\frac{3}{2}}\Big|_9^{11} = \frac{2}{9}\left(u^{\frac{3}{2}}\right)\Big|_9^{11} = \frac{2}{9}(11^{\frac{3}{2}} - 9^{\frac{3}{2}}) = \frac{2}{9}(11\sqrt{11} - 27).$$

§5.4 Area and Volume

1. $A = \int_{-1}^2 (-x^2 + 4) \, dx = \left(-\frac{1}{3}x^3 + 4x\right)\Big|_{-1}^2 = \left[-\frac{1}{3}\cdot 2^3 + 4(2)\right] - \left[-\frac{1}{3}\cdot(-1)^3 + 4(-1)\right]$

$= -\frac{8}{3} + 8 - \frac{1}{3} + 4 = -3 + 12 = 9.$

3. $A = \int_0^1 \frac{x}{(x^2+1)^2} \, dx$. Let $u = x^2 + 1$, then $du = 2xdx \Rightarrow \frac{1}{2}du = xdx$. When $x = 0, u = 1$; when $x = -1, u = (-1)^2 + 1 = 2$:

$$A = \frac{1}{2}\int_1^2 \frac{du}{u^2} = \frac{1}{2}\int_1^2 u^{-2} \, du = \frac{1}{2}\frac{u^{-1}}{-1}\Big|_1^2 = -\frac{1}{2}\cdot\frac{1}{u}\Big|_1^2 = -\frac{1}{2}\left(\frac{1}{2} - 1\right) = \frac{1}{4}.$$

5.

Intersection of $y = x$ and $y = x^2 - 2$:

$$x^2 - 2 = x \Rightarrow x^2 - x - 2 = 0 \Rightarrow (x-2)(x+1) = 0 \Rightarrow x = 2, -1.$$

$$A = \int_{-1}^2 [x - (x^2 - 2)] \, dx = \left(\frac{1}{2}x^2 - \frac{1}{3}x^3 + 2x\right)\Big|_{-1}^2$$

$$= \left[\frac{1}{2}\cdot 2^2 - \frac{1}{3}\cdot 2^3 + 2\cdot 2\right] - \left[\frac{1}{2}(-1)^2 - \frac{1}{3}(-1)^3 + 2(-1)\right]$$

$$= 2 - \frac{8}{3} + 4 - \frac{1}{2} - \frac{1}{3} + 2 = 8 - 3 - \frac{1}{2} = 5 - \frac{1}{2} = \frac{9}{2}.$$

7. $f(x) - g(x) = x^2 + x - (-x^2 + 1) = 2x^2 + x - 1 = (2x - 1)(x + 1) = 0 \Rightarrow x = \frac{1}{2}, -1.$

Sign $f - g$: $\underset{-1 \quad \frac{1}{2}}{+ \overset{c}{\bullet} - \overset{c}{\bullet} +}$ $\Rightarrow$

$$A = \int_a^b |f(x) - g(x)|\, dx = \int_{-1}^{\frac{1}{2}} -(2x^2 + x - 1)\, dx = -\left(2 \cdot \frac{x^3}{3} + \frac{x^2}{2} - x\right)\Big|_{-1}^{\frac{1}{2}}$$

$$= -\left\{\left[\frac{2}{3}(\tfrac{1}{2})^3 + \frac{1}{2}(\tfrac{1}{2})^2 - \frac{1}{2}\right] - \left[-\frac{2}{3} + \frac{1}{2} + 1\right]\right\} = -\left(\frac{1}{12} + \frac{1}{8} - \frac{1}{2} + \frac{2}{3} - \frac{3}{2}\right)$$

$$= -\left(-2 + \frac{2 + 3 + 16}{24}\right) = 2 - \frac{21}{24} = 2 - \frac{7}{8} = \frac{9}{8}.$$

9. $f(x) - g(x) = x^3 + 1 - (x^3 + x^2) = 1 - x^2 = (1 - x)(1 + x) = 0 \Rightarrow x = \pm 1.$

Sign $f - g$: $\underset{-1 \quad 1}{- \overset{c}{\bullet} + \overset{c}{\bullet} -}$ $\Rightarrow$

$$A = \int_a^b |f(x) - g(x)|\, dx = \int_{-1}^1 [x^3 + 1 - (x^3 + x^2)]\, dx = \int_{-1}^1 (1 - x^2)\, dx$$

$$= \left(x - \frac{1}{3}x^3\right)\Big|_{-1}^1 = (1 - \tfrac{1}{3}) - (-1 + \tfrac{1}{3}) = 2 - \frac{2}{3} = \frac{4}{3}.$$

11.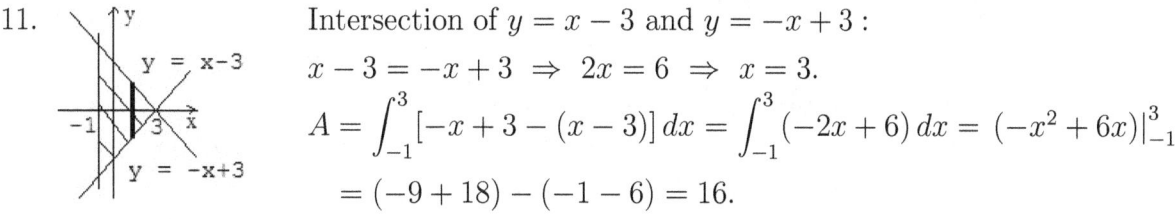

Intersection of $y = x - 3$ and $y = -x + 3$:

$x - 3 = -x + 3 \Rightarrow 2x = 6 \Rightarrow x = 3.$

$$A = \int_{-1}^3 [-x + 3 - (x - 3)]\, dx = \int_{-1}^3 (-2x + 6)\, dx = (-x^2 + 6x)\Big|_{-1}^3$$

$$= (-9 + 18) - (-1 - 6) = 16.$$

13.

Intersection of $y = x$ and $y = -x + 2$:

$x = -x + 2 \Rightarrow 2x = 2 \Rightarrow x = 1.$

$$A = \int_0^1 (-x + 2 - x)\, dx + \int_1^2 [x - (-x + 2)]\, dx$$

$$= \int_0^1 (2 - 2x)\, dx + \int_1^2 (2x - 2)\, dx = (2x - x^2)\Big|_0^1 + (x^2 - 2x)\Big|_1^2$$

$$= 2 - 1 + (4 - 4) - (1 - 2) = 2.$$

15.

Intersection of $x = y^2$ and $x = -y + 2$:

$y^2 = -y + 2 \Rightarrow y^2 + y - 2 = 0 \Rightarrow (y + 2)(y - 1) = 0 \Rightarrow y = -2, 1.$

$$A = \int_0^1 (-y + 2 - y^2)\, dy = \left(-\frac{1}{2}y^2 + 2y - \frac{1}{3}y^3\right)\Big|_0^1 = -\frac{1}{2} + 2 - \frac{1}{3}$$

$$= 2 - \frac{5}{6} = \frac{7}{6}.$$

17.

Intersection of $y = -x + 2$ and $y = x$:
$$-x + 2 = x \Rightarrow 2x = 2 \Rightarrow x = 1$$

Intersection of $y = -x + 2$ and $y = -\frac{x}{2}$:
$$-x + 2 = -\frac{x}{2} \Rightarrow \frac{x}{2} = 2 \Rightarrow x = 4$$

$$A = \int_0^1 \left[x - \left(-\frac{x}{2}\right)\right] dx + \int_1^4 \left[-x + 2 - \left(-\frac{x}{2}\right)\right] dx = \int_0^1 \frac{3}{2}x \, dx + \int_1^4 \left(-\frac{1}{2}x + 2\right) dx$$

$$= \frac{3}{4}x^2 \Big|_0^1 + \left(-\frac{1}{4}x^2 + 2x\right)\Big|_1^4 = \frac{3}{4} + \left[-4 + 8 - \left(-\frac{1}{4} + 2\right)\right] = \frac{3}{4} + 4 + \frac{1}{4} - 2 = 3.$$

19.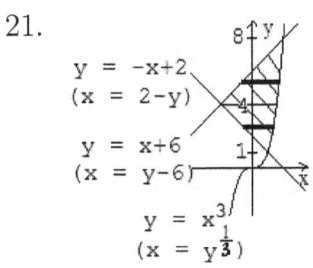

Intersection of $y = \frac{1}{x^2}$ and $y = x$: $\frac{1}{x^2} = x \Rightarrow x^3 = 1 \Rightarrow x = 1$.

$$A = \int_1^2 \left(x - \frac{1}{x^2}\right) dx = \int_1^2 \left(x - x^{-2}\right) dx = \left(\frac{1}{2}x^2 - \frac{x^{-1}}{-1}\right)\Big|_1^2$$

$$= \left(\frac{1}{2}x^2 + \frac{1}{x}\right)\Big|_1^2 = 2 + \frac{1}{2} - \left(\frac{1}{2} + 1\right) = 1.$$

21.

y = -x+2
(x = 2-y)

y = x+6
(x = y-6)

y = x³
(x = y^{\frac{1}{3}})

Intersection of $y = -x + 2$ and $y = x^3$:
$$-x + 2 = x^3 \Rightarrow x^3 + x - 2 = 0 \Rightarrow x = 1 \Rightarrow y = 1.$$

Intersection of $y = x + 6$ and $y = -x + 2$:
$$x + 6 = -x + 2 \Rightarrow 2x = -4 \Rightarrow x = -2 \Rightarrow y = -2 + 6 = 4.$$

Intersection of $y = x + 6$ and x^3 :
$$x^3 = x + 6 \Rightarrow x^3 - x - 6 = 0 \Rightarrow x = 2 \Rightarrow y = 2^3 = 8.$$

$A = A_1 + A_2$:

$$A_1 = \int_1^4 \left[y^{\frac{1}{3}} - (2 - y)\right] dy$$

$$= \left[\frac{3}{4}y^{\frac{4}{3}} - 2y + \frac{1}{2}y^2\right]\Big|_1^4$$

$$= \left[\frac{3}{4}(4)^{\frac{4}{3}} - 8 + 8\right] - \left[\frac{3}{4} - 2 + \frac{1}{2}\right]$$

$$= 3 \cdot 4^{\frac{1}{3}} + \frac{3}{4}$$

$$A_2 = \int_4^8 \left[y^{\frac{1}{3}} - (y - 6)\right] dy$$

$$= \left[\frac{3}{4}y^{\frac{4}{3}} - \frac{1}{2}y^2 + 6y\right]\Big|_4^8$$

$$= \left[\frac{3}{4}(8)^{\frac{4}{3}} - 32 + 48\right] - \left[\frac{3}{4} \cdot 4^{\frac{4}{3}} - 8 + 24\right]$$

$$= (12 + \cancel{16}) - (3 \cdot 4^{\frac{1}{3}} + \cancel{16})$$

$$= 12 - 3 \cdot 4^{\frac{1}{3}}$$

Thus $A = 3 \cdot 4^{\frac{1}{3}} + \frac{3}{4} + 12 - 3 \cdot 4^{\frac{1}{3}} = \frac{51}{4}$.

23.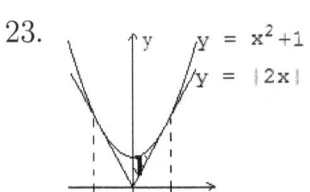

By symmetry, the area A is twice the area in the first quadrant, where $|2x| = 2x$. Intersection of $y = x^2 + 1$ and $y = 2x$ in QI:
$$x^2 + 1 = 2x \Rightarrow x^2 - 2x + 1 = 0 \Rightarrow (x - 1)^2 = 0 \Rightarrow x = 1.$$

$$A = 2\int_0^1 (x^2 + 1 - 2x)dx = 2\left(\frac{1}{3}x^3 + x - x^2\right)\Big|_0^1 = 2\left(\frac{1}{3} + 1 - 1\right) = \frac{2}{3}.$$

25.

Intersection of $y = -x$ and $y = \sqrt{2}$: $-x = \sqrt{2} \Rightarrow x = -\sqrt{2}$.

Intersection of $y = x\sqrt{x^2+1}$ and $y = \sqrt{2}$: ($y > 0$, so $x > 0$)

$x\sqrt{x^2+1} = \sqrt{2} \Rightarrow x^2(x^2+1) = 2 \Rightarrow x^4 + x^2 - 2 = 0 \Rightarrow$

$(x^2-1)(x^2+2) = 0 \Rightarrow x = 1$.

$A = A_1 + A_2$:

$A_1 = \int_{-\sqrt{2}}^{0} \left[\sqrt{2} - (-x)\right] dx = \left(\sqrt{2}x + \tfrac{1}{2}x^2\right)\Big|_{-\sqrt{2}}^{0} = 0 - (-2+1) = 1.$

$A_2 = \int_0^1 \left[\sqrt{2} - x\sqrt{x^2+1}\right] dx = \int_0^1 \sqrt{2}\,dx - \int_0^1 x\sqrt{x^2+1}\,dx$

$\int_0^1 \sqrt{2}\,dx = \sqrt{2}x\Big|_0^1 = \sqrt{2}.$

$\int_0^1 x\sqrt{x^2+1}\,dx$: Let $u = x^2+1 \Rightarrow du = 2x\,dx \Rightarrow x\,dx = \tfrac{1}{2}du$. When $x = 0, u = 1$; when $x = 1, u = 2 \Rightarrow$

$\int_0^1 x\sqrt{x^2+1}\,dx = \tfrac{1}{2}\int_1^2 u^{\frac{1}{2}}\,du = \tfrac{1}{2}\cdot\tfrac{2}{3}u^{\frac{3}{2}}\Big|_1^2 = \tfrac{1}{3}(2^{\frac{3}{2}} - 1) = \tfrac{1}{3}(2\sqrt{2} - 1)$

Thus $A_2 = \sqrt{2} - \tfrac{2}{3}\sqrt{2} + \tfrac{1}{3} = \tfrac{1}{3}\sqrt{2} + \tfrac{1}{3}.$

Therefore $A = 1 + \tfrac{1}{3}\sqrt{2} + \tfrac{1}{3} = \tfrac{1}{3}(4 + \sqrt{2}).$

27.

Intersection of $y = \sin x$ and $y = \cos x$ in $\left[\tfrac{\pi}{4}, \tfrac{\pi}{2}\right]$:

$\sin x = \cos x \Rightarrow \tan x = 1 \Rightarrow x = \tfrac{\pi}{4}.$

$A = \int_{\pi/4}^{\pi/2} (\sin x - \cos x)dx = (-\cos x - \sin x)\Big|_{\pi/4}^{\pi/2}$

$= -\cos\tfrac{\pi}{2} - \sin\tfrac{\pi}{2} - \left(-\cos\tfrac{\pi}{4} - \sin\tfrac{\pi}{4}\right) = 0 - 1 + \tfrac{1}{\sqrt{2}} + \tfrac{1}{\sqrt{2}}$

$= -1 + \tfrac{2}{\sqrt{2}}\cdot\tfrac{\sqrt{2}}{\sqrt{2}} = -1 + \sqrt{2}.$

29. $V = \pi\int_a^b[f(x)]^2 dx = \pi\int_1^3 x^2\,dx = \tfrac{\pi}{3}(x^3)\Big|_1^3 = \tfrac{\pi}{3}(3^3 - 1^3) = \tfrac{26\pi}{3}.$

31. $V = \pi\int_a^b[f(x)]^2 dx = \pi\int_0^1(x^2+1)^2\,dx = \pi\int_0^1(x^4+2x^2+1)dx = \pi\left(\tfrac{1}{5}x^5 + \tfrac{2}{3}x^3 + x\right)\Big|_0^1 dx$

$= \pi\left(\tfrac{1}{5} + \tfrac{2}{3} + 1\right) = \tfrac{\pi}{15}(3 + 10 + 15) = \tfrac{28\pi}{15}.$

33. $V = \pi\int_a^b[f(x)]^2 dx = \pi\int_0^1(-x^2+x)^2 dx = \pi\int_0^1(x^4 - 2x^3 + x^2)dx = \pi\left(\tfrac{1}{5}x^5 - \tfrac{2}{4}x^4 + \tfrac{1}{3}x^3\right)\Big|_0^1$

$= \pi\left(\tfrac{1}{5} - \tfrac{1}{2} + \tfrac{1}{3}\right) = \tfrac{\pi}{30}(6 - 15 + 10) = \tfrac{\pi}{30}.$

35. $V = \pi\int_a^b[f(x)]^2 dx = \pi\int_0^4(\sqrt{x})^2\,dx = \pi\int_0^4 x\,dx = \pi\cdot\tfrac{1}{2}x^2\Big|_0^4 = \tfrac{\pi}{2}(4^2) = 8\pi.$

37.

$V = \pi \int_{-r}^{r} (\sqrt{r^2 - x^2})^2 dx = \pi \int_{-r}^{r} (r^2 - x^2)\, dx = \pi \left(r^2 x - \frac{1}{3}x^3 \right)\Big|_{-r}^{r}$

$= \pi \left[\left(r^3 - \frac{1}{3}r^3 \right) - \left(-r^3 + \frac{1}{3}r^3 \right) \right] = \pi r^3 \left(2 - \frac{2}{3} \right) = \frac{4}{3}\pi r^3.$

39.

Intersection of $y = x^4$ and $y = x$: $x^4 = x \Rightarrow x^4 - x = 0 \Rightarrow$
$x(x^3 - 1) = 0 \Rightarrow x = 0, 1.$

$V = \pi \int_a^b ([f(x)]^2 - [g(x)]^2) dx = \pi \int_0^1 [(x)^2 - (x^4)^2] dx$

$= \pi \int_0^1 (x^2 - x^8) dx = \pi \left(\frac{1}{3}x^3 - \frac{1}{9}x^9 \right)\Big|_0^1 = \pi \left(\frac{1}{3} - \frac{1}{9} \right) = \frac{2\pi}{9}.$

41.

Intersection of $y = x^4 + 1$ and $y = -x^2 + 3$:
$x^4 + 1 = -x^2 + 3 \Rightarrow x^4 + x^2 - 2 = 0 \Rightarrow (x^2 + 2)(x^2 - 1) = 0 \Rightarrow$
$x^2 = 1 \Rightarrow x = \pm 1.$

By symmetry, the volume is twice the volume generated by the region in the first quadrant. Thus

$V = \pi \int_a^b ([f(x)]^2 - [g(x)]^2) dx = 2\pi \int_0^1 [(-x^2 + 3)^2 - (x^4 + 1)^2] dx$

$= 2\pi \int_0^1 (x^4 - 6x^2 + 9 - x^8 - 2x^4 - 1) dx = 2\pi \int_0^1 (-x^8 - x^4 - 6x^2 + 8) dx$

$= 2\pi \left(-\frac{1}{9}x^9 - \frac{1}{5}x^5 - 6 \cdot \frac{1}{3}x^3 + 8x \right)\Big|_0^1 = 2\pi \left(-\frac{1}{9} - \frac{1}{5} - 2 + 8 \right) = \frac{2\pi}{45}(-5 - 9 + 270) = \frac{512\pi}{45}.$

43.

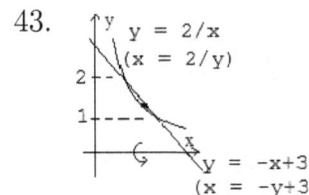

Intersection of $x = \frac{2}{y}$ and $x = -y + 3$:
$\frac{2}{y} = -y + 3 \Rightarrow 2 = -y^2 + 3y \Rightarrow y^2 - 3y + 2 = 0 \Rightarrow$
$(y - 1)(y - 2) = 0 \Rightarrow y = 1, 2.$

$V = 2\pi \int_a^b y[f(y) - g(y)] dy = 2\pi \int_1^2 y \left(-y + 3 - \frac{2}{y} \right) dy = 2\pi \int_1^2 \left(-y^2 + 3y - 2 \right) dy$

$= 2\pi \left(-\frac{1}{3}y^3 + 3 \cdot \frac{1}{2}y^2 - 2y \right)\Big|_1^2 = 2\pi \left[\left(-\frac{8}{3} + 6 - 4 \right) - \left(-\frac{1}{3} + \frac{3}{2} - 2 \right) \right]$

$= 2\pi \left(-\frac{7}{3} + 2 + \frac{1}{2} \right) = \frac{2\pi}{6}(-14 + 12 + 3) = \frac{\pi}{3}.$

45.

Intersection of $y = x^2$ and $y = x + 2$:
$x^2 = x + 2 \Rightarrow x^2 - x - 2 = 0 \Rightarrow (x - 2)(x + 1) = 0 \Rightarrow$
$x = 2, -1.$

$V = \pi \int_a^b ([f(x)]^2 - [g(x)]^2) dx = \pi \int_{-1}^2 [(x + 2)^2 - (x^2)^2]\, dx = \pi \int_{-1}^2 [x^2 + 4x + 4 - x^4]\, dx$

$= \pi \left(\frac{1}{3}x^3 + 4 \cdot \frac{1}{2}x^2 + 4x - \frac{1}{5}x^5 \right)\Big|_{-1}^2 = \pi \left[\left(\frac{8}{3} + 8 + 8 - \frac{32}{5} \right) - \left(-\frac{1}{3} + 2 - 4 + \frac{1}{5} \right) \right]$

$= \pi \left(3 + 16 - \frac{33}{5} + 2 \right) = \pi \left(21 - \frac{33}{5} \right) = \frac{\pi}{5}(105 - 33) = \frac{72\pi}{5}.$

47. As the graph of $f(x) = x^2 + x + 1$ is not readily obtained, and not necessary, we just determine where the graphs of $y = f(x)$ and $y = x + 2$ intersect: $x^2 + x + 1 = x + 2$ $\Rightarrow$ $x^2 = 1$ $\Rightarrow$ $x = \pm 1$. At, for example, $x = 0$, $f(0) = 1$ and $y = x + 2 = 0 + 2 = 2$, which means the graph of f is the lower graph. Hence

$$V = \pi \int_a^b ([f(x)]^2 - [g(x)]^2)dx = \pi \int_{-1}^1 [(x+2)^2 - (x^2+x+1)^2]\,dx$$
$$= \pi \int_{-1}^1 [x^2 + 4x + 4 - (x^4 + x^3 + x^2 + x^3 + x^2 + x + x^2 + x + 1)]\,dx$$
$$= \pi \int_{-1}^1 [-x^4 - 2x^3 - 2x^2 + 2x + 3]\,dx = \pi \left(-\tfrac{1}{5}x^5 - 2\cdot\tfrac{1}{4}x^4 - 2\cdot\tfrac{1}{3}x^3 + x^2 + 3x\right)\Big|_{-1}^1$$
$$= \pi \left[\left(-\tfrac{1}{5} - \tfrac{1}{2} - \tfrac{2}{3} + 1 + 3\right) - \left(\tfrac{1}{5} - \tfrac{1}{2} + \tfrac{2}{3} + 1 - 3\right)\right]$$
$$= \pi \left(-\tfrac{2}{5} - \tfrac{4}{3} + 6\right) = \tfrac{\pi}{15}(-6 - 20 + 90) = \tfrac{64\pi}{15}.$$

49.

y = -x

y = x+4

y = 2x+3

Intersection of $y = -x$ and $y = x + 4$:
$$-x = x + 4 \Rightarrow 2x = -4 \Rightarrow x = -2.$$
Intersection of $y = 2x + 3$ and $y = -x$:
$$2x + 3 = -x \Rightarrow 3x = -3 \Rightarrow x = -1.$$
Intersection of $y = 2x+3$ and $y = x+4$: $2x+3 = x+4 \Rightarrow x = 1$.

$$V = V_1 + V_2: \quad V = \pi \int_a^b ([f(x)]^2 - [g(x)]^2)dx \Rightarrow$$
$$V_1 = \pi \int_{-2}^{-1}[(x+4)^2 - (-x)^2]dx = \pi \int_{-2}^{-1}[x^2 + 8x + 16 - x^2]dx = \pi\,(4x^2 + 16x)|_{-2}^{-1}$$
$$= \pi[(4 - 16) - (16 - 32)] = 4\pi.$$
$$V_2 = \pi \int_{-1}^1[(x+4)^2 - (2x+3)^2]dx = \pi \int_{-1}^1[(x^2 + 8x + 16) - (4x^2 + 12x + 9)]dx$$
$$= \pi \int_{-1}^1(-3x^2 - 4x + 7)dx = \pi\,(-x^3 - 2x^2 + 7x)|_{-1}^1$$
$$= \pi[(-1 - 2 + 7) - (1 - 2 - 7)] = 12\pi.$$
Therefore $V = 4\pi + 12\pi = 16\pi$.

51.

y = √sin x

y = x/2

With a graphing calculator, we graphed $f(x) = \sqrt{\sin x}$, and the two given lines. Thus, $V = \pi \int_a^b ([f(x)]^2 - [g(x)]^2)dx \Rightarrow$

$$V = \pi \int_0^{\frac{\pi}{2}} \left[(\sqrt{\sin x})^2 - \left(\tfrac{x}{2}\right)^2\right] dx = \pi \int_0^{\frac{\pi}{2}} \left(\sin x - \tfrac{1}{4}x^2\right) dx = \pi \left(-\cos x - \tfrac{1}{4}\cdot\tfrac{1}{3}x^3\right)\Big|_0^{\frac{\pi}{2}}$$
$$= \pi \left[\left(-\cos\tfrac{\pi}{2} - \tfrac{1}{12}\cdot(\tfrac{\pi}{2})^3\right) - (-\cos 0 - 0)\right] = \pi \left(-\tfrac{\pi^3}{96} + 1\right) = \tfrac{\pi}{96}(96 - \pi^3).$$

53.

y = x⁴

y = x

Intersection of $y = x^4$ and $y = x$: $x = 0, 1$, from Exercise 39.
$$V = 2\pi \int_a^b x[f(x) - g(x)]dx = 2\pi \int_0^1 x(x - x^4)dx = 2\pi \int_0^1 (x^2 - x^5)dx$$
$$= 2\pi \left(\tfrac{1}{3}x^3 - \tfrac{1}{6}x^6\right)\Big|_0^1 = 2\pi \left(\tfrac{1}{3} - \tfrac{1}{6}\right) = \tfrac{\pi}{3}.$$

55.

Intersection of $y = \frac{2}{x}$ and $y = -x + 3$:

$$\frac{2}{x} = -x + 3 \;\Rightarrow\; 2 = -x^2 + 3x \;\Rightarrow\; x^2 - 3x + 2 = 0 \;\Rightarrow$$

$$(x-2)(x-1) = 0 \;\Rightarrow\; x = 1, 2.$$

$$V = 2\pi \int_a^b x[f(x) - g(x)]dx = 2\pi \int_1^2 x\left(-x + 3 - \frac{2}{x}\right)dx = 2\pi \int_1^2 (-x^2 + 3x - 2)dx$$

$$= 2\pi \left(-\tfrac{1}{3}x^3 + 3 \cdot \tfrac{1}{2}x^2 - 2x\right)\Big|_1^2 = 2\pi\left[\left(-\tfrac{8}{3} + 6 - 4\right) - \left(-\tfrac{1}{3} + \tfrac{3}{2} - 2\right)\right] = 2\pi\left(-\tfrac{7}{3} + 4 - \tfrac{3}{2}\right)$$

$$= \tfrac{2\pi}{6}(-14 + 24 - 9) = \tfrac{\pi}{3}.$$

57.

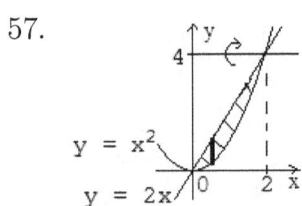

In the general formula for rotation about the x-axis in the washer method, we have to replace $f(x)$ and $g(x)$, the radii of the washers, by $4 - f(x)$ and $4 - g(x)$, the radii of the washers in this case:

$$V = \pi \int_a^b ([4 - f(x)]^2 - [4 - g(x)]^2)dx.$$

Intersection of $y = x^2$ and $y = 2x$:

$$x^2 = 2x \;\Rightarrow\; x^2 - 2x = 0 \;\Rightarrow\; x(x-2) = 0 \;\Rightarrow\; x = 0, 2.$$

$$V = \pi \int_0^2 \left[\left(4 - x^2\right)^2 - (4 - 2x)^2\right]dx = \pi \int_0^2 \left[\cancel{16} - 8x^2 + x^4 - \left(\cancel{16} - 16x + 4x^2\right)\right]dx$$

$$= \pi \int_0^2 (x^4 - 12x^2 + 16x)dx = \pi\left(\tfrac{1}{5}x^5 - 12 \cdot \tfrac{1}{3}x^3 + 16 \cdot \tfrac{1}{2}x^2\right)\Big|_0^2 = \pi\left(\tfrac{32}{5} - \cancel{32} + \cancel{32}\right) = \tfrac{32\pi}{5}.$$

59.

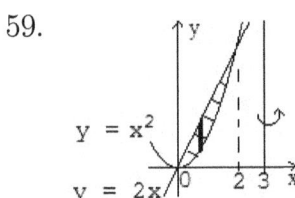

In the general formula for rotation about the y-axis in the shell method, we have to replace x, the radius of the shell, with $3 - x$, the radius of the shell in this case:

$$V = 2\pi \int_a^b (3 - x)[f(x) - g(x)]dx.$$

From Exercise 57, the limits of integration are 0 and 2.

$$V = 2\pi \int_0^2 (3 - x)(2x - x^2)dx = 2\pi \int_0^2 (6x - 5x^2 + x^3)dx = 2\pi \left(6 \cdot \tfrac{1}{2}x^2 - 5 \cdot \tfrac{1}{3}x^3 + \tfrac{1}{4}x^4\right)\Big|_0^2$$

$$= 2\pi \left(12 - \tfrac{40}{3} + 4\right) = \tfrac{2\pi}{3}(48 - 40) = \tfrac{16\pi}{3}.$$

61.

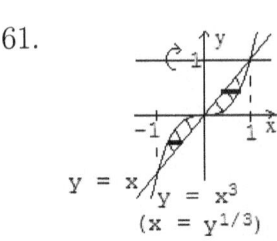

In the general formula for rotation about the x-axis in the shell method, we have to replace y, the radius of the shell, with $1 - y$, the radius of the shell in this case:

$$V = 2\pi \int_a^b (1 - y)[f(y) - g(y)]dy.$$

Intersection of $y = x^3$ and $y = x$: $x^3 = x \;\Rightarrow\; x^3 - x = 0 \;\Rightarrow$
$x(x^2 - 1) = 0 \;\Rightarrow\; x = 0, \pm 1.$

$V = V_1 + V_2$:

$V_1 = 2\pi \int_0^1 (1-y)(y^{\frac{1}{3}} - y)dy = 2\pi \int_0^1 (y^{\frac{1}{3}} - y - y^{\frac{4}{3}} + y^2)dy = 2\pi \left(\frac{3}{4}y^{\frac{4}{3}} - \frac{1}{2}y^2 - \frac{3}{7}y^{\frac{7}{3}} + \frac{1}{3}y^3 \right)\Big|_0^1$

$= 2\pi \left(\frac{3}{4} - \frac{1}{2} - \frac{3}{7} + \frac{1}{3} \right) = \frac{2\pi}{84}(63 - 42 - 36 + 28) = \frac{13\pi}{42}$.

$V_2 = 2\pi \int_{-1}^0 (1-y)(y - y^{\frac{1}{3}})dy = -2\pi \left(\frac{3}{4}y^{\frac{4}{3}} - \frac{1}{2}y^2 - \frac{3}{7}y^{\frac{7}{3}} + \frac{1}{3}y^3 \right)\Big|_{-1}^0 = 2\pi \left(\frac{3}{4} - \frac{1}{2} + \frac{3}{7} - \frac{1}{3} \right)$

$= \frac{2\pi}{84}(63 - 42 + 36 - 28) = \frac{29\pi}{42}$.

Therefore $V = \frac{13\pi}{42} + \frac{29\pi}{42} = \pi$.

63.

In the general formula for rotation about the y-axis in the washer method, we have to replace $f(y)$ and $g(y)$, the radii of the washers, by $1 - f(y)$ and $1 - g(y)$, the radii of the washers in this case:

$$V = \pi \int_a^b ([1 - f(y)]^2 - [1 - g(y)]^2)dy.$$

Intersection of $x = y$ and $x = -\frac{1}{2}y$: $y = -\frac{1}{2}y \Rightarrow \frac{3}{2}y = 0 \Rightarrow y = 0$.

Intersection of $x = y$ and $x = -y + 2$: $y = -y + 2 \Rightarrow 2y = 2 \Rightarrow y = 1$.

Intersection of $x = -y + 2$ and $x = -\frac{1}{2}y$: $-y + 2 = -\frac{1}{2}y \Rightarrow -\frac{1}{2}y = -2 \Rightarrow y = 4$.

$V = V_1 + V_2$:

$V_1 = \pi \int_0^1 \left[\left(1 - \left(-\frac{1}{2}y\right)\right)^2 - (1-y)^2 \right] dy = \pi \int_0^1 \left[\cancel{1} + y + \frac{1}{4}y^2 - (\cancel{1} - 2y + y^2) \right] dy$

$= \pi \int_0^1 \left(3y - \frac{3}{4}y^2 \right) dy = \pi \left(3 \cdot \frac{1}{2}y^2 - \frac{3}{4} \cdot \frac{1}{3}y^3 \right)\Big|_0^1 = \pi \left(\frac{3}{2} - \frac{1}{4} \right) = \frac{5\pi}{4}$.

$V_2 = \pi \int_1^4 \left[\left(1 - \left(-\frac{1}{2}y\right)\right)^2 - (1 - (-y+2))^2 \right] dy = \pi \int_1^4 \left[\left(1 - \left(-\frac{1}{2}y\right)\right)^2 - (y-1)^2 \right] dy$

$= \pi \left(3 \cdot \frac{1}{2}y^2 - \frac{3}{4} \cdot \frac{1}{3}y^3 \right)\Big|_1^4 = \pi \left[\frac{3}{2} \cdot 16 - 16 - \left(\frac{3}{2} - \frac{1}{4} \right) \right] = \pi \left(24 - 16 - \frac{5}{4} \right) = \frac{27\pi}{4}$.

Therefore $V = \frac{5\pi}{4} + \frac{27\pi}{4} = 8\pi$.

65.

By Shells: $V = 2\pi \int_a^b x[f(x) - g(x)]dx$

Intersection of $y = x^2$ and $y = -x + 6$: $x^2 = -x + 6 \Rightarrow$
$x^2 + x - 6 = 0 \Rightarrow (x+3)(x-2) = 0 \Rightarrow x = -3, 2$.

$V = 2\pi \int_1^2 x(-x + 6 - x^2)dx = 2\pi \int_1^2 (-x^2 + 6x - x^3)dx$

$= 2\pi \left(-\frac{1}{3}x^3 + 6 \cdot \frac{1}{2}x^2 - \frac{1}{4}x^4 \right)\Big|_1^2$

$= 2\pi \left[-\frac{8}{3} + 12 - 4 - \left(-\frac{1}{3} + 3 - \frac{1}{4} \right) \right] = 2\pi \left(-\frac{7}{3} + 5 + \frac{1}{4} \right)$

$= \frac{2\pi}{12}(-28 + 60 + 3) = \frac{35\pi}{6}$.

By Washers: $V = \pi \int_a^b ([f(y)]^2 - [g(y)]^2) dy$

When $x = 1, y = x^2 = 1$; when $x = 2, y = x^2 = 4$;

when $x = 1, y = -x + 6 = 5$.

$V = V_1 + V_2$:

$V_1 = \pi \int_4^5 [(-y+6)^2 - 1^2] dy = \pi \int_4^5 (y^2 - 12y + 35) dy = \pi \left(\frac{1}{3}y^3 - 12 \cdot \frac{1}{2}y^2 + 35y \right) \Big|_4^5$

$= \pi \left[\frac{125}{3} - 150 + 175 - \left(\frac{64}{3} - 96 + 140 \right) \right] = \pi \left(\frac{61}{3} + 25 - 44 \right) = \frac{\pi}{3}(61 - 57) = \frac{4\pi}{3}$.

$V_2 = \pi \int_1^4 [(\sqrt{y})^2 - 1^2] dy = \pi \int_1^4 (y - 1) dy = \pi \left(\frac{1}{2}y^2 - y \right) \Big|_1^4 = \pi \left[(8 - 4) - \left(\frac{1}{2} - 1 \right) \right] = \frac{9\pi}{2}$.

Therefore $V = \frac{4\pi}{3} + \frac{9\pi}{2} = \frac{\pi}{6}(8 + 27) = \frac{35\pi}{6}$.

67.

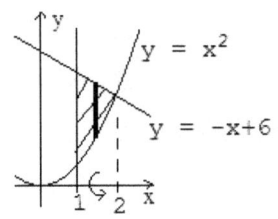

Same region as in Exercise 65, but this time rotation is about the x-axis.

By Shells: $V = 2\pi \int_a^b y[f(y) - g(y)] dy$

$V = V_1 + V_2$:

$V_1 = 2\pi \int_4^5 y(-y + 6 - 1) dy = 2\pi \int_4^5 (-y^2 + 5y) dy = 2\pi \left(-\frac{1}{3}y^3 + 5 \cdot \frac{1}{2}y^2 \right) \Big|_4^5$

$= 2\pi \left[-\frac{125}{3} + \frac{125}{2} - \left(-\frac{64}{3} + 40 \right) \right] = 2\pi \left(-\frac{61}{3} + \frac{125}{2} - 40 \right) = \frac{2\pi}{6}(-122 + 375 - 240)$

$= \frac{13\pi}{3}$.

$V_2 = 2\pi \int_1^4 y(\sqrt{y} - 1) dy = 2\pi \int_1^4 \left(y^{\frac{3}{2}} - y \right) dy = 2\pi \left(\frac{2}{5}y^{\frac{5}{2}} - \frac{1}{2}y^2 \right) \Big|_1^4$

$= 2\pi \left[\frac{2}{5} \cdot 4^{\frac{5}{2}} - 8 - \left(\frac{2}{5} - \frac{1}{2} \right) \right] = 2\pi \left(\frac{64}{5} - 8 - \frac{2}{5} + \frac{1}{2} \right) = \frac{2\pi}{10}(128 - 80 - 4 + 5) = \frac{49\pi}{5}$.

Therefore $V = \frac{13\pi}{3} + \frac{49\pi}{5} = \frac{\pi}{15}(65 + 147) = \frac{212\pi}{15}$.

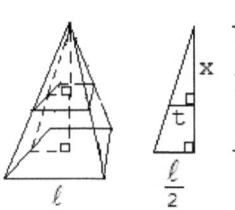

By Washers: $V = \pi \int_a^b ([f(x)]^2 - [g(x)]^2) dx$

$V = \pi \int_1^2 [(-x+6)^2 - (x^2)^2] dx = \pi \int_1^2 (x^2 - 12x + 36 - x^4) dx$

$= \pi \left(\frac{1}{3}x^3 - 12 \cdot \frac{1}{2}x^2 + 36x - \frac{1}{5}x^5 \right) \Big|_1^2$

$= \pi \left[\frac{8}{3} - 24 + 72 - \frac{32}{5} - \left(\frac{1}{3} - 6 + 36 - \frac{1}{5} \right) \right]$

$= \pi \left(\frac{7}{3} + 48 - \frac{31}{5} - 30 \right) = \frac{\pi}{15}(35 + 270 - 93) = \frac{212\pi}{15}$.

69.

From the similar triangles: $\frac{x}{t} = \frac{h}{\frac{\ell}{2}} \Rightarrow t = \frac{x\ell}{2h}$.

The cross-sectional area is $(2t)^2 = 4t^2 = 4 \left(\frac{x\ell}{2h} \right)^2 = \frac{x^2\ell^2}{h^2} \Rightarrow$

$V = \int_0^h \left(\frac{x^2\ell^2}{h^2} \right) dx = \frac{\ell^2}{h^2} \cdot \frac{1}{3}x^3 \Big|_0^h = \frac{\ell^2}{3h^2} \cdot h^3 = \frac{1}{3}\ell^2 h$.

71. The length of the second leg of the right triangle is $\sqrt{r^2 - x^2}$, so the length of one side of the square (the chord shown) is twice this amount. Thus the cross-sectional area $A = (2\sqrt{r^2 - x^2})^2 = 4(r^2 - x^2)$. By symmetry,

$$V = 2\int_0^r 4(r^2 - x^2)dx = 8\left(r^2 x - \tfrac{1}{3}x^3\right)\Big|_0^r = 8\left(r^3 - \tfrac{1}{3}r^3\right) = \tfrac{16}{3}r^3.$$

73. 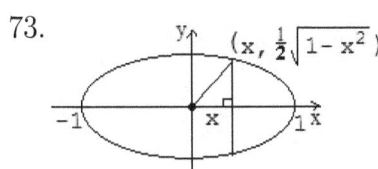 From $x^2 + 4y^2 = 1$, then

$$4y^2 = 1 - x^2 \Rightarrow y^2 = \tfrac{1}{4}(1 - x^2) \Rightarrow y = \pm\tfrac{1}{2}\sqrt{1 - x^2}.$$

The length of one side of the square (the chord shown) is $2y$. Thus the cross-sectional area $A = (2y)^2 = 1 - x^2$.

Therefore, by symmetry, the volume $V = 2\int_0^1 (1 - x^2)dx = 2\left(x - \tfrac{1}{3}x^3\right)\Big|_0^1 = 2\left(1 - \tfrac{1}{3}\right) = \tfrac{4}{3}$.

§5.5 Additional Applications

1.
```
fnInt(√(1+4X²),X
,1,5)
         24.40
```
$L = \int_a^b \sqrt{1 + (y')^2}\, dx : y = x^2 \Rightarrow y' = 2x \Rightarrow$

$$L = \int_1^5 \sqrt{1 + 4x^2}\, dx \approx 24.40.$$

3.
```
fnInt(√(1+1/(2X-
5)),X,4,9)
          5.35
```
$L = \int_a^b \sqrt{1 + (y')^2}\, dx : y = \sqrt{2x - 5} = (2x - 5)^{\frac{1}{2}} \Rightarrow$

$$y' = \tfrac{1}{2}(2x - 5)^{-\frac{1}{2}} \cdot 2 = \frac{1}{\sqrt{2x - 5}} \Rightarrow$$

$$L = \int_4^9 \sqrt{1 + \frac{1}{2x - 5}}\, dx \approx 5.35.$$

5.
```
fnInt(√(1+(6X+4)
²),X, -2,2)
         27.27
```
$L = \int_a^b \sqrt{1 + (y')^2}\, dx : y = 3x^2 + 4x - 5 \Rightarrow$

$$y' = 6x + 4 \Rightarrow$$

$$L = \int_{-2}^2 \sqrt{1 + (6x + 4)^2}\, dx \approx 27.27.$$

7. $f(x) = x^{\frac{3}{2}} \Rightarrow f'(x) = \tfrac{3}{2}x^{\frac{1}{2}} \Rightarrow [f'(x)]^2 = \tfrac{9}{4}x \Rightarrow L = \int_a^b \sqrt{1 + (y')^2}\, dx = \int_0^8 \sqrt{1 + \tfrac{9}{4}x}\, dx$.

Let $u = 1 + \tfrac{9}{4}x$, then $du = \tfrac{9}{4}dx \Rightarrow \tfrac{4}{9}du = dx$. When $x = 0, u = 1$; when $x = 8, u = 19$.

Then $L = \tfrac{4}{9}\int_1^{19} u^{\frac{1}{2}}du = \tfrac{4}{9} \cdot \tfrac{2}{3} u^{\frac{3}{2}}\Big|_1^{19} = \tfrac{8}{27}\left(19^{\frac{3}{2}} - 1\right) = \tfrac{8}{27}(19\sqrt{19} - 1)$.

9. $y = \frac{1}{3}x^3 + \frac{1}{4}x^{-1} \Rightarrow y' = \frac{1}{3}(3x^2) + \frac{1}{4}(-x^{-2}) = x^2 - \frac{1}{4x^2} \Rightarrow (y')^2 = x^4 - \frac{1}{2} + \frac{1}{16x^4} \Rightarrow$

$1 + (y')^2 = 1 + x^4 - \frac{1}{2} + \frac{1}{16x^4} = x^4 + \frac{1}{2} + \frac{1}{16x^4} = \left(x^2 + \frac{1}{4x^2}\right)^2 \Rightarrow \sqrt{1+(y')^2} = x^2 + \frac{1}{4x^2} \Rightarrow$

$L = \int_a^b \sqrt{1+(y')^2}\, dx = \int_1^2 \left(x^2 + \frac{1}{4x^2}\right) dx = \int_1^2 \left(x^2 + \frac{1}{4}x^{-2}\right) dx = \left[\frac{1}{3}x^3 + \frac{1}{4}\left(\frac{x^{-1}}{-1}\right)\right]\Big|_1^2$

$= \left(\frac{1}{3}x^3 - \frac{1}{4x}\right)\Big|_1^2 = \frac{8}{3} - \frac{1}{8} - \left(\frac{1}{3} - \frac{1}{4}\right) = \frac{7}{3} + \frac{1}{8} = \frac{59}{24}.$

11. $y = \int_1^x \sqrt{t^2-1}\, dt \Rightarrow y' = \sqrt{x^2-1} \Rightarrow 1+(y')^2 = 1 + (x^2-1) = x^2 \Rightarrow$

$L = \int_a^b \sqrt{1+(y')^2}\, dx = \int_1^2 \sqrt{x^2}\, dx = \int_1^2 x\, dx = \frac{1}{2}x^2\Big|_1^2 = \frac{1}{2}(4-1) = \frac{3}{2}.$

13. The length is 4 times the length of that part of the ellipse lying in the first quadrant:

In QI, $\frac{x^2}{a^2} + \frac{y^2}{b^2} = 1 \Rightarrow y = b\sqrt{1 - \frac{x^2}{a^2}} = \frac{b}{a}(a^2-x^2)^{\frac{1}{2}} \Rightarrow y' = \frac{b}{a} \cdot \frac{1}{2}(a^2-x^2)^{-\frac{1}{2}}(-2x) \Rightarrow$

$y' = -\frac{bx}{a\sqrt{a^2-x^2}} \Rightarrow (y')^2 = \frac{b^2x^2}{a^2(a^2-x^2)} \Rightarrow L = 4\int_0^a \sqrt{1 + \frac{b^2x^2}{a^2(a^2-x^2)}}\, dx.$

15. On the one hand, the length of the top half of the unit circle, being half the circumference is: $L = \frac{1}{2}(2\pi \cdot 1) = \pi$. On the other hand, we can compute the arc length of the top half of the unit circle from Definition 5.7: $y = \sqrt{1-x^2} = (1-x^2)^{\frac{1}{2}} \Rightarrow$

$y' = \frac{1}{2}(1-x^2)^{-\frac{1}{2}}(-2x) = -\frac{x}{\sqrt{1-x^2}} \Rightarrow (y')^2 = \frac{x^2}{1-x^2} \Rightarrow 1+(y')^2 = 1 + \frac{x^2}{1-x^2} = \frac{1}{1-x^2}.$

Therefore, $L = \int_a^b \sqrt{1+[f'(x)]^2}\, dx = \int_{-1}^1 \sqrt{\frac{1}{1-x^2}}\, dx = \int_{-1}^1 \frac{1}{\sqrt{1-x^2}}\, dx.$

17. Since, 200 cm. equals 2 m, the force $f(2) = 25 = 2k \Rightarrow k = \frac{25}{2} \Rightarrow f(x) = \frac{25}{2}x.$

(a) Since, 100 cm. equals 1 m, $W = \int_0^1 \frac{25}{2}x\, dx = \frac{25}{2} \cdot \frac{1}{2}x^2\Big|_0^1 = \frac{25}{4}$ J.

(b) Since, 50 cm. equals $\frac{1}{2}$ m, $W = \int_0^{\frac{1}{2}} \frac{25}{2}x\, dx = \frac{25}{2} \cdot \frac{1}{2}x^2\Big|_0^{\frac{1}{2}} = \frac{25}{4} \cdot \frac{1}{4} = \frac{25}{16}$ J.

(c) $10 = \frac{25}{2}x \Rightarrow x = \frac{20}{25} = \frac{4}{5}$ m $= \frac{4}{5}$m $\cdot \frac{100 \text{ cm}}{1 \text{ m}} = 80$ cm.

19. Let ℓ be the natural length of the spring, and suppose $f(x) = kx$. Then

$\int_{2-\ell}^{3-\ell} kx\, dx = k \cdot \frac{1}{2}x^2\Big|_{2-\ell}^{3-\ell}$ | $\frac{1}{2}\int_{3-\ell}^{4-\ell} kx\, dx = \frac{1}{2}k \cdot \frac{1}{2}x^2\Big|_{3-\ell}^{4-\ell}$

$= \frac{1}{2}k[(3-\ell)^2 - (2-\ell)^2]$ | $= \frac{1}{4}k[(4-\ell)^2 - (3-\ell)^2]$

$= \frac{1}{2}k[9 - 6\ell + \ell^2 - (4 - 4\ell + \ell^2)]$ | $= \frac{1}{4}k[16 - 8\ell + \ell^2 - (9 - 6\ell + \ell^2)]$

$= \frac{1}{2}k(5 - 2\ell)$ | $= \frac{1}{4}k(7 - 2\ell)$

So, $\frac{1}{2}k(5-2\ell) = \frac{1}{4}k(7-2\ell) \Rightarrow 2(5-2\ell) = 7 - 2\ell \Rightarrow 10 - 4\ell = 7 - 2\ell \Rightarrow 3 = 2\ell \Rightarrow \ell = \frac{3}{2}$ ft.

21. Let ℓ be the natural length of the spring, and suppose $f(x) = kx$. Then

$$\int_{\ell-1}^{\ell-\frac{3}{4}} kx\,dx = k \cdot \tfrac{1}{2}x^2 \Big|_{\ell-1}^{\ell-\frac{3}{4}} \qquad\qquad 2\int_{1-\ell}^{2-\ell} kx\,dx = 2k \cdot \tfrac{1}{2}x^2 \Big|_{1-\ell}^{2-\ell}$$

$$= \tfrac{1}{2}k[(\ell-\tfrac{3}{4})^2 - (\ell-1)^2] \qquad\qquad = k[(2-\ell)^2 - (1-\ell)^2]$$

$$= \tfrac{1}{2}k[\ell^2 - \tfrac{3}{2}\ell + \tfrac{9}{16} - (\ell^2 - 2\ell + 1)] \qquad = k[4 - 4\ell + \ell^2 - (1 - 2\ell + \ell^2)]$$

$$= \tfrac{1}{2}k\left(\tfrac{1}{2}\ell - \tfrac{7}{16}\right) \qquad\qquad = k(3 - 2\ell)$$

Thus, $\tfrac{1}{2}k\left(\tfrac{1}{2}\ell - \tfrac{7}{16}\right) = k(3 - 2\ell) \Rightarrow \tfrac{1}{2}\ell - \tfrac{7}{16} = 2(3 - 2\ell) \Rightarrow 8\ell - 7 = 32(3 - 2\ell) \Rightarrow$

$72\ell = 103 \Rightarrow \ell = \tfrac{103}{72}$ m.

23. Given $W = \int_0^a kx\,dx$, we are to determine $\int_a^{2a} kx\,dx$:

$$W = \int_0^a kx\,dx = k \cdot \tfrac{1}{2}x^2 \Big|_0^a = \tfrac{1}{2}ka^2.$$

$$\int_a^{2a} kx\,dx = k \cdot \tfrac{1}{2}x^2 \Big|_a^{2a} = \tfrac{1}{2}k[(2a)^2 - a^2] = \tfrac{1}{2}k(3a^2) = 3 \cdot \tfrac{1}{2}ka^2 = 3W.$$

25. First, convert 6 oz./in.^{3} to lbs/ft^3: $\dfrac{6 \text{ oz}}{\text{in.}^3} = \dfrac{6 \text{ oz}}{\text{in}^3} \cdot \dfrac{12^3 \text{in.}^3}{1^3 \text{ ft}^3} \cdot \dfrac{1 \text{ lb}}{16\text{oz}} = \dfrac{6(12^3)}{16} \dfrac{\text{lb}}{\text{ft}^3} = 648 \dfrac{\text{lb}}{\text{ft}^3}$.

(a) Like the solution of CYU 5.28, with the units now pounds and feet, the approximate work to lift the disk to the top of the tank is $\Delta W = \pi r^2 \Delta x (648)x$; and again, $r = 1 - \tfrac{x}{3}$.

Thus, $W = 648\pi \int_0^3 xr^2\,dx = 648\pi \int_0^3 x(1 - \tfrac{x}{3})^2 dx = 648\pi \int_0^3 (x - \tfrac{2}{3}x^2 + \tfrac{x^3}{9})\,dx$

$$= 648\pi \left(\tfrac{1}{2}x^2 - \tfrac{2}{9}x^3 + \tfrac{x^4}{36}\right)\Big|_0^3 = 648\pi \left(\tfrac{9}{2} - 6 + \tfrac{9}{4}\right) = 486\pi \text{ ft-lbs.}$$

(b) $W = 648\pi \int_0^3 (x+1)(1 - \tfrac{x}{3})^2 dx = 648\pi \int_0^3 (\tfrac{x^3}{9} - \tfrac{5x^2}{9} + \tfrac{x}{3} + 1)\,dx$

$$= 648\pi \left(\tfrac{x^4}{36} - \tfrac{5x^3}{27} + \tfrac{x^2}{6} + x\right)\Big|_0^3 = 648\pi \left(\tfrac{9}{4} - 5 + \tfrac{3}{2} + 3\right) = 1{,}134\pi \text{ ft-lbs.}$$

(c) First determine the value of r corresponding to half the volume of the cone, in order to get the upper limit of integration for x: As the volume of a cone is $V = \tfrac{1}{3}\pi r^2 h$, the volume of our full cone is $\tfrac{1}{3}\pi(1^2)(3) = \pi \Rightarrow \tfrac{1}{2}\pi = \tfrac{1}{3}\pi r^2(3r) \Rightarrow r^3 = \tfrac{1}{2} \Rightarrow r = \tfrac{1}{2^{\frac{1}{3}}}$. As $r = 1 - \tfrac{x}{3}$, $x = 3(1-r) = 3 - \tfrac{3}{2^{\frac{1}{3}}} = 3 - 3 \cdot 2^{-\frac{1}{3}}$. Then from (a),

$$W = 648\pi \int_0^{3-3\cdot 2^{-\frac{1}{3}}} (x - \tfrac{2}{3}x^2 + \tfrac{x^3}{9})\,dx = 648\pi \left(\tfrac{1}{2}x^2 - \tfrac{2}{9}x^3 + \tfrac{x^4}{36}\right)\Big|_0^{3-3\cdot 2^{-\frac{1}{3}}}$$

$$\approx 648(0.143)\pi \approx 92.66\pi \text{ ft-lbs.}$$

27. Convert the weight of the rope to lbs/ft: $\dfrac{4 \text{ oz}}{\text{ft}} = \dfrac{4 \text{ oz}}{\text{ft}} \cdot \dfrac{1 \text{ lb}}{16 \text{ oz}} = \tfrac{1}{4}$ lb/ft.

As $\Delta W = \tfrac{1}{4}\Delta x(30 - x)$, $W = \int_0^{25} \tfrac{1}{4}(30 - x)\,dx = \tfrac{1}{4}\left(30x - \tfrac{1}{2}x^2\right)\Big|_0^{25} = \tfrac{1}{4}\left(750 - \tfrac{1}{2}\cdot 625\right)$

$$= \tfrac{1}{8}(1500 - 625) = \tfrac{875}{8} \text{ ft-lbs.}$$

29. The work done raising the cable and the bucket 20 ft is the sum of the work done in raising the cable, W_c, and the work done in raising the bucket, W_b:

$$W_b = 50 \text{ lbs} \cdot 20 \text{ ft} = 1000 \text{ ft-lbs.}$$
$$W_c = \int_0^{20} 2x \, dx = x^2 \big|_0^{20} = 20^2 = 400 \text{ ft-lbs.}$$

The total work done is $1000 + 400 = 1400$ ft-lbs.

31. $\bar{x} = \dfrac{10(-5) + 15(3)}{10 + 15} = \dfrac{-5}{25} = -\dfrac{1}{5}.$

33. Let w be the weight sought, then

$$\bar{x} = 0 = \frac{10(-5) + 15(3) + w(7)}{10 + 15 + w} = \frac{-5 + 7w}{25 + w} \;\Rightarrow\; -5 + 7w = 0 \;\Rightarrow\; w = \frac{5}{7} \text{ lbs.}$$

35. $\bar{x} = \dfrac{\displaystyle\int_0^{10} x\left(1 + \frac{x}{10}\right) dx}{\displaystyle\int_0^{10} \left(1 + \frac{x}{10}\right) dx} = \dfrac{\displaystyle\int_0^{10} \left(x + \frac{1}{10}x^2\right) dx}{\displaystyle\int_0^{10} \left(1 + \frac{x}{10}\right) dx} = \dfrac{\left(\frac{1}{2}x^2 + \frac{1}{10}\cdot\frac{1}{3}x^3\right)\big|_0^{10}}{\left(x + \frac{1}{10}\cdot\frac{1}{2}x^2\right)\big|_0^{10}}$

$= \dfrac{50 + \frac{100}{3}}{10 + 5} = \dfrac{250}{45} = \dfrac{50}{9}$ ft.

37. $\bar{x} = \dfrac{10(1) + 20(-2) + 4(-1)}{10 + 20 + 4} = \dfrac{10 - 40 - 4}{34} = -1$

$\bar{y} = \dfrac{10(3) + 20(2) + 4(8)}{10 + 20 + 4} = \dfrac{30 + 40 + 32}{34} = \dfrac{102}{34} = 3 \;\Rightarrow\; (\bar{x}, \bar{y}) = (-1, 3).$

39. Let (a, b) be the location of the 5 lb weight:

$$\bar{x} = 0 = \frac{10(1) + 20(-2) + 4(-1) + 5a}{10 + 20 + 4 + 5} = \frac{-34 + 5a}{39} \;\Rightarrow\; 5a - 34 = 0 \;\Rightarrow\; a = \frac{34}{5}$$

$$\bar{y} = 0 = \frac{10(3) + 20(2) + 4(8) + 5b}{10 + 20 + 4 + 5} = \frac{102 + 5b}{39} \;\Rightarrow\; 5b + 102 = 0 \;\Rightarrow\; b = -\frac{102}{5}$$

The position is $\left(\frac{34}{5}, -\frac{102}{5}\right).$

41. $\bar{x} = \dfrac{\displaystyle\int_0^1 x(x^2) dx}{\displaystyle\int_0^1 x^2 dx} = \dfrac{\frac{1}{4}x^4 \big|_0^1}{\frac{1}{3}x^3 \big|_0^1} = \dfrac{\frac{1}{4}}{\frac{1}{3}} = \dfrac{3}{4}.$

$\bar{y} = \dfrac{\frac{1}{2}\displaystyle\int_0^1 \left[(x^2)^2 - 0^2\right] dx}{\displaystyle\int_0^1 (x^2 - 0) dx} = \dfrac{\left(\frac{1}{2}\cdot\frac{1}{5}x^5\right)\big|_0^1}{\frac{1}{3}} = \dfrac{\frac{1}{10}}{\frac{1}{3}} = \dfrac{3}{10} \;\Rightarrow\; (\bar{x}, \bar{y}) = \left(\frac{3}{4}, \frac{3}{10}\right).$

43.

$$\bar{x} = \frac{\int_0^1 x\left(x - x^2\right)dx}{\int_0^1 \left(x - x^2\right)dx} = \frac{\int_0^1 \left(x^2 - x^3\right)dx}{\int_0^1 \left(x - x^2\right)dx} = \frac{\left(\frac{1}{3}x^3 - \frac{1}{4}x^4\right)\big|_0^1}{\left(\frac{1}{2}x^2 - \frac{1}{3}x^3\right)\big|_0^1}$$

$$= \frac{\frac{1}{3} - \frac{1}{4}}{\frac{1}{2} - \frac{1}{3}} = \frac{\frac{1}{12}}{\frac{1}{6}} = \frac{1}{2}$$

$$\bar{y} = \frac{\frac{1}{2}\int_0^1 \left[x^2 - (x^2)^2\right]dx}{\int_0^1 \left(x - x^2\right)dx} = \frac{\frac{1}{2}\left(\frac{1}{3}x^3 - \frac{1}{5}x^5\right)\big|_0^1}{\frac{1}{6}} = 3\left(\frac{1}{3} - \frac{1}{5}\right) = \frac{2}{5}$$

Therefore, $(\bar{x}, \bar{y}) = \left(\frac{1}{2}, \frac{2}{5}\right).$

45.

Intersection of $y = 2\sqrt{x}$ and $y = x$:

$$2\sqrt{x} = x \;\Rightarrow\; 4x = x^2 \;\Rightarrow\; x^2 - 4x = 0 \;\Rightarrow\; x(x-4) = 0 \;\Rightarrow\; x = 0, 4.$$

$$\bar{x} = \frac{\int_0^4 x\left(2x^{\frac{1}{2}} - x\right)dx}{\int_0^4 \left(2x^{\frac{1}{2}} - x\right)dx} = \frac{\int_0^4 \left(2x^{\frac{3}{2}} - x^2\right)dx}{\int_0^4 \left(2x^{\frac{1}{2}} - x\right)dx} = \frac{\left(2\cdot\frac{2}{5}x^{\frac{5}{2}} - \frac{1}{3}x^3\right)\big|_0^4}{\left(2\cdot\frac{2}{3}x^{\frac{3}{2}} - \frac{1}{2}x^2\right)\big|_0^4}$$

$$= \frac{\frac{4}{5}\cdot 4^{\frac{5}{2}} - \frac{64}{3}}{\frac{4}{3}\cdot 4^{\frac{3}{2}} - 8} = \frac{\frac{128}{5} - \frac{64}{3}}{\frac{8}{3}}\cdot\frac{15}{15} = \frac{3(128) - 5(64)}{5(8)} = \frac{384 - 320}{40} = \frac{64}{40} = \frac{8}{5}$$

$$\bar{y} = \frac{\frac{1}{2}\int_0^4\left[\left(2x^{\frac{1}{2}}\right)^2 - (x)^2\right]dx}{\int_0^4 \left(2x^{\frac{1}{2}} - x\right)dx} = \frac{\frac{1}{2}\int_0^4 \left(4x - x^2\right)dx}{\frac{8}{3}} = \frac{3}{16}\left(2x^2 - \frac{1}{3}x^3\right)\big|_0^4$$

$$= \frac{3}{16}\left(32 - \frac{64}{3}\right) = 3\left(2 - \frac{4}{3}\right) = 6 - 4 = 2$$

Therefore, $(\bar{x}, \bar{y}) = \left(\frac{8}{5}, 2\right).$

47. Let L be the strip width, from similar triangles:

$$\frac{y-1}{L} = \frac{4}{3} \;\Rightarrow\; L = \frac{3}{4}(y-1) \;\Rightarrow$$

$$F = w\int_1^5 \left[y\cdot\frac{3}{4}(y-1)\right]dy = \frac{3w}{4}\int_1^5 (y^2 - y)dy = \frac{3w}{4}\left(\frac{1}{3}y^3 - \frac{1}{2}y^2\right)\big|_1^5$$

$$= \frac{3w}{4}\left[\frac{125}{3} - \frac{25}{2} - \left(\frac{1}{3} - \frac{1}{2}\right)\right] = \frac{3w}{4}\left(\frac{124}{3} - 12\right) = w(31 - 9) = 22w \text{ lbs.}$$

49. Let L be the strip width. In the lower figure, the side of the large triangle is: $\sqrt{6^2 - 3^2} = \sqrt{27} = 3\sqrt{3}$. From similar triangles:

$$\frac{L/2}{3\sqrt{3} - (y-2)} = \frac{3}{3\sqrt{3}} \Rightarrow L = \frac{2}{\sqrt{3}}\left(3\sqrt{3} - (y-2)\right) \Rightarrow$$

$$F = w \int_2^{2+3\sqrt{3}} \left[y \cdot \frac{2}{\sqrt{3}}\left(3\sqrt{3} - (y-2)\right)\right] dy$$

$$= \frac{2w}{\sqrt{3}} \int_2^{2+3\sqrt{3}} [(3\sqrt{3}+2)y - y^2] dy = \frac{2w}{\sqrt{3}}\left[\left(\frac{3\sqrt{3}+2}{2}(y^2) - \frac{1}{3}y^3\right)\right]\Big|_2^{2+3\sqrt{3}}$$

$$= \frac{2w}{\sqrt{3}}\left[\frac{3\sqrt{3}+2}{2}\left(2+3\sqrt{3}\right)^2 - \frac{1}{3}\left(2+3\sqrt{3}\right)^3 - \left(2(3\sqrt{3}+2) - \frac{8}{3}\right)\right]$$

$$= \frac{2w}{\sqrt{3}}\left[\frac{1}{2}(2+3\sqrt{3})^3 - \frac{1}{3}(2+3\sqrt{3})^3 - 6\sqrt{3} - 4 + \frac{8}{3}\right]$$

$$= \frac{2w}{6\sqrt{3}}\left[(2+3\sqrt{3})^3 - 36\sqrt{3} - 8\right] = \frac{w}{3\sqrt{3}}\left[170 + 117\sqrt{3} - 36\sqrt{3} - 8\right]$$

$$= \frac{\sqrt{3}w}{9}(162 + 81\sqrt{3}) = \sqrt{3}w(18 + 9\sqrt{3}) = (27 + 18\sqrt{3})w \text{ lbs.}$$

51. Let L be the strip width. In the lower figure, from similar triangles:

$$\frac{10-y}{k} = \frac{6}{2} \Rightarrow k = \frac{1}{3}(10-y) \Rightarrow$$

$$L = 2k + 12 = \frac{2}{3}(10-y) + 12 = \frac{20 - 2y + 36}{3} = \frac{56 - 2y}{3} \Rightarrow$$

$$F = w\int_4^{10} y\left(\frac{56-2y}{3}\right) dy = \frac{w}{3}\int_4^{10} (56y - 2y^2)\, dy = \frac{w}{3}\left(56 \cdot \frac{1}{2}y^2 - \frac{2}{3}y^3\right)\Big|_4^{10}$$

$$= \frac{w}{3}\left[2800 - \frac{2}{3}(1000) - \left(28 \cdot 16 - \frac{2}{3}\cdot 64\right)\right]$$

$$= \frac{w}{9}(8400 - 2000 - 1344 + 128) = \frac{w}{9}(5184) = 576w \text{ lbs.}$$

53. 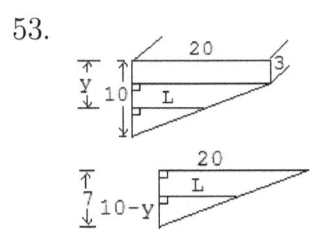 $F = F_1 + F_2$, where F_1 is the force to a depth of 3 feet, and F_2 is the force from 3 feet to 10 feet:

Let L be the strip width. For F_1, L is constant at 20 ft. For F_2, referring to the lower figure, from similar triangles:

$$\frac{10-y}{L} = \frac{7}{20} \Rightarrow 7L = 20(10-y) \Rightarrow L = \frac{20}{7}(10-y).$$

Therefore:

$$F_1 = 62.4 \int_0^3 y(20)\, dy, \text{ and } F_2 = 62.4\int_3^{10} y\left[\frac{20}{7}(10-y)\right] dy. \text{ Thus,}$$

$$F = 62.4\int_0^3 20y\, dy + 62.4\left(\frac{20}{7}\right)\int_3^{10}(10y - y^2)\, dy = 62.4\,(10y^2)\Big|_0^3 + 6.24\left(\frac{20}{7}\right)\left(5y^2 - \frac{1}{3}y^3\right)\Big|_3^{10}$$

$$= 62.4\left[90 + \frac{20}{7}\left(500 - \frac{1000}{3} - (45-9)\right)\right] = 62.4\left(90 + \frac{10,000}{7} - \frac{20,000}{21} - \frac{720}{7}\right)$$

$$= \frac{62.4}{21}(1890 + 30,000 - 20,000 - 2160) = \frac{62.4}{21}(9,730) = 28,912 \text{ lbs.}$$

CHAPTER 6
Additional Transcendental Functions

§6.1 The Natural Logarithmic Function

1. $h'(x) = (x^3 \ln x)' = x^3(\ln x)' + (x^3)' \ln x = x^3 \left(\frac{1}{x}\right) + 3x^2 \ln x = x^2 + 3x^2 \ln x.$

3. $h'(x) = [\ln x^2 + (\ln x)^2]' = \frac{1}{x^2} \cdot 2x + 2 \ln x \cdot \frac{1}{x} = \frac{2}{x} + \frac{2 \ln x}{x}.$

5. $g'(x) = (\sin x \ln x)' = (\sin x)(\ln x)' + (\sin x)' \ln x = \sin x \cdot \frac{1}{x} + \cos x \ln x = \frac{\sin x}{x} + \cos x \ln x.$

7. $\frac{d}{dx} f(x) = \frac{d}{dx} \ln(\cos x) = \frac{1}{\cos x} \cdot - \sin x = -\frac{\sin x}{\cos x} = -\tan x.$

9. $\frac{d}{dx} f(x) = \frac{d}{dx} \frac{\sin x}{\ln x} = \frac{\ln x \cdot \left(\frac{d}{dx} \sin x\right) - \sin x \cdot \frac{d}{dx} \ln x}{(\ln x)^2} = \frac{\ln x \cos x - \sin x \cdot \frac{1}{x}}{(\ln x)^2} = \frac{x \ln x \cos x - \sin x}{x(\ln x)^2}.$

11. $\frac{d}{dx} f(x) = \frac{d}{dx}[\ln(x^2 + 1)]^2 = 2 \ln(x^2 + 1) \cdot \frac{1}{x^2 + 1} \cdot 2x = \frac{4x \ln(x^2 + 1)}{x^2 + 1}.$

13. $\frac{d}{dx} h(x) = \frac{d}{dx} \sqrt{x \ln x} = \frac{d}{dx}(x \ln x)^{\frac{1}{2}} = \frac{1}{2}(x \ln x)^{-\frac{1}{2}} \cdot \frac{d}{dx}(x \ln x) = \frac{1}{2\sqrt{x \ln x}} \left(x \cdot \frac{1}{x} + \ln x\right)$
$= \frac{1 + \ln x}{2\sqrt{x \ln x}}.$

15. $f'(x) = [\ln(\sec x)]' = \frac{1}{\sec x} \cdot (\sec x)' = \frac{1}{\sec x} \sec x \tan x = \tan x.$

17. $f'(x) = [\sin(\ln x^2)]' = \cos(\ln x^2) \cdot (\ln x^2)' = \cos(\ln x^2) \cdot \frac{1}{x^2} \cdot 2x = \frac{2 \cos(\ln x^2)}{x}.$

19. $\frac{dy}{dx} = \frac{d}{dx} \ln x = \frac{1}{x} \Rightarrow \frac{d^2 y}{dx^2} = \frac{d}{dx}\left(\frac{1}{x}\right) = \frac{d}{dx} x^{-1} = -x^{-2} = -\frac{1}{x^2}.$

21. $\frac{dy}{dx} = \frac{d}{dx}\left(\frac{x}{\ln x}\right) = \frac{\ln x \cdot 1 - x \cdot \frac{1}{x}}{(\ln x)^2} = \frac{\ln x - 1}{(\ln x)^2} \Rightarrow$

$\frac{d^2 y}{dx^2} = \frac{d}{dx}\left[\frac{\ln x - 1}{(\ln x)^2}\right] = \frac{(\ln x)^2 \cdot \frac{1}{x} - (\ln x - 1) \cdot 2 \ln x \cdot \frac{1}{x}}{(\ln x)^4} = \frac{\ln x (\ln x - 2 \ln x + 2)}{x(\ln x)^{4\;3}} = \frac{2 - \ln x}{x(\ln x)^3}.$

23. From Exercise 13 above, $\frac{dy}{dx} = \frac{1 + \ln x}{2\sqrt{x \ln x}} \Rightarrow$

$\frac{d^2 y}{dx^2} = \frac{d}{dx}\left(\frac{1 + \ln x}{2\sqrt{x \ln x}}\right) = \frac{2(x \ln x)^{\frac{1}{2}} \cdot \frac{1}{x} - (1 + \ln x) \cdot 2 \cdot \frac{1}{2}(x \ln x)^{-\frac{1}{2}} \cdot \left(x \cdot \frac{1}{x} + \ln x\right)}{4x \ln x} \cdot \frac{(x \ln x)^{\frac{1}{2}}}{(x \ln x)^{\frac{1}{2}}}$

$= \frac{2 \ln x - (1 + \ln x)^2}{4(x \ln x)^{\frac{3}{2}}} = \frac{2 \ln x - [1 + 2 \ln x + (\ln x)^2]}{4(x \ln x)^{\frac{3}{2}}} = -\frac{1 + (\ln x)^2}{4(x \ln x)^{\frac{3}{2}}}.$

25. For $f(x) = 3x^2 + x$ and $g(x) = \ln x$, $f'(x) = 6x + 1$ and $g'(x) = \frac{1}{x}$:

(a) $(g \circ f)'(x) = g'[f(x)]f'(x) = \frac{1}{f(x)} \cdot (6x + 1) = \frac{6x + 1}{3x^2 + x}$.

(b) $(f \circ g)'(x) = f'[g(x)]g'(x) = (6 \ln x + 1) \cdot \frac{1}{x} = \frac{6 \ln x + 1}{x}$.

(c) $(f \circ f)'(x) = f'[f(x)]f'(x) = [6(3x^2 + x) + 1] \cdot (6x + 1) = (18x^2 + 6x + 1)(6x + 1)$
$$= 108x^3 + 18x^2 + 36x^2 + 6x + 6x + 1 = 108x^3 + 54x^2 + 12x + 1.$$

(d) $(g \circ g)'(x) = g'[g(x)]g'(x) = \frac{1}{\ln x} \cdot \frac{1}{x} = \frac{1}{x \ln x}$.

27.   The slope of the tangent line is the first derivative y' at $x = 1$: $y' = 2 \cdot \frac{1}{x}$, at $x = 1, y' = 2$. When $x = 1, y = 2 \ln 1 = 0 \Rightarrow$ the equation of the tangent line is $y - 0 = 2(x - 1) \Rightarrow y = 2x - 2$.

29. The slope of the tangent line is the first derivative y' at $x = \frac{\pi}{2}$:
$y' = [\ln(\sin x)]' = \frac{1}{\sin x} \cdot \cos x = \cot x$. At $x = \frac{\pi}{2}, y' = \cot \frac{\pi}{2} = 0$. When $x = \frac{\pi}{2}$,
$y = \ln(\sin \frac{\pi}{2}) = \ln 1 = 0 \Rightarrow$ the equation of the tangent line is $y = 0$.

31.   Let the point to be found be (a, b). If the tangent line is to pass through the origin, its equation has the form $y = mx$. The slope of the tangent line is $y' = (\ln x)' = \frac{1}{x}$, so at $(a, b), y' = m = \frac{1}{a} \Rightarrow$ the equation of the tangent line is $y = \frac{1}{a}x$. Since the point lies on the graph, $b = \ln a$. Since the point also lies on the tangent line: $b = \frac{1}{a} \cdot a = 1 \Rightarrow 1 = b = \ln a \Rightarrow \ln a = 1 \Rightarrow a = e$. The point is $(e, 1)$.

33. Let $u = x^3 + 2$. Then $du = 3x^2 dx \Rightarrow x^2 dx = \frac{1}{3} du \Rightarrow$
$$\int \frac{x^2}{x^3 + 2} dx = \frac{1}{3} \int \frac{1}{u} du = \frac{1}{3} \ln |u| + C = \frac{1}{3} \ln |x^3 + 2| + C.$$

35. Let $u = \ln x^2 = 2 \ln |x|$. Then $du = 2 \cdot \frac{1}{x} dx \Rightarrow \frac{1}{x} dx = \frac{1}{2} du \Rightarrow$
$$\int \frac{dx}{x \ln x^2} = \frac{1}{2} \int \frac{1}{u} du = \frac{1}{2} \ln |u| + C = \frac{1}{2} \ln |\ln x^2| + C = \frac{1}{2} \ln |2 \ln x| + C$$
$$= \frac{1}{2} (\ln 2 + \ln |\ln x|) + C = \frac{1}{2} \ln 2 + \frac{1}{2} \ln |\ln x| + C = \frac{1}{2} \ln |\ln x| + C.$$

37. Let $u = \ln x$. Then $du = \frac{1}{x} dx \Rightarrow$
$$\int \frac{\cos(\ln x)}{x} dx = \int \cos u \, du = \sin u + C = \sin(\ln x) + C.$$

39. Let $u = \sqrt{x} = x^{\frac{1}{2}}$. Then $du = \frac{1}{2} x^{-\frac{1}{2}} dx = \frac{1}{2} \cdot \frac{1}{\sqrt{x}} dx \Rightarrow \frac{1}{\sqrt{x}} dx = 2 \, du \Rightarrow$
$$\int \frac{\cot \sqrt{x}}{\sqrt{x}} dx = 2 \int \cot u \, du = -2 \ln |\csc u| + C = -2 \ln |\csc \sqrt{x}| + C.$$

41. $\displaystyle\int_{\frac{\pi}{6}}^{\frac{\pi}{4}} \cot x\, dx = -\left(\ln|\csc x|\right)\Big|_{\frac{\pi}{6}}^{\frac{\pi}{4}} = -\left(\ln\left|\csc\frac{\pi}{4}\right| - \ln\left|\csc\frac{\pi}{6}\right|\right) = -\left(\ln\left|\frac{1}{\sin\frac{\pi}{4}}\right| - \ln\left|\frac{1}{\sin\frac{\pi}{6}}\right|\right)$

$$= -(\ln\sqrt{2} - \ln 2) = -(\ln 2^{\frac{1}{2}} - \ln 2) = -\left(\tfrac{1}{2}\ln 2 - \ln 2\right) = \tfrac{1}{2}\ln 2.$$

43. Let $u = x^2 + 1$. Then $du = 2x\,dx \;\Rightarrow\; x\,dx = \tfrac{1}{2}du \;\Rightarrow$

$$f(x) = \int \frac{x}{x^2+1}\,dx = \tfrac{1}{2}\int \frac{1}{u}\,du = \tfrac{1}{2}\ln|u| + C = \tfrac{1}{2}\ln(x^2+1) + C.$$

As $f(0) = 2$, $\; 2 = \tfrac{1}{2}\ln(0^2+1) + C \;\Rightarrow\; C = 2 \;\Rightarrow\; f(x) = \tfrac{1}{2}\ln(x^2+1) + 2.$

45. Let $u = \ln\frac{1}{x} = \ln x^{-1} = -\ln x$, as $x > 0$ for the logarithm to be defined. Then $du = -\frac{1}{x}dx \;\Rightarrow\; \frac{1}{x}dx = -du \;\Rightarrow$

$$f(x) = \int \frac{1}{x}\ln\frac{1}{x}\,dx = -\int u\,du = -\tfrac{1}{2}u^2 + C = -\tfrac{1}{2}\left(\ln\tfrac{1}{x}\right)^2 + C.$$

As $f(e) = 1$, $\; 1 = -\tfrac{1}{2}\left(\ln\tfrac{1}{e}\right)^2 + C = -\tfrac{1}{2}(\ln 1 - \ln e)^2 + C = -\tfrac{1}{2} + C \;\Rightarrow\; C = \tfrac{3}{2} \;\Rightarrow$

$$f(x) = -\tfrac{1}{2}\left(\ln\tfrac{1}{x}\right)^2 + \tfrac{3}{2}.$$

47. The function $f(x) = -\ln(-x-5)$ is only defined for $-x - 5 > 0 \;\Rightarrow\; -x > 5 \;\Rightarrow\; x < -5$. There is no y-intercept, and the x intercept is: $0 = -\ln(-x-5) \;\Rightarrow\; -x - 5 = 1 \;\Rightarrow\; x = -6$. There is a vertical asymptote: $x = -5$.

$f'(x) = -\frac{1}{-x-5}\cdot -1 = -\frac{1}{x+5} > 0$ on its domain,
so the graph is always increasing.

$f''(x) = [-(x+5)^{-1}]' = (x+5)^{-2}\cdot 1 = \frac{1}{(x+5)^2} > 0,$
so the graph is always concave up.

49.

$A = \displaystyle\int_1^e \left[\tfrac{1}{x} - (-x)\right] dx = \left(\ln|x| + \tfrac{1}{2}x^2\right)\Big|_1^e = \ln e + \tfrac{1}{2}e^2 - \left(\ln 1 + \tfrac{1}{2}\right)$

$$= 1 + \tfrac{1}{2}e^2 - \tfrac{1}{2} = \tfrac{1}{2} + \tfrac{1}{2}e^2 = \tfrac{1}{2}(1 + e^2).$$

51.

Intersection of $y = x^2$ and $y = \frac{1}{\sqrt{x}}$:

$$x^2 = \frac{1}{\sqrt{x}} \;\Rightarrow\; x^{\frac{5}{2}} = 1 \;\Rightarrow\; x^5 = 1 \;\Rightarrow\; x = 1.$$

$$V = \pi\int_1^e \left[(x^2)^2 - \left(\frac{1}{\sqrt{x}}\right)^2\right] dx = \pi\int_1^e \left[x^4 - \tfrac{1}{x}\right] dx$$

$$= \pi\left[\tfrac{1}{5}x^5 - \ln|x|\right]\Big|_1^e = \pi\left[\tfrac{1}{5}e^5 - \ln e - \left(\tfrac{1}{5} - \ln 1\right)\right]$$

$$= \pi\left(\tfrac{1}{5}e^5 - 1 - \tfrac{1}{5}\right) = \pi\left(\tfrac{1}{5}e^5 - \tfrac{6}{5}\right) = \tfrac{\pi}{5}(e^5 - 6).$$

53. The work done is $\int_2^9 \frac{1}{x}\, dx = (\ln|x|)\Big|_2^9 = \ln 9 - \ln 2 = \ln\frac{9}{2}$.

55. $v' = \left(\frac{\ln t}{t}\right)' = \dfrac{t\cdot\frac{1}{t} - \ln t\cdot 1}{t^2} = \dfrac{1-\ln t}{t^2} = 0 \Rightarrow \ln t = 1 \Rightarrow t = e$. As we can see, for $1 < t < e$, $v' > 0$ and for $t > e$, $v' < 0 \Rightarrow$ the maximum value of v occurs when $t = e$.

57.   Since $(\ln x^r)' = \frac{1}{x^r}\cdot rx^{r-1} = \frac{r}{x}$, and $(r\ln x)' = r\cdot\frac{1}{x} = \frac{r}{x}$, the two functions can only differ by a constant (Theorem 5.1): $\ln x^r = r\ln x + C$. Evaluating this equation at $x = 1$: $\ln 1 = r\ln 1 + C \Rightarrow C = 0 \Rightarrow \ln x^r = r\ln x$.

§6.2 The Natural Exponential Function

1. $f'(x) = (e^{2x})' = e^{2x}\cdot(2x)' = e^{2x}\cdot 2 = 2e^{2x}$.

3. $f'(x) = (x^3 e^x)' = (x^3)' e^x + x^3 (e^x)' = 3x^2 e^x + x^3(e^x) = x^2 e^x (3+x)$.

5. $g'(x) = (e^{2x} - 2e^x)' = e^{2x}\cdot(2x)' - 2e^x = 2e^{2x} - 2e^x = 2e^x(e^x - 1)$.

7. $g'(x) = \left(\dfrac{e^{2x}}{2x}\right)' = \dfrac{2x(e^{2x}\cdot 2) - e^{2x}\cdot 2}{(2x)^2} = \dfrac{2e^{2x}(2x-1)}{4x^2} = \dfrac{(2x-1)e^{2x}}{2x^2}$.

9. $g'(x) = [(x+e^x)^5]' = 5(x+e^x)^4\cdot(x+e^x)' = 5(x+e^x)^4(1+e^x)$.

11. $f'(x) = [(x^2+e^2)^5]' = 5(x^2+e^2)^4\cdot(x^2+e^2)' = 5(x^2+e^2)^4(2x) = 10x(x^2+e^2)^4$.

13. $g'(x) = (e^{\sin x})' = e^{\sin x}\cdot(\sin x)' = e^{\sin x}\cos x$.

15. $f'(x) = (e^x\sin x^2)' = e^x(\sin x^2)' + (e^x)'\sin x^2 = e^x(\cos x^2\cdot 2x) + e^x\sin x^2$
    $= e^x(2x\cos x^2 + \sin x^2)$.

17. $g'(x) = \left(\dfrac{\tan x^2}{e^{x^2}}\right)' = \dfrac{e^{x^2}(\tan x^2)' - \tan x^2\cdot(e^{x^2})'}{(e^{x^2})^2} = \dfrac{e^{x^2}(\sec^2 x^2\cdot 2x) - \tan x^2(e^{x^2}\cdot 2x)}{(e^{x^2})^2}$
    $= \dfrac{2x\,e^{x^2}(\sec^2 x^2 - \tan x^2)}{(e^{x^2})^2} = \dfrac{2x(\sec^2 x^2 - \tan x^2)}{e^{x^2}}$.

19. $f'(x) = [\ln(x^2+e^{x^2})]' = \dfrac{1}{x^2+e^{x^2}}\cdot(x^2+e^{x^2})' = \dfrac{1}{x^2+e^{x^2}}[2x + (e^{x^2}\cdot 2x)] = \dfrac{2x(1+e^{x^2})}{x^2+e^{x^2}}$.

21. $f'(x) = (e^{e^x})' = e^{e^x}\cdot(e^x)' = e^{e^x}\cdot e^x = e^{x+e^x}$.

23. $(xe^y + \ln y - x^2)' = (1)' \Rightarrow (x)'e^y + x(e^y)' + \frac{1}{y} \cdot y' - 2x = 0 \Rightarrow e^y + x \cdot e^y \cdot y' + \frac{1}{y}(y') - 2x = 0$

$\Rightarrow \left(xe^y + \frac{1}{y}\right)y' = 2x - e^y \Rightarrow \left(\frac{xye^y + 1}{y}\right)y' = 2x - e^y \Rightarrow y' = \frac{y(2x - e^y)}{xye^y + 1}.$

25. $y' = (xe^x)' = x(e^x)' + x'e^x = xe^x + e^x = (x+1)e^x \Rightarrow$

$y'' = [(x+1)e^x]' = (x+1)'e^x + (x+1)(e^x)' = e^x + (x+1)e^x = (x+2)e^x.$

27. $y = \ln xe^{x^2} = \ln x + \ln e^{x^2} = \ln x + x^2 \Rightarrow y' = (\ln x + x^2)' = \frac{1}{x} + 2x \Rightarrow$

$y'' = (x^{-1} + 2x)' = -x^{-2} + 2 = -\frac{1}{x^2} + 2.$

29. $y' = (e^{2x} \cos 3x)' = (e^{2x})' \cos 3x + e^{2x}(\cos 3x)' = [e^{2x} \cdot (2x)'] \cos 3x + e^{2x}[-\sin 3x \cdot (3x)']$

$= 2e^{2x} \cos 3x - 3e^{2x} \sin 3x = e^{2x}(2 \cos 3x - 3 \sin 3x)$

$y'' = (e^{2x})'(2 \cos 3x - 3 \sin 3x) + e^{2x}(2 \cos 3x - 3 \sin 3x)'$

$= 2e^{2x}(2 \cos 3x - 3 \sin 3x) + e^{2x}(-2 \sin 3x \cdot 3 - 3 \cos 3x \cdot 3)$

$= e^{2x}(4 \cos 3x - 6 \sin 3x - 6 \sin 3x - 9 \cos 3x) = -e^{2x}(5 \cos 3x + 12 \sin 3x).$

31. $f(x) = 3x^2 + x \Rightarrow f'(x) = 6x + 1$ and $g(x) = e^x = g'(x)$:

(a) $(g \circ f)'(x) = g'[f(x)]f'(x) = e^{3x^2+x}(6x + 1) = (6x + 1)e^{3x^2+x}.$

(b) $(f \circ g)'(x) = f'[g(x)]g'(x) = (6e^x + 1)e^x.$

(c) $(f \circ f)'(x) = f'[f(x)]f'(x) = [6(3x^2 + x) + 1](6x + 1) = (18x^2 + 6x + 1)(6x + 1)$

$= 108x^3 + 18x^2 + 36x^2 + 6x + 6x + 1 = 108x^3 + 54x^2 + 12x + 1.$

(d) $(g \circ g)'(x) = g'[g(x)]g'(x) = e^{e^x}e^x = e^{e^x+x}.$

33. The slope of the tangent line at $x = 2$ is $y'(2)$, where $y = e^{x^2}$. Then $y' = e^{x^2} \cdot 2x = 2xe^{x^2} \Rightarrow y'(2) = 2(2)e^{2^2} = 4e^4$. When $x = 2$, $y = e^{2^2} = e^4$. Thus the equation of the tangent line is $y - e^4 = 4e^4(x - 2) \Rightarrow y = 4e^4x - 7e^4$.

35. The slope of the tangent line at $x = \frac{\pi}{2}$ is $y'\left(\frac{\pi}{2}\right)$, where $y = e^x \sin x$. Then $y' = (e^x)' \sin x + e^x(\sin x)' = e^x \sin x + e^x \cos x = e^x(\sin x + \cos x) \Rightarrow$ $y'\left(\frac{\pi}{2}\right) = e^{\frac{\pi}{2}}\left(\sin \frac{\pi}{2} + \cos \frac{\pi}{2}\right) = e^{\frac{\pi}{2}}(1 + 0) = e^{\frac{\pi}{2}}$. When $x = \frac{\pi}{2}$, $y = e^{\frac{\pi}{2}} \sin \frac{\pi}{2} = e^{\frac{\pi}{2}}$. Thus the equation of the tangent line is $y - e^{\frac{\pi}{2}} = e^{\frac{\pi}{2}}\left(x - \frac{\pi}{2}\right) \Rightarrow y = e^{\frac{\pi}{2}}x + e^{\frac{\pi}{2}}\left(1 - \frac{\pi}{2}\right).$

37. The slope of the tangent line is $f'(x) = (e^{x^2})' = e^{x^2} \cdot 2x = 2xe^{x^2}$. To solve the equation: $f'(x) = f(x) \Rightarrow 2xe^{x^2} = e^{x^2} \Rightarrow 2x = 1 \Rightarrow x = \frac{1}{2} \Rightarrow y = e^{\left(\frac{1}{2}\right)^2} = e^{\frac{1}{4}}$. The only such point is $\left(\frac{1}{2}, e^{\frac{1}{4}}\right).$

39. Let $u = x^3$. Then $du = 3x^2 dx \implies x^2 dx = \frac{1}{3} du$:

$\int x^2 e^{x^3} dx = \frac{1}{3} \int e^u du = \frac{1}{3} e^u + C = \frac{1}{3} e^{x^3} + C.$

41. Let $u = \frac{1}{x} = x^{-1}$. Then $du = -x^{-2} dx = -\frac{1}{x^2} dx \implies \frac{1}{x^2} dx = -du$:

$\int \frac{e^{1/x}}{x^2} dx = -\int e^u du = -e^u + C = -e^{1/x} + C.$

43. $\int \frac{e^x + 1}{e^x} dx = \int (1 + e^{-x}) dx = x - e^{-x} + C.$

45. Let $u = e^x$. Then $du = e^x dx$:

$\int e^x \cos e^x dx = \int \cos u \, du = \sin u + C = \sin e^x + C.$

47. Let $u = e^x - e^{-x}$. Then $du = (e^x - e^{-x} \cdot -1) dx = (e^x + e^{-x}) dx$:

$\int \frac{(e^x + e^{-x})}{(e^x - e^{-x})} dx = \int \frac{1}{u} du = \ln |u| + C = \ln |e^x - e^{-x}| + C.$

49. Let $u = \sin x$. Then $du = \cos x \, dx$. When $x = 0$, $u = \sin 0 = 0$; when $x = \frac{\pi}{4}$, $u = \sin \frac{\pi}{4} = \frac{1}{\sqrt{2}}$:

$\int_0^{\frac{\pi}{4}} e^{\sin x} \cos x \, dx = \int_0^{\frac{1}{\sqrt{2}}} e^u \, du = e^u \Big|_0^{\frac{1}{\sqrt{2}}} = e^{\frac{1}{\sqrt{2}}} - e^0 = e^{\frac{1}{\sqrt{2}}} - 1.$

51. Let $u = x^2$. Then $du = 2x \, dx \implies x \, dx = \frac{1}{2} du$:

$f(x) = \int x e^{x^2} dx = \frac{1}{2} \int e^u du = \frac{1}{2} e^{x^2} + C$

As $f(0) = 2$, $2 = \frac{1}{2} e^{0^2} + C = \frac{1}{2} + C \implies C = \frac{3}{2} \implies f(x) = \frac{1}{2} e^{x^2} + \frac{3}{2}.$

53. Let $u = 2x$. Then $du = 2 \, dx \implies dx = \frac{1}{2} du$:

$f'(x) = \int e^{2x} dx = \frac{1}{2} \int e^u du = \frac{1}{2} e^u + C = \frac{1}{2} e^{2x} + C$

As $f'(0) = 1$, $1 = \frac{1}{2} e^0 + C = \frac{1}{2} + C \implies C = \frac{1}{2} \implies f'(x) = \frac{1}{2} e^{2x} + \frac{1}{2} = \frac{1}{2} (e^{2x} + 1).$

$f(x) = \frac{1}{2} \int (e^{2x} + 1) dx = \frac{1}{2} [\int e^{2x} dx + \int dx] = \frac{1}{2} [\frac{1}{2} e^{2x} + x] + C = \frac{1}{4} e^{2x} + \frac{1}{2} x + C$

As $f(0) = 1$, $1 = \frac{1}{4} e^0 + \frac{1}{2} (0) + C = \frac{1}{4} + C \implies C = \frac{3}{4} \implies f(x) = \frac{1}{4} e^{2x} + \frac{1}{2} x + \frac{3}{4}.$

55. $y = Ae^{-x} + Bxe^{-x} = (A + Bx)e^{-x} \Rightarrow$

$y' = (A+Bx)'e^{-x} + (A+Bx)(e^{-x})' = Be^{-x} + (A+Bx)e^{-x}(-1) = Be^{-x} - (A+Bx)e^{-x}$

$y'' = [Be^{-x} - (A+Bx)e^{-x}]' = Be^{-x}(-1) - [Be^{-x} - (A+Bx)e^{-x}]$

$= -2Be^{-x} + (A+Bx)e^{-x}$

Then, $y'' + 2y' + y = -2Be^{-x} + (A+Bx)e^{-x} + 2[Be^{-x} - (A+Bx)e^{-x}] + (A+Bx)e^{-x}$

$= e^{-x}[-2B + (A+Bx) + 2B - 2(A+Bx) + (A+Bx)] = 0.$

57. $y = e^{ax} \Rightarrow y' = e^{ax} \cdot a = ae^{ax} \Rightarrow y'' = (ae^{ax})' = a(e^{ax} \cdot a) = a^2 e^{ax}.$ Then

$y'' - 5y' + 6y = 0 \Rightarrow a^2 e^{ax} - 5(ae^{ax}) + 6e^{ax} = 0 \Rightarrow e^{ax}(a^2 - 5a + 6) = 0 \Rightarrow$

$0 = a^2 - 5a + 6 = (a - 2)(a - 3) \Rightarrow a = 2$ or 3.

59. The graph of the function $f(x) = xe^x$ passes through the origin: $f(0) = 0$.

$f'(x) = x(e^x)' + (x)'e^x = xe^x + e^x = (x + 1)e^x$

$f'(x) = 0 \Rightarrow x = -1 \Rightarrow y = -e^{-1} = -\dfrac{1}{e}$

$f''(x) = [(x + 1)e^x]' = (x + 1)'e^x + (x + 1)(e^x)' = e^x + (x + 1)e^x = (x + 2)e^x$

$f''(x) = 0 \Rightarrow x = -2 \Rightarrow y = -2e^{-2} = -\dfrac{2}{e^2}$

As $f' < 0$ for $x < -1$, the graph is decreasing there. As $f' > 0$ for $x > -1$, the graph is increasing there. Thus there is a minimum at $(-1, -\frac{1}{e})$.

As $f'' < 0$ for $x < -2$, the graph is concave down there. As $f'' > 0$ for $x > -2$, the graph is concave up there. Thus there is an inflection point at $(2, -\frac{2}{e^2})$.

Since the graph is decreasing and concave down for $x < -2$, with no x-intercept there, the x-axis is a horizontal asymptote for the graph. Clearly, as x tends to $+\infty$, so does $f(x)$.

61. The graph of the function $f(x) = \frac{\ln x}{x}$ is only defined for $x > 0$, so there is no y-intercept. $f(x) = 0 \Rightarrow \ln x = 0 \Rightarrow x = 1$ is the only x-intercept.

$f'(x) = \dfrac{x(\ln x)' - (\ln x)(x)'}{x^2} = \dfrac{x \cdot \frac{1}{x} - \ln x}{x^2} = \dfrac{1 - \ln x}{x^2}$

$f'(x) = 0 \Rightarrow \ln x = 1 \Rightarrow x = e \Rightarrow y = \dfrac{\ln e}{e} = \dfrac{1}{e}$

$f''(x) = \left(\dfrac{1 - \ln x}{x^2}\right)' = \dfrac{x^2(1 - \ln x)' - (1 - \ln x)(x^2)'}{(x^2)^2} = \dfrac{x^2(-\frac{1}{x}) - (1 - \ln x)(2x)}{x^4}$

$= \dfrac{x[-1 - 2 + 2\ln x]}{x^{\cancel{4}3}} = \dfrac{2\ln x - 3}{x^3}$

$f''(x) = 0 \Rightarrow 2\ln x - 3 = 0 \Rightarrow \ln x = \dfrac{3}{2} \Rightarrow x = e^{\ln x} = e^{\frac{3}{2}} \Rightarrow y = \dfrac{\ln e^{\frac{3}{2}}}{e^{\frac{3}{2}}} = \dfrac{3}{2e^{\frac{3}{2}}}$

As $f' > 0$ for $x < e$, the graph is increasing there. As $f' < 0$ for $x > e$, the graph is decreasing there. Thus there is a maximum at $(e, \frac{1}{e})$.

As $f'' < 0$ for $0 < x < e^{\frac{3}{2}}$, the graph is concave down there. As $f'' > 0$ for $x > e^{\frac{3}{2}}$, the graph is concave up there. Thus there is an inflection point at $(e^{\frac{3}{2}}, \frac{3}{2e^{\frac{3}{2}}})$.

Since the graph is decreasing and concave up for $x > e^{\frac{3}{2}}$, with no x-intercept there, the positive x-axis is a horizontal asymptote for the graph. Clearly, as x tends to 0, $f(x)$ tends to $-\infty$, so the y-axis is a vertical asymptote for the graph.

63. $A = \int_0^{\ln 2} (e^{2x} - e^{-2x}) \, dx$. We could use the substitution method to integrate each exponential function, but instead note that as $(e^{2x})' = e^{2x} \cdot 2 = 2e^{2x}$, an antiderivative of e^{2x} is $\frac{e^{2x}}{2}$. Similarly, an antiderivative of e^{-2x} is $\frac{e^{-2x}}{-2}$. Thus

$$\int_0^{\ln 2} (e^{2x} - e^{-2x}) \, dx = \left(\frac{e^{2x}}{2} - \frac{e^{-2x}}{-2} \right)\Big|_0^{\ln 2} = \frac{1}{2} \left(e^{2x} + e^{-2x} \right)\big|_0^{\ln 2} = \frac{1}{2}[e^{2\ln 2} + e^{-2\ln 2} - (1+1)]$$

$$= \frac{1}{2}(e^{\ln 2^2} + e^{\ln 2^{-2}} - 2) = \frac{1}{2}\left(4 + \frac{1}{4} - 2\right) = \frac{9}{8}.$$

65. The intersection of $y = e^x$ and $y = e$ is: $e^x = e \Rightarrow x = 1$. By the washer method:

$$V = \pi \int_0^1 [e^2 - (e^x)^2] \, dx = \pi \int_0^1 (e^2 - e^{2x}) \, dx = \pi \left(e^2 x - \frac{e^{2x}}{2} \right)\Big|_0^1$$

$$= \pi \left[e^2 - \frac{e^2}{2} - \left(-\frac{1}{2} \right) \right] = \pi \left(\frac{e^2}{2} + \frac{1}{2} \right) = \frac{\pi}{2}(e^2 + 1).$$

67. First graph the curve $y = f(x) = e^{-x^2}$: The graph lies above the x-axis, as $e^{-x^2} > 0$, and is symmetric wrt the y-axis, since f is an even function. As x tends to $\pm\infty$, $f(x)$ approaches $0 \Rightarrow$ the x-axis is a horizontal asymptote for the graph.
   $f'(x) = e^{-x^2} \cdot -2x = -2xe^{-x^2}$: so $f' > 0$ for $x < 0$, and $f' < 0$ when $x > 0 \Rightarrow$ there is a maximum at $(0, 1)$.
   $f''(x) = -2[x'e^{-x^2} + x(e^{-x^2})'] = -2[e^{-x^2} + x \cdot -2xe^{-x^2}] = -2e^{-x^2}(1 - 2x^2)$
   $= -2e^{-x^2}(1 - \sqrt{2}x)(1 + \sqrt{2}x)$. Sign f'':

There are two points of inflection, at $x = \pm\frac{1}{\sqrt{2}}$.

The area of the rectangle pictured in the figure is $2x$ wide by e^{-x^2} high, so its area is $A = 2xe^{-x^2}$, which equals $-f'(x)$. The maximum area occurs where $A' = 0$, but $A' = -f'' = 0$ for $x > 0$ at $x = \frac{1}{\sqrt{2}}$. The area is indeed maximal there, from the sign chart for f'' (remembering that $A' = -f''$). By symmetry, the other vertex of the rectangle is located at $x = -\frac{1}{\sqrt{2}}$, which means that the upper two vertices are located at the points of inflection of the graph of f.

69. $A(t) = A_0 e^{rt}$, where here $r = 0.04 \Rightarrow A(t) = A_0 e^{0.04t}$. We seek A_0 such that $A(5) = 10,000$:
$$10,000 = A_0 e^{(0.04)(5)} = A_0 e^{0.2} \Rightarrow A_0 = \frac{10,000}{e^{0.2}} \approx \$8,187.31.$$

71.   Let $P(t)$ denote the population t years after 1990, so $P(0) = P_0 = 5.28$, and $P(t) = P_0 e^{kt} = 5.28\,e^{kt}$. The year 2004 corresponds to $t = 14 \;\Rightarrow\; P(14) = 6.37 \;\Rightarrow\;$

$6.37 = 5.28\,e^{14k} \;\Rightarrow\; \ln\left(\frac{6.37}{5.28}\right) = 14k \;\Rightarrow\; k = \frac{1}{14}\ln\left(\frac{6.37}{5.28}\right) \;\Rightarrow\; P(t) = 5.28\,e^{\left(\frac{1}{14}\ln\frac{6.37}{5.28}\right)t}$

(a) The year 2010 corresponds to $t = 20 \;\Rightarrow\; P(20) = 5.28\,e^{\left(\frac{1}{14}\ln\frac{6.37}{5.28}\right)20} \approx 6.9$ billion.

(b) To determine t such that $P(t) = 9$:

$9 = 5.28\,e^{\left(\frac{1}{14}\ln\frac{6.37}{5.28}\right)t} \;\Rightarrow\; \frac{9}{5.28} = e^{\left(\frac{1}{14}\ln\frac{6.37}{5.28}\right)t} \;\Rightarrow\; \ln\frac{9}{5.28} = \left(\frac{1}{14}\ln\frac{6.37}{5.28}\right)t \;\Rightarrow\; t = \frac{\ln\frac{9}{5.28}}{\left(\frac{1}{14}\ln\frac{6.37}{5.28}\right)} \approx 40$

This is about 40 years after 1990, so in about the year 2030.

73. As the natural logarithmic function is one-to-one, we need only show $\ln(e^a)^b = \ln e^{ab}$. From $\ln e^x = x$ and $\ln x^r = r\ln x$: $\quad \ln(e^a)^b = b\ln e^a = b\cdot a = ab$, and, $\ln e^{ab} = ab$.

75.   Let $g(x) = e^{-x}f(x)$. Then $g'(x) = e^{-x}f'(x) + (e^{-x}\cdot -1)f(x) = e^{-x}f'(x) - e^{-x}f(x) = 0$, since $f'(x) = f(x)$. By Theorem 4.3(c), the function $g(x)$ is constant, call it c. Then $g(x) = e^{-x}f(x) = c \;\Rightarrow\; f(x) = \dfrac{c}{e^{-x}} = ce^x$.

§6.3 a^x and $\log_a x$

1. $f'(x) = (5^{2x})' = 5^{2x}\ln 5 \cdot (2x)' = (2\ln 5)\,5^{2x}$.

3. $f'(x) = (x^3\,3^x)' = (x^3)'\,3^x + x^3(3^x)' = (3x^2)\,3^x + x^3(3^x \ln 3) = x^2\,3^x(3 + x\ln 3)$.

5. $g(x) = \dfrac{2^{2x}}{2x} = \dfrac{2^{2x-1}}{x} \;\Rightarrow\;$

$g'(x) = \dfrac{x(2^{2x-1})' - 2^{2x-1}(x)'}{x^2} = \dfrac{x[(2^{2x-1}\ln 2)(2x-1)'] - 2^{2x-1}}{x^2} = \dfrac{2^{2x-1}[(2\ln 2)x - 1]}{x^2}$.

7. $g'(x) = (5^{\sin x})' = 5^{\sin x}\ln 5 \cdot (\sin x)' = 5^{\sin x}\ln 5\cos x = (\ln 5)\,5^{\sin x}\cos x$.

9. $f'(x) = (5^x \sin x^2)' = 5^x(\sin x^2)' + (5^x)'\sin x^2 = 5^x(\cos x^2 \cdot 2x) + (5^x \ln 5)\sin x^2$
$= 5^x[2x\cos x^2 + (\ln 5)\sin x^2]$.

11. $g'(x) = \left(\dfrac{\ln x}{3^x}\right)' = \dfrac{3^x(\ln x)' - (\ln x)(3^x)'}{(3^x)^2} = \dfrac{3^x\cdot\frac{1}{x} - (\ln x)(3^x \ln 3)}{(3^x)^2} = \dfrac{3^x[1 - (\ln 3)x\ln x]}{x(3^x)^2}$
$= \dfrac{1 - (\ln 3)x\ln x}{x3^x}$.

13. $f'(x) = (2^x \log_2 x)' = 2^x (\log_2 x)' + (2^x)' \log_2 x = 2^x \cdot \dfrac{1}{x \ln 2} + (2^x \ln 2) \log_2 x$

$\qquad = 2^x \left[\dfrac{1}{x \ln 2} + (\ln 2) \log_2 x \right].$

15. $f'(x) = [\ln(\log_2 x)]' = \dfrac{1}{\log_2 x} \cdot (\log_2 x)' = \dfrac{1}{\log_2 x} \cdot \dfrac{1}{x \ln 2} = \dfrac{1}{(\ln 2)x \log_2 x}.$

17. $f'(x) = [\log_2(\log_2 x)]' = \dfrac{1}{(\log_2 x) \ln 2} \cdot (\log_2 x)' = \dfrac{1}{(\log_2 x) \ln 2} \cdot \dfrac{1}{x \ln 2} = \dfrac{1}{(\ln 2)^2 x \log_2 x}.$

19. $f'(x) = [(3x)^x]' = (e^{x \ln 3x})' = e^{x \ln 3x}(x \ln 3x)' = (3x)^x [x(\ln 3x)' + (x)' \ln 3x]$

$\qquad = (3x)^x [x \cdot \dfrac{1}{3x}(3) + \ln 3x] = (3x)^x (1 + \ln 3x).$

21.[1] $\displaystyle\int x 5^{x^2}\, dx = \dfrac{1}{2 \ln 5} \int du = \dfrac{1}{2 \ln 5} u + C = \dfrac{5^{x^2}}{2 \ln 5} + C = \dfrac{5^{x^2}}{\ln 5^2} + C = \dfrac{5^{x^2}}{\ln 25} + C.$

$$\boxed{\begin{aligned} u &= 5^{x^2} \\ du &= (5^{x^2} \ln 5)(x^2)'dx = (5^{x^2} \ln 5)(2x)dx \;\Rightarrow\; x5^{x^2} dx = \dfrac{1}{2 \ln 5} du \end{aligned}}$$

23.[1] $\displaystyle\int \dfrac{5^{1/x}}{x^2}\, dx = -\int 5^u du = -\dfrac{5^u}{\ln 5} + C = -\dfrac{5^{1/x}}{\ln 5} + C.$

$$\boxed{\begin{aligned} u &= \dfrac{1}{x} = x^{-1} \\ du &= -x^{-2}dx = -\dfrac{1}{x^2}\, dx \;\Rightarrow\; \dfrac{1}{x^2}\, dx = -du \end{aligned}}$$

25.[1] $\displaystyle\int_1^{\sqrt{2}} x\, 2^{x^2}\, dx = \dfrac{1}{2}\int_1^2 2^u\, du = \dfrac{1}{2 \ln 2}(2^u)\Big|_1^2 = \dfrac{1}{2 \ln 2}(2^2 - 2^1) = \dfrac{2}{2 \ln 2} = \dfrac{1}{\ln 2}.$

$$\boxed{\begin{aligned} u &= x^2, & x = 1 &\Rightarrow u = 1 \\ du &= 2x\, dx, & x = \sqrt{2} &\Rightarrow u = 2 \end{aligned}}$$

27. $A(t) = A_0 e^{kt}$. A half-life of $H \Rightarrow A(H) = \frac{1}{2}A_0 \Rightarrow \frac{1}{2}A_0 = A_0 e^{kH} \Rightarrow \ln \frac{1}{2} = \ln e^{kH} = kH$

$\Rightarrow k = \frac{1}{H} \ln \frac{1}{2} = \frac{1}{H}(\ln 1 - \ln 2) = -\frac{\ln 2}{H}$. Therefore $e^{kt} = e^{-\frac{\ln 2}{H}t} = (e^{\ln 2})^{-t/H} = 2^{-t/H}$. Then $A(t) = A_0 e^{kt} = A_0 \cdot 2^{-t/H}$.

29. From "an average experienced typist can type 73 words/min." we conclude $a = 73 \Rightarrow$ $L(t) = 73(1 - b^t)$. From "it book Bill 9 days to learn to type 30 words/min.", we have $30 = 73(1 - b^9)$ which we can solve for b: $1 - b^9 = \frac{30}{73} \Rightarrow b^9 = 1 - \frac{30}{73} = \frac{43}{73} \Rightarrow b = \left(\frac{43}{73}\right)^{\frac{1}{9}}$. Then $L(t) = 73\left[1 - \left(\frac{43}{73}\right)^{\frac{1}{9}t}\right]$. We are to find t such that $L(t) = 60$:

---

[1]Instead of substituting for the exponent in an exponential integrand, one can substitute for the entire exponential function, as in Exercise 21.

$60 = 73 \left[1 - \left(\frac{43}{73} \right)^{\frac{1}{9}t} \right] \Rightarrow 1 - \left(\frac{43}{73} \right)^{\frac{1}{9}t} = \frac{60}{73} \Rightarrow \left(\frac{43}{73} \right)^{\frac{1}{9}t} = 1 - \frac{60}{73} = \frac{13}{73}$. Applying the natural logarithmic function to both sides of the equation:

$$\ln \left(\frac{43}{73} \right)^{\frac{1}{9}t} = \ln \frac{13}{73} \Rightarrow \frac{1}{9}t \ln \frac{43}{73} = \ln \frac{13}{73} \Rightarrow t = \frac{9 \ln(13/73)}{\ln(43/73)} \approx 29.3 \text{ days.}$$

31. As $a^{\log_a x} = x \Rightarrow \log_b a^{\log_a x} = \log_b x \Rightarrow \log_a x \log_b a = \log_b x \Rightarrow \log_a x = \frac{\log_b x}{\log_b a}$.

33. $L = 10 \log \frac{I}{10^{-12}} = 10(\log I - \log 10^{-12}) = 10(\log I + 12)$.

(a) $90 = 10(\log I + 12) \Rightarrow 9 = \log I + 12 \Rightarrow \log I = 9 - 12 = -3 \Rightarrow I = 10^{-3} \text{ Watts/m}^2$.

(b) $30 = 10(\log I + 12) \Rightarrow 3 = \log I + 12 \Rightarrow \log I = 3 - 12 = -9 \Rightarrow I = 10^{-9} \text{ Watts/m}^2$.

(c) $10 = 10(\log I + 12) \Rightarrow 1 = \log I + 12 \Rightarrow \log I = 1 - 12 = -11 \Rightarrow I = 10^{-11} \text{ Watts/m}^2$.

35. Let x be the number of students required. First, determine the intensity of one average whisper: $20 = 10 \log \frac{I}{I_0} \Rightarrow 2 = \log \frac{I}{I_0}$. Apply the exponential function base 10:

$10^2 = 10^{\log(I/I_0)} \Rightarrow 10^2 = \frac{I}{I_0} \Rightarrow I = 100 I_0$. Then the intensity of x students whispering is $I = 100 x I_0$. Thus we need to solve the following equation for x:

$$60 = 10 \log \frac{100 x I_0}{I_0} \Rightarrow 6 = \log 100x \Rightarrow 10^6 = 10^{\log 100x} = 100x \Rightarrow x = 10^4 = 10,000 \text{ students.}$$

37. Let I_T denote the intensity of the earthquake in Turkey, and I_C that of California. Then $7.4 = \log \frac{I_T}{I_0} \Rightarrow 10^{7.4} = 10^{\log(I_T/I_0)} \Rightarrow 10^{7.4} = \frac{I_T}{I_0} \Rightarrow I_T = 10^{7.4} I_0$.

$$5 = \log \frac{I_C}{I_0} \Rightarrow 10^5 = 10^{\log(I_C/I_0)} \Rightarrow 10^5 = \frac{I_C}{I_0} \Rightarrow I_C = 10^5 I_0.$$

Then, $\frac{I_T}{I_C} = \frac{10^{7.4} I_0}{10^5 I_0} = 10^{2.4} \approx 251$. The earthquake in Turkey was about 251 times as intense as the California one.

39. $\text{pH} = -\log[H^+] = -\log(6.31 \times 10^{-9}) = -[\log 6.31 + \log 10^{-9}] = -[\log 6.31 - 9]$
$\qquad = -\log 6.31 + 9 \approx 8.2$.

§6.4 Inverse Trigonometric Functions

1. $f(x) = \sin^{-1}(\ln x)$: The logarithm is only defined for $x > 0$. The inverse sine requires $-1 \leq \ln x \leq 1 \Rightarrow e^{-1} \leq e^{\ln x} \leq e^1 \Rightarrow \frac{1}{e} \leq x \leq e$. Thus x has to be positive and range between $\frac{1}{e}$ and $e \Rightarrow D_f = \left[\frac{1}{e}, e \right]$.

3. $f(x) = e^{\tan^{-1} x}$: As the domain of the exponential function is all reals. The domain of f is the domain of the inverse tangent function, which is also all reals. Therefore $D_f = (-\infty, \infty)$.

5. $f(x) = \sin(\sin^{-1} x)$: The domain of the sine function is all reals, so the domain of f is the domain of the inverse sine function. $D_f = [-1, 1]$.

7. $f'(x) = [\sin^{-1}(x^2)]' = \dfrac{1}{\sqrt{1 - (x^2)^2}} \cdot (x^2)' = \dfrac{2x}{\sqrt{1 - x^4}}$.

9. $f(x) = [\cos^{-1}(x^2)]' = -\dfrac{1}{\sqrt{1 - (x^2)^2}} \cdot (x^2)' = -\dfrac{2x}{\sqrt{1 - x^4}}$.

11. $f'(x) = [\tan^{-1}(\cos x)]' = \dfrac{1}{1 + (\cos x)^2} \cdot (\cos x)' = -\dfrac{\sin x}{1 + \cos^2 x}$.

13. $f'(x) = [\sec^{-1}(e^x)]' = \dfrac{1}{e^x \sqrt{(e^x)^2 - 1}} \cdot (e^x)' = \dfrac{e^x}{e^x \sqrt{e^{2x} - 1}} = \dfrac{1}{\sqrt{e^{2x} - 1}}$.

15. $f'(x) = [\sin^{-1}(e^{2x})]' = \dfrac{1}{\sqrt{1 - (e^{2x})^2}} \cdot (e^{2x})' = \dfrac{2e^{2x}}{\sqrt{1 - e^{4x}}}$.

17. $f'(x) = \left[\dfrac{\tan^{-1} x}{x}\right]' = \dfrac{x(\tan^{-1} x)' - (\tan^{-1} x)x'}{x^2} = \dfrac{x \cdot \frac{1}{1 + x^2} - \tan^{-1} x}{x^2}$

$\qquad = \dfrac{x - (1 + x^2)\tan^{-1} x}{x^2(1 + x^2)}$.

19. $f'(x) = \left[\sqrt{\tan^{-1} 2x}\right]' = \left[(\tan^{-1} 2x)^{\frac{1}{2}}\right]' = \frac{1}{2}(\tan^{-1} 2x)^{-\frac{1}{2}}(\tan^{-1} 2x)'$

$\qquad = \dfrac{1}{2\sqrt{\tan^{-1} 2x}} \cdot \dfrac{1}{1 + (2x)^2} \cdot 2 = \dfrac{1}{(1 + 4x^2)\sqrt{\tan^{-1} 2x}}$.

21. $f'(x) = \left[\sin^{-1}\left(\dfrac{x}{x+1}\right)\right]' = \dfrac{1}{\sqrt{1 - \left(\frac{x}{x+1}\right)^2}} \cdot \left(\dfrac{x}{x+1}\right)' = \dfrac{1}{\sqrt{1 - \frac{x^2}{(x+1)^2}}} \cdot \dfrac{(x+1)(x)' - x(x+1)'}{(x+1)^2}$

$\qquad = \dfrac{1}{\sqrt{1 - \frac{x^2}{(x+1)^2}}} \cdot \dfrac{x + 1 - x}{(x+1)^2} = \dfrac{1}{\sqrt{\frac{(x+1)^2 - x^2}{(x+1)^2}} \cdot \sqrt{(x+1)^2}} = \dfrac{1}{|x+1|\sqrt{2x + 1}}$.

23. $f'(x) = [\csc^{-1}(x^2 + 1)]' = -\dfrac{1}{(x^2 + 1)\sqrt{(x^2 + 1)^2 - 1}} \cdot (x^2 + 1)' = -\dfrac{2x}{(x^2 + 1)\sqrt{x^4 + 2x^2}}$

$\qquad = -\dfrac{2x}{(x^2 + 1)\sqrt{x^2(x^2 + 2)}} = -\dfrac{2x}{(x^2 + 1)|x|\sqrt{x^2 + 2}}$.

25. The slope of the tangent line at $x = 0$ is $y'(0)$: $y'(x) = (\sin^{-1} x)' = \dfrac{1}{\sqrt{1 - x^2}} \Rightarrow$ $y'(0) = 1$. When $x = 0$, $y = \sin^{-1} 0 = 0$, so the equation of the tangent line is $y - 0 = 1(x - 0) \Rightarrow y = x$.

27. The slope of the tangent line at $x = 1$ is $y'(1)$:

$y'(x) = [\sec^{-1}(2x)]' = \dfrac{1}{2x\sqrt{(2x)^2 - 1}} \cdot (2x)' = \dfrac{2}{2x\sqrt{(2x)^2 - 1}} = \dfrac{1}{x\sqrt{4x^2 - 1}} \Rightarrow y'(1) = \dfrac{1}{\sqrt{3}}$

When $x = 1$, $y = \sec^{-1} 2 \Rightarrow \sec y = 2$ and $0 \le y < \frac{\pi}{2} \Rightarrow \cos y = \frac{1}{2} \Rightarrow y = \frac{\pi}{3}$.

The tangent line is $y - \frac{\pi}{3} = \frac{1}{\sqrt{3}}(x - 1) \Rightarrow y = \frac{1}{\sqrt{3}}x + \frac{\pi}{3} - \frac{1}{\sqrt{3}} = \frac{1}{\sqrt{3}}x + \frac{\pi - \sqrt{3}}{3}$.

29. $\displaystyle\int \dfrac{dx}{\sqrt{9 - x^2}} = \int \dfrac{dx}{\sqrt{9[1 - (\frac{x}{3})^2]}} = \dfrac{1}{3}\int \dfrac{dx}{\sqrt{1 - (\frac{x}{3})^2}} = \int \dfrac{du}{\sqrt{1 - u^2}} = \sin^{-1} u + C = \sin^{-1}\frac{x}{3} + C$

$$\boxed{u = \tfrac{x}{3} \Rightarrow du = \tfrac{1}{3}\, dx}$$

31. $\displaystyle\int \dfrac{dx}{\sqrt{3 - 4x^2}} = \int \dfrac{dx}{\sqrt{3(1 - \frac{4x^2}{3})}} = \dfrac{1}{\sqrt{3}}\int \dfrac{dx}{\sqrt{1 - (\frac{2x}{\sqrt{3}})^2}} = \dfrac{1}{2}\int \dfrac{du}{\sqrt{1 - u^2}}$

$$\boxed{\begin{array}{c} u = \dfrac{2x}{\sqrt{3}} \\ du = \dfrac{2}{\sqrt{3}}\, dx \Rightarrow \dfrac{1}{\sqrt{3}}\, dx = \dfrac{1}{2}\, du \end{array}}$$

$= \dfrac{1}{2}\sin^{-1} u + C = \dfrac{1}{2}\sin^{-1}\dfrac{2x}{\sqrt{3}} + C$.

33. $\displaystyle\int \dfrac{e^x}{\sqrt{1 - e^{2x}}}\, dx = \int \dfrac{e^x}{\sqrt{1 - (e^x)^2}}\, dx = \int \dfrac{du}{\sqrt{1 - u^2}} = \sin^{-1} u + C = \sin^{-1} e^x + C$.

$$\boxed{u = e^x \Rightarrow du = e^x\, dx}$$

35. $\displaystyle\int \dfrac{dx}{(\sqrt{1 - x^2})\sin^{-1} x} = \int \dfrac{1}{u}\, du = \ln|u| + C = \ln|\sin^{-1} x| + C$.

$$\boxed{u = \sin^{-1} x \Rightarrow du = \dfrac{1}{\sqrt{1 - x^2}}\, dx}$$

37. $\displaystyle\int \dfrac{dx}{\sqrt{x}(1 + x)} = \int \dfrac{dx}{\sqrt{x}[1 + (\sqrt{x})^2]} = 2\int \dfrac{du}{1 + u^2} = 2\tan^{-1} u + C = 2\tan^{-1}\sqrt{x} + C$.

$$\boxed{\begin{array}{c} u = \sqrt{x} = x^{\frac{1}{2}} \\ du = \dfrac{1}{2}x^{-\frac{1}{2}}\, dx = \dfrac{1}{2\sqrt{x}}\, dx \Rightarrow \dfrac{1}{\sqrt{x}}\, dx = 2\, du \end{array}}$$

39. $\displaystyle\int \dfrac{\cos x}{\sin^2 x + 9}\, dx = \int \dfrac{\cos x}{9\left(1 + \frac{\sin^2 x}{9}\right)}\, dx = \dfrac{1}{9}\int \dfrac{\cos x}{1 + \left(\frac{\sin x}{3}\right)^2}\, dx = \dfrac{1}{3}\int \dfrac{du}{1 + u^2}$

$$\boxed{u = \dfrac{\sin x}{3} \Rightarrow du = \dfrac{1}{3}\cos x\, dx}$$

$= \dfrac{1}{3}\tan^{-1} u + C = \dfrac{1}{3}\tan^{-1}\left(\dfrac{\sin x}{3}\right) + C$.

41. $\int_0^{\frac{1}{4}} \frac{dx}{\sqrt{1-4x^2}} = \int_0^{\frac{1}{4}} \frac{dx}{\sqrt{1-(2x)^2}} = \frac{1}{2}\int_0^{\frac{1}{2}} \frac{du}{\sqrt{1-u^2}} = \frac{1}{2}\sin^{-1}u\Big|_0^{\frac{1}{2}} = \frac{1}{2}[\sin^{-1}\frac{1}{2} - \sin^{-1}0]$

$$\boxed{\begin{array}{ll} u = 2x, & x = 0 \Rightarrow u = 0 \\ du = 2\,dx, & x = \frac{1}{4} \Rightarrow u = \frac{1}{2} \end{array}}$$

$$= \frac{1}{2}\left(\frac{\pi}{6} - 0\right) = \frac{\pi}{12}.$$

43. $\int_{\frac{1}{\sqrt{3}}}^{\sqrt{3}} \frac{8}{1+x^2}\,dx = (8\tan^{-1}x)\Big|_{\frac{1}{\sqrt{3}}}^{\sqrt{3}} = 8\left(\tan^{-1}\sqrt{3} - \tan^{-1}\frac{1}{\sqrt{3}}\right) = 8\left(\frac{\pi}{3} - \frac{\pi}{6}\right) = 8\cdot\frac{\pi}{6} = \frac{4\pi}{3}.$

45. $\int_0^{\frac{1}{2}} \frac{\sin^{-1}x}{\sqrt{1-x^2}}\,dx = \int_0^{\frac{\pi}{6}} u\,du = \frac{1}{2}u^2\Big|_0^{\frac{\pi}{6}} = \frac{1}{2}[(\frac{\pi}{6})^2 - 0] = \frac{\pi^2}{72}.$

$$\boxed{\begin{array}{ll} u = \sin^{-1}x, & x = 0 \Rightarrow u = \sin^{-1}0 = 0 \\ du = \frac{1}{\sqrt{1-x^2}}\,dx, & x = \frac{1}{2} \Rightarrow u = \sin^{-1}\frac{1}{2} = \frac{\pi}{6} \end{array}}$$

47. $f(x) = \int \frac{1}{1+3x^2}\,dx = \int \frac{1}{1+(\sqrt{3}x)^2}\,dx = \frac{1}{\sqrt{3}}\int \frac{1}{1+u^2}\,du = \frac{1}{\sqrt{3}}\tan^{-1}u + C$

$$\boxed{\begin{array}{l} u = \sqrt{3}x \\ du = \sqrt{3}\,dx \Rightarrow dx = \frac{1}{\sqrt{3}}\,du \end{array}}$$

$$= \frac{1}{\sqrt{3}}\tan^{-1}\sqrt{3}x + C$$

$f(\frac{1}{\sqrt{3}}) = \frac{1}{\sqrt{3}} \Rightarrow \frac{1}{\sqrt{3}} = \frac{1}{\sqrt{3}}\tan^{-1}(\sqrt{3}\cdot\frac{1}{\sqrt{3}}) + C = \frac{1}{\sqrt{3}}\tan^{-1}1 + C = \frac{1}{\sqrt{3}}\cdot\frac{\pi}{4} + C \Rightarrow$

$C = \frac{1}{\sqrt{3}} - \frac{1}{\sqrt{3}}\cdot\frac{\pi}{4} = \frac{1}{\sqrt{3}}(1 - \frac{\pi}{4}) = \frac{4-\pi}{4\sqrt{3}} \Rightarrow f(x) = \frac{1}{\sqrt{3}}\tan^{-1}\sqrt{3}x + \frac{4-\pi}{4\sqrt{3}}.$

49. $A = \int_0^{3^{\frac{3}{2}}} \frac{x}{x^4+9}\,dx = \int_0^{3^{\frac{3}{2}}} \frac{x}{9\left[1+\left(\frac{x^2}{3}\right)^2\right]}\,dx = \frac{3}{2}\int_0^9 \frac{1}{9(1+u^2)}\,du = \frac{1}{6}\int_0^9 \frac{1}{1+u^2}\,du$

$$\boxed{\begin{array}{ll} u = \frac{x^2}{3}, & x = 0 \Rightarrow u = 0 \\ du = \frac{2}{3}x\,dx, & x = 3^{\frac{3}{2}} \Rightarrow u = \frac{(3^{\frac{3}{2}})^2}{3} = 3^2 = 9 \end{array}}$$

$$= \frac{1}{6}\tan^{-1}u\Big|_0^9 = \frac{1}{6}(\tan^{-1}9 - \tan^{-1}0) = \frac{1}{6}\tan^{-1}9.$$

51. ring

To determine $\dfrac{d\theta}{dt}\Big|_{x=20} = \left(\dfrac{d\theta}{dx} \cdot \dfrac{dx}{dt}\right)\Big|_{x=20}$, where $\dfrac{dx}{dt} = -1$: From the adjacent figure, $\theta = \sin^{-1}\dfrac{12}{x}$,

$$\frac{d\theta}{dx} = \frac{1}{\sqrt{1-\left(\frac{12}{x}\right)^2}} \cdot (12x^{-1})' = \frac{1}{\sqrt{1-\frac{144}{x^2}}} \cdot -12x^{-2} = -\frac{12}{x^2\sqrt{1-\frac{144}{x^2}}}$$

$$\frac{d\theta}{dt}\Big|_{x=20} = -\frac{12}{20^2\sqrt{1-\frac{144}{20^2}}} \cdot (-1) = \frac{12}{400\sqrt{1-\frac{144}{400}}} = \frac{3}{100\sqrt{1-\frac{9}{25}}} = \frac{3}{100\cdot\frac{4}{5}} = \frac{3}{80}.$$

The angle is increasing at the rate of $\dfrac{3}{80}$ radians/sec.

53. To show $\dfrac{d}{dx}\tan^{-1}x = \dfrac{1}{1+x^2}$: Differentiating $\tan(\tan^{-1}x) = x$,

$$\frac{d}{dx}[\tan(\tan^{-1}x)] = \frac{d}{dx}(x) \;\Rightarrow\; \sec^2(\tan^{-1}x)\cdot\frac{d}{dx}\tan^{-1}x = 1 \;\Rightarrow\; \frac{d}{dx}\tan^{-1}x = \frac{1}{\sec^2(\tan^{-1}x)}.$$

Applying the identity $\sec^2 x = 1 + \tan^2 x$:

$$\frac{d}{dx}\tan^{-1}x = \frac{1}{1+\tan^2(\tan^{-1}x)} = \frac{1}{1+[\tan(\tan^{-1}x)]^2} = \frac{1}{1+x^2}.$$

55. To show $\dfrac{d}{dx}\sec^{-1}x = \dfrac{1}{x\sqrt{x^2-1}}$: Differentiating $\sec(\sec^{-1}x) = x$,

$$\frac{d}{dx}[\sec(\sec^{-1}x)] = \frac{d}{dx}(x) \;\Rightarrow\; \sec(\sec^{-1}x)\cdot\tan(\sec^{-1}x)\cdot\frac{d}{dx}\sec^{-1}x = 1 \;\Rightarrow$$

$$\frac{d}{dx}\sec^{-1}x = \frac{1}{\sec(\sec^{-1}x)\tan(\sec^{-1}x)} = \frac{1}{x\tan(\sec^{-1}x)}$$

Applying the identity $1 + \tan^2 x = \sec^2 x$:

$1+\tan^2(\sec^{-1}x) = \sec^2(\sec^{-1}x) = x^2 \;\Rightarrow\; \tan^2(\sec^{-1}x) = x^2-1 \;\Rightarrow\; \tan(\sec^{-1}x) = \sqrt{x^2-1}$ since the tangent is positive in QI and QIII, the range of the inverse secant. Therefore,

$$\frac{d}{dx}\sec^{-1}x = \frac{1}{x\sqrt{x^2-1}}.$$

57. Let a be the positive square root of a^2.

(a) By substitution: $\displaystyle\int \frac{dx}{\sqrt{a^2-x^2}} = \int \frac{dx}{\sqrt{a^2\left[1-\left(\frac{x}{a}\right)^2\right]}} = \int \frac{dx}{a\sqrt{1-\left(\frac{x}{a}\right)^2}} = \int \frac{du}{\sqrt{1-u^2}}$

$$\boxed{\begin{array}{l} u = \dfrac{x}{a} \\[4pt] du = \dfrac{1}{a}\,dx \end{array}}$$

$$= \sin^{-1}u + C = \sin^{-1}\frac{x}{a} + C.$$

(b) By differentiating:

$$\left[\sin^{-1}\frac{x}{a}\right]' = \frac{1}{\sqrt{1-\left(\frac{x}{a}\right)^2}} \cdot \left(\frac{x}{a}\right)' = \frac{1}{\sqrt{1-\frac{x^2}{a^2}}} \cdot \frac{1}{a} = \frac{1}{\sqrt{1-\frac{x^2}{a^2}}} \cdot \frac{1}{\sqrt{a^2}} = \frac{1}{\sqrt{a^2-x^2}}.$$

59. Let a be the positive square root of a^2.

(a) By substitution: $\displaystyle \int \frac{dx}{a^2 + (x+b)^2} = \int \frac{dx}{a^2\left[1 + \left(\frac{x+b}{a}\right)^2\right]} = \frac{1}{a}\int \frac{du}{1+u^2} = \frac{1}{a}\tan^{-1}u + C$

$$\boxed{\begin{array}{c} u = \dfrac{x+b}{a} \\ du = \dfrac{1}{a}\,dx \end{array}}$$

$$= \frac{1}{a}\tan^{-1}\left(\frac{x+b}{a}\right) + C.$$

(b) By differentiating:

$$\left[\frac{1}{a}\tan^{-1}\left(\frac{x+b}{a}\right)\right]' = \frac{1}{a}\cdot\frac{1}{1+\left(\frac{x+b}{a}\right)^2}\cdot\left(\frac{x+b}{a}\right)' = \frac{1}{a^2}\cdot\frac{1}{1 + \frac{(x+b)^2}{a^2}} = \frac{1}{a^2 + (x+b)^2}.$$

61. $f'(x) = (\sin^{-1}x)' = \frac{1}{\sqrt{1-x^2}} \Rightarrow f'(c) = \frac{1}{\sqrt{1-c^2}}$, while $\frac{f(1)-f(0)}{1-0} = \sin^{-1}1 - \sin^{-1}0 = \frac{\pi}{2} - 0 = \frac{\pi}{2}$. We want c such that $\frac{1}{\sqrt{1-c^2}} = \frac{\pi}{2} \Rightarrow \sqrt{1-c^2} = \frac{2}{\pi} \Rightarrow 1 - c^2 = \frac{4}{\pi^2} \Rightarrow c^2 = 1 - \frac{4}{\pi^2} \Rightarrow$ as $c > 0$, $c = \sqrt{1 - \frac{4}{\pi^2}} = \sqrt{\frac{\pi^2-4}{\pi^2}} = \frac{\sqrt{\pi^2-4}}{\pi} \approx 0.77$, which lies in $(0,1)$.

63. $f'(x) = (\tan^{-1}x)' = \frac{1}{1+x^2} \Rightarrow f'(c) = \frac{1}{1+c^2}$, while $\frac{f(1)-f(0)}{1-0} = \tan^{-1}1 - \tan^{-1}0 = \frac{\pi}{4}$. We want c such that $\frac{1}{1+c^2} = \frac{\pi}{4} \Rightarrow 1 + c^2 = \frac{4}{\pi} \Rightarrow c^2 = \frac{4}{\pi} - 1 \Rightarrow c = \sqrt{\frac{4}{\pi} - 1} = \sqrt{\frac{4-\pi}{\pi}} \approx 0.52$, which lies in $(0,1)$.

CHAPTER 7
Techniques of Integration

§7.1 Integration by Parts

1. $\int u\,dv = uv - \int v\,du$
$\int xe^{-x}\,dx = -xe^{-x} + \int e^{-x}\,dx = -xe^{-x} - e^{-x} + C = -e^{-x}(x+1) + C.$

$$\boxed{\begin{array}{ll} u = x & dv = e^{-x}\,dx \\ du = dx & v = -e^{-x} \end{array}}$$

3. $\int u\,dv = uv - \int v\,du$
$\int x\sin 3x\,dx = -\frac{1}{3}x\cos 3x + \frac{1}{3}\int \cos 3x\,dx = -\frac{1}{3}x\cos 3x + \frac{1}{9}\sin 3x + C.$

$$\boxed{\begin{array}{ll} u = x & dv = \sin 3x\,dx \\ du = dx & v = -\frac{1}{3}\cos 3x \end{array}}$$

5. $\int u\,dv = uv - \int v\,du$
$\int x\sin ax\,dx = -\frac{1}{a}x\cos ax + \frac{1}{a}\int \cos ax\,dx = -\frac{1}{a}x\cos ax + \frac{1}{a^2}\sin ax + C.$

$$\boxed{\begin{array}{ll} u = x & dv = \sin ax\,dx \\ du = dx & v = -\frac{1}{a}\cos ax \end{array}}$$

7. $\int u\,dv = uv - \int v\,du$
$\int x^2\ln x\,dx = \frac{1}{3}x^3\ln x - \frac{1}{3}\int x^2\,dx = \frac{1}{3}x^3\ln x - \frac{1}{9}x^3 + C = \frac{1}{9}x^3(3\ln x - 1) + C.$

$$\boxed{\begin{array}{ll} u = \ln x & dv = x^2\,dx \\ du = \frac{1}{x}\,dx & v = \frac{1}{3}x^3 \end{array}}$$

9. $\int u\,dv = uv - \int v\,du$
$\int x^2 e^{3x}\,dx = \frac{1}{3}x^2 e^{3x} - \frac{2}{3}\int xe^{3x}\,dx = \frac{1}{3}x^2 e^{3x} - \frac{2}{3}\left[\frac{1}{3}xe^{3x} - \frac{1}{3}\int e^{3x}\,dx\right]$

$$\boxed{\begin{array}{ll} u = x^2 & dv = e^{3x}\,dx \\ du = 2x\,dx & v = \frac{1}{3}e^{3x} \end{array}} \qquad \boxed{\begin{array}{ll} u = x & dv = e^{3x}\,dx \\ du = dx & v = \frac{1}{3}e^{3x} \end{array}}$$

$$= \frac{1}{3}x^2 e^{3x} - \frac{2}{9}xe^{3x} + \frac{2}{9}\cdot\frac{e^{3x}}{3} + C = \frac{1}{27}e^{3x}(9x^2 - 6x + 2) + C.$$

11.   We could use integration by parts here, with $u = x^2$ and $dv = \dfrac{x}{\sqrt{1+x^2}}$ leading to the answer: $\frac{1}{3}\sqrt{1+x^2}(x^2 - 2) + C$. Instead we apply the simpler substitution method, and arrive at an equivalent answer.

$$\int \frac{x^3}{\sqrt{1+x^2}}\,dx = \tfrac{1}{2}\int \frac{u-1}{\sqrt{u}}\,du = \tfrac{1}{2}\int(u^{\frac{1}{2}} - u^{-\frac{1}{2}})\,du = \tfrac{1}{2}\left(\tfrac{2}{3}u^{\frac{3}{2}} - 2u^{\frac{1}{2}}\right) + C$$

$$
\boxed{\begin{array}{l}
u = 1 + x^2 \;\Rightarrow\; x^2 = u - 1 \\[4pt]
du = 2x\,dx \;\Rightarrow\; x\,dx = \tfrac{1}{2}\,du \;\Rightarrow\; x^3\,dx = \tfrac{1}{2}(u-1)\,du
\end{array}}
$$

$$= \tfrac{1}{3}(1+x^2)^{\frac{3}{2}} - \sqrt{1+x^2} + C \quad \left[= \tfrac{1}{3}\sqrt{1+x^2}(1+x^2-3) + C\right].$$

13. $\int u\,dv = uv - \int v\,du$

$\int x^2 e^{-x}dx = -x^2 e^{-x} + 2\int xe^{-x}dx = -x^2 e^{-x} + 2\left[-xe^{-x} + \int e^{-x}\,dx\right]$

$$
\boxed{\begin{array}{ll}
u = x^2 & dv = e^{-x}\,dx \\[3pt]
du = 2x\,dx & v = -e^{-x}
\end{array}}
\qquad
\boxed{\begin{array}{ll}
u = x & dv = e^{-x}\,dx \\[3pt]
du = dx & v = -e^{-x}
\end{array}}
$$

$$= -x^2 e^{-x} - 2xe^{-x} - 2e^{-x} + C = -e^{-x}(x^2 + 2x + 2) + C.$$

15. $\int u\,dv = uv - \int v\,du$

$\int \sqrt{x}\,\ln x\,dx = \tfrac{2}{3}x^{\frac{3}{2}}\ln x - \tfrac{2}{3}\int x^{\frac{3}{2}}\cdot\tfrac{1}{x}\,dx = \tfrac{2}{3}x^{\frac{3}{2}}\ln x - \tfrac{2}{3}\cdot\tfrac{2}{3}x^{\frac{3}{2}} + C = \tfrac{2}{3}x^{\frac{3}{2}}\ln x - \tfrac{4}{9}x^{\frac{3}{2}} + C$

$$
\boxed{\begin{array}{ll}
u = \ln x & dv = x^{\frac{1}{2}}\,dx \\[3pt]
du = \tfrac{1}{x}\,dx & v = \tfrac{2}{3}x^{\frac{3}{2}}
\end{array}}
$$

$$= \tfrac{2}{9}x^{\frac{3}{2}}(3\ln x - 2) + C.$$

17. $\int u\,dv = uv - \int v\,du$

$\int \ln\tfrac{1}{x}\,dx = x\ln\tfrac{1}{x} + \int x\cdot\tfrac{1}{x}\,dx = x\ln\tfrac{1}{x} + x + C = x\left(1 + \ln\tfrac{1}{x}\right) + C.$

$$
\boxed{\begin{array}{ll}
u = \ln x^{-1} & dv = dx \\[3pt]
du = \dfrac{1}{x^{-1}}\cdot -x^{-2}\,dx = -\tfrac{1}{x}\,dx & v = x
\end{array}}
$$

19. $\int u\,dv = uv - \int v\,du$

$\int x\ln(x+c)\,dx = \tfrac{1}{2}x^2\ln(x+c) - \tfrac{1}{2}\int \frac{x^2}{x+c}\,dx$

$$
\boxed{\begin{array}{ll}
u = \ln(x+c) & dv = x\,dx \\[3pt]
du = \dfrac{1}{x+c}\,dx & v = \tfrac{1}{2}x^2
\end{array}}
\qquad
\boxed{\begin{array}{l}
w = x + c \;\Rightarrow\; x^2 = (w-c)^2 \\[3pt]
dw = dx
\end{array}}
$$

$\int \frac{x^2}{x+c}\,dx = \int \frac{(w-c)^2}{w}\,dw = \int \frac{w^2 - 2cw + c^2}{w}\,dw = \int\left(w - 2c + \frac{c^2}{w}\right)dw$

$$= \tfrac{1}{2}w^2 - 2cw + c^2\ln|w| + C = \tfrac{1}{2}(x+c)^2 - 2c(x+c) + c^2\ln(x+c) + C$$

$$= \tfrac{1}{2}x^2 + cx + \tfrac{1}{2}c^2 - 2cx - 2c^2 + c^2\ln(x+c) + C = \tfrac{1}{2}x^2 - cx + c^2\ln(x+c) + C.$$

(absorbing the constant terms into C)

Thus,

$$\int x \ln(x+c)\, dx = \tfrac{1}{2}x^2 \ln(x+c) - \tfrac{1}{2}\left[\tfrac{1}{2}x^2 - cx + c^2 \ln(x+c)\right] + C$$
$$= \tfrac{1}{2}x^2 \ln(x+c) - \tfrac{1}{4}x^2 + \tfrac{1}{2}cx - \tfrac{1}{2}c^2 \ln(x+c) + C$$
$$= \tfrac{1}{2}(x^2 - c^2)\ln(x+c) - \tfrac{1}{4}x^2 + \tfrac{1}{2}cx + C.$$

21. $\int u\, dv = uv - \int v\, du$

$\int \cos(\ln x)\, dx = x\cos(\ln x) + \int \sin(\ln x)\, dx = x\cos(\ln x) + [x\sin(\ln x) - \int \cos(\ln x)\, dx]$

$u = \cos(\ln x)$	$dv = dx$
$du = -\sin(\ln x)\cdot\dfrac{1}{x}\, dx$	$v = x$

$u = \sin(\ln x)$	$dv = dx$
$du = \cos(\ln x)\cdot\dfrac{1}{x}\, dx$	$v = x$

From $\int \cos(\ln x)\, dx = x\cos(\ln x) + x\sin(\ln x) - \int \cos(\ln x)\, dx$

$\quad 2\int \cos(\ln x)\, dx = x\cos(\ln x) + x\sin(\ln x) + C$

$\quad \int \cos(\ln x)\, dx = \tfrac{1}{2}x[\cos(\ln x) + \sin(\ln x)] + C.$

23. $\int u\, dv = uv - \int v\, du$

$\int \cos x \ln(\sin x)\, dx = \sin x \ln(\sin x) - \int \cos x\, dx = \sin x \ln(\sin x) - \sin x + C$

$u = \ln(\sin x)$	$dv = \cos x\, dx$
$du = \dfrac{1}{\sin x}\cdot \cos x\, dx$	$v = \sin x$

$$= (\sin x)[\ln(\sin x) - 1] + C.$$

25. $\int u\, dv = uv - \int v\, du$

$\int (x^2 - 5x)e^x\, dx = (x^2 - 5x)e^x - \int (2x - 5)e^x\, dx = (x^2 - 5x)e^x - [(2x - 5)e^x - 2\int e^x\, dx]$

$u = x^2 - 5x$	$dv = e^x\, dx$
$du = (2x - 5)\, dx$	$v = e^x$

$u = 2x - 5$	$dv = e^x\, dx$
$du = 2\, dx$	$v = e^x$

$$= (x^2 - 5x)e^x - (2x - 5)e^x + 2e^x + C = e^x(x^2 - 5x - 2x + 5 + 2) + C$$
$$= e^x(x^2 - 7x + 7) + C.$$

27. $\int x\tan^2 x\, dx = \int x(\sec^2 x - 1)\, dx = \int x\sec^2 x\, dx - \int x\, dx = \int x\sec^2 x\, dx - \tfrac{1}{2}x^2$

$\qquad\qquad \int u\, dv = uv - \int v\, du$

$\quad \int x\sec^2 x\, dx = x\tan x - \int \tan x\, dx = x\tan x - \ln|\sec x| + C \;\Rightarrow$

$u = x$	$dv = \sec^2 x\, dx$
$du = dx$	$v = \tan x$

$\int x\tan^2 x\, dx = x\tan x - \ln|\sec x| - \tfrac{1}{2}x^2 + C.$

29. $\int \cos^3 x \, dx = \int \cos^2 x \cos x \, dx = \int (1 - \sin^2 x) \cos x \, dx = \int \cos x \, dx - \int \sin^2 x \cos x \, dx$

$= \sin x - \int \sin^2 x \cos x \, dx = \sin x - \int u^2 \, du = \sin x - \frac{1}{3} u^3 + C$

$$\boxed{\begin{array}{l} u = \sin x \\ du = \cos x \, dx \end{array}}$$

$= \sin x - \frac{1}{3} \sin^3 x + C.$

[OR - by the Reduction formula with $n = 3$:

$\int \cos^3 x \, dx = \frac{\sin x}{3} \cos^2 x + \frac{2}{3} \int \cos x \, dx = \frac{\sin x}{3} \cos^2 x + \frac{2}{3} \sin x + C = \frac{\sin x}{3} (\cos^2 x + 2) + C.$]

31. $\displaystyle \int \frac{x^2}{(1 + x^2)^2} \, dx = \int \left[x \cdot \frac{x}{(1 + x^2)^2} \right] \, dx$

$$\boxed{\begin{array}{ll} u = x & dv = \dfrac{x}{(1 + x^2)^2} \, dx \\[2mm] du = dx & v = -\dfrac{1}{2(1 + x^2)} \quad \text{(see below)} \end{array}}$$

$v : \displaystyle \int \frac{x}{(1 + x^2)^2} \, dx = \frac{1}{2} \int \frac{dt}{t^2} = \frac{1}{2} \int t^{-2} \, dt = -\frac{1}{2} t^{-1} + C = -\frac{1}{2(1 + x^2)} + C$

$$\boxed{\begin{array}{l} t = 1 + x^2 \\ dt = 2x \, dx \implies x \, dx = \frac{1}{2} dt \end{array}}$$

Then, $\int u \, dv = uv - \int v \, du$

$\displaystyle \int \frac{x^2}{(1 + x^2)^2} \, dx = x \cdot -\frac{1}{2(1 + x^2)} + \frac{1}{2} \int \frac{1}{1 + x^2} \, dx = -\frac{x}{2(1 + x^2)} + \frac{1}{2} \tan^{-1} x + C.$

33. $\int u \, dv = uv - \int v \, du$

$\displaystyle \int x^2 \tan^{-1} x \, dx = \frac{1}{3} x^3 \tan^{-1} x - \frac{1}{3} \int \frac{x^3}{1 + x^2} \, dx = \frac{1}{3} x^3 \tan^{-1} x - \frac{1}{3} \int x^2 \cdot \frac{x}{1 + x^2} \, dx$

$$\boxed{\begin{array}{ll} u = \tan^{-1} x & dv = x^2 \, dx \\[2mm] du = \dfrac{1}{1 + x^2} \, dx & v = \dfrac{1}{3} x^3 \end{array}} \qquad \boxed{\begin{array}{l} w = 1 + x^2 \implies x^2 = w - 1 \\ dw = 2x \, dx \implies x \, dx = \frac{1}{2} dw \end{array}}$$

$\displaystyle \int x^2 \cdot \frac{x}{1 + x^2} \, dx = \frac{1}{2} \int \frac{w - 1}{w} \, dw = \frac{1}{2} \int \left(1 - \frac{1}{w} \right) \, dw = \frac{1}{2} (w - \ln |w|) + C$

$= \frac{1}{2} [\cancel{1} + x^2 - \ln(1 + x^2)] + C \qquad \text{(after absorbing the ``}\frac{1}{2} \cdot 1\text{'' into ``}C\text{'')}$

$= \frac{1}{2} [x^2 - \ln(1 + x^2)] + C$

Therefore, $\displaystyle \int x^2 \tan^{-1} x \, dx = \frac{1}{3} x^3 \tan^{-1} x - \frac{1}{6} [x^2 - \ln(1 + x^2)] + C$

$= \frac{1}{6} [2x^3 \tan^{-1} x - x^2 + \ln(1 + x^2)] + C.$

35. $\int u\,dv = uv - \int v\,du$

$\int \sin 3x \cos 2x\,dx = \frac{1}{2}\sin 3x \sin 2x - \frac{3}{2}\int \cos 3x \sin 2x\,dx$

$u = \sin 3x$	$dv = \cos 2x\,dx$	$u = \cos 3x$	$dv = \sin 2x\,dx$
$du = 3\cos 3x\,dx$	$v = \frac{1}{2}\sin 2x$	$du = -3\sin 3x\,dx$	$v = -\frac{1}{2}\cos 2x$

$\int \cos 3x \sin 2x\,dx = -\frac{1}{2}\cos 3x \cos 2x - \frac{3}{2}\int \sin 3x \cos 2x\,dx$

Thus, $\int \sin 3x \cos 2x\,dx = \frac{1}{2}\sin 3x \sin 2x - \frac{3}{2}\left[-\frac{1}{2}\cos 3x \cos 2x - \frac{3}{2}\int \sin 3x \cos 2x\,dx\right]$

$$= \frac{1}{2}\sin 3x \sin 2x + \frac{3}{4}\cos 3x \cos 2x + \frac{9}{4}\int \sin 3x \cos 2x\,dx$$

$\left(1 - \frac{9}{4}\right)\int \sin 3x \cos 2x\,dx = \frac{1}{2}\sin 3x \sin 2x + \frac{3}{4}\cos 3x \cos 2x + C$

$$\int \sin 3x \cos 2x\,dx = -\frac{4}{5}\left[\frac{1}{2}\sin 3x \sin 2x + \frac{3}{4}\cos 3x \cos 2x\right] + C$$

$$= -\frac{2}{5}\sin 3x \sin 2x - \frac{3}{5}\cos 3x \cos 2x + C$$

37. $\int u\,dv = uv - \int v\,du$

$\int_0^1 x^2 e^x\,dx = x^2 e^x\big|_0^1 - 2\int_0^1 xe^x\,dx = e - 2\left[xe^x\big|_0^1 - \int_0^1 e^x\,dx\right] = e - 2e + 2e^x\big|_0^1$

$u = x^2$	$dv = e^x\,dx$	$u = x$	$dv = e^x\,dx$
$du = 2x\,dx$	$v = e^x$	$du = dx$	$v = e^x$

$$= -e + 2(e - 1) = e - 2.$$

39. $\int u\,dv = uv - \int v\,du$

$\int_0^1 x^2 e^{-3x}\,dx = -\frac{1}{3}x^2 e^{-3x}\Big|_0^1 + \frac{2}{3}\int_0^1 xe^{-3x}\,dx = -\frac{1}{3}e^{-3} + \frac{2}{3}\left[-\frac{x}{3}e^{-3x}\Big|_0^1 + \frac{1}{3}\int_0^1 e^{-3x}\,dx\right]$

$u = x^2$	$dv = e^{-3x}\,dx$	$u = x$	$dv = e^{-3x}\,dx$
$du = 2x\,dx$	$v = -\frac{1}{3}e^{-3x}$	$du = dx$	$v = -\frac{1}{3}e^{-3x}$

$$= -\frac{1}{3}e^{-3} + \frac{2}{3}\left(-\frac{1}{3}e^{-3} - \frac{1}{9}e^{-3x}\Big|_0^1\right) = -\frac{1}{3}e^{-3} - \frac{2}{9}e^{-3} - \frac{2}{27}(e^{-3} - 1)$$

$$= \frac{1}{27}(-9e^{-3} - 6e^{-3} - 2e^{-3} + 2) = \frac{1}{27}\left(2 - \frac{17}{e^3}\right).$$

41. $\int u\,dv = uv - \int v\,du$

$\int_0^1 \tan^{-1} x\,dx = x\tan^{-1} x\big|_0^1 - \int_0^1 \frac{x}{1 + x^2}\,dx = \tan^{-1} 1 - \frac{1}{2}\int_1^2 \frac{dw}{w} = \frac{\pi}{4} - \frac{1}{2}\ln|w|\Big|_1^2$

$u = \tan^{-1} x$	$dv = dx$	$w = 1 + x^2, \ x = 0 \Rightarrow w = 1$
$du = \dfrac{1}{1 + x^2}\,dx$	$v = x$	$dw = 2x\,dx, \quad x = 1 \Rightarrow w = 2$
		$\frac{1}{2}dw = x\,dx$

$$= \frac{\pi}{4} - \frac{1}{2}(\ln 2 - \ln 1) = \frac{\pi}{4} - \frac{1}{2}\ln 2 = \frac{1}{4}(\pi - 2\ln 2).$$

43. $\int u \, dv = uv - \int v \, du$

$$\int_{-2}^{2} \ln(x+3) \, dx = x \ln(x+3) \Big|_{-2}^{2} - \int_{-2}^{2} \frac{x}{x+3} \, dx = 2 \ln 5 - (-2 \ln 1) - \int_{1}^{5} \frac{w-3}{w} \, dw$$

| $u = \ln(x+3) \quad dv = dx$ | $w = x+3, \quad x = -2 \Rightarrow w = 1$ |
| $du = \frac{1}{x+3} \, dx \quad v = x$ | $dw = dx, \quad x = 2 \Rightarrow w = 5$ |

$$\int_{1}^{5} \frac{w-3}{w} \, dw = \int_{1}^{5} \left(1 - \frac{3}{w}\right) dw = (w - 3 \ln|w|) \Big|_{1}^{5} = 5 - 3 \ln 5 - (1 - 3 \ln 1) = 4 - 3 \ln 5$$

Therefore, $\int_{-2}^{2} \ln(x+3) \, dx = 2 \ln 5 - (4 - 3 \ln 5) = 5 \ln 5 - 4$.

45. $\int u \, dv = uv - \int v \, du$

$$\int_{0}^{\frac{\pi}{2}} x \sin 4x \, dx = -\frac{1}{4} x \cos 4x \Big|_{0}^{\frac{\pi}{2}} + \frac{1}{4} \int_{0}^{\frac{\pi}{2}} \cos 4x \, dx = -\frac{1}{4} \cdot \frac{\pi}{2} \cos 2\pi + \frac{1}{4} \cdot \frac{\sin 4x}{4} \Big|_{0}^{\frac{\pi}{2}}$$

| $u = x \quad dv = \sin 4x \, dx$ |
| $du = dx \quad v = -\frac{1}{4} \cos 4x$ |

$$= -\frac{\pi}{8} + \frac{1}{16}(\sin 2\pi - \sin 0) = -\frac{\pi}{8}.$$

47. $\int u \, dv = uv - \int v \, du$

$$\int_{0}^{1} \frac{x^3}{\sqrt{x^2+1}} \, dx = x^2 \sqrt{x^2+1} \Big|_{0}^{1} - \int_{0}^{1} 2x \sqrt{x^2+1} \, dx = \sqrt{2} - \int_{0}^{1} 2x \sqrt{x^2+1} \, dx$$

| $u = x^2 \quad dv = \frac{x}{\sqrt{x^2+1}} \, dx$ | $s = x^2 + 1, \quad x = 0 \Rightarrow s = 1$ |
| $du = 2x \, dx \quad v = \sqrt{x^2+1}$ (see below) | $ds = 2x \, dx, \quad x = 1 \Rightarrow s = 2$ |

$$v : \int \frac{x}{\sqrt{x^2+1}} \, dx = \frac{1}{2} \int \frac{dt}{\sqrt{t}} = \frac{1}{2} \int t^{-\frac{1}{2}} \, dt = t^{\frac{1}{2}} + C = \sqrt{x^2+1} + C$$

| $t = x^2 + 1$ |
| $dt = 2x \, dx \Rightarrow x \, dx = \frac{1}{2} dt$ |

$$\int_{0}^{1} 2x \sqrt{x^2+1} \, dx = \int_{1}^{2} \sqrt{s} \, ds = \int_{1}^{2} s^{\frac{1}{2}} \, ds = \frac{2}{3} s^{\frac{3}{2}} \Big|_{1}^{2} = \frac{2}{3}(2^{\frac{3}{2}} - 1) = \frac{2}{3}(2\sqrt{2} - 1)$$

Therefore, $\int_{0}^{1} \frac{x^3}{\sqrt{x^2+1}} \, dx = \sqrt{2} - \frac{2}{3}(2\sqrt{2} - 1) = \frac{1}{3}(3\sqrt{2} - 4\sqrt{2} + 2) = \frac{1}{3}(2 - \sqrt{2})$.

49.   In $[0, 3\pi]$, $f(x) = x \sin x = 0 \Rightarrow x = 0, \pi, 2\pi, 3\pi$. As $x \geq 0$ in $[0, 3\pi]$, the sign of f is the same as the sign of the sine function which is positive in $(0, \pi)$ and $(2\pi, 3\pi)$ and negative in $(\pi, 2\pi)$. Consequently,
$A = \int_{0}^{3\pi} |f(x)| dx = \int_{0}^{\pi} f(x) dx - \int_{\pi}^{2\pi} f(x) dx + \int_{2\pi}^{3\pi} f(x) dx$
$\qquad = \int_{0}^{\pi} x \sin x \, dx - \int_{\pi}^{2\pi} x \sin x \, dx + \int_{2\pi}^{3\pi} x \sin x \, dx$

Now, $\int u\,dv = uv - \int v\,du$

$\int x \sin x\,dx = -x\cos x + \int \cos x\,dx = -x\cos x + \sin x + C$

$u = x$	$dv = \sin x$
$du = dx$	$v = -\cos x\,dx$

So, $A = (-x\cos x + \sin x)\Big|_0^\pi - (-x\cos x + \sin x)\Big|_\pi^{2\pi} + (-x\cos x + \sin x)\Big|_{2\pi}^{3\pi}$ (as $\sin n\pi = 0$)

$= -\pi\cos\pi + (2\pi\cos 2\pi - \pi\cos\pi) + (-3\pi\cos 3\pi + 2\pi\cos 2\pi)$

$= \pi + 2\pi + \pi + 3\pi + 2\pi = 9\pi.$

51.

$y = \ln x$

$\int u\,dv = uv - \int v\,du$

$V = \pi\int_1^e (\ln x)^2\,dx = \pi\left[x(\ln x)^2\Big|_1^e - 2\int_1^e (\ln x)\cdot\frac{1}{x}\cdot x\,dx\right] \Rightarrow$

$u = (\ln x)^2$	$dv = dx$
$du = 2(\ln x)\cdot\frac{1}{x}\,dx$	$v = x$

$u = \ln x$	$dv = dx$
$du = \frac{1}{x}\,dx$	$v = x$

$V = \pi\left[e(\ln e)^2 - (\ln 1)^2 - 2\int_1^e \ln x\,dx\right] = \pi e - 2\pi\left[x\ln x\Big|_1^e - \int_1^e dx\right]$

$= \pi e - 2\pi\left(e\ln e - \ln 1 - x\Big|_1^e\right) = \pi e - 2\pi[e - (e-1)] = \pi e - 2\pi = \pi(e-2).$

53. Distance traveled, $D = \int_a^b |v(t)|\,dt$. Since $v(t) = te^t \ge 0$ on $[0,3]$,

$\int u\,dv = uv - \int v\,du$

$D = \int_0^3 te^t\,dt = te^t\Big|_0^3 - \int_0^3 e^t\,dt = 3e^3 - e^t\Big|_0^3 = 3e^3 - (e^3 - 1) = 2e^3 + 1$ feet.

$u = t$	$dv = e^t\,dt$
$du = dt$	$v = e^t$

55. $\int u\,dv = uv - \int v\,du$

$\int x^n e^x\,dx = x^n e^x - n\int x^{n-1}e^x\,dx.$

$u = x^n$	$dv = e^x\,dx$
$du = nx^{n-1}\,dx$	$v = e^x$

57. $\int u\,dv = uv - \int v\,du$

$\int \sec^n x\,dx = \int(\sec^{n-2}x)\sec^2 x\,dx = (\sec^{n-2}x)\tan x - (n-2)\int(\sec^{n-2}x)\tan^2 x\,dx$

$\sec^2 x - 1$

$u = \sec^{n-2}x$	$dv = \sec^2 x\,dx$
$du = (n-2)(\sec^{n-3}x)\sec x\tan x\,dx$ $= (n-2)(\sec^{n-2}x)\tan x\,dx$	$v = \tan x$

$= (\sec^{n-2}x)\tan x - (n-2)\int(\sec^n x - \sec^{n-2}x)\,dx$

$= (\sec^{n-2}x)\tan x - (n-2)\int\sec^n x\,dx + (n-2)\int(\sec^{n-2}x)\,dx$, next, combine integrals of $\sec^n x$: $\underbrace{[1 + (n-2)]}_{n-1}\int\sec^n x\,dx = (\sec^{n-2}x)\tan x + (n-2)\int(\sec^{n-2}x)\,dx$

$\int\sec^n x\,dx = \frac{(\sec^{n-2}x)\tan x}{n-1} + \frac{n-2}{n-1}\int(\sec^{n-2}x)\,dx.$

59. $\qquad \int u\,dv = uv - \int v\,du$

$$\int x^m (\ln x)^n dx = \frac{x^{m+1}(\ln x)^n}{m+1} - \int \frac{x^{m\not{+}1}}{m+1} \cdot \frac{n(\ln x)^{n-1}}{\not{x}}\,dx$$

$u = (\ln x)^n$	$dv = x^m\,dx$
$du = n(\ln x)^{n-1} \cdot \frac{1}{x}\,dx$	$v = \dfrac{x^{m+1}}{m+1}$

$$= \frac{x^{m+1}(\ln x)^n}{m+1} - \frac{n}{m+1}\int x^m (\ln x)^{n-1}\,dx.$$

61. For the integration by parts with $u = \ln(x + \sqrt{x^2+a^2}) = \ln[x + (x^2+a^2)^{\frac{1}{2}}]$, we will need the derivative of u:

$$u' = \frac{1}{x + (x^2+a^2)^{\frac{1}{2}}}\left[1 + \frac{1}{\not2}(x^2+a^2)^{-\frac{1}{2}} \cdot \not{2}x\right] = \frac{1 + \dfrac{x}{\sqrt{x^2+a^2}}}{x + \sqrt{x^2+a^2}} \cdot \frac{\sqrt{x^2+a^2}}{\sqrt{x^2+a^2}}$$

$$= \frac{\sqrt{x^2+a^2} \not{+} x}{(x \not{+} \sqrt{x^2+a^2})\sqrt{x^2+a^2}} = \frac{1}{\sqrt{x^2+a^2}}$$

Then, $\qquad \int u\,dv = uv - \int v\,du$

$$\int \ln(x + \sqrt{x^2+a^2})\,dx = x\ln(x + \sqrt{x^2+a^2}) - \int \frac{x}{\sqrt{x^2+a^2}}\,dx$$

$u = \ln(x + \sqrt{x^2+a^2})$	$dv = dx$
$du = \dfrac{1}{\sqrt{x^2+a^2}}\,dx$	$v = x$

$w = x^2 + a^2$
$dw = 2x\,dx \;\Rightarrow\; x\,dx = \frac{1}{2}\,dw$

$$= x\ln(x + \sqrt{x^2+a^2}) - \frac{1}{2}\int w^{-\frac{1}{2}}\,dw = x\ln(x + \sqrt{x^2+a^2}) - w^{\frac{1}{2}} + C$$

$$= x\ln(x + \sqrt{x^2+a^2}) - \sqrt{x^2+a^2} + C.$$

63. $\qquad \int u\,dv = uv - \int v\,du$

$$\int e^{ax}\sin bx\,dx = \frac{1}{a}e^{ax}\sin bx - \frac{b}{a}\int e^{ax}\cos bx\,dx$$

$u = \sin bx$	$dv = e^{ax}\,dx$
$du = b\cos bx\,dx$	$v = \frac{1}{a}e^{ax}$

$u = \cos bx$	$dv = e^{ax}\,dx$
$du = -b\sin bx\,dx$	$v = \frac{1}{a}e^{ax}$

Since $\quad \int e^{ax}\cos bx\,dx = \frac{1}{a}e^{ax}\cos bx + \frac{b}{a}\int e^{ax}\sin bx\,dx$,

$$\int e^{ax}\sin bx\,dx = \frac{1}{a}e^{ax}\sin bx - \frac{b}{a}\left[\frac{1}{a}e^{ax}\cos bx + \frac{b}{a}\int e^{ax}\sin bx\,dx\right]$$

$$= \frac{1}{a}e^{ax}\sin bx - \frac{b}{a^2}e^{ax}\cos bx - \frac{b^2}{a^2}\int e^{ax}\sin bx\,dx$$

$$\left(1 + \frac{b^2}{a^2}\right)\int e^{ax}\sin bx\,dx = \frac{e^{ax}}{a}\left(\sin bx - \frac{b}{a}\cos bx\right) + C \;\Rightarrow$$

$$(a^2 + b^2)\int e^{ax}\sin bx\,dx = e^{ax}(a\sin bx - b\cos bx) + C \;\Rightarrow$$

$$\int e^{ax}\sin bx\,dx = \frac{e^{ax}}{a^2+b^2}(a\sin bx - b\cos bx) + C.$$

65. $\;u(v+C) - \int(v+C)\,du = uv + Cu - \int v\,du - \int C\,du = uv + Cu - \int v\,du - Cu = uv - \int v\,du.$

§7.2 Completing the Square and Partial Fractions

1. $\displaystyle\int \frac{dx}{x^2 + 2x + 5} = \int \frac{dx}{\left[x^2 + 2x + \left(\frac{2}{2}\right)^2\right] + 4} = \int \frac{dx}{(x+1)^2 + 4} = \int \frac{dx}{4\left[1 + \left(\frac{x+1}{2}\right)^2\right]} = \frac{1}{2}\int \frac{du}{1 + u^2}$

$$= \frac{1}{2}\tan^{-1} u + C = \frac{1}{2}\tan^{-1}\left(\frac{x+1}{2}\right) + C.$$

$$\boxed{\begin{array}{l} u = \dfrac{x+1}{2} \\ du = \dfrac{1}{2}dx \Rightarrow \dfrac{1}{4}dx = \dfrac{1}{2}du \end{array}}$$

3. $\displaystyle\int \frac{dx}{x^2 + 6x + 25} = \int \frac{dx}{\left[x^2 + 6x + \left(\frac{6}{2}\right)^2\right] + 16} = \int \frac{dx}{(x+3)^2 + 16} = \int \frac{dx}{16\left[1 + \left(\frac{x+3}{4}\right)^2\right]}$

$$\boxed{\begin{array}{l} u = \dfrac{x+3}{4} \\ du = \dfrac{1}{4}dx \Rightarrow \dfrac{1}{16}dx = \dfrac{1}{4}du \end{array}}$$

$$= \frac{1}{4}\int \frac{du}{1 + u^2} = \frac{1}{4}\tan^{-1} u + C = \frac{1}{4}\tan^{-1}\left(\frac{x+3}{4}\right) + C.$$

5. As $x^2 - 6x = x^2 - 6x + \left(\frac{6}{2}\right)^2 - 9 = (x-3)^2 - 9$,

$$\int \frac{dx}{\sqrt{6x - x^2}} = \int \frac{dx}{\sqrt{9 - (x-3)^2}} = \int \frac{dx}{\sqrt{9\left[1 - \left(\frac{x-3}{3}\right)^2\right]}} = \frac{1}{3}\int \frac{dx}{\sqrt{1 - \left(\frac{x-3}{3}\right)^2}} = \int \frac{du}{\sqrt{1 - u^2}}$$

$$= \sin^{-1} u + C = \sin^{-1}\left(\frac{x-3}{3}\right) + C.$$

$$\boxed{\begin{array}{l} u = \dfrac{x-3}{3} \\ du = \dfrac{1}{3}dx \end{array}}$$

7. As $4x^2 + 8x + 29 = 4\left[x^2 + 2x + \left(\frac{2}{2}\right)^2\right] + 25 = 4(x+1)^2 + 25$,

$$\int \frac{dx}{4x^2 + 8x + 29} = \int \frac{dx}{4(x+1)^2 + 25} = \frac{1}{25}\int \frac{dx}{1 + \left[\frac{2(x+1)}{5}\right]^2} = \frac{1}{25}\cdot\frac{5}{2}\int \frac{du}{1 + u^2} = \frac{1}{10}\tan^{-1} u + C$$

$$= \frac{1}{10}\tan^{-1}\left(\frac{2x+2}{5}\right) + C.$$

$$\boxed{\begin{array}{l} u = \dfrac{2(x+1)}{5} \\ du = \dfrac{2}{5}dx \Rightarrow dx = \dfrac{5}{2}du \end{array}}$$

9. As $x^2 + 3x + 5 = x^2 + 3x + \left(\frac{3}{2}\right)^2 - \left(\frac{3}{2}\right)^2 + 5 = \left(x + \frac{3}{2}\right)^2 + \frac{11}{4}$,

$$\int \frac{dx}{x^2 + 3x + 5} = \frac{dx}{(x + \frac{3}{2})^2 + \frac{11}{4}} = \int \frac{1}{\frac{11}{4}\left[1 + \left(\frac{x + \frac{3}{2}}{\frac{\sqrt{11}}{2}}\right)^2\right]}dx = \frac{4}{11}\int \frac{1}{1 + \left(\frac{2x+3}{\sqrt{11}}\right)^2}dx$$

$$\boxed{\begin{array}{l} u = \dfrac{2x+3}{\sqrt{11}} \\ du = \dfrac{2}{\sqrt{11}}dx \Rightarrow dx = \dfrac{\sqrt{11}}{2}du \end{array}}$$

$$= \frac{4}{11}\cdot\frac{\sqrt{11}}{2}\int \frac{1}{1 + u^2}du = \frac{2}{\sqrt{11}}\tan^{-1} u + C = \frac{2}{\sqrt{11}}\tan^{-1}\left(\frac{2x+3}{\sqrt{11}}\right) + C.$$

11. As $\sqrt{3 + 2x - x^2} = \sqrt{-(x^2 - 2x + 1)^2 + 1 + 3} = \sqrt{4 - (x-1)^2}$,

$$\int \frac{x}{\sqrt{3 + 2x - x^2}}\, dx = \int \frac{x}{\sqrt{4 - (x-1)^2}}\, dx = \int \frac{x}{\sqrt{4[1 - (\frac{x-1}{2})^2]}}\, dx = \frac{1}{2} \cdot 2 \int \frac{2u + 1}{\sqrt{1 - u^2}}\, du$$

$$\boxed{\begin{array}{l} u = \frac{x-1}{2} \;\Rightarrow\; x = 2u + 1 \\ du = \frac{1}{2}\, dx \;\Rightarrow\; dx = 2\, du \end{array}}$$

$$= \int \frac{2u}{\sqrt{1 - u^2}}\, du \;\searrow\; + \int \frac{1}{\sqrt{1 - u^2}}\, du = -\;\nearrow\; \int v^{-\frac{1}{2}}\, dv + \sin^{-1} u$$

$$\boxed{\begin{array}{l} v = 1 - u^2 \\ dv = -2u\, du \;\Rightarrow\; 2u\, du = -dv \end{array}}$$

$$\int v^{-\frac{1}{2}}\, dv = 2v^{\frac{1}{2}} + C = 2\sqrt{1 - u^2} + C = \sqrt{4 - (x-1)^2} + C = \sqrt{3 + 2x - x^2} + C$$

Therefore, $\int \dfrac{x}{\sqrt{3 + 2x - x^2}}\, dx = -\sqrt{3 + 2x - x^2} + \sin^{-1}(\frac{x-1}{2}) + C.$

13. As $-5 + 8x - 4x^2 = -4[x^2 - 2x + (\frac{2}{2})^2] + 4 - 5 = -4(x-1)^2 - 1$,

$$\int_{\frac{3}{2}}^{1 + \frac{\sqrt{3}}{2}} \frac{dx}{-5 + 8x - 4x^2} = \int_{\frac{3}{2}}^{1 + \frac{\sqrt{3}}{2}} \frac{dx}{-4(x-1)^2 - 1} = -\int_{\frac{3}{2}}^{1 + \frac{\sqrt{3}}{2}} \frac{dx}{1 + [2(x-1)]^2} = -\frac{1}{2} \int_{1}^{\sqrt{3}} \frac{du}{1 + u^2}$$

$$\boxed{\begin{array}{ll} u = 2(x-1), & x = \frac{3}{2} \;\Rightarrow\; u = 1 \\ du = 2\, dx, & x = 1 + \frac{\sqrt{3}}{2} \;\Rightarrow\; u = \sqrt{3} \\ dx = \frac{1}{2}\, du & \end{array}}$$

$$= -\frac{1}{2} \tan^{-1} u \Big|_{1}^{\sqrt{3}} = -\frac{1}{2}[\tan^{-1}\sqrt{3} - \tan^{-1} 1] = -\frac{1}{2}[\frac{\pi}{3} - \frac{\pi}{4}] = -\frac{\pi}{24}.$$

15. As $4x - x^2 = -[x^2 - 4x + 2^2] + 4 = -(x-2)^2 + 4$,

$$\int_{1}^{2} \frac{dx}{\sqrt{4x - x^2}} = \int_{1}^{2} \frac{dx}{\sqrt{4 - (x-2)^2}} = \int_{1}^{2} \frac{dx}{\sqrt{4[1 - (\frac{x-2}{2})^2]}} = \frac{1}{2} \int_{1}^{2} \frac{dx}{\sqrt{1 - (\frac{x-2}{2})^2}} = \int_{-\frac{1}{2}}^{0} \frac{du}{\sqrt{1 - u^2}}$$

$$\boxed{\begin{array}{ll} u = \frac{x-2}{2}, & x = 1 \;\Rightarrow\; u = -\frac{1}{2} \\ du = \frac{1}{2}\, dx, & x = 2 \;\Rightarrow\; u = 0 \end{array}}$$

$$= \sin^{-1} u \Big|_{-\frac{1}{2}}^{0} = \sin^{-1} 0 - \sin^{-1}(-\frac{1}{2}) = 0 - (-\frac{\pi}{6}) = \frac{\pi}{6}.$$

****

NOTE: $\int \frac{dx}{x + a}$, with a a constant, arises frequently in the following exercises:

Let $u = x + a$, then $du = dx$ and $\int \frac{dx}{x + a} = \int \frac{du}{u} = \ln|u| + C = \ln|x + a| + C.$

17. $\displaystyle\int\frac{dx}{x^2-x-2}=\int\frac{dx}{(x-2)(x+1)}$:

$$\frac{1}{(x-2)(x+1)}=\frac{A}{x-2}+\frac{B}{x+1}$$

$$1=A(x+1)+B(x-2)$$

$$x=-1:1=B(-1-2)\ \Rightarrow\ B=-\tfrac{1}{3}$$

$$x=2:1=A(2+1)\ \Rightarrow\ A=\tfrac{1}{3}$$

$$\int\frac{dx}{x^2-x-2}=\tfrac{1}{3}\int\frac{dx}{x-2}-\tfrac{1}{3}\int\frac{dx}{x+1}$$

$$=\tfrac{1}{3}\ln|x-2|-\tfrac{1}{3}\ln|x+1|+C$$

$$=\tfrac{1}{3}\ln\left|\frac{x-2}{x+1}\right|+C.$$

19. $\displaystyle\int\frac{x}{x^2-3x+2}\,dx=\int\frac{x}{(x-2)(x-1)}\,dx$:

$$\frac{x}{(x-2)(x-1)}=\frac{A}{x-2}+\frac{B}{x-1}$$

$$x=A(x-1)+B(x-2)$$

$$x=1:1=B(1-2)\ \Rightarrow\ B=-1$$

$$x=2:2=A(2-1)\ \Rightarrow\ A=2$$

$$\int\frac{x}{x^2-3x+2}\,dx=2\int\frac{dx}{x-2}-\int\frac{dx}{x-1}$$

$$=2\ln|x-2|-\ln|x-1|+C$$

$$=\ln\frac{(x-2)^2}{|x-1|}+C.$$

21.

$$\frac{x^2}{(x-1)^2(x+1)}=\frac{A}{x-1}+\frac{B}{(x-1)^2}+\frac{C}{x+1}$$

$$x^2=A(x-1)(x+1)+B(x+1)+C(x-1)^2$$

$$x=1:1=B(1+1)\ \Rightarrow\ B=\tfrac{1}{2}$$

$$x=-1:1=C(-1-1)^2\ \Rightarrow\ C=\tfrac{1}{4}$$

$$x=0:0=-A+\tfrac{1}{2}+\tfrac{1}{4}\ \Rightarrow\ A=\tfrac{3}{4}$$

$$\int\frac{x^2}{(x-1)^2(x+1)}\,dx=\tfrac{3}{4}\int\frac{dx}{x-1}+\tfrac{1}{2}\int\frac{dx}{(x-1)^2}+\tfrac{1}{4}\int\frac{dx}{x+1}$$

$$\boxed{\begin{array}{c}u=x-1\\du=dx\end{array}}$$

$$\tfrac{1}{2}\int\frac{dx}{(x-1)^2}=\tfrac{1}{2}\int u^{-2}\,du=\tfrac{1}{2}\cdot\frac{u^{-1}}{-1}+C=-\frac{1}{2u}+C=-\frac{1}{2x-2}+C\ \Rightarrow$$

$$\int\frac{x^2}{(x-1)^2(x+1)}\,dx=\tfrac{3}{4}\ln|x-1|-\frac{1}{2x-2}+\tfrac{1}{4}\ln|x+1|+C=\tfrac{1}{4}\ln|(x-1)^3(x+1)|-\frac{1}{2x-2}+C.$$

23. $\displaystyle\int \frac{x}{6x^2 - x - 2}\,dx = \int \frac{x}{(2x+1)(3x-2)}\,dx$:

$$\frac{x}{(2x+1)(3x-2)} = \frac{A}{2x+1} + \frac{B}{3x-2}$$

$$x = A(3x-2) + B(2x+1)$$

$$x = \tfrac{2}{3}: \quad \tfrac{2}{3} = B(\tfrac{4}{3}+1) = \tfrac{7}{3}B \;\Rightarrow\; 7B = 2 \;\Rightarrow\; B = \tfrac{2}{7}$$

$$x = -\tfrac{1}{2}: \quad -\tfrac{1}{2} = A(-\tfrac{3}{2}-2) = -\tfrac{7}{2}A \;\Rightarrow\; 7A = 1 \;\Rightarrow\; A = \tfrac{1}{7}$$

$$\int \frac{x}{6x^2 - x - 2}\,dx = \tfrac{1}{7}\int \frac{dx}{2x+1} + \tfrac{2}{7}\int \frac{dx}{3x-2} = \tfrac{1}{14}\int \frac{du}{u} + \tfrac{2}{21}\int \frac{dv}{v} = \tfrac{1}{14}\ln|u| + \tfrac{2}{21}\ln|v| + C$$

$$\boxed{\begin{array}{l} u = 2x+1 \\ du = 2\,dx \\ dx = \tfrac{1}{2}\,du \end{array}} \qquad \boxed{\begin{array}{l} v = 3x-2 \\ dv = 3\,dx \\ dx = \tfrac{1}{3}\,dv \end{array}}$$

$$= \tfrac{1}{14}\ln|2x+1| + \tfrac{2}{21}\ln|3x-2| + C.$$

25. $\displaystyle\int \frac{7x+3}{x^3 - 2x^2 - 3x}\,dx = \int \frac{7x+3}{x(x^2 - 2x - 3)}\,dx = \int \frac{7x+3}{x(x-3)(x+1)}\,dx$:

$$\frac{7x+3}{x(x-3)(x+1)} = \frac{A}{x} + \frac{B}{x-3} + \frac{C}{x+1}$$

$$7x + 3 = A(x-3)(x+1) + Bx(x+1) + Cx(x-3)$$

$$x = 3: \quad 7(3)+3 = B\cdot 3(3+1) \;\Rightarrow\; 24 = 12B \;\Rightarrow\; B = 2$$

$$x = -1: \quad 7(-1)+3 = C(-1)(-1-3) \;\Rightarrow\; -4 = 4C \;\Rightarrow\; C = -1$$

$$x = 0: \quad 7(0)+3 = A(0-3)(0+1) \;\Rightarrow\; 3 = -3A \;\Rightarrow\; A = -1$$

$$\int \frac{7x+3}{x^3 - 2x^2 - 3x}\,dx = -\int \frac{dx}{x} + 2\int \frac{dx}{x-3} - \int \frac{dx}{x+1} = -\ln|x| + 2\ln|x-3| - \ln|x+1| + C$$

$$= \ln\left|\frac{(x-3)^2}{x(x+1)}\right| + C.$$

27.

$$\frac{5x^2 + 18x - 1}{(x+4)^2(x-3)} = \frac{A}{x+4} + \frac{B}{(x+4)^2} + \frac{C}{x-3}$$

$$5x^2 + 18x - 1 = A(x+4)(x-3) + B(x-3) + C(x+4)^2$$

$$x = -4: \quad 5(-4)^2 + 18(-4) - 1 = B(-4-3) \;\Rightarrow\; 80 - 72 - 1 = -7B \;\Rightarrow\; B = -1$$

$$x = 3: \quad 5(3)^2 + 18(3) - 1 = C(3+4)^2 \;\Rightarrow\; 45 + 54 - 1 = 49C \;\Rightarrow\; C = 2$$

$$x = 0: \quad 5(0)^2 + 18(0) - 1 = -12A - (-3) + 2(16) \;\Rightarrow\; 12A = 36 \;\Rightarrow\; A = 3$$

$$\int \frac{5x^2 + 18x - 1}{(x+4)^2(x-3)}\, dx = 3\int \frac{dx}{x+4} - \int \frac{dx}{(x+4)^2} + 2\int \frac{dx}{x-3}$$

$$\boxed{\begin{array}{c} u = x + 4 \\ du = dx \end{array}}$$

$$\int \frac{dx}{(x+4)^2} = \int u^{-2}\, du = \frac{u^{-1}}{-1} + C = -\frac{1}{u} + C = -\frac{1}{x+4} + C \Rightarrow$$

$$\int \frac{5x^2 + 18x - 1}{(x+4)^2(x-3)}\, dx = 3\ln|x+4| + \frac{1}{x+4} + 2\ln|x-3| + C = \ln\left|(x+4)^3(x-3)^2\right| + \frac{1}{x+4} + C.$$

29.
$$\frac{4x^2 + x + 2}{x^3(x+2)} = \frac{A}{x} + \frac{B}{x^2} + \frac{C}{x^3} + \frac{D}{x+2}$$

$$(*)\ 4x^2 + x + 2 = Ax^2(x+2) + Bx(x+2) + C(x+2) + Dx^3$$

$$x = 0: \qquad\qquad 2 = C(0+2) \Rightarrow C = 1$$

$$x = -2:\ 4(-2)^2 - 2 + 2 = D(-2)^3 \Rightarrow 16 = -8D \Rightarrow D = -2$$

Substituting these values for C and D in $(*)$, we then equate coefficients of like powers of x:

$$4x^2 + x + 2 = Ax^3 + 2Ax^2 + Bx^2 + 2Bx + x + 2 - 2x^3$$
$$= (A-2)x^3 + (2A+B)x^2 + (2B+1)x + 2$$

$$x^3:\ 0 = A - 2 \Rightarrow A = 2$$
$$x^2:\ 4 = 2A + B \Rightarrow 4 = 4 + B \Rightarrow B = 0$$

$$\int \frac{4x^2 + x + 2}{x^3(x+2)}\, dx = 2\int \frac{dx}{x} + \int x^{-3}\, dx - 2\int \frac{dx}{x+2} = 2\ln|x| + \frac{x^{-2}}{-2} - 2\ln|x+2| + C$$

$$= \ln\left[\frac{x^2}{(x+2)^2}\right] - \frac{1}{2x^2} + C.$$

31. $\dfrac{x^3 - x}{(x^2+1)^2} = \dfrac{Ax + B}{x^2+1} + \dfrac{Cx + D}{(x^2+1)^2}$

$$x^3 - x = (Ax+B)(x^2+1) + Cx + D$$
$$= Ax^3 + Ax + Bx^2 + B + Cx + D$$
$$= Ax^3 + Bx^2 + (A+C)x + (B+D)$$

Equating coefficients of like powers of x:

$$x^3:\ 1 = A$$
$$x^2:\ 0 = B$$
$$x:\ -1 = A + C \Rightarrow -1 = 1 + C$$
$$\Rightarrow C - -2$$
$$\text{const}:\ 0 = B + D \Rightarrow 0 = 0 + D$$
$$\Rightarrow D = 0$$

$$\int \frac{x^3 - x}{(x^2+1)^2}\, dx = \int \frac{x}{x^2+1}\, dx - \int \frac{2x}{(x^2+1)^2}\, dx = \frac{1}{2}\int \frac{du}{u} - \int u^{-2}\, du = \frac{1}{2}\ln|u| + \frac{1}{u} + C$$

$$\boxed{\begin{array}{c} u = x^2 + 1 \\ du = 2x\, dx \end{array}}$$

$$= \frac{1}{2}\ln(x^2 + 1) + \frac{1}{x^2+1} + C.$$

33. $\dfrac{x-1}{x(x^2+1)} = \dfrac{A}{x} + \dfrac{Bx+C}{x^2+1}$

Equating coefficients of like powers of x:

$$x-1 = A(x^2+1) + (Bx+C)x$$
$$= Ax^2 + A + Bx^2 + Cx$$
$$= (A+B)x^2 + Cx + A$$

$\text{const}: -1 = A$

$x: \quad 1 = C$

$x^2: \quad 0 = A+B \Rightarrow 0 = -1+B \Rightarrow B = 1$

$$\int \frac{x-1}{x(x^2+1)}\,dx = -\int \frac{dx}{x} + \int \frac{x+1}{x^2+1}\,dx = -\ln|x| + \int \frac{x}{x^2+1}\,dx + \int \frac{1}{x^2+1}\,dx$$

$$\boxed{\begin{array}{l} u = x^2+1 \\ du = 2x\,dx \Rightarrow x\,dx = \frac{1}{2}\,du \end{array}}$$

$$= -\ln|x| + \frac{1}{2}\int \frac{du}{u} + \tan^{-1} x = -\ln|x| + \frac{1}{2}\ln(x^2+1) + \tan^{-1} x + C.$$

35. $\dfrac{7x^3 - 3x^2 + 9x - 6}{x^4 + 3x^2 + 2} = \dfrac{7x^3 - 3x^2 + 9x - 6}{(x^2+2)(x^2+1)} = \dfrac{Ax+B}{x^2+2} + \dfrac{Cx+D}{x^2+1}$

$$7x^3 - 3x^2 + 9x - 6 = (Ax+B)(x^2+1) + (Cx+D)(x^2+2)$$
$$= Ax^3 + Ax + Bx^2 + B + Cx^3 + 2Cx + Dx^2 + 2D$$
$$= (A+C)x^3 + (B+D)x^2 + (A+2C)x + (B+2D)$$

Equating coefficients of like powers of x:

$\left.\begin{array}{l} x^3: (1) \quad 7 = A+C \\ x^2: (2) \; -3 = B+D \\ x: (3) \quad 9 = A+2C \\ \text{const}: (4) \; -6 = B+2D \end{array}\right\} \rightarrow$

$(3)-(1): \quad C = 9 - 7 = 2$

$(4)-(2): \quad D = -6 - (-3) = -3$

From $(1): \quad A = 7 - C = 7 - 2 = 5$

From $(2): \quad B = -3 - D = -3 - (-3) = 0$

$$\int \frac{7x^3 - 3x^2 + 9x - 6}{x^4 + 3x^2 + 2}\,dx = \int \frac{5x}{x^2+2}\,dx + \int \frac{2x-3}{x^2+1}\,dx = \frac{5}{2}\int \frac{du}{u} + \int \frac{2x}{x^2+1}\,dx - 3\int \frac{1}{1+x^2}\,dx$$

$$\boxed{\begin{array}{l} u = x^2+2 \\ du = 2x\,dx \Rightarrow 5x\,dx = \frac{5}{2}\,du \end{array}} \qquad \boxed{\begin{array}{l} v = x^2+1 \\ dv = 2x\,dx \end{array}}$$

$$= \frac{5}{2}\ln|u| + \int \frac{dv}{v} - 3\tan^{-1} x = \frac{5}{2}\ln(x^2+2) + \ln(x^2+1) - 3\tan^{-1} x + C.$$

37. The first step is to divide the denominator of the integrand into the numerator, resulting in: $\dfrac{x^2+2}{x^2+2x} = 1 + \dfrac{-2x+2}{x^2+2x}$. Next, obtain the partial fraction expansion of the fractional term:

$$\frac{-2x+2}{x^2+2x} = \frac{-2x+2}{x(x+2)} = \frac{A}{x} + \frac{B}{x+2}$$

$$-2x+2 = A(x+2) + Bx$$

$$x = -2: \quad -2(-2) + 2 = -2B \Rightarrow 6 = -2B \Rightarrow B = -3$$

$$x = 0: \quad 2 = 2A \Rightarrow A = 1$$

$$\int \frac{x^2+2}{x^2+2x}\,dx = \int 1\,dx + \int \frac{dx}{x} - 3\int \frac{dx}{x+2} = x + \ln|x| - 3\ln|x+2| + C$$

$$= x + \ln\left|\frac{x}{(x+2)^3}\right| + C.$$

39. The first step is to divide the denominator of the integrand into the numerator, resulting in: $\dfrac{x^3 - 3x^2 + 2x - 3}{x^2+1} = x - 3 + \dfrac{x}{x^2+1}$.

$$\int \frac{x^3 - 3x^2 + 2x - 3}{x^2+1}\,dx = \int (x-3)\,dx + \int \frac{x}{x^2+1}\,dx = -\frac{1}{2}x^2 - 3x + \frac{1}{2}\int \frac{du}{u}$$

$$\boxed{\begin{array}{l} u = x^2 + 1 \\ du = 2x\,dx \ \Rightarrow \ x\,dx = \frac{1}{2}\,du \end{array}}$$

$$= -\frac{1}{2}x^2 - 3x + \frac{1}{2}\ln|u| + C = -\frac{1}{2}x^2 - 3x + \frac{1}{2}\ln(x^2+1) + C.$$

41. Let $u = \sin x$, then $du = \cos x\,dx$ and $\displaystyle\int \frac{\cos x}{\sin^2 x + \sin x - 6}\,dx = \int \frac{du}{u^2 + u - 6}$

$$\frac{1}{u^2 + u - 6} = \frac{1}{(u+3)(u-2)} = \frac{A}{u+3} + \frac{B}{u-2}$$

$$1 = A(u-2) + B(u+3)$$

$$u = 2 : 1 = 5B \ \Rightarrow \ B = \frac{1}{5}$$

$$u = -3 : 1 = -5A \ \Rightarrow \ A = -\frac{1}{5}$$

$$\int \frac{du}{u^2 + u - 6} = -\frac{1}{5}\int \frac{du}{u+3} + \frac{1}{5}\int \frac{du}{u-2} = -\frac{1}{5}\ln|u+3| + \frac{1}{5}\ln|u-2| + C$$

$$= -\frac{1}{5}\ln|\sin x + 3| + \frac{1}{5}\ln|\sin x - 2| + C = \frac{1}{5}\ln\left|\frac{\sin x - 2}{\sin x + 3}\right| + C.$$

43. Let $u = e^x$, then $du = e^x\,dx$ and $\displaystyle\int \frac{e^x}{e^{2x} - 1}\,dx = \int \frac{du}{u^2 - 1} = \int \frac{du}{(u-1)(u+1)}$

$$\frac{1}{(u-1)(u+1)} = \frac{A}{u-1} + \frac{B}{u+1}$$

$$1 = A(u+1) + B(u-1)$$

$$u = 1 : 1 = 2A \ \Rightarrow \ A = \frac{1}{2}$$

$$u = -1 : 1 = -2B \ \Rightarrow \ B = -\frac{1}{2}$$

$$\int \frac{e^x}{e^{2x} - 1}\,dx = \frac{1}{2}\int \frac{du}{u-1} - \frac{1}{2}\int \frac{du}{u+1} = \frac{1}{2}\ln|u-1| - \frac{1}{2}\ln|u+1| + C$$

$$= \frac{1}{2}\ln|e^x - 1| - \frac{1}{2}\ln|e^x + 1| + C = \frac{1}{2}\ln\left|\frac{e^x - 1}{e^x + 1}\right| + C.$$

45. $\displaystyle\int_0^1 \frac{dx}{x^2-9} = \int_0^1 \frac{dx}{(x-3)(x+3)}$

$$\frac{1}{(x-3)(x+3)} = \frac{A}{x-3} + \frac{B}{x+3}$$

$$1 = A(x+3) + B(x-3)$$

$x = -3 : 1 = -6B \;\Rightarrow\; B = -\frac{1}{6}$

$x = 3 : 1 = 6A \;\Rightarrow\; A = \frac{1}{6}$

$\displaystyle\int_0^1 \frac{dx}{x^2-9} = \frac{1}{6}\int_0^1 \frac{1}{x-3}\,dx - \frac{1}{6}\int_0^1 \frac{1}{x+3}\,dx = \frac{1}{6}\ln|x-3|\Big|_0^1 - \frac{1}{6}\ln|x+3|\Big|_0^1$

$\qquad = \frac{1}{6}(\ln 2 - \ln 3) - \frac{1}{6}(\ln 4 - \ln 3) = \frac{1}{6}(\ln 2 - \ln 4) = \frac{1}{6}\ln\frac{2}{4} = \frac{1}{6}\ln\frac{1}{2}$

$\qquad = \frac{1}{6}(\ln 1 - \ln 2) = -\frac{1}{6}\ln 2.$

47. $\displaystyle\int_1^2 \frac{22}{6x^2+5x-4}\,dx = \int_1^2 \frac{22}{(2x-1)(3x+4)}\,dx$

$$\frac{22}{(2x-1)(3x+4)} = \frac{A}{2x-1} + \frac{B}{3x+4}$$

$$22 = A(3x+4) + B(2x-1)$$

$x = -\frac{4}{3} : \; 22 = B(-\frac{8}{3}-1) \;\Rightarrow\; -\left(\frac{11}{3}\right)B = 22 \;\Rightarrow\; B = -6$

$x = \frac{1}{2} : \; 22 = A(\frac{3}{2}+4) \;\Rightarrow\; \left(\frac{11}{2}\right)A = 22 \;\Rightarrow\; A = 4$

$\displaystyle\int_1^2 \frac{22}{(2x-1)(3x+4)}\,dx = 4\int_1^2 \frac{1}{2x-1}\,dx - 6\int_1^2 \frac{1}{3x+4}\,dx$

$\qquad = 4\cdot\frac{1}{2}\ln|2x-1|\,\Big|_1^2 - 6\cdot\frac{1}{3}\ln|3x+4|\,\Big|_1^2$

$\qquad = 2(\ln 3 - \ln 1) - 2(\ln 10 - \ln 7) = \ln 9 - \ln 100 + \ln 49 = \ln(4.41).$

49. $\displaystyle\int_{-\sqrt{3}}^1 \frac{2x^3+4x^2+2x-3}{x^4+x^2}\,dx = \int_{-\sqrt{3}}^1 \frac{2x^3+4x^2+2x-3}{x^2(x^2+1)}\,dx$

$\displaystyle\frac{2x^3+4x^2+2x-3}{x^2(x^2+1)} = \frac{A}{x} + \frac{B}{x^2} + \frac{Cx+D}{x^2+1}$

$2x^3 + 4x^2 + 2x - 3$

$\qquad = Ax(x^2+1) + B(x^2+1) + (Cx+D)x^2$

$\qquad = Ax^3 + Ax + Bx^2 + B + Cx^3 + Dx^2$

$\qquad = (A+C)x^3 + (B+D)x^2 + Ax + B$

Equating coefficients of like powers of x:

const: $-3 = B$

$\quad x: \quad 2 = A$

$\quad x^2: \quad 4 = B+D = -3+D \Rightarrow D = 7$

$\quad x^3: \quad 2 = A+C = 2+C \Rightarrow C = 0$

$\displaystyle\int_{-\sqrt{3}}^1 \frac{2x^3+4x^2+2x-3}{x^4+x^2}\,dx = 2\int_{-\sqrt{3}}^1 \frac{dx}{x} - 3\int_{-\sqrt{3}}^1 \frac{dx}{x^2} + 7\int_{-\sqrt{3}}^1 \frac{dx}{x^2+1}$

$\qquad = 2\ln|x|\,\Big|_{-\sqrt{3}}^1 + \frac{3}{x}\,\Big|_{-\sqrt{3}}^1 + 7\tan^{-1}x\,\Big|_{-\sqrt{3}}^1$

$\qquad = 2(\ln 1 - \ln\sqrt{3}) + 3(1 + \frac{1}{\sqrt{3}}) + 7[\tan^{-1}1 - \tan^{-1}(-\sqrt{3})]$

$\qquad = -\ln 3 + 3 + \sqrt{3} + 7[\frac{\pi}{4} + \frac{\pi}{3}] = 3 + \sqrt{3} - \ln 3 + \frac{49\pi}{12}.$

51. $\displaystyle\int_1^2 \frac{2x^2 - x + 2}{x^3 + x}\, dx = \int_1^2 \frac{2x^2 - x - 2}{x(x^2 + 1)}\, dx$

$$\frac{2x^2 - x + 2}{x(x^2 + 1)} = \frac{A}{x} + \frac{Bx + C}{x^2 + 1}$$

Equating coefficients of like powers of x:

const: $2 = A$

$x: -1 = C$

$x^2: \quad 2 = A + B = 2 + B \implies B = 0$

$$2x^2 - x + 2 = A(x^2 + 1) + (Bx + C)x$$
$$= Ax^2 + A + Bx^2 + Cx$$
$$= (A + B)x^2 + Cx + A$$

$$\int_1^2 \frac{2x^2 - x + 2}{x^3 + x}\, dx = 2\int_1^2 \frac{dx}{x} - \int_1^2 \frac{dx}{1 + x^2} = 2\ln|x| \Big|_1^2 - \tan^{-1} x \Big|_1^2$$

$$= 2(\ln 2 - \ln 1) - (\tan^{-1} 2 - \tan^{-1} 1) = 2\ln 2 - \tan^{-1} 2 + \tfrac{\pi}{4}$$

$$= \ln 4 - \tan^{-1} 2 + \tfrac{\pi}{4}.$$

53. $\displaystyle f(x) = \int \frac{dx}{x^2 - 3x + 2} = \frac{dx}{(x - 2)(x - 1)}, \quad f(3) = 0.$

$$\frac{1}{(x - 2)(x - 1)} = \frac{A}{x - 2} + \frac{B}{x - 1}$$

$$1 = A(x - 1) + B(x - 2)$$

$$x = 1: 1 = B(1 - 2) \implies -B = 1 \implies B = -1$$

$$x = 2: 1 = A(2 - 1) \implies A = 1$$

$$f(x) = \int \frac{dx}{x - 2} - \int \frac{dx}{x - 1} = \ln|x - 2| - \ln|x - 1| + C$$

$$f(3) = 0 \implies \ln 1 - \ln 2 + C = 0 \implies C = \ln 2 \implies$$

$$f(x) = \ln|x - 2| - \ln|x - 1| + \ln 2 = \ln\left|\frac{2(x - 2)}{x - 1}\right|.$$

55. Area $\displaystyle= \int_3^5 \frac{7x - 5}{4x^2 - 7x - 2}\, dx = \int_3^5 \frac{7x - 5}{(4x + 1)(x - 2)}\, dx$

$$\frac{7x - 5}{(4x + 1)(x - 2)} = \frac{A}{4x + 1} + \frac{B}{x - 2}$$

$$7x - 5 = A(x - 2) + B(4x + 1)$$

$$x = 2: 7(2) - 5 = B(9) \implies 9B = 9 \implies B = 1$$

$$x = -\tfrac{1}{4}: \; -\tfrac{7}{4} - 5 = A(-\tfrac{1}{4}) - 2) \implies -27 = -9A \implies A = 3$$

$$\text{Area } = 3\int_3^5 \frac{dx}{4x + 1} + \int_3^5 \frac{dx}{x - 2} = \tfrac{3}{4}\ln|4x + 1| \Big|_3^5 + \ln|x - 2| \Big|_3^5$$

$$= \tfrac{3}{4}(\ln 21 - \ln 13) + (\ln 3 - \ln 1) = \tfrac{3}{4}\ln\tfrac{21}{13} + \ln 3.$$

57. $f(x) = \dfrac{x}{x^2 - 4} > 0$ on $[3, 5] \Rightarrow V = \pi \displaystyle\int_3^5 \left[\dfrac{x}{(x-2)(x+2)}\right]^2 dx = \pi \displaystyle\int_3^5 \dfrac{x^2}{(x-2)^2(x+2)^2} dx$

$$\dfrac{x^2}{(x-2)^2(x+2)^2} = \dfrac{A}{x-2} + \dfrac{B}{(x-2)^2} + \dfrac{C}{x+2} + \dfrac{D}{(x+2)^2}$$

$$x^2 = A(x-2)(x+2)^2 + B(x+2)^2 + C(x+2)(x-2)^2 + D(x-2)^2$$

$$x = 2 : 4 = B(4)^2 \Rightarrow B = \tfrac{1}{4}$$

$$x = -2 : 4 = 16D \Rightarrow D = \tfrac{1}{4}$$

$$x = 0 : 0 = -8A + 1 + 8C + 1 \Rightarrow 8(-A + C) = -2 \Rightarrow -A + C = -\tfrac{1}{4} \quad (*)$$

$$x = 1 : 1 = A(-1)(9) + \tfrac{9}{4} + 3C + \tfrac{1}{4} \Rightarrow -9A + 3C = -\tfrac{3}{2} \Rightarrow -3A + C = -\tfrac{1}{2} \quad (**)$$

Subtracting equations $(*)-(**)$: $-A - (-3A) = \tfrac{1}{4} - -\tfrac{1}{2} \Rightarrow 2A = \tfrac{1}{4} \Rightarrow A = \tfrac{1}{8}$.

From $(*)$: $-\tfrac{1}{8} + C = -\tfrac{1}{4} \Rightarrow C = -\tfrac{1}{4} + \tfrac{1}{8} = -\tfrac{1}{8}$.

Therefore, $V = \pi \left[\dfrac{1}{8}\displaystyle\int_3^5 \dfrac{1}{x-2} dx + \dfrac{1}{4}\displaystyle\int_3^5 \dfrac{1}{(x-2)^2} dx - \dfrac{1}{8}\displaystyle\int_3^5 \dfrac{1}{x+2} dx + \dfrac{1}{4}\displaystyle\int_3^5 \dfrac{1}{(x+2)^2} dx\right]$

$$\dfrac{1}{8}\int_3^5 \dfrac{1}{x-2} dx = \tfrac{1}{8} \ln|x-2| \ \Big|_3^5 = \tfrac{1}{8}(\ln 3 - \cancel{\ln 1}) = \tfrac{1}{8} \ln 3$$

$$\dfrac{1}{4}\int_3^5 \dfrac{1}{(x-2)^2} dx = \tfrac{1}{4}\int_1^3 u^{-2} du = \tfrac{1}{4} \cdot -\tfrac{1}{u} \ \Big|_1^3 = -\tfrac{1}{4}(\tfrac{1}{3} - 1) = -\tfrac{1}{4}(-\tfrac{2}{3}) = \tfrac{1}{6}$$

$$\boxed{\begin{array}{l} u = x - 2, \ x = 3 \Rightarrow u = 1 \\ du = dx, \qquad x = 5 \Rightarrow u = 3 \end{array}}$$

$$-\dfrac{1}{8}\int_3^5 \dfrac{1}{x+2} dx = -\tfrac{1}{8} \ln|x+2| \ \Big|_3^5 = -\tfrac{1}{8}(\ln 7 - \ln 5)$$

$$\dfrac{1}{4}\int_3^5 \dfrac{1}{(x+2)^2} dx = \tfrac{1}{4}\int_5^7 u^{-2} du = \tfrac{1}{4} \cdot -\tfrac{1}{u} \ \Big|_5^7 = -\tfrac{1}{4}(\tfrac{1}{7} - \tfrac{1}{5}) = -\tfrac{1}{4}(-\tfrac{2}{35}) = \tfrac{1}{70}$$

$$\boxed{\begin{array}{l} u = x + 2, \ x = 3 \Rightarrow u = 5 \\ du = dx, \qquad x = 5 \Rightarrow u = 7 \end{array}}$$

So, $V = \pi \left[\tfrac{1}{8} \ln 3 + \tfrac{1}{6} - \tfrac{1}{8}(\ln 7 - \ln 5) + \tfrac{1}{70}\right] = \pi \left[\tfrac{76}{420} + \tfrac{1}{8} \ln \tfrac{3 \cdot 5}{7}\right] = \pi \left[\tfrac{19}{105} + \tfrac{1}{8} \ln \tfrac{15}{7}\right].$

59. $\displaystyle\int \dfrac{dx}{x^2 - a^2} = \int \dfrac{dx}{(x-a)(x+a)}$:

$$\dfrac{1}{(x-a)(x+a)} = \dfrac{A}{x-a} + \dfrac{B}{x+a}$$

$$1 = A(x+a) + B(x-a)$$

$$x = -a : 1 = B(-a-a) \Rightarrow B = -\dfrac{1}{2a}$$

$$x = a : 1 = A(a+a) \Rightarrow A = \dfrac{1}{2a}$$

$$\int \dfrac{dx}{x^2 - a^2} = \dfrac{1}{2a}\int \dfrac{dx}{x-a} - \dfrac{1}{2a}\int \dfrac{dx}{x+a}$$

$$= \dfrac{1}{2a} \ln|x-a| - \dfrac{1}{2a} \ln|x+a| + C$$

$$= \dfrac{1}{2a} \ln\left|\dfrac{x-a}{x+a}\right| + C.$$

§7.3 Powers of Trigonometric Functions and Trigonometric Substitution

1. $\int \sin^2 x \, dx = \frac{1}{2} \int (1 - \cos 2x) \, dx = \frac{1}{2}[x - \frac{1}{2}\sin 2x] + C = \frac{1}{2}x - \frac{1}{4}\sin 2x + C.$

3. $\int \sin^3 3x \, dx = \int \sin^2 3x \sin 3x \, dx = \int (1 - \cos^2 3x) \sin 3x \, dx = \int \sin 3x \, dx - \int \cos^2 3x \sin 3x \, dx$

$$\boxed{\begin{aligned} u &= \cos 3x \\ du &= -3\sin 3x \, dx \\ -\sin 3x &= \frac{1}{3} du \end{aligned}}$$

$$= -\frac{1}{3}\cos 3x + \frac{1}{3}\int u^2 \, du = -\frac{1}{3}\cos 3x + \frac{1}{3} \cdot \frac{u^3}{3} + C = -\frac{1}{3}\cos 3x + \frac{1}{9}\cos^3 3x + C.$$

5. $\displaystyle \int \cos^4 2x \, dx = \int (\cos^2 2x)(\cos^2 2x) \, dx = \int (\cos^2 2x)(1 - \sin^2 2x) \, dx$

$$= \int \cos^2 2x \, dx - \int \sin^2 2x \cos^2 2x \, dx = \int \frac{1 + \cos 4x}{2} \, dx - \int \left(\frac{\sin 4x}{2}\right)^2 dx$$

$$= \int \left(\frac{1}{2} + \frac{1}{2}\cos 4x\right) dx - \frac{1}{4} \int \frac{1 - \cos 8x}{2} \, dx$$

$$= \frac{1}{2}x + \frac{1}{2} \cdot \frac{1}{4}\sin 4x - \frac{1}{8}x + \frac{1}{8} \cdot \frac{1}{8}\sin 8x + C = \frac{3}{8}x + \frac{1}{8}\sin 4x + \frac{1}{64}\sin 8x + C.$$

7. $\displaystyle \int \sin^2 x \cos^4 x = \int (\sin x \cos x)^2 \cos^2 x \, dx = \int \left(\frac{\sin 2x}{2}\right)^2 \left(\frac{1 + \cos 2x}{2}\right) dx$

$$= \frac{1}{8}\left[\int \sin^2 2x \, dx + \int \sin^2 2x \cos 2x \, dx\right]$$

$$\boxed{\begin{aligned} u &= \sin 2x \\ du &= 2\cos 2x \, dx \ \Rightarrow \ \cos 2x \, dx = \frac{1}{2} du \end{aligned}}$$

$$= \frac{1}{8}\left[\int \left(\frac{1 - \cos 4x}{2}\right) dx + \frac{1}{2}\int u^2 \, du\right] = \frac{1}{16}x - \frac{1}{16} \cdot \frac{\sin 4x}{4} + \frac{1}{16} \cdot \frac{u^3}{3} + C$$

$$= \frac{1}{16}x - \frac{1}{64}\sin 4x + \frac{1}{48}\sin^3 2x + C.$$

9. $\int \tan^3 x \, dx = \int (\tan x)(\tan^2 x) \, dx = \int (\tan x)(\sec^2 x - 1) \, dx = \int \tan x \sec^2 x \, dx - \int \tan x \, dx$

$$\boxed{\begin{aligned} u &= \tan x \\ du &= \sec^2 x \, dx \end{aligned}}$$

$$= \int u \, du - \ln|\sec x| + C = \frac{1}{2}u^2 - \ln|\sec x| + C = \frac{1}{2}\tan^2 x - \ln|\sec x| + C.$$

11. $\int \csc^4 2x\, dx = \int (\csc^2 2x)(\csc^2 2x)\, dx = \int \csc^2 2x(1 + \cot^2 2x)\, dx$

$\quad = \int \csc^2 2x\, dx + \cot^2 2x \csc^2 2x\, dx = -\frac{1}{2}\cot 2x - \frac{1}{2}\int u^2\, du$

$$\boxed{\begin{array}{l} u = \cot 2x \\[4pt] du = -2\csc^2 2x\, dx \;\Rightarrow\; \csc^2 2x\, dx = -\frac{1}{2}\, du \end{array}}$$

$\quad = -\frac{1}{2}\cot 2x - \frac{1}{2}\cdot\frac{u^3}{3} + C = -\frac{1}{2}\cot 2x - \frac{1}{6}\cot^3 2x + C.$

13. Two possible solutions:

$$\int \frac{\sec^4 x}{\cot^3 x}\, dx = \int \tan^3 x \sec^2 x \sec^2 x\, dx = \tan^3 x(1 + \tan^2 x)\sec^2 x\, dx$$

$$= \int (\tan^3 x + \tan^5 x)\sec^2 x\, dx = \int (u^3 + u^5)\, du = \frac{1}{4}u^4 + \frac{1}{6}u^6 + C$$

$$\boxed{\begin{array}{l} u = \tan x \\[4pt] du = \sec^2 x\, dx \end{array}}$$

$$= \frac{1}{4}\tan^4 x + \frac{1}{6}\tan^6 x + C.$$

OR

$$\int \frac{\sec^4 x}{\cot^3 x}\, dx = \int \sec^4 x \tan^3 x\, dx = \int \sec^3 x \tan^2 x(\sec x \tan x)\, dx$$

$$= \int \sec^3 x(\sec^2 x - 1)(\sec x \tan x)\, dx = \int (\sec^5 x - \sec^3 x)(\sec x \tan x)\, dx$$

$$\boxed{\begin{array}{l} u = \sec x \\[4pt] du = \sec x \tan x\, dx \end{array}}$$

$$= \int (u^5 - u^3)\, du = \frac{1}{6}u^6 - \frac{1}{4}u^4 + C = \frac{1}{6}\sec^6 x - \frac{1}{4}\sec^4 x + C.$$

15. Two possible solutions:

$$\int \tan 5x \sec^4 5x\, dx = \int \sec^3 5x(\sec 5x \tan 5x)\, dx = \frac{1}{5}\int u^3\, du = \frac{1}{5}\cdot\frac{u^4}{4} + C$$

$$\boxed{\begin{array}{l} u = \sec 5x \\[4pt] du = 5\sec 5x \tan 5x\, dx \;\Rightarrow\; \sec 5x \tan 5x\, dx = \frac{1}{5}\, du \end{array}}$$

$$= \frac{1}{20}\sec^4 5x + C.$$

OR

$$\int \tan 5x \sec^4 5x\, dx = \int \tan 5x \sec^2 5x \sec^2 5x\, dx = \int \tan 5x(1 + \tan^2 5x)\sec^2 5x\, dx$$

$$= \int (\tan 5x + \tan^3 5x)\sec^2 5x\, dx = \frac{1}{5}\int (u + u^3)\, du$$

$$\boxed{\begin{array}{l} u = \tan 5x \\[4pt] du = 5\sec^2 5x\, dx \;\Rightarrow\; \sec^2 5x\, dx = \frac{1}{5}\, du \end{array}}$$

$$= \frac{1}{5}\left[\frac{1}{2}u^2 + \frac{1}{4}u^4\right] + C = \frac{1}{10}\tan^2 5x + \frac{1}{20}\tan^4 5x + C.$$

17. $\quad\quad\quad\quad \int u\,dv = uv - \int v\,du$

$\int x\sec x\tan x\,dx = x\sec x - \int \sec x\,dx = x\sec x - \ln|\sec x + \tan x| + C.$

$$\begin{array}{ll} u = x & dv = \sec x\tan x\,dx \\ du = dx & v = \sec x \end{array}$$

19. $\displaystyle\int \frac{\tan^5 x}{\cos x}\,dx = \int (\tan^2 x)^2 \tan x\sec x\,dx = \int (\sec^2 x - 1)^2 \sec x\tan x\,dx$

$\quad\quad = \displaystyle\int (\sec^4 x - 2\sec^2 x + 1)\sec x\tan x\,dx = \int (u^4 - 2u^2 + 1)\,du$

$$\begin{array}{l} u = \sec x \\ du = \sec x\tan x\,dx \end{array}$$

$\quad\quad = \frac{1}{5}u^5 - \frac{2}{3}u^3 + u + C = \frac{1}{5}\sec^5 x - \frac{2}{3}\sec^3 x + \sec x + C.$

21. $\int \cot^3 x\csc^3 x\,dx = \int \cot^2 x\csc^2 x\csc x\cot x\,dx = \int (\csc^2 x - 1)\csc^2 x\csc x\cot x\,dx$

$\quad\quad = \int (\csc^4 x - \csc^2 x)\csc x\cot x\,dx = -\int (u^4 - u^2)\,du = -\frac{1}{5}u^5 + \frac{1}{3}u^3 + C$

$$\begin{array}{l} u = \csc x \\ du = -\csc x\cot x\,dx \end{array}$$

$\quad\quad = -\frac{1}{5}\csc^5 x + \frac{1}{3}\csc^3 x + C.$

23. $\displaystyle\int \frac{1 - \tan^2 x}{\sec^2 x}\,dx = \int (\cos^2 x)\left(1 - \frac{\sin^2 x}{\cos^2 x}\right)\,dx = \int (\cos^2 x - \sin^2 x)\,dx = \int \cos 2x\,dx$

$\quad\quad = \frac{1}{2}\sin 2x + C.$

25. $\int \sqrt{\tan x}\sec^4 x\,dx = \int (\tan x)^{\frac{1}{2}}\sec^2 x\sec^2 x\,dx = \int (\tan x)^{\frac{1}{2}}(1 + \tan^2 x)\sec^2 x\,dx$

$\quad\quad = \int [(\tan x)^{\frac{1}{2}} + (\tan x)^{\frac{5}{2}}]\sec^2 x\,dx = \int (u^{\frac{1}{2}} + u^{\frac{5}{2}})\,du = \frac{2}{3}u^{\frac{3}{2}} + \frac{2}{7}u^{\frac{7}{2}} + C$

$$\begin{array}{l} u = \tan x \\ du = \sec^2 x\,dx \end{array}$$

$\quad\quad = \frac{2}{3}(\tan x)^{\frac{3}{2}} + \frac{2}{7}(\tan x)^{\frac{7}{2}} + C.$

27. $\int_0^{\frac{\pi}{4}} \cos^3 x\,dx = \int_0^{\frac{\pi}{4}} \cos^2 x\cos x\,dx = \int_0^{\frac{\pi}{4}} (1-\sin^2 x)\cos x\,dx = \int_0^{\frac{\pi}{4}} \cos x\,dx - \int_0^{\frac{\pi}{4}} \sin^2 x\cos x\,dx$

$$\begin{array}{ll} u = \sin x & x = 0 \;\Rightarrow\; u = 0 \\ du = \cos x\,dx & x = \frac{\pi}{4} \;\Rightarrow\; u = \dfrac{1}{\sqrt{2}} \end{array}$$

$\quad\quad = \sin x\Big|_0^{\frac{\pi}{4}} - \int_0^{\frac{1}{\sqrt{2}}} u^2\,du = \sin\frac{\pi}{4} - \sin 0 - \frac{1}{3}u^3\Big|_0^{\frac{1}{\sqrt{2}}} = \frac{1}{\sqrt{2}} - \frac{1}{3}\left(\frac{1}{\sqrt{2}}\right)^3$

$\quad\quad = \frac{1}{\sqrt{2}} - \frac{1}{6}\cdot\frac{1}{\sqrt{2}} = \frac{5}{6\sqrt{2}}\cdot\frac{\sqrt{2}}{\sqrt{2}} = \frac{5\sqrt{2}}{12}.$

29. $\int_0^{\frac{\pi}{3}} \sin^4 3x \cos^3 3x \, dx = \int_0^{\frac{\pi}{3}} \sin^4 3x \cos^2 3x \cos 3x \, dx = \int_0^{\frac{\pi}{3}} \sin^4 3x (1 - \sin^2 3x) \cos 3x \, dx$

$$= \int_0^{\frac{\pi}{3}} \sin^4 3x \cos 3x \, dx - \int_0^{\frac{\pi}{3}} \sin^6 3x \cos 3x \, dx$$

$$\boxed{\begin{array}{ll} u = \sin 3x & x = 0 \Rightarrow u = 0 \\ du = 3\cos 3x \, dx & x = \frac{\pi}{3} \Rightarrow u = \sin \pi = 0 \end{array}}$$

As $\int_0^0 = 0$, $\int_0^{\frac{\pi}{3}} \sin^4 3x \cos^3 3x \, dx = 0$.

31. $\int_{\frac{\pi}{6}}^{\frac{\pi}{2}} \cot^2 x \, dx = \int_{\frac{\pi}{6}}^{\frac{\pi}{2}} (\csc^2 x - 1) \, dx = (-\cot x - x) \Big|_{\frac{\pi}{6}}^{\frac{\pi}{2}} = -\cot \frac{\pi}{2} - \frac{\pi}{2} - (-\cot \frac{\pi}{6} - \frac{\pi}{6})$

$$= -\frac{\pi}{2} + \sqrt{3} + \frac{\pi}{6} = -\frac{\pi}{3} + \sqrt{3}.$$

33. $\int_0^{2\pi} \cos^4 x \, dx = \int_0^{2\pi} (\cos^2 x)^2 \, dx = \int_0^{2\pi} \left[\frac{1 + \cos 2x}{2} \right]^2 dx$

$$= \frac{1}{4} \int_0^{2\pi} (1 + 2\cos 2x + \cos^2 2x) \, dx = \frac{1}{4} \left[(x + \sin 2x) \Big|_0^{2\pi} + \int_0^{2\pi} \frac{1 + \cos 4x}{2} \right]$$

$$= \frac{1}{4} \left[2\pi + \sin 4\pi + \frac{1}{2} \left(x + \frac{\sin 4x}{4} \right) \Big|_0^{2\pi} \right] = \frac{\pi}{2} + \frac{1}{8} \left(2\pi + \frac{\sin 8\pi}{4} \right) = \frac{\pi}{2} + \frac{\pi}{4}$$

$$= \frac{3\pi}{4}.$$

35. $\int_0^{\frac{\pi}{3}} 10 \sec^6 x \, dx = 10 \int_0^{\frac{\pi}{3}} (\sec^2 x)^2 \sec^2 x \, dx = 10 \int_0^{\frac{\pi}{3}} (1 + \tan^2 x)^2 \sec^2 x \, dx$

$$= 10 \int_0^{\frac{\pi}{3}} (1 + 2\tan^2 x + \tan^4 x) \sec^2 x \, dx = 10 \int_0^{\sqrt{3}} (1 + 2u^2 + u^4) \, du$$

$$\boxed{\begin{array}{ll} u = \tan x & x = 0 \Rightarrow u = 0 \\ du = \sec^2 x \, dx & x = \frac{\pi}{3} \Rightarrow u = \tan \frac{\pi}{3} = \sqrt{3} \end{array}}$$

$$= 10 \left[u + \frac{2}{3} u^3 + \frac{1}{5} u^5 \right] \Big|_0^{\sqrt{3}} = 10 [\sqrt{3} + \frac{2}{3}(\sqrt{3})^3 + \frac{1}{5}(\sqrt{3})^5]$$

$$= 10 \left(\sqrt{3} + 2\sqrt{3} + \frac{9\sqrt{3}}{5} \right) = 10\sqrt{3} + 20\sqrt{3} + 18\sqrt{3} = 48\sqrt{3}.$$

37. $\int \frac{x^2}{\sqrt{1 - x^2}} \, dx = \int \frac{\sin^2 \theta \cos \theta}{\cos \theta} \, d\theta = \int \frac{1 - \cos 2\theta}{2} \, d\theta = \frac{1}{2} \left(\theta - \frac{\sin 2\theta}{2} \right) + C$

$$\boxed{\begin{array}{l} x = \sin \theta \Rightarrow 1 - x^2 = 1 - \sin^2 \theta = \cos^2 \theta \\ dx = \cos \theta \, d\theta \end{array}}$$

$$= \frac{1}{2} [\theta - \sin \theta \cos \theta] + C = \frac{1}{2} (\sin^{-1} x - x\sqrt{1 - x^2}) + C.$$

39. $\int \frac{x}{\sqrt{9 - x^2}} \, dx = \int \frac{(3\sin \theta)(3\cos \theta)}{3\cos \theta} \, d\theta = 3 \int \sin \theta \, d\theta = -3\cos \theta + C$

$$\boxed{\begin{array}{l} x = 3\sin \theta \Rightarrow 9 - x^2 = 9 - 9\sin^2 \theta = 9\cos^2 \theta \\ dx = 3\cos \theta \, d\theta \end{array}}$$

$$= -3 \cdot \frac{\sqrt{9 - x^2}}{3} + C = -\sqrt{9 - x^2} + C.$$

41. $\displaystyle\int \frac{dx}{x^2\sqrt{25-x^2}} = \int \frac{5\cos\theta\,d\theta}{(25\sin^2\theta)(5\cos\theta)} = \frac{1}{25}\int \csc^2\theta\,d\theta = \frac{1}{25}(-\cot\theta)+C$

$$\boxed{\begin{array}{l} x=5\sin\theta \;\Rightarrow\; 25-x^2 = 25-25\sin^2\theta = 25\cos^2\theta \\ dx = 5\cos\theta\,d\theta \end{array}}$$

$$= -\frac{1}{25}\cdot\frac{\sqrt{25-x^2}}{x}+C = -\frac{\sqrt{25-x^2}}{25x}+C.$$

43. $\displaystyle\int \frac{dx}{x\sqrt{4x^2+9}} = \frac{3}{2}\int \frac{\sec^2\theta\,d\theta}{(\frac{3}{2}\tan\theta)(3\sec\theta)} = \frac{1}{3}\int \frac{\sec\theta\,d\theta}{\tan\theta} = \frac{1}{3}\int \frac{\frac{1}{\cos\theta}}{\frac{\sin\theta}{\cos\theta}}\,d\theta = \frac{1}{3}\int \frac{1}{\sin\theta}\,d\theta$

$$\boxed{\begin{array}{l} 4x^2 = 9\tan^2\theta \;\Rightarrow\; 4x^2+9 = 9\tan^2\theta+9 = 9\sec^2\theta \\ 2x = 3\tan\theta \;\Rightarrow\; x=\frac{3}{2}\tan\theta \;\Rightarrow\; dx=\frac{3}{2}\sec^2\theta\,d\theta \end{array}}$$

$$= \frac{1}{3}\int \csc\theta\,d\theta = -\frac{1}{3}\ln|\csc\theta+\cot\theta|+C = -\frac{1}{3}\ln\left|\frac{\sqrt{4x^2+9}}{2x}+\frac{3}{2x}\right|+C$$

$$= -\frac{1}{3}\ln\left|\frac{\sqrt{4x^2+9}+3}{2x}\right|+C.$$

45. $\displaystyle\int \frac{dx}{x^2\sqrt{4x^2-9}} = \int \frac{\frac{3}{2}\sec\theta\tan\theta\,d\theta}{\frac{9}{4}\sec^2\theta\cdot 3\tan\theta} = \frac{1}{2}\cdot\frac{4}{9}\int \frac{d\theta}{\sec\theta} = \frac{2}{9}\int \cos\theta\,d\theta = \frac{2}{9}\sin\theta+C$

$$\boxed{\begin{array}{l} 4x^2 = 9\sec^2\theta \;\Rightarrow\; 4x^2-9 = 9\sec^2\theta-9 = 9\tan^2\theta \\ 2x = 3\sec\theta \;\Rightarrow\; x=\frac{3}{2}\sec\theta \;\Rightarrow\; dx=\frac{3}{2}\sec\theta\tan\theta\,d\theta \end{array}}$$

$$= \frac{2}{9}\frac{\sqrt{4x^2-9}}{2x}+C = \frac{\sqrt{4x^2-9}}{9x}+C.$$

47. $\displaystyle\int \frac{dx}{(9x^2-1)^{\frac{3}{2}}} = \frac{1}{3}\int \frac{\sec\theta\tan\theta\,d\theta}{\tan^3\theta} = \frac{1}{3}\int \frac{\sec\theta}{\tan^2\theta}\,d\theta = \frac{1}{3}\int \left(\frac{1}{\cos\theta}\cdot\frac{\cos^2\theta}{\sin^2\theta}\right)d\theta$

$$\boxed{\begin{array}{l} 9x^2 = \sec^2\theta \;\Rightarrow\; 9x^2-1 = \sec^2\theta-1 = \tan^2\theta \\ 3x = \sec\theta \;\Rightarrow\; x=\frac{1}{3}\sec\theta \;\Rightarrow\; dx=\frac{1}{3}\sec\theta\tan\theta\,d\theta \end{array}}$$

$$= \frac{1}{3}\int \frac{\cos\theta}{\sin^2\theta}\,d\theta = \frac{1}{3}\int u^{-2}\,du = \frac{1}{3}\cdot\frac{u^{-1}}{-1}+C = -\frac{1}{3u}+C = -\frac{1}{3\sin\theta}+C$$

$$\boxed{\begin{array}{l} u = \sin\theta \\ du = \cos\theta\,d\theta \end{array}}$$

$$= -\frac{3x}{3\sqrt{9x^2-1}}+C = -\frac{x}{\sqrt{9x^2-1}}+C$$

49. $\displaystyle\int \frac{x^3}{\sqrt{2-x^2}}\,dx = \int \frac{2\sqrt{2}\sin^3\theta \cdot \sqrt{2}\cos\theta\,d\theta}{\sqrt{2}\cos\theta} = 2\sqrt{2}\int \sin^3\theta\,d\theta = 2\sqrt{2}\int \sin^2\theta\sin\theta\,d\theta$

$$\boxed{\begin{array}{l} x^2 = 2\sin^2\theta \;\Rightarrow\; 2-x^2 = 2-2\sin^2\theta = 2\cos^2\theta \\ x = \sqrt{2}\sin\theta \;\Rightarrow\; dx = \sqrt{2}\cos\theta\,d\theta \end{array}}$$

$$= 2\sqrt{2}\int (1-\cos^2\theta)\sin\theta\,d\theta = -2\sqrt{2}\int (1-u^2)\,du = -2\sqrt{2}\left(u - \tfrac{1}{3}u^3\right) + C$$

$$\boxed{\begin{array}{l} u = \cos\theta \\ du = -\sin\theta\,d\theta \end{array}}$$

$$= -2\sqrt{2}\left(\cos\theta - \tfrac{1}{3}\cos^3\theta\right) + C = -2\sqrt{2}\left[\frac{\sqrt{2-x^2}}{\sqrt{2}} - \frac{1}{3}\cdot\frac{(2-x^2)^{\frac{3}{2}}}{2\sqrt{2}}\right] + C$$

$$= -2\sqrt{2-x^2} + \tfrac{1}{3}(2-x^2)^{\frac{3}{2}} + C.$$

51. $\displaystyle\int \frac{\sqrt{x^2+16}}{x^4}\,dx = \int \frac{4\sec\theta\cdot 4\sec^2\theta\,d\theta}{16\cdot 16\tan^4\theta} = \frac{1}{16}\int \frac{\sec^3\theta}{\tan^4\theta}\,d\theta = \frac{1}{16}\int \frac{1}{\cos^3\theta}\cdot\frac{\cos^4\theta}{\sin^4\theta}\,d\theta$

$$\boxed{\begin{array}{l} x^2 = 16\tan^2\theta \;\Rightarrow\; x^2+16 = 16\tan^2\theta + 16 = 16\sec^2\theta \\ x = 4\tan\theta \;\Rightarrow\; dx = 4\sec^2\theta\,d\theta \end{array}}$$

$$= \frac{1}{16}\int \frac{\cos\theta}{\sin^4\theta}\,d\theta = \frac{1}{16}\int u^{-4}\,du = \frac{1}{16}\cdot\frac{u^{-3}}{-3} + C = -\frac{1}{48u^3} + C$$

$$\boxed{\begin{array}{l} u = \sin\theta \\ du = \cos\theta\,d\theta \end{array}}$$

$$= -\frac{1}{48\sin^3\theta} + C = -\frac{1}{48\left(\dfrac{x}{\sqrt{x^2+16}}\right)^3} + C = -\frac{(x^2+16)^{\frac{3}{2}}}{48x^3} + C.$$

53. $\displaystyle\int \sqrt{1-4x^2}\,dx = \int \cos\theta\cdot\tfrac{1}{2}\cos\theta\,d\theta = \tfrac{1}{2}\int \cos^2\theta\,d\theta = \tfrac{1}{2}\int\left(\frac{1+\cos 2\theta}{2}\right)d\theta = \tfrac{1}{4}\int(1+\cos 2\theta)\,d\theta$

$$\boxed{\begin{array}{l} 4x^2 = \sin^2\theta \;\Rightarrow\; 1-4x^2 = 1-\sin^2\theta = \cos^2\theta \\ 2x = \sin\theta \;\Rightarrow\; x = \tfrac{1}{2}\sin\theta \;\Rightarrow\; dx = \tfrac{1}{2}\cos\theta\,d\theta \end{array}}$$

$$= \tfrac{1}{4}\left(\theta + \frac{\sin 2\theta}{2}\right) + C = \tfrac{1}{4}\left(\theta + \frac{2\sin\theta\cos\theta}{2}\right) + C = \tfrac{1}{4}\left(\sin^{-1} 2x + 2x\sqrt{1-4x^2}\right) + C$$

$$= \tfrac{1}{4}\sin^{-1} 2x + \tfrac{1}{2}x\sqrt{1-4x^2} + C.$$

55. $\int e^{2x}\sqrt{1-e^{2x}}\,dx = \int \sin^2\theta\cos\theta\cdot\dfrac{\cos\theta}{\sin\theta}\,d\theta = \int \cos^2\theta\sin\theta\,d\theta$

$$\boxed{\begin{aligned}&e^{2x}=\sin^2\theta \;\Rightarrow\; 1-e^{2x}=1-\sin^2\theta=\cos^2\theta\\ &e^x=\sin\theta \;\Rightarrow\; x=\ln e^x=\ln\sin\theta \;\Rightarrow\; dx=\frac{1}{\sin\theta}\cdot\cos\theta\,d\theta\end{aligned}}$$

$$= -\int u^2\,du = -\tfrac{1}{3}u^3 + C = -\tfrac{1}{3}\cos^3\theta + C = -\tfrac{1}{3}(1-e^{2x})^{\frac{3}{2}} + C.$$

$$\boxed{u=\cos\theta,\ du=-\sin\theta\,d\theta}$$

57. $21+4x-x^2 = -(x^2-4x+4)+4+21 = 25-(x-2)^2 \;\Rightarrow$

$\int\sqrt{21+4x-x^2}\,dx = \int\sqrt{25-(x-2)^2}\,dx = \int 5\cos\theta\cdot 5\cos\theta\,d\theta = 25\int\cos^2\theta\,d\theta$

$$\boxed{\begin{aligned}&(x-2)^2=25\sin^2\theta \;\Rightarrow\; 25-(x-2)^2=25-25\sin^2\theta=25\cos^2\theta\\ &x-2=5\sin\theta \;\Rightarrow\; x=2+5\sin\theta \;\Rightarrow\; dx=5\cos\theta\,d\theta\end{aligned}}$$

$$= 25\int\left(\frac{1+\cos 2\theta}{2}\right)d\theta = \frac{25}{2}[\theta + \tfrac{1}{2}\sin 2\theta] + C = \frac{25}{2}[\theta + \tfrac{1}{2}(2\sin\theta\cos\theta)]$$

$$= \frac{25}{2}\left[\sin^{-1}\left(\frac{x-2}{5}\right) + \left(\frac{x-2}{5}\right)\left(\frac{\sqrt{25-(x-2)^2}}{5}\right)\right] + C$$

$$= \frac{25}{2}\sin^{-1}\left(\frac{x-2}{5}\right) + \left(\frac{x-2}{2}\right)\sqrt{21+4x-x^2} + C.$$

59. $x^2+x+1 = [x^2+x+(\tfrac{1}{2})^2] - \tfrac{1}{4}+1 = (x+\tfrac{1}{2})^2 + \tfrac{3}{4} = \left(\frac{2x+1}{2}\right)^2 + \tfrac{3}{4} = \tfrac{1}{4}[(2x+1)^2+3]$

$$\Rightarrow \sqrt{x^2+x+1} = \tfrac{1}{2}\sqrt{(2x+1)^2+3} = \tfrac{1}{2}\sqrt{3\left[1+\left(\frac{2x+1}{\sqrt{3}}\right)^2\right]}$$

$$\int\frac{x}{\sqrt{x^2+x+1}}\,dx = \int\frac{x}{\frac{\sqrt{3}}{2}\sqrt{1+\left(\frac{2x+1}{\sqrt{3}}\right)^2}}\,dx = \int\frac{\frac{\sqrt{3}\tan\theta-1}{2}\cdot\frac{\sqrt{3}}{2}\sec^2\theta\,d\theta}{\frac{\sqrt{3}}{2}\sec\theta}$$

$$\boxed{\begin{aligned}&\frac{2x+1}{\sqrt{3}}=\tan\theta \;\Rightarrow\; 2x+1=\sqrt{3}\tan\theta \;\Rightarrow\\ &x=\frac{\sqrt{3}\tan\theta-1}{2} \;\Rightarrow\; dx=\frac{\sqrt{3}}{2}\sec^2\theta\,d\theta\end{aligned}}$$

$$= \int\frac{\sqrt{3}}{2}\tan\theta\sec\theta\,d\theta - \frac{1}{2}\int\sec\theta\,d\theta$$

$$= \frac{\sqrt{3}}{2}\sec\theta - \frac{1}{2}\ln|\sec\theta+\tan\theta| + C$$

$$= \frac{\sqrt{3}}{2}\cdot\frac{2\sqrt{x^2+x+1}}{\sqrt{3}} - \frac{1}{2}\ln\left|\frac{\sqrt{x^2+x+1}}{\sqrt{3}} + \frac{2x+1}{\sqrt{3}}\right| + C$$

$$= \sqrt{x^2+x+1} - \frac{1}{2}(\ln|\sqrt{x^2+x+1}+2x+1| - \ln\sqrt{3}) + C$$

$$= \sqrt{x^2+x+1} - \frac{1}{2}\ln|\sqrt{x^2+x+1}+2x+1| + C$$

(after absorbing the constant term into C)

61. First divide the denominator into the numerator:

$$\int \frac{2x^3 - 8x^2 + 20x - 5}{x^2 - 4x + 8} \, dx = \int \left(2x + \frac{4x - 5}{x^2 - 4x + 8}\right) dx = x^2 + \int \frac{4x - 5}{x^2 - 4x + 8} \, dx.$$

Then complete the square in the denominator:

$$x^2 - 4x + 8 = x^2 - 4x + 4 + 4 = (x - 2)^2 + 4 = 4\left[1 + \left(\tfrac{x-2}{2}\right)^2\right]$$

$$\int \frac{4x - 5}{x^2 - 4x + 8} \, dx = \int \frac{4x - 5}{4\left[1 + \left(\frac{x-2}{2}\right)^2\right]} \, dx = \int \frac{4(2\tan\theta + 2) - 5}{\cancel{4}\sec^2\theta \,_2} \cdot \cancel{2}\sec^2\theta \, d\theta$$

$$\boxed{\frac{x-2}{2} = \tan\theta \;\Rightarrow\; x = 2\tan\theta + 2 \;\Rightarrow\; dx = 2\sec^2\theta \, d\theta}$$

(triangle: hypotenuse $\sqrt{(x-2)^2 + 4} = x^2 - 4x + 8$, opposite $x - 2$, adjacent 2, angle θ)

$$= \tfrac{1}{2}\int (8\tan\theta + 3) \, d\theta = 4\ln|\sec\theta| + \tfrac{3}{2}\theta + C$$

$$= 4\ln\left(\frac{\sqrt{x^2 - 4x + 8}}{2}\right) + \tfrac{3}{2}\tan^{-1}\left(\tfrac{x-2}{2}\right) + C$$

$$= 4[\ln(x^2 - 4x + 8)^{\frac{1}{2}} - \cancel{\ln 2}] + \tfrac{3}{2}\tan^{-1}\left(\tfrac{x-2}{2}\right) + C$$

$$= 2\ln(x^2 - 4x + 8) + \tfrac{3}{2}\tan^{-1}\left(\tfrac{x-2}{2}\right) + C$$

(after absorbing the constant term into C)

Therefore, $\displaystyle\int \frac{2x^3 - 8x^2 + 20x - 5}{x^2 - 4x + 8} \, dx = x^2 + 2\ln(x^2 - 4x + 8) + \tfrac{3}{2}\tan^{-1}\left(\tfrac{x-2}{2}\right) + C.$

63. $\displaystyle\int_{2\sqrt{2}}^{4} \frac{4}{\sqrt{x^2 - 4}} \, dx = 4\int_{\frac{\pi}{4}}^{\frac{\pi}{3}} \frac{2\sec\theta\tan\theta \, d\theta}{2\tan\theta} = 4\int_{\frac{\pi}{4}}^{\frac{\pi}{3}} \sec\theta \, d\theta = 4\ln|\sec\theta + \tan\theta|\,\Big|_{\frac{\pi}{4}}^{\frac{\pi}{3}}$

$$\boxed{\begin{aligned} x^2 &= 4\sec^2\theta \;\Rightarrow\; x = 2\sec\theta \;\Rightarrow\; dx = 2\sec\theta\tan\theta \, d\theta \\ x &= 2\sqrt{2} \Rightarrow 2\sec\theta = 2\sqrt{2} \Rightarrow \cos\theta = \tfrac{1}{\sqrt{2}} \Rightarrow \theta = \tfrac{\pi}{4} \\ x &= 4 \Rightarrow 2\sec\theta = 4 \Rightarrow \cos\theta = \tfrac{1}{2} \Rightarrow \theta = \tfrac{\pi}{3} \end{aligned}}$$

(triangles: $30°$-$60°$ with sides 2, $\sqrt{3}$, 1; $45°$-$45°$ with sides $\sqrt{2}$, 1, 1)

$$= 4\left[\ln\left|\sec\tfrac{\pi}{3} + \tan\tfrac{\pi}{3}\right| - \ln\left|\sec\tfrac{\pi}{4} + \tan\tfrac{\pi}{4}\right|\right]$$

$$= 4[\ln|2 + \sqrt{3}| - \ln|\sqrt{2} + 1|] = 4\ln\left(\frac{2 + \sqrt{3}}{1 + \sqrt{2}}\right).$$

65. $\displaystyle\int_0^1 \frac{3}{\sqrt{16 + 9x^2}} \, dx = \cancel{3} \cdot \frac{\cancel{4}}{\cancel{3}}\int_0^{\tan^{-1}\frac{3}{4}} \frac{\sec^{\cancel{2}}\theta \, d\theta}{\cancel{4}\sec\theta} = \int_0^{\tan^{-1}\frac{3}{4}} \sec\theta \, d\theta = \ln|\sec\theta + \tan\theta|\,\Big|_0^{\tan^{-1}\frac{3}{4}}$

$$\boxed{\begin{aligned} 9x^2 &= 16\tan^2\theta \;\Rightarrow\; x = \tfrac{4}{3}\tan\theta \;\Rightarrow\; dx = \tfrac{4}{3}\sec^2\theta \, d\theta \\ x &= 0 \Rightarrow \tfrac{4}{3}\tan\theta = 0 \Rightarrow \tan\theta = 0 \Rightarrow \theta = 0 \\ x &= 1 \Rightarrow \tfrac{4}{3}\tan\theta = 1 \Rightarrow \tan\theta = \tfrac{3}{4} \Rightarrow \theta = \tan^{-1}\tfrac{3}{4} \end{aligned}}$$

(triangle: hypotenuse 5, opposite 3, adjacent 4, angle θ)

$$= \ln\left|\sec\left(\tan^{-1}\tfrac{3}{4}\right) + \tan\left(\tan^{-1}\tfrac{3}{4}\right)\right| - \ln|\sec 0 + \tan 0|$$

$$= \ln\left|\tfrac{5}{4} + \tfrac{3}{4}\right| - \ln|1 + 0| = \ln 2.$$

67. $\displaystyle\int_{1}^{\frac{4}{3}} \frac{10}{(25x^2-16)^{\frac{3}{2}}}\,dx = 10\int_{\cos^{-1}\frac{4}{5}}^{\cos^{-1}\frac{3}{5}} \frac{\frac{4}{5}\sec\theta\tan\theta\,d\theta}{(4\tan\theta)^3} = \frac{1}{8}\int_{\cos^{-1}\frac{4}{5}}^{\cos^{-1}\frac{3}{5}} \frac{\sec\theta}{\tan^2\theta}\,d\theta$

$$\boxed{\begin{array}{l} 25x^2 = 16\sec^2\theta \;\Rightarrow\; x = \frac{4}{5}\sec\theta \;\Rightarrow\; dx = \frac{4}{5}\sec\theta\tan\theta\,d\theta \\[4pt] x = 1 \;\Rightarrow\; \frac{4}{5}\sec\theta = 1 \;\Rightarrow\; \cos\theta = \frac{4}{5} \;\Rightarrow\; \theta = \cos^{-1}\frac{4}{5} \\[4pt] x = \frac{4}{3} \;\Rightarrow\; \frac{4}{5}\sec\theta = \frac{4}{3} \;\Rightarrow\; \cos\theta = \frac{3}{5} \;\Rightarrow\; \theta = \cos^{-1}\frac{3}{5} \end{array}}$$

$\theta = \cos^{-1}\frac{3}{5}$ $\theta = \cos^{-1}\frac{4}{5}$

$$= \frac{1}{8}\int_{\cos^{-1}\frac{4}{5}}^{\cos^{-1}\frac{3}{5}} \left(\frac{1}{\cos\theta}\cdot\frac{\cos^2\theta}{\sin^2\theta}\right) d\theta = \frac{1}{8}\int_{\cos^{-1}\frac{4}{5}}^{\cos^{-1}\frac{3}{5}} \frac{\cos\theta}{\sin^2\theta}\,d\theta$$

$$\boxed{\begin{array}{ll} u = \sin\theta, & \theta = \cos^{-1}\frac{4}{5} \;\Rightarrow\; u = \sin(\cos^{-1}\frac{4}{5}) = \frac{3}{5} \\[4pt] du = \cos\theta\,d\theta, & \theta = \cos^{-1}\frac{3}{5} \;\Rightarrow\; u = \sin(\cos^{-1}\frac{3}{5}) = \frac{4}{5} \end{array}}$$

$$= \frac{1}{8}\int_{\frac{3}{5}}^{\frac{4}{5}} u^{-2}\,du = \frac{1}{8}\cdot\frac{u^{-1}}{-1}\Big|_{\frac{3}{5}}^{\frac{4}{5}} = -\frac{1}{8}\left(\frac{1}{u}\right)\Big|_{\frac{3}{5}}^{\frac{4}{5}} = -\frac{1}{8}\left(\frac{1}{\frac{4}{5}} - \frac{1}{\frac{3}{5}}\right)$$

$$= -\frac{1}{8}\left(\frac{5}{4} - \frac{5}{3}\right) = -\frac{5}{8}\left(-\frac{1}{12}\right) = \frac{5}{96}.$$

69. Since $y = \dfrac{\sqrt{9-x^2}}{3} \;\Rightarrow\; x^2 + 9y^2 = 9$, is the top half of the ellipse centered at the origin, the area enclosed is:

$$A = \int_{-3}^{3} \frac{\sqrt{9-x^2}}{3}\,dx = \int_{-\frac{\pi}{2}}^{\frac{\pi}{2}} \frac{\cancel{3}\cos\theta\cdot 3\cos\theta\,d\theta}{\cancel{3}} = 3\int_{-\frac{\pi}{2}}^{\frac{\pi}{2}}\cos^2\theta\,d\theta = \frac{3}{2}\int_{-\frac{\pi}{2}}^{\frac{\pi}{2}}(1+\cos 2\theta)\,d\theta$$

$$\boxed{\begin{array}{ll} x^2 = 9\sin^2\theta \;\Rightarrow\; x = 3\sin\theta, & x = -3 \;\Rightarrow\; 3\sin\theta = -3 \;\Rightarrow\; \sin\theta = -1 \;\Rightarrow\; \theta = -\frac{\pi}{2} \\[4pt] dx = 3\cos\theta\,d\theta, & x = 3 \;\Rightarrow\; 3\sin\theta = 3 \;\Rightarrow\; \sin\theta = 1 \;\Rightarrow\; \theta = \frac{\pi}{2} \end{array}}$$

$$= \frac{3}{2}\left[\theta + \frac{1}{2}\sin 2\theta\right]\Big|_{-\frac{\pi}{2}}^{\frac{\pi}{2}} = \frac{3}{2}\left[\frac{\pi}{2} + \frac{1}{2}\sin\pi - \left(-\frac{\pi}{2} + \frac{1}{2}\sin(-\pi)\right)\right] = \frac{3\pi}{2}.$$

71.

$A = \int_0^{2\pi} |\sin x - \sin^2 x|\,dx$

$= \int_0^{\pi}(\sin x - \sin^2 x)\,dx + \int_{\pi}^{2\pi}(\sin^2 x - \sin x)\,dx$

$= -\cos x\Big|_0^{\pi} - \frac{1}{2}\int_0^{\pi}(1-\cos 2x)\,dx + \frac{1}{2}\int_{\pi}^{2\pi}(1-\cos 2x)\,dx + \cos x\Big|_{\pi}^{2\pi}$

$= -\cos\pi + \cos 0 - \frac{1}{2}\left(x - \frac{1}{2}\sin 2x\right)\Big|_0^{\pi} + \frac{1}{2}\left(x - \frac{1}{2}\sin 2x\right)\Big|_{\pi}^{2\pi}$

$\hspace{8cm} + \cos 2\pi - \cos\pi$

$= 1 + 1 - \frac{\pi}{2} + \frac{1}{2}(2\pi - \pi) + 1 + 1 = 4.$

73. $V = \pi \int_0^2 \left(\dfrac{4}{x^2 + 4} \right)^2 dx = 16\pi \int_0^2 \dfrac{1}{(x^2 + 4)^2}\, dx = 16\pi \int_0^{\frac{\pi}{4}} \dfrac{2 \sec^2 \theta}{16 \sec^4 \theta}\, d\theta = 2\pi \int_0^{\frac{\pi}{4}} \cos^2 \theta\, d\theta$

$$\begin{array}{c} x^2 = 4\tan^2 \theta \;\Rightarrow\; x = 2\tan\theta \;\Rightarrow\; dx = 2\sec^2\theta\, d\theta \\ x = 0 \;\Rightarrow\; 2\tan\theta = 0 \;\Rightarrow\; \tan\theta = 0 \;\Rightarrow\; \theta = 0 \\ x = 2 \;\Rightarrow\; 2\tan\theta = 2 \;\Rightarrow\; \tan\theta = 1 \;\Rightarrow\; \theta = \frac{\pi}{4} \end{array}$$

$= \pi \int_0^{\frac{\pi}{4}} (1 + \cos 2\theta)\, d\theta = \pi \left[\theta + \dfrac{\sin 2\theta}{2} \right] \Big|_0^{\frac{\pi}{4}} = \pi \left[\dfrac{\pi}{4} + \dfrac{\sin \frac{\pi}{2}}{2} \right] = \dfrac{\pi^2}{4} + \dfrac{\pi}{2}.$

75.

By symmetry, $V = 2 \cdot \pi \int_0^r \left[(R + \sqrt{r^2 - x^2})^2 - (R - \sqrt{r^2 - x^2})^2 \right] dx$

$$\begin{array}{ll} x = r\sin\theta, & x = 0 \;\Rightarrow\; r\sin\theta = 0 \;\Rightarrow\; \theta = 0 \\ dx = r\cos\theta\, d\theta, & x = r \;\Rightarrow\; r\sin\theta = r \;\Rightarrow\; \sin\theta = 1 \;\Rightarrow\; \theta = \frac{\pi}{2} \end{array}$$

$V = 2\pi \int_0^{\frac{\pi}{2}} \left[(R + r\cos\theta)^2 - (R - r\cos\theta)^2 \right] r\cos\theta\, d\theta$

$= 2\pi \int_0^{\frac{\pi}{2}} \left[(R^2 + 2Rr\cos\theta + r^2\cos^2\theta) - (R^2 - 2Rr\cos\theta + r^2\cos^2\theta) \right] r\cos\theta\, d\theta$

$= 2\pi \int_0^{\frac{\pi}{2}} 4Rr\cos\theta \cdot r\cos\theta\, d\theta = 8\pi Rr^2 \int_0^{\frac{\pi}{2}} \cos^2\theta\, d\theta = 4\pi Rr^2 \int_0^{\frac{\pi}{2}} (1 + \cos 2\theta)\, d\theta$

$= 4\pi Rr^2 [\theta + \tfrac{1}{2}\sin 2\theta] \Big|_0^{\frac{\pi}{2}} = 4\pi Rr^2 \cdot \dfrac{\pi}{2} = 2\pi^2 Rr^2.$

77. $\int \tan^n x \sec^m x\, dx = \int \tan x \tan^{n-1} x \sec x \sec^{m-1} x\, dx = \int \sec^{m-1} x \tan^{n-1} x \sec x \tan x\, dx.$ It remains to show that $\tan^{n-1} x = (\sec^2 x - 1)^{\frac{n-1}{2}}$: As n is an odd positive integer, $n - 1$ is even, say $n - 1 = 2k$. From $\tan^2 x = \sec^2 x - 1$, then $\tan^{n-1} x = \tan^{2k} x = (\tan^2 x)^k = (\sec^2 x - 1)^k = (\sec^2 x - 1)^{\frac{n-1}{2}}.$

§7.4 A Hodgepodge of integrals

1. $\int \dfrac{x^2 - 3x + 2}{x^2}\, dx = \int (1 - \tfrac{3}{x} + 2x^{-2})\, dx = x - 3\ln|x| - 2x^{-1} + C = x - 3\ln|x| - \tfrac{2}{x} + C.$

3. First, divide the denominator into the numerator:

$\int \dfrac{x^2}{x^2 + 2x + 5}\, dx = \int \left(1 - \dfrac{2x + 5}{x^2 + 2x + 5} \right) dx = x - \int \dfrac{2x + 5}{x^2 + 2x + 5}\, dx$

Then complete the square in the denominator:

$x^2 + 2x + 5 = x^2 + 2x + 1 + 4 = (x + 1)^2 + 4 = 4 \left[1 + \left(\dfrac{x+1}{2} \right)^2 \right]$

$$\int \frac{2x+5}{x^2+2x+5}\,dx = \frac{1}{4}\int \frac{2x+5}{1+(\frac{x+1}{2})^2}\,dx = \frac{1}{4}\int \frac{2(2\tan\theta-1)+5}{\sec^2\theta}\cdot 2\sec^2\theta\,d\theta$$

$$\boxed{\begin{array}{c} \frac{x+1}{2} = \tan\theta \;\Rightarrow\; x = 2\tan\theta - 1 \\ dx = 2\sec^2\theta\,d\theta \end{array}}$$

$$= \frac{1}{2}\int (4\tan\theta + 3)\,d\theta = 2\ln|\sec\theta| + \frac{3}{2}\theta + C$$

$$= 2\ln\frac{\sqrt{x^2+2x+5}}{2} + \frac{3}{2}\tan^{-1}(\frac{x+1}{2}) + C$$

$$= \ln(\sqrt{x^2+2x+5})^2 - 2\ln 2 + \frac{3}{2}\tan^{-1}(\frac{x+1}{2}) + C$$

$$= \ln(x^2+2x+5) + \frac{3}{2}\tan^{-1}(\frac{x+1}{2}) + C \quad \text{(after absorbing } -2\ln 2 \text{ into } C)$$

Therefore, $\int \frac{x^2}{x^2+2x+5}\,dx = x - \ln(x^2+2x+5) - \frac{3}{2}\tan^{-1}(\frac{x+1}{2}) + C.$

5. $\int \frac{dx}{1+3x^2} = \int \frac{dx}{1+(\sqrt{3}x)^2} = \frac{1}{\sqrt{3}}\int \frac{du}{1+u^2} = \frac{1}{\sqrt{3}}\tan^{-1}u + C = \frac{1}{\sqrt{3}}\tan^{-1}\sqrt{3}x + C.$

$$\boxed{u = \sqrt{3}x,\; du = \sqrt{3}dx \;\Rightarrow\; dx = \frac{1}{\sqrt{3}}du}$$

7.
$$\int u\,dv = uv - \int v\,du$$
$$\int x^5 e^{-x^3}\,dx = \int x^3(x^2 e^{-x^3}\,dx) = -\frac{1}{3}x^3 e^{-x^3} + \int x^2 e^{-x^3}\,dx = -\frac{1}{3}x^3 e^{-x^3} - \frac{1}{3}e^{-x^3} + C$$

$$\boxed{\begin{array}{cc} u = x^3 & dv = x^2 e^{-x^3}\,dx \\ du = 3x^2\,dx & v = \frac{e^{-x^3}}{-3} \end{array}}$$

$$= -\frac{1}{3}e^{-x^3}(x^3+1) + C.$$

9. $\int \frac{e^x}{\sqrt{1-e^{2x}}}\,dx = \int \frac{du}{\sqrt{1-u^2}} = \sin^{-1}u + C = \sin^{-1}e^x + C.$

$$\boxed{u = e^x,\; du = e^x\,dx}$$

11.
$$\int u\,dv = uv - \int v\,du \qquad\qquad \int u\,dv = uv - \int v\,du$$
$$\int \sin(\ln 2x)\,dx = x\sin(\ln 2x) - \int \cos(\ln 2x)\,dx \qquad \int \cos(\ln 2x)\,dx = x\cos(\ln 2x) + \int \sin(\ln 2x)\,dx$$

$$\boxed{\begin{array}{cc} u = \sin(\ln 2x) & dv = dx \\ du = \cos(\ln 2x)\cdot\frac{1}{2x}\cdot 2\,dx & v = x \end{array}} \qquad \boxed{\begin{array}{cc} u = \cos(\ln 2x) & dv = dx \\ du = -\sin(\ln 2x)\cdot\frac{1}{2x}\cdot 2\,dx & v = x \end{array}}$$

Thus, $\int \sin(\ln 2x)\,dx = x\sin(\ln 2x) - [x\cos(\ln 2x) + \int \sin(\ln 2x)\,dx]$
$$= x\sin(\ln 2x) - x\cos(\ln 2x) - \int \sin(\ln 2x)\,dx$$
$$2\int \sin(\ln 2x)\,dx = x\sin(\ln 2x) - x\cos(\ln 2x) + C$$
$$\int \sin(\ln 2x)\,dx = \frac{x}{2}[\sin(\ln 2x) - \cos(\ln 2x)] + C.$$

13. $\int \dfrac{e^x}{e^{2x}+e^x-2}\,dx = \int \dfrac{du}{u^2+u-2} = \int \dfrac{du}{(u-1)(u+2)} = \int \dfrac{\frac{1}{3}}{u-1}\,du + \int \dfrac{-\frac{1}{3}}{u+2}\,du$

$$\boxed{\begin{array}{l} u = e^x \\ du = e^x\,dx \end{array}}$$

$$\boxed{\begin{array}{c} \dfrac{1}{(u-1)(u+2)} = \dfrac{A}{u-1} + \dfrac{B}{u+2} \\ 1 = A(u+2) + B(u-1) \\ x=1 : 1 = 3A \;\Rightarrow\; A = \frac{1}{3} \\ x=-2 : 1 = -3B \;\Rightarrow\; B = -\frac{1}{3} \end{array}}$$

$$= \tfrac{1}{3}\ln|u-1| - \tfrac{1}{3}\ln|u+2| + C = \tfrac{1}{3}\ln\left|\dfrac{u-1}{u+2}\right| + C = \tfrac{1}{3}\ln\dfrac{|e^x-1|}{e^x+2} + C.$$

15. $\int (x+2)\sqrt{x-5}\,dx = \int (u^2+7)u \cdot 2u\,du = 2\int(u^4+7u^2)\,du = 2\left(\tfrac{1}{5}u^5 + \tfrac{7}{3}u^3\right) + C$

$$\boxed{\begin{array}{l} u = \sqrt{x-5} \;\Rightarrow\; u^2 = x-5 \;\Rightarrow\; x = u^2+5 \\ dx = 2u\,du \end{array}}$$

$$= 2u^3\left(\tfrac{1}{5}u^2 + \tfrac{7}{3}\right) + C = 2(x-5)^{\frac{3}{2}}\left[\tfrac{1}{5}(x-5) + \tfrac{7}{3}\right] + C$$

$$= 2(x-5)^{\frac{3}{2}}\left(\dfrac{3x-15+35}{15}\right) + C = \tfrac{2}{15}(x-5)^{\frac{3}{2}}(3x+20) + C.$$

17. $\int \dfrac{\sqrt{x}}{1-\sqrt[3]{x}}\,dx = \int \dfrac{u^3 \cdot 6u^5\,du}{1-u^2} = 6\int \dfrac{u^8}{1-u^2}\,du$

$$\boxed{\begin{array}{l} u = x^{\frac{1}{6}} \;\Rightarrow\; u^6 = x \\ dx = 6u^5\,du \end{array}}$$

$$\begin{array}{r} -u^6-u^4-u^2-1 \\ \hline -u^2+1\,\overline{)\,u^8} \\ \underline{u^8-u^6} \\ u^6 \\ \underline{u^6-u^4} \\ u^4 \\ \underline{u^4-u^2} \\ u^2 \\ \underline{u^2-1} \\ 1 \end{array}$$

$$= 6\int\left(-u^6 - u^4 - u^2 - 1 + \dfrac{1}{1-u^2}\right)du$$

$$= -6\left(\tfrac{1}{7}u^7 + \tfrac{1}{5}u^5 + \tfrac{1}{3}u^3 + u\right) + 6\int \dfrac{du}{1-u^2}$$

$$\boxed{\begin{array}{l} u = \sin\theta \\ du = \cos\theta\,d\theta \end{array}}$$

$$\int \dfrac{du}{1-u^2} = \int \dfrac{\cos\theta\,d\theta}{\cos^2\theta} = \int \sec\theta\,d\theta = \ln|\sec\theta + \tan\theta| + C$$

$$= \ln\left|\dfrac{1}{\sqrt{1-u^2}} + \dfrac{u}{\sqrt{1-u^2}}\right| + C = \ln|1+u| - \tfrac{1}{2}\ln|(1+u)(1-u)| + C$$

$$= \ln|1+u| - \tfrac{1}{2}\ln|1+u| - \tfrac{1}{2}\ln|1-u| + C = \tfrac{1}{2}\ln\left|\dfrac{1+u}{1-u}\right| + C$$

Thus, $\int \dfrac{\sqrt{x}}{1-\sqrt[3]{x}}\,dx = -6\left(\tfrac{1}{7}u^7 + \tfrac{1}{5}u^5 + \tfrac{1}{3}u^3 + u\right) + 3\ln\left|\dfrac{1+u}{1-u}\right| + C$

$$= -\tfrac{6}{7}x^{\frac{7}{6}} - \tfrac{6}{5}x^{\frac{5}{6}} - 2\sqrt{x} - 6x^{\frac{1}{6}} + 3\ln\left|\dfrac{1+x^{\frac{1}{6}}}{1-x^{\frac{1}{6}}}\right| + C.$$

19. $3x - x^2 = -\left[x^2 - 3x + \left(\frac{3}{2}\right)^2\right] + \frac{9}{4} = \frac{9}{4} - \left(x - \frac{3}{2}\right)^2 \Rightarrow$

$$\int \frac{dx}{\sqrt{3x - x^2}} = \int \frac{dx}{\sqrt{\frac{9}{4} - (x - \frac{3}{2})^2}} = \frac{3}{2} \int \frac{\cos\theta \, d\theta}{\frac{3}{2}\cos\theta} = \int d\theta = \theta + C = \sin^{-1}\left(\frac{2}{3}x - 1\right) + C.$$

$$\boxed{\begin{array}{c}\left(x - \frac{3}{2}\right)^2 = \frac{9}{4}\sin^2\theta \Rightarrow x = \frac{3}{2} + \frac{3}{2}\sin\theta = \frac{3}{2}(1 + \sin\theta) \\ dx = \frac{3}{2}\cos\theta \, d\theta\end{array}}$$

21. $\int \frac{\sqrt{x^2 - 1}}{x} dx = \int \frac{\tan\theta \cdot \sec\theta \tan\theta \, d\theta}{\sec\theta} = \int \tan^2\theta \, d\theta = \int (\sec^2\theta - 1) \, d\theta = \tan\theta - \theta + C$

$$\boxed{\begin{array}{c}x = \sec\theta \\ dx = \sec\theta\tan\theta \, d\theta\end{array}}$$

$$= \sqrt{x^2 - 1} - \tan^{-1}\sqrt{x^2 - 1} + C.$$

23. $\int \frac{x^2}{(x^2 + 9)^{\frac{3}{2}}} dx = \int \frac{9\tan^2\theta \cdot 3\sec^2\theta \, d\theta}{(3\sec\theta)^3} = \int \frac{\tan^2\theta}{\sec\theta} d\theta = \int \frac{\sec^2\theta - 1}{\sec\theta} d\theta$

$$\boxed{\begin{array}{c}x^2 = 9\tan^2\theta \Rightarrow x = 3\tan\theta \\ dx = 3\sec^2\theta \, d\theta\end{array}}$$

$$= \int \sec\theta \, d\theta - \int \cos\theta \, d\theta = \ln|\sec\theta + \tan\theta| - \sin\theta + C$$

$$= \ln\left|\frac{\sqrt{x^2 + 9}}{3} + \frac{x}{3}\right| - \frac{x}{\sqrt{x^2 + 9}} + C$$

$$= \ln|\sqrt{x^2 + 9} + x| - \ln 3 - \frac{x}{\sqrt{x^2 + 9}} + C \quad (\text{absorbing} - \ln 3 \text{ into } C)$$

$$= \ln|\sqrt{x^2 + 9} + x| - \frac{x}{\sqrt{x^2 + 9}} + C.$$

25. $x^2 - 2x + 10 = x^2 - 2x + 1 + 9 = (x - 1)^2 + 9 \Rightarrow$

$$\int \frac{dx}{(x^2 - 2x + 10)^{\frac{3}{2}}} = \int \frac{dx}{[(x - 1)^2 + 9]^{\frac{3}{2}}} = \int \frac{3\sec^2\theta \, d\theta}{(3\sec\theta)^3} = \frac{1}{9}\int \frac{d\theta}{\sec\theta} = \frac{1}{9}\int \cos\theta \, d\theta$$

$$\boxed{\begin{array}{c}(x - 1)^2 = 9\tan^2\theta \Rightarrow x = 1 + 3\tan\theta \\ dx = 3\sec^2\theta \, d\theta\end{array}}$$

$$= \frac{1}{9}\sin\theta \mid C = \frac{x - 1}{9\sqrt{x^2 - 2x + 10}} + C.$$

27. $4x^2 + 8x + 29 = 4(x^2 + 2x) + 29 = 4(x^2 + 2x + 1) - 4 + 29 = 4(x + 1)^2 + 25 \Rightarrow$

$$\int \frac{dx}{4x^2 + 8x + 29} = \int \frac{dx}{4(x+1)^2 + 25} = \int \frac{\frac{5}{2}\sec^2\theta\,d\theta}{25\sec^2\theta} = \frac{1}{10}\int d\theta = \frac{1}{10}\theta + C$$

$$\boxed{\begin{array}{c} 4(x+1)^2 = 25\tan^2\theta \Rightarrow 2(x+1) = 5\tan\theta \\ x = \frac{5}{2}\tan\theta - 1 \Rightarrow dx = \frac{5}{2}\sec^2\theta\,d\theta \end{array}}$$

$$= \frac{1}{10}\tan^{-1}\left(\frac{2x+2}{5}\right) + C.$$

29. $\int u\,dv = uv - \int v\,du$

$$\int \frac{\tan^{-1} x}{x^2}\,dx = -\frac{\tan^{-1} x}{x} + \int \frac{dx}{x(1+x^2)}$$

$$\boxed{\begin{array}{cc} u = \tan^{-1} x & dv = x^{-2}\,dx \\ du = \dfrac{1}{1+x^2}\,dx & v = -\dfrac{1}{x} \end{array}}$$

$$\frac{1}{x(1+x^2)} = \frac{A}{x} + \frac{Bx+C}{1+x^2}$$

$$1 = A(1+x^2) + (Bx+C)x$$

$$= A + Ax^2 + Bx^2 + Cx$$

$$= (A+B)x^2 + Cx + A$$

$$\text{const: } 1 = A$$

$$x : 0 = C$$

$$x^2 : 0 = A + B \Rightarrow 1 + B = 0 \Rightarrow B = -1$$

$$\int \frac{dx}{x(1+x^2)} = \int \frac{1}{x}\,dx - \int \frac{x}{1+x^2}\,dx = \ln|x| - \frac{1}{2}\int \frac{du}{u} = \ln|x| - \frac{1}{2}\ln|u| + C$$

$$\boxed{\begin{array}{c} u = 1 + x^2 \\ du = 2x\,dx \Rightarrow x\,dx = \frac{1}{2}\,du \end{array}}$$

$$= \ln|x| - \frac{1}{2}\ln(1+x^2) + C = \ln\frac{|x|}{\sqrt{1+x^2}} + C$$

Thus, $$\int \frac{\tan^{-1} x}{x^2}\,dx = -\frac{\tan^{-1} x}{x} + \ln\frac{|x|}{\sqrt{1+x^2}} + C$$

31. $\int u\,dv = uv - \int v\,du$

$\int \sec^3 x\,dx = \sec x\tan x - \int \sec x\tan^2 x\,dx = \sec x\tan x - \int \sec x(\sec^2 x - 1)\,dx$

$$\boxed{\begin{array}{cc} u = \sec x & dv = \sec^2 x\,dx \\ du = \sec x\tan x\,dx & v = \tan x \end{array}}$$

$$= \sec x\tan x - \int \sec^3 x\,dx + \int \sec x\,dx$$

$$2\int \sec^3 x\,dx = \sec x\tan x + \ln|\sec x + \tan x| + C$$

$$\int \sec^3 x\,dx = \frac{1}{2}\left[\sec x\tan x + \ln|\sec x + \tan x|\right] + C.$$

33. $\displaystyle\int \frac{1+\sin x}{1-\sin x}\,dx = \int \frac{1+\sin x}{1-\sin x}\cdot\frac{1+\sin x}{1+\sin x}\,dx = \int \frac{1+2\sin x+\sin^2 x}{1-\sin^2 x}\,dx$

$$= \int \frac{1+2\sin x + 1 - \cos^2 x}{\cos^2 x}\,dx = \int \frac{2}{\cos^2 x}\,dx + \int \frac{2\sin x}{\cos^2 x}\,dx - \int dx$$

$$= 2\int \sec^2 x\,dx + 2\int \sec x\tan x\,dx - x$$

$$= 2\tan x + 2\sec x - x + C.$$

35. $\displaystyle\int \frac{\sin x\cos 2x}{\sin x+\sec x}\,dx = \int \frac{\sin x\cos 2x}{\sin x + \frac{1}{\cos x}}\,dx = \int \frac{\sin x\cos x\cos 2x}{\sin x\cos x+1}\,dx = \int \frac{\frac{1}{2}\sin 2x\cos 2x}{\frac{1}{2}\sin 2x+1}\,dx$

$$= \int \frac{\sin 2x\cos 2x}{\sin 2x+2}\,dx = \int \frac{(u-2)\cdot\frac{1}{2}\,du}{u} = \int\left(\tfrac{1}{2}-\tfrac{1}{u}\right)du$$

$$\boxed{\begin{array}{l} u = \sin 2x + 2 \\ du = 2\cos 2x\,dx \;\Rightarrow\; \tfrac{1}{2}\,du = \cos 2x\,dx \end{array}}$$

$$= \tfrac{1}{2}u - \ln|u| + C = \tfrac{1}{2}(\sin 2x+2) - \ln|\sin 2x + 2| + C$$

$$= \tfrac{1}{2}(2\sin x\cos x) + \cancel{1} - \ln|2(\sin x\cos x+1)| + C$$

$$= \sin x\cos x - \cancel{\ln 2} - \ln|\sin x\cos x+1| + C \quad (\text{absorbing} -\ln 2\ \text{into } C)$$

$$= \sin x\cos x - \ln|\sin x\cos x+1| + C.$$

37. $\displaystyle\int \frac{\sec x}{2\tan x+\sec x-1}\,dx = \int \frac{\frac{1}{\cos x}}{2\cdot\frac{\sin x}{\cos x}+\frac{1}{\cos x}-1}\,dx = \int \frac{dx}{2\sin x-\cos x+1}.$

Let $u = \tan\frac{x}{2}$:

$$= \int \frac{\frac{2\,du}{u^2+1}}{\frac{4u}{u^2+1}-\frac{1-u^2}{u^2+1}+1} = \int \frac{2\,du}{4u-\cancel{1}+u^2+u^2+\cancel{1}} = \int \frac{du}{u^2+2u}$$

Apply the method of partial fractions:

$$\frac{1}{u^2+2u} = \frac{1}{u(u+2)} = \frac{A}{u}+\frac{B}{u+2}$$

$$1 = A(u+2) + Bu$$

$$u = 0 : 1 = 2A \;\Rightarrow\; A = \tfrac{1}{2}$$

$$u = -2 : 1 = -2B \;\Rightarrow\; B = -\tfrac{1}{2}$$

so that $\displaystyle\int \frac{\sec x}{2\tan x+\sec x-1}\,dx = \frac{1}{2}\left[\int \frac{du}{u}-\int \frac{du}{u+2}\right] = \tfrac{1}{2}\bigl[\ln|u|-\ln|u+2|\bigr]+C$

$$= \tfrac{1}{2}\ln\left|\frac{u}{u+2}\right| + C = \tfrac{1}{2}\ln\left|\frac{\tan\frac{x}{2}}{\tan\frac{x}{2}+2}\right| + C.$$

39. $\int_1^e \dfrac{dx}{x\sqrt{1+(\ln x)^2}} = \int_0^1 \dfrac{du}{\sqrt{1+u^2}} = \int_0^{\frac{\pi}{4}} \dfrac{\sec^2\theta\, d\theta}{\sec\theta} = \int_0^{\frac{\pi}{4}} \sec\theta\, d\theta = \ln|\sec\theta + \tan\theta|\,\Big|_0^{\frac{\pi}{4}}$

$$\boxed{\begin{array}{l} u = \ln x, \quad x = 1 \Rightarrow u = 0 \\ du = \frac{1}{x}\,dx, \;\; x = e \Rightarrow u = 1 \end{array}} \qquad \boxed{\begin{array}{ll} u = \tan\theta, & u = 0 \Rightarrow \tan\theta = 0 \Rightarrow \theta = 0 \\ du = \sec^2\theta\, d\theta, & u = 1 \Rightarrow \tan\theta = 1 \Rightarrow \theta = \frac{\pi}{4} \end{array}}$$

$$= \ln\left|\sec\frac{\pi}{4} + \tan\frac{\pi}{4}\right| - \ln|\sec 0 + \tan 0| = \ln(\sqrt{2}+1) - \ln 1 = \ln(\sqrt{2}+1).$$

41. $\int_0^1 \dfrac{dx}{x^3 + x^2 + x + 1} = \int_0^1 \dfrac{dx}{(x+1)(x^2+1)}.$ Apply the method of partial fractions:

$$\frac{1}{(x+1)(x^2+1)} = \frac{A}{x+1} + \frac{Bx+C}{x^2+1}$$

$$1 = A(x^2+1) + (Bx+C)(x+1)$$

$$x = -1 : 1 = 2A \;\Rightarrow\; A = \tfrac{1}{2}$$

$$x = 0 : 1 = A + C = \tfrac{1}{2} + C \;\Rightarrow\; C = \tfrac{1}{2}$$

$$x = 1 : 1 = 2A + 2B + 2C = 1 + 2B + 1 \;\Rightarrow\; 2B = -1 \;\Rightarrow\; B = -\tfrac{1}{2}$$

$$\int_0^1 \frac{dx}{x^3 + x^2 + x + 1} = \int_0^1 \left[\frac{\frac{1}{2}}{x+1} + \frac{-\frac{1}{2}x + \frac{1}{2}}{x^2+1}\right] dx$$

$$= \frac{1}{2}\left[\int_0^1 \frac{1}{x+1}\,dx - \int_0^1 \frac{x}{x^2+1}\,dx + \int_0^1 \frac{1}{1+x^2}\,dx\right]$$

$$\boxed{\begin{array}{l} u = x^2 + 1, \;\; x = 0 \;\Rightarrow\; u = 1 \\ du = 2x\,dx, \;\;\; x = 1 \;\Rightarrow\; u = 2 \end{array}}$$

$$= \frac{1}{2}\left[\ln|x+1|\,\Big|_0^1 - \frac{1}{2}\int_1^2 \frac{du}{u} + \tan^{-1}x\,\Big|_0^1\right]$$

$$= \frac{1}{2}\left[(\ln 2 - \ln 1) - \frac{1}{2}\ln|u|\,\Big|_1^2 + \tan^{-1}1 - \tan^{-1}0\right]$$

$$= \frac{1}{2}\left[\ln 2 - \frac{1}{2}(\ln 2 - \ln 1) + \frac{\pi}{4}\right] = \frac{1}{4}\ln 2 + \frac{\pi}{8}.$$

43. $\int_0^{\frac{\pi}{2}} \dfrac{dx}{2 + \cos x} = \int_0^1 \dfrac{\frac{2\,du}{u^2+1}}{2 + \frac{1-u^2}{u^2+1}} = \int_0^1 \dfrac{2\,du}{2u^2 + 2 + 1 - u^2} = \int_0^1 \dfrac{2\,du}{u^2 + 3} = \int_0^{\frac{\pi}{6}} \dfrac{2\sqrt{3}\sec^2\theta\, d\theta}{3\sec^2\theta}$

$$\boxed{\begin{array}{l} u = \tan\frac{x}{2} \\ x = 0 \;\Rightarrow\; u = \tan 0 = 0 \\ x = \frac{\pi}{2} \;\Rightarrow\; u = \tan\frac{\pi}{4} = 1 \end{array}} \qquad \boxed{\begin{array}{l} u^2 = 3\tan^2\theta \;\Rightarrow\; u = \sqrt{3}\tan\theta, u = 0 \;\Rightarrow\; \theta = 0 \\ du = \sqrt{3}\sec^2\theta, u = 1 \;\Rightarrow\; \tan\theta = \frac{1}{\sqrt{3}} \;\Rightarrow\; \theta = \frac{\pi}{6} \end{array}}$$

$$= \frac{2\sqrt{3}}{3}\int_0^{\frac{\pi}{6}} d\theta = \frac{2\sqrt{3}}{3}\theta\,\Big|_0^{\frac{\pi}{6}} = \frac{2\sqrt{3}}{3} \cdot \frac{\pi}{6} = \frac{\sqrt{3}\pi}{9}.$$

45. $\int u \, dv = uv - \int v \, du$

$\int_0^2 xe^{2x} \, dx = \frac{1}{2}xe^{2x}\Big|_0^2 - \frac{1}{2}\int_0^2 e^{2x} \, dx = \frac{1}{2}\cdot 2e^4 - \frac{1}{2}\left(\frac{e^{2x}}{2}\right)\Big|_0^2 = e^4 - \frac{1}{4}(e^4 - e^0) = \frac{3}{4}e^4 + \frac{1}{4} = \frac{1}{4}(3e^4 + 1).$

$$\boxed{\begin{array}{ll} u = x & dv = e^{2x}\,dx \\ du = dx & v = \frac{1}{2}e^{2x} \end{array}}$$

47. $\int u \, dv = uv - \int v \, du$

$\int_0^{\frac{\pi}{2}} x \sin 4x \, dx = -\frac{1}{4}x\cos 4x\Big|_0^{\frac{\pi}{2}} + \frac{1}{4}\int_0^{\frac{\pi}{2}} \cos 4x \, dx = -\frac{\pi}{8}\cos 2\pi + \frac{1}{4}\left(\frac{\sin 4x}{4}\right)\Big|_0^{\frac{\pi}{2}}$

$$\boxed{\begin{array}{ll} u = x & dv = \sin 4x \, dx \\ du = dx & v = -\frac{1}{4}\cos 4x \end{array}}$$

$$= -\frac{\pi}{8} + \frac{1}{16}(\sin 2\pi - \sin 0) = -\frac{\pi}{8}.$$

49. $\int_0^{\frac{\pi}{6}} \tan^2 2x \, dx = \int_0^{\frac{\pi}{6}} (\sec^2 2x - 1) \, dx = \left[\frac{1}{2}\tan 2x - x\right]_0^{\frac{\pi}{6}} = \frac{1}{2}\tan\frac{\pi}{3} - \frac{\pi}{6}$

$$= \frac{\sqrt{3}}{2} - \frac{\pi}{6} = \frac{1}{6}(3\sqrt{3} - \pi).$$

51. $\displaystyle\int_{\sqrt{2}}^2 \frac{dx}{x^2\sqrt{x^2-1}} = \int_{\frac{\pi}{4}}^{\frac{\pi}{3}} \frac{\sec\theta\tan\theta\,d\theta}{\sec^2\theta\tan\theta} = \int_{\frac{\pi}{4}}^{\frac{\pi}{3}} \cos\theta\,d\theta = \sin\theta\Big|_{\frac{\pi}{4}}^{\frac{\pi}{3}} = \sin\frac{\pi}{3} - \sin\frac{\pi}{4} = \frac{\sqrt{3}}{2} - \frac{1}{\sqrt{2}}$

$$\boxed{\begin{array}{ll} x = \sec\theta & x = \sqrt{2} \Rightarrow \sec\theta = \sqrt{2} \Rightarrow \cos\theta = \frac{1}{\sqrt{2}} \Rightarrow \theta = \frac{\pi}{4} \\ dx = \sec\theta\tan\theta\,d\theta & x = 2 \Rightarrow \sec\theta = 2 \Rightarrow \cos\theta = \frac{1}{2} \Rightarrow \theta = \frac{\pi}{3} \end{array}}$$

$$= \frac{\sqrt{3}}{2} - \frac{\sqrt{2}}{2} = \frac{1}{2}(\sqrt{3} - \sqrt{2}).$$

53. $\displaystyle\int_{\frac{\pi}{2}}^{\frac{3\pi}{4}} \frac{dx}{1 - \cos x} = \int_{\frac{\pi}{2}}^{\frac{3\pi}{4}} \frac{1}{1 - \cos x} \cdot \frac{1 + \cos x}{1 + \cos x} \, dx = \int_{\frac{\pi}{2}}^{\frac{3\pi}{4}} \frac{1 + \cos x}{1 - \cos^2 x} \, dx = \int_{\frac{\pi}{2}}^{\frac{3\pi}{4}} \frac{1 + \cos x}{\sin^2 x} \, dx$

$$= \int_{\frac{\pi}{2}}^{\frac{3\pi}{4}} (\csc^2 x + \csc x \cot x) \, dx = [-\cot x - \csc x]\Big|_{\frac{\pi}{2}}^{\frac{3\pi}{4}}$$

$$= -\cot\frac{3\pi}{4} - \csc\frac{3\pi}{4} - \left(-\cot\frac{\pi}{2} - \csc\frac{\pi}{2}\right)$$

$$= -(-1) - \sqrt{2} - (0 - 1) = 1 - \sqrt{2} + 1 = 2 - \sqrt{2}.$$

55. $\displaystyle\int_{-\frac{\pi}{10}}^0 \sin 2x \cos 3x \, dx = \frac{1}{2}\int_{-\frac{\pi}{10}}^0 [\sin(2x + 3x) + \sin(2x - 3x)] \, dx = \frac{1}{2}\int_{-\frac{\pi}{10}}^0 [\sin 5x - \sin x] \, dx$

$$= \frac{1}{2}\left[-\frac{1}{5}\cos 5x + \cos x\right]\Big|_{-\frac{\pi}{10}}^0$$

$$- \frac{1}{2}\left\{-\frac{1}{5}\cos 0 + \cos 0 - \left[-\frac{1}{5}\cos\left(-\frac{\pi}{2}\right) + \cos\left(-\frac{\pi}{10}\right)\right]\right\}$$

$$= \frac{1}{2}\left(\frac{4}{5} - \cos\frac{\pi}{10}\right) = \frac{2}{5} - \frac{1}{2}\cos\frac{\pi}{10} \qquad \text{(See note at the end of this section.)}$$

57. $\int_0^{\frac{\pi}{3}} \sin^3 3x \, dx = \int_0^{\frac{\pi}{3}} \sin^2 3x \sin 3x \, dx = \int_0^{\frac{\pi}{3}} (1 - \cos^2 3x) \sin 3x \, dx$

$$= \int_0^{\frac{\pi}{3}} \sin 3x \, dx - \int_0^{\frac{\pi}{3}} \cos^2 3x \sin 3x \, dx = -\frac{1}{3} \cos 3x \Big|_0^{\frac{\pi}{3}} + \frac{1}{3} \int_1^{-1} u^2 \, du$$

$$\boxed{\begin{array}{ll} u = \cos 3x & x = 0 \Rightarrow u = \cos 0 = 1 \\ du = -3 \sin 3x \, dx & x = \frac{\pi}{3} \Rightarrow u = \cos \pi = -1 \\ -\frac{1}{3} du = \sin 3x \, dx & \end{array}}$$

$$= -\frac{1}{3}(\cos \pi - \cos 0) + \frac{1}{9} u^3 \Big|_1^{-1} = -\frac{1}{3}(-1 - 1) + \frac{1}{9}(-1 - 1) = \frac{2}{3} - \frac{2}{9} = \frac{4}{9}.$$

59. $\int_{-1}^0 x(x+1)^{\frac{1}{3}} \, dx = \int_0^1 (u^3 - 1)u \cdot 3u^2 \, du = 3 \int_0^1 (u^6 - u^3) \, du = 3 \left(\frac{1}{7} u^7 - \frac{1}{4} u^4 \right) \Big|_0^1$

$$\boxed{\begin{array}{l} u = (x+1)^{\frac{1}{3}} \Rightarrow u^3 = x + 1 \Rightarrow x = u^3 - 1 \Rightarrow dx = 3u^2 \, du \\ x = -1 \Rightarrow u = 0, \ x = 0 \Rightarrow u = 1 \end{array}}$$

$$= 3 \left(\frac{1}{7} - \frac{1}{4} \right) = 3 \left(-\frac{3}{28} \right) = -\frac{9}{28}.$$

61. $\int_0^{\frac{\pi}{2}} \frac{\cos x}{2 - \cos x} \, dx = \int_0^1 \frac{\frac{1-u^2}{u^2+1} \cdot \frac{2 \, du}{u^2+1}}{2 - \frac{1-u^2}{u^2+1}} = \int_0^1 \frac{2 - 2u^2}{(3u^2 + 1)(u^2 + 1)} \, du$

$$\boxed{\begin{array}{l} u = \tan \frac{x}{2}, \quad x = 0 \Rightarrow u = \tan 0 = 0 \\ x = \frac{\pi}{2} \Rightarrow u = \tan \frac{\pi}{4} = 1 \end{array}}$$

Apply the method of partial fractions:
$$\frac{2 - 2u^2}{(3u^2 + 1)(u^2 + 1)} = \frac{Au + B}{3u^2 + 1} + \frac{Cu + D}{u^2 + 1}$$
$$2 - 2u^2 = (Au + B)(u^2 + 1) + (Cu + D)(3u^2 + 1)$$
$$= Au^3 + Au + Bu^2 + B + 3Cu^3 + Cu + 3Du^2 + D$$
$$= (A + 3C)u^3 + (B + 3D)u^2 + (A + C)u + (B + D)$$
$$u^3 : (1): \quad 0 = A + 3C$$
$$u^2 : (2): -2 = B + 3D$$
$$u : (3): \quad 0 = A + C \Rightarrow (1) - (3): 2C = 0 \Rightarrow C = 0$$
$$const : (4): \quad 2 = B + D \Rightarrow (2) - (4): 2D = -4 \Rightarrow D = -2$$

From (1), $0 = A + 0 \Rightarrow A = 0$, and from (4), $2 = B - 2 \Rightarrow B = 4$, so that:

$$\int_0^{\frac{\pi}{2}} \frac{\cos x}{2 - \cos x} \, dx = \int_0^1 \frac{4}{3u^2 + 1} du - 2 \int_0^1 \frac{1}{1 + u^2} \, du = \frac{4}{\sqrt{3}} \int_0^{\frac{\pi}{3}} \frac{\sec^2 \theta \, d\theta}{\sec^2 \theta} - 2 \tan^{-1} u \Big|_0^1$$

$$\boxed{\begin{array}{ll} 3u^2 = \tan^2 \theta \Rightarrow u = \frac{1}{\sqrt{3}} \tan \theta & u = 0 \Rightarrow \tan \theta = 0 \Rightarrow \theta = 0 \\ du = \frac{1}{\sqrt{3}} \sec^2 \theta \, d\theta & u = 1 \Rightarrow \tan \theta = \sqrt{3} \Rightarrow \theta = \frac{\pi}{3} \end{array}}$$

$$= \frac{4}{\sqrt{3}} \theta \Big|_0^{\frac{\pi}{3}} - 2(\tan^{-1} 1 - \tan^{-1} 0) = \frac{4\pi}{3\sqrt{3}} - 2 \cdot \frac{\pi}{4} = \frac{4\sqrt{3}\pi}{9} - \frac{\pi}{2} = \frac{\pi}{18}(8\sqrt{3} - 9).$$

63. (a) $\cos(A-B) - \cos(A+B) = (\cos A \cos B + \sin A \sin B) - (\cos A \cos B - \sin A \sin B)$
$$= \cancel{\cos A \cos B} + \sin A \sin B - \cancel{\cos A \cos B} + \sin A \sin B$$
$$= 2\sin A \sin B$$

(b) $\sin(A+B) + \sin(A-B) = (\sin A \cos B + \cos A \sin B) + (\sin A \cos B - \cos A \sin B)$
$$= \sin A \cos B + \cancel{\cos A \sin B} + \sin A \cos B - \cancel{\cos A \sin B}$$
$$= 2\sin A \cos B$$

(c) $\cos(A-B) + \cos(A+B) = (\cos A \cos B + \sin A \sin B) + (\cos A \cos B - \sin A \sin B)$
$$= \cos A \cos B + \cancel{\sin A \sin B} + \cos A \cos B - \cancel{\sin A \sin B}$$
$$= 2\cos A \cos B.$$

Note re Exercise 55 above: The exact value of $\cos\frac{\pi}{10} = \cos 18° = \frac{1}{4}\sqrt{10 + 2\sqrt{5}}$, can be obtained in several ways, two of which appear below. Therefore, the answer in Exercise 55 is: $\int_{-\frac{\pi}{10}}^{0} \sin 2x \cos 3x\, dx = \frac{2}{5} - \frac{1}{8}\sqrt{10 + 2\sqrt{5}}$.

(1) The isosceles triangle appearing on the left is known as "The golden triangle":

From the right triangle, $\sin 18° = \dfrac{\frac{1}{2}}{\frac{1+\sqrt{5}}{2}} = \dfrac{1}{1+\sqrt{5}} \cdot \dfrac{1-\sqrt{5}}{1-\sqrt{5}} = \dfrac{1-\sqrt{5}}{-4} = \frac{1}{4}(\sqrt{5}-1) \Rightarrow$

$$\cos 18° = \sqrt{1 - \sin^2 18°} = \sqrt{1 - \left[\frac{1}{4}(\sqrt{5}-1)\right]^2} = \sqrt{1 - \frac{1}{16}(5 - 2\sqrt{5}+1)}$$

$$= \sqrt{1 - \frac{1}{16}(6 - 2\sqrt{5})} = \sqrt{1 - \frac{1}{8}(3 - \sqrt{5})} = \sqrt{\frac{8 - 3 + \sqrt{5}}{8}} = \sqrt{\frac{5+\sqrt{5}}{8} \cdot \frac{2}{2}} = \frac{1}{4}\sqrt{10 + 2\sqrt{5}}$$

(2) Begin by noting that $\cos\frac{2\pi}{5} = -\cos(\pi - \frac{2\pi}{5})$, and then apply double angle formulas:

$$\cos\frac{2\pi}{5} = -\cos(\pi - \frac{2\pi}{5}) = -\cos\frac{3\pi}{5} = -\cos(\frac{\pi}{5} + \frac{2\pi}{5})$$

$$\cos 2\left(\frac{\pi}{5}\right) = 2\cos^2\frac{\pi}{5} - 1 = -[\cos\frac{\pi}{5}\cos\frac{2\pi}{5} - \sin\frac{\pi}{5}\sin\frac{2\pi}{5}]$$

$$= [-\cos\frac{\pi}{5}][2\cos^2\frac{\pi}{5} - 1] + [\sin\frac{\pi}{5}][2\sin\frac{\pi}{5}\cos\frac{\pi}{5}]$$

$$= -2\cos^3\frac{\pi}{5} + \cos\frac{\pi}{5} + 2[1 - \cos^2\frac{\pi}{5}]\cos\frac{\pi}{5}$$

Let $x = \cos\frac{\pi}{5}$. Then $0 < x < 1$ and:
$$2x^2 - 1 = -2x^3 + x + 2x - 2x^3 = -4x^3 + 3x$$

$$4x^3 + 2x^2 - 3x - 1 = 0 \quad \text{(note that } -1 \text{ is a root, so } x + 1 \text{ is a factor)}$$

$$(x+1)(4x^2 - 2x - 1) = 0$$

$$x = -1 \text{ or } x = \frac{-(-2) \pm \sqrt{4 - 4(-4)}}{8} = \frac{1 \pm \sqrt{5}}{4}$$

As $x > 0$, $x = \frac{1+\sqrt{5}}{4} = \cos\frac{\pi}{5}$. By the half-angle formula:

$$\cos\frac{\pi}{10} = \sqrt{\frac{1 + \cos\frac{\pi}{5}}{2}} = \sqrt{\frac{1 + \frac{1+\sqrt{5}}{4}}{2}} = \sqrt{\frac{5+\sqrt{5}}{8} \cdot \frac{2}{2}} = \frac{1}{4}\sqrt{10 + 2\sqrt{5}}.$$

CHAPTER 8
L'Hôpital's Rule and Improper Integrals

§8.1 L'Hôpital's Rule

1. $\lim\limits_{x \to 2} x^3 - 5x + 2 = 2^3 - 5 \cdot 2 + 2 = 0$, and $\lim\limits_{x \to 2} x^4 + 6x^2 - 40 = 2^4 + 6 \cdot 4 - 40 = 0 \Rightarrow$

   Form $\frac{0}{0}$: $\lim\limits_{x \to 2} \dfrac{x^3 - 5x + 2}{x^4 + 6x^2 - 40} = \lim\limits_{x \to 2} \dfrac{3x^2 - 5}{4x^3 + 12x} = \dfrac{3 \cdot 4 - 5}{4 \cdot 8 + 24} = \dfrac{7}{56} = \dfrac{1}{8}$.

3. Form $\frac{\infty}{\infty}$: $\lim\limits_{x \to \infty} \dfrac{x^2 - 5x + 2}{x^4 + 6x^2 - 40} = \lim\limits_{x \to \infty} \dfrac{2x - 5}{4x^3 + 12x} = \lim\limits_{x \to \infty} \dfrac{2}{12x^2 + 12} = 0$.

5. Form $\frac{0}{0}$: $\lim\limits_{x \to 0} \dfrac{1 - \cos x}{x} = \lim\limits_{x \to 0} \dfrac{\sin x}{1} = \dfrac{\sin 0}{1} = \dfrac{0}{1} = 0$.

7. Form $\frac{0}{0}$: $\lim\limits_{x \to \infty} \dfrac{\ln\left(1 + \frac{1}{x}\right)}{\frac{1}{x}} = \lim\limits_{x \to \infty} \dfrac{\frac{1}{1 + \frac{1}{x}} \cdot -x^{-2}}{-x^{-2}} = \lim\limits_{x \to \infty} \dfrac{1}{1 + \frac{1}{x}} = 1$.

9. Form $\frac{\infty}{\infty}$: $\lim\limits_{x \to \infty} \dfrac{\ln x}{\sqrt{x}} = \lim\limits_{x \to \infty} \dfrac{\frac{1}{x}}{\frac{1}{2\sqrt{x}}} = \lim\limits_{x \to \infty} \dfrac{2\sqrt{x}}{x} = \lim\limits_{x \to \infty} \dfrac{2}{\sqrt{x}} = 0$.

11. Form $\frac{\infty}{\infty}$: $\lim\limits_{x \to 1} \dfrac{\frac{1}{x - 1}}{\frac{1}{\ln x}} = \lim\limits_{x \to 1} \dfrac{\ln x}{x - 1} = \lim\limits_{x \to 1} \dfrac{\frac{1}{x}}{1} = \lim\limits_{x \to 1} \dfrac{1}{x} = 1$.

13. Form $\frac{\infty}{\infty}$: $\lim\limits_{x \to 0^+} \dfrac{\ln x}{\frac{1}{x}} = \lim\limits_{x \to 0^+} \dfrac{\frac{1}{x}}{-\frac{1}{x^2}} = \lim\limits_{x \to 0^+} (-x) = 0$.

15. Form $\frac{0}{0}$: $\lim\limits_{x \to \frac{\pi}{2}} \dfrac{\sin 2x}{\pi - 2x} = \lim\limits_{x \to \frac{\pi}{2}} \dfrac{2 \cos 2x}{-2} = \dfrac{2 \cos \pi}{-2} = \dfrac{-2}{-2} = 1$.

17. Form $\frac{0}{0}$: $\lim\limits_{x \to \frac{\pi}{2}} \dfrac{1 + \cos 2x}{1 - \sin x} = \lim\limits_{x \to \frac{\pi}{2}} \dfrac{\cancel{-}2 \sin 2x}{\cancel{-} \cos x} = \lim\limits_{x \to \frac{\pi}{2}} \dfrac{4 \cos 2x}{-\sin x} = \dfrac{4 \cos \pi}{-\sin \frac{\pi}{2}} = \dfrac{-4}{-1} = 4$.

19. Form $\frac{0}{0}$: $\lim\limits_{x \to \infty} \dfrac{\sin \frac{3}{x}}{\sin \frac{9}{x}} = \lim\limits_{x \to \infty} \dfrac{(\cos \frac{3}{x})(-\frac{3}{x^2})}{(\cos \frac{9}{x})(-\frac{9}{x^2})} = \lim\limits_{x \to \infty} \dfrac{1}{3} \cdot \dfrac{\cos \frac{3}{x}}{\cos \frac{9}{x}} = \dfrac{1}{3} \cdot 1 = \dfrac{1}{3}$.

21. Form $\frac{0}{0}$: $\displaystyle\lim_{x \to 0} \frac{x^2 + 2x - 2e^x + 2}{\sin x - x} = \lim_{x \to 0} \frac{2x + 2 - 2e^x}{\cos x - 1} = \lim_{x \to 0} \frac{2 - 2e^x}{-\sin x} = \lim_{x \to 0} \frac{\not{-}2e^x}{\not{-}\cos x}$

$$= \frac{2e^0}{\cos 0} = \frac{2}{1} = 2.$$

23. Form $\frac{0}{0}$: $\displaystyle\lim_{x \to 0} \frac{\tan x - \sin x}{x^3} = \lim_{x \to 0} \frac{\sec^2 x - \cos x}{3x^2} = \lim_{x \to 0} \frac{2\sec^2 x \tan x + \sin x}{6x}$

$$= \lim_{x \to 0} \frac{2[\sec^2 x \cdot \sec^2 x + \tan x \cdot 2\sec^2 x \tan x] + \cos x}{6}$$

$$= \lim_{x \to 0} \frac{2\sec^4 x + 4\sec^2 x \tan^2 x + \cos x}{6}$$

$$= \frac{2\sec^4 0 + 4\sec^2 0 \tan^2 0 + \cos 0}{6} = \frac{2 + 0 + 1}{6} = \frac{1}{2}.$$

25. Form $\frac{0}{0}$: $\displaystyle\lim_{x \to 0} \frac{\sin x - x\cos x}{x - \sin x} = \lim_{x \to 0} \frac{\cos x - [-x\sin x + \cos x]}{1 - \cos x} = \lim_{x \to 0} \frac{x\sin x}{1 - \cos x}$

$$= \lim_{x \to 0} \frac{x\cos x + \sin x}{\sin x} = \lim_{x \to 0} \frac{-x\sin x + \cos x + \cos x}{\cos x}$$

$$= \lim_{x \to 0} \frac{0 + 2\cos 0}{\cos 0} = 2.$$

27. $\displaystyle\lim_{x \to 1^-} \ln(1 - x) = -\infty$, and $\displaystyle\lim_{x \to 1^-} \cos \pi x = \cos \pi = -1$, so l'Hôpital's Rule does not apply, and $\displaystyle\lim_{x \to 1^-} \frac{\ln(1 - x)}{\cos \pi x} = \infty.$

29. Form $\frac{0}{0}$: $\displaystyle\lim_{x \to 0} \frac{a^x - b^x}{x} = \lim_{x \to 0} \frac{a^x \ln a - b^x \ln b}{1} = a^0 \ln a - b^0 \ln b = \ln a - \ln b = \ln \frac{a}{b}.$

31. Form $\frac{0}{0}$: $\displaystyle\lim_{x \to 0} \frac{x - \sin^{-1} x}{\sin^3 x} = \lim_{x \to 0} \frac{1 - \dfrac{1}{\sqrt{1 - x^2}}}{3\sin^2 x \cos x}$

$$= \lim_{x \to 0} \frac{\frac{1}{\not{2}}(1 - x^2)^{-\frac{3}{2}}(-\not{2}x)}{3[(\sin^2 x)(-\sin x) + (\cos x)(2\sin x \cos x)]}$$

$$= \lim_{x \to 0} \frac{\dfrac{-x}{(1 - x^2)^{\frac{3}{2}}}}{3[-\sin^3 x + 2\sin x \cos^2 x]}$$

$$= \lim_{x \to 0} \frac{\dfrac{(1 - x^2)^{\frac{3}{2}}(-1) - (-x) \cdot \frac{3}{2}(1 - x^2)^{\frac{1}{2}}(-2x)}{(1 - x^2)^3}}{3[-3\sin^2 x \cos x + 2(\sin x \cdot 2\cos x(-\sin x) + \cos^3 x)]}$$

$$= \frac{-1 - 0}{3[0 + 2(0 + 1)]} = -\frac{1}{6}.$$

33. Form $\frac{0}{0}$: $\displaystyle\lim_{x\to 0}\frac{x-\tan x}{\sin x-x}=\lim_{x\to 0}\frac{1-\sec^2 x}{\cos x-1}=\lim_{x\to 0}\frac{\not{-}2\sec^2 x\tan x}{\not{-}\sin x}$

$$=\lim_{x\to 0}\frac{2\cdot\dfrac{1}{\cos^2}\cdot\dfrac{\sin x}{\cos x}}{\sin x}=\lim_{x\to 0}\frac{2}{\cos^3 x}=\frac{2}{1}=2.$$

35. Form $\frac{\infty}{\infty}$: $\displaystyle\lim_{x\to\infty}\frac{\ln(\ln x)}{\ln x}=\lim_{x\to\infty}\frac{\dfrac{1}{\ln x}\cdot\dfrac{1}{x}}{\dfrac{1}{x}}=\lim_{x\to\infty}\frac{1}{\ln x}=0.$

37. Form $\frac{\infty}{\infty}$: $\displaystyle\lim_{x\to 0^+}\frac{\ln(x^2-x)}{\ln x}=\lim_{x\to 0^+}\frac{\dfrac{1}{x^2-x}\cdot(2x-1)}{\dfrac{1}{x}}=\lim_{x\to 0^+}\frac{\not{x}(2x-1)}{\not{x}(x-1)}=\frac{-1}{-1}=1.$

39. Form $\infty\cdot 0$: $\displaystyle\lim_{x\to\infty}x^3 e^{-x^2}=\lim_{x\to\infty}\frac{x^3}{e^{x^2}}=\lim_{x\to\infty}\frac{3x^2}{2xe^{x^2}}=\lim_{x\to\infty}\frac{6x}{2[e^{x^2}+xe^{x^2}\cdot 2x]}$

$$=\lim_{x\to\infty}\frac{6x}{2e^{x^2}(1+2x^2)}=\lim_{x\to\infty}\frac{6}{2e^{x^2}(4x)+(1+2x^2)\cdot 2e^{x^2}\cdot 2x}$$

$$=\lim_{x\to\infty}\frac{6}{4xe^{x^2}(2+1+2x^2)}=0.$$

41. Form $\infty\cdot 0$: $\displaystyle\lim_{x\to\frac{\pi}{2}}\tan x\cdot\ln(\sin x)=\lim_{x\to\frac{\pi}{2}}\frac{\ln(\sin x)}{\cot x}=\lim_{x\to\frac{\pi}{2}}\frac{\dfrac{1}{\sin x}\cdot\cos x}{-\csc^2 x}=\lim_{x\to\frac{\pi}{2}}\frac{\dfrac{\cos x}{\sin x}}{-\dfrac{1}{\sin^2 x}}$

$$=\lim_{x\to\frac{\pi}{2}}-\sin x\cos x=-1\cdot 0=0.$$

43. Form $\infty-\infty$: $\displaystyle\lim_{x\to 1}\left(\frac{1}{\ln x}-\frac{1}{x-1}\right)=\lim_{x\to 1}\frac{x-1-\ln x}{(x-1)\ln x}=\lim_{x\to 1}\frac{1-\dfrac{1}{x}}{(x-1)\cdot\dfrac{1}{x}+\ln x}$

$$=\lim_{x\to 1}\frac{x-1}{x-1+x\ln x}=\lim_{x\to 1}\frac{1}{1+x\cdot\dfrac{1}{x}+\ln x}$$

$$=\frac{1}{1+1+0}=\frac{1}{2}.$$

45. Form $\infty-\infty$: $\displaystyle\lim_{x\to\infty}[\ln 2x-\ln(x+1)]=\lim_{x\to\infty}[\ln 2+\ln x-\ln(x+1)]$

$$=\lim_{x\to\infty}\left[\ln 2+\ln\frac{x}{x+1}\right]$$

Form $\frac{\infty}{\infty}$: $\displaystyle\lim_{x\to\infty}\frac{x}{x+1}=\lim_{x\to\infty}\frac{1}{1}=1\ \Rightarrow\ \lim_{x\to\infty}\ln\frac{x}{x+1}=\ln 1=0\ \Rightarrow$

$$\lim_{x\to\infty}[\ln 2x-\ln(x+1)]=\ln 2+0=\ln 2.$$

47. Form 0^0 : $\lim\limits_{x\to 0+}(\sin x)^{\sin x}$. Let $y=(\sin x)^{\sin x}$, then $\ln y = \ln(\sin x)^{\sin x} = \sin x \ln(\sin x)$. Now consider $\lim\limits_{x\to 0+}\sin x \ln(\sin x)$: Form $0\cdot\infty$:

$$\lim_{x\to 0+}\sin x\ln(\sin x) = \lim_{x\to 0+}\frac{\ln(\sin x)}{\csc x} = \lim_{x\to 0+}\frac{\frac{1}{\sin x}\cdot\cos x}{-\csc x\cot x}$$

$$= \lim_{x\to 0+}\frac{\frac{\cos x}{\sin x}}{-\frac{1}{\sin x}\cdot\frac{\cos x}{\sin x}} = \lim_{x\to 0+}-\sin x = 0,$$

so $\lim\limits_{x\to 0+}\ln y = 0 \Rightarrow \lim\limits_{x\to 0+}y = \lim\limits_{x\to 0+}e^{\ln y} = e^0 = 1$, i.e. $\lim\limits_{x\to 0+}(\sin x)^{\sin x}=1$.

49. Form 1^∞ : $\lim\limits_{x\to\infty}\left(1+\frac{1}{x}\right)^x$. Let $y=\left(1+\frac{1}{x}\right)^x$, then $\ln y = x\ln\left(1+\frac{1}{x}\right)$. Now consider $\lim\limits_{x\to\infty}\ln y = \lim\limits_{x\to\infty}x\ln\left(1+\frac{1}{x}\right)$:

Form $\infty\cdot 0$: $\lim\limits_{x\to\infty}x\ln\left(1+\frac{1}{x}\right) = \lim\limits_{x\to\infty}\frac{\ln\left(1+\frac{1}{x}\right)}{\frac{1}{x}} = \lim\limits_{x\to\infty}\frac{\frac{1}{1+\frac{1}{x}}\cdot-\frac{1}{x^2}}{-\frac{1}{x^2}} = \lim\limits_{x\to\infty}\frac{1}{1+\frac{1}{x}}=1,$

so $\lim\limits_{x\to\infty}\ln y = 1 \Rightarrow \lim\limits_{x\to\infty}y = \lim\limits_{x\to\infty}e^{\ln y} = e^1 = e$, i.e. $\lim\limits_{x\to\infty}\left(1+\frac{1}{x}\right)^x = e$.

51. Form 1^∞ : $\lim\limits_{x\to 0}(\cos x)^{1/x^2}$. Let $y=(\cos x)^{1/x^2}$, then $\ln y = \frac{1}{x^2}\ln\cos x = \frac{\ln\cos x}{x^2}$. Now consider $\lim\limits_{x\to 0}\ln y = \lim\limits_{x\to 0}\frac{\ln\cos x}{x^2}$:

Form $\frac{0}{0}$: $\lim\limits_{x\to 0}\frac{\ln\cos x}{x^2} = \lim\limits_{x\to 0}\frac{\frac{1}{\cos x}\cdot-\sin x}{2x} = \lim\limits_{x\to 0}\frac{-\tan x}{2x} = \lim\limits_{x\to 0}\frac{-\sec^2 x}{2} = -\frac{1}{2},$

so $\lim\limits_{x\to 0}\ln y = -\frac{1}{2} \Rightarrow \lim\limits_{x\to 0}y = \lim\limits_{x\to 0}e^{\ln y} = e^{-\frac{1}{2}} = \frac{1}{\sqrt{e}}.$

53. Form $\frac{\infty}{\infty}$: $\lim\limits_{x\to\infty}\frac{\int_0^x e^{t^2}dt}{x} = \lim\limits_{x\to\infty}\frac{e^{x^2}}{1} = \infty.$

55. (1) Form $\frac{\infty}{\infty}$: $\lim\limits_{x\to\infty}\frac{e^x+x}{xe^x} = \lim\limits_{x\to\infty}\frac{e^x+1}{xe^x+e^x} = \lim\limits_{x\to\infty}\frac{e^x}{(xe^x+e^x)+e^x}$

$$= \lim_{x\to\infty}\frac{\cancel{e^x}\,1}{\cancel{e^x}(x+2)} = \lim_{x\to\infty}\frac{1}{x+2} = 0.$$

(2) $\lim\limits_{x\to-\infty}\frac{e^x+x}{xe^x} = \lim\limits_{x\to-\infty}\frac{e^x+x}{xe^x}\cdot\frac{e^{-x}}{e^{-x}} = \lim\limits_{x\to-\infty}\frac{1+xe^{-x}}{x}$. Let $t=-x$, then

$$\lim_{x\to-\infty}\frac{1+xe^{-x}}{x} = \lim_{t\to\infty}\frac{1-te^t}{-t}, \text{ a limit of the form } \frac{\infty}{\infty}:$$

$$\lim_{t\to\infty}\frac{1-te^t}{-t} = \lim_{t\to\infty}\frac{\cancel{-}(te^t+e^t)}{\cancel{-}1} = \lim_{t\to\infty}e^t(t+1) = \infty.$$

57. $\lim\limits_{x \to 3} x^3 - 2x^2 - 3x = 3^3 - 2(3^2) - 3(3) = 27 - 18 - 9 = 0$, and

$\lim\limits_{x \to 3} x^2 + 2x - 15 = 3^2 + 2(3) - 15 = 9 + 6 - 15 = 0 \Rightarrow$

Form $\frac{0}{0}$: $\lim\limits_{x \to 3} \dfrac{x^3 - 2x^2 - 3x}{x^2 + 2x - 15} = \lim\limits_{x \to 3} \dfrac{3x^2 - 4x - 3}{2x + 2} = \dfrac{3(3^2) - 4(3) - 3}{2(3) + 2} = \dfrac{27 - 12 - 3}{6 + 2} = \dfrac{12}{8} = \dfrac{3}{2}$.

59. Form $\frac{0}{0}$: $\lim\limits_{x \to 0} \dfrac{\sqrt{x+1} - 1}{x} = \lim\limits_{x \to 0} \dfrac{\frac{1}{2\sqrt{x+1}}}{1} = \dfrac{1}{2\sqrt{0} + 1} = \dfrac{1}{2}$.

61. Form $\frac{0}{0}$: $\lim\limits_{h \to 0} \dfrac{\sqrt{9+h} - 3}{h} = \lim\limits_{h \to 0} \dfrac{\frac{1}{2\sqrt{9+h}}}{1} = \dfrac{1}{2\sqrt{9} + 0} = \dfrac{1}{6}$.

63. Form $\frac{0}{0}$: $\lim\limits_{x \to 0} \dfrac{\cos ax - 1}{x^2} = \lim\limits_{x \to 0} \dfrac{-a \sin ax}{2x} = \lim\limits_{x \to 0} \dfrac{-a^2 \cos ax}{2} = -\dfrac{a^2}{2} = -8 \Rightarrow$
$a^2 = 16 \Rightarrow a = \pm 4$.

65. Using the facts that $(e^x)' = e^x$ and that the nth derivative of x^n equals $n!$ (Exercise 63, page 88), we apply l'Hôpital's Rule n times to go from $\lim\limits_{x \to \infty} \dfrac{e^x}{x^n}$ to $\lim\limits_{x \to \infty} \dfrac{e^x}{n!} = \infty$.

§8.2 Improper Integrals

1. $\displaystyle\int_1^\infty \dfrac{1}{x^2}\,dx$ converges by Theorem 8.3, as $p = 2 > 1$, and the value of the integral is $\dfrac{1}{p-1} = \dfrac{1}{2-1} = 1$ (converges).

3.
$$\int u\,dv = uv - \int v\,du$$
$\displaystyle\int_{-\infty}^0 xe^x\,dx = \lim\limits_{t \to -\infty} \int_t^0 xe^x\,dx = \lim\limits_{t \to -\infty} \left[xe^x \big|_t^0 - \int_t^0 e^x\,dx \right] = \lim\limits_{t \to -\infty} \left[-te^t - e^x \big|_t^0 \right]$

$$\boxed{\begin{array}{ll} u = x & dv = e^x\,dx \\ du = dx & v = e^x \end{array}}$$

$= \lim\limits_{t \to -\infty} \left[-te^t - (1 - e^t) \right]$. Since $\lim\limits_{t \to -\infty} te^t = \lim\limits_{t \to -\infty} \dfrac{t}{e^{-t}} = \lim\limits_{t \to -\infty} \dfrac{1}{-e^{-t}} = 0$,
$\displaystyle\int_{-\infty}^0 xe^x\,dx = 0 - 1 + 0 = -1$ (converges).

5. $\displaystyle\int_{-\infty}^0 \dfrac{1}{(2x-1)^3}\,dx = \lim\limits_{t \to -\infty} \int_t^0 \dfrac{1}{(2x-1)^3}\,dx = \lim\limits_{t \to -\infty} \dfrac{1}{2} \int_{2t-1}^{-1} u^{-3}\,du = \lim\limits_{t \to -\infty} -\dfrac{1}{4} \cdot \dfrac{1}{u^2} \bigg|_{2t-1}^{-1}$

$$\boxed{\begin{array}{ll} u = 2x - 1, & x = t \Rightarrow u = 2t - 1 \\ du = 2\,dx, & x = 0 \Rightarrow u = -1 \end{array}}$$

$= \lim\limits_{t \to -\infty} -\dfrac{1}{4}\left[1 - \dfrac{1}{(2t-1)^2}\right] = -\dfrac{1}{4}$ (converges).

7. $\displaystyle\int_0^\infty \frac{1}{4+x^2}\,dx = \lim_{t\to\infty}\int_0^t \frac{1}{4+x^2}\,dx$

$\displaystyle\int \frac{1}{4+x^2}\,dx = \frac{1}{4}\int \frac{1}{1+(\frac{x}{2})^2}\,dx = \frac{1}{2}\int \frac{du}{1+u^2} = \frac{1}{2}\tan^{-1}u + C = \frac{1}{2}\tan^{-1}\frac{x}{2} + C$

$$\boxed{u = \tfrac{x}{2},\ du = \tfrac{1}{2}\,dx}$$

$\displaystyle\int_0^\infty \frac{1}{4+x^2}\,dx = \lim_{t\to\infty}\ \frac{1}{2}\tan^{-1}\frac{x}{2}\Big|_0^t = \lim_{t\to\infty}\ \frac{1}{2}[\tan^{-1}\frac{t}{2}-\tan^{-1}0] = \frac{1}{2}\cdot\frac{\pi}{2} = \frac{\pi}{4}$ (converges).

9. $\displaystyle\int_0^\infty \cos\pi x\,dx = \lim_{t\to\infty}\int_0^t \cos\pi x\,dx = \lim_{t\to\infty}\frac{1}{\pi}\sin\pi x\Big|_0^t = \lim_{t\to\infty}\frac{1}{\pi}[\sin\pi t - \sin 0]$ does not exist, as the sine function continues to vary from -1 to 1, forever: (diverges).

11. $\displaystyle\int_0^\infty \frac{x}{(1+x^2)^2}\,dx = \lim_{t\to\infty}\int_0^t \frac{x}{(1+x^2)^2}\,dx$

$\displaystyle\int \frac{x}{(1+x^2)^2}\,dx = \frac{1}{2}\int u^{-2}\,du = -\frac{1}{2u} + C = -\frac{1}{2(1+x^2)} + C$

$$\boxed{u = 1+x^2,\ du = 2x\,dx}$$

$\displaystyle\int_0^\infty \frac{x}{(1+x^2)^2}\,dx = \lim_{t\to\infty}\ -\frac{1}{2(1+x^2)}\Big|_0^t = \lim_{t\to\infty}\ -\frac{1}{2}\left[\frac{1}{1+t^2} - 1\right] = \frac{1}{2}$ (converges).

13. $\displaystyle\int_{-\infty}^\infty xe^{-x^2}\,dx = \int_{-\infty}^0 xe^{-x^2}\,dx + \int_0^\infty xe^{-x^2}\,dx = \lim_{t\to-\infty}\int_t^0 xe^{-x^2}\,dx + \lim_{t\to\infty}\int_0^t xe^{-x^2}\,dx$

$\displaystyle\int xe^{-x^2}\,dx = -\frac{1}{2}\int e^u\,du = -\frac{1}{2}e^u + C = -\frac{1}{2}e^{-x^2} + C$

$$\boxed{u = -x^2,\ du = -2x\,dx}$$

$\displaystyle\int_{-\infty}^\infty xe^{-x^2}\,dx = \lim_{t\to-\infty}\ -\frac{1}{2}e^{-x^2}\Big|_t^0 + \lim_{t\to\infty}\ -\frac{1}{2}e^{-x^2}\Big|_0^t = \lim_{t\to-\infty}\ -\frac{1}{2}(1-e^{-t^2}) + \lim_{t\to\infty}\ -\frac{1}{2}(e^{-t^2}-1)$

$\displaystyle\qquad\qquad = -\frac{1}{2} + \frac{1}{2} = 0$ (converges).

15. $\displaystyle\int_0^\infty \frac{6x^2+8}{(x^2+1)(x^2+2)}\,dx = \lim_{t\to\infty}\int_0^t \frac{6x^2+8}{(x^2+1)(x^2+2)}\,dx$. Apply the method of Partial Fractions:

$$\frac{6x^2+8}{(x^2+1)(x^2+2)} = \frac{Ax+B}{x^2+1} + \frac{Cx+D}{x^2+2}$$

$$6x^2+8 = (Ax+B)(x^2+2) + (Cx+D)(x^2+1)$$

$$= Ax^3 + Bx^2 + 2Ax + 2B + Cx^3 + Dx^2 + Cx + D$$

$$= (A+C)x^3 + (B+D)x^2 + (2A+C)x + (2B+D)$$

$$x^3 : (1) : \quad 0 = A + C$$
$$x^2 : (2) : \quad 6 = B + D$$
$$x : (3) : \quad 0 = 2A + C \quad \text{From (3)--(1):} \quad A = 0 \Rightarrow \text{From (1):} \quad C = 0$$
$$\text{const} : (4) : \quad 8 = 2B + D \quad \text{From (4)--(2):} \quad B = 2 \Rightarrow \text{From (2):} \quad D = 4$$

$$\int \frac{6x^2 + 8}{(x^2 + 1)(x^2 + 2)} \, dx = \int \frac{2}{x^2 + 1} + \int \frac{4}{x^2 + 2} = 2\tan^{-1} x + 2 \int \frac{1}{1 + (\frac{x}{\sqrt{2}})^2} \, dx$$

$$\boxed{u = \frac{x}{\sqrt{2}}, \ du = \frac{1}{\sqrt{2}} \, dx}$$

$$= 2\tan^{-1} x + 2 \int \frac{\sqrt{2} \, du}{1 + u^2} = 2\tan^{-1} x + 2\sqrt{2} \tan^{-1} \frac{x}{\sqrt{2}} + C$$

$$\int_0^\infty \frac{6x^2 + 8}{(x^2 + 1)(x^2 + 2)} \, dx = \lim_{t \to \infty} \left[2\tan^{-1} x + 2\sqrt{2} \tan^{-1} \frac{x}{\sqrt{2}} \right]\Big|_0^t$$

$$= \lim_{t \to \infty} \left[2\tan^{-1} t + 2\sqrt{2} \tan^{-1} \frac{t}{\sqrt{2}} - (2\tan^{-1} 0 + 2\sqrt{2} \tan^{-1} 0) \right]$$

$$= 2 \cdot \frac{\pi}{2} + 2\sqrt{2} \cdot \frac{\pi}{2} = \pi(1 + \sqrt{2}) \quad \text{(converges)}.$$

17. $\displaystyle \int_e^\infty \frac{\ln x}{x} \, dx = \lim_{t \to \infty} \int_e^t \frac{\ln x}{x} \, dx = \lim_{t \to \infty} \int_1^{\ln t} u \, du = \lim_{t \to \infty} \frac{1}{2} u^2 \Big|_1^{\ln t} = \lim_{t \to \infty} \frac{1}{2} [(\ln t)^2 - 1]$

$$\boxed{\begin{array}{l} u = \ln x, \ x = e \Rightarrow u = 1 \\ du = \frac{1}{x} \, dx, \ x = t \Rightarrow u = \ln t \end{array}}$$

Since $\ln t$ becomes infinite, the integral diverges.

19. $\displaystyle \int_0^8 x^{-\frac{1}{3}} \, dx = \lim_{t \to 0^+} \int_t^8 x^{-\frac{1}{3}} \, dx = \lim_{t \to 0^+} \frac{3}{2} x^{\frac{2}{3}} \Big|_t^8 = \lim_{t \to 0^+} \frac{3}{2} (8^{\frac{2}{3}} - t^{\frac{2}{3}}) = \frac{3}{2} (4) = 6$ (converges).

21. $\displaystyle \int_{-\infty}^\infty \frac{16}{9x^4 + 10x^2 + 1} \, dx = \int_{-\infty}^\infty \frac{16}{(9x^2 + 1)(x^2 + 1)} \, dx$

$$= \lim_{t \to -\infty} \int_t^0 \frac{16}{(9x^2 + 1)(x^2 + 1)} \, dx + \lim_{t \to \infty} \int_0^t \frac{16}{(9x^2 + 1)(x^2 + 1)} \, dx.$$

Apply the method of Partial Fractions:

$$\frac{16}{(9x^2 + 1)(x^2 + 1)} = \frac{Ax + B}{9x^2 + 1} + \frac{Cx + D}{x^2 + 1}$$

$$16 = (Ax + B)(x^2 + 1) + (Cx + D)(9x^2 + 1)$$
$$= Ax^3 + Bx^2 + Ax + B + 9Cx^3 + 9Dx^2 + Cx + D$$
$$= (A + 9C)x^3 + (B + 9D)x^2 + (A + C)x + (B + D)$$

$$x^3 : (1) : \quad 0 = A + 9C$$
$$x^2 : (2) : \quad 0 = B + 9D$$
$$x : (3) : \quad 0 = A + C \quad \text{From (1)--(3):} \quad C = 0 \Rightarrow \text{From (1):} \quad A = 0$$
$$\text{const} : (4) : \quad 16 = B + D \quad \text{From (2)--(4):} \quad 8D = -16 \Rightarrow D = -2 \Rightarrow \text{From (4):} \quad B = 18$$

Thus,

$$\frac{16}{(9x^2+1)(x^2+1)} = \frac{18}{9x^2+1} - \frac{2}{x^2+1}$$

$$\int \frac{18}{9x^2+1}\,dx = 18\int \frac{dx}{1+(3x)^2} = 6\int \frac{du}{1+u^2} = 6\tan^{-1}u + C = 6\tan^{-1}3x + C$$

$$\boxed{u = 3x,\ du = 3\,dx}$$

$$\int \frac{2}{x^2+1}\,dx = 2\int \frac{dx}{1+x^2} = 2\tan^{-1}x + C. \quad \text{Putting it all together:}$$

$$\int_{-\infty}^{\infty} \frac{16}{9x^4+10x^2+1}\,dx = \lim_{t\to -\infty}\left[6\tan^{-1}3x - 2\tan^{-1}x\right]\Big|_t^0 + \lim_{t\to\infty}\left[6\tan^{-1}3x - 2\tan^{-1}x\right]\Big|_0^t$$

$$= \lim_{t\to -\infty}\left[0 - (6\tan^{-1}3t - 2\tan^{-1}t)\right] + \lim_{t\to\infty}\left[6\tan^{-1}3t - 2\tan^{-1}t - 0\right]$$

$$= \left[-6(-\tfrac{\pi}{2}) + 2(-\tfrac{\pi}{2})\right] + \left[6(\tfrac{\pi}{2}) - 2(\tfrac{\pi}{2})\right] = 3\pi - \pi + 3\pi - \pi = 4\pi$$

(converges).

23. $\int_0^{\frac{\pi}{2}} \tan x\,dx = \lim\limits_{t\to \frac{\pi}{2}^-}\int_0^t \tan x\,dx = \lim\limits_{t\to \frac{\pi}{2}^-} \ln|\sec x|\big|_0^t = \lim\limits_{t\to \frac{\pi}{2}^-}\left[\ln|\sec t| - \ln|\sec 0|\right]$

As $\sec t$ becomes infinite, and therefore $\ln|\sec t|$ becomes infinite, the integral diverges.

25. $\int_{-3}^{-1} 8(x+1)^{-\frac{1}{5}}\,dx = \lim\limits_{t\to -1^-}\int_{-3}^t 8(x+1)^{-\frac{1}{5}}\,dx = \lim\limits_{t\to -1^-} 10(x+1)^{\frac{4}{5}}\Big|_{-3}^t$

$$= \lim_{t\to -1^-} 10\left[(t+1)^{\frac{4}{5}} - (-2)^{\frac{4}{5}}\right] = -10\cdot 2^{\frac{4}{5}} \quad \text{(converges).}$$

27. $\int_{-5}^{-1} 3(x+3)^{-\frac{2}{5}}\,dx = \int_{-5}^{-3} 3(x+3)^{-\frac{2}{5}}\,dx + \int_{-3}^{-1} 3(x+3)^{-\frac{2}{5}}\,dx$

$$= \lim_{t\to -3^-}\int_{-5}^t 3(x+3)^{-\frac{2}{5}}\,dx + \lim_{t\to -3^+}\int_t^{-1} 3(x+3)^{-\frac{2}{5}}\,dx$$

Now, $\int 3(x+3)^{-\frac{2}{5}}\,dx = 3\cdot\frac{5}{3}(x+3)^{\frac{3}{5}} + C = 5(x+3)^{\frac{3}{5}} + C$

Therefore,

$$\int_{-5}^{-1} 3(x+3)^{-\frac{2}{5}}\,dx = \lim_{t\to -3^-} 5(x+3)^{\frac{3}{5}}\Big|_{-5}^t + \lim_{t\to -3^+} 5(x+3)^{\frac{3}{5}}\Big|_t^{-1}$$

$$= \lim_{t\to -3^-} 5\left[(t+3)^{\frac{3}{5}} - (-2)^{\frac{3}{5}}\right] + \lim_{t\to -3^+} 5\left[2^{\frac{3}{5}} - (t+3)^{\frac{3}{5}}\right]$$

$$= 5\cdot 2^{\frac{3}{5}} + 5\cdot 2^{\frac{3}{5}} = 10\cdot 2^{\frac{3}{5}}$$

(converges).

29. $\int_2^{\infty} \frac{x+3}{(x-1)(x^2+1)}\,dx = \lim\limits_{t\to\infty}\int_2^t \frac{x+3}{(x-1)(x^2+1)}\,dx.$

Apply the method of Partial Fractions:

$$\frac{x+3}{(x-1)(x^2+1)} = \frac{A}{x-1} + \frac{Bx+C}{x^2+1}$$

$$x+3 = A(x^2+1) + (Bx+C)(x-1)$$

$$x=1: \ 4 = 2A \ \Rightarrow \ A = 2 \ \Rightarrow$$

$$x+3 = 2x^2 + 2 + Bx^2 - Bx + Cx - C \ \Rightarrow$$

$$-2x^2 + x + 1 = Bx^2 + (C-B)x - C$$

$$x^2: \ -2 = B$$

$$\text{const}: \quad 1 = -C \ \Rightarrow \ C = -1$$

Thus,

$$\frac{x+3}{(x-1)(x^2+1)} = \frac{2}{x-1} - \frac{2x+1}{x^2+1}$$

$$\int \frac{2}{x-1}\,dx = 2\ln|x-1| + C = \ln(x-1)^2 + C, \text{ and } \int \frac{2x+1}{x^2+1}\,dx = \int \frac{2x}{x^2+1}\,dx + \int \frac{1}{x^2+1}\,dx$$

$$\int \frac{2x}{x^2+1}\,dx = \int \frac{1}{u}\,du = \ln|u| + C = \ln(x^2+1) + C$$

$$\boxed{u = x^2 + 1, \ du = 2x\,dx}$$

Consequently,

$$\int_2^\infty \frac{x+3}{(x-1)(x^2+1)}\,dx = \lim_{t\to\infty} \left\{ \ln(x-1)^2 - \ln(x^2+1) - \tan^{-1}x \right\}\Big|_2^t$$

$$= \lim_{t\to\infty} \left\{ \ln\left[\frac{(x-1)^2}{x^2+1}\right] - \tan^{-1}x \right\}\Big|_2^t$$

$$= \lim_{t\to\infty} \left\{ \ln\left[\frac{(t-1)^2}{t^2+1}\right] - \tan^{-1}t - [\ln\tfrac{1}{5} - \tan^{-1}2] \right\}$$

Form $\frac{\infty}{\infty}: \ \lim_{t\to\infty} \frac{(t-1)^2}{t^2+1} = \lim_{t\to\infty} \frac{2(t-1)}{2t} = \lim_{t\to\infty} \frac{2}{2} = 1 \ \Rightarrow \ \lim_{t\to\infty} \ln\left[\frac{(t-1)^2}{t^2+1}\right] = \ln 1 = 0 \ \Rightarrow$

$$\int_2^\infty \frac{x+3}{(x-1)(x^2+1)}\,dx = -\tfrac{\pi}{2} - \ln\tfrac{1}{5} + \tan^{-1}2 = -\tfrac{\pi}{2} + \ln 5 + \tan^{-1}2 \ \text{ (converges)}.$$

31. $\displaystyle \int_0^1 \frac{1}{\sqrt{x}}\,dx = \lim_{t\to 0^+} \int_t^1 x^{-\frac{1}{2}}\,dx = \lim_{t\to 0^+} 2x^{\frac{1}{2}}\Big|_t^1 = \lim_{t\to 0^+} 2(1-\sqrt{t}) = 2 \ \text{ (converges)}.$

33. $\displaystyle \int_1^2 \frac{1}{\sqrt[3]{x-2}}\,dx - \lim_{t\to 2^-} \int_1^t (x-2)^{-\frac{1}{3}}\,dx = \lim_{t\to 2^-} \tfrac{3}{2}(x-2)^{\frac{2}{3}}\Big|_1^t = \lim_{t\to 2^-} \tfrac{3}{2}[(t-2)^{\frac{2}{3}} - (-1)^{\frac{2}{3}}]$

$$= -\tfrac{3}{2} \ \text{ (converges)}.$$

35. $\displaystyle \int_0^{\frac{\pi}{2}} \sec x\,dx = \lim_{t\to\frac{\pi}{2}^-} \int_0^t \sec x\,dx = \lim_{t\to\frac{\pi}{2}^-} \ln|\sec x + \tan x|\,\big|_0^t$

$$= \lim_{t\to\frac{\pi}{2}^-} [\ln|\sec t + \tan t| - \ln|\sec 0 + \tan 0|] = \lim_{t\to\frac{\pi}{2}^-} \ln\left|\frac{1+\sin t}{\cos t}\right|$$

which becomes infinite as the cosine tends to 0, so the integral diverges.

37. $\displaystyle\int_0^\infty \frac{1}{\sqrt{x}(x+4)}\,dx = \int_0^1 \frac{1}{\sqrt{x}(x+4)}\,dx + \int_1^\infty \frac{1}{\sqrt{x}(x+4)}\,dx$

$$= \lim_{t\to 0^+} \int_t^1 \frac{1}{\sqrt{x}(x+4)}\,dx + \lim_{t\to\infty} \int_1^t \frac{1}{\sqrt{x}(x+4)}\,dx$$

$$\int \frac{1}{\sqrt{x}(x+4)}\,dx = \int \frac{2\,\cancel{u}\,du}{\cancel{u}(u^2+4)} = 2\int \frac{du}{u^2+4} = \frac{1}{2}\int \frac{du}{1+(\frac{u}{2})^2} = \int \frac{dv}{1+v^2} = \tan^{-1} v + C$$

$$\boxed{\begin{array}{l} u = \sqrt{x} \;\Rightarrow\; x = u^2 \\ dx = 2u\,du \end{array}} \qquad\qquad \boxed{\begin{array}{l} v = \dfrac{u}{2} \\ dv = \dfrac{1}{2}\,du \end{array}}$$

$$= \tan^{-1}\frac{u}{2} + C = \tan^{-1}\frac{\sqrt{x}}{2} + C$$

Therefore,
$$\int_0^\infty \frac{1}{\sqrt{x}(x+4)}\,dx = \lim_{t\to 0^+} \tan^{-1}\frac{\sqrt{x}}{2}\Big|_t^1 + \lim_{t\to\infty} \tan^{-1}\frac{\sqrt{x}}{2}\Big|_1^t$$

$$= \lim_{t\to 0^+}\left[\tan^{-1}\frac{1}{2} - \tan^{-1}\frac{\sqrt{t}}{2}\right] + \lim_{t\to\infty}\left[\tan^{-1}\frac{\sqrt{t}}{2} - \tan^{-1}\frac{1}{2}\right]$$

$$= \tan^{-1}\frac{1}{2} + \frac{\pi}{2} - \tan^{-1}\frac{1}{2} = \frac{\pi}{2} \quad \text{(converges)}.$$

39.
$$\int u\,dv = uv - \int v\,du$$
$$\int_0^1 x\ln x\,dx = \lim_{t\to 0^+}\int_t^1 x\ln x\,dx = \lim_{t\to 0^+}\left[\frac{1}{2}x^2\ln x\Big|_t^1 - \frac{1}{2}\int_t^1 x\,dx\right]$$

$$\boxed{\begin{array}{ll} u = \ln x & dv = x\,dx \\ du = \dfrac{1}{x}\,dx & v = \dfrac{1}{2}x^2 \end{array}}$$

$$= \lim_{t\to 0^+}\left[\frac{1}{2}(\ln 1 - t^2\ln t) - \frac{1}{2}\cdot\frac{x^2}{2}\Big|_t^1\right] = \lim_{t\to 0^+}\left[-\frac{1}{2}t^2\ln t - \frac{1}{4}(1 - t^2)\right]$$

Form $0\cdot\infty$: $\displaystyle\lim_{t\to 0^+} t^2\ln t = \lim_{t\to 0^+}\frac{\ln t}{t^{-2}} = \lim_{t\to 0^+}\frac{\frac{1}{t}}{\frac{-2}{t^3}} = \lim_{t\to 0^+}\frac{t^2}{-2} = 0 \;\Rightarrow\; \int_0^1 x\ln x\,dx = -\frac{1}{4}$

(converges).

41. $\displaystyle\int_1^{16} \frac{1}{\sqrt{x}}\left(\frac{1}{2-\sqrt{x}}\right)^{\frac{1}{3}} dx = \int_1^4 \frac{1}{\sqrt{x}}\left(\frac{1}{2-\sqrt{x}}\right)^{\frac{1}{3}} dx + \int_4^{16} \frac{1}{\sqrt{x}}\left(\frac{1}{2-\sqrt{x}}\right)^{\frac{1}{3}} dx$

$$= \lim_{t\to 4^-}\int_1^t \frac{1}{\sqrt{x}}\left(\frac{1}{2-\sqrt{x}}\right)^{\frac{1}{3}} dx + \lim_{t\to 4^+}\int_t^{16} \frac{1}{\sqrt{x}}\left(\frac{1}{2-\sqrt{x}}\right)^{\frac{1}{3}} dx$$

$$\int \frac{1}{\sqrt{x}}\left(\frac{1}{2-\sqrt{x}}\right)^{\frac{1}{3}} dx = -2\int u^{-\frac{1}{3}}\,du = -3u^{\frac{2}{3}} + C = -3(2-\sqrt{x})^{\frac{2}{3}} + C$$

$$\boxed{\begin{array}{l} u = 2 - \sqrt{x} \\ du = -\dfrac{1}{2\sqrt{x}}\,dx \end{array}}$$

Therefore,

$$\int_1^{16} \frac{1}{\sqrt{x}}\left(\frac{1}{2-\sqrt{x}}\right)^{\frac{1}{3}} dx = \lim_{t\to 4^-} -3(2-\sqrt{x})^{\frac{2}{3}}\Big|_1^t + \lim_{t\to 4^+} -3(2-\sqrt{x})^{\frac{2}{3}}\Big|_t^{16}$$
$$= \lim_{t\to 4^-} -3[(2-\sqrt{t})^{\frac{2}{3}}-1] + \lim_{t\to 4^+} -3[(-2)^{\frac{2}{3}}-(2-\sqrt{t})^{\frac{2}{3}}]$$
$$= 3 - 3\cdot 2^{\frac{2}{3}} = 3(1-2^{\frac{2}{3}}) \quad (\text{converges}).$$

43. If $a \geq 0$, $\int_0^1 x^a\,dx = \frac{x^{a+1}}{a+1}\Big|_0^1 = \frac{1}{a+1}(1-0) = \frac{1}{a+1}$, and the integral converges.

If $a < 0$ but $a \neq -1$, then

$$\int_0^1 x^a\,dx = \lim_{t\to 0^+}\int_t^1 x^a\,dx = \lim_{t\to 0^+}\frac{x^{a+1}}{a+1}\Big|_t^1 = \lim_{t\to 0^+}\frac{1}{a+1}(1-t^{a+1})$$

If $-1 < a < 0$, then $a+1 > 0$ and the integral converges; and its value is $\frac{1}{a+1}$.

If $a < -1$, then $a+1 < 0$ and the integral diverges.

If $a = -1$, $\int_0^1 x^a\,dx = \lim_{t\to 0^+}\int_t^1 \frac{1}{x}\,dx = \lim_{t\to 0^+}\ln|x|\Big|_t^1 = \lim_{t\to 0^+}[\ln 1 - \ln|t|]$, which becomes infinite, so the integral diverges.

Therefore, the integral converges only when $a > -1$.

45. $$\int_0^\infty \left(\frac{x}{x^2+1} - \frac{a}{3x+1}\right) dx = \lim_{t\to\infty}\int_0^t \left(\frac{x}{x^2+1} - \frac{a}{3x+1}\right) dx$$
$$= \lim_{t\to\infty}\left\{\left[\tfrac{1}{2}\ln(x^2+1) - \tfrac{a}{3}\ln|3x+1|\right]\Big|_0^t\right\}$$
$$= \lim_{t\to\infty}\ln\left[\frac{(x^2+1)^{\frac{1}{2}}}{(3x+1)^{\frac{a}{3}}}\right]\Big|_0^t$$
$$= \lim_{t\to\infty}\left\{\ln\left[\frac{(t^2+1)^{\frac{1}{2}}}{(3t+1)^{\frac{a}{3}}}\right] - \ln 1\right\} = \lim_{t\to\infty}\ln\left[\frac{(t^2+1)^{\frac{1}{2}}}{(3t+1)^{\frac{a}{3}}}\right]$$

Consider $\lim_{t\to\infty}\frac{(t^2+1)^{\frac{1}{2}}}{(3t+1)^{\frac{a}{3}}}$. If $a \leq 0$, the limit is infinite, so that the limit of the logarithm is also infinite, and the integral diverges. So, $a > 0$, and we can apply l'Hôpital's Rule:

$$\lim_{t\to\infty}\frac{(t^2+1)^{\frac{1}{2}}}{(3t+1)^{\frac{a}{3}}} = \lim_{t\to\infty}\frac{\frac{1}{2}(t^2+1)^{-\frac{1}{2}}(2t)}{\frac{a}{3}(3t+1)^{\frac{a}{3}-1}(3)} = \lim_{t\to\infty}\frac{\frac{t}{\sqrt{t^2+1}}}{a(3t+1)^{\frac{a}{3}-1}} \quad (*)$$

The limit of the numerator of $(*)$ is 1:
$$\lim_{t\to\infty}\frac{t}{\sqrt{t^2+1}} = \lim_{t\to\infty}\frac{1}{\sqrt{\frac{t^2+1}{t^2}}} = \lim_{t\to\infty}\frac{1}{\sqrt{1+\frac{1}{t^2}}} = 1$$

Now, the denominator of $(*)$:

If $\frac{a}{3}-1 < 0$, i.e. $a < 3$, the limit in $(*)$ is infinite, and the integral diverges.

If $\frac{a}{3}-1 = 0$, i.e. $a = 3$, expression $(*)$ becomes $\lim_{t\to\infty}\frac{t}{3\sqrt{t^2+1}} = 1\cdot\frac{1}{3} = \frac{1}{3}$, and the limit of the logarithm is $\ln \frac{1}{3}$, and the integral converges.

If $\frac{a}{3} - 1 > 0$, i.e. $a > 3$, the limit in (*) is 0, and the limit of the logarithm is infinite. The integral diverges.

Conclusion: The integral only converges when $a = 3$.

47. $\int_0^1 x^n (\ln x)^2 \, dx = \lim\limits_{t \to 0^+} \int_t^1 x^n (\ln x)^2 \, dx$ (*)

If $n = -1$, then (*) $= \lim\limits_{t \to 0^+} \frac{1}{3}(\ln x)^3 \Big|_t^1 = \lim\limits_{t \to 0^+} \frac{1}{3}[(\ln 1)^3 - (\ln t)^3]$ which is infinite, as the logarithm becomes infinite. Thus the integral diverges in this case.

So, consider $n \neq -1$. Then, by integration by parts:
$$\int u \, dv = uv - \int v \, du$$

$$\int_t^1 x^n (\ln x)^2 \, dx = \frac{x^{n+1}}{n+1}(\ln x)^2 \Big|_t^1 - 2\int_t^1 (\ln x)\frac{x^n}{n+1}\, dx = -\frac{t^{n+1}}{n+1}(\ln t)^2 - \frac{2}{n+1}\int_t^1 (\ln x)x^n \, dx$$

$u = (\ln x)^2$	$dv = x^n \, dx$
$du = (2\ln x) \cdot \frac{1}{x}\, dx$	$v = \frac{x^{n+1}}{n+1}$

Now, $\int_t^1 (\ln x)x^n \, dx = \frac{x^{n+1}}{n+1}(\ln x)\Big|_t^1 - \int_t^1 \frac{x^n}{n+1}\, dx = -\frac{t^{n+1}}{n+1}(\ln t) - \frac{1}{(n+1)^2}x^{n+1}\Big|_t^1$

$u = \ln x$	$dv = x^n \, dx$
$du = \frac{1}{x}\, dx$	$v = \frac{x^{n+1}}{n+1}$

$$= -\frac{t^{n+1}}{n+1}(\ln t) - \frac{1}{(n+1)^2}(1 - t^{n+1}) \Rightarrow$$

$$\int_t^1 x^n (\ln x)^2 \, dx = -\frac{t^{n+1}}{n+1}(\ln t)^2 - \frac{2}{(n+1)}\left[-\frac{t^{n+1}}{n+1}(\ln t) - \frac{1}{(n+1)^2}(1 - t^{n+1})\right]$$

$$= -\frac{t^{n+1}}{n+1}(\ln t)^2 + \frac{2t^{n+1}(\ln t)}{(n+1)^2} + \frac{2}{(n+1)^3}(1 - t^{n+1})$$

The convergence of the given integral depends on the limits of these three terms (we can ignore the constants in each term):

If $n + 1 < 0$, each of the three terms tends to $-\infty$ as $t \to 0^+$, so the integral diverges.

If $n + 1 > 0$:

$\lim\limits_{t \to 0^+}(1 - t^{n+1}) = 1$ (limit exists).

Form $0 \cdot \infty$: $\lim\limits_{t \to 0^+} t^{n+1}(\ln t) = \lim\limits_{t \to 0^+} \frac{\ln t}{t^{-(n+1)}} = \lim\limits_{t \to 0^+} \frac{\frac{1}{t}}{-(n+1)t^{-(n+1)-1}}$

$$= \lim\limits_{t \to 0^+} \frac{1}{-(n+1)t^{-(n+1)}} = \lim\limits_{t \to 0^+} \frac{t^{n+1}}{-(n+1)} = 0 \text{ (limit exists).}$$

Form $0 \cdot \infty$: $\lim\limits_{t \to 0^+} t^{n+1}(\ln t)^2 = \lim\limits_{t \to 0^+} \frac{(\ln t)^2}{t^{-(n+1)}} = \lim\limits_{t \to 0^+} \frac{2(\ln t)(\frac{1}{t})}{-(n+1)t^{-(n+1)-1}}$

$$= \lim\limits_{t \to 0^+} \frac{2\ln t}{-(n+1)t^{-(n+1)}} = \lim\limits_{t \to 0^+} \frac{\frac{2}{t}}{(n+1)^2 t^{-(n+1)-1}}$$

$$= \lim\limits_{t \to 0^+} \frac{2t^{(n+1)}}{(n+1)^2} = 0 \text{ (limit exists).}$$

Conclusion: Integral converges for $n + 1 > 0$, i.e. for $n > -1$.

49. $\int_0^\infty \sin x \, dx = \lim_{t \to \infty} \int_0^t \sin x \, dx = \lim_{t \to \infty} (-\cos x)\big|_0^t = \lim_{t \to \infty} - [\cos t - \cos 0]$

$= \lim_{t \to \infty} [-\cos t + 1]$ d.n.e., as the cosine continues to vary from -1 to 1,

forever. The integral diverges.

$\int_{-\infty}^0 \sin x \, dx = \lim_{t \to -\infty} \int_t^0 \sin x \, dx = \lim_{t \to -\infty} (-\cos x)\big|_t^0 = \lim_{t \to -\infty} - [\cos 0 - \cos t]$

$= \lim_{t \to -\infty} [-1 + \cos t]$ d.n.e., as the cosine continues to vary from -1 to 1, forever.

The integral diverges.

$\lim_{t \to \infty} \int_{-t}^t \sin x \, dx = \lim_{t \to \infty} (-\cos x)\big|_{-t}^t = \lim_{t \to \infty} - [\cos t - \cos(-t)] = 0$, as $\cos t = \cos(-t)$ and

$\lim_{t \to \infty} 0 = 0$.

51.

$A = \int_0^\infty e^{-x} dx = \lim_{t \to \infty} \int_0^t e^{-x} \, dx = \lim_{t \to \infty} -e^{-x}\big|_0^t = \lim_{t \to \infty} - (e^{-t} - e^0)$

$= \lim_{t \to \infty} (-\frac{1}{e^t} + 1) = 1.$

53.

$A = \int_0^1 x^{-\frac{1}{4}} dx = \lim_{t \to 0^+} \int_t^1 x^{-\frac{1}{4}} \, dx = \lim_{t \to 0^+} \frac{4}{3} x^{\frac{3}{4}}\big|_t^1 = \lim_{t \to 0^+} \frac{4}{3}(1 - t^{\frac{3}{4}}) = \frac{4}{3}.$

55.

$V = 2\pi \int_0^\infty x e^{-x} \, dx = 2\pi \lim_{t \to \infty} \int_0^t x e^{-x} \, dx$

$u = x$	$dv = e^{-x} \, dx$
$du = dx$	$v = -e^{-x}$

$= 2\pi \lim_{t \to \infty} \left[-x e^{-x}\big|_0^t + \int_0^t e^{-x} \, dx \right] = 2\pi \lim_{t \to \infty} \left[-t e^{-t} - e^{-x}\big|_0^t \right]$

$= 2\pi \lim_{t \to \infty} \left[-t e^{-t} - (e^{-t} - e^0) \right] = 2\pi \lim_{t \to \infty} \left[-(t+1) e^{-t} + 1 \right]$

Since $\lim_{t \to \infty} (t+1) e^{-t} = \lim_{t \to \infty} \frac{t+1}{e^t} = \lim_{t \to \infty} \frac{1}{e^t} = 0,$ the volume $V = 2\pi.$

57.

$V = \pi \int_1^\infty (y^{-4})^2 \, dy - \pi \lim_{t \to \infty} \int_1^t y^{-8} \, dy = \pi \lim_{t \to \infty} \frac{y^{-7}}{-7}\bigg|_1^t$

$= \pi \lim_{t \to \infty} -\frac{1}{7}(\frac{1}{t^7} - 1) = \frac{\pi}{7}.$

CHAPTER 9
Sequences and Series

§9.1 Sequences

1. In the sequence $\left(\frac{1}{2}, \frac{2}{3}, \frac{3}{4}, \frac{4}{5}, \ldots\right)$, the numerators of the fractions are the positive integers, starting at 1, and the denominators are the positive integers starting at 2, so a formula for the nth term is $\frac{n}{n+1}$.

3. In the sequence $\left(\frac{1}{2}, -\frac{4}{5}, \frac{9}{8}, -\frac{16}{11}, \ldots\right)$ the numerators of the fractions are the squares of the positive integers, so n^2, while the denominators start at 2 and increase by 3: $2 + 3(n-1) = 2 + 3n - 3 = 3n - 1$, and the sign alternates starting with '+', so $(-1)^{n+1}$. A formula for the nth term is $\frac{(-1)^{n+1}n^2}{3n-1}$.

5. (a) In $a_n = 1 + \frac{1}{\sqrt{2n}}$, $2n \to \infty \Rightarrow \frac{1}{\sqrt{2n}} \to 0 \Rightarrow 1 + \frac{1}{\sqrt{2n}} \to 1$, so $L = 1$.

(b) We want N such that

$$n > N \Rightarrow \left|1 + \frac{1}{\sqrt{2n}} - 1\right| < \frac{1}{10}$$

i.e: $n > N \Rightarrow \frac{1}{\sqrt{2n}} < \frac{1}{10}$

i.e: $n > N \Rightarrow \sqrt{2n} > 10$

i.e: $n > N \Rightarrow 2n > 100$

i.e: $n > N \Rightarrow n > 50$

The smallest N is therefore 50.

(c) We want N such that

$$n > N \Rightarrow \left|1 + \frac{1}{\sqrt{2n}} - 1\right| < \frac{1}{100}$$

i.e: $n > N \Rightarrow \frac{1}{\sqrt{2n}} < \frac{1}{100}$

i.e: $n > N \Rightarrow \sqrt{2n} > 100$

i.e: $n > N \Rightarrow 2n > 10,000$

i.e: $n > N \Rightarrow n > 5,000$

The smallest N is therefore 5,000.

(d) We want N such that

$$n > N \Rightarrow \left|1 + \frac{1}{\sqrt{2n}} - 1\right| < \epsilon$$

i.e: $n > N \Rightarrow \frac{1}{\sqrt{2n}} < \epsilon$

i.e: $n > N \Rightarrow \sqrt{2n} > \frac{1}{\epsilon}$

i.e: $n > N \Rightarrow 2n > \frac{1}{\epsilon^2}$

i.e: $n > N \Rightarrow n > \frac{1}{2\epsilon^2}$

The smallest N is therefore the smallest integer greater than or equal to $\frac{1}{2\epsilon^2}$.

7. $\frac{2n+1}{5n} = \frac{2 + \frac{1}{n}}{5} \to \frac{2}{5}$, as $\frac{1}{n} \to 0 \Rightarrow L = \frac{2}{5}$.

(b) We want N such that

$$n > N \Rightarrow \left| \frac{2n+1}{5n} - \frac{2}{5} \right| < \frac{1}{10}$$

i.e: $n > N \Rightarrow \left| \frac{2n+1-2n}{5n} \right| < \frac{1}{10}$

i.e: $n > N \Rightarrow \frac{1}{5n} < \frac{1}{10}$

i.e: $n > N \Rightarrow 5n > 10$

i.e: $n > N \Rightarrow n > 2$

The smallest N is therefore 2.

(c) We want N such that

$$n > N \Rightarrow \left| \frac{2n+1}{5n} - \frac{2}{5} \right| < \frac{1}{100}$$

i.e: $n > N \Rightarrow \left| \frac{2n+1-2n}{5n} \right| < \frac{1}{100}$

i.e: $n > N \Rightarrow \frac{1}{5n} < \frac{1}{100}$

i.e: $n > N \Rightarrow 5n > 100$

i.e: $n > N \Rightarrow n > 20$

The smallest N is therefore 20.

(d) We want N such that

$$n > N \Rightarrow \left| \frac{2n+1}{5n} - \frac{2}{5} \right| < \epsilon$$

i.e: $n > N \Rightarrow \left| \frac{2n+1-2n}{5n} \right| < \epsilon$

i.e: $n > N \Rightarrow \frac{1}{5n} < \epsilon$

i.e: $n > N \Rightarrow 5n > \frac{1}{\epsilon}$

i.e: $n > N \Rightarrow n > \frac{1}{5\epsilon}$

The smallest N is therefore the smallest integer greater than or equal to $\frac{1}{5\epsilon}$.

9. The sequence of terms $a_n = \frac{n}{10^{100}}$ is unbounded, and therefore divergent: Let M be any large positive number, then $a_n = \frac{n}{10^{100}} > M$ for all $n > 10^{100} M$. (See CYU 9.7: If a sequence converges, then it is bounded.)

11. For $n > 100$, $2n = n + n > 100 + n \Rightarrow a_n = \frac{n}{\sqrt{n+100}} > \frac{n}{\sqrt{2n}} = \frac{\sqrt{n}}{\sqrt{2}} \to \infty$ as $n \to \infty$. Thus, the sequence (a_n) is unbounded and therefore divergent. (See CYU 9.7: If a sequence converges, then it is bounded.)

13. $a_n = \frac{4}{n} \to 0$ as $n \to \infty$. The sequence converges to 0.

15. As $n \to \infty$, $a_n = 5 - \frac{1}{n} \to 5$, since $\frac{1}{n} \to 0$. The sequence converges to 5.

17. $a_n = 1 + (-1)^n \Rightarrow (a_n) = (0, 2, 0, 2, 0, 2, \dots)$ diverges, as no one number is approached as $n \to \infty$.

19. $(a_n) = \left(\frac{(-1)^n n}{n+1} \right)$ diverges: $\lim_{n \to \infty} \frac{n}{n+1} = \lim_{n \to \infty} \frac{1}{1 + \frac{1}{n}} = 1$ since $\frac{1}{n} \to 0$, and consecutive terms have opposite signs. The distance between consecutive terms tends to 2, so no one number is approached.

21. By l'Hôpital's Rule, $\lim\limits_{x \to \infty} \dfrac{x}{x^2+1} = \lim\limits_{x \to \infty} \dfrac{1}{2x} = 0 \Rightarrow a_n = \dfrac{n}{n^2+1} \to 0$. The sequence converges to 0.

23. $a_n = \dfrac{(n+1)!}{n!} = \dfrac{(n+1)n!}{n!} = n+1 \to \infty$. The sequence diverges as it is unbounded (See CYU 9.7.)

25. $a_n = \cos n\pi = (-1)^n \Rightarrow (a_n) = (-1, 1, -1, 1, \ldots)$. The sequence diverges as no one number is approached.

27. $a_n = \dfrac{\sin \frac{n\pi}{3}}{\cos \frac{n\pi}{3}} = \tan \dfrac{n\pi}{3} = (\tan \frac{\pi}{3}, \tan \frac{2\pi}{3}, \tan \pi, \tan \frac{4\pi}{3}, \tan \frac{5\pi}{3}, \tan 2\pi, \tan \frac{\pi}{3}, \tan \frac{2\pi}{3}, \ldots)$
$$= (\sqrt{3}, -\sqrt{3}, 0, \sqrt{3}, -\sqrt{3}, 0, \ldots)$$
The sequence diverges, as the three values keep repeating, and no one number is approached.

29. Let $b_n = 5 - \left(\frac{1}{2}\right)^n$ and $c_n = 5 + \left(\frac{1}{2}\right)^n$. Since $\left(\frac{1}{2}\right)^n \to 0$, as $\left|\frac{1}{2}\right| < 1$, both (b_n) and (c_n) converge to 5. For all n, $b_n \le a_n = 5 + \left(-\frac{1}{2}\right)^n \le c_n$, so (a_n) converges to 5.

31. $\dfrac{n + |\cos n|}{n} = 1 + \dfrac{|\cos n|}{n}$. Since $0 \le \dfrac{|\cos n|}{n} \le \dfrac{1}{n}$ for all n, and $\lim\limits_{n \to \infty} 0 = \lim\limits_{n \to \infty} \dfrac{1}{n} = 0$, $\lim\limits_{n \to \infty} \dfrac{|\cos n|}{n} = 0 \Rightarrow \lim\limits_{n \to \infty} 1 + \dfrac{|\cos n|}{n} = 1$.

33. As $0 \le \left(\frac{1}{n}\right)^2 \le \frac{1}{n}$ for all n, and $\lim\limits_{n \to \infty} 0 = \lim\limits_{n \to \infty} \dfrac{1}{n} = 0$, $\lim\limits_{n \to \infty}\left(\frac{1}{n}\right)^2 = 0 \Rightarrow \lim\limits_{n \to \infty}\left[4 + \left(\frac{1}{n}\right)^2\right] = 4 \Rightarrow \lim\limits_{n \to \infty} \sqrt{4 + \left(\frac{1}{n}\right)^2} = \sqrt{4} = 2$.

35. For all n, $\dfrac{3n - \frac{1}{3}}{5n} \le \dfrac{3n + (-\frac{1}{3})^n}{5n} \le \dfrac{3n + \frac{1}{3}}{5n}$, and $\lim\limits_{n \to \infty} \dfrac{3n - \frac{1}{3}}{5n} = \lim\limits_{n \to \infty} \dfrac{3n + \frac{1}{3}}{5n} = \dfrac{3}{5}$, since, by l'Hôpital's Rule, $\lim\limits_{x \to \infty} \dfrac{3x \pm \frac{1}{3}}{5x} = \lim\limits_{x \to \infty} \dfrac{3}{5} = \dfrac{3}{5}$. Thus $\lim\limits_{n \to \infty} \dfrac{3n + (-\frac{1}{3})^n}{5n} = \dfrac{3}{5}$.

37. $\lim\limits_{n \to \infty} \dfrac{5n\pi}{4n - 1} = \lim\limits_{n \to \infty} \dfrac{5\pi}{4 - \frac{1}{n}} = \dfrac{5\pi}{4}$, since $\dfrac{1}{n} \to 0$. Therefore, $\lim\limits_{n \to \infty} \left|\sin \dfrac{5n\pi}{4n - 1}\right| = \left|\sin \dfrac{5\pi}{4}\right| = \left|-\dfrac{1}{\sqrt{2}}\right| = \dfrac{1}{\sqrt{2}}$.

39. $\lim\limits_{n \to \infty} \dfrac{en + 1}{n - 1} = \lim\limits_{n \to \infty} \dfrac{e + \frac{1}{n}}{1 - \frac{1}{n}} = e$, since $\dfrac{1}{n} \to 0$. Therefore, $\lim\limits_{n \to \infty} \ln \dfrac{en + 1}{n - 1} = \ln e = 1$.

41. $a_n = \ln n^2 - \ln 5n^2 = \ln \dfrac{n^2}{5n^2} = \ln \dfrac{1}{5}$, a constant. Thus $\lim\limits_{n \to \infty} a_n = \ln \dfrac{1}{5}$.

43. $\dfrac{4n}{\sqrt{n^2 + 1}} = \dfrac{4}{\sqrt{\frac{n^2+1}{n^2}}} = \dfrac{4}{\sqrt{1 + \frac{1}{n^2}}} \to 4$, as $\dfrac{1}{n^2} \to 0$. Therefore, $a_n = \sqrt{\dfrac{4n}{\sqrt{n^2 + 1}}} \to \sqrt{4} = 2$.

45. $\lim\limits_{x\to\infty}\dfrac{e^x}{x^3}=\lim\limits_{x\to\infty}\dfrac{e^x}{3x^2}=\lim\limits_{x\to\infty}\dfrac{e^x}{6x}=\lim\limits_{x\to\infty}\dfrac{e^x}{6}=\infty$, so the sequence $(a_n)=\left(\dfrac{e^n}{n^3}\right)$ diverges.

47. $\lim\limits_{x\to\infty}\dfrac{(\ln x)^2}{x}=\lim\limits_{x\to\infty}\dfrac{2\ln x\cdot\frac1x}{1}=\lim\limits_{x\to\infty}\dfrac{2\ln x}{x}=\lim\limits_{x\to\infty}\dfrac{2\cdot\frac1x}{1}=\lim\limits_{x\to\infty}\dfrac2x=0$. Therefore, the

sequence $(a_n)=\left(\dfrac{(\ln n)^2}{n}\right)$ converges to 0.

49. $\lim\limits_{x\to\infty}\dfrac{x^2\sin\frac1x}{2x-1}=\lim\limits_{x\to\infty}\dfrac{\sin\frac1x}{\frac2x-\frac1{x^2}}=\lim\limits_{x\to\infty}\dfrac{(\cos\frac1x)(-x^{-2})}{2(-x^{-2})+2(x^{-3})}\cdot\dfrac{x^2}{x^2}=\lim\limits_{x\to\infty}\dfrac{-\cos\frac1x}{-2+\frac2x}$

$=\dfrac{-\cos 0}{-2}=\dfrac12.$ Thus, $\left(\dfrac{n^2\sin\frac1n}{2n-1}\right)$ converges to $\dfrac12.$

51. $a_n=n-\sqrt{n^2-n}=n\left(1-\dfrac{\sqrt{n^2-n}}{n}\right)=\dfrac{1-\sqrt{\frac{n^2-n}{n^2}}}{\frac1n}=\dfrac{1-\sqrt{1-\frac1n}}{\frac1n}$

$\lim\limits_{x\to\infty}\dfrac{1-(1-x^{-1})^{\frac12}}{x^{-1}}=\lim\limits_{x\to\infty}\dfrac{-\frac12(1-\frac1x)^{-\frac12}x^{-2}}{-x^{-2}}=\lim\limits_{x\to\infty}\dfrac{1}{2\sqrt{1-\frac1x}}=\dfrac12\Rightarrow(a_n)\to\dfrac12.$

53. $f(x)=\dfrac{x}{2x+1}\Rightarrow f'(x)=\dfrac{(2x+1)\cdot1-x\cdot2}{(2x+1)^2}=\dfrac{1}{(2x+1)^2}>0\Rightarrow$ the sequence

$(a_n)=\left(\dfrac{n}{2n+1}\right)$ is an increasing sequence.

55. $f(x)=\dfrac{x}{2^x}\Rightarrow f'(x)=\dfrac{2^x\cdot1-x\cdot2^x\ln2}{(2^x)^2}=\dfrac{2^x(1-x\ln2)}{(2^x)^2}=\dfrac{1-x\ln2}{2^x}.$

For all x sufficiently large, $1-x\ln2<0\Rightarrow f'(x)<0\Rightarrow$ the sequence $(a_n)=\left(\dfrac{n}{2^n}\right)$ is a decreasing sequence.

57. $f(x)=\dfrac{-x}{2x+1}\Rightarrow f'(x)=\dfrac{(2x+1)(-1)+x(2)}{(2x+1)^2}=\dfrac{-2x-1+2x}{(2x+1)^2}=-\dfrac{1}{(2x+1)^2}<0\Rightarrow$ the

sequence $(a_n)=\left(\dfrac{-n}{2n+1}\right)$ is a decreasing sequence.

59. $f(x)=\dfrac{\ln(x+1)}{x+1}\Rightarrow f'(x)=\dfrac{(x+1)\cdot\frac1{x+1}-\ln(x+1)\cdot1}{(x+1)^2}=\dfrac{1-\ln(x+1)}{(x+1)^2}.$ For all x suffi-

ciently large, $1-\ln(x+1)<0\Rightarrow f'(x)<0\Rightarrow$ the sequence $(a_n)=\left(\dfrac{\ln(n+1)}{n+1}\right)$ is a decreasing sequence.

61. $f(x)=\dfrac{\ln x}{x}\Rightarrow f'(x)=\dfrac{x\cdot\frac1x-\ln x}{x^2}=\dfrac{1-\ln x}{x^2}.$ For $x>e,\ 1-\ln x<0\Rightarrow f'(x)<0\Rightarrow$

the sequence $(a_n)=\left(\dfrac{\ln n}{n}\right)$ is a decreasing sequence, hence a monotone sequence. Being decreasing it is bounded above, and it is bounded below by 0, so it is a bounded sequence. Since it is monotone and bounded, the sequence converges.

63. We show that the sequence $(a_n) = (-1, 1, -1, 1, \ldots)$ diverges, by demonstrating that no fixed but *arbitrary* real number L can be the limit of the sequence:

Let N be *any* positive integer, and let $\epsilon = \frac{1}{2}$. Since any two numbers in the interval $(L - \frac{1}{2}, L + \frac{1}{2})$ are less than one unit apart, both a_{N+1} and a_{N+2} cannot be contained in this interval, as one of the numbers is -1 while the other is 1. This shows that no N "works" for $\epsilon = \frac{1}{2}$, and that, consequently the *arbitrarily* chosen number L cannot be a limit of the sequence.

65. One possible solution is the sequence $(-1, 1, -1, 1, \ldots)$, since each of the subsequences $(-1, -1, -1, \ldots)$ and $(1, 1, 1, \ldots)$ of constants converges to that constant.

67. Given $\epsilon > 0$, we want N first of all to be so large that $n > N \Rightarrow rn + 1 \neq 0$, i.e., that $n > \left| \frac{1}{r} \right|$. Then we want:

$$n > N \Rightarrow \left| \frac{n}{rn+1} - \frac{1}{r} \right| < \epsilon$$

$$\text{i.e: } n > N \Rightarrow \left| \frac{nr - (rn+1)}{r(rn+1)} \right| < \epsilon$$

$$\text{i.e: } n > N \Rightarrow \frac{1}{|r(rn+1)|} < \epsilon$$

$$\text{i.e: } n > N \Rightarrow |r(rn+1)| > \frac{1}{\epsilon}$$

$$\text{i.e: } n > N \Rightarrow |rn+1| > \frac{1}{\epsilon|r|}$$

The Triangle Inequality, $|a| \pm b| \geq |a| - |b|$, $\Rightarrow |rn+1| \geq |rn| - |1| = n|r| - 1 \Rightarrow$

want: $n > N \Rightarrow n|r| - 1 > \frac{1}{\epsilon|r|}$ i.e.: $n > N \Rightarrow n > \dfrac{\frac{1}{\epsilon|r|} + 1}{|r|} = \dfrac{1 + \epsilon|r|}{\epsilon r^2}$

Thus, let N be a positive integer larger than both $\left| \frac{1}{r} \right|$ and $\frac{1 + \epsilon|r|}{\epsilon r^2}$.

69. Given any $\epsilon > 0$, we seek an N such that $\big| |a_n| - |L| \big| < \epsilon$. As $\lim\limits_{x \to \infty} a_n = L$, there is an N such that $|a_n - L| < \epsilon$. The desired result follows from the (alternate) Triangle Inequality, as $|a_n - L| \geq \big| |a_n| - |L| \big|$.

§9.2 Series

1. In the sum $\frac{1}{3} + \frac{1}{5} + \frac{1}{7} + \frac{1}{9}$, the numerators are all 1 and the denominators are positive odd integers starting at 3 and ending at 9. An odd integer is denoted either as $2n - 1$ or $2n + 1$. Since we want to start n at 1, we use $2n + 1$. Thus the expression is $\sum\limits_{n=1}^{4} \frac{1}{2n+1}$ or, starting at 0, $\sum\limits_{n=0}^{3} \frac{1}{2(n+1)+1} = \sum\limits_{n=0}^{3} \frac{1}{2n+3}$.

3. $1 + 10 + 100 + 1000 + 10000 = \sum\limits_{n=0}^{4} 10^n$, or starting at 1, $\sum\limits_{n=1}^{5} 10^{n-1}$.

5. In the series $-\frac{5}{2} + \frac{10}{4} - \frac{15}{8} + \frac{20}{16} - \ldots$, the numerators are multiples of 5 starting at 5, so $5n$, and the denominators are powers of 2, starting at the first power, so 2^n. The alternating sign can be accommodated by $(-1)^n$. Putting it all together: $\sum_{n=1}^{\infty} \frac{(-1)^n 5n}{2^n}$, or starting at 0, $\sum_{n=0}^{\infty} \frac{(-1)^{n+1} 5(n+1)}{2^{n+1}} = \sum_{n=0}^{\infty} \frac{(-1)^{n+1}(5n+5)}{2^{n+1}}.$

7. (a) $\sum_{n=1}^{\infty} \left(\frac{1}{5}\right)^{n-1}$ is a geometric series with $a = 1$ and $r = \frac{1}{5}$. Since $|r| < 1$, the series converges, and its sum is $\frac{a}{1-r} = \frac{1}{1-\frac{1}{5}} = \frac{1}{\frac{4}{5}} = \frac{5}{4}.$

(b) $\sum_{n=1}^{\infty} \left(\frac{1}{5}\right)^{n} = \sum_{n=1}^{\infty} \frac{1}{5}\left(\frac{1}{5}\right)^{n-1}$ is a convergent geometric series with $a = \frac{1}{5}$ and $r = \frac{1}{5}$, so

its sum is $\frac{a}{1-r} = \frac{\frac{1}{5}}{1-\frac{1}{5}} = \frac{\frac{1}{5}}{\frac{4}{5}} = \frac{1}{4}.$

(c) $\sum_{n=0}^{\infty} \left(\frac{1}{5}\right)^{n} = \sum_{n=1}^{\infty} \left(\frac{1}{5}\right)^{n-1}$, whose sum is $\frac{5}{4}$, from (a).

9. $\sum_{n=1}^{\infty} \left(\frac{5}{7}\right)^{n-1}$ is a convergent geometric series, since $r = \frac{5}{7}$ and $|r| < 1$. As $a = 1$, the sum of the series is $\frac{a}{1-r} = \frac{1}{1-\frac{5}{7}} = \frac{1}{\frac{2}{7}} = \frac{7}{2}.$

11. $\sum_{n=0}^{\infty} \frac{5}{100^n} = \sum_{n=0}^{\infty} 5\left(\frac{1}{100}\right)^{n}$ which is a convergent geometric series, as $r = \frac{1}{100}$ and $|r| < 1$. With $a = 5$, the sum of the series is $\frac{a}{1-r} = \frac{5}{1-\frac{1}{100}} = \frac{5}{\frac{99}{100}} = \frac{500}{99}.$

13. The series $\sum_{n=1}^{\infty} \frac{2}{3^{n-1}} = \sum_{n=1}^{\infty} 2\left(\frac{1}{3}\right)^{n-1}$ is a convergent geometric series with $a = 2$ and $r = \frac{1}{3}$. Its sum is $\frac{a}{1-r} = \frac{2}{1-\frac{1}{3}} = \frac{2}{\frac{2}{3}} = 3.$

15. The series $\sum_{n=1}^{\infty} \frac{n^n}{n!}$ diverges by the Divergence Test, as $\lim_{n \to \infty} \frac{n^n}{n!} \neq 0$, because

$$\frac{n^n}{n!} = \frac{\not{n} \cdot n \cdot n \cdots n}{\not{n}(n-1)(n-2)(n-3)\cdots(2)(1)} = \frac{n}{n-1} \cdot \frac{n}{n-2} \cdot \frac{n}{n-3} \cdots \frac{n}{2} \cdot \frac{n}{1} > 1$$

as each term in the product is greater than 1, if $n > 1$.

17. The series $\sum_{n=0}^{\infty} \frac{1}{\sin\frac{n\pi}{4}}$ diverges, since for any N and $n > N$ there are undefined terms whenever n is a multiple of 4, as the sine function is then 0.

19. $\sum_{n=1}^{\infty} \frac{3^{n-1}}{9^n} = \sum_{n=1}^{\infty} \frac{1}{9}\left(\frac{3}{9}\right)^{n-1} = \sum_{n=1}^{\infty} \frac{1}{9}\left(\frac{1}{3}\right)^{n-1}$ is a convergent geometric series with $a = \frac{1}{9}$

and $r = \frac{1}{3}$. The sum of the series is $\frac{a}{1-r} = \frac{\frac{1}{9}}{1-\frac{1}{3}} = \frac{\frac{1}{9}}{\frac{2}{3}} = \frac{1}{9} \cdot \frac{3}{2} = \frac{1}{6}$.

21. $\sum_{n=1}^{\infty} \left[\frac{3}{2^n} + \left(\frac{1}{4}\right)^n\right] = \sum_{n=1}^{\infty} \frac{3}{2^n} + \sum_{n=1}^{\infty} \left(\frac{1}{4}\right)^n$, as each is a convergent geometric series.

In $\sum_{n=1}^{\infty} \frac{3}{2^n} = \sum_{n=1}^{\infty} \frac{3}{2}\left(\frac{1}{2}\right)^{n-1}$, $a = \frac{3}{2}$ and $r = \frac{1}{2}$ so that the sum of this series is

$$\frac{a}{1-r} = \frac{\frac{3}{2}}{1-\frac{1}{2}} = \frac{\frac{3}{2}}{\frac{1}{2}} = 3.$$

In $\sum_{n=1}^{\infty} \left(\frac{1}{4}\right)^n = \sum_{n=1}^{\infty} \frac{1}{4}\left(\frac{1}{4}\right)^{n-1}$, $a = \frac{1}{4}$ and $r = \frac{1}{4}$ so that the sum of this series is

$$\frac{a}{1-r} = \frac{\frac{1}{4}}{1-\frac{1}{4}} = \frac{\frac{1}{4}}{\frac{3}{4}} = \frac{1}{3}.$$

Therefore, the series $\sum_{n=1}^{\infty} \left[\frac{3}{2^n} + \left(\frac{1}{4}\right)^n\right]$ converges to $3 + \frac{1}{3} = \frac{10}{3}$.

23. The series $\frac{1}{3} - \frac{3}{2} + \frac{1}{9} - \frac{3}{4} + \frac{1}{27} - \frac{3}{8} + \frac{1}{3^4} - \frac{3}{2^4} + \cdots = \sum_{n=1}^{\infty} \left[\left(\frac{1}{3}\right)^n - \frac{3}{2^n}\right] = \sum_{n=1}^{\infty} \left(\frac{1}{3}\right)^n - \sum_{n=1}^{\infty} \frac{3}{2^n}$,

the difference of two convergent geometric series.

In $\sum_{n=1}^{\infty} \left(\frac{1}{3}\right)^n = \sum_{n=1}^{\infty} \frac{1}{3}\left(\frac{1}{3}\right)^{n-1}$, $a = \frac{1}{3}$ and $r = \frac{1}{3}$ so that the sum of this series is

$$\frac{a}{1-r} = \frac{\frac{1}{3}}{1-\frac{1}{3}} = \frac{\frac{1}{3}}{\frac{2}{3}} = \frac{1}{2}.$$

In $\sum_{n=1}^{\infty} \frac{3}{2^n} = \sum_{n=1}^{\infty} \frac{3}{2}\left(\frac{1}{2}\right)^{n-1}$, $a = \frac{3}{2}$ and $r = \frac{1}{2}$ so that the sum of this series is

$$\frac{a}{1-r} = \frac{\frac{3}{2}}{1-\frac{1}{2}} = \frac{\frac{3}{2}}{\frac{1}{2}} = 3.$$

Therefore, the given series converges to $\frac{1}{2} - 3 = -\frac{5}{2}$.

25. Let $f(x) = \frac{x}{x^2+1}$, and consider: (i) $\lim_{x \to \infty} f(x) = \lim_{x \to \infty} \frac{x}{x^2+1} - \lim_{x \to \infty} \frac{1}{2x} - 0 \to$
$\lim_{n \to \infty} \frac{n}{n^2+1} = 0$.

(ii) $f'(x) = \frac{(x^2+1) - x(2x)}{(x^2+1)^2} = \frac{1-x^2}{(x^2+1)^2} < 0$ for $x > 1$, so the sequence $\left(\frac{n}{n^2+1}\right)$ is de-

creasing. Thus, by the Alternating Series Theorem, the series $\sum_{n=0}^{\infty} (-1)^{n-1}\frac{n}{n^2+1}$ converges.

27.   As $n \to \infty, \ln n \to \infty \Rightarrow \frac{1}{\ln n} \to 0$. Since the logarithmic function is an increasing function, $\ln(n+1) > \ln n$ so that for all $n \geq 2$, $\frac{1}{\ln n} > \frac{1}{\ln(n+1)}$, proving that the sequence $(a_n) = \left(\frac{1}{\ln n}\right)$ is a decreasing sequence. Thus, by the Alternating Series Theorem, the series $\sum_{n=2}^{\infty} (-1)^n \frac{1}{\ln n}$ converges.

29.   Since $\lim_{x \to \infty} \frac{x}{\ln x} = \lim_{x \to \infty} \frac{1}{\frac{1}{x}} = \lim_{x \to \infty} x = \infty$, $\lim_{n \to \infty} \left| \frac{(-1)^n n}{\ln n} \right| \neq 0 \Rightarrow \sum_{n=2}^{\infty} (-1)^n \frac{n}{\ln n}$ diverges, by the Divergence Test.

31.   The series $\sum_{n=1}^{\infty} (-1)^n \frac{1}{\sin^2 n}$ diverges as its terms do not tend to 0, because the sine function being periodic implies its square is also, and the values of $\sin^2 n$ keep varying between 0 and 1 forever.

33. $a_n = \frac{1}{n}$ and we want N such that $a_{N+1} < .0001$. Now, $.0001 = \frac{1}{10,000}$, so we want $\frac{1}{N+1} < \frac{1}{10,000} \Rightarrow N+1 > 10,000 \Rightarrow N = 10,000$.

35. $a_n = \frac{1}{(2n)!}$ and we want N such that $a_{N+1} < .0001$, i.e. $\frac{1}{(2(N+1))!} < .0001 \Rightarrow \frac{1}{(2N+2)!} < .0001$. Now $\frac{1}{6!} \approx 0.0014$ and $\frac{1}{8!} \approx 0.0000248$, so $2N+2 = 8 \Rightarrow N = 3$.

37. $a_n = \frac{1}{\sqrt{n}}$ and we want N such that $a_{N+1} < .0001$, i.e. $\frac{1}{\sqrt{N+1}} < 10^{-4} \Rightarrow \frac{1}{N+1} < 10^{-8}$ $\Rightarrow N+1 > 10^8 \Rightarrow N = 10^8$.

39. $\sum_{n=1}^{\infty} \left(\frac{1}{\sqrt{n}} - \frac{1}{\sqrt{n+1}} \right) \Rightarrow s_4 = \left(1 - \frac{1}{\sqrt{2}}\right) + \left(\frac{1}{\sqrt{2}} - \frac{1}{\sqrt{3}}\right) + \left(\frac{1}{\sqrt{3}} - \frac{1}{\sqrt{4}}\right) + \left(\frac{1}{\sqrt{4}} - \frac{1}{\sqrt{5}}\right) = 1 - \frac{1}{\sqrt{5}}$
$s_n = \left(1 - \frac{1}{\sqrt{2}}\right) + \left(\frac{1}{\sqrt{2}} - \frac{1}{\sqrt{3}}\right) + \cdots + \left(\frac{1}{\sqrt{n-1}} - \frac{1}{\sqrt{n}}\right) + \left(\frac{1}{\sqrt{n}} - \frac{1}{\sqrt{n+1}}\right) = 1 - \frac{1}{\sqrt{n+1}}$. The sum of the series is $\lim_{n \to \infty} s_n = \lim_{n \to \infty} 1 - \frac{1}{\sqrt{n+1}} = 1$, as $\frac{1}{\sqrt{n+1}} \to 0$ as $n \to \infty$.

41. $\sum_{n=1}^{\infty} \left(\frac{1}{\ln(n+2)} - \frac{1}{\ln(n+1)} \right) \Rightarrow$
$s_4 = \left(\frac{1}{\ln 3} - \frac{1}{\ln 2}\right) + \left(\frac{1}{\ln 4} - \frac{1}{\ln 3}\right) + \left(\frac{1}{\ln 5} - \frac{1}{\ln 4}\right) + \left(\frac{1}{\ln 6} - \frac{1}{\ln 5}\right) = \frac{1}{\ln 6} - \frac{1}{\ln 2}$
$s_n = \left(\frac{1}{\ln 3} - \frac{1}{\ln 2}\right) + \left(\frac{1}{\ln 4} - \frac{1}{\ln 3}\right) + \cdots + \left(\frac{1}{\ln(n+1)} - \frac{1}{\ln n}\right) + \left(\frac{1}{\ln(n+2)} - \frac{1}{\ln(n+1)}\right)$
$= \frac{1}{\ln(n+2)} - \frac{1}{\ln 2}$
The sum of the series is $\lim_{n \to \infty} s_n = \lim_{n \to \infty} \left(\frac{1}{\ln(n+2)} - \frac{1}{\ln 2} \right) = -\frac{1}{\ln 2}$.

43. First rewrite a_n by expressing the product as a sum, using the method of partial fractions:

$$\frac{3}{(2x-1)(2x+1)} = \frac{A}{2x-1} + \frac{B}{2x+1}$$

$$3 = A(2x+1) + B(2x-1)$$

$$x = \tfrac{1}{2} : 3 = 2A \ \Rightarrow \ A = \tfrac{3}{2}$$

$$x = -\tfrac{1}{2} : 3 = -2B \ \Rightarrow \ B = -\tfrac{3}{2}$$

So, $\displaystyle\sum_{n=1}^{\infty} \frac{3}{(2n-1)(2n+1)} = \sum_{n=1}^{\infty} \left(\frac{\frac{3}{2}}{2n-1} - \frac{\frac{3}{2}}{2n+1} \right) = \sum_{n=1}^{\infty} \frac{3}{2} \left(\frac{1}{2n-1} - \frac{1}{2n+1} \right)$

$$s_4 = \frac{3}{2}\left[\left(1 - \tfrac{1}{3}\right) + \left(\tfrac{1}{3} - \tfrac{1}{5}\right) + \left(\tfrac{1}{5} - \tfrac{1}{7}\right) + \left(\tfrac{1}{7} - \tfrac{1}{9}\right) \right] = \frac{3}{2}\left(1 - \tfrac{1}{9}\right) = \frac{3}{2} \cdot \frac{8}{9} = \frac{4}{3}.$$

$$s_n = \frac{3}{2}\left[\left(1 - \tfrac{1}{3}\right) + \left(\tfrac{1}{3} - \tfrac{1}{5}\right) + \cdots + \left(\tfrac{1}{2n-1} - \tfrac{1}{2n+1}\right) \right] = \frac{3}{2}\left[1 - \tfrac{1}{2n+1} \right].$$

The sum of the series is $\displaystyle\lim_{n\to\infty} s_n = \lim_{n\to\infty} \frac{3}{2}\left[1 - \tfrac{1}{2n+1} \right] = \frac{3}{2}$, as $\tfrac{1}{2n+1} \to 0$ as $n \to \infty$.

45. (a) Let s_{n_a}, s_{n_b} and s_n denote the partial sums of $\displaystyle\sum_{n=1}^{\infty} a_n, \sum_{n=1}^{\infty} b_n$, and $\displaystyle\sum_{n=1}^{\infty} (a_n - b_n)$, respectively.

$$s_n = (a_1 - b_1) + (a_2 - b_2) + \cdots + (a_n - b_n) = (a_1 + \cdots + a_n) - (b_1 + \cdots + b_n) = s_{n_a} - s_{n_b} \Rightarrow$$

$$\sum_{n=1}^{\infty} (a_n - b_n) = \lim_{n\to\infty} s_n = \lim_{n_a\to\infty} s_{n_a} - \lim_{n_b\to\infty} s_{n_b} = \sum_{n=1}^{\infty} a_n - \sum_{n=1}^{\infty} b_n.$$

(b) Let s_{n_a} and s_n denote the partial sums of $\displaystyle\sum_{n=1}^{\infty} a_n$ and $\displaystyle\sum_{n=1}^{\infty} c a_n$, respectively.

$$s_n = c a_1 + c a_2 + \cdots + c a_n = c(a_1 + a_2 + \cdots + a_n) = c s_{n_a} \to c \sum_{n=1}^{\infty} a_n.$$

47. (a) Proof by contradiction: Suppose that $\displaystyle\sum_{n=1}^{\infty} (a_n + b_n)$ converges. By Exercise 45(a), if two series converge, so does the series whose terms are the difference of the terms of the two series. Thus, $\displaystyle\sum_{n=1}^{\infty} b_n = \sum_{n=1}^{\infty} [(a_n + b_n) - a_n]$ converges, a contradiction. Thus $\displaystyle\sum_{n=1}^{\infty} (a_n + b_n)$ diverges.

(b) One possible answer: (i) Let $a_n = 1$ and $b_n = -1$ for all n. Then both $\displaystyle\sum_{n=1}^{\infty} a_n$ and $\displaystyle\sum_{n=1}^{\infty} b_n$ diverge, as their terms do not tend to 0 (Divergence Test), but $\displaystyle\sum_{n=1}^{\infty} (a_n + b_n) = \sum_{n=1}^{\infty} 0 = 0$ converges.

(ii) One possible answer: (i) Let $a_n = 1$ and $b_n = 2$ for all n. Then both $\displaystyle\sum_{n=1}^{\infty} a_n$ and $\displaystyle\sum_{n=1}^{\infty} b_n$ diverge, as their terms do not tend to 0 (Divergence Test), and $\displaystyle\sum_{n=1}^{\infty} (a_n + b_n) = \sum_{n=1}^{\infty} 3$ also diverges, by the Divergence Test.

49. The total vertical distance is the sum of the initial distance 60 feet plus an infinite sum of twice the 2/3 length of the previous height. We need the factor of 2, because the ball rises and then falls the same distance, each time:

$$60 + 2\left\{\tfrac{2}{3}(60) + \tfrac{2}{3}\left[\tfrac{2}{3}(60)\right] + \cdots\right\} = 60 + 2\sum_{n=1}^{\infty} 60 \left(\tfrac{2}{3}\right)^{n} = 60 + 2\sum_{n=1}^{\infty} 60 \left(\tfrac{2}{3}\right)\left(\tfrac{2}{3}\right)^{n-1}$$

$$= 60 + 2\sum_{n=1}^{\infty} 40 \left(\tfrac{2}{3}\right)^{n-1} = 60 + 2\left[\frac{40}{1 - \tfrac{2}{3}}\right]$$

$$= 60 + 2(120) = 300 \text{ feet.}$$

51.   Just note that the length of the side of the "next" square is the length of the side of the present square multiplied by one-half $\sqrt{2}$. The one-half because of the midpoint process and the $\sqrt{2}$ because that new side is the hypotenuse of a 45 degree right triangle. Thus, the total area is:

$$4^2 + \left(4 \cdot \tfrac{1}{2}\sqrt{2}\right)^2 + \left(4 \cdot \tfrac{1}{2}\sqrt{2} \cdot \tfrac{1}{2}\sqrt{2}\right)^2 + \cdots = \sum_{n=0}^{\infty} 4^2 \left(\frac{\sqrt{2}}{2}\right)^{2n} = \sum_{n=0}^{\infty} 16 \left(\tfrac{1}{2}\right)^{n} = \sum_{n=1}^{\infty} 16 \left(\tfrac{1}{2}\right)^{n-1}$$

$$= \frac{16}{1 - \tfrac{1}{2}} = 32 \text{ sq. ft.}$$

§9.3 Series of Positive Terms

1.   Let $f(x) = \dfrac{x}{x^2+1}$, then f is continuous and positive for $x \geq 1$, and $f'(x) = \dfrac{x^2 + 1 - x(2x)}{(x^2+1)^2} =$ $\dfrac{1-x^2}{(x^2+1)^2} < 0$ for $x > 1 \Rightarrow f$ is decreasing for $x > 1$, so f satisfies the hypotheses of the Integral Test for $x \geq 2$.

$$\int \frac{x}{x^2+1}\, dx = \tfrac{1}{2}\int \frac{du}{u} = \tfrac{1}{2}\ln|u| + C = \tfrac{1}{2}\ln(x^2+1) + C \Rightarrow$$

$$\boxed{u = x^2 + 1, \ du = 2x\,dx}$$

$$\int_2^{\infty} \frac{x}{x^2+1}\, dx = \lim_{t\to\infty} \int_2^{t} \frac{x}{x^2+1}\, dx = \lim_{t\to\infty} \tfrac{1}{2}\ln(x^2+1)\Big|_2^{t} = \lim_{t\to\infty} \tfrac{1}{2}\left[\ln(t^2+1) - \ln 3\right] = \infty.$$

Since the integral diverges, so does the series $\sum \dfrac{n}{n^2+1}$, by the Integral Test.

3.   Let $f(x) = \dfrac{x}{e^x}$, then f is continuous and positive for $x \geq 1$, and $f'(x) = \dfrac{e^x - xe^x}{(e^x)^2} =$ $\dfrac{1-x}{e^x} < 0$ for $x > 1 \Rightarrow f$ is decreasing for $x > 1$, so f satisfies the hypotheses of the Integral Test for $x \geq 2$.

$$\int \frac{x}{e^x}\, dx = \int xe^{-x}\, dx = -xe^{-x} + \int e^{-x}\, dx = -xe^{-x} - e^{-x} + C = -e^{-x}(1+x) + C \Rightarrow$$

$$\boxed{\begin{array}{ll} u = x & dv = e^{-x}\, dx \\ du = dx & v = -e^{-x} \end{array}}$$

$$\int_2^\infty \frac{x}{e^x}\,dx = \lim_{t\to\infty} \int_2^t xe^{-x}\,dx = \lim_{t\to\infty} -e^{-x}(1+x)\big|_2^t = \lim_{t\to\infty}\left[-e^{-t}(1+t)+e^{-2}(3)\right]$$

Now, $\displaystyle\lim_{t\to\infty} -e^{-t}(1+t) = \lim_{t\to\infty}\frac{1+t}{e^t} = \lim_{t\to\infty}\frac{1}{e^t} = 0 \Rightarrow \int_2^\infty \frac{x}{e^x}\,dx = \frac{3}{e^2}$. Since the integral converges, so does the series $\sum \frac{n}{e^n}$ converge, by the Integral Test.

5. Since $\dfrac{6}{6^n+3} < \dfrac{6}{6^n} = \dfrac{1}{6^{n-1}} = \left(\dfrac{1}{6}\right)^{n-1}$ and $\sum \left(\dfrac{1}{6}\right)^{n-1}$ is a convergent geometric series, $\sum \dfrac{6}{6^n+3}$ converges, by the Comparison Test.

7. Compare $\sum \dfrac{1}{n^2+5n}$ with $\sum \dfrac{1}{n^2}$: $\displaystyle\lim_{n\to\infty}\frac{\frac{1}{n^2+5n}}{\frac{1}{n^2}} = \lim_{n\to\infty}\frac{n^2}{n^2+5n} = \lim_{n\to\infty}\frac{1}{1+\frac{5}{n}} = 1 > 0$.

As $\sum \dfrac{1}{n^2}$ is a convergent p-series ($p = 2 > 1$), $\sum \dfrac{1}{n^2+5n}$ also converges, by the Limit Comparison Test.

9. Compare $\sum \dfrac{5\sqrt{n}+9}{2n^2+3}$ with $\sum \dfrac{\sqrt{n}}{n^2} = \sum \dfrac{1}{n^{\frac{3}{2}}}$:

$$\lim_{n\to\infty}\frac{\frac{5\sqrt{n}+9}{2n^2+3}}{\frac{1}{n^{\frac{3}{2}}}} = \lim_{n\to\infty}\frac{(5n^{\frac{1}{2}}+9)n^{\frac{3}{2}}}{2n^2+3} = \lim_{n\to\infty}\frac{5n^2+9n^{\frac{3}{2}}}{2n^2+3} = \lim_{n\to\infty}\frac{5+\frac{9}{n^{\frac{1}{2}}}}{2+\frac{3}{n^2}} = \frac{5}{2} > 0.$$ As $\sum \dfrac{1}{n^{\frac{3}{2}}}$ is a

convergent p-series ($p = \frac{3}{2} > 1$), $\sum \dfrac{5\sqrt{n}+9}{2n^2+3}$ also converges, by the Limit Comparison Test.

11. $\displaystyle\lim_{n\to\infty}\frac{a_{n+1}}{a_n} = \lim_{n\to\infty}\frac{\frac{2^{n+1}}{(n+1)!}}{\frac{2^n}{n!}} = \lim_{n\to\infty}\frac{2^{n+1}n!}{(n+1)!2^n} = \lim_{n\to\infty}\frac{2}{n+1} = 0 < 1 \Rightarrow \sum \dfrac{2^n}{n!}$

converges, by the Ratio Test.

13. $\displaystyle\lim_{n\to\infty} a_n^{\frac{1}{n}} = \lim_{n\to\infty}\left[\frac{(3n+5)^n}{(2n-1)^n}\right]^{1/n} = \lim_{n\to\infty}\frac{3n+5}{2n-1} = \lim_{n\to\infty}\frac{3+\frac{5}{n}}{2-\frac{1}{n}} = \frac{3}{2} > 1 \Rightarrow \sum \dfrac{(3n+5)^n}{(2n-1)^n}$

diverges, by the Root Test.

15. $\displaystyle\lim_{n\to\infty} a_n^{\frac{1}{n}} = \lim_{n\to\infty}\left[\frac{3^n}{n^{10}}\right]^{1/n} = \lim_{n\to\infty}\frac{3}{(n^{1/n})^{10}} = 3$, as $n^{1/n} \to 1$. As $3 > 1$, the se-

ries $\sum \dfrac{3^n}{n^{10}}$ diverges, by the Root Test.

17. Compare $\sum \dfrac{1}{\sqrt{n(n-1)}}$ with $\sum \dfrac{1}{\sqrt{n^2}} = \sum \dfrac{1}{n}$. As $0 < n(n-1) < n^2$, then

$0 < \sqrt{n(n-1)} < n \Rightarrow \dfrac{1}{\sqrt{n(n-1)}} > \dfrac{1}{n}$. The series $\sum \dfrac{1}{n}$ is the divergent harmonic series.

By the Comparison Test, $\sum \dfrac{1}{\sqrt{n(n-1)}}$ diverges.

19. Compare $\sum \dfrac{1}{n\sqrt{n^2-1}}$ with $\sum \dfrac{1}{n\sqrt{n^2}} = \sum \dfrac{1}{n^2}$:

$$\lim_{n\to\infty} \frac{\frac{1}{n\sqrt{n^2-1}}}{\frac{1}{n^2}} = \lim_{n\to\infty} \frac{n^2}{n\sqrt{n^2-1}} = \lim_{n\to\infty} \frac{n}{\sqrt{n^2-1}} = \lim_{n\to\infty} \frac{1}{\sqrt{\frac{n^2-1}{n^2}}} = \lim_{n\to\infty} \frac{1}{\sqrt{1-\frac{1}{n^2}}} = 1 > 0,$$

as $\dfrac{1}{n^2} \to 0$. By the Limit Comparison Test, as $\sum \dfrac{1}{n^2}$ is a convergent p-series ($p = 2 > 1$), so does $\sum \dfrac{1}{n\sqrt{n^2-1}}$ converge.

21. Try the Integral Test, since we can integrate $f(x) = \dfrac{1}{x\ln x}$. For $x \geq 2$, $f(x)$ is continuous and positive and clearly decreasing, so f satisfies the hypotheses of the Integral Test.

$$\int \frac{1}{x\ln x}\,dx = \int \frac{1}{u}\,du = \ln|u| + C = \ln|\ln x| + C \Rightarrow$$

$$\boxed{u = \ln x, \ du = \frac{1}{x}\,dx}$$

$$\int_2^\infty \frac{1}{x\ln x}\,dx = \lim_{t\to\infty} \ln|\ln x|\Big|_2^t = \lim_{t\to\infty}[\ln(\ln t) - \ln(\ln 2)] = \infty \Rightarrow$$

the integral diverges. Therefore, by the Integral Test, the series $\sum \dfrac{1}{n\ln n}$ also diverges.

23. Since $\sum \dfrac{1}{(n+1)\ln(n+1)}$ is essentially the same as the divergent series $\sum \dfrac{1}{n\ln n}$ of Exercise 21, and since $\dfrac{1}{n\ln(n+1)} > \dfrac{1}{(n+1)\ln(n+1)}$, as $n < n+1$, the series $\sum \dfrac{1}{n\ln(n+1)}$ also diverges, by the Comparison Test.

25. $\lim_{n\to\infty} a_n^{\frac{1}{n}} = \lim_{n\to\infty} \left(\dfrac{2^n}{n^4}\right)^{\frac{1}{n}} = \lim_{n\to\infty} \dfrac{2}{(n^{\frac{1}{n}})^4} = \dfrac{2}{1} = 2 > 1 \Rightarrow$ the series $\sum \dfrac{2^n}{n^4}$ diverges, by the Root Test.

27. $\lim_{n\to\infty} \dfrac{a_{n+1}}{a_n} = \lim_{n\to\infty} \dfrac{\frac{5^{n+1}}{(n+1)!}}{\frac{5^n}{n!}} = \lim_{n\to\infty} \dfrac{5^{n+1}n!}{(n+1)!5^n} = \lim_{n\to\infty} \dfrac{5}{n+1} = 0 < 1 \Rightarrow$ the series $\sum \dfrac{5^n}{n!}$ converges, by the Ratio Test.

29. $\lim_{n\to\infty} a_n^{\frac{1}{n}} = \lim_{n\to\infty} (ne^{-2n})^{\frac{1}{n}} = \lim_{n\to\infty} \left[\dfrac{n}{(e^2)^n}\right]^{\frac{1}{n}} = \lim_{n\to\infty} \dfrac{n^{\frac{1}{n}}}{e^2} = \dfrac{1}{e^2} < 1 \Rightarrow$ the series $\sum ne^{-2n}$ converges, by the Root Test.

31. $\displaystyle\lim_{n\to\infty}\frac{a_{n+1}}{a_n}=\lim_{n\to\infty}\frac{\frac{[(n+1)!]^2}{[2(n+1)]!}}{\frac{(n!)^2}{(2n)!}}=\lim_{n\to\infty}\frac{[(n+1)!]^2(2n)!}{(2n+2)!(n!)^2}=\lim_{n\to\infty}\frac{(n+1)^2}{(2n+2)(2n+1)}=\lim_{n\to\infty}\frac{n^2+2n+1}{4n^2+6n+2}$

$\displaystyle=\lim_{n\to\infty}\frac{1+\frac{2}{n}+\frac{1}{n^2}}{4+\frac{6}{n}+\frac{2}{n^2}}=\frac14<1 \Rightarrow$ the series $\displaystyle\sum\frac{(n!)^2}{(2n)!}$ converges, by the Ratio Test.

33. Let $f(x)=\frac{\ln x}{x^2}$. Then f is continuous and positive for $x\ge 2$, and $f'(x)=\dfrac{x^2\cdot\frac{1}{x}-(\ln x)(2x)}{x^4}=$

$\dfrac{1-2\ln x}{x^3}<0$ for $x\ge 3$, so the hypotheses of the Integral Test are satisfied.

$\displaystyle\int\frac{\ln x}{x^2}\,dx=-\frac{\ln x}{x}+\int x^{-2}\,dx=-\frac{\ln x}{x}-\frac1x+C \Rightarrow$

$$\boxed{\begin{array}{ll} u=\ln x & dv=x^{-2}\,dx\\[4pt] du=\frac1x\,dx & v=-x^{-1} \end{array}}$$

$\displaystyle\int_3^\infty\frac{\ln x}{x^2}\,dx=\lim_{t\to\infty}\int_3^t\frac{\ln x}{x^2}\,dx=\lim_{t\to\infty}\left[-\frac{\ln x}{x}-\frac1x\right]_3^t=\lim_{t\to\infty}\left[-\frac{\ln t}{t}-\frac1t+\frac{\ln 3}{3}+\frac13\right]=\frac{\ln 3}{3}+\frac13$

since, by l'Hôpital's Rule, $\displaystyle\lim_{t\to\infty}\frac{\ln t}{t}=\lim_{t\to\infty}\frac{\frac1t}{1}=\lim_{t\to\infty}\frac1t=0$. Since the integral converges, so

does the series $\displaystyle\sum\frac{\ln n}{n^2}$, by the Integral Test.

35. $\displaystyle\lim_{n\to\infty}\frac{a_{n+1}}{a_n}=\lim_{n\to\infty}\frac{\frac{\ln(n+1)}{e^{n+1}}}{\frac{\ln n}{e^n}}=\lim_{n\to\infty}\frac{\ln(n+1)e^n}{e^{n+1}\ln n}=\lim_{n\to\infty}\frac{\ln(n+1)}{e\ln n}=\frac1e<1,$

since, by l'Hôpital's Rule, $\displaystyle\lim_{x\to\infty}\frac{\ln(x+1)}{e\ln x}=\lim_{x\to\infty}\frac{\frac{1}{x+1}}{\frac{e}{x}}=\lim_{x\to\infty}\frac{x}{e(x+1)}=\lim_{x\to\infty}\frac1e=\frac1e.$

Therefore, the series $\displaystyle\sum\frac{\ln n}{e^n}$ converges, by the Ratio Test.

37. Let $f(x)=\ln\left(\dfrac{x^2+1}{x^2}\right)=\ln\left(1+\dfrac{1}{x^2}\right)$. Then f is continuous and positive for $x\ge 1$,

and $f'(x)=\dfrac{1}{1+\frac{1}{x^2}}\cdot(-2x^{-3})=-\dfrac{2}{x^3\left(1+\frac{1}{x^2}\right)}=-\dfrac{2}{x(x^2+1)}<0$ for $x\ge 1$ which means that f

is decreasing there. Thus f satisfies the hypotheses of the Integral Test there.

$\int\ln(1+x^{-2})\,dx-x\ln(1+x^{-2})+2\int\dfrac{dx}{1+x^2}-x\ln(1+x^{-2})+2\tan^{-1}x+C \Rightarrow$

$$\boxed{\begin{array}{ll} u=\ln(1+x^{-2}) & dv=dx\\[4pt] du=-\dfrac{2}{x(x^2+1)}\,dx & v=x \end{array}}$$

$\int_1^\infty\ln\left(1+\dfrac{1}{x^2}\right)dx=\lim_{t\to\infty}\int_1^t\ln\left(1+\dfrac{1}{x^2}\right)dx=\lim_{t\to\infty}\left[x\ln(1+x^{-2})+2\tan^{-1}x\right]\Big|_1^t$

$\qquad\qquad=\lim_{t\to\infty}\left[t\ln(1+t^{-2})+2\tan^{-1}t-\ln 2-2\tan^{-1}1\right]=2\cdot\dfrac{\pi}{2}-\ln 2-2\cdot\dfrac{\pi}{4},$

because, by l'Hôpital's Rule,

$$\lim_{t\to\infty} t\ln(1+t^{-2}) = \lim_{t\to\infty} \frac{\ln(1+t^{-2})}{t^{-1}} = \lim_{t\to\infty} \frac{\frac{1}{1+t^{-2}}\cdot(-2t^{-3})}{-t^{-2}}\cdot\frac{t^4}{t^4} = \lim_{t\to\infty} \frac{2t}{t^2+1} = \lim_{t\to\infty} \frac{2}{2t} = 0.$$

Since the integral converges, so does the series $\sum \ln\left(\dfrac{n^2+1}{n^2}\right)$, by the Integral Test.

39.   Compare $\sum \dfrac{n+1}{(n+2)2^n}$ with the convergent geometric series $\sum \dfrac{1}{2^n} = \sum \left(\dfrac{1}{2}\right)^n$. As $0 < \dfrac{n+1}{(n+2)2^n} < \left(\dfrac{1}{2}\right)^n$, since $n+1 < n+2 \Rightarrow \dfrac{n+1}{n+2} < 1$, the series $\sum \dfrac{n+1}{(n+2)2^n}$ converges, by the Comparison Test.

41. $\lim_{n\to\infty} \dfrac{a_{n+1}}{a_n} = \lim_{n\to\infty} \dfrac{\dfrac{(n+1)!}{(n+1)2^{n+1}}}{\dfrac{n!}{n2^n}} = \lim_{n\to\infty} \dfrac{(n+1)!n2^n}{(n+1)2^{n+1}n!} = \lim_{n\to\infty} \dfrac{n}{2} = \infty \Rightarrow$ the series $\sum \dfrac{n!}{n2^n}$ diverges, by the Ratio Test.

43. $\lim_{n\to\infty} \dfrac{a_{n+1}}{a_n} = \lim_{n\to\infty} \dfrac{\dfrac{2(n+1)-1}{2^{n+1}}}{\dfrac{2n-1}{2^n}} = \lim_{n\to\infty} \dfrac{(2n+1)2^n}{2^{n+1}(2n-1)} = \lim_{n\to\infty} \dfrac{2n+1}{4n-2} = \lim_{n\to\infty} \dfrac{2+\frac{1}{n}}{4-\frac{2}{n}} = \dfrac{2}{4} = \dfrac{1}{2} < 1$. The series $\sum \dfrac{2n-1}{2^n}$ converges, by the Ratio Test.

45. $\lim_{n\to\infty} a_n^{\frac{1}{n}} = \lim_{n\to\infty} \left[\left(\dfrac{n^2+10n}{2n^2+5}\right)^n\right]^{\frac{1}{n}} = \lim_{n\to\infty} \dfrac{n^2+10n}{2n^2+5} = \lim_{n\to\infty} \dfrac{1+\frac{10}{n}}{2+\frac{5}{n^2}} = \dfrac{1}{2} < 1 \Rightarrow$ the series $\sum \left(\dfrac{n^2+10n}{2n^2+5}\right)^n$ converges, by the Root Test.

47. $\lim_{n\to\infty} \dfrac{a_{n+1}}{a_n} = \lim_{n\to\infty} \dfrac{\dfrac{[2(n+1)]!}{(n+1)!2^{n+1}}}{\dfrac{(2n)!}{n!2^n}} = \lim_{n\to\infty} \dfrac{(2n+2)!n!2^n}{(n+1)!2^{n+1}(2n)!} = \lim_{n\to\infty} \dfrac{(2n\cancel{+}2)(2n+1)}{\cancel{2}(n\cancel{+}1)}$

$= \lim_{n\to\infty} (2n+1) = \infty \Rightarrow$ the series $\sum \dfrac{(2n)!}{n!2^n}$ diverges, by the Ratio Test.

49.   Suspecting that the terms of the series do not converge to zero, we check the limit, with l'Hôpital's Rule:

$$\lim_{x\to\infty} \frac{2^x}{1+(\ln x)^2} = \lim_{x\to\infty} \frac{2^x\ln 2}{2\ln x\cdot\frac{1}{x}} = \lim_{x\to\infty} \frac{x2^x\ln 2}{2\ln x} = \lim_{x\to\infty} \frac{(\ln 2)[x2^x\ln 2 + 2^x]}{2\cdot\frac{1}{x}} = \infty \Rightarrow$$

the series $\sum \dfrac{2^n}{1+(\ln n)^2}$ diverges, by the Divergence Test.

51. Let $f(x) = \dfrac{1}{(x+1)[\ln(x+1)]^2}$. Then f is continuous, positive, and decreasing for $x \geq 1$, so f satisfies the hypotheses of the Integral Test.

$$\int \frac{1}{(x+1)[\ln(x+1)]^2}\,dx = \int \frac{du}{u^2} = \int u^{-2}\,du = -u^{-1} + C = -\frac{1}{\ln(x+1)} + C \Rightarrow$$

$$\boxed{u = \ln(x+1), \ du = \frac{1}{x+1}\,dx}$$

$$\int_1^\infty \frac{1}{(x+1)[\ln(x+1)]^2}\,dx = \lim_{t\to\infty} \int_1^t \frac{1}{(x+1)[\ln(x+1)]^2}\,dx = \lim_{t\to\infty} -\frac{1}{\ln(x+1)}\Big|_1^t$$

$$= \lim_{t\to\infty} \left[-\frac{1}{\ln(t+1)} + \frac{1}{\ln 2} \right] = \frac{1}{\ln 2}$$

Since the integral converges, the series $\displaystyle\sum \frac{1}{(n+1)[\ln(n+1)]^2}$ converges, by the Integral Test.

53. By Exercise 51, $\displaystyle\sum \frac{1}{(n+1)[\ln(n+1)]^2}$ which is essentially the same as $\displaystyle\sum \frac{1}{n(\ln n)^2}$ converges. The series $\displaystyle\sum \frac{1}{n^2}$ is a convergent p-series ($p = 2 > 1$). By Theorem 9.10, $\displaystyle\sum \left[\frac{1}{n(\ln n)^2} - \frac{1}{n^2} \right]$ also converges.

55. $\displaystyle\sum \frac{1}{n^{3/2}}$ is a convergent p-series ($p = \frac{3}{2} > 1$). Next, consider $\displaystyle\sum \frac{e^n}{(1+e^n)^2}$: Let $f(x) = \dfrac{e^x}{(1+e^x)^2}$. Then f is continuous and positive for $x \geq 1$, and

$$f'(x) = \frac{(1+e^x)^2 e^x - e^x \cdot 2(1+e^x)e^x}{(1+e^x)^4} = \frac{e^x[(1+e^x) - 2e^x]}{(1+e^x)^3} = \frac{e^x(1-e^x)}{(1+e^x)^3} < 0 \text{ for } x \geq 1.$$

Therefore, f satisfies the hypotheses of the Integral Test.

$$\int \frac{e^x}{(1+e^x)^2}\,dx = \int \frac{du}{u^2} = \int u^{-2}\,du = -u^{-1} + C = -\frac{1}{1+e^x} + C \Rightarrow$$

$$\boxed{u = 1 + e^x, \ du = e^x\,dx}$$

$$\int_1^\infty \frac{e^x}{(1+e^x)^2}\,dx = \lim_{t\to\infty} \int_1^t \frac{e^x}{(1+e^x)^2}\,dx = \lim_{t\to\infty} -\frac{1}{1+e^x}\Big|_1^t = \lim_{t\to\infty} \left[-\frac{1}{1+e^t} + \frac{1}{1+e} \right] = \frac{1}{1+e}$$

Since the integral converges, the series $\displaystyle\sum \frac{e^n}{(1+e^n)^2}$ converges, by the Integral Test.

By Theorem 9.10, as both $\displaystyle\sum \frac{1}{n^{3/2}}$ and $\displaystyle\sum \frac{e^n}{(1+e^n)^2}$ converge, so does $\displaystyle\sum \left[\frac{1}{n^{3/2}} + \frac{e^n}{(1+e^n)^2} \right]$.

57. $\displaystyle\sum \frac{\sqrt{n}+5}{n^2} = \sum \frac{\sqrt{n}}{n^2} + \sum \frac{5}{n^2}$ converges, because each of the two series converges:

$$\sum \frac{\sqrt{n}}{n^2} = \sum \frac{1}{n^{3/2}} \text{ is a convergent } p\text{-series } \left(p = \frac{3}{2} > 1\right), \text{ and}$$

$$\sum \frac{5}{n^2} = 5 \sum \frac{1}{n^2} \text{ converges, as } \sum \frac{1}{n^2} \text{ is a convergent } p\text{-series } (p = 2 > 1).$$

Now consider the series $\displaystyle\sum \frac{3n+1}{n2^n}$:

$$\lim_{n\to\infty}\frac{a_{n+1}}{a_n}=\lim_{n\to\infty}\frac{\dfrac{3(n+1)+1}{(n+1)2^{n+1}}}{\dfrac{3n+1}{n2^n}}=\lim_{n\to\infty}\frac{(3n+4)n2^n}{(n+1)2^{n+1}(3n+1)}=\lim_{n\to\infty}\frac{3n^2+4n}{6n^2+8n+2}$$

$$=\lim_{n\to\infty}\frac{3+\dfrac{4}{n}}{6+\dfrac{8}{n}+\dfrac{2}{n^2}}=\frac{3}{6}=\frac{1}{2}<1\;\Rightarrow\;\text{the series converges, by the Ratio Test.}$$

Therefore, the given series $\sum\left[\dfrac{\sqrt{n}+5}{n^2}+\dfrac{3n+1}{n2^n}\right]$, which is the sum of two convergent series, converges, by Theorem 9.10.

59. To prove that the series $\sum\dfrac{1}{n(\ln n)^p}$ converges if and only if $p>1$.

Case 1: If $p=0$, then the series is the divergent harmonic series.

Case 2: If $p=1$, that series was proven in Exercise 21 to be divergent.

Case 3: If $p\neq0$ and $p\neq1$, let $f(x)=\dfrac{1}{x(\ln x)^p}$. Then f is continuous and positive for $x\geq2$,

and $f'(x)=-[x(\ln x)^p]^{-2}\cdot\left[xp(\ln x)^{p-1}\cdot\dfrac{1}{x}+(\ln x)^p\right]=-\dfrac{[(\ln x)^{p-1}(p+\ln x)]}{x^2(\ln x)^{2p}}<0$ for x suf-

ficiently large $(\ln x>|p|)$, say $x>N$, so that f is decreasing for such x. The function f satisfies the hypotheses of the Integral Test for $x>N$.

$$\int\frac{dx}{x(\ln x)^p}=\int u^{-p}\,du=\frac{u^{-p+1}}{-p+1}+C=\frac{1}{(1-p)(\ln x)^{p-1}}+C\;\Rightarrow$$

$$\boxed{u=\ln x,\ du=\frac{1}{x}\,dx}$$

$$\int_N^\infty\frac{1}{x(\ln x)^p}\,dx=\lim_{t\to\infty}\int_N^t\frac{1}{x(\ln x)^p}\,dx=\lim_{t\to\infty}\frac{1}{(1-p)(\ln x)^{p-1}}\Big|_N^t$$

$$=\lim_{t\to\infty}\left[\frac{1}{(1-p)(\ln t)^{p-1}}-\frac{1}{(1-p)(\ln N)^{p-1}}\right]$$

If $p<1$, the limit is infinite and the integral and hence the series diverges, by the Integral Test.

If $p>1$, the limit exists and is finite, and the integral and hence the series converges, by the Integral Test.

61. No, this does not violate the Integral Test, as f is not a decreasing function:

$f'(x)=2(\sin\pi x)(\cos\pi x)\pi-\dfrac{2}{x^3}=\pi\sin2\pi x-\dfrac{2}{x^3}$ is not negative for all x sufficiently large, since $\sin2\pi x$ is a periodic function which forever varies between -1 and 1 (so $\pi\sin2\pi x$ varies between $-\pi$ and π) as $x\to\infty$, while $\dfrac{2}{x^3}\to0$.

$$\int_1^\infty\left(\sin^2\pi x+\frac{1}{x^2}\right)dx=\lim_{t\to\infty}\int_1^t\left(\sin^2\pi x+\frac{1}{x^2}\right)dx=\lim_{t\to\infty}\int_1^t\left(\frac{1-\cos2\pi x}{2}-x^{-2}\right)dx$$

$$=\lim_{t\to\infty}\left(\frac{1}{2}x-\frac{1}{4\pi}\sin2\pi x+\frac{1}{x}\right)\Big|_1^t=\lim_{t\to\infty}\left(\frac{1}{2}t-\frac{1}{4\pi}\sin2\pi t+\frac{1}{t}-\frac{1}{2}-1\right)$$

is infinite, so the integral diverges.

The series $\sum\left(\sin^2\pi n+\dfrac{1}{n^2}\right)$ converges, as $\sin^2\pi n=0$ for all integers n, and $\sum\dfrac{1}{n^2}$ is a convergent p-series ($p=2>1$).

63.   True: As $a_n > 0$, so is $\ln(1 + a_n) > 0$. Given $\sum \ln(1 + a_n)$ converges, $\ln(1 + a_n) \to 0$
$\Rightarrow 1 + a_n \to 1 \Rightarrow a_n \to 0$ as $n \to \infty$. Thus,

$$\lim_{n \to \infty} \frac{\ln(1 + a_n)}{a_n} = \lim_{x \to 0} \frac{\ln(1 + x)}{x} = \lim_{x \to 0} \frac{\frac{1}{1+x}}{1} = \lim_{x \to 0} \frac{1}{1+x} = 1 \Rightarrow \text{the series } \sum a_n \text{ also}$$

converges, by the Limit Comparison Test.

65.   True: Since $a_n > 0$ and $\sum a_n$ converges, $\lim_{n \to \infty} a_n = 0 \Rightarrow$ there exists an N such
that for all $n > N$, $0 < a_n < 1 \Rightarrow 0 < a_n{}^2 < a_n$ for $n > N$. Therefore, by the Comparison
Test, $\sum a_n{}^2$ converges.

§9.4 Absolute and Conditional Convergence

1. $\sum (-1)^n \frac{1}{\sqrt{n}}$ is not absolutely convergent, as $\sum |(-1)^n \frac{1}{\sqrt{n}}| = \sum \frac{1}{n^{\frac{1}{2}}}$ which is a divergent

p-series ($p = \frac{1}{2} < 1$). The given alternating series does converge by the Alternating Series
Theorem, as $a_n = \frac{1}{\sqrt{n}}$ is positive and decreasing to 0 as $n \to \infty$. Since the series converges,
but not absolutely, it is conditionally convergent.

3. $\sum (-1)^n \frac{1}{\ln n}$ is not absolutely convergent, since, as we will show, $\sum |(-1)^n \frac{1}{\ln n}| = \sum \frac{1}{\ln n}$
diverges by the Comparison Test, comparing it with the divergent harmonic series $\sum \frac{1}{n}$:

Since $\lim_{x \to \infty} \frac{\ln x}{x} = \lim_{x \to \infty} \frac{\frac{1}{x}}{1} = \lim_{x \to \infty} \frac{1}{x} = 0$, for x sufficiently large, $0 < \frac{\ln x}{x} < 1 \Rightarrow 0 < \ln x < x$.
Thus, for n sufficiently large, $0 < \ln n < n \Rightarrow \frac{1}{\ln n} > \frac{1}{n}$.

The given alternating series does converge by the Alternating Series Theorem, as $a_n = \frac{1}{\ln n}$
is positive and decreasing to 0 as $n \to \infty$. Since the series converges, but not absolutely, it
is conditionally convergent.

5.   Try the Ratio Test: $\lim_{n \to \infty} \left| \frac{a_{n+1}}{a_n} \right| = \lim_{n \to \infty} \frac{\frac{(n+1)^{50}}{(n+1)!}}{\frac{n^{50}}{n!}} = \lim_{n \to \infty} \frac{(n+1)^{50} n!}{n^{50} (n+1)!}$

$$= \lim_{n \to \infty} \left(\frac{n+1}{n} \right)^{50} \left(\frac{1}{n+1} \right) = \lim_{n \to \infty} \left(1 + \frac{1}{n} \right)^{50} \left(\frac{1}{n+1} \right)$$

$$= 1^{50} \cdot 0 = 0 < 1$$

The given series, $\sum (-1)^n \frac{n^{50}}{n!}$ converges absolutely.

7. $\sum (-1)^n \frac{1+n}{n^2}$ is not absolutely convergent, by the Comparison Test, since, $a_n = \frac{1+n}{n^2} >$
$\frac{n}{n^2} = \frac{1}{n}$ and $\sum \frac{1}{n}$ is the divergent harmonic series. Try the Alternating Series Theorem:
$a_n > 0$, is it decreasing?

$$\frac{1 + (n+1)}{(n+1)^2} \overset{?}{<} \frac{1+n}{n^2}$$

$$(n+2)(n^2) \overset{?}{<} (n+1)^3$$

$$n^3 + 2n^2 \overset{?}{<} n^3 + 3n^2 + 3n + 1$$

yes!, and $\lim\limits_{n \to \infty} \frac{1+n}{n^2} = \lim\limits_{n \to \infty} \left(\frac{1}{n^2} + \frac{1}{n}\right) = 0$. By the Alternating Series Theorem, the given series converges. Therefore it converges conditionally.

9. The series $\sum \left|(-1)^n \frac{1}{n(\ln n)^2}\right| = \sum \frac{1}{n(\ln n)^2}$, and the series $\sum \frac{1}{(n+1)[\ln(n+1)]^2}$ agree in all but possibly a finite number of terms. In Section 9.3, #51, we showed that the latter series converges. Therefore the given series is absolutely convergent.

11. By Theorem 3.5, page 90, $\lim\limits_{x \to 0} \frac{\sin x}{x} = 1$. Thus, for small $\theta > 0$, $0 < \frac{\sin \theta}{\theta} < \frac{3}{2}$ $\Rightarrow$ $\sin \theta < \left(\frac{3}{2}\right)\theta$. Consequently, for sufficiently large n, $0 < \sin\left(\frac{1}{n}\right) < \left(\frac{3}{2}\right)\frac{1}{n}$ which implies that $\frac{\sin\left(\frac{1}{n}\right)}{n} < \left(\frac{3}{2}\right) \cdot \frac{\frac{1}{n}}{n} = \left(\frac{3}{2}\right)\frac{1}{n^2}$. The series $\sum \frac{1}{n^2}$ is a convergent p-series $(p = 2 > 1)$ $\Rightarrow$ $\left(\frac{3}{2}\right)\sum \frac{1}{n^2}$ converges. By the Comparison Test, $\sum \left|(-1)^n \frac{\sin\left(\frac{1}{n}\right)}{n}\right| = \sum \frac{\sin\left(\frac{1}{n}\right)}{n}$ converges, which means that the given alternating series is absolutely convergent.

13. $\lim\limits_{n \to \infty} |a_n|^{\frac{1}{n}} = \lim\limits_{n \to \infty} \frac{n^2 + 1}{2n^2 + 1} = \lim\limits_{n \to \infty} \frac{1 + \frac{1}{n^2}}{2 + \frac{1}{n^2}} = \frac{1}{2} < 1$. By the Root Test, the series $\sum (-1)^n \left(\frac{n^2 + 1}{2n^2 + 1}\right)^n$ is absolutely convergent.

15. The series $\sum (-1)^n \frac{e^n}{n}$ is divergent, by the Divergence Test, as, by l'Hôpital's Rule, $\lim\limits_{x \to \infty} \frac{e^x}{x} = \lim\limits_{x \to \infty} \frac{e^x}{1} = \infty$ $\Rightarrow$ $\lim\limits_{n \to \infty} |a_n| = \lim\limits_{n \to \infty} \frac{e^n}{n} \neq 0$ $\Rightarrow$ $\lim\limits_{n \to \infty} a_n \neq 0$.

17. Since $-\frac{\pi}{2} < \tan^{-1} x < \frac{\pi}{2}$, $\left|(-1)^n \frac{\tan^{-1} n}{n^2}\right| < \frac{\frac{\pi}{2}}{n^2}$. The series $\sum \frac{1}{n^2}$ is a convergent p-series $(p = 2 > 1)$, which implies that $\sum \frac{\frac{\pi}{2}}{n^2}$ is also convergent (Theorem 9.10). By the Comparison Test, therefore, the series $\sum (-1)^n \frac{\tan^{-1} n}{n^2}$ is absolutely convergent.

19. The series $\sum (-1)^n 3^{\frac{1}{n}}$ diverges by the Divergence Test, since:

$$\lim\limits_{n \to \infty} |a_n| = \lim\limits_{n \to \infty} |(-1)^n 3^{\frac{1}{n}}| = \lim\limits_{n \to \infty} 3^{\frac{1}{n}} = \lim\limits_{n \to \infty} e^{\ln 3^{\frac{1}{n}}} = \lim\limits_{n \to \infty} e^{\frac{1}{n}\ln 3} = e^{\lim\limits_{n \to \infty} \frac{1}{n}\ln 3}$$

$$= e^0 = 1 \neq 0 \Rightarrow \lim\limits_{n \to \infty} a_n \neq 0.$$

21. $\sum \left| (-1)^n \frac{1}{n \ln n} \right| = \sum \frac{1}{n \ln n}$ which was shown to diverge in Exercise 21 of section 9.3, above. Since $\frac{1}{n \ln n} > 0$ and decreasing to 0 as $n \to \infty$, the given series converges, by the Alternating Series Theorem. Thus, the given series is conditionally convergent.

23. $\sum (-1)^n \frac{\ln n}{n}$ is not absolutely convergent, as $\sum \frac{\ln n}{n}$ diverges, by the Comparison Test, because $\frac{\ln n}{n} > \frac{1}{n}$ and $\sum \frac{1}{n}$ is the divergent harmonic series. Try the Alternating Series Theorem: Let $f(x) = \frac{\ln x}{x}$. Then $f'(x) = \frac{x \cdot \frac{1}{x} - \ln x}{x^2} = \frac{1 - \ln x}{x^2} < 0$ for $x > e$, so f is a decreasing function for such x. By l'Hôpital's Rule, $\lim\limits_{x \to \infty} \frac{\ln x}{x} = \lim\limits_{x \to \infty} \frac{\frac{1}{x}}{1} = \lim\limits_{x \to \infty} \frac{1}{x} = 0 \Rightarrow f$ tends to 0 as $x \to \infty$. By the Alternating Series Theorem, the given series converges. Thus it converges conditionally.

25. First note that $\left| \frac{\cos(\frac{n\pi}{4})}{n!} \right| \le \frac{1}{n!}$. Use the Ratio Test to see that the series $\sum \frac{1}{n!}$ converges:

$$\lim_{n \to \infty} \frac{a_{n+1}}{a_n} = \lim_{n \to \infty} \frac{\frac{1}{(n+1)!}}{\frac{1}{n!}} = \lim_{n \to \infty} \frac{n!}{(n+1)!} = \lim_{n \to \infty} \frac{1}{n+1} = 0 < 1.$$

By the Comparison Test, $\sum \frac{\cos(\frac{n\pi}{4})}{n!}$ converges absolutely.

27. Consider absolute convergence first: Try the Integral Test: $f(x) = \frac{1}{x\sqrt{\ln x}} > 0$ and is decreasing. Consider $\int_2^\infty \frac{dx}{x(\ln x)^{\frac{1}{2}}}$:

$$\int \frac{dx}{x(\ln x)^{\frac{1}{2}}} = \int u^{-\frac{1}{2}} du = 2u^{\frac{1}{2}} + C = 2\sqrt{\ln x} + C$$

$$\boxed{u = \ln x, \ du = \frac{1}{x} dx}$$

Thus, $\int_2^\infty \frac{dx}{x(\ln x)^{\frac{1}{2}}} = \lim\limits_{t \to \infty} \int_2^t \frac{dx}{x(\ln x)^{\frac{1}{2}}} = \lim\limits_{t \to \infty} 2\sqrt{\ln x} \Big|_2^t = \lim\limits_{t \to \infty} (2\sqrt{\ln t} - 2\sqrt{\ln 2}) = \infty.$

By the Integral Test, as the integral diverges, the series $\sum \frac{1}{n\sqrt{\ln n}}$ also diverges, so the given alternating series does not converge absolutely.

Consider conditional convergence: Try the Alternating Series Theorem: $a_n = \frac{1}{n\sqrt{\ln n}}$ is positive and decreasing to 0 as $n \to \infty$. Therefore, the series $\sum (-1)^n \frac{1}{n\sqrt{\ln n}}$ converges, and thus converges conditionally.

29. $\sum (-1)^n \frac{n + \ln n}{n^{\frac{3}{2}}}$ is not absolutely convergent: $\left| (-1)^n \frac{n + \ln n}{n^{\frac{3}{2}}} \right| = \frac{n + \ln n}{n^{\frac{3}{2}}} > \frac{n}{n^{\frac{3}{2}}} = \frac{1}{n^{\frac{1}{2}}}$ and $\sum \frac{1}{n^{\frac{1}{2}}}$ is a divergent p-series ($p = \frac{1}{2} < 1$). By the Comparison Test, $\sum \frac{n + \ln n}{n^{\frac{3}{2}}}$ also diverges.

Is it conditionally convergent? Consider the Alternating Series Theorem:

$a_n = \frac{n + \ln n}{n^{\frac{3}{2}}} > 0$, is it decreasing?

$$\left[\frac{x + \ln x}{x^{\frac{3}{2}}}\right]' = \frac{x^{\frac{3}{2}}(1 + \frac{1}{x}) - (x + \ln x)\frac{3}{2}x^{\frac{1}{2}}}{x^3} = \frac{x + 1 - \frac{3}{2}(x + \ln x)}{x^{\frac{5}{2}}} = \frac{-\frac{1}{2}x - \ln x + 1}{x^{\frac{5}{2}}} < 0$$

for $x > 2$, so (a_n) is decreasing for $n > 2$. Is the limit 0? By l'Hôpital's Rule, $\lim\limits_{x \to \infty} \frac{x + \ln x}{x^{\frac{3}{2}}} =$

$\lim\limits_{x \to \infty} \frac{1 + \frac{1}{x}}{\frac{3}{2}x^{\frac{1}{2}}} = 0 \Rightarrow \lim\limits_{n \to \infty} a_n = 0$ - yes! By the Alternating Series Theorem, the given series

converges, and therefore converges conditionally.

31. Try the Ratio Test: $\lim\limits_{n \to \infty}\left|\frac{a_{n+1}}{a_n}\right| = \lim\limits_{n \to \infty} \frac{\frac{[2(n+1)]!}{(n+1)2^{n+1}(n+1)!}}{\frac{(2n)!}{n2^n n!}} = \lim\limits_{n \to \infty} \frac{(2n+2)!n2^n n!}{(n+1)2^{n+1}(n+1)!(2n)!}$

$= \lim\limits_{n \to \infty} \frac{(2n\!\!\not{+}\!2)(2n+1)n}{(n\!\!\not{+}\!1)\,\not{2}(n+1)} = \lim\limits_{n \to \infty} \frac{2n^2 + n}{n+1} = \lim\limits_{n \to \infty} \frac{2 + \frac{1}{n}}{\frac{1}{n} + \frac{1}{n^2}} = \infty$. The given alternating series

diverges, by the Ratio Test.

33. $(\sqrt{n + \sqrt{n}} - \sqrt{n}) \cdot \frac{\sqrt{n + \sqrt{n}} + \sqrt{n}}{\sqrt{n + \sqrt{n}} + \sqrt{n}} = \frac{\not{n} + \sqrt{n} - \not{n}}{\sqrt{n + \sqrt{n}} + \sqrt{n}} = \frac{1}{\sqrt{\frac{n + \sqrt{n}}{n}} + 1} = \frac{1}{\sqrt{1 + \frac{1}{\sqrt{n}}} + 1} \to \frac{1}{2}$

as $n \to \infty$. As $|a_n| \not\to 0$ as $n \to \infty$, $a_n \not\to 0$ as $n \to \infty$, so that the given alternating series $\sum(-1)^n(\sqrt{n + \sqrt{n}} - \sqrt{n})$ diverges by the Divergence Test.

35. If $\sum|a_n|$ converges, then so must $\sum a_n$ (Theorem 9.20). Consequently, if $\sum a_n$ diverges, then so must $\sum|a_n|$.

37.   (a) As $\sum|a_n|$ and $\sum|b_n|$ converge, so does $\sum(|a_n| + |b_n|)$ (Theorem 9.10). By the Triangle Inequality, $|a_n + b_n| \leq |a_n| + |b_n|$ which implies that $\sum|a_n + b_n|$ converges by the Comparison Test. Thus $\sum(a_n + b_n)$ converges absolutely.
   (b) No: For example, let $a_n = 1$ and $b_n = -1$ for all n. Then, $\sum(a_n + b_n) = \sum 0$ is absolutely convergent, but neither $\sum|a_n|$ nor $\sum|b_n|$ converges.
   (c) No, see (b).
   (d) Yes: $\sum|a_n|$ converges, implies $|a_n| \to 0$ as $n \to \infty$, which means that for some N and all $n > N$, $|a_n| < 1$. Thus $|a_n b_n| = |a_n||b_n| < |b_n|$ for all $n > N$. By the Comparison Test, as $\sum|b_n|$ converges, so does $\sum|a_n b_n|$.
   (e) No: For example, let $a_n = b_n = \frac{1}{n}$, then $\sum|a_n b_n| = \sum\frac{1}{n^2}$ which is a convergent p-series ($p = 2 > 1$), while neither $\sum a_n$ nor $\sum b_n$ converge, as each is the divergent harmonic series.
   (f) No, see (e).

39.   (a) Let M be such that $|y_n| \leq M$ for all n. Since $|a_n y_n| \leq M|a_n|$ and the convergence of $\sum|a_n|$ implies the convergence of $\sum M|a_n|$, by the Comparison Test, $\sum|a_n y_n|$ converges.

(b) Let $a_n = (-1)^n \frac{1}{n}$. Then $\sum a_n$ is the conditionally convergent alternating harmonic series. Let $y_n = (-1)^n$. Then $|y_n| = 1$, is a bounded sequence, but $\sum a_n y_n = \sum (-1)^n (-1)^n \frac{1}{n} = \sum \frac{1}{n}$ [as $(-1)^n(-1)^n = (-1)^{2n} = 1$ for all n] is the divergent harmonic series.

§9.5 Power Series

1. $\lim\limits_{n \to \infty} \left| \frac{a_{n+1}}{a_n} \right| = \lim\limits_{n \to \infty} \left| \frac{\frac{x^{n+1}}{(n+1)+1}}{\frac{x^n}{n+1}} \right| = \lim\limits_{n \to \infty} \frac{|x|^{n+1}(n+1)}{|x|^n(n+2)} = \lim\limits_{n \to \infty} \frac{|x|(1+\frac{1}{n})}{1+\frac{2}{n}} = |x| \Rightarrow$ the

series $\sum \frac{x^n}{n+1}$ converges for $|x| < 1$ and diverges for $|x| > 1$ by the Ratio Test. The radius of convergence is $R = 1$. As the center of the power series is at $a = 0$, the interval of convergence to be considered is $(-1, 1)$. Checking the endpoints:

At $x = 1$: $\sum \frac{1}{n+1}$ which is the divergent harmonic series.

At $x = -1$: $\sum \frac{(-1)^n}{n+1}$ is the alternating harmonic series, which converges.
Therefore, the interval of convergence is $[-1, 1)$.

3. $\lim\limits_{n \to \infty} |a_n^{\frac{1}{n}}| = \lim\limits_{n \to \infty} \left| \left(\frac{nx^n}{2^n} \right)^{\frac{1}{n}} \right| = \lim\limits_{n \to \infty} \frac{n^{\frac{1}{n}}|x|}{2} = \frac{|x|}{2}$, as $n^{\frac{1}{n}} \to 1$, as $n \to \infty$. Thus the

series converges for $\frac{|x|}{2} < 1$ and diverges for $\frac{|x|}{2} > 1$, i.e it converges for $|x| < 2$ and diverges for $|x| > 2$ by the Root Test. The radius of convergence is $R = 2$. As the center of the power series is at $a = 0$, the interval of convergence to be considered is $(-2, 2)$. Checking the endpoints:

At $x = 2$: $\sum \frac{n2^n}{2^n} = \sum n$, which diverges by the Divergence Test, as $n \to \infty$, not zero.

At $x = -2$: $\sum \frac{n(-2)^n}{2^n} = \sum (-1)^n n$, which diverges by the Divergence Test, as $|(-1)^n n| \to \infty$, not zero.
Therefore, the interval of convergence is $(-2, 2)$.

5. $\lim\limits_{n \to \infty} \left| \frac{a_{n+1}}{a_n} \right| = \lim\limits_{n \to \infty} \left| \frac{\frac{(n+1)[(n+1)+1]x^{n+1}}{5^{n+1}}}{\frac{n(n+1)x^n}{5^n}} \right| = \lim\limits_{n \to \infty} \frac{(n+1)(n+2)|x|^{n+1}5^n}{5^{n+1}n(n+1)|x|^n} = \lim\limits_{n \to \infty} \frac{n+2}{5n}|x| =$

$\lim\limits_{n \to \infty} \frac{1+\frac{2}{n}}{5}|x| = \frac{|x|}{5}$. Thus the series converges for $\frac{|x|}{5} < 1$ and diverges for $\frac{|x|}{5} > 1$, i.e. it converges for $|x| < 5$ and diverges for $|x| > 5$ by the Ratio Test. The radius of convergence is $R = 5$. As the center of the power series is at $a = 0$, the interval of convergence to be considered is $(-5, 5)$. Checking the endpoints:

At $x = 5$: $\sum \frac{n(n+1)5^n}{5^n} = \sum n(n+1)$ diverges by the Divergence Test, as its terms do not tend to 0.

At $x = -5$: $\sum \frac{n(n+1)(-5)^n}{5^n} = \sum (-1)^n n(n+1)$ also diverges by the Divergence Test, as its terms do not tend to 0.

Therefore, the interval of convergence is $(-5, 5)$.

7. $\lim_{n \to \infty} |a_n^{\frac{1}{n}}| = \lim_{n \to \infty} \left| \frac{x+4}{2} \right| = \left| \frac{x+4}{2} \right|$. By the Root Test, the series converges for $\frac{|x+4|}{2} < 1$ and diverges for $\frac{|x+4|}{2} > 1$, i.e. it converges for $|x+4| < 2$ and diverges for $|x+4| > 2 \Rightarrow$ the radius of convergence is $R = 2$. As the center of the power series is at $a = -4$, the interval of convergence to be considered is $(-4-2, -4+2) = (-6, -2)$. Checking the endpoints:

At $x = -2 : \sum \frac{2^n}{2^n} = \sum 1$ diverges by the Divergence Test, as its terms do not tend to 0.

At $x = -6 : \sum \frac{(-2)^n}{2^n} = \sum (-1)^n$ diverges by the Divergence Test, as its terms do not tend to 0.

Thus, the interval of convergence is $(-6, -2)$.

9. $\lim_{n \to \infty} \left| \frac{a_{n+1}}{a_n} \right| = \lim_{n \to \infty} \left| \dfrac{\dfrac{(n+1)2^{n+1}(x-1)^{n+1}}{(n+1)+1}}{\dfrac{n2^n(x-1)^n}{n+1}} \right| = \lim_{n \to \infty} \frac{(n+1)2^{n+1}|x-1|^{n+1}(n+1)}{(n+2)n2^n|x-1|^n} = $

$\lim_{n \to \infty} \frac{(n+1)^2 2|x-1|}{(n+2)n} = \lim_{n \to \infty} \frac{2n^2 + 4n + 2}{n^2 + 2n}|x-1| = \lim_{n \to \infty} \frac{2 + \frac{4}{n} + \frac{2}{n^2}}{1 + \frac{2}{n}}|x-1| = 2|x-1|$. By the Ratio Test, the series converges for $2|x-1| < 1$ and diverges for $2|x-1| > 1$, i.e. it converges for $|x-1| < \frac{1}{2}$ and diverges for $|x-1| > \frac{1}{2} \Rightarrow$ the radius of convergence is $R = \frac{1}{2}$. As the center of the power series is $a = 1$, the interval of convergence to be considered is $(1 - \frac{1}{2}, 1 + \frac{1}{2}) = (\frac{1}{2}, \frac{3}{2})$. Checking the endpoints:

At $x = \frac{1}{2} : \sum \frac{n2^n(-\frac{1}{2})^n}{n+1} = \sum (-1)^n \frac{n}{n+1}$ which diverges by the Divergence Test, as $\frac{n}{n+1} \to 1 \neq 0$.

At $x = \frac{3}{2} : \sum \frac{n2^n(\frac{1}{2})^n}{n+1} = \sum \frac{n}{n+1}$ which diverges by the Divergence Test, as $\frac{n}{n+1} \to 1 \neq 0$.

Thus, the interval of convergence is $(\frac{1}{2}, \frac{3}{2})$.

11. $\lim_{n \to \infty} \left| \frac{a_{n+1}}{a_n} \right| = \lim_{n \to \infty} \left| \dfrac{\dfrac{(x+4)^{n+1}}{(n+1)[(n+1)+1]}}{\dfrac{(x+4)^n}{n(n+1)}} \right| = \lim_{n \to \infty} \frac{|x+4|^{n+1}n(n+1)}{(n+1)(n+2)|x+4|^n} = \lim_{n \to \infty} \frac{n}{n+2}|x+4|$

$= \lim_{n \to \infty} \frac{1}{1 + \frac{2}{n}}|x+4| = |x+4|$. By the Ratio Test, the series converges for $|x+4| < 1$ and diverges for $|x+4| > 1 \Rightarrow$ the radius of convergence is $R = 1$. As the center of the power series is $a = -4$, the interval of convergence to be considered is $(-4-1, -4+1) = (-5, -3)$. Checking the endpoints:

At $x = -3 : \sum \frac{1}{n(n+1)}$ converges by the Comparison Test, comparing it with the convergent p-series $\sum \frac{1}{n^2}$, as $\frac{1}{n(n+1)} = \frac{1}{n^2 + n} < \frac{1}{n^2}$.

At $x = -5 : \sum \frac{(-1)^n}{n(n+1)}$ was shown to be absolutely convergent in the previous step. Thus, the interval of convergence is $[-5, -3]$.

13. $\lim_{n \to \infty} |a_n^{\frac{1}{n}}| = \lim_{n \to \infty} \left| \left[\frac{(-1)^{n-1} x^n}{\sqrt{n}} \right]^{\frac{1}{n}} \right| = \lim_{n \to \infty} \frac{|x|}{(n^{\frac{1}{n}})^{\frac{1}{2}}} = |x|$, as $n^{\frac{1}{n}} \to 1$, as $n \to \infty$.

By the Root Test, the series converges for $|x| < 1$ and diverges for $|x| > 1 \Rightarrow$ the radius of convergence is $R = 1$. As the center of the power series is $a = 0$, the interval of convergence to be considered is $(-1, 1)$. Checking the endpoints:

At $x = -1 : \sum \frac{(-1)^{n-1}(-1)^n}{\sqrt{n}} = \sum \frac{(-1)^{2n-1}}{\sqrt{n}} = -\sum \frac{1}{n^{\frac{1}{2}}}$ which diverges, because it is just a constant times a divergent p-series $(p = \frac{1}{2} < 1)$.

At $x = 1 : \sum \frac{(-1)^{n-1}}{\sqrt{n}}$ converges by the Alternating Series Theorem, as $a_n = \frac{1}{\sqrt{n}}$ is positive and decreasing to 0 as $n \to \infty$.
Therefore, the interval of convergence is $(-1, 1]$.

15. $\lim_{n \to \infty} \left| \frac{a_{n+1}}{a_n} \right| = \lim_{n \to \infty} \left| \frac{\frac{x^{n+1}}{1+(n+1)^2}}{\frac{x^n}{1+n^2}} \right| = \lim_{n \to \infty} \frac{|x|^{n+1}(1+n^2)}{|x|^n(n^2+2n+2)} = \lim_{n \to \infty} \frac{\frac{1}{n^2}+1}{1+\frac{2}{n}+\frac{2}{n^2}} |x| = |x|$.

By the Ratio Test, the series converges for $|x| < 1$ and diverges for $|x| > 1 \Rightarrow$ the radius of convergence is $R = 1$. As the center of the power series is $a = 0$, the interval of convergence to be considered is $(-1, 1)$. Checking the endpoints:

At $x = 1 : \sum \frac{1}{1+n^2}$ converges by the Comparison Test, comparing it with the convergent p-series $\sum \frac{1}{n^2}$ $(p = 2 > 1)$, as $\frac{1}{1+n^2} < \frac{1}{n^2}$.

At $x = -1 : \sum \frac{(-1)^n}{1+n^2}$ is absolutely convergent (see above).
Consequently, the interval of convergence is $[-1, 1]$.

17. From Example 9.20(a): $\frac{1}{1-x} = \sum_{n=0}^{\infty} x^n$ converges for $|x| < 1$, then $\frac{1}{1+x} = \frac{1}{1-(-x)} = \sum_{n=0}^{\infty} (-x)^n = \sum_{n=0}^{\infty} (-1)^n x^n$ converges for $|-x| < 1 \Rightarrow |x| < 1$, so that the radius of convergence is $R = 1$, and the interval of convergence is $(-1, 1)$.

19. From Example 9.20(a): $\frac{1}{1-x} = \sum_{n=0}^{\infty} x^n$ converges for $|x| < 1$. So we rewrite $f(x) = \frac{2}{3-x}$ in the form of a constant times $\frac{1}{1 - \text{something}}$, and then apply this example series:
$$\frac{2}{3-x} = \frac{2}{3(1-\frac{x}{3})} = \frac{2}{3} \cdot \frac{1}{1-\frac{x}{3}} = \frac{2}{3} \sum_{n=0}^{\infty} \left(\frac{x}{3}\right)^n = 2 \sum_{n=0}^{\infty} \frac{x^n}{3^{n+1}}$$
and the series converges for $|\frac{x}{3}| < 1$, i.e. for $|x| < 3 \Rightarrow$ the radius of convergence is $R = 3$ and the interval of convergence is $(-3, 3)$.

21. From Example 9.20(a): $\frac{1}{1-x} = \sum_{n=0}^{\infty} x^n$ converges for $|x| < 1$. So we rewrite $f(x) =$

$\frac{x}{9+x^2} = x\left(\frac{1}{9+x^2}\right)$ in the form of x times a constant multiple of $\frac{1}{1-\text{something}}$, and then apply this example series:

$$\frac{1}{9+x^2} = \frac{1}{9(1+\frac{x^2}{9})} = \frac{1}{9} \cdot \frac{1}{1-(-\frac{x^2}{9})} = \frac{1}{9}\sum_{n=0}^{\infty}\left(-\frac{x^2}{9}\right)^n = \sum_{n=0}^{\infty}\frac{(-1)^n x^{2n}}{9^{n+1}}$$

Thus, $f(x) = x\sum_{n=0}^{\infty}\frac{(-1)^n x^{2n}}{9^{n+1}} = \sum_{n=0}^{\infty}\frac{(-1)^n x^{2n+1}}{9^{n+1}}$. The series converges for $|\frac{x^2}{9}| < 1$, i.e. for $|x^2| < 9 \Rightarrow |x| < 3 \Rightarrow$ the radius of convergence is $R = 3$ and the interval of convergence is $(-3, 3)$.

23. From Example 9.20(d): $\ln(1-x) = -\sum_{n=1}^{\infty}\frac{x^n}{n}$ converges for $|x| < 1$. So we rewrite $f(x) = \ln(5-x)$ in terms of $\ln(1-\text{something})$: $\ln(5-x) = \ln[5(1-\frac{x}{5})] = \ln 5 + \ln(1-\frac{x}{5})$.

From the example, $\ln(1-\frac{x}{5}) = -\sum_{n=1}^{\infty}\frac{(\frac{x}{5})^n}{n} = -\sum_{n=1}^{\infty}\frac{x^n}{n5^n}$. Thus, $f(x) = \ln 5 - \sum_{n=1}^{\infty}\frac{x^n}{n5^n}$, and the series converges for $|\frac{x}{5}| < 1 \Rightarrow |x| < 5 \Rightarrow$ the radius of convergence is $R = 5$ and the interval of convergence is $(-5, 5)$.

25. $f(x) = \frac{3}{x^2-x-2} = \frac{3}{(x-2)(x+1)} = \frac{A}{x-2} + \frac{B}{x+1}$

$$3 = A(x+1) + B(x-2)$$
$$x = -1 : 3 = -3B \Rightarrow B = -1$$
$$x = 2 : 3 = 3A \Rightarrow A = 1$$

So, $f(x) = \frac{1}{x-2} - \frac{1}{x+1}$. We can obtain a power series for each of these expressions, from Example 9.20(a): $\frac{1}{1-x} = \sum_{n=0}^{\infty} x^n$ converges for $|x| < 1$. Thus,

$$\frac{1}{x-2} = -\frac{1}{2-x} = -\frac{1}{2(1-\frac{x}{2})} = -\frac{1}{2} \cdot \frac{1}{1-\frac{x}{2}} = -\frac{1}{2}\sum_{n=0}^{\infty}\left(\frac{x}{2}\right)^n = -\sum_{n=0}^{\infty}\frac{x^n}{2^{n+1}},$$ converging for

$|\frac{x}{2}| < 1 \Rightarrow |x| < 2 \Rightarrow R = 2.$

$$\frac{1}{x+1} = \frac{1}{1-(-x)} = \sum_{n=0}^{\infty}(-x)^n = \sum_{n=0}^{\infty}(-1)^n x^n,$$ converging for $|-x| < 1 \Rightarrow |x| < 1 \Rightarrow R = 1.$

Hence, $f(x) = -\sum_{n=0}^{\infty}\frac{x^n}{2^{n+1}} - \sum_{n=0}^{\infty}(-1)^n x^n = \sum_{n=0}^{\infty}(-1)^{n+1} x^n - \sum_{n=0}^{\infty}\frac{x^n}{2^{n+1}}$, with a radius of convergence equal to the smaller of the radii of the two series, $R = 1$, and the interval of convergence is $(-1, 1)$.

27. From Example 9.20(a): $\frac{1}{1-x} = \sum\limits_{n=0}^{\infty} x^n$ converges for $|x| < 1$. Thus, $\frac{1}{1-2x} = \sum\limits_{n=0}^{\infty} (2x)^n =$

$\sum\limits_{n=0}^{\infty} 2^n x^n$ converging for $|2x| < 1 \Rightarrow |x| < \frac{1}{2}$. As $[\ln(1-2x)]' = \frac{1}{1-2x}(-2)$,

$$\ln(1-2x) = -2\int \frac{dx}{1-2x} = -2\int \sum\limits_{n=0}^{\infty} 2^n x^n \, dx = -2\sum\limits_{n=0}^{\infty} 2^n \int x^n \, dx = -\sum\limits_{n=0}^{\infty} 2^{n+1} \frac{x^{n+1}}{n+1} + C$$

Evaluating at $x = 0$: $\ln 1 = 0 + C \Rightarrow C = \ln 1 = 0 \Rightarrow$

$$\ln(1-2x) = -\sum\limits_{n=0}^{\infty} 2^{n+1} \frac{x^{n+1}}{n+1} = -\sum\limits_{n=1}^{\infty} \frac{2^n x^n}{n}$$

with a radius of convergence of $R = \frac{1}{2}$ and an interval of convergence of $\left(-\frac{1}{2}, \frac{1}{2}\right)$.

29. From Example 9.20(a): $\frac{1}{1-x} = \sum\limits_{n=0}^{\infty} x^n$ converges for $|x| < 1$. Thus, $\frac{1}{1-x^2} = \sum\limits_{n=0}^{\infty} (x^2)^n =$

$\sum\limits_{n=0}^{\infty} x^{2n}$ converging for $|x^2| < 1 \Rightarrow |x| < 1$. As $[\ln(1-x^2)]' = \frac{1}{1-x^2}(-2x)$,

$$\ln(1-x^2) = \int \left(-2x \sum\limits_{n=0}^{\infty} x^{2n} \right) dx = -2\sum\limits_{n=0}^{\infty} \left(\int x^{2n+1} dx \right) = -2\sum\limits_{n=0}^{\infty} \frac{x^{2n+2}}{2n+2} + C$$

Evaluating at $x = 0$: $\ln 1 = 0 + C \Rightarrow C = \ln 1 = 0 \Rightarrow$

$$\ln(1-x^2) = -2\sum\limits_{n=0}^{\infty} \frac{x^{2n+2}}{2n+2} = -2\sum\limits_{n=1}^{\infty} \frac{x^{2n}}{2n} = -\sum\limits_{n=1}^{\infty} \frac{x^{2n}}{n}$$

with a radius of convergence of $R = 1$ and an interval of convergence of $(-1, 1)$.

31.   As $\tan^{-1} x = \sum\limits_{n=0}^{\infty} (-1)^n \frac{x^{2n+1}}{2n+1}$, for $|x| < 1$, then $\tan^{-1} x^2 = \sum\limits_{n=0}^{\infty} (-1)^n \frac{(x^2)^{2n+1}}{2n+1} =$

$\sum\limits_{n=0}^{\infty} (-1)^n \frac{x^{4n+2}}{2n+1}$, for $|x^2| < 1 \Rightarrow |x| < 1$. Therefore, the radius of convergence is $R = 1$, and

the interval of convergence is $(-1, 1)$.

33. $f(x) = \sum\limits_{n=0}^{\infty} \frac{x^n}{n!} \Rightarrow f'(x) = \sum\limits_{n=1}^{\infty} \frac{nx^{n-1}}{n!} - \sum\limits_{n=1}^{\infty} \frac{x^{n-1}}{(n-1)!} \Rightarrow f''(x) = \sum\limits_{n=2}^{\infty} \frac{(n-1)x^{n-2}}{(n-1)!} =$

$\sum\limits_{n=2}^{\infty} \frac{x^{n-2}}{(n-2)!} = \sum\limits_{n=0}^{\infty} \frac{x^n}{n!} = f(x)$. Hence, $f''(x) - f(x) = 0$.

35. $\displaystyle\lim_{n\to\infty}\left|\frac{a_{n+1}}{a_n}\right| = \lim_{n\to\infty}\left|\frac{\dfrac{(n+1+s)!|x|^{n+1}}{(n+1)!(n+1+t)!}}{\dfrac{(n+s)!|x|^n}{n!(n+t)!}}\right| = \lim_{n\to\infty}\frac{(n+s+1)!n!(n+t)!}{(n+1)!(n+1+t)!(n+s)!}|x|$

$\displaystyle = \lim_{n\to\infty}\frac{(n+s+1)}{(n+1)(n+t+1)}|x| = \lim_{n\to\infty}\frac{(n+s+1)}{n^2+(t+2)n+(t+1)}|x|$

$\displaystyle = \lim_{n\to\infty}\frac{\dfrac{1}{n}+\dfrac{s+1}{n^2}}{1+\dfrac{t+2}{n}+\dfrac{t+1}{n^2}}|x| = 0\cdot|x| = 0 < 1.$

By the Ratio Test, the series converges for all x, which means that $R = \infty$.

37. $\displaystyle\lim_{n\to\infty}\left|a_n^{\frac{1}{n}}\right| = \lim_{n\to\infty}\left|(c_nx^n)^{\frac{1}{n}}\right| = \lim_{n\to\infty}|c_n|^{\frac{1}{n}}|x| = |x|\left(\lim_{n\to\infty}|c_n|^{\frac{1}{n}}\right) = |x|L.$ By the Root Test, the series converges for $|x|L < 1 \Rightarrow |x| < \dfrac{1}{L} \Rightarrow$ the radius of convergence is $R = \dfrac{1}{L}$.

§9.6 Taylor Series

1.

$$\begin{aligned} f(x) &= \ln(1+x) \\ f'(x) &= \frac{1}{1+x} = (1+x)^{-1} \\ f''(x) &= -(1+x)^{-2} \\ f'''(x) &= 2(1+x)^{-3} \\ f^{(4)}(x) &= -2\cdot3(1+x)^{-4} \\ f^{(5)}(x) &= 2\cdot3\cdot4(1+x)^{-5} \\ &= \ldots \\ f^{(n)}(x) &= (-1)^{n+1}(n-1)!(1+x)^{-n}, \ n\geq 1 \end{aligned}$$

$$\begin{aligned} f(0) &= \ln 1 = 0 \\ f'(0) &= 1 \\ f''(0) &= -1 \\ f'''(0) &= 2 \\ f^{(4)}(0) &= -2\cdot3 \\ f^{(5)}(0) &= 2\cdot3\cdot4 \\ &= \ldots \\ f^{(n)}(0) &= (-1)^{n+1}(n-1)!, \ n\geq 1 \end{aligned}$$

The Maclaurin series is then $\displaystyle\sum_{n=0}^{\infty}\frac{f^{(n)}(0)}{n!}x^n = \sum_{n=1}^{\infty}\frac{(-1)^{n+1}(n-1)!}{n!}x^n = \sum_{n=1}^{\infty}\frac{(-1)^{n+1}x^n}{n}$. Apply the Ratio Test to determine the radius of convergence:

$$\left|\frac{a_{n+1}}{a_n}\right| = \left|\frac{\dfrac{x^{n+1}}{n+1}}{\dfrac{x^n}{n}}\right| = \frac{n}{n+1}|x| \to |x|$$

Thus the series converges for $|x| < 1$ and diverges for $|x| > 1$, so that the radius of convergence $R = 1$.

3.
$$f(x) = \frac{1}{1+x} = (1+x)^{-1}$$
$$f'(x) = -(1+x)^{-2}$$
$$f''(x) = 2(1+x)^{-3}$$
$$f'''(x) = -2\cdot 3(1+x)^{-4}$$
$$f^{(4)}(x) = 2\cdot 3\cdot 4(1+x)^{-5}$$
$$= \ldots$$
$$f^{(n)}(x) = (-1)^n n!(1+x)^{-(n+1)}$$

$$f(0) = 1$$
$$f'(0) = -1$$
$$f''(0) = 2$$
$$f'''(0) = -2\cdot 3$$
$$f^{(4)}(0) = 2\cdot 3\cdot 4$$
$$= \ldots$$
$$f^{(n)}(0) = (-1)^n n!$$

The Maclaurin series is then $\displaystyle\sum_{n=0}^{\infty}\frac{f^{(n)}(0)}{n!}x^n = \sum_{n=0}^{\infty}\frac{(-1)^n n!}{n!}x^n = \sum_{n=0}^{\infty}(-1)^n x^n$. Apply

the Ratio Test to determine the radius of convergence:

$$\left|\frac{a_{n+1}}{a_n}\right| = \left|\frac{x^{n+1}}{x^n}\right| = |x| \to |x|$$

Thus the series converges for $|x| < 1$ and diverges for $|x| > 1$, so that the radius of convergence $R = 1$.

5.
$$f(x) = \frac{1}{1-3x} = (1-3x)^{-1}$$
$$f'(x) = -(1-3x)^{-2}(-3)$$
$$f''(x) = 2(1-3x)^{-3}(-3)^2$$
$$f'''(x) = -2\cdot 3(1-3x)^{-4}(-3)^3$$
$$f^{(4)}(x) = 2\cdot 3\cdot 4(1-3x)^{-5}(-3)^4$$
$$= \ldots$$
$$f^{(n)}(x) = n!(1-3x)^{-(n+1)}3^n$$

$$f(0) = 1$$
$$f'(0) = 3$$
$$f''(0) = 2(3^2)$$
$$f'''(0) = 2\cdot 3(3^3)$$
$$f^{(4)}(0) = 2\cdot 3\cdot 4(3^4)$$
$$= \ldots$$
$$f^{(n)}(0) = n!(3^n)$$

The Maclaurin series is then $\displaystyle\sum_{n=0}^{\infty}\frac{f^{(n)}(0)}{n!}x^n = \sum_{n=0}^{\infty}\frac{n!(3^n)}{n!}x^n = \sum_{n=0}^{\infty}3^n x^n$. Apply

the Ratio Test to determine the radius of convergence:

$$\left|\frac{a_{n+1}}{a_n}\right| = \left|\frac{3^{n+1}x^{n+1}}{3^n x^n}\right| = 3|x| \to 3|x|$$

Thus the series converges for $3|x| < 1 \Rightarrow |x| < \frac{1}{3}$ and diverges for $3|x| > 1 \Rightarrow |x| > \frac{1}{3}$, so that the radius of convergence $R = \frac{1}{3}$.

7.
$$f(x) = \frac{x}{e^x} = xe^{-x}$$
$$f'(x) = x(-e^{-x}) + e^{-x} = e^{-x}(1-x)$$
$$f''(x) = e^{-x}(-1) + (1-x)(-e^{-x}) = e^{-x}(x-2)$$
$$f'''(x) = e^{-x}(1) + (x-2)(-e^{-x}) = e^{-x}(3-x)$$
$$f^{(4)}(x) = e^{-x}(-1) + (3-x)(-e^{-x}) = e^{-x}(x-4)$$
$$= \ldots$$
$$f^{(n)}(x) = e^{-x}(-1)^n(x-n), \quad n \geq 1$$

$$f(0) = 0$$
$$f'(0) = 1$$
$$f''(0) = -2$$
$$f'''(0) = 3$$
$$f^{(4)}(0) = -4$$
$$= \ldots$$
$$f^{(n)}(0) = (-1)^{n+1}n, \quad n \geq 1$$

The Maclaurin series is then $\sum_{n=0}^{\infty} \frac{f^{(n)}(0)}{n!}x^n = \sum_{n=1}^{\infty} \frac{(-1)^{n+1}n}{n!}x^n = \sum_{n=1}^{\infty} \frac{(-1)^{n+1}}{(n-1)!}x^n = \sum_{n=0}^{\infty} \frac{(-1)^n}{n!}x^{n+1}$.

Apply the Ratio Test to determine the radius of convergence:

$$\left|\frac{a_{n+1}}{a_n}\right| = \left|\frac{\frac{x^{n+2}}{(n+1)!}}{\frac{x^{n+1}}{n!}}\right| = \frac{|x|}{n+1} \to 0$$

Thus the series converges for all x, so that the radius of convergence $R = \infty$.

9.
$$f(x) = 100\cos\pi x$$
$$f'(x) = -100\pi\sin\pi x$$
$$f''(x) = -100\pi^2\cos\pi x$$
$$f'''(x) = 100\pi^3\sin\pi x$$
$$f^{(4)}(x) = 100\pi^4\cos\pi x$$
$$= \ldots$$
$$f^{(2n+1)}(x) = (-1)^{n+1}100\pi^{2n+1}\sin\pi x$$
$$f^{(2n)}(x) = (-1)^n100\pi^{2n}\cos\pi x$$

$$f(0) = 100$$
$$f'(0) = 0$$
$$f''(0) = -100\pi^2$$
$$f'''(0) = 0$$
$$f^{(4)}(0) = 100\pi^4$$
$$= \ldots$$
$$f^{(2n+1)}(0) = 0$$
$$f^{(2n)}(0) = (-1)^n100\pi^{2n}$$

The Maclaurin series is then $\sum_{n=0}^{\infty} \frac{f^{(n)}(0)}{n!}x^n = \sum_{n=0}^{\infty} \frac{(-1)^n100\pi^{2n}}{(2n)!}x^{2n}$. Apply the Ratio Test to determine the radius of convergence:

$$\left|\frac{a_{n+1}}{a_n}\right| = \left|\frac{\frac{100\pi^{2(n+1)}x^{2(n+1)}}{[2(n+1)]!}}{\frac{100\pi^{2n}x^{2n}}{(2n)!}}\right| = \frac{\pi^2|x|^2}{(2n+2)(2n+1)} \to 0$$

Thus the series converges for all x, so that the radius of convergence $R = \infty$.

11. $f(x) = e^x = f^{(n)}(x) \Rightarrow f^{(n)}(1) = e^1 = e$. Hence the Taylor series is $\sum\limits_{n=0}^{\infty} \dfrac{f^{(n)}(1)}{n!}(x-1)^n =$

$\sum\limits_{n=0}^{\infty} \dfrac{e(x-1)^n}{n!}$. Apply the Ratio Test to determine the radius of convergence:

$$\left|\frac{a_{n+1}}{a_n}\right| = \left|\frac{\dfrac{e(x-1)^{n+1}}{(n+1)!}}{\dfrac{e(x-1)^n}{n!}}\right| = \frac{|x-1|}{n+1} \to 0 \text{ as } n \to \infty. \text{ Therefore the series converges}$$

for all x, and the radius of convergence is $R = \infty$.

13.

$$
\begin{aligned}
f(x) &= \cos x \\
f'(x) &= -\sin x \\
f''(x) &= -\cos x \\
f'''(x) &= \sin x \\
f^{(4)}(x) &= \cos x \\
&= \dots \\
f^{(2n+1)}(x) &= (-1)^{n+1}\sin x \\
f^{(2n)}(x) &= (-1)^n \cos x
\end{aligned}
\qquad
\begin{aligned}
f\left(\tfrac{\pi}{6}\right) &= \cos\tfrac{\pi}{6} = \tfrac{\sqrt{3}}{2} \\
f'\left(\tfrac{\pi}{6}\right) &= -\sin\tfrac{\pi}{6} = -\tfrac{1}{2} \\
f''\left(\tfrac{\pi}{6}\right) &= -\cos\tfrac{\pi}{6} = -\tfrac{\sqrt{3}}{2} \\
f'''\left(\tfrac{\pi}{6}\right) &= \sin\tfrac{\pi}{6} = \tfrac{1}{2} \\
f^{(4)}\left(\tfrac{\pi}{6}\right) &= \cos\tfrac{\pi}{6} = \tfrac{\sqrt{3}}{2} \\
&= \dots \\
f^{(2n+1)}\left(\tfrac{\pi}{6}\right) &= (-1)^{n+1}\tfrac{1}{2} \\
f^{(2n)}\left(\tfrac{\pi}{6}\right) &= (-1)^n \tfrac{\sqrt{3}}{2}
\end{aligned}
$$

Hence the Taylor series is:

$$\frac{\sqrt{3}}{2} - \frac{1}{2}\left(x - \frac{\pi}{6}\right) - \frac{\frac{\sqrt{3}}{2}\left(x - \frac{\pi}{6}\right)^2}{2!} + \frac{\frac{1}{2}\left(x - \frac{\pi}{6}\right)^3}{3!} + \frac{\frac{\sqrt{3}}{2}\left(x - \frac{\pi}{6}\right)^4}{4!} - \frac{\frac{1}{2}\left(x - \frac{\pi}{6}\right)^5}{5!} - \frac{\frac{\sqrt{3}}{2}\left(x - \frac{\pi}{6}\right)^6}{6!} +$$

$$\frac{\frac{1}{2}\left(x - \frac{\pi}{6}\right)^7}{7!} + \cdots. \quad \text{Now apply the Ratio test:}$$

$$\left|\frac{a_{2n}}{a_{2n-1}}\right| = \left|\frac{\dfrac{\frac{\sqrt{3}}{2}\left(x - \frac{\pi}{6}\right)^{2n}}{(2n)!}}{\dfrac{\frac{1}{2}\left(x - \frac{\pi}{6}\right)^{2n-1}}{(2n-1)!}}\right| = \frac{\sqrt{3}\left|x - \frac{\pi}{6}\right|}{2n} \to 0 \text{ for all } x, \text{ and}$$

$$\left|\frac{a_{2n+1}}{a_{2n}}\right| = \left|\frac{\dfrac{\frac{1}{2}\left(x - \frac{\pi}{6}\right)^{2n+1}}{(2n+1)!}}{\dfrac{\frac{\sqrt{3}}{2}\left(x - \frac{\pi}{6}\right)^{2n}}{(2n)!}}\right| = \frac{\left|x - \frac{\pi}{6}\right|}{\sqrt{3}(2n+1)} \to 0 \text{ for all } x,$$

so the series converges absolutely for all x, $R = \infty$, and we can write the series as:

$$\sum_{n=0}^{\infty} \frac{(-1)^{n+1}\left(x - \frac{\pi}{6}\right)^{2n+1}}{2(2n+1)!} + \sum_{n=0}^{\infty} \frac{(-1)^n \sqrt{3}\left(x - \frac{\pi}{6}\right)^{2n}}{2(2n)!}.$$

15.
$$f(x) = \ln x$$
$$f'(x) = \frac{1}{x} = x^{-1}$$
$$f''(x) = -x^{-2}$$
$$f'''(x) = 2x^{-3}$$
$$f^{(4)}(x) = -2 \cdot 3x^{-4}$$
$$= \ldots$$
$$f^{(n)}(x) = (-1)^{n+1}(n-1)!x^{-n}, \ n \geq 1$$

$$f(1) = 0$$
$$f'(1) = 1$$
$$f''(1) = -1$$
$$f'''(1) = 2$$
$$f^{(4)}(1) = -2 \cdot 3$$
$$= \ldots$$
$$f^{(n)}(1) = (-1)^{n+1}(n-1)!, \ n \geq 1$$

Hence the Taylor series is:

$$\sum_{n=0}^{\infty} \frac{f^{(n)}(1)}{n!}(x-1)^n = \sum_{n=1}^{\infty} \frac{(-1)^{n+1}(n-1)!}{n!}(x-1)^n = \sum_{n=1}^{\infty} \frac{(-1)^{n+1}(x-1)^n}{n}.$$ Apply the Ratio

Test to determine the radius of convergence:

$$\left| \frac{a_{n+1}}{a_n} \right| = \left| \frac{\frac{(x-1)^{n+1}}{n+1}}{\frac{(x-1)^n}{n}} \right| = \left(\frac{n}{n+1} \right)|x-1| = \left[\frac{1}{1+\frac{1}{n}} \right]|x-1| \to |x-1| \ \Rightarrow \ \text{the}$$

series converges for $|x-1| < 1$ and diverges for $|x-1| > 1$ $\Rightarrow$ the radius of convergence $R = 1$.

17.
$$f(x) = \sin \pi x$$
$$f'(x) = \pi \cos \pi x$$
$$f''(x) = -\pi^2 \sin \pi x$$
$$f'''(x) = -\pi^3 \cos \pi x$$
$$f^{(4)}(x) = \pi^4 \sin \pi x$$
$$= \ldots$$
$$f^{(2n+1)}(x) = (-1)^n \pi^{2n+1} \cos \pi x$$
$$f^{(2n)}(x) = (-1)^n \pi^{2n} \sin \pi x$$

$$f(\tfrac{1}{2}) = \sin \tfrac{\pi}{2} = 1$$
$$f'(\tfrac{1}{2}) = \pi \cos \tfrac{\pi}{2} = 0$$
$$f''(\tfrac{1}{2}) = -\pi^2 \sin \tfrac{\pi}{2} = -\pi^2$$
$$f'''(\tfrac{1}{2}) = -\pi^3 \cos \tfrac{\pi}{2} = 0$$
$$f^{(4)}(\tfrac{1}{2}) = \pi^4 \sin \tfrac{\pi}{2} = \pi^4$$
$$= \ldots$$
$$f^{(2n+1)}(\tfrac{1}{2}) = 0$$
$$f^{(2n)}(\tfrac{1}{2}) = (-1)^n \pi^{2n}$$

Hence the Taylor series is:

$$\sum_{n=0}^{\infty} \frac{f^{(n)}(\frac{1}{2})}{n!}(x-\tfrac{1}{2})^n = \sum_{n=0}^{\infty} \frac{(-1)^n \pi^{2n}}{(2n)!}(x-\tfrac{1}{2})^{2n}.$$ Apply the Ratio Test to determine the radius

of convergence:

$$\left| \frac{a_{n+1}}{a_n} \right| = \left| \frac{\frac{\pi^{2(n+1)}(x-\frac{1}{2})^{2(n+1)}}{[2(n+1)]!}}{\frac{\pi^{2n}(x-\frac{1}{2})^{2n}}{(2n)!}} \right| = \frac{\pi^2|x-\frac{1}{2}|^2}{(2n+2)(2n+1)} \to 0, \text{ as } n \to \infty.$$

Therefore the series converges for all x, and the radius of convergence is $R = \infty$.

19. $f(x) = \dfrac{1}{\sqrt{x}} = x^{-\frac{1}{2}}$

$f'(x) = -\dfrac{1}{2}x^{-\frac{3}{2}}$

$f''(x) = -\dfrac{1}{2}\left(-\dfrac{3}{2}\right)x^{-\frac{5}{2}}$

$f'''(x) = -\dfrac{1}{2}\left(-\dfrac{3}{2}\right)\left(-\dfrac{5}{2}\right)x^{-\frac{7}{2}}$

$= \ldots$ so, for $n \geq 1$,

$f^{(n)}(x) = \dfrac{(-1)^n[1 \cdot 3 \cdot 5 \cdots (2n-1)]}{2^n}x^{-\left(\frac{2n+1}{2}\right)}$

$f(9) = 9^{-\frac{1}{2}} = \dfrac{1}{3}$

$f'(9) = -\dfrac{1}{2} \cdot 9^{-\frac{3}{2}} = -\dfrac{1}{2}\left(\dfrac{1}{3^3}\right)$

$f''(9) = -\dfrac{1}{2}\left(-\dfrac{3}{2}\right)\left(9^{-\frac{5}{2}}\right) = -\dfrac{1}{2}\left(-\dfrac{3}{2}\right)\left(\dfrac{1}{3^5}\right)$

$f'''(9) = -\dfrac{1}{2}\left(-\dfrac{3}{2}\right)\left(-\dfrac{5}{2}\right)\left(\dfrac{1}{3^7}\right)$

$= \ldots$ so, for $n \geq 1$,

$f^{(n)}(9) = \dfrac{(-1)^n[1 \cdot 3 \cdot 5 \cdots (2n-1)]}{2^n}\left(\dfrac{1}{3^{2n+1}}\right)$

Hence the Taylor series is:

$$\sum_{n=0}^{\infty}\dfrac{f^{(n)}(9)}{n!}(x-9)^n = \dfrac{1}{3} + \sum_{n=1}^{\infty}\dfrac{(-1)^n[1 \cdot 3 \cdot 5 \cdots (2n-1)]}{2^n\,3^{2n+1}\,n!}(x-9)^n.$$ Apply the Ratio Test to determine the radius of convergence:

$$\left|\dfrac{a_{n+1}}{a_n}\right| = \left|\dfrac{\dfrac{[1 \cdot 3 \cdot 5 \cdots (2n-1)(2n+1)](x-9)^{n+1}}{2^{n+1}3^{2(n+1)+1}(n+1)!}}{\dfrac{[1 \cdot 3 \cdot 5 \cdots (2n-1)](x-9)^n}{2^n\,3^{2n+1}\,n!}}\right| = \dfrac{2n+1}{2(3^2)(n+1)}|x-9| = \dfrac{2+\frac{1}{n}}{18(1+\frac{1}{n})}|x-9| \rightarrow$$

$\dfrac{1}{9}|x-9|$, as $n \rightarrow \infty$. Therefore, the series converges for $\dfrac{1}{9}|x-9| < 1 \Rightarrow |x-9| < 9$ and diverges for $|x-9| > 9 \Rightarrow$ the radius of convergence $R = 9$.

21. In $e^x = \displaystyle\sum_{n=0}^{\infty}\dfrac{x^n}{n!}$ which converges for all x, substitute $x-1$ for x. Then $e^{x-1} = \displaystyle\sum_{n=0}^{\infty}\dfrac{(x-1)^n}{n!}$.

But $e^{x-1} = (e^x)(e^{-1}) \Rightarrow e^x = e\displaystyle\sum_{n=0}^{\infty}\dfrac{(x-1)^n}{n!} = \displaystyle\sum_{n=0}^{\infty}\dfrac{e(x-1)^n}{n!}$. The series converges for all $x \Rightarrow$ the radius of convergence $R = \infty$.

23. $f(x) = \dfrac{1}{1+x^2} = \dfrac{1}{1-(-x^2)}$. In $\dfrac{1}{1-x} = \displaystyle\sum_{n=0}^{\infty}x^n$ which converges for $|x| < 1$, substitute $-x^2$

for x. Then $\dfrac{1}{1+x^2} = \displaystyle\sum_{n=0}^{\infty}(-x^2)^n = \displaystyle\sum_{n=0}^{\infty}(-1)^n x^{2n}$ which converges for $|-x^2| < 1 \Rightarrow |x| < 1 \Rightarrow$ the radius of convergence $R = 1$.

25. As in Exercise 23 above, substituting $-x^2$ for x in $\dfrac{1}{1-x} = \displaystyle\sum_{n=0}^{\infty}x^n$ yields $\dfrac{1}{1+x^2} = \displaystyle\sum_{n=0}^{\infty}(-1)^n x^{2n}$ which converges for $|x| < 1$. Integrate:

$$\tan^{-1}x = \int\dfrac{1}{1+x^2}\,dx = \int\left(\sum_{n=0}^{\infty}(-1)^n x^{2n}\right)dx = \sum_{n=0}^{\infty}(-1)^n\int x^{2n}dx = \sum_{n=0}^{\infty}(-1)^n\dfrac{x^{2n+1}}{2n+1}+C,$$

converging for $|x| < 1 \Rightarrow$ the radius of convergence $R = 1$. Evaluating at $x = 0$:

$0 = \tan^{-1}0 = C \Rightarrow C = 0 \Rightarrow \tan^{-1}x = \displaystyle\sum_{n=0}^{\infty}(-1)^n\dfrac{x^{2n+1}}{2n+1}$.

27. $f(x) = \dfrac{1}{b+x} = \dfrac{1}{(a+b)+(x-a)} = \dfrac{1}{a+b}\left[\dfrac{1}{1+\frac{x-a}{a+b}}\right] = \dfrac{1}{a+b}\left[\dfrac{1}{1-\left(-\frac{x-a}{a+b}\right)}\right].$ From

$\dfrac{1}{1-x} = \displaystyle\sum_{n=0}^{\infty} x^n$ which converges for $|x| < 1$, substitute $-\dfrac{x-a}{a+b}$ for x. Then

$\dfrac{1}{1-\left(-\frac{x-a}{a+b}\right)} = \displaystyle\sum_{n=0}^{\infty}\left(-\dfrac{x-a}{a+b}\right)^n = \displaystyle\sum_{n=0}^{\infty}\dfrac{(-1)^n(x-a)^n}{(a+b)^n} \Rightarrow f(x) = \left(\dfrac{1}{a+b}\right)\displaystyle\sum_{n=0}^{\infty}\dfrac{(-1)^n(x-a)^n}{(a+b)^n} =$

$\displaystyle\sum_{n=0}^{\infty}\dfrac{(-1)^n(x-a)^n}{(a+b)^{n+1}}$ with the series converging for $\left|-\dfrac{x-a}{a+b}\right| < 1 \Rightarrow |x-a| < |a+b| \Rightarrow$ the

radius of convergence $R = |a+b|$.

29. Since $f(x) = \dfrac{1}{1+x} = \dfrac{1}{1-(-x)}$, in $\dfrac{1}{1-x} = \displaystyle\sum_{n=0}^{\infty} x^n$ which converges for $|x| < 1$, sub-

stitute $-x$ for x. Then $f(x) = \dfrac{1}{1+x} = \displaystyle\sum_{n=0}^{\infty}(-x)^n = \displaystyle\sum_{n=0}^{\infty}(-1)^n x^n$, and the series converges for

$|-x| < 1 \Rightarrow |x| < 1 \Rightarrow$ the radius of convergence $R = 1$.

31. As $f(x) = \dfrac{x}{e^x} = xe^{-x}$, begin by substituting $-x$ for x in $e^x = \displaystyle\sum_{n=0}^{\infty}\dfrac{x^n}{n!}$ which converges for

all x. Then $e^{-x} = \displaystyle\sum_{n=0}^{\infty}\dfrac{(-x)^n}{n!} = \displaystyle\sum_{n=0}^{\infty}\dfrac{(-1)^n x^n}{n!}$, and $f(x) = xe^{-x} = x\displaystyle\sum_{n=0}^{\infty}\dfrac{(-1)^n x^n}{n!} = \displaystyle\sum_{n=0}^{\infty}\dfrac{(-1)^n x^{n+1}}{n!}$.
The series converges for all $x \Rightarrow$ the radius of convergence $R = \infty$.

33. Substituting $x-\pi$ for x in $\cos x = \displaystyle\sum_{n=0}^{\infty}\dfrac{(-1)^n x^{2n}}{(2n)!}$ which converges for all x, we have

$\cos(x-\pi) = \displaystyle\sum_{n=0}^{\infty}\dfrac{(-1)^n(x-\pi)^{2n}}{(2n)!}$. But $\cos(x-\pi) = \cos x\cos\pi + \sin x\sin\pi = -\cos x$, as $\sin\pi = 0$

and $\cos\pi = -1$. Consequently, $-\cos x = \displaystyle\sum_{n=0}^{\infty}\dfrac{(-1)^n(x-\pi)^{2n}}{(2n)!} \Rightarrow \cos x = \displaystyle\sum_{n=0}^{\infty}\dfrac{(-1)^{n+1}(x-\pi)^{2n}}{(2n)!}$.
The series converges for all $x \Rightarrow$ the radius of convergence is $R = \infty$.

35. The series $\displaystyle\sum_{n=0}^{\infty}\dfrac{(-1)^n x^{2n+1}}{(2n+1)!}$ converges to $\sin x$, for all x . Multiplying by x^2 we have:

$f(x) = x^2\sin x = x^2\displaystyle\sum_{n=0}^{\infty}\dfrac{(-1)^n x^{2n+1}}{(2n+1)!} = \displaystyle\sum_{n=0}^{\infty}\dfrac{(-1)^n x^{2n+3}}{(2n+1)!}$, and the series converges for all x, so the
radius of convergence is $R = \infty$.

37. $f(x) = \dfrac{1}{(2+x)^3} = \dfrac{1}{[2(1+\frac{x}{2})]^3} = \dfrac{1}{2^3}(1+\frac{x}{2})^{-3}$. Apply the binomial series:

$(1+x)^r = \displaystyle\sum_{k=0}^{\infty}\binom{r}{k}x^k$ with $r = -3$ and x replaced with $\frac{x}{2}$ to obtain the expansion

$$(1 + \tfrac{x}{2})^{-3} = \sum_{k=0}^{\infty} \binom{-3}{k} \left(\tfrac{x}{2}\right)^k = \sum_{k=0}^{\infty} \binom{-3}{k} \frac{x^k}{2^k} \Rightarrow f(x) = \frac{1}{2^3} \sum_{k=0}^{\infty} \binom{-3}{k} \frac{x^k}{2^k} = \sum_{k=0}^{\infty} \binom{-3}{k} \frac{x^k}{2^{k+3}}.$$

Since the binomial series converges for $|x| < 1$, the series representation of f converges for $|\tfrac{x}{2}| < 1 \Rightarrow |x| < 2$, so the radius of convergence $R = 2$.

39. $f(x) = \sin x = \sum_{n=0}^{\infty} \frac{(-1)^n x^{2n+1}}{(2n+1)!}$. On $[0, \pi]$, $|x| < \pi$ and $|f^{(n)}(c)| = |\sin x|$ or $|\cos x|$

both of which are at most 1. By Taylor's inequality, the error $|E_N(x)| \le \frac{M|x - a|^{N+1}}{(N+1)!}$ which

in this case becomes $|E_N(x)| \le \frac{\pi^{N+1}}{(N+1)!}$. Checking values, we find that $\frac{\pi^{14}}{14!} \approx 0.000104 >$

0.0001 but, $\frac{\pi^{15}}{15!} \approx 0.000021 < 0.0001$. So we need a polynomial of degree 14, namely:

$x - \frac{x^3}{3!} + \frac{x^5}{5!} - \frac{x^7}{7!} + \frac{x^9}{9!} - \frac{x^{11}}{11!} + \frac{x^{13}}{13!} + 0x^{14} = x - \frac{x^3}{3!} + \frac{x^5}{5!} - \frac{x^7}{7!} + \frac{x^9}{9!} - \frac{x^{11}}{11!} + \frac{x^{13}}{13!}$. Hence we need 7 terms of the series.

[Note: As the power series for $\sin x$ is an alternating series, by Theorem 9.12 on page 340, the error $|E_N(x)| \le a_{N+1}$. By the above, on $[0, \pi]$, $a_{15} = \frac{x^{15}}{15!} \le \frac{\pi^{15}}{15!}$ is the first term less than 0.0001, so $\frac{x^{13}}{13!}$ is the last term in the required polynomial – as we found above.]

41. From Exercise 23 above, $\frac{1}{x^2 + 1} = \sum_{n=0}^{\infty} (-1)^n x^{2n}$. As it would be difficult to use Taylor's

inequality to estimate the error, as that requires computing the (n+1)st derivative, instead, observing that the power series is an alternating series, we apply Theorem 9.12 on page 340, which gives the error $|E_N(x)| \le a_{N+1}$. On $[-\frac{1}{2}, \frac{1}{2}]$, $a_{N+1} = (x^2)^{N+1} \le [(\frac{1}{2})^2]^{N+1} = (\frac{1}{4})^{N+1}$.

Checking values, we find that $(\frac{1}{4})^6 \approx 0.0002 > 0.0001$ but, $(\frac{1}{4})^7 \approx 0.00006 < 0.0001$.

Therefore, the required polynomial ends with the power $(x^2)^6 = x^{12}$, and is therefore $1 - x^2 + x^4 - x^6 + x^8 - x^{10} + x^{12}$. Hence 7 terms are required.

43.

$f(x) = \sin x$ on $[0, \pi]$ with the first N terms of its Maclaurin series for N = 1,2,3,and 4.

45.

$f(x) = e^x$ on $[-1, 2]$ with the first N terms of its Maclaurin series for N = 1,2,3,and 4.

47. (a) From $\sin x = \sum_{n=0}^{\infty} \frac{(-1)^n x^{2n+1}}{(2n+1)!}$, we get $\sin x^2 = \sum_{n=0}^{\infty} \frac{(-1)^n (x^2)^{2n+1}}{(2n+1)!} = \sum_{n=0}^{\infty} \frac{(-1)^n x^{4n+2}}{(2n+1)!}$.

(b) $\int \sin x^2 \, dx = \int \left[\sum_{n=0}^{\infty} \frac{(-1)^n x^{4n+2}}{(2n+1)!} \right] = \sum_{n=0}^{\infty} \left[\frac{(-1)^n}{(2n+1)!} \int x^{4n+2} \, dx \right]$

$$= \sum_{n=0}^{\infty} \frac{(-1)^n x^{4n+3}}{(2n+1)!(4n+3)} + C$$

(c) On $[0,1]$, the series is an alternating series, so by Theorem 9.12 on page 340, the error $|E_N(x)| \leq a_{N+1}$, where $a_n = \frac{x^{4n+3}}{(2n+1)!(4n+3)}$. Now $a_{N+1} = \frac{x^{4(N+1)+3}}{[2(N+1)+1]![4(N+1)+3]} = \frac{x^{4N+7}}{(2N+3)!(4N+7)} \leq \frac{1}{(2N+3)!(4N+7)}$ on $[0,1]$. Checking values, we find that $a_1 \leq \frac{1}{3!7} = \frac{1}{42} \approx 0.023 > 0.001$ and $a_2 \leq \frac{1}{5!11} \approx 0.0007 < 0.001$, so the polynomial ends with a_1, i.e. with $x^7 : \frac{x^3}{3} - \frac{x^7}{3!7} = \frac{x^3}{3} - \frac{x^7}{42}$.

The value of the integral of $\sin x$ from 0 to 1 is about $\left(\frac{x^3}{3} - \frac{x^7}{42} \right) \Big|_0^1 = \frac{1}{3} - \frac{1}{42} \approx 0.3095$

(d) The polynomial $p(x) = \frac{x^3}{3} - \frac{x^7}{42}$ in (c) above satisfies the requirements here. Approximating the integral over $[0,1]$ to within 0.001 means it approximates the integral over any subinterval to at least that same degree of accuracy.

49. As suggested, we first show that $\frac{d^k (x-a)^n}{dx^k}$ is (1) 0 if $n < k$, (2) $k!$ if $n = k$, and (3) $n(n-1) \cdots (n-k+1)(x-a)^{n-k}$, i.e. $n(n-1) \cdots [n-(k-1)](x-a)^{n-k}$ if $n > k$:

(1) Every time you differentiate $(x-a)^n$ the power of $x-a$ is reduced by 1, so the nth derivative is a constant and higher order derivatives are therefore 0.

(2) We can show that the nth derivative of $(x-a)^n$ is $n!$, by the Principle of Mathematical Induction. The first derivative of $x-a$ is 1 and $1!=1$. Assume $\frac{d^k (x-a)^k}{dx^k} = k!$. We have only to show that $\frac{d^{k+1} (x-a)^{k+1}}{dx^{k+1}} = (k+1)!$

$$\frac{d^{k+1} (x-a)^{k+1}}{dx^{k+1}} = \frac{d^k}{dx^k} \left[\frac{d}{dx}(x-a)^{k+1} \right] = \frac{d^k (k+1)(x-a)^k}{dx^k} = (k+1) \frac{d^k (x-a)^k}{dx^k} = (k+1)k!$$

$$= (k+1)!.$$

(3) Again, use Mathematical Induction to show that if $n > k$, then $\frac{d^k (x-a)^n}{dx^k} = n(n-1) \cdots [n-(k-1)](x-a)^{n-k}$. If $k=1$, $[(x-a)^n]' = n(x-a)^{n-1}$ – check, and $[(x-a)^n]'' = n(n-1)(x-a)^{n-2}$ – check. Now assume $\frac{d^K}{dx^K}(x-a)^n = n(n-1) \cdots [n-(K-1)](x-a)^{n-K}$ for some $k = K < n$. We are to show that

$$\frac{d^{K+1}}{dx^{K+1}}(x-a)^n = n(n-1) \cdots (n-K)(x-a)^{n-(K+1)}.$$

Indeed, $\dfrac{d^{K+1}}{dx^{K+1}}(x-a)^n = \dfrac{d}{dx}\left[\dfrac{d^K}{dx^K}(x-a)^n\right] = \dfrac{d}{dx}\left[n(n-1)\cdots[n-(K-1)](x-a)^{n-K}\right]$

$$= \{n(n-1)\cdots[n-(K-1)]\}\cdot(n-K)(x-a)^{(n-K)-1}$$

$$= n(n-1)\cdots(n-K)(x-a)^{n-(K+1)}.$$

Lastly, as suggested, consider the following:

$$f(x) = \sum_{n=0}^{\infty}c_n(x-a)^n = \sum_{n=0}^{k-1}c_n(x-a)^n + c_k(x-a)^k + \sum_{n=k+1}^{\infty}c_n(x-a)^n$$

Then $f^{(k)}(x) = \dfrac{d^k}{dx^k}\left(\sum_{n=0}^{k-1}c_n(x-a)^n + c_k(x-a)^k + \sum_{n=k+1}^{\infty}c_n(x-a)^n\right)$

$$= \sum_{n=0}^{k-1}c_n\dfrac{d^k}{dx^k}(x-a)^n + c_k\dfrac{d^k}{dx^k}(x-a)^k + \sum_{n=k+1}^{\infty}c_n\dfrac{d^k}{dx^k}(x-a)^n$$

$$= \qquad 0 \qquad\qquad + c_k\,k! \qquad\qquad + \sum_{n=k+1}^{\infty}c_n\,n(n-1)\cdots[n-(k-1)](x-a)^{n-k}$$

Evaluating at $x=a$, every term of the last series drops out because it contains $(x-a)$ to some power, and we have $f^{(k)}(a) = c_k\,k! \Rightarrow c_k = \dfrac{f^{(k)}(a)}{k!}$.

51. First, to prove that the Binomial series converges for $|x|<1$, apply the Ratio Test:

$$\left|\dfrac{a_{k+1}}{a_k}\right| = \left|\dfrac{\dfrac{r(r-1)\cdots(r-k)x^{k+1}}{(k+1)!}}{\dfrac{r(r-1)\cdots(r-k+1)x^k}{k!}}\right| = \left|\dfrac{r(r-1)\cdots(r-k+1)(r-k)x}{r(r-1)\cdots(r-k+1)(k+1)}\right| = \left|\dfrac{(r-k)x}{k+1}\right| = \left|\dfrac{\frac{r}{k}-1}{1+\frac{1}{k}}\right||x| \to$$

$|x|$ as $k\to\infty$. Therefore, the series converges for $|x|<1$, by the Ratio Test.

Next, we will prove that the series $g(x) = \sum_{k=0}^{\infty}\binom{r}{k}x^k$ converges to $(1+x)^r$:

We will show that $h(x) = (1+x)^{-r}g(x) = 1$, by proving that $h'(x) = 0$ which means that $h(x)$ is constant. Then $h(x) = h(0) = g(0) = 1 \Rightarrow g(x) = (1+x)^r$.

Now, $h'(x) = 0$ when $h'(x) = -r(1+x)^{-r-1}g(x) + (1+x)^{-r}g'(x) = 0$, that is, if $g'(x) = r(1+x)^{-r-1}g(x)(1+x)^r = r(1+x)^{-1}g(x) = \dfrac{rg(x)}{1+x}$. So it remains to show that $g'(x) = \dfrac{rg(x)}{1+x}$, i.e. $(1+x)g'(x) = rg(x)$:

$$(1+x)g'(x) = (1+x)\sum_{k=1}^{\infty}\dfrac{r(r-1)\cdots(r-k+1)}{k!}kx^{k-1}$$

$$= 1\cdot\sum_{k=1}^{\infty}\dfrac{r(r-1)\cdots(r-k+1)}{k!}kx^{k-1} + x\cdot\sum_{k=1}^{\infty}\dfrac{r(r-1)\cdots(r-k+1)}{k!}kx^{k-1}$$

$$= \sum_{k=1}^{\infty}\dfrac{r(r-1)\cdots(r-k+1)}{k!}kx^{k-1} + \sum_{k=1}^{\infty}\dfrac{r(r-1)\cdots(r-k+1)}{k!}kx^k$$

Rewrite the first series, lowering the index by 1, and in the second series, add 0 to the sum, by lowering the index by 1. Then the two series can be combined into one series:

$$= \sum_{k=0}^{\infty} \frac{r(r-1)\cdots(r-k)}{(k\!\!\!/+1)!\ k!}(k\!\!\!/+1)x^k + \sum_{k=0}^{\infty} \frac{r(r-1)\cdots(r-k+1)}{k!}kx^k$$

$$= \sum_{k=0}^{\infty} \frac{[r(r-1)\cdots(r-k+1)](r-k)}{k!}x^k + \sum_{k=0}^{\infty} \frac{[r(r-1)\cdots(r-k+1)](k)}{k!}x^k$$

$$= \sum_{k=0}^{\infty} \frac{r(r-1)\cdots(r-k+1)[(r-k\!\!\!/)+k\!\!\!/]}{k!}x^k = rg(x)$$

CHAPTER 10
Parametrization of Curves and Polar Coordinates

§10.1 Parametrization of Curves

1. $x = \sqrt{t} + 1, y = t^{\frac{3}{2}} \Rightarrow \sqrt{t} = x - 1 \Rightarrow y = (x-1)^3$ (see leftmost figure). At $t = 1, x = 2$, and $y = 1$. As t, increases, so do x and y. Since $t \geq 1$, we only want the part of the curve to the right of the point $(2,1)$, including that point (see rightmost figure).

3. From the identity $1 + \tan^2 t = \sec^2 t \Rightarrow \sec^2 t - \tan^2 t = 1$. As $x = \sec t$, and $y = \tan t$, then $x^2 - y^2 = 1$ whose graph is a hyperbola (see leftmost figure). As $-\frac{\pi}{2} < t < \frac{\pi}{2}, \cos t$ varies between 0 and 1, so $\sec t$ varies between 1 and $\infty \Rightarrow x \geq 1$ and only the right half of the hyperbola is traversed. As t increases so does y which gives the orientation shown in the rightmost figure.

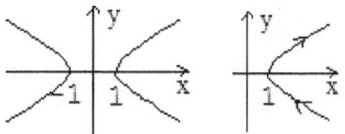

5. Apply the identity $\sin^2 t + \cos^2 t = 1$ to $x = 3\sin t - 2$ and $y = 2\cos t$ to get: $\frac{(x+2)^2}{3^2} + \frac{y^2}{2^2} = 1$, an ellipse centered at $(-2, 0)$ (see figure). With $0 \leq t \leq 2\pi$, we check points to determine the orientation: At $t = 0, (x, y) = (-2, 2)$. At $t = \frac{\pi}{2}, (x, y) = (1, 0)$. At $t = \pi, (x, y) = (-2, -2)$. At $t = \frac{3\pi}{2}, (x, y) = (-5, 0)$, and at $t = 2\pi, (x, y) = (-2, 2)$ again (see figure).

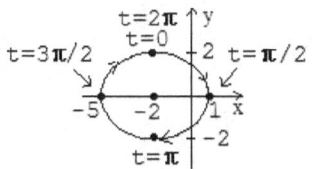

7. $x = t^3, y = t^2 \Rightarrow t = x^{\frac{1}{3}} \Rightarrow y = (x^{\frac{1}{3}})^2 = x^{\frac{2}{3}}$ $y(0) = 0$, and $y' = \frac{2}{3}x^{-\frac{1}{3}}$ is negative for $x < 0$ and positive for $x > 0$, so the curve is decreasing for $x < 0$ and increasing for $x > 0$. $y'' = -\frac{2}{9}x^{-\frac{4}{3}} < 0$ for $x \neq 0$, which means the curve is always concave down. As t varies from $-\infty$ to ∞, x increases from $-\infty$ to ∞ which gives the orientation in the figure.

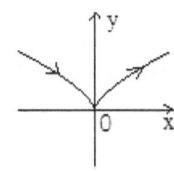

9. $x = t^2, y = \sqrt{t^4 + 1} \Rightarrow y = \sqrt{x^2 + 1} \Rightarrow y^2 = x^2 + 1, y \geq 0$. This is the top half of the hyperbola $y^2 - x^2 = 1$ (see leftmost figure). When $t = 0, x = 0$ and $y = 1$. Then as t increases so do both x and y, giving us just the right half of the upper portion of the hyperbola, and the orientation shown in the rightmost figure.

11. Applying the identity $\cos 2t = 1 - 2\sin^2 t$ to $x = \cos 2t$, $y = \sin t$, we have $x = 1 - 2y^2$ whose graph is a parabola opening to the left, with vertex at $(1,0)$. As $-\frac{\pi}{2} < t < \frac{\pi}{2}$, we check some points: At $t = -\frac{\pi}{2}$, $x = \cos(-\pi) = -1$ and $y = \sin(-\frac{\pi}{2}) = -1$. At $t = 0$, $x = \cos 0 = 1$ and $y = \sin 0 = 0$. Finally, at $t = \frac{\pi}{2}$, $x = \cos(\pi) = -1$ and $y = \sin(\frac{\pi}{2}) = 1$. See figure.

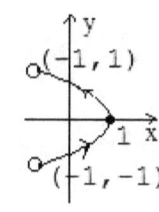

13. We seek the tangent line to the curve parametrized by $x = t^4 + 1$, $y = t^3 + t$ at $t = -1$. Finding the slope of the tangent line: $\dfrac{dy}{dx} = \dfrac{\frac{dy}{dt}}{\frac{dx}{dt}} = \dfrac{3t^2 + 1}{4t^3} \Rightarrow \dfrac{dy}{dx}\Big|_{t=-1} = \dfrac{4}{-4} = -1$.

At the point when $t = -1$, $x = 2$ and $y = -2$, so the equation of the tangent line at that point is $y - (-2) = -1(x - 2) \Rightarrow y = -x$.

15. The tangent line to the curve parametrized by $x = 3\cos t$, $y = 4\sin t$ has slope given by: $\dfrac{dy}{dx} = \dfrac{\frac{dy}{dt}}{\frac{dx}{dt}} = \dfrac{4\cos t}{-3\sin t}$. At $t = \frac{\pi}{4}$, $\dfrac{dy}{dx}\Big|_{t=\frac{\pi}{4}} = -\dfrac{4\cos\frac{\pi}{4}}{3\sin\frac{\pi}{4}} = -\dfrac{4}{3}$. When $t = \frac{\pi}{4}$, $x = 3\cos\frac{\pi}{4} = \dfrac{3}{\sqrt{2}} = \dfrac{3\sqrt{2}}{2}$ and $y = 4\sin\frac{\pi}{4} = \dfrac{4}{\sqrt{2}} = \dfrac{4\sqrt{2}}{2} = 2\sqrt{2}$. Thus the equation of the tangent line is $y - 2\sqrt{2} = -\frac{4}{3}(x - \frac{3\sqrt{2}}{2}) \Rightarrow y = -\frac{4}{3}x + 4\sqrt{2}$.

17. The tangent line to the curve parametrized by $x = t^2 - 1$, $y = 2e^t$ has slope given by: $\dfrac{dy}{dx} = \dfrac{\frac{dy}{dt}}{\frac{dx}{dt}} = \dfrac{2e^t}{2t} = \dfrac{e^t}{t}$. At $t = 2$, $\dfrac{dy}{dx}\Big|_{t=2} = \frac{1}{2}e^2$. When $t = 2$, $x = 3$ and $y = 2e^2$. Thus the equation of the tangent line is $y - 2e^2 = \frac{1}{2}e^2(x - 3) \Rightarrow y = \frac{1}{2}e^2 x + \frac{1}{2}e^2$.

19. $x = 3t^2$, $y = t^3 - 4t \Rightarrow \dfrac{dy}{dx} = \dfrac{\frac{dy}{dt}}{\frac{dx}{dt}} = \dfrac{3t^2 - 4}{6t}$. A vertical tangent occurs where $6t = 0 \Rightarrow t = 0 \Rightarrow x = y = 0$ i.e., at the point $(0,0)$. A horizontal tangent occurs where $3t^2 - 4 = 0 \Rightarrow t = \pm\frac{2}{\sqrt{3}}$. There, $x = 3(\frac{4}{3}) = 4$ and $y = \pm(\frac{8}{3\sqrt{3}} - \frac{8}{\sqrt{3}}) = \pm(\frac{16}{3\sqrt{3}}) = \pm(\frac{16\sqrt{3}}{9})$. Thus a horizontal tangent occurs at the points $(4, \pm\frac{16\sqrt{3}}{9})$.

21. $x = 2\cos t$, $y = \sin 2t$. Since $\cos t$ is periodic of period 2π, and $\sin 2t$ has period π, we can restrict t to the interval $[0, 2\pi)$. $\dfrac{dy}{dx} = \dfrac{\frac{dy}{dt}}{\frac{dx}{dt}} = \dfrac{2\cos 2t}{-2\sin t} = -\dfrac{\cos 2t}{\sin t}$. A vertical tangent occurs where $\sin t = 0 \Rightarrow t = 0, \pi$. When $t = 0$, $x = 2\cos 0 = 2$ and $y = \sin 0 = 0 \Rightarrow (2, 0)$. When $t = \pi$, $x = 2\cos\pi = -2$ and $y = \sin 2\pi = 0 \Rightarrow (-2, 0)$. So, vertical tangents occur at $(\pm 2, 0)$. A horizontal tangent happens when $\cos 2t = 0 \Rightarrow 2t = \frac{\pi}{2} + n\pi \Rightarrow t = \frac{\pi}{4} + \frac{n\pi}{2}$.

In $[0, 2\pi]$, $t = \frac{\pi}{4}, \frac{3\pi}{4}, \frac{5\pi}{4}, \frac{7\pi}{4}$:

$$t = \frac{\pi}{4} \;:\;\; x = 2\cos\frac{\pi}{4} = \sqrt{2}, \qquad y = \sin\frac{\pi}{2} = 1 \;\Rightarrow\;\; (\sqrt{2}, 1)$$

$$t = \frac{3\pi}{4} \;:\;\; x = 2\cos\frac{3\pi}{4} = -\sqrt{2}, \quad y = \sin\frac{3\pi}{2} = -1 \;\Rightarrow\;\; (-\sqrt{2}, -1)$$

$$t = \frac{5\pi}{4} \;:\;\; x = 2\cos\frac{5\pi}{4} = -\sqrt{2}, \quad y = \sin\frac{5\pi}{2} = 1 \;\Rightarrow\;\; (-\sqrt{2}, 1)$$

$$t = \frac{7\pi}{4} \;:\;\; x = 2\cos\frac{7\pi}{4} = \sqrt{2}, \qquad y = \sin\frac{7\pi}{2} = -1 \;\Rightarrow\;\; (\sqrt{2}, -1)$$

Thus, a horizontal tangent occurs at $(\pm\sqrt{2}, -1)$ and $(\pm\sqrt{2}, 1)$.

23.   For $x = \frac{1}{t}$, $y = t^2 + 3$, $\dfrac{dy}{dx} = \dfrac{\frac{dy}{dt}}{\frac{dx}{dt}} = \dfrac{2t}{-t^{-2}} = -\dfrac{2t}{\frac{1}{t^2}}$, so there are no vertical tangents, as

$\frac{1}{t^2}$ is never 0. Since $t \neq 0$ (else x is undefined), there are also no horizontal tangents.

25.   For $x = t^2 + 4$, $y = t^3 + t^2$, $\dfrac{dy}{dx} = \dfrac{\frac{dy}{dt}}{\frac{dx}{dt}} = \dfrac{3t^2 + 2t}{2t} = \dfrac{3t + 2}{2}$ for $t \neq 0$. The derivative is

positive for $t > -\frac{2}{3}$ $\Rightarrow$ the curve is increasing there. Since $\dfrac{d^2y}{dx^2} = \dfrac{\frac{d}{dt}\left(\frac{dy}{dx}\right)}{\frac{dx}{dt}} = \dfrac{\frac{3}{2}}{2t} = \dfrac{3}{4t} > 0$

for $t > 0$, the curve is concave up for $t > 0$.

27.   For $x = t - e^t$, $y = t + e^{-t}$, $\dfrac{dy}{dx} = \dfrac{\frac{dy}{dt}}{\frac{dx}{dt}} = \dfrac{1 - e^{-t}}{1 - e^t} \cdot \dfrac{e^t}{e^t} = \dfrac{e^t - 1}{e^t(1 - e^t)} = -\dfrac{1}{e^t} = -e^{-t} > 0$

never, as the exponential function is always positive. So the curve is never increasing. Since

$\dfrac{d^2y}{dx^2} = \dfrac{\frac{d}{dt}\left(\frac{dy}{dx}\right)}{\frac{dx}{dt}} = \dfrac{\frac{d}{dt}(-e^{-t})}{1 - e^t} = \dfrac{e^{-t}}{1 - e^t} > 0$ when $e^t < 1$ $\Rightarrow$ $\ln e^t < \ln 1$, i.e. $t < 0$: The curve is

concave up when $t < 0$.

29. $x = 3t^2 + 1$, $y = 2t^2 + 4$ $\Rightarrow$ $t^2 = \frac{x-1}{3}$ $\Rightarrow$ $y = 2\left(\frac{x-1}{3}\right) + 4 = \frac{2}{3}x + \left(4 - \frac{2}{3}\right) = \frac{2}{3}x + \frac{10}{3}$.
This is the equation of a line of slope $\frac{2}{3}$ and y–intercept $\frac{10}{3}$. But $-\infty < t < \infty$ $\Rightarrow$ $x \geq 1$
and $y \geq 4$, so the graph is only a half-line terminating at the point (1,4).
As t increases from $-\infty$ to 0, the curve is traversed from right to
left terminating at the point (1,4) (see leftmost figure), and as
t increases from 0 to ∞, the curve is traversed from (1,4) back
along the half-line to the right (see rightmost figure). There are
no max/min or inflection points.

31. $x = \frac{t^2}{2}$, $y = \frac{t^3}{2} - 6t$ for $0 \leq t \leq 4$ $\Rightarrow$ $t = \sqrt{2x}$ $\Rightarrow$ $y = \frac{1}{2}(2x)^{\frac{3}{2}} - 6(2x)^{\frac{1}{2}} = \sqrt{2x}(x - 6)$.

The derivative $\dfrac{dy}{dx} = \dfrac{\frac{dy}{dt}}{\frac{dx}{dt}} = \dfrac{\frac{3}{2}t^2 - 6}{t} = 0$ when $\frac{3}{2}t^2 = 6$ $\Rightarrow$ $t^2 = 4$ $\Rightarrow$ $t = 2$, as $t \geq 0$. This is the point $(2, -8)$. Since x increases as t increases, and the derivative is negative just to the left of $t = 2$ and positive just to the right, there is a local minimum at $(2, -8)$.

The second derivative $\dfrac{d^2y}{dx^2} = \dfrac{\frac{d}{dt}\left(\frac{dy}{dx}\right)}{\frac{dx}{dt}} = \dfrac{\frac{d}{dt}\left(\frac{3}{2}t - 6t^{-1}\right)}{\frac{dx}{dt}} =$

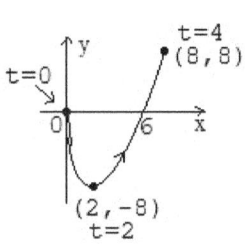

$\dfrac{\frac{3}{2} + 6t^{-2}}{t} = \dfrac{3t^2 + 12}{2t^3} > 0$ always as $t \geq 0$, so the curve is concave

up throughout its domain: $0 \leq t \leq 4$. Checking the endpoints of the curve: $t = 0 \Rightarrow (x, y) = (0, 0)$ and $t = 4 \Rightarrow (x, y) = (8, 8)$, which gives the orientation shown in the figure.

33. The length of the curve $L = \displaystyle\int_0^1 \sqrt{\left(\frac{dx}{dt}\right)^2 + \left(\frac{dy}{dt}\right)^2}\, dt$. As $x = 3t^2 + 1$, $y = 2t^3 + 4$, then

$\dfrac{dx}{dt} = 6t$ and $\dfrac{dy}{dt} = 6t^2 \Rightarrow L = \displaystyle\int_0^1 \sqrt{(6t)^2 + (6t^2)^2}\, dt = \int_0^1 \sqrt{36t^2 + 36t^4}\, dt = 6\int_0^1 t\sqrt{1 + t^2}\, dt$

$$\boxed{\begin{array}{l} u = 1 + t^2, \quad t = 0 \;\Rightarrow\; u = 1 \\ du = 2t\, dt, \quad\quad t = 1 \;\Rightarrow\; u = 2 \end{array}}$$

$$= \frac{6}{2}\int_1^2 u^{\frac{1}{2}}\, du = 3 \cdot \frac{2}{3}u^{\frac{3}{2}}\Big|_1^2 = 2[2^{\frac{3}{2}} - 1] = 2[2\sqrt{2} - 1] = 4\sqrt{2} - 2.$$

35.   The length of the curve $L = \displaystyle\int_0^\pi \sqrt{\left(\frac{dx}{dt}\right)^2 + \left(\frac{dy}{dt}\right)^2}\, dt$. As $x = 3\cos t - \cos 3t$, and

$y = 3\sin t - \sin 3t$, then $\dfrac{dx}{dt} = -3\sin t + 3\sin 3t$ and $\dfrac{dy}{dt} = 3\cos t - 3\cos 3t \Rightarrow$

$$\begin{aligned} \left(\frac{dx}{dt}\right)^2 + \left(\frac{dy}{dt}\right)^2 &= 9[\sin^2 t - 2\sin t \sin 3t + \sin^2 3t] + 9[\cos^2 t - 2\cos t \cos 3t + \cos^2 3t] \\ &= 9[1 - 2(\cos t \cos 3t + \sin t \sin 3t) + 1] = 18[1 - \cos(t - 3t)] \\ &= 18(1 - \cos 2t) = 18[1 - (1 - 2\sin^2 t)] = 36\sin^2 t \end{aligned}$$

Therefore, $L = \displaystyle\int_0^\pi \sqrt{36\sin^2 t}\, dt = 6\int_0^\pi \sin t\, dt = -\,6\cos t\Big|_0^\pi = -6(-1 - 1) = 12.$

37.   The length of the curve $L = \displaystyle\int_0^2 \sqrt{\left(\frac{dx}{dt}\right)^2 + \left(\frac{dy}{dt}\right)^2}\, dt$. As $x = t - t^2$, and $y = 3t^{\frac{3}{2}}$, then

$\dfrac{dx}{dt} = 1 - 2t$ and $\dfrac{dy}{dt} = \dfrac{9}{2}t^{\frac{1}{2}} \Rightarrow L = \displaystyle\int_0^2 \sqrt{(1 - 2t)^2 + \frac{81}{4}t}\, dt = \int_0^2 \sqrt{1 - 4t + 4t^2 + \frac{81}{4}t}\, dt =$

$\displaystyle\int_0^2 \sqrt{1 + \left(\frac{65}{4}\right)t + 4t^2}\, dt.$

39. As $x = 2\cos t$ and $y = \sin t$, then $\dfrac{dx}{dt} = -2\sin t$ and $\dfrac{dy}{dt} = \cos t$. Thus,

$L = \displaystyle\int_0^{2\pi} \sqrt{\left(\frac{dx}{dt}\right)^2 + \left(\frac{dy}{dt}\right)^2}\, dt = \int_0^{2\pi} \sqrt{4\sin^2 t + \cos^2 t}\, dt.$

§10.2 Polar Coordinates

1. $(r, \theta) = (2, 0) \Rightarrow x = r \cos \theta = 2 \cos 0 = 2$ and $y = r \sin \theta = 2 \sin 0 = 0$. The rectangular coordinates of $(2,0)$ are again $(2,0)$.

3. $(r, \theta) = (-3, \pi) \Rightarrow x = r \cos \theta = -3 \cos \pi = 3$ and $y = r \sin \theta = -3 \sin \pi = 0$. The rectangular coordinates are $(3,0)$.

5. $(r, \theta) = (-2, -\frac{\pi}{3}) \Rightarrow x = r \cos \theta = -2 \cos(-\frac{\pi}{3}) = -2 \cdot \frac{1}{2} = -1$ and $y = r \sin \theta = -2 \sin(-\frac{\pi}{3}) = 2 \sin \frac{\pi}{3} = 2 \cdot \frac{\sqrt{3}}{2} = \sqrt{3}$. The rectangular coordinates are $(-1, \sqrt{3})$.

7. $(r, \theta) = (-2, \frac{3\pi}{4}) \Rightarrow x = r \cos \theta = -2 \cos \frac{3\pi}{4} = -2(-\cos \frac{\pi}{4}) = 2 \cdot \frac{1}{\sqrt{2}} = \sqrt{2}$ and $y = r \sin \theta = -2 \sin(\frac{3\pi}{4}) = -2 \sin \frac{\pi}{4} = -2 \cdot \frac{1}{\sqrt{2}} = -\sqrt{2}$. The rectangular coordinates are $(\sqrt{2}, -\sqrt{2})$.

9. $(x, y) = (4, -4) \Rightarrow r^2 = x^2 + y^2 = 16 + 16 = 32 \Rightarrow r = \pm\sqrt{32} = \pm 4\sqrt{2}$. As $\tan \theta = \frac{-4}{4} = -1$, the reference angle $\theta_r = \frac{\pi}{4}$. Since $(4, -4)$ lies in QIV, when $r = 4\sqrt{2} > 0, \theta = \frac{7\pi}{4}$, so one pair of suitable coordinates is $(4\sqrt{2}, \frac{7\pi}{4})$. The other pair is $(-4\sqrt{2}, \frac{7\pi}{4} - \pi) = (-4\sqrt{2}, \frac{3\pi}{4})$.

11. $(x, y) = (0, 1) \Rightarrow r^2 = x^2 + y^2 = 1 \Rightarrow r = \pm 1$. Clearly, when $r = 1 > 0, \theta = \frac{\pi}{2}$, so one pair of suitable coordinates is $(1, \frac{\pi}{2})$. The other pair is $(-1, \frac{\pi}{2} + \pi) = (-1, \frac{3\pi}{2})$.

13. $(x, y) = (-1, \sqrt{3}) \Rightarrow r^2 = x^2 + y^2 = 1 + 3 = 4 \Rightarrow r = \pm 2$. As $\tan \theta = \frac{\sqrt{3}}{-1} = -\sqrt{3}$, the reference angle $\theta_r = \frac{\pi}{3}$. Since $(-1, \sqrt{3})$ lies in QII, when $r = 2 > 0, \theta = \frac{2\pi}{3}$, and all polar coordinates are $(2, \frac{2\pi}{3} + 2k\pi)$. When $r = -2$, all polar coordinates are $(-2, \frac{2\pi}{3} + \pi + 2k\pi) = (-2, \frac{5\pi}{3} + 2k\pi)$.

15. $(x, y) = (4\sqrt{3}, 4) \Rightarrow r^2 = x^2 + y^2 = (4\sqrt{3})^2 + 4^2 = 48 + 16 = 64 \Rightarrow r = \pm 8$. As $\tan \theta = \frac{4}{4\sqrt{3}} = \frac{1}{\sqrt{3}}$, the reference angle $\theta_r = \frac{\pi}{6}$. Since $(4\sqrt{3}, 4)$ lies in QI, when $r = 8 > 0, \theta = \frac{\pi}{6}$, and all polar coordinates are $(8, \frac{\pi}{6} + 2k\pi)$. When $r = -8$, all polar coordinates are $(-8, \frac{\pi}{6} + \pi + 2k\pi) = (-8, \frac{7\pi}{6} + 2k\pi)$.

17. Since $x^2 + y^2 = r^2$, $x^2 + y^2 = 9 \Rightarrow r^2 = 9 \Rightarrow r = 3$ is a rectangular equation.

19. From $x = r \cos \theta$, $y = r \sin \theta$, $x = -y^2 \Rightarrow r \cos \theta = -(r \sin \theta)^2 \Rightarrow \cos \theta = -r \sin^2 \theta \Rightarrow r = -\frac{\cos \theta}{\sin^2 \theta} = -(\frac{\cos \theta}{\sin \theta})(\frac{1}{\sin \theta}) = -\cot \theta \csc \theta$.

21. From $x = r\cos\theta$, $y = r\sin\theta$, $x^2 + y^2 + 4x = 0 \Rightarrow r^2 + 4r\cos\theta = 0 \Rightarrow r = -4\cos\theta$

23. $r = 4 \Rightarrow r^2 = 16 \Rightarrow x^2 + y^2 = 16$.

25. $r = 3\sin\theta \Rightarrow r^2 = 3r\sin\theta \Rightarrow x^2 + y^2 = 3y$, i.e. $x^2 + y^2 - 3y = 0$.

27. $r = \csc\theta \Rightarrow r = \dfrac{1}{\sin\theta} \Rightarrow r\sin\theta = 1 \Rightarrow y = 1$.

29. $r = 4$ is the polar equation of a circle of radius 4 centered at the origin.

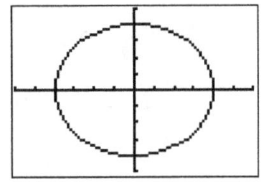

31. $r = 6\sin\theta$: As θ increases from 0 to $\frac{\pi}{2}$, r increases from 0 to 6. As θ increases from $\frac{\pi}{2}$ to π, r decreases from 6 to 0. This is the polar equation of a circle centered on the y-axis of radius $\frac{6}{2} = 3$.

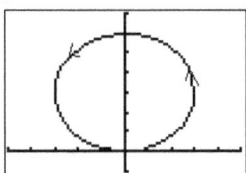

33. $r = 2 - 2\cos\theta$: When $\theta = 0, r = 0$. As θ increases from 0 to $\frac{\pi}{2}$, $\cos\theta$ decreases from 1 to 0, so r increases from 0 to 2. As θ increases from $\frac{\pi}{2}$ to π, $\cos\theta$ decreases from 0 to -1, so r increases from 2 to 4. As θ increases from π to $\frac{3\pi}{2}$, $\cos\theta$ increases from -1 to 0, so r decreases from 4 to 2. As θ increases from $\frac{3\pi}{2}$ to 2π, $\cos\theta$ increases from 0 to 1, so r decreases from 2 to 0. This is the graph of a cardioid.

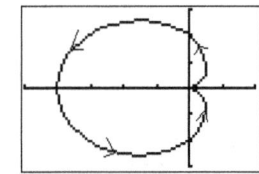

35. $r = 4\sin 3\theta$: This is a 3-leaved rose. We first determine when $r = 0$ and when r is a maximum. $r = 0 \Rightarrow \sin 3\theta = 0 \Rightarrow 3\theta = n\pi \Rightarrow \theta = \frac{n\pi}{3} \Rightarrow \theta = 0, \frac{\pi}{3}, \frac{2\pi}{3}, \pi, \frac{4\pi}{3}, \frac{5\pi}{3}, 2\pi$. $|r|$ is a maximum of 4, (1) when $\sin 3\theta = 1 \Rightarrow 3\theta = \frac{\pi}{2} + 2n\pi \Rightarrow \theta = \frac{\pi}{6} + \frac{2n\pi}{3} = (4n+1)\frac{\pi}{6} \Rightarrow \theta = \frac{\pi}{6}, \frac{5\pi}{6}, \dots$ or (2) when $\sin 3\theta = -1 \Rightarrow 3\theta = \frac{3\pi}{2} + 2n\pi \Rightarrow \theta = \frac{\pi}{2} + \frac{2n\pi}{3} = (4n+3)\frac{\pi}{6} \Rightarrow \theta = \frac{\pi}{2}$, etc.

As θ increases from 0 to $\frac{\pi}{3}$, r increases from 0 to its maximum of 4 at $\theta = \frac{\pi}{6}$, and then decreases back to 0, which traces out a leaf in QI. As θ increases from $\frac{\pi}{3}$ to $\frac{2\pi}{3}$, r is negative; it decreases from 0 to a minimum of -4 at $\frac{\pi}{2}$ and then it increases back to 0, which traces out a leaf in QIII-QIV with its maximum on the negative y-axis $[(r,\theta) = (-4, \frac{\pi}{2}) = (4, \frac{3\pi}{2})]$. Finally, as θ increases from $\frac{2\pi}{3}$ to π, r increases from 0 to its maximum of 4 when $\theta = \frac{5\pi}{6}$ and then back down to 0, which yields the third leaf, in QII.

37. $r = 2 + 4\cos\theta$: $r = 0$ when $4\cos\theta = -2 \Rightarrow \cos\theta = -\frac{1}{2} \Rightarrow \theta = \frac{2\pi}{3}$ and $\frac{4\pi}{3}$. As θ increases from 0 to $\frac{\pi}{2}$, $4\cos\theta$ decreases from 4 to 0, so r decreases from 6 to 2 (the part of the outer curve in QI). As θ increases from $\frac{\pi}{2}$ to $\frac{2\pi}{3}$, r continues to decrease from 2 to 0 (the part of the curve in QII). As θ increases from $\frac{2\pi}{3}$ to π, r becomes negative and decreases from 0 to -2, forming the lower part of the inner loop. As θ increases from π to $\frac{4\pi}{3}$, r is still negative, but increasing from -2 to 0, forming the upper part of the inner loop. As θ increases from $\frac{4\pi}{3}$ to $\frac{3\pi}{2}$, r increases from 0 to 2 (the part of the curve in QIII). Finally, as θ increases from $\frac{3\pi}{2}$ to 2π, r increases from 2 to 6 (the part of the outer curve in QIV).

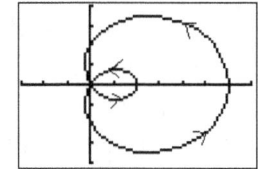

39. $r = \frac{3}{2} + \cos\theta$: Since $-1 \leq \cos\theta \leq 1$, r is always positive. As θ increases from 0 to $\frac{\pi}{2}$, $\cos\theta$ decreases from 1 to 0, so r decreases from $\frac{5}{2}$ to $\frac{3}{2}$. As θ increases from $\frac{\pi}{2}$ to π, $\cos\theta$ decreases from 0 to -1, which means that r decreases from $\frac{3}{2}$ to $\frac{1}{2}$. As θ increases from π to $\frac{3\pi}{2}$, $\cos\theta$ increases from -1 to 0, so that r increases from $\frac{1}{2}$ to $\frac{3}{2}$. Finally, as θ increases from $\frac{3\pi}{2}$ to 2π, $\cos\theta$ increases from 0 to 1, so that r increases from $\frac{3}{2}$ to $\frac{5}{2}$.

41. $r = 2\tan\theta$: First we note that, $r = 2\left(\frac{\sin\theta}{\cos\theta}\right) \Rightarrow r\cos\theta = 2\sin\theta$ for $\cos\theta \neq 0$, i.e. $x = 2\sin\theta$. Since $\sin\theta$ varies between -1 and 1, then x varies between -2 and 2.

As θ increases from 0 to $\frac{\pi}{2}$, r increases from 0 to ∞ (the part of the curve in QI). As θ increases from $\frac{\pi}{2}$ to π, $\tan\theta < 0 \Rightarrow r < 0$ and increasing from $-\infty$ to 0 (the part of the curve in QIV). As θ increases from π to $\frac{3\pi}{2}$, r increases from 0 to ∞ (the part of the curve in QIII). Finally, as θ increases from $\frac{3\pi}{2}$ to 2π, $r < 0$ and increasing from $-\infty$ to 0 (the part of the curve in QII).

43. $r^2 = 9\sin 2\theta$: The left side of the equation is never negative, so there is no curve corresponding to $\sin 2\theta < 0$, i.e. when $\pi < 2\theta < 2\pi \Rightarrow \frac{\pi}{2} < \theta < \pi$ which is QII, and when $3\pi < 2\theta < 4\pi \Rightarrow \frac{3\pi}{2} < \theta < 2\pi$ which is QIV. Also, $|r|$ is a maximum of 3 when $\sin 2\theta = 1 \Rightarrow 2\theta = \frac{\pi}{2} + 2n\pi \Rightarrow \theta = \frac{\pi}{4} + n\pi$.

As θ increases from 0 to $\frac{\pi}{4}$, 2θ increases from 0 to $\frac{\pi}{2}$, so r^2 increases from 0 to 9, which gives rise to 2 branches of the curve; one corresponding to r increasing from 0 to 3 (the lower half of the loop in QI), and the other corresponding to $r < 0$ decreasing from 0 to -3 (the upper half of the loop in QIII). As θ increases from $\frac{\pi}{4}$ to $\frac{\pi}{2}$, 2θ increases from $\frac{\pi}{2}$ to π, so r^2

decreases from 9 to 0, giving rise again to 2 branches; one corresponding to r decreasing from 3 to 0 (upper half of loop in QI), and the other corresponding to $r < 0$ increasing from -3 to 0 (lower half of loop in QIII).

45. $r = 2\theta, 0 \le \theta \le 3\pi$: The graph is part of a spiral. As θ increases, so does r. The graph starts at the origin, and ends at $(6\pi, 3\pi)$. We plotted a few additional points.

θ	0	$\frac{\pi}{2}$	π	$\frac{3\pi}{2}$	2π	$\frac{5\pi}{2}$	3π
r	0	π	2π	3π	4π	5π	6π

47. $r = 2\sin\theta = f(\theta) \Rightarrow f'(\theta) = 2\cos\theta$

$$\frac{dy}{dx} = \frac{f'(\theta)\sin\theta + f(\theta)\cos\theta}{f'(\theta)\cos\theta - f(\theta)\sin\theta} = \frac{2\cos\theta\sin\theta + 2\sin\theta\cos\theta}{2\cos^2\theta - 2\sin^2\theta}$$

$$= \frac{\cancel{2}[2\sin\theta\cos\theta]}{\cancel{2}(\cos^2\theta - \sin^2\theta)} = \frac{\sin 2\theta}{\cos 2\theta} = \tan 2\theta$$

$$\left.\frac{dy}{dx}\right|_{\theta = \frac{\pi}{6}} = \tan 2(\tfrac{\pi}{6}) = \tan\tfrac{\pi}{3} = \sqrt{3}.$$

49. $r = \cos 2\theta = f(\theta) \Rightarrow f'(\theta) = -2\sin 2\theta$

$$\frac{dy}{dx} = \frac{f'(\theta)\sin\theta + f(\theta)\cos\theta}{f'(\theta)\cos\theta - f(\theta)\sin\theta} = \frac{-2\sin 2\theta\sin\theta + \cos 2\theta\cos\theta}{-2\sin 2\theta\cos\theta - \cos 2\theta\sin\theta}$$

$$\left.\frac{dy}{dx}\right|_{\theta = \frac{\pi}{4}} = \frac{-2\sin\frac{\pi}{2}\sin\frac{\pi}{4} + \cos\frac{\cancel{\pi}}{2}\cos\frac{\pi}{4}}{-2\sin\frac{\pi}{2}\cos\frac{\pi}{4} - \cos\frac{\cancel{\pi}}{2}\sin\frac{\pi}{4}} = \tan\tfrac{\pi}{4} = 1.$$

51. $r = \tfrac{1}{\theta} = \theta^{-1} = f(\theta) \Rightarrow f'(\theta) = -\theta^{-2} = -\tfrac{1}{\theta^2}$

$$\frac{dy}{dx} = \frac{f'(\theta)\sin\theta + f(\theta)\cos\theta}{f'(\theta)\cos\theta - f(\theta)\sin\theta} = \frac{-\tfrac{1}{\theta^2}\sin\theta + \tfrac{1}{\theta}\cos\theta}{-\tfrac{1}{\theta^2}\cos\theta - \tfrac{1}{\theta}\sin\theta} \cdot \frac{\theta^2}{\theta^2} = \frac{-\sin\theta + \theta\cos\theta}{-\cos\theta - \theta\sin\theta}$$

$$\left.\frac{dy}{dx}\right|_{\theta = \pi} = \frac{-\sin\cancel{\pi} + \pi\cos\pi}{-\cos\pi - \pi\sin\cancel{\pi}} = -\pi.$$

53. $r = 4 + 4\cos\theta = f(\theta) \Rightarrow f'(\theta) = -4\sin\theta$

$$\frac{dy}{dx} = \frac{f'(\theta)\sin\theta + f(\theta)\cos\theta}{f'(\theta)\cos\theta - f(\theta)\sin\theta} = \frac{-4\sin^2\theta + (4 + 4\cos\theta)\cos\theta}{-4\sin\theta\cos\theta - (4 + 4\cos\theta)\sin\theta}$$

$$= \frac{-\sin^2\theta + \cos\theta + \cos^2\theta}{-2\sin\theta\cos\theta - \sin\theta} = \frac{-(1 - \cos^2\theta) + \cos\theta + \cos^2\theta}{-\sin\theta(2\cos\theta + 1)}$$

$\frac{dy}{dx} = 0 \Rightarrow 1-\cos^2\theta = \cos\theta+\cos^2\theta \Rightarrow 2\cos^2\theta+\cos\theta-1 = 0 \Rightarrow (2\cos\theta-1)(\cos\theta+1) = 0 \Rightarrow$
$\cos\theta = \frac{1}{2}$ or $\cos\theta = -1 \Rightarrow \theta = \frac{\pi}{3}, \frac{5\pi}{3}$, or π which makes the denominator of $\frac{dy}{dx}$ zero, so
there are only two max/min points; when $\cos\theta = \frac{1}{2} \Rightarrow r = 4 + 4\cos\theta = 4 + 4(\frac{1}{2}) = 6$. As

$x = r\cos\theta, x = 6(\frac{1}{2}) = 3$ for both points. $y = 6\sin\frac{\pi}{3} = 6(\frac{\sqrt{3}}{2}) = 3\sqrt{3}$

and $y = 6\sin\frac{5\pi}{3} = 6(-\frac{\sqrt{3}}{2}) = -3\sqrt{3}$. Thus the two points have (x, y)
coordinates $(3, 3\sqrt{3})$ and $(3, -3\sqrt{3})$. Since the graph is the cardioid in
the figure, $(3, 3\sqrt{3})$ is a maximum point and $(3, -3\sqrt{3})$ is a minimum.

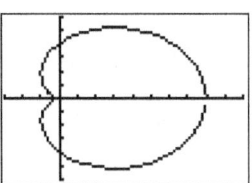

55. $r = \cos^2\theta = f(\theta) \Rightarrow f'(\theta) = 2\cos\theta(-\sin\theta) = -2\sin\theta\cos\theta$

$$\frac{dy}{dx} = \frac{f'(\theta)\sin\theta + f(\theta)\cos\theta}{f'(\theta)\cos\theta - f(\theta)\sin\theta} = \frac{-2\sin\theta\cos\theta\sin\theta + \cos^2\theta\cos\theta}{-2\sin\theta\cos\theta\cos\theta - \cos^2\theta\sin\theta}$$
$$= \frac{(\cos\theta)(-2\sin^2\theta + \cos^2\theta)}{-3\sin\theta\cos^2\theta}$$

$\frac{dy}{dx} = 0 \Rightarrow 2\sin^2\theta = \cos^2\theta \Rightarrow \tan^2\theta = \frac{1}{2} \Rightarrow \tan\theta = \pm\frac{1}{\sqrt{2}} \Rightarrow \tan\theta_r = \frac{1}{\sqrt{2}}$:

Depending on the quadrant, $\cos\theta = \pm\cos\theta_r$, and from the triangle then , $r = \cos^2\theta = \frac{2}{3}$.

In QI:	In QII:	In QIII:	In QIV:
$\cos\theta = \frac{\sqrt{2}}{\sqrt{3}} = \frac{\sqrt{6}}{3}$	$\cos\theta = -\frac{\sqrt{2}}{\sqrt{3}} = -\frac{\sqrt{6}}{3}$	$\cos\theta = -\frac{\sqrt{2}}{\sqrt{3}} = -\frac{\sqrt{6}}{3}$	$\cos\theta = \frac{\sqrt{2}}{\sqrt{3}} = \frac{\sqrt{6}}{3}$
$\sin\theta = \frac{1}{\sqrt{3}} = \frac{\sqrt{3}}{3}$	$\sin\theta = \frac{1}{\sqrt{3}} = \frac{\sqrt{3}}{3}$	$\sin\theta = -\frac{1}{\sqrt{3}} = -\frac{\sqrt{3}}{3}$	$\sin\theta = -\frac{1}{\sqrt{3}} = -\frac{\sqrt{3}}{3}$
$x = r\cos\theta = \frac{2}{3}\cdot\frac{\sqrt{6}}{3}$	$x = r\cos\theta = \frac{2}{3}\cdot-\frac{\sqrt{6}}{3}$	$x = r\cos\theta = \frac{2}{3}\cdot-\frac{\sqrt{6}}{3}$	$x = r\cos\theta = \frac{2}{3}\cdot\frac{\sqrt{6}}{3}$
$= \frac{2\sqrt{6}}{9}$	$= -\frac{2\sqrt{6}}{9}$	$= -\frac{2\sqrt{6}}{9}$	$= \frac{2\sqrt{6}}{9}$
$y = r\sin\theta = \frac{2}{3}\cdot\frac{\sqrt{3}}{3}$	$y = r\sin\theta = \frac{2}{3}\cdot\frac{\sqrt{3}}{3}$	$y = r\sin\theta = \frac{2}{3}\cdot-\frac{\sqrt{3}}{3}$	$y = r\sin\theta = \frac{2}{3}\cdot-\frac{\sqrt{3}}{3}$
$= \frac{2\sqrt{3}}{9}$	$= \frac{2\sqrt{3}}{9}$	$= -\frac{2\sqrt{3}}{9}$	$= -\frac{2\sqrt{3}}{9}$
$(x, y) = (\frac{2\sqrt{6}}{9}, \frac{2\sqrt{3}}{9})$	$(x, y) = (-\frac{2\sqrt{6}}{9}, \frac{2\sqrt{3}}{9})$	$(x, y) = (-\frac{2\sqrt{6}}{9}, -\frac{2\sqrt{3}}{9})$	$(x, y) = (\frac{2\sqrt{6}}{9}, -\frac{2\sqrt{3}}{9})$

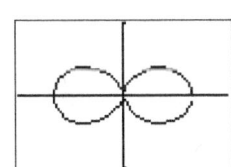

From the figure, we see that $(\pm\frac{2\sqrt{6}}{9}, \frac{2\sqrt{3}}{9})$ are maximum
points, and $(\pm\frac{2\sqrt{6}}{9}, -\frac{2\sqrt{3}}{9})$ are minimum points.

57. $r = 1 + 2\cos\theta = f(\theta) \;\Rightarrow\; f'(\theta) = -2\sin\theta$

$$\frac{dy}{dx} = \frac{f'(\theta)\sin\theta + f(\theta)\cos\theta}{f'(\theta)\cos\theta - f(\theta)\sin\theta} = \frac{-2\sin^2\theta + (1 + 2\cos\theta)\cos\theta}{-2\sin\theta\cos\theta - (1 + 2\cos\theta)\sin\theta}$$

$$= \frac{-2(1 - \cos^2\theta) + \cos\theta + 2\cos^2\theta}{-\sin\theta(4\cos\theta + 1)} = \frac{-2 + 4\cos^2\theta + \cos\theta}{-\sin\theta(4\cos\theta + 1)}$$

$\frac{dy}{dx} = 0 \;\Rightarrow\; 4\cos^2\theta + \cos\theta - 2 = 0 \;\Rightarrow\; \cos\theta = \frac{-1 \pm \sqrt{1 - 4(-8)}}{8} = \frac{-1 \pm \sqrt{33}}{8}$

If $\cos\theta = \frac{-1 + \sqrt{33}}{8}$, then $\theta \approx 53.6°$ or $360° - 53.6° = 306.4°$. If $\cos\theta = \frac{-1 - \sqrt{33}}{8}$, then

$\theta \approx 147.5°$ or $360° - 147.5° = 212.5°$. Now $r = 1 + 2\cos\theta \;\Rightarrow\; r = 1 + 2(\frac{-1 \pm \sqrt{33}}{8}) =$

$\frac{3 \pm \sqrt{33}}{4}$ Therefore, there are 4 max/min points, with polar coordinates $(\frac{3 + \sqrt{33}}{4}, 53.6°)$,

$(\frac{3 + \sqrt{33}}{4}, 306.4°)$, $(\frac{3 - \sqrt{33}}{4}, 147.5°)$, and $(\frac{3 - \sqrt{33}}{4}, 212.5°)$.

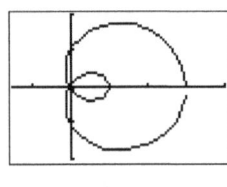

From the figure, we see that $\left(\frac{3 + \sqrt{33}}{4}, 53.6°\right)$ (at top of curve) and

$\left(\frac{3 - \sqrt{33}}{4}, 212.5°\right)$ (at top of inner loop) are maximum points, and

$\left(\frac{3 + \sqrt{33}}{4}, 306.4°\right)$ (at bottom of curve) and $\left(\frac{3 - \sqrt{33}}{4}, 147.5°\right)$ (at

bottom of inner loop) are minimum points.

59.

$r = \sin\theta$ and $r = -\cos\theta$ are two circles. From the figure we see that one point of intersection is the origin. To find the other point, we solve the equation $\sin\theta = -\cos\theta \;\Rightarrow\; \tan\theta = -1 \;\Rightarrow\; \tan\theta_r = 1 \;\Rightarrow\; \theta_r = \frac{\pi}{4}$. As $\tan\theta < 0, \theta = \frac{3\pi}{4}$ or $\frac{7\pi}{4}$. When $\theta = \frac{3\pi}{4}, r = \sin\frac{3\pi}{4} = \sin\frac{\pi}{4} = \frac{1}{\sqrt{2}}$, and we have the polar coordinates of the other

point $(\frac{1}{\sqrt{2}}, \frac{3\pi}{4})$. [As you can verify, using $\theta = \frac{7\pi}{4}$ yields the point $(-\frac{1}{\sqrt{2}}, \frac{7\pi}{4})$ which is the same

point in QII.] The rectangular coordinates are: $x = r\cos\theta = \frac{1}{\sqrt{2}}\cos\frac{3\pi}{4} = \frac{1}{\sqrt{2}} \cdot -\frac{1}{\sqrt{2}} = -\frac{1}{2}$

and $y = r\sin\theta = \frac{1}{\sqrt{2}} \cdot \frac{1}{\sqrt{2}} = \frac{1}{2} \;\Rightarrow\; (-\frac{1}{2}, \frac{1}{2})$.

61.

$r = 1 - \cos\theta$ is a cardioid, and $r = \cos\theta$ is a circle. From the figure we see that they intersect at the origin and two additional, symmetric points, so we can find the top point and use the symmetry to get the other one. To find the top point of intersection, we solve the equation $1 - \cos\theta = \cos\theta \;\Rightarrow\; 2\cos\theta = 1 \;\Rightarrow\; \cos\theta = \frac{1}{2} \;\Rightarrow\; \theta = \frac{\pi}{3}$.

From $r = \cos\theta$, we have $r = \frac{1}{2}$. So the polar coordinates of the top point of intersec-

tion are $(\frac{1}{2}, \frac{\pi}{3})$. As $x = r\cos\theta$, we find $x = \frac{1}{2} \cdot \frac{1}{2} = \frac{1}{4}$, and as $y = r\sin\theta$, we get

$y = \frac{1}{2}\sin\frac{\pi}{3} = \frac{1}{2} \cdot \frac{\sqrt{3}}{2} = \frac{\sqrt{3}}{4}$. So the top point has rectangular coordinates $(\frac{1}{4}, \frac{\sqrt{3}}{4})$. By

symmetry, the bottom point is $(\frac{1}{4}, -\frac{\sqrt{3}}{4})$.

63. $r = a \sin\theta + b\cos\theta \Rightarrow r^2 = ar\sin\theta + br\cos\theta \Rightarrow x^2 + y^2 = ay + bx \Rightarrow$
$x^2 - bx + \frac{b^2}{4} + y^2 - ay + \frac{a^2}{4} = 0 + \frac{b^2}{4} + \frac{a^2}{4} \Rightarrow (x - \frac{b}{2})^2 + (y - \frac{a}{2})^2 = \frac{a^2 + b^2}{4}$ which is
the equation of a circle centered at $(\frac{b}{2}, \frac{a}{2})$ of radius $\frac{1}{2}\sqrt{a^2 + b^2}$.

65.   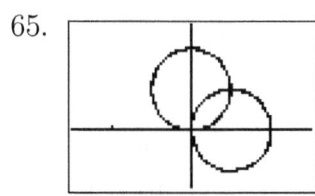 From the figure, we see that the two circles $r = \cos\theta$ and $r = \sin\theta$
intersect at the origin and one other point: $\cos\theta = \sin\theta \Rightarrow \tan\theta = 1 \Rightarrow \theta = \frac{\pi}{4} \Rightarrow r = \cos\frac{\pi}{4} = \frac{1}{\sqrt{2}}$ giving the polar coordinates of
the second point of intersection as $(\frac{1}{\sqrt{2}}, \frac{\pi}{4})$. Let's find the slope of
each curve: $\dfrac{dy}{dx} = \dfrac{f'(\theta)\sin\theta + f(\theta)\cos\theta}{f'(\theta)\cos\theta - f(\theta)\sin\theta}$

For $r = \cos\theta = f(\theta) \Rightarrow f'(\theta) = -\sin\theta$:
$$\frac{dy}{dx} = \frac{-\sin^2\theta + \cos^2\theta}{-\sin\theta\cos\theta - \cos\theta\sin\theta} = \frac{-\sin^2\theta + \cos^2\theta}{-2\sin\theta\cos\theta} = -\frac{\cos 2\theta}{\sin 2\theta}$$
For $r = \sin\theta = f(\theta) \Rightarrow f'(\theta) = \cos\theta$:
$$\frac{dy}{dx} = \frac{\cos\theta\sin\theta + \sin\theta\cos\theta}{\cos^2\theta - \sin^2\theta} = \frac{2\sin\theta\cos\theta}{\cos^2\theta - \sin^2\theta} = \frac{\sin 2\theta}{\cos 2\theta}$$
The slopes are negative reciprocals of one another! To confirm: At $(0,0)$, one slope is 0
(horizontal tangent), and the other is undefined (vertical tangent)–check. At $(\frac{1}{\sqrt{2}}, \frac{\pi}{4})$, one
slope is $\dfrac{\sin\frac{\pi}{2}}{\cos\frac{\pi}{2}}$ which is undefined (vertical tangent), and the other is $-\dfrac{\cos\frac{\pi}{2}}{\sin\frac{\pi}{2}} = 0$ (horizontal
tangent)–check. The circles intersect at right angles.

§10.3 Area and Length

1.   The curve $r = 1 + \cos\theta$ is a cardioid. r has its maximum value
of 2 when $\cos\theta = 1$, at $\theta = 0$. As θ increases from 0 to $\frac{\pi}{2}$, $\cos\theta$
decreases from 1 to 0, so r decreases from 2 to 1. As θ increases from
$\frac{\pi}{2}$ to π, $\cos\theta$ continues to decrease to -1, which causes r to decrease
to 0. We use the symmetry of the cardioid to complete the sketch
(see figure).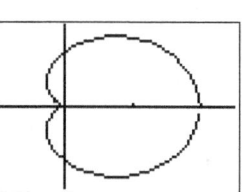

The area enclosed by the cardioid $r = 1 + \cos\theta$ is:
$$A = \frac{1}{2}\int_0^{2\pi} r^2\, d\theta = \frac{1}{2}\int_0^{2\pi}(1 + \cos\theta)^2 d\theta = \frac{1}{2}\int_0^{2\pi}(1 + 2\cos\theta + \cos^2\theta)d\theta$$
$$= \frac{1}{2}\int_0^{2\pi}(1 + 2\cos\theta + \frac{1 + \cos 2\theta}{2})d\theta = \frac{1}{2}\left[\theta + 2\sin\theta + \frac{\theta}{2} + \frac{\sin 2\theta}{4}\right]\Big|_0^{2\pi}$$
$$= \frac{1}{2}\left[2\pi + 2\sin 2\pi + \pi + \frac{\sin 4\pi}{4}\right] = \frac{3\pi}{2}$$

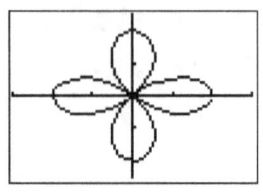

3. The curve $r = 2\cos 2\theta$ is a 4-leaved rose: r has a maximum value of 2 when $\cos 2\theta = 1 \Rightarrow \theta = 0$ (among other values), and $r = 0$ when $2\theta = \frac{\pi}{2} \Rightarrow \theta = \frac{\pi}{4}$ (for one). That completes half a leaf. As θ increases from $\frac{\pi}{4}$ to $\frac{3\pi}{4}$, 2θ increases from $\frac{\pi}{2}$ to $\frac{3\pi}{2}$, and the cosine goes from 0 to -1 and back to 0, tracing out the bottom leaf. As θ increases from $\frac{3\pi}{4}$ to $\frac{5\pi}{4}$, 2θ increases from $\frac{3\pi}{2}$ to $\frac{5\pi}{2}$ and $r > 0$, tracing out the leftmost leaf. As θ increases from $\frac{5\pi}{4}$ to $\frac{7\pi}{4}$, 2θ increases from $\frac{5\pi}{2}$ to $\frac{7\pi}{2}$, and the cosine goes from 0 to -1 and back to 0, tracing out the top leaf. Finally, as θ increases from $\frac{7\pi}{4}$ to 2π, 2θ increases from $\frac{7\pi}{2}$ to 4π, the cosine goes from 0 to 1, completing the bottom half of the rightmost leaf, and we get the graph shown in the figure.

The area enclosed by the rose is :

$$A = \frac{1}{2}\int_\alpha^\beta r^2\, d\theta = \frac{1}{2}\int_0^{2\pi} 4\cos^2 2\theta\, d\theta = 2\int_0^{2\pi}\left(\frac{1+\cos 4\theta}{2}\right)d\theta$$

$$= \left[\theta + \frac{\sin 4\theta}{4}\right]\Big|_0^{2\pi} = 2\pi$$

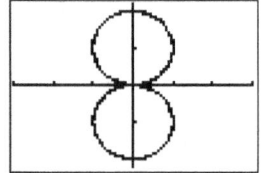

5. The curve $r^2 = 4\sin^2\theta$ is composed of 2 circles: $r = \pm 2\sin\theta$. Sketching $r = 2\sin\theta$: As θ increases from 0 to $\frac{\pi}{2}$, r increases from 0 to 2. As θ increases from $\frac{\pi}{2}$ to π, r decreases from 2 to 0. The circle is centered on the positive y-axis with a radius of $\frac{2}{2} = 1$ and center at $(0,1)$.

Sketching $r = -2\sin\theta$: As θ increases from 0 to $\frac{\pi}{2}$, r is negative and decreases from 0 to -2 which sketches half the circle in QIII. As θ increases from $\frac{\pi}{2}$ to π, r is still negative and increases from -2 to 0 which sketches the other half of the circle below the x-axis, in QIV. Again a circle, but this time centered on the negative y-axis of radius $\frac{2}{2} = 1$ and center at $(0,-1)$.

(We know the area of each circle is $\pi r^2 = \pi(1^2) = \pi$ so the total area is 2π, but we'll verify it by integrating.)

By symmetry, the area enclosed is twice the area of the upper circle. (Note that the circle is traced out once as θ goes from 0 to π (not 2π):

$$A = \frac{1}{2}\int_\alpha^\beta r^2\, d\theta = 2\cdot\frac{1}{2}\int_0^\pi (2\sin\theta)^2\, d\theta = 4\int_0^\pi \sin^2\theta\, d\theta = 4\int_0^\pi\left(\frac{1-\cos 2\theta}{2}\right)d\theta$$

$$= 2\left[\theta - \frac{\sin 2\theta}{2}\right]\Big|_0^\pi = 2\pi.$$

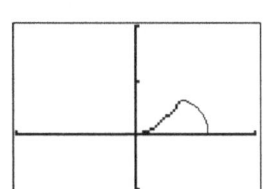

7. From $r = \theta^2, 0 \le \theta \le \frac{\pi}{4}$, we see that $r \ge 0$ and as θ increases from 0 to $\frac{\pi}{4}$, r increases from 0 to $\left(\frac{\pi}{4}\right)^2$. The graph is contained in the first quadrant.

$$A = \frac{1}{2}\int_0^{\pi/4}(\theta^2)^2 d\theta = \frac{1}{2}\int_0^{\pi/4}\theta^4\, d\theta = \frac{1}{2}\cdot\frac{\theta^5}{5}\Big|_0^{\pi/4} = \frac{1}{10}\left(\frac{\pi}{4}\right)^5.$$

9. From $r = \tan 2\theta, 0 \le \theta \le \frac{\pi}{8}$, we see that $r \ge 0$ and as θ increases from 0 to $\frac{\pi}{8}, r$ increases from 0 to $\tan \frac{\pi}{4} = 1$.

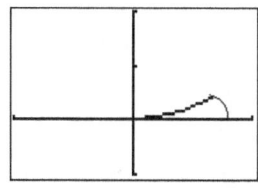

$$A = \frac{1}{2} \int_0^{\pi/8} (\tan 2\theta)^2 d\theta = \frac{1}{2} \int_0^{\pi/8} \tan^2 2\theta \, d\theta = \frac{1}{2} \int_0^{\pi/8} (\sec^2 2\theta - 1) \, d\theta$$

$$= \frac{1}{2} \left(\frac{\tan 2\theta}{2} - \theta \right)\Big|_0^{\pi/8} = \frac{1}{2} \left(\frac{\tan \frac{\pi}{4}}{2} - \frac{\pi}{8} \right) = \frac{1}{2}(\frac{1}{2} - \frac{\pi}{8}) = \frac{1}{16}(4 - \pi).$$

11. From $r = \sqrt{1 - \cos\theta}, \frac{\pi}{2} \le \theta \le \pi$, we see that $r > 0$ and as θ increases from $\frac{\pi}{2}$ to π, r increases from 1 to $\sqrt{2}$.

$$A = \frac{1}{2} \int_{\pi/2}^\pi (\sqrt{1 - \cos\theta})^2 d\theta = \frac{1}{2} \int_{\pi/2}^\pi (1 - \cos\theta) \, d\theta = \frac{1}{2} (\theta - \sin\theta)|_{\pi/2}^\pi$$

$$= \frac{1}{2}[\pi - \sin\pi - (\frac{\pi}{2} - \sin\frac{\pi}{2})] = \frac{1}{2}(\frac{\pi}{2} + 1) = \frac{1}{4}(\pi + 2).$$

13. The curve $r = 2$ is a circle centered at the origin, of radius 2. The curve $r = 2(1 - \cos\theta)$ is a cardioid, with $r \ge 0$ reaching its maximum value of 4 when $\theta = \pi$, and when θ is $\frac{\pi}{2}$ or $\frac{3\pi}{2}, r = 2$; when $\theta = 0$ or 2π then $r = 0$. Thus we see that the curves intersect at the two symmetric points $(0, \pm 2)$. Using the symmetry, we double the

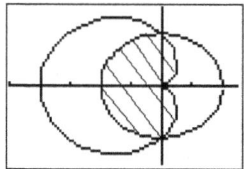

area inside the cardioid from 0 to $\frac{\pi}{2}$ (the shaded area in QI) plus the area of the quarter circle (shaded area in QII) which we know is $\frac{1}{4}(\pi \cdot 2^2) = \pi$:

$$A = 2\left\{ \frac{1}{2} \int_0^{\pi/2} [2(1 - \cos\theta)]^2 d\theta + \pi \right\} = 4 \int_0^{\pi/2} (1 - 2\cos\theta + \cos^2\theta) d\theta + 2\pi$$

$$= 4 [\theta - 2\sin\theta]\Big|_0^{\pi/2} + 4 \int_0^{\pi/2} \frac{1 + \cos 2\theta}{2} d\theta + 2\pi = 4(\frac{\pi}{2} - 2\sin\frac{\pi}{2}) + 2 (\theta + \frac{\sin 2\theta}{2})\Big|_0^{\pi/2} + 2\pi$$

$$= 2\pi - 8 + 2(\frac{\pi}{2}) + 2\pi = 5\pi - 8.$$

15. The curves $r = \cos\theta$ and $r = \sin\theta$ are both circles of radius $\frac{1}{2}$, the latter one is centered on the positive y-axis ($\sin\frac{\pi}{2} = 1$), and the former one on the positive x-axis ($\cos 0 = 1$). First determine where the circles intersect besides at the origin, somewhere in QI: $\cos\theta = \sin\theta \Rightarrow \tan\theta = 1 \Rightarrow \theta = \frac{\pi}{4}$. Clearly the shaded area is

symmetric about the half-line $\theta = \frac{\pi}{4}$, so we will determine the area of the lower half of the shaded region and double the answer:

$$A = 2 \cdot \frac{1}{2} \int_0^{\pi/4} \sin^2\theta \, d\theta = \int_0^{\pi/4} \frac{1 + \cos 2\theta}{2} d\theta = \frac{1}{2} (\theta - \frac{\sin 2\theta}{2})\Big|_0^{\pi/4} = \frac{1}{2}(\frac{\pi}{4} - \frac{\sin\frac{\pi}{2}}{2})$$

$$= \frac{1}{2}(\frac{\pi}{4} - \frac{1}{2}) = \frac{\pi - 2}{8}.$$

17. The curve $r = -4\sin\theta$ is a circle centered on the negative y-axis, of radius 2: As θ varies from 0 to $\frac{\pi}{2}$, r goes from 0 to -4. As θ varies from $\frac{\pi}{2}$ to π, r increases from -4 back to 0.

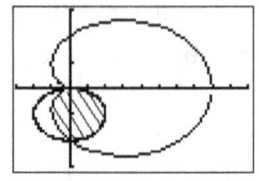

The curve $r = 4(1 + \cos\theta)$ is a cardioid: $r = 0$ when $\cos\theta = -1$, at $\theta = \pi$, and r reaches its maximum value of 8, when $\cos\theta = 1$, at $\theta = 0$. As θ increases from 0 to $\frac{\pi}{2}$, $\cos\theta$ decreases from 1 to 0, so r decreases from 8 to 4. As θ increases from $\frac{\pi}{2}$ to π, $\cos\theta$ continues to decrease from 0 to -1, so r decreases from 4 to 0. The other half of the cardioid we get by symmetry about the x-axis.

We see that the two curves intersect at the origin and at $(0, -4)$. The area in common is the area inside the cardioid from $\theta = \pi$ to $\theta = \frac{3\pi}{2}$ plus the area in QIV inside the circle. The latter area is half the area of the circle, namely $\frac{1}{2}\pi(2^2) = 2\pi$:

$$A = \frac{1}{2}\int_\pi^{3\pi/2} [4(1 + \cos\theta)]^2 d\theta + 2\pi = 8\int_\pi^{3\pi/2}(1 + 2\cos\theta + \cos^2\theta)d\theta + 2\pi$$

$$= 8\left[\theta + 2\sin\theta\right]\Big|_\pi^{3\pi/2} + 8\int_\pi^{3\pi/2}\frac{1 + \cos 2\theta}{2}d\theta + 2\pi$$

$$= 8[(\tfrac{3\pi}{2} + 2\sin\tfrac{3\pi}{2}) - (\pi + 2\sin\pi)] + 8\left(\tfrac{\theta}{2} + \tfrac{\sin 2\theta}{4}\right)\Big|_\pi^{3\pi/2} + 2\pi$$

$$= 12\pi - 16 - 8\pi + 8(\tfrac{3\pi}{4} - \tfrac{\pi}{2}) + 2\pi = 6\pi - 16 + 2\pi = 8\pi - 16.$$

19. The curves $r^2 = \sin 2\theta$ and $r^2 = \cos 2\theta$ are two lemniscates: $r^2 = \sin 2\theta$: The left side of the equation is never negative, so there is no curve corresponding to $\sin 2\theta < 0$, i.e. when $\pi < 2\theta < 2\pi$ $\Rightarrow$ $\frac{\pi}{2} < \theta < \pi$ which is QII, and when $3\pi < 2\theta < 4\pi$ $\Rightarrow$ $\frac{3\pi}{2} < \theta < 2\pi$ which is QIV. Also, $|r|$ is a maximum of 1 when $\sin 2\theta = 1$ $\Rightarrow$ $2\theta = \frac{\pi}{2} + 2n\pi$ $\Rightarrow$ $\theta = \frac{\pi}{4} + n\pi$.

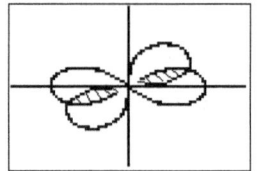

As θ increases from 0 to $\frac{\pi}{4}$, 2θ increases from 0 to $\frac{\pi}{2}$, so r^2 increases from 0 to 1, which gives rise to 2 branches of the curve; one corresponding to r increasing from 0 to 1 (the lower half of the loop in QI), and the other corresponding to $r < 0$ decreasing from 0 to -1 (the upper half of the loop in QIII). As θ increases from $\frac{\pi}{4}$ to $\frac{\pi}{2}$, 2θ increases from $\frac{\pi}{2}$ to π, so r^2 decreases from 1 to 0, giving rise again to 2 branches; one corresponding to r decreasing from 1 to 0 (upper half of loop in QI), and the other corresponding to $r < 0$ increasing from -1 to 0 (lower half of loop in QIII).

Similarly, we get the graph of the other lemniscate. Suffice it here to find where the maximum of $|r|$ occurs, to position the graph: r^2 is largest when $\cos 2\theta = 1$ $\Rightarrow$ $2\theta = 0$ $\Rightarrow$ $\theta = 0$. Using the symmetry of the shaded regions, we determine the area in QI, and double it. First we need to find where the curves intersect in QI, besides at the origin: Setting $\sin 2\theta = \cos 2\theta$ $\Rightarrow$ $\tan 2\theta = 1$ $\Rightarrow$ $2\theta = \frac{\pi}{4}$ $\Rightarrow$ $\theta = \frac{\pi}{8}$. Clearly, the area in QI is symmetric about the half-line through $\theta = \frac{\pi}{8}$, so we find the area below and double it:

$$A = 2 \cdot 2 \cdot \frac{1}{2}\int_0^{\pi/8} \sin^2 2\theta\, d\theta = 2\int_0^{\pi/8}\frac{1 - \cos 2\theta}{2}d\theta = \left(\theta - \frac{\sin 2\theta}{2}\right)\Big|_0^{\pi/8} = \frac{\pi}{8} - \frac{\sin\frac{\pi}{4}}{2}$$

$$= \frac{\pi}{8} - \frac{\sqrt{2}}{4} = \frac{\pi - 2\sqrt{2}}{8}.$$

21. Using symmetry, we will find and double the shaded area in QI. First determine where the cardioid and the circle intersect in QI: $4(1 + \cos \theta) = 6 \Rightarrow 4 \cos \theta = 2 \Rightarrow \cos \theta = \frac{1}{2} \Rightarrow \theta = \frac{\pi}{3}$:. The required area is the area inside the cardioid minus the area inside the circle:

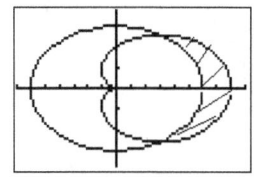

$$A = 2 \cdot \frac{1}{2} \int_0^{\pi/3} \left\{ [4(1 + \cos \theta)]^2 - 6^2 \right\} d\theta = \int_0^{\pi/3} [16(1 + 2 \cos \theta + \cos^2 \theta) - 36] \, d\theta$$

$$= \int_0^{\pi/3} \left[16 + 32 \cos \theta + 16 \left(\frac{1 + \cos 2\theta}{2} \right) - 36 \right] d\theta = \int_0^{\pi/3} [32 \cos \theta + 8 \cos 2\theta - 12] \, d\theta$$

$$= (32 \sin \theta + 4 \sin 2\theta - 12\theta) \Big|_0^{\pi/3} = 32 \sin \frac{\pi}{3} + 4 \sin \frac{2\pi}{3} - 12(\frac{\pi}{3}) = 32(\frac{\sqrt{3}}{2}) + 4(\frac{\sqrt{3}}{2}) - 4\pi$$

$$= 18\sqrt{3} - 4\pi.$$

23. We first find the two values of θ that determine the beginning and end of the formation of the inner loop. They correspond to the values of θ that make $r = 0$ for $0 \le \theta < 2\pi$: $1 - 2 \sin \theta = 0 \Rightarrow \sin \theta = \frac{1}{2} \Rightarrow \theta = \frac{\pi}{6}$ and $\frac{5\pi}{6}$.

$$A = \frac{1}{2} \int_{\pi/6}^{5\pi/6} (1 - 2 \sin \theta)^2 d\theta = \frac{1}{2} \int_{\pi/6}^{5\pi/6} (1 - 4 \sin \theta + 4 \sin^2 \theta) d\theta$$

$$= \frac{1}{2} (\theta + 4 \cos \theta) \Big|_{\pi/6}^{5\pi/6} + 2 \int_{\pi/6}^{5\pi/6} \frac{1 - \cos 2\theta}{2} \, d\theta$$

$$= \frac{1}{2} \left[\frac{5\pi}{6} + 4 \cos \frac{5\pi}{6} - (\frac{\pi}{6} + 4 \cos \frac{\pi}{6}) \right] + (\theta - \frac{1}{2} \sin 2\theta) \Big|_{\pi/6}^{5\pi/6}$$

$$= \frac{5\pi}{12} + 2(-\frac{\sqrt{3}}{2}) - \frac{\pi}{12} - 2(\frac{\sqrt{3}}{2}) + \frac{5\pi}{6} - \frac{1}{2} \sin \frac{5\pi}{3} - \frac{\pi}{6} + \frac{1}{2} \sin \frac{\pi}{3}$$

$$= \pi - 2\sqrt{3} + \frac{\sqrt{3}}{2} = \frac{2\pi - 3\sqrt{3}}{2}.$$

25. The area of the region between the two limaçons $r = 4 + \cos \theta$ and $r = 2 + \cos \theta$ is the area inside the outer one minus the area inside the inner one:

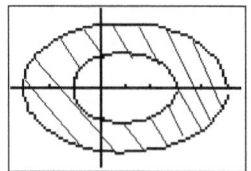

$$A = \frac{1}{2} \int_0^{2\pi} [(4 + \cos \theta)^2 - (2 + \cos \theta)^2] d\theta$$

$$= \frac{1}{2} \int_0^{2\pi} [16 + 8 \cos \theta + \cos^2 \theta - 4 - 4 \cos \theta - \cos^2 \theta] d\theta = \frac{1}{2} \int_0^{2\pi} (12 + 4 \cos \theta) d\theta$$

$$= \frac{1}{2} (12\theta + 4 \sin \theta) \Big|_0^{2\pi} = 6(2\pi) = 12\pi.$$

27. The area of the region outside the circle $r = 2$ and inside the cardioid $r = 4(1 + \cos \theta)$ is symmetric wrt the x-axis, so we will find the area above the x-axis, and double it. First determine where the two curves intersect (in QII): $2 = 4(1 + \cos \theta) \Rightarrow 1 + \cos \theta = \frac{1}{2} \Rightarrow \cos \theta = -\frac{1}{2} \Rightarrow \theta = \frac{2\pi}{3}$. The area is then twice the area inside the the cardioid minus the area inside the circle between $\theta = 0$ and $\theta = \frac{2\pi}{3}$:

$$A = 2 \cdot \frac{1}{2} \int_0^{2\pi/3} \left\{ [4(1 + \cos\theta)]^2 - 2^2 \right\} d\theta = \int_0^{2\pi/3} (16 + 32\cos\theta + 16\cos^2\theta - 4)d\theta$$

$$= \int_0^{2\pi/3} [12 + 32\cos\theta + 8(1 + \cos 2\theta)]d\theta = \int_0^{2\pi/3} (20 + 32\cos\theta + 8\cos 2\theta)d\theta$$

$$= (20\theta + 32\sin\theta + 4\sin 2\theta)\Big|_0^{2\pi/3} = 20(\tfrac{2\pi}{3}) + 32\sin\tfrac{2\pi}{3} + 4\sin\tfrac{4\pi}{3}$$

$$= \tfrac{40\pi}{3} + 32(\tfrac{\sqrt{3}}{2}) + 4(-\tfrac{\sqrt{3}}{2}) = \tfrac{40\pi}{3} + 14\sqrt{3} = \tfrac{40\pi + 42\sqrt{3}}{3}.$$

29. Using the symmetry of the shaded region, its area is 4 times the area shaded in QI. Now determine where the curves intersect in QI: $r^2 = 4\cos 2\theta = (\sqrt{2})^2 \Rightarrow \cos 2\theta = \frac{1}{2} \Rightarrow 2\theta = \frac{\pi}{3} \Rightarrow \theta = \frac{\pi}{6}$. Thus the desired area is 4 times the area inside the lemniscate minus the area inside the circle between $\theta = 0$ and $\theta = \frac{\pi}{6}$:

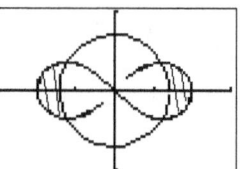

$$A = 4 \cdot \frac{1}{2} \int_0^{\pi/6} [4\cos 2\theta - (\sqrt{2})^2]d\theta = 2\left[2\sin 2\theta - 2\theta\right]\Big|_0^{\pi/6} = 2[2\sin\tfrac{\pi}{3} - \tfrac{\pi}{3}]$$

$$= 2[2 \cdot \tfrac{\sqrt{3}}{2} - \tfrac{\pi}{3}] = 2\sqrt{3} - \tfrac{2\pi}{3} = \tfrac{6\sqrt{3} - 2\pi}{3}.$$

31. The length $L = \int_\alpha^\beta \sqrt{r^2 + \left(\frac{dr}{d\theta}\right)^2}\, d\theta$. Here $r = 1 + \cos\theta \Rightarrow \frac{dr}{d\theta} = -\sin\theta$ and traversing the cardioid once means θ varies from 0 to 2π, but we will use the symmetry of the cardioid to integrate from 0 to π and double the answer:

$$L = 2\int_0^\pi \sqrt{(1 + \cos\theta)^2 + (-\sin\theta)^2}\, d\theta = 2\int_0^\pi \sqrt{1 + 2\cos\theta + (\cos^2\theta + \sin^2\theta)}\, d\theta$$

$$= 2\int_0^\pi \sqrt{2 + 2\cos\theta}\, d\theta = 2\sqrt{2}\int_0^\pi \sqrt{1 + \cos\theta}\, d\theta = 2\sqrt{2}\int_0^\pi \sqrt{2\cos^2\tfrac{\theta}{2}}\, d\theta$$

$$= 4\int_0^\pi \sqrt{\cos^2\tfrac{\theta}{2}}\, d\theta = 4\int_0^\pi \cos\tfrac{\theta}{2}\, d\theta = 8\sin\tfrac{\theta}{2}\Big|_0^\pi = 8\sin\tfrac{\pi}{2} = 8.$$

[Note: Had we integrated from 0 to 2π, then $\int_0^{2\pi} \sqrt{\cos^2\tfrac{\theta}{2}}\, d\theta = \int_0^{2\pi} \left|\cos\tfrac{\theta}{2}\right| d\theta \neq \int_0^{2\pi} \cos\tfrac{\theta}{2}\, d\theta$, since $\cos\tfrac{\theta}{2}$ is negative when $\pi \leq \theta \leq 2\pi$, as $\tfrac{\pi}{2} \leq \tfrac{\theta}{2} \leq \pi$ and the cosine is negative in QII.

Then, $\int_0^{2\pi} \left|\cos\tfrac{\theta}{2}\right| d\theta = \int_0^\pi \cos\tfrac{\theta}{2}\, d\theta + \int_\pi^{2\pi} (-\cos\tfrac{\theta}{2})\, d\theta.$

Moral: Use symmetry whenever possible.]

33. The length $L = \int_\alpha^\beta \sqrt{r^2 + \left(\frac{dr}{d\theta}\right)^2}\, d\theta$. Here $r = \cos^3\tfrac{\theta}{3}$ and $0 \leq \theta \leq \tfrac{\pi}{4}$. Then $\frac{dr}{d\theta} = 3\left[\cos^2\tfrac{\theta}{3}\right]\left[-\sin\tfrac{\theta}{3}\right]\left(\tfrac{1}{3}\right) = -\cos^2\tfrac{\theta}{3}\sin\tfrac{\theta}{3} \Rightarrow :$

$$L = \int_0^{\pi/4} \sqrt{\cos^6\tfrac{\theta}{3} + \cos^4\tfrac{\theta}{3}\sin^2\tfrac{\theta}{3}}\, d\theta = \int_0^{\pi/4} \sqrt{\cos^6\tfrac{\theta}{3} + \cos^4\tfrac{\theta}{3}\left(1 - \cos^2\tfrac{\theta}{3}\right)}\, d\theta$$

$$= \int_0^{\pi/4} \sqrt{\cos^6\tfrac{\theta}{3} + \cos^4\tfrac{\theta}{3} - \cos^6\tfrac{\theta}{3}}\, d\theta = \int_0^{\pi/4} \sqrt{\cos^4\tfrac{\theta}{3}}\, d\theta = \int_0^{\pi/4} \cos^2\tfrac{\theta}{3}\, d\theta$$

$$= \frac{1}{2}\int_0^{\pi/4} \left(1 + \cos\tfrac{2\theta}{3}\right) d\theta = \frac{1}{2}\left[\theta + \tfrac{3}{2}\sin\tfrac{2\theta}{3}\right]\Big|_0^{\pi/4} = \frac{1}{2}[\tfrac{\pi}{4} + \tfrac{3}{2}\sin\tfrac{\pi}{6}] = \tfrac{\pi}{8} + \tfrac{3}{4} \cdot \tfrac{1}{2} = \tfrac{\pi + 3}{8}.$$

35. The length $L = \int_\alpha^\beta \sqrt{r^2 + \left(\frac{dr}{d\theta}\right)^2}\, d\theta$. Here $r = \dfrac{6}{1 + \cos\theta}$, $0 \le \theta \le \frac{\pi}{2}$ $\Rightarrow$

$$\frac{dr}{d\theta} = \frac{d}{d\theta}\, 6(1 + \cos\theta)^{-1} = 6(-1)(1 + \cos\theta)^{-2}(-\sin\theta) = \frac{6\sin\theta}{(1 + \cos\theta)^2} \Rightarrow :$$

$$L = \int_0^{\pi/2} \sqrt{\frac{36}{(1 + \cos\theta)^2} + \frac{36\sin^2\theta}{(1 + \cos\theta)^4}}\, d\theta = 6\int_0^{\pi/2} \sqrt{\frac{(1 + \cos\theta)^2 + \sin^2\theta}{(1 + \cos\theta)^4}}\, d\theta$$

$$= 6\int_0^{\pi/2} \frac{\sqrt{2 + 2\cos\theta}}{(1 + \cos\theta)^2} = 6\int_0^{\pi/2} \frac{\sqrt{2}(1 + \cos\theta)^{\frac{1}{2}}}{(1 + \cos\theta)^2} = 6\sqrt{2}\int_0^{\pi/2} \frac{d\theta}{(1 + \cos\theta)^{\frac{3}{2}}}$$

$$= 6\sqrt{2}\int_0^{\pi/2} \frac{d\theta}{\left[2\cos^2\frac{\theta}{2}\right]^{\frac{3}{2}}} = 3\int_0^{\pi/2} \frac{d\theta}{\cos^3\frac{\theta}{2}} = 3\int_0^{\pi/2} \sec^3\frac{\theta}{2}\, d\theta = 6\int_0^{\pi/4} \sec^3 u\, du$$

$$\boxed{\begin{array}{ll} u = \frac{\theta}{2}, & \theta = 0 \Rightarrow u = 0 \\ du = \frac{1}{2}d\theta, & \theta = \frac{\pi}{2} \Rightarrow u = \frac{\pi}{4} \end{array}}$$

At this point, rather than invoking the reduction formula in Exercise 57 of Section 7.1, we will determine $\int \sec^3 x\, dx$, by integrating by parts, and then combining the two integrals of secant cubed: $\qquad (\int u\, dv = uv - \int v\, du)$

$$\int \sec^3 x\, dx = \int \sec x \sec^2 x\, dx = \sec x \tan x - \int \tan^2 x \sec x\, dx$$

$$\boxed{\begin{array}{ll} u = \sec x, & dv = \sec^2 x\, dx \\ du = \sec x \tan x\, dx, & v = \tan x \end{array}}$$

$$= \sec x \tan x - \int (\sec^2 x - 1)\sec x\, dx = \sec x \tan x - \int \sec^3 x\, dx + \int \sec x\, dx$$

$$2\int \sec^3 x\, dx = \sec x \tan x + \int \sec x\, dx = \sec x \tan x + \ln|\sec x + \tan x| + C$$

$$\int \sec^3 x\, dx = \tfrac{1}{2} \sec x \tan x + \tfrac{1}{2} \ln|\sec x + \tan x| + C$$

Returning to determining the length:

$$L = 6\int_0^{\pi/4} \sec^3 u\, du = 6\left[\tfrac{1}{2} \sec u \tan u + \tfrac{1}{2} \ln|\sec u + \tan u|\right]\Big|_0^{\pi/4}$$

$$= 3\sec\tfrac{\pi}{4}\tan\tfrac{\pi}{4} + 3\ln|\sec\tfrac{\pi}{4} + \tan\tfrac{\pi}{4}| - (3\ln|\sec 0|) = 3[\sqrt{2} + \ln(\sqrt{2} + 1)].$$

37. The length $L = \int_\alpha^\beta \sqrt{r^2 + \left(\frac{dr}{d\theta}\right)^2}\, d\theta$. Here $r = 4\cos 2\theta$ $\Rightarrow$ $\frac{dr}{d\theta} = 4(-\sin 2\theta)(2) = -8\sin 2\theta$. The 4-leaved rose is traversed once as θ varies from 0 to 2π. Therefore:

$$L = \int_0^{2\pi} \sqrt{(4\cos 2\theta)^2 + (-8\sin 2\theta)^2}\, d\theta$$

$$= \int_0^{2\pi} \sqrt{16\cos^2 2\theta + 64\sin^2 2\theta}\, d\theta \approx 38.75.$$

CHAPTER 11
Functions of Several Variables

§11.1 Limits and Continuity

1. As the denominator of $f(x,y) = \dfrac{xy - 100}{x^2 + 9}$ is defined and never zero, the domain is all of $\Re^2$.

3. Since the tangent is undefined at odd multiples of $\frac{\pi}{2}$, the domain of $f(x,y) = \tan xy$ is $\left\{(x,y) \mid xy \text{ is not an odd multiple of } \frac{\pi}{2}\right\}$.

5. The rational function $f(x,y,z) = \dfrac{1}{x + y + z}$ is undefined where its denominator is zero. Thus the domain of f is $\{(x,y,z) \mid x + y + z \neq 0\}$.

7. For given $\epsilon > 0$, we seek $\delta > 0$ such that
$$0 < \|(x,y) - (1,1)\| < \delta \implies |(2x + y) - 3| < \epsilon$$
As $|2x + y - 3| = |2(x - 1) + (y - 1)| \leq 2|x - 1| + |y - 1|$, we find a $\delta > 0$ for which $2|x - 1| < \frac{\epsilon}{2}$ (i.e. $|x - 1| < \frac{\epsilon}{4}$) and $|y - 1| < \frac{\epsilon}{2}$. Such a δ is $\delta = \frac{\epsilon}{4}$, since if $\|(x,y) - (1,1)\| = \|(x - 1, y - 1)\| < \frac{\epsilon}{4}$, then both x and y are at most $\frac{\epsilon}{4}$ units from 1 so that:
$$|2x + y - 3| = |2(x - 1) + (y - 1)| \leq 2|x - 1| + |y - 1| < 2\left(\tfrac{\epsilon}{4}\right) + \tfrac{\epsilon}{4} = \tfrac{3}{4}\epsilon < \epsilon.$$

9. For given $\epsilon > 0$, we seek $\delta > 0$ such that
$$0 < \|(x,y) - (2,1)\| < \delta \implies \left|\frac{x^2 - 4xy + 4y^2}{x - 2y}\right| < \epsilon$$
As $\left|\dfrac{x^2 - 4xy + 4y^2}{x - 2y}\right| = \left|\dfrac{(x - 2y)^2}{x - 2y}\right| = |x - 2y| = |(x - 2) - 2(y - 1)| \leq |x - 2| + 2|y - 1|$, we find a $\delta > 0$ for which $|x - 2| < \frac{\epsilon}{2}$ and $2|y - 1| < \frac{\epsilon}{2}$ (i.e. $|y - 1| < \frac{\epsilon}{4}$). Such a δ is $\delta = \frac{\epsilon}{4}$, since if $\|(x,y) - (2,1)\| = \|(x - 2, y - 1)\| < \frac{\epsilon}{4}$, then x is at most $\frac{\epsilon}{4}$ units from 2 and y is at most $\frac{\epsilon}{4}$ units from 1 so that:
$$\left|\frac{x^2 - 4xy + 4y^2}{x - 2y}\right| = |x - 2y| = |(x - 2) - 2(y - 1)| \leq |x - 2| + 2|y - 1| < \tfrac{\epsilon}{4} + 2\left(\tfrac{\epsilon}{4}\right) = \tfrac{3}{4}\epsilon < \epsilon.$$

11. As $\lim\limits_{x \to 0} x \sin x = 0 \cdot 0 = 0$, $\lim\limits_{(x,y) \to (0,\pi)} \dfrac{y + x \sin x}{y^2} = \dfrac{\pi + 0}{\pi^2} = \dfrac{1}{\pi}$.

13. $\lim\limits_{(x,y) \to (2,2)} \dfrac{x^4 - y^4}{x^2 - y^2} = \lim\limits_{(x,y) \to (2,2)} \dfrac{(x^2 - y^2)(x^2 + y^2)}{x^2 - y^2} = \lim\limits_{(x,y) \to (2,2)} (x^2 + y^2) = 2^2 + 2^2 = 8.$

15. $\lim\limits_{(x,y) \to (1,e)} \ln xy = \ln(1 \cdot e) = \ln e = 1.$

17. $$\lim_{(x,y)\to(1,2)}\frac{x^2+2xy+y^2-9}{x+y-3}=\lim_{(x,y)\to(1,2)}\frac{(x+y)^2-3^2}{(x+y)-3}$$
$$=\lim_{(x,y)\to(1,2)}\frac{[(x+y)\not=3][(x+y)+3]}{(x+y)\not=3}$$
$$=\lim_{(x,y)\to(1,2)}(x+y+3)=1+2+3=6.$$

19. $$\lim_{(x,y,z)\to(1,1,1)}\frac{zx-zy}{x-y}=\lim_{(x,y,z)\to(1,1,1)}\frac{z(x\not=y)}{x\not=y}=\lim_{(x,y,z)\to(1,1,1)}z=1.$$

21. $f(x,y)=\dfrac{x^4-y^4}{x^2-y^2}$ is discontinuous only where it is not defined, which is where its

denominator $x^2-y^2=0\Rightarrow(x-y)(x+y)=0\Rightarrow y=\pm x$, so the set of discontinuities is
$\{(x,\pm x)|x\in\Re\}$.

23. $f(x,y)=\dfrac{x^2e^{xy}}{\ln e^x}$ is discontinuous only where it is not defined, which is where its

denominator is 0. As $\ln e^x=x$, f is discontinuous only at $x=0$, so the set of discontinuities
is $\{(0,y)|y\in\Re\}$.

25. $f(x,y,z)=\dfrac{zx-zy}{x-y}$ is discontinuous only where it is not defined, which is where its

denominator $x-y=0\Rightarrow y=x$, so the set of discontinuities is $\{(x,x,z)|x,z\in\Re\}$.

27.   For $x\ne y$, $f(x,y)=\dfrac{x^2-y^2}{x-y}=\dfrac{(x\not=y)(x+y)}{x\not=y}=x+y$, and for $x=y$, $f(x,y)=x+y$.
Therefore $f(x,y)=x+y$ for all x and y, and so f is continuous everywhere. The set of its
discontinuities is the emply set, $\emptyset$.

29.   There are many possibilities. Realizing that $\dfrac{1}{x^2+y^2}$ is only discontinuous where
it is not defined, which is where its denominator is zero, namely at $(0,0)$, the function
$f(x,y)=\dfrac{1}{x^2+(y-1)^2}$ is only discontinuous at $(0,1)$.

31.   There are many possibilities. Realizing that $\dfrac{1}{x^2+y^2+z^2}$ is only discontinuous where
it is not defined, which is where its denominator is zero, namely at $(0,0,0)$, the function
$f(x,y,z)=\dfrac{1}{x^2+(y-1)^2+z^2}$ is only discontinuous at $(0,1,0)$.

33.   Let $\epsilon>0$ be given. We seek $\delta>0$ such that
$$0<\|(x,y)-(x_0,y_0)\|<\delta\Rightarrow|(f-g)(x,y)-(L-M)|<\epsilon\quad(*)$$
Now, $|(f-g)(x,y)-(L-M)|=|f(x,y)-g(x,y)-L+M|=|[f(x,y)-L]-[g(x,y)-M]|$
and $|[f(x,y)-L]-[g(x,y)-M]|\le|f(x,y)-L|+|g(x,y)-M|$. Since the limit of f is L

and the limit of g is M, there exist $\delta_1 > 0$ and $\delta_2 > 0$ such that

$$0 < \|(x, y) - (x_0, y_0)\| < \delta_1 \;\Rightarrow\; |f(x, y) - L| < \tfrac{\epsilon}{2} \text{ and}$$
$$0 < \|(x, y) - (x_0, y_0)\| < \delta_2 \;\Rightarrow\; |g(x, y) - M| < \tfrac{\epsilon}{2}$$

It follows that for δ the smaller of δ_1 and δ_2, (*) holds.

35. To prove: $\displaystyle\lim_{(x,y)\to(x_0,y_0)} (f + g)(x, y) = (f + g)(x_0, y_0)$.

By definition, $(f + g)(x, y) = f(x, y) + g(x, y)$. Then,

$$\lim_{(x,y)\to(x_0,y_0)} (f + g)(x, y) = \lim_{(x,y)\to(x_0,y_0)} [f(x, y) + g(x, y)]$$

$$= \lim_{(x,y)\to(x_0,y_0)} f(x, y) + \lim_{(x,y)\to(x_0,y_0)} g(x, y) \text{ , by Theorem 11.1(a)}$$

$$= f(x_0, y_0) \quad + \quad g(x_0, y_0),$$

since f and g are continuous at (x_0, y_0),

$$= (f + g)(x_0, y_0), \text{ by definition.}$$

37. We are to show that: $\displaystyle\lim_{(x,y)\to(x_0,y_0)} \left(\frac{f}{g}\right)(x, y) = \left(\frac{f}{g}\right)(x_0, y_0), \text{ provided } g(x_0, y_0) \neq 0.$

By definition, $\left(\dfrac{f}{g}\right)(x, y) = \dfrac{f(x, y)}{g(x, y)}$. Then,

$$\lim_{(x,y)\to(x_0,y_0)} \left(\frac{f}{g}\right)(x, y) = \lim_{(x,y)\to(x_0,y_0)} \frac{f(x, y)}{g(x, y)} = \frac{\displaystyle\lim_{(x,y)\to(x_0,y_0)} f(x, y)}{\displaystyle\lim_{(x,y)\to(x_0,y_0)} g(x, y)}$$

by Theorem 11.1(d), provided $\displaystyle\lim_{(x,y)\to(x_0,y_0)} g(x, y) \neq 0$.

Since both f and g are continuous at (x_0, y_0),

$$\frac{\displaystyle\lim_{(x,y)\to(x_0,y_0)} f(x, y)}{\displaystyle\lim_{(x,y)\to(x_0,y_0)} g(x, y)} = \frac{f(x_0, y_0)}{g(x_0, y_0)} = \left(\frac{f}{g}\right)(x_0, y_0), \text{ by definition, and}$$

because g is continuous at (x_0, y_0), the condition $\displaystyle\lim_{(x,y)\to(x_0,y_0)} g(x, y) \neq 0$ is just $g(x_0, y_0) \neq 0$.

39. We are to show that if $f(x, y) = x$, then $\displaystyle\lim_{(x,y)\to(x_0,y_0)} f(x, y) - x_0$.

Define two functions, $a(x) = x$ and $b(y) = 1$. Each is a continuous function of one variable. By Theorem 11.4, their product $f(x, y) = a(x)b(y) = x \cdot 1 = x$ is a continuous function of two variables. Thus, $\displaystyle\lim_{(x,y)\to(x_0,y_0)} f(x, y) = f(x_0, y_0) = x_0$.

41. To prove: $\displaystyle\lim_{(x,y)\to(x_0,y_0)} c = c$. For given $\epsilon > 0$, we are to find a $\delta > 0$, such that:

$$0 < \|(x, y) - (x_0, y_0)\| < \delta \Rightarrow |c - c| < \epsilon. \text{ As } |c - c| = 0 < \epsilon \text{ always, any } \delta \text{ 'works.'}$$

§11.2 Graphing Functions of Two Variables

1. Find and plot the x-, y-, and z-intercepts: Set $y = z = 0$ in $z = -4x - 2y + 2 \Rightarrow x = \frac{1}{2}$; set $x = z = 0 \Rightarrow y = 1$; set $x = y = 0 \Rightarrow z = 2$. Then connect them as in the figure.

3. Since the three points $(0,0,0), (1,0,2)$, and $(0,2,5)$, lie on the plane $z = ax + by + c$, their coordinates satisfy the equation of the plane, leading us to a system of 3 equations in 3 unknowns, which are readily solved:

$$0 = a \cdot 0 + b \cdot 0 + c \Rightarrow c = 0$$
$$2 = a \cdot 1 + b \cdot 0 \Rightarrow a = 2$$
$$5 = a \cdot 0 + b \cdot 2 \Rightarrow b = \frac{5}{2}$$

Thus the equation of the plane containing the given points is $z = 2x + \frac{5}{2}y \Rightarrow 4x + 5y - 2z = 0$.

5. Since z does not occur in the equation, $x + 2y = 1$, this is the equation of the same line no matter what the value of z. This line has x-intercept 1 and y-intercept $\frac{1}{2}$. Just raise or lower this line to various values of z to see that the result is a plane parallel to the z axis.

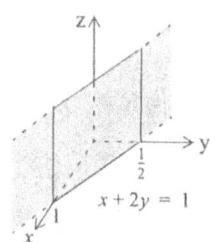

7. In the yz-plane, $z = -y^2$ is the equation of a parabola through the origin, opening downward. Repeat this all along the x-axis (as there is no x in the equation) and we have a parabolic cylinder, through the origin.

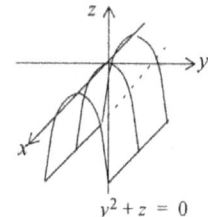

9. In the xy-plane, $x^2 - 2y^2 = 1$ is the equation of a hyperbola centered at the origin. When $y = 0, x = \pm 1$. As z is missing from the equation, the traces are all the same hyperbola at different levels of z. This is a hyperbolic cylinder.

11. The equation $z = x^2 + y^2$ is of the form of an elliptic paraboloid,

$\dfrac{z}{c} = \dfrac{x^2}{a^2} + \dfrac{y^2}{b^2}$, with $a = b = c = 1$. Since $a = b$, it is a circular

paraboloid. The traces z equals a positive constant are circles of
increasing radius as the constant increases.

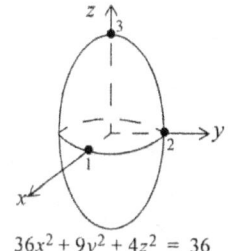

13. The equation $36x^2 + 9y^2 + 4z^2 = 36 \Rightarrow \dfrac{x^2}{1^2} + \dfrac{y^2}{2^2} + \dfrac{z^2}{3^2} = 1$

is of the form $\dfrac{x^2}{a^2} + \dfrac{y^2}{b^2} + \dfrac{z^2}{c^2} = 1$. Thus, this is an ellipsoid with

x–intercept, y–intercept, z–intercept, resp., 1,2,3.

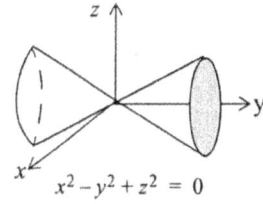

15. The equation $x^2 - y^2 + z^2 = 0 \Rightarrow \dfrac{y^2}{1^2} = \dfrac{x^2}{1^2} + \dfrac{z^2}{1^2}$

is of the form $\dfrac{y^2}{c^2} = \dfrac{x^2}{a^2} + \dfrac{z^2}{b^2}$. As $a = b$, this is a circular
cone. The traces y equals a constant are circles of increasing
radius as the constant increases in magnitude.

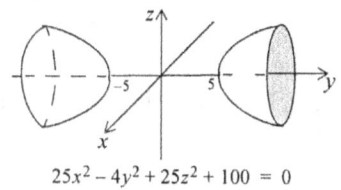

17. The equation $25x^2 - 4y^2 + 25z^2 + 100 = 0 \Rightarrow$

$-\dfrac{x^2}{2^2} + \dfrac{y^2}{5^2} - \dfrac{z^2}{2^2} = 1$ is of the form $-\dfrac{x^2}{a^2} + \dfrac{y^2}{b^2} - \dfrac{z^2}{c^2} = 1$.

It is a hyperboloid of two sheets. When $x = z = 0, y = \pm 5$.

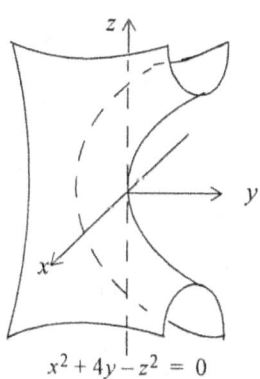

19. The equation $x^2 + 4y - z^2 = 0 \Rightarrow \dfrac{y}{-1} = \dfrac{x^2}{2^2} - \dfrac{z^2}{2^2}$ is

of the form $\dfrac{y}{c} = \dfrac{x^2}{a^2} - \dfrac{z^2}{b^2}$. It is a hyperbolic paraboloid.

When $x = 0$, $y = \frac{1}{4}z^2$ is a parabola in the yz-plane opening

in the positive y direction, and when $z = 0$, $y = -\frac{1}{4}x^2$

is a parabola in the xy-plane opening in the negative y
direction.

21. The equation $x^2 + y^2 - 4z^2 = 4 \Rightarrow \dfrac{x^2}{2^2} + \dfrac{y^2}{2^2} - \dfrac{z^2}{1^2} = 1$

is of the form $\dfrac{x^2}{a^2} + \dfrac{y^2}{b^2} - \dfrac{z^2}{c^2} = 1$. It is a hyperboloid of one sheet. The traces z equals a constant are circles of increasing radius as the magnitude of the constant increases.

$x^2 + y^2 - 4z^2 = 4$

23.   The equation $25x^2 - 4y^2 + 25z^2 = 100 \Rightarrow \dfrac{x^2}{2^2} - \dfrac{y^2}{5^2} + \dfrac{z^2}{2^2} = 1$ is of the form $\dfrac{x^2}{a^2} - \dfrac{y^2}{b^2} + \dfrac{z^2}{c^2} = 1$. It is a hyperboloid of one sheet. The traces y equals a constant are circles of increasing radius as the magnitude of the constant increases.

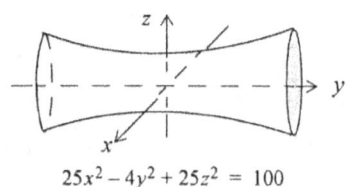

$25x^2 - 4y^2 + 25z^2 = 100$

25.   The equation $9x^2 - 4y^2 = 36z \Rightarrow \dfrac{z}{1} = \dfrac{x^2}{2^2} - \dfrac{y^2}{3^2}$ is of the form $\dfrac{z}{c} = \dfrac{x^2}{a^2} - \dfrac{y^2}{b^2}$. It is a hyperbolic paraboloid. When $y = 0$, $z = \frac{1}{4}x^2$ is a parabola in the xz-plane opening upward, and when $x = 0$, $z = -\frac{1}{9}y^2$ is a parabola in the yz-plane opening downward.

$9x^2 - 4y^2 = 36z$

§11.3 Double Integrals

1. (a) $V = \displaystyle\int_0^1 \int_1^2 xy^2\,dy\,dx = \int_0^1 \left[\tfrac{1}{3}xy^3\right]_{y=1}^{y=2} dx = \tfrac{1}{3}\int_0^1 x(8-1)\,dx = \tfrac{7}{3}\int_0^1 x\,dx = \tfrac{7}{3}\left[\tfrac{1}{2}x^2\right]_0^1$

   $= \tfrac{7}{6}(1-0) = \tfrac{7}{6}.$

   (b) $V = \displaystyle\int_1^2 \int_0^1 xy^2\,dx\,dy = \int_1^2 \left[\tfrac{1}{2}x^2y^2\right]_{x=0}^{x=1} dy = \tfrac{1}{2}\int_1^2 y^2(1-0)\,dy = \tfrac{1}{2}\left[\tfrac{1}{3}y^3\right]_1^2$

   $= \tfrac{1}{6}(2^3 - 1) = \tfrac{7}{6}.$

3. (a) $V = \displaystyle\int_1^2 \int_3^5 (y - 2x)\,dy\,dx = \int_1^2 \left[\tfrac{1}{2}y^2 - 2xy\right]_{y=3}^{y=5} dx = \int_1^2 \left[\left(\tfrac{25}{2} - 10x\right) - \left(\tfrac{9}{2} - 6x\right)\right] dx$

   $= \displaystyle\int_1^2 (8 - 4x)\,dx = \left[8x - 2x^2\right]_1^2 = 16 - 8 - (8 - 2) = 2.$

   (b) $V = \displaystyle\int_3^5 \int_1^2 (y - 2x)\,dx\,dy = \int_3^5 \left[xy - x^2\right]_{x=1}^{x=2} dy = \int_3^5 \left[(2y - 4) - (y - 1)\right] dy$

   $= \displaystyle\int_3^5 (y - 3)\,dy = \left[\tfrac{1}{2}y^2 - 3y\right]_3^5 = \tfrac{25}{2} - 15 - \left(\tfrac{9}{2} - 9\right) = 8 - 6 = 2.$

5. (a) $V = \int_0^{\pi/2} \int_0^{\pi/2} (\sin x + \cos y) dy dx = \int_0^{\pi/2} \Big[y \sin x + \sin y \Big]_{y=0}^{y=\pi/2} dx$

$= \int_0^{\pi/2} \left(\frac{\pi}{2} \sin x + \sin \frac{\pi}{2} \right) dx = \int_0^{\pi/2} \left(\frac{\pi}{2} \sin x + 1 \right) dx = \left[-\frac{\pi}{2} \cos x + x \right]_0^{\pi/2}$

$= -\frac{\pi}{2} \cos \frac{\pi}{2} + \frac{\pi}{2} - \left(-\frac{\pi}{2} \cos 0 + 0 \right) = \frac{\pi}{2} + \frac{\pi}{2} = \pi.$

(b) $V = \int_0^{\pi/2} \int_0^{\pi/2} (\sin x + \cos y) dx dy = \int_0^{\pi/2} \Big[-\cos x + x \cos y \Big]_{x=0}^{x=\pi/2} dy$

$= \int_0^{\pi/2} \left[-\cos \frac{\pi}{2} + \frac{\pi}{2} \cos y - (-\cos 0 + 0) \right] dy = \int_0^{\pi/2} \left(\frac{\pi}{2} \cos y + 1 \right) dy$

$= \left[\frac{\pi}{2} \sin y + y \right]_0^{\pi/2} = \frac{\pi}{2} \sin \frac{\pi}{2} + \frac{\pi}{2} = \frac{\pi}{2} + \frac{\pi}{2} = \pi.$

7. (a)

Intersection of the two curves: $x^2 = 2x \Rightarrow x(x - 2) = 0 \Rightarrow x = 0, 2$ (see figure).

$V = \int_0^2 \int_{x^2}^{2x} 3xy^2 dy dx = \int_0^2 \Big[xy^3 \Big]_{y=x^2}^{y=2x} dx = \int_0^2 \Big[x(2x)^3 - x(x^2)^3 \Big] dx$

$= \int_0^2 \left(8x^4 - x^7 \right) dx = \left[\frac{8}{5} x^5 - \frac{1}{8} x^8 \right]_0^2 = \frac{8(32)}{5} - 32 = \frac{96}{5}.$

(b)

$V = \int_0^4 \int_{y/2}^{\sqrt{y}} 3xy^2 dx dy = \int_0^4 \Big[\frac{3}{2} x^2 y^2 \Big]_{x=y/2}^{x=\sqrt{y}} dy = \frac{3}{2} \int_0^4 \left(y^3 - \frac{1}{4} y^4 \right) dy$

$= \frac{3}{2} \left[\frac{1}{4} y^4 - \frac{1}{20} y^5 \right]_0^4 = \frac{3}{2} \left[64 - \frac{16(16)}{5} \right] = 96 - \frac{3(8)(16)}{5} = \frac{480 - 384}{5}$

$= \frac{96}{5}.$

9.

$V = \int_0^1 \int_0^{1-x} (1 - x - y) dy dx = \int_0^1 \Big[y - xy - \frac{1}{2} y^2 \Big]_{y=0}^{y=1-x} dx$

$= \int_0^1 \Big[(1 - x) - x(1 - x) - \frac{1}{2}(1 - x)^2 \Big] dx$

$= \int_0^1 \left(1 - x - x + x^2 - \frac{1}{2} + x - \frac{1}{2} x^2 \right) dx = \int_0^1 \left(\frac{1}{2} - x + \frac{1}{2} x^2 \right) dx$

$= \left[\frac{1}{2} x - \frac{1}{2} x^2 + \frac{1}{6} x^3 \right]_0^1 = \frac{1}{2} - \frac{1}{2} + \frac{1}{6} = \frac{1}{6}.$

11.

Equation of the line joining $(1,2)$ to $(4,1)$: Slope $m = \frac{2 - 1}{1 - 4} = -\frac{1}{3} \Rightarrow y - 2 = -\frac{1}{3}(x - 1) \Rightarrow$ $y = -\frac{1}{3} x + \frac{7}{3}$. (see figure).

$V = \int_1^4 \int_1^{-\frac{1}{3}x+\frac{7}{3}} xy \, dy dx = \int_1^4 \Big[\frac{1}{2} xy^2 \Big]_{y=1}^{y=-\frac{1}{3}x+\frac{7}{3}} dx = \frac{1}{2} \int_1^4 \left[x \left(-\frac{1}{3} x + \frac{7}{3} \right)^2 - x \right] dx$

$= \frac{1}{2} \int_1^4 \left(\frac{x^3}{9} - \frac{14}{9} x^2 + \frac{49}{9} x - x \right) dx = \frac{1}{18} \int_1^4 \left(x^3 - 14x^2 + 40x \right) dx$

$= \frac{1}{18} \left[\frac{x^4}{4} - \frac{14x^3}{3} + 20x^2 \right]_1^4 = \frac{1}{18} \left[64 - \frac{14}{3}(64) + 20(16) - \left(\frac{1}{4} - \frac{14}{3} + 20 \right) \right] = \frac{1}{18} \left(\frac{279}{4} \right) = \frac{31}{8}.$

13.

$$V = \int_{-1}^{1} \int_{0}^{1-y^2} (2x + y + 2)\, dx\, dy = \int_{-1}^{1} \left[x^2 + xy + 2x \right]_{x=0}^{x=1-y^2} dy$$

$$= \int_{-1}^{1} \left[(1 - y^2)^2 + y(1 - y^2) + 2(1 - y^2) \right] dy$$

$$= \int_{-1}^{1} \left(1 - 2y^2 + y^4 + y - y^3 + 2 - 2y^2 \right) dy = \int_{-1}^{1} \left(3 - 4y^2 + y^4 - y^3 + y \right) dy$$

$$= \left[3y - \tfrac{4}{3}y^3 + \tfrac{1}{5}y^5 - \tfrac{1}{4}y^4 + \tfrac{1}{2}y^2 \right]_{-1}^{1} = 3 - \tfrac{4}{3} + \tfrac{1}{5} - \tfrac{1}{4} + \tfrac{1}{2} - \left(-3 + \tfrac{4}{3} - \tfrac{1}{5} - \tfrac{1}{4} + \tfrac{1}{2} \right)$$

$$= 6 - \tfrac{8}{3} + \tfrac{2}{5} = \tfrac{90 - 40 + 6}{15} = \tfrac{56}{15}.$$

15.
$$\int_{-1}^{1} \int_{-x^2}^{x^2} (x^2 - y)\, dy\, dx = \int_{-1}^{1} \left[x^2 y - \tfrac{1}{2}y^2 \right]_{y=-x^2}^{y=x^2} dx = \int_{-1}^{1} \left[x^4 - \tfrac{1}{2}x^4 - \left(-x^4 - \tfrac{1}{2}x^4 \right) \right] dx$$

$$= \int_{-1}^{1} 2x^4\, dx = \left[\tfrac{2}{5}x^5 \right]_{-1}^{1} = \tfrac{2}{5}[1 - (-1)] = \tfrac{4}{5}.$$

17.
$$\int_{0}^{1} \int_{-1}^{y} (xy - y^3)\, dx\, dy = \int_{0}^{1} \left[\tfrac{1}{2}x^2 y - xy^3 \right]_{x=-1}^{x=y} dy = \int_{0}^{1} \left[\tfrac{1}{2}y^3 - y^4 - \left(\tfrac{1}{2}y + y^3 \right) \right] dy$$

$$= \int_{0}^{1} \left(-\tfrac{1}{2}y^3 - y^4 - \tfrac{1}{2}y \right) dy = \left[-\tfrac{1}{8}y^4 - \tfrac{1}{5}y^5 - \tfrac{1}{4}y^2 \right]_{0}^{1}$$

$$= -\tfrac{1}{8} - \tfrac{1}{5} - \tfrac{1}{4} = \tfrac{-5 - 8 - 10}{40} = -\tfrac{23}{40}.$$

19.
$$\int_{0}^{2} \int_{0}^{\sqrt{4-x^2}} (x + 2y)\, dy\, dx = \int_{0}^{2} \left[xy + y^2 \right]_{y=0}^{y=\sqrt{4-x^2}} dx = \int_{0}^{2} x\sqrt{4 - x^2}\, dx + \int_{0}^{2} (4 - x^2)\, dx$$

$$= -\tfrac{1}{2} \int_{4}^{0} u^{\frac{1}{2}}\, du + \left[4x - \tfrac{1}{3}x^3 \right]_{0}^{2} \qquad \boxed{\begin{array}{l|l} u = 4 - x^2 & x = 0 \Rightarrow u = 4 \\ du = -2x\, dx & x = 2 \Rightarrow u = 0 \end{array}}$$

$$= -\tfrac{1}{2} \left[\tfrac{2}{3}u^{\frac{3}{2}} \right]_{4}^{0} + 8 - \tfrac{8}{3} = -\tfrac{1}{3}(0 - 8) + \tfrac{16}{3} = 8.$$

21.
$$\int_{0}^{2} \int_{0}^{x/2} e^{x^2}\, dy\, dx = \int_{0}^{2} \left[y e^{x^2} \right]_{y=0}^{y=x/2} dx = \int_{0}^{2} \tfrac{1}{2}x e^{x^2}\, dx = \tfrac{1}{4} \int_{1}^{e^4} du$$

$$= \tfrac{1}{4} \left[u \right]_{1}^{e^4} = \tfrac{1}{4}(e^4 - 1). \qquad \boxed{\begin{array}{l|l} u = e^{x^2} & x = 0 \Rightarrow u = 1 \\ du = 2x e^{x^2}\, dx & x = 2 \Rightarrow u = e^4 \end{array}}$$

23.
$$\int_{1}^{2} \int_{y}^{\sqrt{2y}} xy^2\, dx\, dy = \int_{1}^{2} \left[\tfrac{1}{2}x^2 y^2 \right]_{x=y}^{x=\sqrt{2y}} dy = \int_{1}^{2} \left(y^3 - \tfrac{1}{2}y^4 \right) dy = \left[\tfrac{1}{4}y^4 - \tfrac{1}{10}y^5 \right]_{1}^{2}$$

$$= 4 - \tfrac{1}{10}(32) - \left(\tfrac{1}{4} - \tfrac{1}{10} \right) = \tfrac{80 - 64 - 5 + 2}{20} = \tfrac{13}{20}.$$

25. $\displaystyle\int_0^{\sqrt{\pi}}\int_0^{x^2} x\sin y\, dy dx = \int_0^{\sqrt{\pi}}\Big[-x\cos y\Big]_{y=0}^{y=x^2} dx = \int_0^{\sqrt{\pi}}\Big[-x\cos x^2 - (-x\cos 0)\Big]\, dx$

$$= -\int_0^{\sqrt{\pi}} x\cos x^2\, dx + \Big[\tfrac{1}{2}x^2\Big]_0^{\sqrt{\pi}} = -\tfrac{1}{2}\int_0^{\pi}\cos u\, du + \tfrac{\pi}{2}$$

$u = x^2$	$x = 0 \Rightarrow u = 0$
$du = 2x dx$	$x = \sqrt{\pi} \Rightarrow u = \pi$

$$= -\tfrac{1}{2}\Big[\sin u\Big]_0^{\pi} + \tfrac{\pi}{2} = -\tfrac{1}{2}(\sin\pi - \sin 0) + \tfrac{\pi}{2} = \tfrac{\pi}{2}.$$

27. 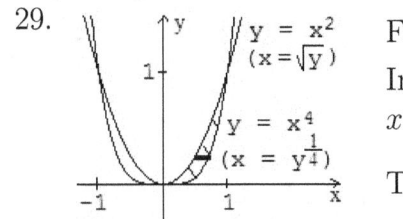 From the given double integral: $0 \le y \le 1$, $-3 \le x \le 2$.

Then: $\displaystyle\int_{-3}^{2}\int_0^1 h(x,y)\, dy dx = \int_0^1\int_{-3}^2 h(x,y)\, dx dy.$

29. 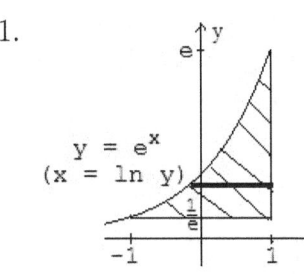 From the given double integral: $x^4 \le y \le x^2$, $0 \le x \le 1$.

Intersection of the two curves: $x^2 = x^4 \Rightarrow x^2(x^2 - 1) = 0 \Rightarrow x = 0, \pm 1.$

Then: $\displaystyle\int_0^1\int_{x^4}^{x^2} h(x,y)\, dy dx = \int_0^1\int_{\sqrt{y}}^{y^{\frac{1}{4}}} h(x,y)\, dx dy.$

31. 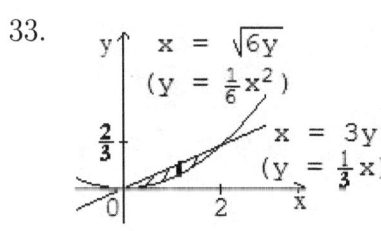 From the given double integral: $1/e \le y \le e^x$, $-1 \le x \le 1$.

Then: $\displaystyle\int_{-1}^1\int_{1/e}^{e^x} h(x,y)\, dy dx = \int_{1/e}^{e}\int_{\ln y}^{1} h(x,y)\, dx dy.$

33. From the given double integral: $3y \le x \le \sqrt{6y}$, $0 \le y \le \tfrac{2}{3}$.

Intersection of the two curves: $3y = \sqrt{6y} \Rightarrow 9y^2 = 6y \Rightarrow$

$3y^2 - 2y = 0 \Rightarrow y(3y-2) = 0 \Rightarrow y = 0, \tfrac{2}{3} \Rightarrow x = 3y = 0, 2,$ respectively.

Then: $\displaystyle\int_0^{2/3}\int_{3y}^{\sqrt{6y}} h(x,y)\, dx dy = \int_0^2\int_{x^2/6}^{x/3} h(x,y)\, dy dx.$

35. From the given double integral:

$$-\sqrt{9-y^2} \le x \le \sqrt{9-y^2}, \quad 0 \le y \le 3.$$

Then: $\displaystyle\int_0^3\int_{-\sqrt{9-y^2}}^{\sqrt{9-y^2}} h(x,y)\, dx dy = \int_{-3}^3\int_0^{\sqrt{9-x^2}} h(x,y)\, dy dx.$

37. From the given double integral:

$1 - y \leq x \leq 1 + y$, and $0 \leq y \leq 1$.

Then:

$$\int_0^1 \int_{1-y}^{1+y} h(x,y)\,dx\,dy = \int_0^1 \int_{-x+1}^1 h(x,y)\,dy\,dx + \int_1^2 \int_{x-1}^1 h(x,y)\,dy\,dx$$

39. From the given double integral: $\dfrac{y}{2} \leq x \leq 2, \quad 0 \leq y \leq 4$.

Then:

$$\int_0^4 \int_{y/2}^2 e^{x^2}\,dx\,dy = \int_0^2 \int_0^{2x} e^{x^2}\,dy\,dx = \int_0^2 \left[y e^{x^2} \right]_{y=0}^{y=2x} dx = \int_0^2 2x e^{x^2}\,dx$$

$$= \int_0^4 e^u\,du = \left[e^u \right]_0^4 = e^4 - e^0$$

$$= e^4 - 1.$$

$$\boxed{\begin{aligned} u &= x^2 \\ du &= 2x\,dx \\ x = 0 &\Rightarrow u = 0 \\ x = 2 &\Rightarrow u = 4 \end{aligned}}$$

41. From the given double integral: $x \leq y \leq 1, \quad 0 \leq x \leq 1$.

Then:

$$\int_0^1 \int_x^1 \sin y^2\,dy\,dx = \int_0^1 \int_0^y \sin y^2\,dx\,dy = \int_0^1 \left[x \sin y^2 \right]_{x=0}^{x=y} dy = \int_0^1 y \sin y^2\,dy$$

$$= \frac{1}{2} \int_0^1 \sin u\,du = \frac{1}{2} \left[-\cos u \right]_0^1$$

$$= -\frac{1}{2}(\cos 1 - \cos 0)$$

$$= \frac{1}{2}(1 - \cos 1).$$

$$\boxed{\begin{aligned} u &= y^2 \\ du &= 2y\,dy \\ y = 0 &\Rightarrow u = 0 \\ y = 1 &\Rightarrow u = 1 \end{aligned}}$$

43.

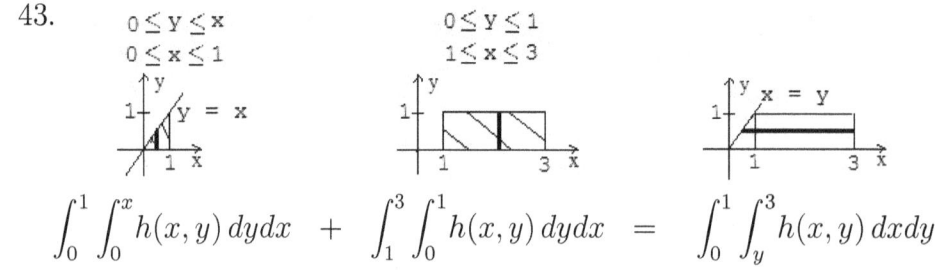

$$\int_0^1 \int_0^x h(x,y)\,dy\,dx \;+\; \int_1^3 \int_0^1 h(x,y)\,dy\,dx \;=\; \int_0^1 \int_y^3 h(x,y)\,dx\,dy.$$

45.

$$\int_1^2 \int_1^y h(x,y)\,dxdy \;+\; \int_2^4 \int_{y/2}^y h(x,y)\,dxdy \;+\; \int_4^8 \int_{y/2}^4 h(x,y)\,dxdy \;=\; \int_1^4 \int_x^{2x} h(x,y)\,dydx.$$

47.

$$M = \iint_R xy\,dA = \int_0^1 \int_0^{x^3} xy\,dydx = \int_0^1 \left[\tfrac{1}{2}xy^2\right]_{y=0}^{y=x^3} dx = \tfrac{1}{2}\int_0^1 x^7\,dx$$

$$= \tfrac{1}{2}\left[\tfrac{1}{8}x^8\right]_0^1 = \tfrac{1}{16}.$$

49.

$$M = \iint_R (6x+6y+6)\,dA = \int_0^1 \int_{2x}^{x+1} (6x+6y+6)\,dydx$$

$$= \int_0^1 \left[6xy + 3y^2 + 6y\right]_{y=2x}^{y=x+1} dx$$

$$= \int_0^1 \left[6x(x+1) + 3(x+1)^2 + 6(x+1) - (12x^2 + 12x^2 + 12x)\right] dx$$

$$= \int_0^1 \left(6x^2 + 6x + 3x^2 + 6x + 3 + 6x + 6 - 24x^2 - 12x\right) dx$$

$$= \int_0^1 (-15x^2 + 6x + 9)dx = \left[-5x^3 + 3x^2 + 9x\right]_0^1 = -5 + 3 + 9 = 7.$$

51.

Intersection of the two curves:

$$2x - 4 = \sqrt{x^2 - 4} \;\Rightarrow\; 4x^2 - 16x + 16 = x^2 - 4 \;\Rightarrow$$

$$3x^2 - 16x + 20 = 0 \;\Rightarrow\; (3x-10)(x-2) = 0 \;\Rightarrow\; x = \tfrac{10}{3},\; 2$$

(see figure).

$$M = \iint_R 4x\,dA = \int_2^{10/3}\int_{2x-4}^{\sqrt{x^2-4}} 4x\,dydx = \int_2^{10/3}\left[4xy\right]_{y=2x-4}^{y=\sqrt{x^2-4}} dx = \int_2^{10/3} 4x\left[\sqrt{x^2-4} - 2x + 4\right] dx$$

$$= \int_2^{10/3} 4x\sqrt{x^2-4}\,dx + \int_2^{10/3}(-8x^2 + 16x)\,dx = \int_0^{64/9} 2u^{\frac{1}{2}}du + \left[-\tfrac{8}{3}x^3 + 8x^2\right]_2^{10/3}$$

$$\boxed{\begin{array}{l} u = x^2 - 4 \\ du = 2x\,dx \\ x = 2 \Rightarrow u = 0 \\ x = \tfrac{10}{3} \Rightarrow u = \tfrac{64}{9} \end{array}} \qquad \begin{array}{l} = \left[\tfrac{4}{3}u^{\frac{3}{2}}\right]_0^{64/9} - \tfrac{8}{3}\left(\tfrac{10}{3}\right)^3 + 8\left(\tfrac{10}{3}\right)^2 - \left(-\tfrac{64}{3} + 32\right) \\[2mm] = \tfrac{4}{3}\left(\tfrac{8}{3}\right)^3 - \tfrac{8000}{81} + \tfrac{800}{9} - \tfrac{32}{3} = \tfrac{2048 - 8000 + 7200 - 864}{81} \\[2mm] = \tfrac{384}{81} = \tfrac{128}{27} \end{array}$$

Alternatively, integrating first wrt x then wrt y:

$$M = \int_0^{8/3} \int_{\sqrt{4+y^2}}^{(y+4)/2} 4x\, dxdy = \int_0^{8/3} \left[2x^2\right]_{x=\sqrt{4+y^2}}^{x=(y+4)/2} dy$$

$$= 2\int_0^{8/3} \left[\tfrac{1}{4}(y+4)^2 - (4+y^2)\right] dy$$

$$= 2\int_0^{8/3} \left(\tfrac{1}{4}y^2 + 2y + 4 - 4 - y^2\right) dy = 2\int_0^{8/3} \left(-\tfrac{3}{4}y^2 + 2y\right) dy$$

$$= 2\left[-\tfrac{1}{4}y^3 + y^2\right]_0^{8/3} = 2\left[-\tfrac{1}{4}\left(\tfrac{512}{27}\right) + \tfrac{64}{9}\right] = 2\left(\tfrac{-128+192}{27}\right) = \tfrac{128}{27}.$$

53.

$$M = \iint_R |\cos x|\, dA = \int_0^\pi \int_0^{\sin x} |\cos x|\, dydx$$

$$= \int_0^{\pi/2} \int_0^{\sin x} \cos x\, dydx + \int_{\pi/2}^{\pi} \int_0^{\sin x} [-\cos x]\, dydx$$

$$= \int_0^{\pi/2} \left[y\cos x\right]_{y=0}^{y=\sin x} dx - \int_{\pi/2}^{\pi} \left[y\cos x\right]_{y=0}^{y=\sin x} dx$$

$$= \int_0^{\pi/2} \sin x \cos x\, dx - \int_{\pi/2}^{\pi} \sin x \cos x\, dx$$

$$= \left[\tfrac{1}{2}\sin^2 x\right]_0^{\pi/2} - \left[\tfrac{1}{2}\sin^2 x\right]_{\pi/2}^{\pi} = \tfrac{1}{2}\sin^2\tfrac{\pi}{2} + \tfrac{1}{2}\sin^2\tfrac{\pi}{2}$$

$$= \tfrac{1}{2} + \tfrac{1}{2} = 1.$$

55.

$$M = \iint_R (6x + 6y + 6)\, dA = \int_0^3 \int_0^2 (6x + 6y + 6)\, dydx$$

$$= \int_0^3 \left[6xy + 3y^2 + 6y\right]_{y=0}^{y=2} dx = \int_0^3 (12x + 12 + 12)dx$$

$$= \int_0^3 (12x + 24)dx = \left[6x^2 + 24x\right]_0^3 = 54 + 72 = 126$$

$$M_y = \int_0^3 \int_0^2 x(6x + 6y + 6)\, dydx = \int_0^3 \int_0^2 (6x^2 + 6xy + 6x)\, dydx$$

$$= \int_0^3 \left[6x^2y + 3xy^2 + 6xy\right]_{y=0}^{y=2} dx = \int_0^3 (12x^2 + 12x + 12x)dx$$

$$= \int_0^3 (12x^2 + 24x)dx = \left[4x^3 + 12x^2\right]_0^3 = 4(27) + 12(9) = 108 + 108 = 216$$

Thus $\bar{x} = \dfrac{M_y}{M} = \dfrac{216}{126} = \dfrac{12}{7}$

$$M_x = \int_0^3 \int_0^2 y(6x + 6y + 6)\, dydx = \int_0^3 \int_0^2 (6xy + 6y^2 + 6y)\, dydx$$

$$= \int_0^3 \left[3xy^2 + 2y^3 + 3y^2\right]_{y=0}^{y=2} dx = \int_0^3 (12x + 16 + 12)dx$$

$$= \int_0^3 (12x + 28)dx = \left[6x^2 + 28x\right]_0^3 = 54 + 84 = 138$$

Thus $\bar{y} = \dfrac{M_x}{M} = \dfrac{138}{126} = \dfrac{23}{21}$ $\Rightarrow$ $(\bar{x}, \bar{y}) = \left(\dfrac{12}{7}, \dfrac{23}{21}\right).$

57.

$$M = \iint\limits_R (x+y)\,dA = \int_0^2 \int_{x/2}^{-x+3} (x+y)\,dy\,dx = \int_0^2 \left[xy + \tfrac{1}{2}y^2\right]_{y=x/2}^{y=-x+3} dx$$

$$= \int_0^2 \left[x(-x+3) + \tfrac{1}{2}(-x+3)^2 - (\tfrac{1}{2}x^2 + \tfrac{1}{8}x^2)\right] dx$$

$$= \int_0^2 \left(-x^2 + 3x + \tfrac{1}{2}x^2 - 3x + \tfrac{9}{2} - \tfrac{5}{8}x^2\right) dx = \int_0^2 \left(-\tfrac{9}{8}x^2 + \tfrac{9}{2}\right) dx$$

$$= \left[-\tfrac{3}{8}x^3 + \tfrac{9}{2}x\right]_0^2 = -3 + 9 = 6$$

$$M_y = \int_0^2 \int_{x/2}^{-x+3} x(x+y)\,dy\,dx = \int_0^2 \int_{x/2}^{-x+3} (x^2+xy)\,dy\,dx = \int_0^2 \left[x^2 y + \tfrac{1}{2}xy^2\right]_{y=x/2}^{y=-x+3} dx$$

$$= \int_0^2 \left[x^2(-x+3) + \tfrac{1}{2}x(-x+3)^2 - (\tfrac{1}{2}x^3 + \tfrac{1}{8}x^3)\right] dx = \int_0^2 \left(-\tfrac{9}{8}x^3 + \tfrac{9}{2}x\right) dx$$

$$= \left[-\tfrac{9}{32}x^4 + \tfrac{9}{4}x^2\right]_0^2 = -\tfrac{9}{2} + 9 = \tfrac{9}{2}$$

Thus, $\bar{x} = \dfrac{M_y}{M} = \dfrac{\tfrac{9}{2}}{6} = \dfrac{9}{12} = \dfrac{3}{4}$

$$M_x = \int_0^2 \int_{x/2}^{-x+3} y(x+y)\,dy\,dx = \int_0^2 \int_{x/2}^{-x+3} (xy+y^2)\,dy\,dx = \int_0^2 \left[\tfrac{1}{2}xy^2 + \tfrac{1}{3}y^3\right]_{y=x/2}^{y=-x+3} dx$$

$$= \int_0^2 \left[\tfrac{1}{2}x(-x+3)^2 + \tfrac{1}{3}(-x+3)^3 - (\tfrac{1}{8}x^3 + \tfrac{1}{24}x^3)\right] dx$$

$$= \int_0^2 \left(\tfrac{1}{2}x^3 - 3x^2 + \tfrac{9}{2}x - \tfrac{1}{3}x^3 + 3x^2 - 9x + 9 - \tfrac{1}{6}x^3\right) dx = \int_0^2 \left(-\tfrac{9}{2}x + 9\right) dx$$

$$= \left[-\tfrac{9}{4}x^2 + 9x\right]_0^2 = -9 + 18 = 9$$

Therefore, $\bar{y} = \dfrac{M_x}{M} = \dfrac{9}{6} = \dfrac{3}{2} \Rightarrow (\bar{x}, \bar{y}) = \left(\tfrac{3}{4}, \tfrac{3}{2}\right)$.

59.

$$M = \iint\limits_R (x+y)\,dA = \int_0^1 \int_0^{\sqrt{x}} (x+y)\,dy\,dx = \int_0^1 \left[xy + \tfrac{1}{2}y^2\right]_{y=0}^{y=\sqrt{x}} dx$$

$$= \int_0^1 \left(x^{3/2} + \tfrac{1}{2}x\right) dx = \left[\tfrac{2}{5}x^{5/2} + \tfrac{1}{4}x^2\right]_0^1 = \tfrac{2}{5} + \tfrac{1}{4} = \tfrac{13}{20}$$

$$M_y = \int_0^1 \int_0^{\sqrt{x}} x(x+y)\,dy\,dx = \int_0^1 \int_0^{\sqrt{x}} (x^2+xy)\,dy\,dx = \int_0^1 \left[x^2 y + \tfrac{1}{2}xy^2\right]_{y=0}^{y=\sqrt{x}} dx$$

$$= \int_0^1 \left(x^{5/2} + \tfrac{1}{2}x^2\right) dx = \left[\tfrac{2}{7}x^{7/2} + \tfrac{1}{6}x^3\right]_0^1 = \tfrac{2}{7} + \tfrac{1}{6} = \tfrac{19}{42}$$

Therefore, $\bar{x} = \dfrac{M_y}{M} = \dfrac{\tfrac{19}{42}}{\tfrac{13}{20}} = \dfrac{190}{273}$

$$M_x = \int_0^1 \int_0^{\sqrt{x}} y(x+y)\,dy\,dx = \int_0^1 \int_0^{\sqrt{x}} (xy+y^2)\,dy\,dx = \int_0^1 \left[\tfrac{1}{2}xy^2 + \tfrac{1}{3}y^3\right]_{y=0}^{y=\sqrt{x}} dx$$

$$= \int_0^1 \left(\tfrac{1}{2}x^2 + \tfrac{1}{3}x^{3/2}\right) dx = \left[\tfrac{1}{6}x^3 + \tfrac{2}{15}x^{5/2}\right]_0^1 = \tfrac{1}{6} + \tfrac{2}{15} = \tfrac{9}{30} = \tfrac{3}{10}$$

Therefore, $\bar{y} = \dfrac{M_x}{M} = \dfrac{\tfrac{3}{10}}{\tfrac{13}{20}} = \dfrac{6}{13} \Rightarrow (\bar{x}, \bar{y}) = \left(\dfrac{190}{273}, \dfrac{6}{13}\right)$.

61.

$$M = \int\int_R y\, dA = \int_0^1 \int_0^{e^x} y\, dy dx = \int_0^1 \left[\tfrac{1}{2}y^2\right]_{y=0}^{y=e^x} dx = \tfrac{1}{2}\int_0^1 e^{2x}\, dx = \left[\tfrac{1}{4}e^{2x}\right]_0^1$$

$$= \tfrac{1}{4}(e^2 - e^0) = \tfrac{1}{4}(e^2 - 1)$$

$$M_y = \int_0^1 \int_0^{e^x} xy\, dy dx = \int_0^1 \left[\tfrac{1}{2}xy^2\right]_{y=0}^{y=e^x} dx = \tfrac{1}{2}\int_0^1 xe^{2x}\, dx = \tfrac{1}{2}\left\{\left[\tfrac{1}{2}xe^{2x}\right]_0^1 - \tfrac{1}{2}\int_0^1 e^{2x}\, dx\right\}$$

$$\boxed{\begin{array}{ll} u = x & dv = e^{2x}dx \\ du = dx & v = \tfrac{1}{2}e^{2x} \end{array}}$$

$$= \tfrac{1}{2}\left\{\tfrac{1}{2}e^2 - \left[\tfrac{1}{4}e^{2x}\right]_0^1\right\} = \tfrac{1}{4}e^2 - \tfrac{1}{8}(e^2 - 1)$$

$$= \tfrac{1}{4}e^2 - \tfrac{1}{8}e^2 + \tfrac{1}{8} = \tfrac{1}{8}(e^2 + 1)$$

Thus $\overline{x} = \dfrac{M_y}{M} = \dfrac{\tfrac{1}{8}(e^2 + 1)}{\tfrac{1}{4}(e^2 - 1)} = \dfrac{e^2 + 1}{2(e^2 - 1)}$

$$M_x = \int_0^1 \int_0^{e^x} y^2\, dy dx = \int_0^1 \left[\tfrac{1}{3}y^3\right]_{y=0}^{y=e^x} dx = \tfrac{1}{3}\int_0^1 e^{3x}\, dx = \tfrac{1}{9}\left[e^{3x}\right]_0^1 = \tfrac{1}{9}(e^3 - 1)$$

Thus $\overline{y} = \dfrac{M_x}{M} = \dfrac{\tfrac{1}{9}(e^3 - 1)}{\tfrac{1}{4}(e^2 - 1)} = \dfrac{4(e^3 - 1)}{9(e^2 - 1)} = \dfrac{4(e-1)(e^2 + e + 1)}{9(e-1)(e + 1)} = \dfrac{4(e^2 + e + 1)}{9(e + 1)} \Rightarrow$

$$(\overline{x}, \overline{y}) = \left(\dfrac{e^2 + 1}{2(e^2 - 1)}, \dfrac{4(e^2 + e + 1)}{9(e + 1)}\right).$$

63.

$$M = \int\int_R 8y\, dA = \int_0^\pi \int_0^{\sin x} 8y\, dy dx = \int_0^\pi \left[4y^2\right]_{y=0}^{y=\sin x} dx = 4\int_0^\pi \sin^2 x\, dx$$

$$= 2\int_0^\pi (1 - \cos 2x)dx = 2\left[x - \tfrac{1}{2}\sin 2x\right]_0^\pi = 2\left(\pi - \tfrac{1}{2}\sin 2\pi\right) = 2\pi$$

$$M_y = \int_0^\pi \int_0^{\sin x} 8xy\, dy dx = \int_0^\pi \left[4xy^2\right]_{y=0}^{y=\sin x} dx = 4\int_0^\pi x\sin^2 x\, dx = 2\int_0^\pi x(1 - \cos 2x)\, dx$$

$$= 2\left\{\left[x\left(x - \tfrac{1}{2}\sin 2x\right)\right]_0^\pi - \int_0^\pi \left(x - \tfrac{1}{2}\sin 2x\right)\, dx\right\}$$

$$\boxed{\begin{array}{ll} u = x & dv = (1 - \cos 2x)\, dx \\ du = dx & v = x - \tfrac{1}{2}\sin 2x \end{array}}$$

$$= 2\left\{\pi^2 - 0 - \left[\tfrac{1}{2}x^2 + \tfrac{1}{4}\cos 2x\right]_0^\pi\right\}$$

$$= 2\left[\pi^2 - \left(\tfrac{1}{2}\pi^2 + \tfrac{1}{4}\cos 2\pi\right) + \tfrac{1}{4}\cos 0\right] = 2\left(\tfrac{1}{2}\pi^2\right) = \pi^2$$

Thus $\overline{x} = \dfrac{M_y}{M} = \dfrac{\pi^2}{2\pi} = \dfrac{\pi}{2}$

$$M_x = \int_0^\pi \int_0^{\sin x} 8y^2\, dy dx = \tfrac{8}{3}\int_0^\pi \left[y^3\right]_{y=0}^{y=\sin x} dx = \tfrac{8}{3}\int_0^\pi \sin^3 x\, dx = \tfrac{8}{3}\int_0^\pi (1 - \cos^2 x)\sin x\, dx$$

$$= \tfrac{8}{3}\int_0^\pi (\sin x - \cos^2 x \sin x)dx = \tfrac{8}{3}\left[-\cos x + \tfrac{1}{3}\cos^3 x\right]_0^\pi$$

$$= \tfrac{8}{3}\left[-\cos \pi + \tfrac{1}{3}\cos^3 \pi - \left(-\cos 0 + \tfrac{1}{3}\cos^3 0\right)\right] = \tfrac{8}{3}\left(1 - \tfrac{1}{3} + 1 - \tfrac{1}{3}\right) = \tfrac{8}{3}\cdot\tfrac{4}{3} = \tfrac{32}{9}$$

Thus $\overline{y} = \dfrac{M_x}{M} = \dfrac{\tfrac{32}{9}}{2\pi} = \dfrac{16}{9\pi} \Rightarrow (\overline{x}, \overline{y}) = \left(\dfrac{\pi}{2}, \dfrac{16}{9\pi}\right).$

§11.4 Double Integrals in Polar Coordinates

1. $\displaystyle\int_0^{\pi/4}\int_0^{\sqrt{4\cos 2\theta}} r\,dr\,d\theta = \frac{1}{2}\int_0^{\pi/4}\Big[r^2\Big]_{r=0}^{r=\sqrt{4\cos 2\theta}}\,d\theta = \frac{1}{2}\int_0^{\pi/4} 4\cos 2\theta\,d\theta = \Big[\sin 2\theta\Big]_0^{\pi/4}$

$$= \sin\frac{\pi}{2} = 1.$$

3. $\displaystyle\int_{-\pi/4}^{\pi/4}\int_0^{\cos 2\theta} r\,dr\,d\theta = \frac{1}{2}\int_{-\pi/4}^{\pi/4}\Big[r^2\Big]_{r=0}^{r=\cos 2\theta}\,d\theta = \frac{1}{2}\int_{-\pi/4}^{\pi/4}\cos^2 2\theta\,d\theta = \frac{1}{2}\int_{-\pi/4}^{\pi/4}\left(\frac{1+\cos 4\theta}{2}\right)d\theta$

$$= \frac{1}{4}\Big[\theta + \frac{1}{4}\sin 4\theta\Big]_{-\pi/4}^{\pi/4} = \frac{1}{4}\Big[\frac{\pi}{4} - \big(-\frac{\pi}{4}\big)\Big] = \frac{\pi}{8}.$$

5. $\displaystyle\int_0^{\pi}\int_0^1 e^{r^2} r\,dr\,d\theta = \int_0^{\pi}\int_0^1 \frac{1}{2}e^u\,du\,d\theta = \frac{1}{2}\int_0^{\pi}\Big[e^u\Big]_{u=0}^{u=1}\,d\theta = \frac{1}{2}\int_0^{\pi}(e-1)\,d\theta = \frac{(e-1)\pi}{2}.$

$$\boxed{\begin{array}{ll} u = r^2, & r = 0 \Rightarrow u = 0 \\ du = 2r\,dr, & r = 1 \Rightarrow u = 1 \end{array}}$$

7. $\displaystyle\int_0^{\pi/2}\int_0^{\sin\theta} r\cos\theta\,dr\,d\theta = \frac{1}{2}\int_0^{\pi/2}\Big[r^2\cos\theta\Big]_{r=0}^{r=\sin\theta}\,d\theta = \frac{1}{2}\int_0^{\pi/2}\sin^2\theta\cos\theta\,d\theta = \frac{1}{2}\Big[\frac{1}{3}\sin^3\theta\Big]_0^{\pi/2}$

$$= \frac{1}{6}\sin^3\frac{\pi}{2} = \frac{1}{6}.$$

9. $\displaystyle\int_0^{\pi/6}\int_{\sin\theta}^{\sec\theta} r\,dr\,d\theta = \frac{1}{2}\int_0^{\pi/6}\Big[r^2\Big]_{r=\sin\theta}^{r=\sec\theta}\,d\theta = \frac{1}{2}\int_0^{\pi/6}(\sec^2\theta - \sin^2\theta)d\theta$

$$= \frac{1}{2}\Big[\tan\theta\Big]_0^{\pi/6} - \frac{1}{4}\int_0^{\pi/6}(1-\cos 2\theta)d\theta = \frac{1}{2}\tan\frac{\pi}{6} - \frac{1}{4}\Big[\theta - \frac{1}{2}\sin 2\theta\Big]_0^{\pi/6}$$

$$= \frac{1}{2}\cdot\frac{1}{\sqrt{3}} - \frac{1}{4}\cdot\frac{\pi}{6} + \frac{1}{8}\sin\frac{\pi}{3} = \frac{\sqrt{3}}{6} - \frac{\pi}{24} + \frac{\sqrt{3}}{16} = \frac{8\sqrt{3} - 2\pi + 3\sqrt{3}}{48}$$

$$= \frac{11\sqrt{3} - 2\pi}{48}.$$

11. $\displaystyle\int_0^{2\pi}\int_0^1 r(1-r^2)^{\frac{3}{2}}\,dr\,d\theta = \int_0^{2\pi}\int_1^0 -\frac{1}{2}u^{\frac{3}{2}}\,du\,d\theta = -\frac{1}{2}\int_0^{2\pi}\Big[\frac{2}{5}u^{\frac{5}{2}}\Big]_{u=1}^{u=0}\,d\theta = -\frac{1}{5}\int_0^{2\pi}(0-1)d\theta$

$$\boxed{\begin{array}{ll} u = 1-r^2, & r = 0 \Rightarrow u = 1 \\ du = -2r\,dr, & r = 1 \Rightarrow u = 0 \end{array}} \qquad = \frac{1}{5}(2\pi) = \frac{2\pi}{5}.$$

13. $y = \sqrt{2x - x^2} \Rightarrow y \geq 0$ and $y^2 = 2x - x^2 \Rightarrow x^2 - 2x + 1 + y^2 = 1$
$\Rightarrow (x-1)^2 + y^2 = 1$ which is the top half of a circle of radius 1
centered at $(1,0)$ (see figure at right).

Change to polar coordinates: $x = r\cos\theta$, $y = r\sin\theta \Rightarrow x^2 + y^2 = r^2$,
then, $y = \sqrt{2x - x^2} \Rightarrow y^2 = 2x - x^2 \Rightarrow$
$$r^2 \sin^2\theta = 2r\cos\theta - r^2\cos^2\theta$$
$$r\sin^2\theta = 2\cos\theta - r\cos^2\theta$$
$$r(\sin^2\theta + \cos^2\theta) = 2\cos\theta$$
$$r = 2\cos\theta$$
(see figure at right).

Thus:
$$\int_0^2 \int_0^{\sqrt{2x-x^2}} \sqrt{x^2 + y^2}\, dy\, dx = \int_0^{\pi/2} \int_0^{2\cos\theta} r \cdot r\, dr\, d\theta = \int_0^{\pi/2} \left[\tfrac{1}{3}r^3\right]_{r=0}^{r=2\cos\theta} d\theta$$
$$= \int_0^{\pi/2} \tfrac{1}{3}(2\cos\theta)^3 d\theta = \tfrac{8}{3}\int_0^{\pi/2}(1 - \sin^2\theta)\cos\theta\, d\theta$$
$$= \tfrac{8}{3}\left[\sin\theta - \tfrac{1}{3}\sin^3\theta\right]_0^{\pi/2} = \tfrac{8}{3}(1 - \tfrac{1}{3}) = \tfrac{16}{9}.$$

15. The region over which the integration occurs is defined by $0 \leq x \leq y$
and $0 \leq y \leq 6$ (see figure at right).

Change to polar coordinates: $x = y \Rightarrow r\cos\theta = r\sin\theta \Rightarrow \tan\theta = 1$
$\Rightarrow \theta = \frac{\pi}{4}$, and $y = 6 \Rightarrow r\sin\theta = 6 \Rightarrow r = \frac{6}{\sin\theta}$ (see figure at right).

Thus:
$$\int_0^6 \int_0^y x\, dx\, dy = \int_{\pi/4}^{\pi/2} \int_0^{\frac{6}{\sin\theta}} (r\cos\theta)\, r\, dr\, d\theta = \int_{\pi/4}^{\pi/2} \tfrac{1}{3}\left[r^3\cos\theta\right]_{r=0}^{r=\frac{6}{\sin\theta}} d\theta$$
$$= \tfrac{1}{3}\int_{\pi/4}^{\pi/2} \left(\tfrac{6}{\sin\theta}\right)^3 \cos\theta\, d\theta = 72\int_{\pi/4}^{\pi/2} (\sin\theta)^{-3}\cos\theta\, d\theta = 72\left[\frac{(\sin\theta)^{-2}}{-2}\right]_{\pi/4}^{\pi/2}$$
$$= -36\left(\frac{1}{\sin^2\frac{\pi}{2}} - \frac{1}{\sin^2\frac{\pi}{4}}\right) = -36(1 - 2) = 36.$$

17. The region over which the integration occurs is defined by
$-\sqrt{1 - x^2} \leq y \leq 0$ and $-1 \leq x \leq 0$. As $y = -\sqrt{1 - x^2} \Rightarrow y \leq 0$ and
$y^2 = 1 - x^2 \Rightarrow x^2 + y^2 = 1$, the region is the third quadrant piece of
the unit disk (see figure at right).

Change to polar coordinates: $x^2 + y^2 = r^2 = 1 \Rightarrow r = 1$, and we see
that $\pi \leq \theta \leq \frac{3\pi}{2}$ (see figure at right).

Thus:

$$\int_{-1}^{0}\int_{-\sqrt{1-x^2}}^{0}\frac{2}{1+\sqrt{x^2+y^2}}\,dy\,dx = \int_{\pi}^{3\pi/2}\int_{0}^{1}\frac{2}{1+r}\,r\,dr\,d\theta = 2\int_{\pi}^{3\pi/2}\int_{1}^{2}\frac{u-1}{u}\,du\,d\theta$$

$$\boxed{\begin{array}{ll} u = 1+r, & r = 0 \Rightarrow u = 1 \\ du = dr, & r = 1 \Rightarrow u = 2 \end{array}}$$

$$= 2\int_{\pi}^{3\pi/2}\Big[u-\ln|u|\Big]_{u=1}^{u=2}\,d\theta = 2\int_{\pi}^{3\pi/2}(2-\ln 2-1)\,d\theta$$

$$= 2(1-\ln 2)\left(\frac{3\pi}{2}-\pi\right) = \pi(1-\ln 2).$$

19.   In polar coordinates, $x^2+y^2=r^2$ and R is the unit circle $r=1$ with $0\le\theta\le 2\pi$. Therefore

$$\iint_R e^{-(x^2+y^2)}\,dA = \int_0^{2\pi}\int_0^1 e^{-r^2}r\,dr\,d\theta = -\frac{1}{2}\int_0^{2\pi}\int_0^{-1}e^u\,du\,d\theta = -\frac{1}{2}\int_0^{2\pi}\Big[e^u\Big]_{u=0}^{u=-1}\,d\theta$$

$$\boxed{\begin{array}{ll} u = -r^2, & r = 0 \Rightarrow u = 0 \\ du = -2r\,dr, & r = 1 \Rightarrow u = -1 \end{array}}$$

$$= -\frac{1}{2}\int_0^{2\pi}(e^{-1}-e^0)\,d\theta = -\frac{1}{2}\left(\frac{1}{e}-1\right)\cdot 2\pi = \pi\left(1-\frac{1}{e}\right).$$

21. In polar coordinates, $x^2+y^2=r^2$ and R is the first quadrant portion of the unit circle $r=1$, so that $0\le\theta\le\frac{\pi}{2}$. Therefore

$$\iint_R \sqrt{9-x^2-y^2}\,dA = \int_0^{\pi/2}\int_0^1 \sqrt{9-r^2}\,r\,dr\,d\theta = -\frac{1}{2}\int_0^{\pi/2}\int_9^8 u^{\frac{1}{2}}\,du\,d\theta$$

$$\boxed{\begin{array}{ll} u = 9-r^2, & r = 0 \Rightarrow u = 9 \\ du = -2r\,dr, & r = 1 \Rightarrow u = 8 \end{array}}$$

$$= -\frac{1}{2}\int_0^{\pi/2}\Big[\frac{2}{3}u^{\frac{3}{2}}\Big]_{u=9}^{u=8}\,d\theta = -\frac{1}{3}\int_0^{\pi/2}\left(8^{\frac{3}{2}}-9^{\frac{3}{2}}\right)\,d\theta$$

$$= -\frac{1}{3}\left(8^{\frac{3}{2}}-9^{\frac{3}{2}}\right)\cdot\frac{\pi}{2} = -\frac{\pi}{6}\Big[8\left(2\sqrt{2}\right)-27\Big]$$

$$= \frac{\pi}{6}\left(27-16\sqrt{2}\right).$$

23. $R: x^2+y^2-y=0 \Rightarrow x^2+y^2-y+\left(\frac{1}{2}\right)^2 = \left(\frac{1}{2}\right)^2 \Rightarrow x^2+\left(y-\frac{1}{2}\right)^2 = \left(\frac{1}{2}\right)^2$, which is a circle of radius $\frac{1}{2}$ centered at $\left(0,\frac{1}{2}\right)$. Its polar equation is: $x^2+y^2=y \Rightarrow r^2=r\sin\theta \Rightarrow r=\sin\theta$. The circle is traversed once as θ varies from 0 to π. Therefore

$$\iint_R y\,dA = \int_0^\pi\int_0^{\sin\theta} r\sin\theta\,r\,dr\,d\theta = \int_0^\pi\Big[\frac{1}{3}r^3\sin\theta\Big]_{r=0}^{r=\sin\theta}\,d\theta = \frac{1}{3}\int_0^\pi\sin^4\theta\,d\theta = \frac{1}{3}\int_0^\pi(\sin^2\theta)^2\,d\theta$$

$$= \frac{1}{3}\int_0^\pi\left(\frac{1-\cos 2\theta}{2}\right)^2\,d\theta = \frac{1}{12}\int_0^\pi(1-2\cos 2\theta+\cos^2 2\theta)\,d\theta$$

$$= \frac{1}{12}\Big[\theta-\sin 2\theta\Big]_0^\pi + \frac{1}{12}\int_0^\pi\left(\frac{1+\cos 4\theta}{2}\right)\,d\theta = \frac{\pi}{12} + \frac{1}{24}\Big[\theta+\frac{1}{4}\sin 4\theta\Big]_0^\pi$$

$$= \frac{\pi}{12} + \frac{\pi}{24} = \frac{3\pi}{24} = \frac{\pi}{8}.$$

25. (a) (b) (c)

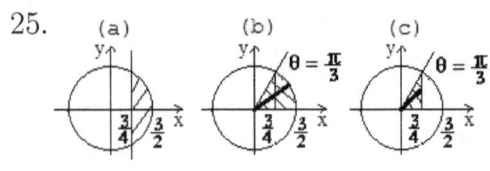

The polar equation of the vertical line is given: $r = \dfrac{3}{4\cos\theta}$. We find the intersection of the line and the circle: $\dfrac{3}{2} = \dfrac{3}{4\cos\theta} \Rightarrow 4\cos\theta = 2 \Rightarrow \cos\theta = \dfrac{1}{2} \Rightarrow \theta = \dfrac{\pi}{3}$.

By symmetry, the shaded area in figure (a) is twice the shaded area in the first quadrant. This area is the area in the whole sector, figure (b), minus the area shown in figure (c). Therefore,

$$A = 2\left[\int_0^{\pi/3}\int_0^{\frac{3}{2}} r\,dr\,d\theta - \int_0^{\pi/3}\int_0^{\frac{3}{4\cos\theta}} r\,dr\,d\theta\right] = 2\left\{\int_0^{\pi/3}\left[\tfrac{1}{2}r^2\right]_{r=0}^{r=\frac{3}{2}} d\theta - \int_0^{\pi/3}\left[\tfrac{1}{2}r^2\right]_{r=0}^{r=\frac{3}{4\cos\theta}} d\theta\right\}$$

$$= \left(\tfrac{3}{2}\right)^2\int_0^{\pi/3} d\theta - \int_0^{\pi/3}\left(\tfrac{3}{4\cos\theta}\right)^2 d\theta = \left(\tfrac{9}{4}\cdot\tfrac{\pi}{3}\right) - \tfrac{9}{16}\int_0^{\pi/3}\sec^2\theta\,d\theta = \tfrac{3\pi}{4} - \tfrac{9}{16}\Big[\tan\theta\Big]_0^{\pi/3}$$

$$= \tfrac{3\pi}{4} - \tfrac{9}{16}\cdot\tan\tfrac{\pi}{3} = \tfrac{3\pi}{4} - \tfrac{9\sqrt{3}}{16} = \tfrac{12\pi - 9\sqrt{3}}{16}.$$

27. (a) (b)

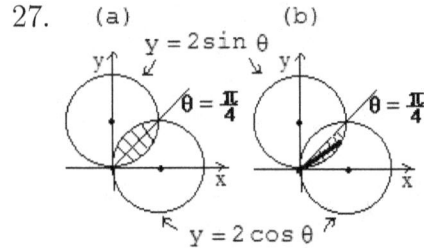

The intersection of the two circles: $2\cos\theta = 2\sin\theta \Rightarrow \tan\theta = 1 \Rightarrow \theta = \tfrac{\pi}{4}$. By symmetry, the area of the region common to the two circles, shown in figure (a), is twice the area of the lower half, shown in figure (b). This region lies entirely inside the circle $r = 2\sin\theta$. Therefore:

$$A = 2\int_0^{\pi/4}\int_0^{2\sin\theta} r\,dr\,d\theta = 2\int_0^{\pi/4}\left[\tfrac{1}{2}r^2\right]_{r=0}^{r=2\sin\theta} d\theta = 4\int_0^{\pi/4}\sin^2\theta\,d\theta = 2\int_0^{\pi/4}(1-\cos 2\theta)\,d\theta$$

$$= 2\left[\theta - \tfrac{1}{2}\sin 2\theta\right]_0^{\pi/4} = 2\left(\tfrac{\pi}{4} - \tfrac{1}{2}\sin\tfrac{\pi}{2}\right) = \tfrac{\pi}{2} - 1.$$

29. (a) (b) (c)

The intersection of the two circles: $3\cos\theta = \tfrac{3}{2} \Rightarrow \cos\theta = \tfrac{1}{2} \Rightarrow \theta = \tfrac{\pi}{3}$. By symmetry, the area of the shaded region in figure (a) is twice the area of the shaded part in the first quadrant. This area is the difference of the area inside the circle $r = 3\cos\theta$, figure (b), and the area inside $r = \tfrac{3}{2}$, figure (c), from $\theta = 0$ to $\theta = \tfrac{\pi}{3}$. Therefore:

$$A = 2\left[\int_0^{\pi/3}\int_0^{3\cos\theta} r\,dr\,d\theta - \int_0^{\pi/3}\int_0^{3/2} r\,dr\,d\theta\right] = 2\left(\int_0^{\pi/3}\left[\tfrac{1}{2}r^2\right]_{r=0}^{r=3\cos\theta} d\theta - \int_0^{\pi/3}\left[\tfrac{1}{2}r^2\right]_{r=0}^{r=3/2} d\theta\right)$$

$$= \int_0^{\pi/3}\left(9\cos^2\theta - \tfrac{9}{4}\right) d\theta = 9\int_0^{\pi/3}\left(\tfrac{1+\cos 2\theta}{2} - \tfrac{1}{4}\right) d\theta = 9\int_0^{\pi/3}\left(\tfrac{1}{4} + \tfrac{1}{2}\cos 2\theta\right) d\theta$$

$$= 9\left[\tfrac{1}{4}\theta + \tfrac{1}{4}\sin 2\theta\right]_0^{\pi/3} = \tfrac{9}{4}\left(\tfrac{\pi}{3} + \sin\tfrac{2\pi}{3}\right) = \tfrac{9}{4}\left(\tfrac{\pi}{3} + \tfrac{\sqrt{3}}{2}\right) = \tfrac{9}{4}\left(\tfrac{2\pi + 3\sqrt{3}}{6}\right) = \tfrac{3}{8}\left(2\pi + 3\sqrt{3}\right).$$

31.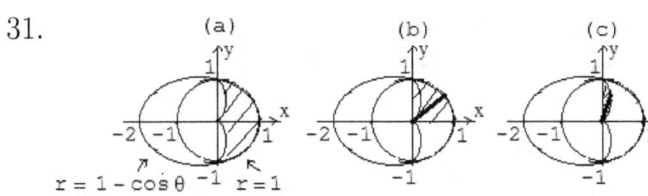

$r = 1 - \cos\theta$ $r = 1$

By symmetry, the area of the shaded region in figure (a) is twice the area of the shaded part in the first quadrant. This area is the difference of the area inside the circle $r = 1$, figure (b), and the area inside $r = 1 - \cos\theta$, figure (c), from $\theta = 0$ to $\theta = \frac{\pi}{2}$. Therefore:

$$A = 2\left[\int_0^{\pi/2}\int_0^1 r\,dr d\theta - \int_0^{\pi/2}\int_0^{1-\cos\theta} r\,dr d\theta\right] = 2\left(\int_0^{\pi/2}\left[\tfrac{1}{2}r^2\right]_{r=0}^{r=1} d\theta - \int_0^{\pi/2}\left[\tfrac{1}{2}r^2\right]_{r=0}^{r=1-\cos\theta} d\theta\right)$$

$$= \int_0^{\pi/2}\left[1 - (1-\cos\theta)^2\right] d\theta = \int_0^{\pi/2}\left(2\cos\theta - \cos^2\theta\right) d\theta = \int_0^{\pi/2}\left[2\cos\theta - \left(\frac{1+\cos 2\theta}{2}\right)\right] d\theta$$

$$= \left[2\sin\theta - \left(\tfrac{1}{2}\theta + \tfrac{1}{4}\sin 2\theta\right)\right]_0^{\pi/2} = 2\sin\frac{\pi}{2} - \frac{1}{2}\cdot\frac{\pi}{2} = 2 - \frac{\pi}{4}.$$

33.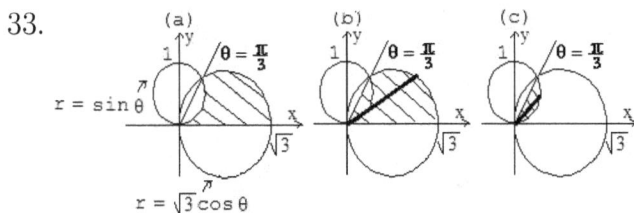

$r = \sin\theta$ $r = \sqrt{3}\cos\theta$

The intersection of the two circles: $\sqrt{3}\cos\theta = \sin\theta \Rightarrow \tan\theta = \sqrt{3} \Rightarrow \theta = \frac{\pi}{3}$. The area is the difference of the area inside the circle $r = \sqrt{3}\cos\theta$, figure (b), and the area inside $r = \sin\theta$, figure (c), from $\theta = 0$ to $\theta = \frac{\pi}{3}$. Therefore:

$$A = \int_0^{\pi/3}\int_0^{\sqrt{3}\cos\theta} r\,dr d\theta - \int_0^{\pi/3}\int_0^{\sin\theta} r\,dr d\theta = \int_0^{\pi/3}\left[\tfrac{1}{2}r^2\right]_{r=0}^{r=\sqrt{3}\cos\theta} d\theta - \int_0^{\pi/3}\left[\tfrac{1}{2}r^2\right]_{r=0}^{r=\sin\theta} d\theta$$

$$= \frac{1}{2}\int_0^{\pi/3}\left(3\cos^2\theta - \sin^2\theta\right) d\theta = \frac{1}{2}\int_0^{\pi/3}\left(3\cdot\frac{1+\cos 2\theta}{2} - \frac{1-\cos 2\theta}{2}\right) d\theta$$

$$= \frac{1}{4}\int_0^{\pi/3}\left(2 + 4\cos 2\theta\right) d\theta = \frac{1}{2}\int_0^{\pi/3}\left(1 + 2\cos 2\theta\right) d\theta = \frac{1}{2}\left[\theta + \sin 2\theta\right]_0^{\pi/3} = \frac{1}{2}\left(\frac{\pi}{3} + \sin\frac{2\pi}{3}\right)$$

$$= \frac{1}{2}\left(\frac{\pi}{3} + \frac{\sqrt{3}}{2}\right) = \frac{2\pi + 3\sqrt{3}}{12}.$$

35. Consider the sphere of radius a centered at the origin: $x^2+y^2+z^2 = a^2$. By symmetry, the volume is twice the volume above the xy-plane, where $z = \sqrt{a^2 - (x^2 + y^2)}$, which in polar coordinates is $z = \sqrt{a^2 - r^2}$. The region of integration is the disk of radius a centered at the origin, in the xy-plane (see figure).

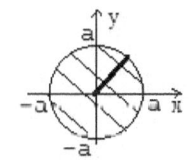

$$V = 2\int_0^{2\pi}\int_0^a \sqrt{a^2 - r^2}\,r\,dr d\theta = -\int_0^{2\pi}\int_{a^2}^0 u^{\frac{1}{2}}\,du d\theta = -\int_0^{2\pi}\left[\frac{2}{3}u^{\frac{3}{2}}\right]_{u=a^2}^{u=0} d\theta$$

$$\boxed{\begin{array}{l} u = a^2 - r^2, \quad r = 0 \Rightarrow u = a^2 \\ du = -2r\,dr, \quad r = a \Rightarrow u = 0 \end{array}}$$

$$= \frac{2}{3}(a^3)\int_0^{2\pi} d\theta = \frac{4}{3}\pi a^3.$$

37.   The region of integration is the interior of the circle, in polar coordinates: $x^2 + y^2 = 4 \Rightarrow r = 2$ (see figure). The equation of the cone in polar coordinates: $z = \sqrt{x^2 + y^2} \Rightarrow z = r$.

$$V = \int_0^{2\pi} \int_0^2 r \cdot r\,dr\,d\theta = \int_0^{2\pi} \left[\tfrac{1}{3}r^3\right]_{r=0}^{r=2} d\theta = \tfrac{8}{3}\int_0^{2\pi} d\theta = \tfrac{16\pi}{3}.$$

39.   The projection of the region onto the xy-plane is the interior of the circle $x^2 + y^2 = 2x \Rightarrow x^2 - 2x + 1 + y^2 = 1 \Rightarrow (x-1)^2 + y^2 = 1$, and its equation in polar coordinates is: $r^2 = 2r\cos\theta \Rightarrow r = 2\cos\theta$ (see figure). The circle is traversed once as θ varies from 0 to π. The equation of the paraboloid in polar coordinates: $z = x^2 + y^2 \Rightarrow z = r^2$. Thus,

$$V = \int_0^{\pi} \int_0^{2\cos\theta} r^2 \cdot r\,dr\,d\theta = \int_0^{\pi} \left[\tfrac{1}{4}r^4\right]_{r=0}^{r=2\cos\theta} d\theta = \tfrac{1}{4}\int_0^{\pi}(2\cos\theta)^4\,d\theta = 4\int_0^{\pi}\cos^4\theta\,d\theta$$

$$= 4\int_0^{\pi}\left(\frac{1+\cos 2\theta}{2}\right)^2 d\theta = \int_0^{\pi}(1 + 2\cos 2\theta + \cos^2 2\theta)d\theta$$

$$= \left[\theta + \sin 2\theta\right]_0^{\pi} + \int_0^{\pi}\left(\frac{1+\cos 4\theta}{2}\right)d\theta = \pi + \left[\tfrac{1}{2}\theta + \tfrac{1}{8}\sin 4\theta\right]_0^{\pi} = \pi + \tfrac{\pi}{2} = \tfrac{3\pi}{2}.$$

41.   The equation of the upper hemisphere of the sphere whose equation is $x^2 + y^2 + z^2 = 4$, is $z = \sqrt{4 - (x^2 + y^2)}$, which in polar coordinates is $z = \sqrt{4 - r^2}$. The region of integration is the interior of the unit circle $x^2 + y^2 = 1$, in polar coordinates: $r = 1$ (see figure).

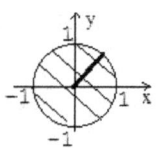

$$V = \int_0^{2\pi} \int_0^1 \sqrt{4 - r^2}\,r\,dr\,d\theta = -\tfrac{1}{2}\int_0^{2\pi}\int_4^3 u^{\frac{1}{2}}\,du\,d\theta = -\tfrac{1}{2}\int_0^{2\pi}\left[\tfrac{2}{3}u^{\frac{3}{2}}\right]_{u=4}^{u=3} d\theta$$

$$\boxed{\begin{array}{ll} u = 4 - r^2, & r = 0 \Rightarrow u = 4 \\ du = -2r\,dr, & r = 1 \Rightarrow u = 3 \end{array}}$$

$$= -\tfrac{1}{3}\left(3\sqrt{3} - 8\right)\int_0^{2\pi} d\theta = \tfrac{2\pi}{3}\left(8 - 3\sqrt{3}\right).$$

43.   The equation of the ellipsoid is $4x^2 + 4y^2 + z^2 = 64$. By symmetry, the volume is twice the volume above the xy-plane, where $z = \sqrt{64 - (4x^2 + 4y^2)} = 2\sqrt{16 - (x^2 + y^2)}$, which in polar coordinates is $z = 2\sqrt{16 - r^2}$. The region of integration is the disk of radius 2 centered at the origin in the xy-plane (see figure).

$$V = 2\int_0^{2\pi}\int_0^2 2\sqrt{16 - r^2}\,r\,dr\,d\theta = -2\int_0^{2\pi}\int_{16}^{12} u^{\frac{1}{2}}\,du\,d\theta = -2\int_0^{2\pi}\left[\tfrac{2}{3}u^{\frac{3}{2}}\right]_{u=16}^{u=12} d\theta$$

$$\boxed{\begin{array}{ll} u = 16 - r^2, & r = 0 \Rightarrow u = 16 \\ du = -2r\,dr, & r = 2 \Rightarrow u = 12 \end{array}}$$

$$= -\tfrac{4}{3}\left(12^{\frac{3}{2}} - 64\right)\int_0^{2\pi} d\theta = \tfrac{8\pi}{3}\left(64 - 24\sqrt{3}\right) = \tfrac{64\pi}{3}\left(8 - 3\sqrt{3}\right).$$

45.

$$M = \int_0^{2\pi} \int_0^{1+\sin\theta} r \cdot r\,dr\,d\theta = \int_0^{2\pi} \left[\tfrac{1}{3}r^3\right]_{r=0}^{r=1+\sin\theta} d\theta = \tfrac{1}{3}\int_0^{2\pi} (1+\sin\theta)^3\,d\theta$$

$$= \tfrac{1}{3}\int_0^{2\pi} (1 + 3\sin\theta + 3\sin^2\theta + \sin^3\theta)\,d\theta$$

$$= \tfrac{1}{3}\int_0^{2\pi} \left[1 + 3\sin\theta + 3\cdot\frac{1-\cos 2\theta}{2} + (1-\cos^2\theta)\sin\theta\right] d\theta$$

$$= \tfrac{1}{3}\int_0^{2\pi} \left(\tfrac{5}{2} + 4\sin\theta - \tfrac{3}{2}\cos 2\theta - \cos^2\theta\sin\theta\right) d\theta$$

$$= \tfrac{1}{3}\left[\tfrac{5}{2}\theta - 4\cos\theta - \tfrac{3}{4}\sin 2\theta + \tfrac{1}{3}\cos^3\theta\right]_0^{2\pi}$$

$$= \tfrac{1}{3}\left[\tfrac{5}{2}\cdot 2\pi - 4 + \tfrac{1}{3} - \left(-4 + \tfrac{1}{3}\right)\right] = \tfrac{5\pi}{3}.$$

47.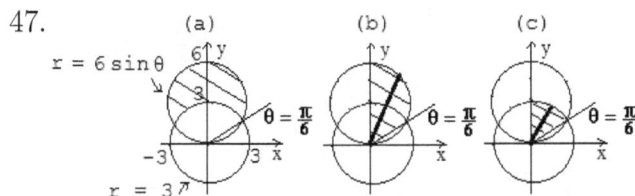

(a) (b) (c)

The intersection of the two circles, figure (a): $3 = 6\sin\theta \Rightarrow \sin\theta = \tfrac{1}{2} \Rightarrow \theta = \tfrac{\pi}{6}$. By the symmetry of the region and the density function, the mass is twice the mass in the first quadrant. The integration over that region is the difference of the integration over the circle $r = 6\sin\theta$, figure (b), and that over the circle $r = 3$, figure (c), from $\theta = \tfrac{\pi}{6}$ to $\theta = \tfrac{\pi}{2}$. Therefore:

$$M = 2\left[\int_{\pi/6}^{\pi/2}\int_0^{6\sin\theta} \tfrac{1}{r}\cdot r\,dr\,d\theta - \int_{\pi/6}^{\pi/2}\int_0^{3} \tfrac{1}{r}\cdot r\,dr\,d\theta\right] = 2\left[\int_{\pi/6}^{\pi/2}(6\sin\theta - 3)\,d\theta\right]$$

$$= 2\left[-6\cos\theta - 3\theta\right]_{\pi/6}^{\pi/2} = 2\left[-6\cos\tfrac{\pi}{2} - 3\cdot\tfrac{\pi}{2} - \left(-6\cos\tfrac{\pi}{6} - 3\cdot\tfrac{\pi}{6}\right)\right] = 2(-\pi + 3\sqrt{3}).$$

49.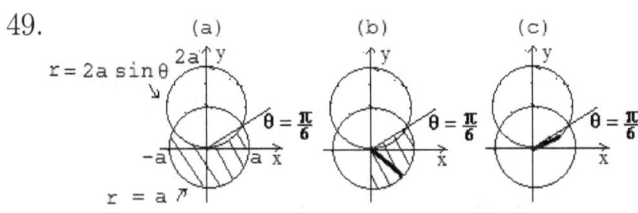

(a) (b) (c)

The intersection of the two circles, figure (a): $a = 2a\sin\theta \Rightarrow \sin\theta = \tfrac{1}{2} \Rightarrow \theta = \tfrac{\pi}{6}$. By the symmetry of the region and the density function, the mass is twice the mass for $x > 0$. The integration over that region is the difference of the integration over the circle $r = a$, from $\theta = -\tfrac{\pi}{2}$ to $\theta = \tfrac{\pi}{6}$, figure (b), and that over the circle $r = 2a\sin\theta$, from $\theta = 0$ to $\theta = \tfrac{\pi}{6}$, figure (c). Therefore:

$$M = 2\left[\int_{-\pi/2}^{\pi/6}\int_0^{a} \tfrac{1}{r}\cdot r\,dr\,d\theta - \int_0^{\pi/6}\int_0^{2a\sin\theta} \tfrac{1}{r}\cdot r\,dr\,d\theta\right] = 2\left[\int_{-\pi/2}^{\pi/6} a\,d\theta - \int_0^{\pi/6} 2a\sin\theta\,d\theta\right]$$

$$= 2a\left\{\tfrac{\pi}{6} - \left(-\tfrac{\pi}{2}\right) + \left[2\cos\theta\right]_0^{\pi/6}\right\} = 2a\left[\tfrac{\pi}{6} + \tfrac{\pi}{2} + 2\left(\cos\tfrac{\pi}{6} - \cos 0\right)\right]$$

$$= 2a\left[\tfrac{2\pi}{3} + \sqrt{3} - 2\right].$$

51. From Exercise #45, $M = \frac{5\pi}{3}$, and

$$M_y = \int_0^{2\pi} \int_0^{1+\sin\theta} (r\cos\theta) r \cdot r \, dr d\theta = \int_0^{2\pi} \left[\tfrac{1}{4} r^4 \cos\theta\right]_{r=0}^{r=1+\sin\theta} d\theta = \tfrac{1}{4} \int_0^{2\pi} (1+\sin\theta)^4 \cos\theta d\theta$$

$$= \tfrac{1}{4} \left[\tfrac{1}{5}(1+\sin\theta)^5\right]_0^{2\pi} = 0 \;\Rightarrow\; \bar{x} = \frac{M_y}{M} = \frac{0}{\frac{5\pi}{3}} = 0$$

(Also clear from the symmetry, that $\bar{x} = 0$.)

$$M_x = \int_0^{2\pi} \int_0^{1+\sin\theta} (r\sin\theta) r \cdot r \, dr d\theta = \int_0^{2\pi} \left[\tfrac{1}{4} r^4 \sin\theta\right]_{r=0}^{r=1+\sin\theta} d\theta = \tfrac{1}{4} \int_0^{2\pi} (1+\sin\theta)^4 \sin\theta d\theta$$

$$= \tfrac{1}{4} \int_0^{2\pi} \left(\sin\theta + 4\sin^2\theta + 6\sin^3\theta + 4\sin^4\theta + \sin^5\theta\right) d\theta$$

$$= \tfrac{1}{4} \int_0^{2\pi} \left[\sin\theta + 4\left(\frac{1-\cos 2\theta}{2}\right) + 6(1-\cos^2\theta)\sin\theta \right.$$
$$\left. + 4\left(\frac{1-\cos 2\theta}{2}\right)^2 + (1-\cos^2\theta)^2 \sin\theta\right] d\theta$$

$$= \tfrac{1}{4} \int_0^{2\pi} (\sin\theta + 2 - 2\cos 2\theta + 6\sin\theta - 6\cos^2\theta\sin\theta + 1 - 2\cos 2\theta + \cos^2 2\theta + \sin\theta$$
$$- 2\cos^2\theta\sin\theta + \cos^4\theta\sin\theta) d\theta$$

$$= \tfrac{1}{4} \int_0^{2\pi} (8\sin\theta + 3 - 4\cos 2\theta - 8\cos^2\theta\sin\theta + \tfrac{1}{2} + \tfrac{1}{2}\cos 4\theta + \cos^4\theta\sin\theta) d\theta$$

$$= \tfrac{1}{4} \left[-8\cos\theta + 3\theta - 2\sin 2\theta - 8\left(-\tfrac{1}{3}\cos^3\theta\right) + \tfrac{1}{2}\theta + \tfrac{1}{8}\sin 4\theta - \tfrac{1}{5}\cos^5\theta\right]_0^{2\pi}$$

$$= \tfrac{1}{4} \left[-8 + 6\pi + \tfrac{8}{3} + \pi - \tfrac{1}{5} - (-8 + \tfrac{8}{3} - \tfrac{1}{5})\right] = \frac{7\pi}{4}$$

Thus,

$$\bar{y} = \frac{M_x}{M} = \frac{\frac{7\pi}{4}}{\frac{5\pi}{3}} = \frac{21}{20} \;\Rightarrow\; (\bar{x}, \bar{y}) = \left(0, \frac{21}{20}\right).$$

53. From Exercise #47, $M = 6\sqrt{3} - 2\pi$, and

$$M_y = \int_{\pi/6}^{5\pi/6} \int_0^{6\sin\theta} (r\cos\theta) \cdot \tfrac{1}{r} \cdot r \, dr d\theta - \int_{\pi/6}^{5\pi/6} \int_0^3 (r\cos\theta) \cdot \tfrac{1}{r} \cdot r \, dr d\theta$$

$$= \int_{\pi/6}^{5\pi/6} \left[\tfrac{1}{2} r^2 \cos\theta\right]_{r=0}^{r=6\sin\theta} d\theta - \int_{\pi/6}^{5\pi/6} \left[\tfrac{1}{2} r^2 \cos\theta\right]_{r=0}^{r=3} d\theta$$

$$= \tfrac{1}{2} \int_{\pi/6}^{5\pi/6} (36\sin^2\theta\cos\theta - 9\cos\theta) d\theta = \tfrac{1}{2} \left[36 \cdot \tfrac{1}{3}\sin^3\theta - 9\sin\theta\right]_{\pi/6}^{5\pi/6}$$

$$= \tfrac{1}{2} \left[12\sin^3 \tfrac{5\pi}{6} - 9\sin \tfrac{5\pi}{6} - \left(12\sin^3 \tfrac{\pi}{6} - 9\sin \tfrac{\pi}{6}\right)\right] = 0 \;\Rightarrow\; \bar{x} = \frac{M_y}{M} = \frac{0}{6\sqrt{3}-2\pi} = 0$$

(Also clear from the symmetry, that $\bar{x} = 0$.)

$$M_x = \int_{\pi/6}^{5\pi/6} \int_0^{6\sin\theta} (r\sin\theta) \cdot \frac{1}{r} \cdot r \, dr d\theta - \int_{\pi/6}^{5\pi/6} \int_0^3 (r\sin\theta) \cdot \frac{1}{r} \cdot r \, dr d\theta$$

$$= \int_{\pi/6}^{5\pi/6} \left[\frac{1}{2}r^2 \sin\theta\right]_{r=0}^{r=6\sin\theta} d\theta - \int_{\pi/6}^{5\pi/6} \left[\frac{1}{2}r^2 \sin\theta\right]_{r=0}^{r=3} d\theta$$

$$= \frac{1}{2}\int_{\pi/6}^{5\pi/6} (36\sin^2\theta \cdot \sin\theta - 9\sin\theta) \, d\theta = \frac{1}{2}\int_{\pi/6}^{5\pi/6} [36(1-\cos^2\theta)\cdot\sin\theta - 9\sin\theta] \, d\theta$$

$$= \frac{1}{2}\int_{\pi/6}^{5\pi/6} (27\sin\theta - 36\cos^2\theta\sin\theta) \, d\theta = \frac{1}{2}\left[-27\cos\theta + 12\cos^3\theta\right]_{\pi/6}^{5\pi/6}$$

$$= \frac{1}{2}\left[-27\cos\frac{5\pi}{6} + 12\cos^3\frac{5\pi}{6} - \left(-27\cos\frac{\pi}{6} + 12\cos^3\frac{\pi}{6}\right)\right]$$

$$= \frac{1}{2}\left[-27\left(-\frac{\sqrt{3}}{2}\right) + 12\left(-\frac{\sqrt{3}}{2}\right)^3 + 27\left(\frac{\sqrt{3}}{2}\right) - 12\left(\frac{\sqrt{3}}{2}\right)^3\right] = \frac{27\sqrt{3}}{2} - \frac{9\sqrt{3}}{2} = 9\sqrt{3}$$

Thus,

$$\bar{y} = \frac{M_x}{M} = \frac{9\sqrt{3}}{6\sqrt{3} - 2\pi} \Rightarrow (\bar{x}, \bar{y}) = \left(0, \frac{9\sqrt{3}}{6\sqrt{3} - 2\pi}\right) \approx (0, 3.794).$$

55.

$$M = \int_{-\pi/4}^{\pi/4} \int_0^{\cos\theta} r \cdot r \, dr d\theta = \int_{-\pi/4}^{\pi/4}\left[\frac{1}{3}r^3\right]_{r=0}^{r=\cos\theta} d\theta = \frac{1}{3}\int_{-\pi/4}^{\pi/4} \cos^3\theta \, d\theta$$

$$= \frac{1}{3}\int_{-\pi/4}^{\pi/4} (1-\sin^2\theta)\cos\theta \, d\theta = \frac{1}{3}\int_{-\pi/4}^{\pi/4}\left(\cos\theta - \sin^2\theta\cos\theta\right) d\theta$$

$$= \frac{1}{3}\left[\sin\theta - \frac{1}{3}\sin^3\theta\right]_{-\pi/4}^{\pi/4} = \frac{1}{3}\left\{\sin\frac{\pi}{4} - \frac{1}{3}\sin^3\frac{\pi}{4} - \left[\sin\left(-\frac{\pi}{4}\right) - \frac{1}{3}\sin^3\left(-\frac{\pi}{4}\right)\right]\right\}$$

$$= \frac{1}{3}\left\{\frac{1}{\sqrt{2}} - \frac{1}{3}\left(\frac{1}{2\sqrt{2}}\right) - \left[-\frac{1}{\sqrt{2}} - \frac{1}{3}\left(-\frac{1}{2\sqrt{2}}\right)\right]\right\}$$

$$= \frac{1}{3}\left[\frac{1}{\sqrt{2}} - \frac{1}{3}\left(\frac{1}{2\sqrt{2}}\right) + \frac{1}{\sqrt{2}} - \frac{1}{3}\left(\frac{1}{2\sqrt{2}}\right)\right] = \frac{1}{3}\left[\frac{2}{\sqrt{2}} - \frac{1}{3\sqrt{2}}\right] = \frac{5}{9\sqrt{2}}$$

$$M_y = \int_{-\pi/4}^{\pi/4}\int_0^{\cos\theta} r\cos\theta \cdot r \cdot r \, dr d\theta = \int_{-\pi/4}^{\pi/4}\left[\frac{1}{4}r^4\cos\theta\right]_{r=0}^{r=\cos\theta} d\theta = \frac{1}{4}\int_{-\pi/4}^{\pi/4}\cos^4\theta \cdot \cos\theta \, d\theta$$

$$= \frac{1}{4}\int_{-\pi/4}^{\pi/4}(1-\sin^2\theta)^2\cos\theta \, d\theta = \frac{1}{4}\int_{-\pi/4}^{\pi/4}(1 - 2\sin^2\theta + \sin^4\theta)\cos\theta d\theta$$

$$= \frac{1}{4}\left[\sin\theta - 2\cdot\frac{1}{3}\sin^3\theta + \frac{1}{5}\sin^5\theta\right]_{-\pi/4}^{\pi/4}$$

$$= \frac{1}{4}\left\{\sin\frac{\pi}{4} - \frac{2}{3}\sin^3\frac{\pi}{4} + \frac{1}{5}\sin^5\frac{\pi}{4} - \left[\sin\left(-\frac{\pi}{4}\right) - \frac{2}{3}\sin^3\left(-\frac{\pi}{4}\right) + \frac{1}{5}\sin^5\left(-\frac{\pi}{4}\right)\right]\right\}$$

$$= \frac{1}{4}\left[\frac{1}{\sqrt{2}} - \frac{2}{3}\cdot\frac{1}{2\sqrt{2}} + \frac{1}{5}\cdot\frac{1}{4\sqrt{2}} - \left(-\frac{1}{\sqrt{2}} + \frac{2}{3}\cdot\frac{1}{2\sqrt{2}} - \frac{1}{5}\cdot\frac{1}{4\sqrt{2}}\right)\right]$$

$$= \frac{1}{4}\left(\frac{2}{\sqrt{2}} - \frac{2}{3\sqrt{2}} + \frac{1}{10\sqrt{2}}\right) = \frac{1}{4}\left(\frac{60 - 20 + 3}{30\sqrt{2}}\right) = \frac{43}{120\sqrt{2}} \Rightarrow \bar{x} = \frac{M_y}{M} = \frac{\frac{43}{120\sqrt{2}}}{\frac{5}{9\sqrt{2}}} = \frac{129}{200} \approx 0.645$$

and,

$$M_x = \int_{-\pi/4}^{\pi/4} \int_0^{\cos\theta} r\sin\theta \cdot r \cdot r\,dr\,d\theta = \int_{-\pi/4}^{\pi/4} \left[\tfrac{1}{4}r^4 \sin\theta\right]_{r=0}^{r=\cos\theta} d\theta = \tfrac{1}{4}\int_{-\pi/4}^{\pi/4} \cos^4\theta \sin\theta\,d\theta$$

$$= \tfrac{1}{4}\left[-\tfrac{1}{5}\cos^5\theta\right]_{-\pi/4}^{\pi/4} = -\tfrac{1}{20}\left[\cos^5\tfrac{\pi}{4} - \cos^5\left(-\tfrac{\pi}{4}\right)\right] = 0 \;\Rightarrow\; \bar{y} = \frac{M_x}{M} = \frac{0}{\frac{5}{9\sqrt{2}}} = 0$$

(Also clear from the symmetry, that $\bar{y} = 0$.)

Hence, $(\bar{x}, \bar{y}) = \left(\frac{129}{200}, 0\right) \approx (0.645, 0)$.

§11.5 Triple Integrals

1.
$$\int_0^1 \int_0^{x^2} \int_{x+y}^{x^2+y^2} dz\,dy\,dx = \int_0^1 \int_0^{x^2} \left[z\right]_{z=x+y}^{z=x^2+y^2} dy\,dx = \int_0^1 \int_0^{x^2} \left[x^2 + y^2 - (x+y)\right] dy\,dx$$

$$= \int_0^1 \left[x^2 y + \tfrac{1}{3}y^3 - xy - \tfrac{1}{2}y^2\right]_{y=0}^{y=x^2} dx = \int_0^1 \left(x^4 + \tfrac{1}{3}x^6 - x^3 - \tfrac{1}{2}x^4\right) dx$$

$$= \left[\tfrac{1}{5}x^5 + \tfrac{1}{21}x^7 - \tfrac{1}{4}x^4 - \tfrac{1}{10}x^5\right]_0^1 = \tfrac{1}{5} + \tfrac{1}{21} - \tfrac{1}{4} - \tfrac{1}{10} = -\tfrac{43}{420}.$$

3.
$$\int_0^1 \int_0^{2-3x} \int_0^{x+y} x\,dz\,dy\,dx = \int_0^1 \int_0^{2-3x} \left[xz\right]_{z=0}^{z=x+y} dy\,dx = \int_0^1 \int_0^{2-3x} x(x+y)\,dy\,dx$$

$$= \int_0^1 \left[x^2 y + \tfrac{1}{2}xy^2\right]_{y=0}^{y=2-3x} dx = \int_0^1 \left[x^2(2-3x) + \tfrac{1}{2}x(2-3x)^2\right] dx$$

$$= \int_0^1 \left(2x^2 - 3x^3 + 2x - 6x^2 + \tfrac{9}{2}x^3\right) dx = \int_0^1 \left(\tfrac{3}{2}x^3 - 4x^2 + 2x\right) dx$$

$$= \left[\tfrac{3}{8}x^4 - \tfrac{4}{3}x^3 + x^2\right]_0^1 = \tfrac{3}{8} - \tfrac{4}{3} + 1 = \frac{9 - 32 + 24}{24} = \tfrac{1}{24}.$$

5.
$$\int_0^3 \int_{-1}^1 \int_0^1 xyz^2\,dz\,dy\,dx = \int_0^3 \int_{-1}^1 \left[\tfrac{1}{3}xyz^3\right]_{z=0}^{z=1} dy\,dx = \tfrac{1}{3}\int_0^3 \int_{-1}^1 xy\,dy\,dx$$

$$= \tfrac{1}{3}\int_0^3 \left[\tfrac{1}{2}xy^2\right]_{y=-1}^{y=1} dx = \tfrac{1}{3}\int_0^3 0\,dx = 0.$$

7.
$$\int_1^2 \int_x^{2x} \int_{y-x}^y (4x - 12z)\,dz\,dy\,dx = \int_1^2 \int_x^{2x} \left[4xz - 6z^2\right]_{z=y-x}^{z=y} dy\,dx$$

$$= \int_1^2 \int_x^{2x} \left\{4xy - 6y^2 - \left[4x(y-x) - 6(y-x)^2\right]\right\} dy\,dx$$

$$= \int_1^2 \int_x^{2x} \left(4xy - 6y^2 - 4xy + 4x^2 + 6y^2 - 12xy + 6x^2\right) dy\,dx$$

$$= \int_1^2 \int_x^{2x} \left(10x^2 - 12xy\right) dy\,dx = \int_1^2 \left[10x^2 y - 6xy^2\right]_{y=x}^{y=2x} dx$$

$$= \int_1^2 [20x^3 - 24x^3 - (10x^3 - 6x^3)]dx = \int_1^2 (-8x^3)dx$$

$$= \left[-2x^4\right]_1^2 = -2(16 - 1) = -30.$$

9.
$$\int_0^1 \int_0^{x^2} \int_0^{xy^3} xy^2z^3 \, dzdydx = \int_0^1 \int_0^{x^2} \left[\frac{1}{4}xy^2z^4\right]_{z=0}^{z=xy^3} dydx = \frac{1}{4}\int_0^1 \int_0^{x^2} xy^2(xy^3)^4 \, dydx$$
$$= \frac{1}{4}\int_0^1 \int_0^{x^2} x^5y^{14} \, dydx = \frac{1}{4}\int_0^1 \left[\frac{1}{15}x^5y^{15}\right]_{y=0}^{y=x^2} dx$$
$$= \frac{1}{60}\int_0^1 x^{35} \, dx = \frac{1}{60}\left[\frac{1}{36}x^{36}\right]_0^1 = \frac{1}{60\cdot36} = \frac{1}{2160}.$$

11.
$$\int_0^1 \int_0^z \int_0^{z^2+y^2} dxdydz = \int_0^1 \int_0^z \left[x\right]_{x=0}^{x=z^2+y^2} dydz = \int_0^1 \int_0^z (z^2+y^2)dydz$$
$$= \int_0^1 \left[z^2y + \frac{1}{3}y^3\right]_{y=0}^{y=z} dz = \int_0^1 \left(z^3 + \frac{1}{3}z^3\right) dz = \frac{4}{3}\int_0^1 z^3 \, dz$$
$$= \left[\frac{1}{3}z^4\right]_0^1 = \frac{1}{3}.$$

13.
$$\int_0^1 \int_0^\pi \int_0^\pi y\sin z \, dxdydz = \int_0^1 \int_0^\pi \left[xy\sin z\right]_{x=0}^{x=\pi} dydz = \pi\int_0^1 \int_0^\pi y\sin z \, dydz$$
$$= \pi\int_0^1 \left[\frac{1}{2}y^2\sin z\right]_{y=0}^{y=\pi} dz = \frac{\pi}{2}\int_0^1 \pi^2\sin z \, dz = \frac{\pi^3}{2}\left[-\cos z\right]_0^1$$
$$= \frac{\pi^3}{2}(-\cos 1 + \cos 0) = \frac{\pi^3}{2}(1 - \cos 1).$$

15.
$$\int_0^1 \int_0^{x^2} \int_0^{x^3y} x^3y^2z \, dzdydx = \int_0^1 \int_0^{x^2} \left[\frac{1}{2}x^3y^2z^2\right]_{z=0}^{z=x^3y} dydx = \frac{1}{2}\int_0^1 \int_0^{x^2} (x^3y^2)(x^3y)^2 \, dydx$$
$$= \frac{1}{2}\int_0^1 \int_0^{x^2} x^9y^4 \, dydx = \frac{1}{2}\int_0^1 \left[\frac{1}{5}x^9y^5\right]_{y=0}^{y=x^2} dx$$
$$= \frac{1}{10}\int_0^1 x^9\cdot x^{10} \, dx = \frac{1}{10}\left[\frac{1}{20}x^{20}\right]_0^1 = \frac{1}{200}.$$

17.
$$\int_0^1 \int_{-x}^x \int_{-x-z}^{x+z} e^{x+y+z} \, dydzdx = \int_0^1 \int_{-x}^x \left[e^{x+y+z}\right]_{y=-x-z}^{y=x+z} dzdx$$
$$= \int_0^1 \int_{-x}^x \left(e^{x+(x+z)+z} - e^{x+(-x-z)+z}\right) dzdx$$
$$= \int_0^1 \int_{-x}^x \left(e^{2x+2z} - 1\right) dzdx = \int_0^1 \left[\frac{1}{2}e^{2x+2z} - z\right]_{z=-x}^{z=x} dx$$
$$= \int_0^1 \left[\frac{1}{2}e^{2x+2x} - x - \left(\frac{1}{2}e^{2x-2x} + x\right)\right] dx$$
$$= \int_0^1 \left(\frac{1}{2}e^{4x} - 2x - \frac{1}{2}\right) dx = \left[\frac{1}{8}e^{4x} - x^2 - \frac{1}{2}x\right]_0^1$$
$$= \frac{1}{8}e^4 - 1 - \frac{1}{2} - \frac{1}{8} = \frac{1}{8}(e^4 - 8 - 4 - 1) = \frac{1}{8}(e^4 - 13).$$

19.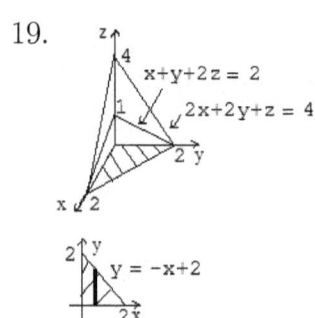

We note that z varies from the plane $z = \frac{2 - x - y}{2}$ up to the plane $z = 4 - 2x - 2y$, see upper figure. The projection of the solid onto the xy-plane yields the region of xy-integration, shown in the lower figure.

$$V = \int_0^2 \int_0^{-x+2} \int_{\frac{2-x-y}{2}}^{4-2x-2y} dz\,dy\,dx = \int_0^2 \int_0^{-x+2} \left[z\right]_{z=\frac{2-x-y}{2}}^{z=4-2x-2y} dy\,dx$$

$$= \int_0^2 \int_0^{-x+2} \left(4 - 2x - 2y - 1 + \tfrac{1}{2}x + \tfrac{1}{2}y\right) dy\,dx$$

$$= \int_0^2 \int_0^{-x+2} \left(3 - \tfrac{3}{2}x - \tfrac{3}{2}y\right) dy\,dx$$

$$= \int_0^2 \left[\left(3 - \tfrac{3}{2}x\right)y - \tfrac{3}{4}y^2\right]_{y=0}^{y=-x+2} dx$$

$$= \int_0^2 \left[\left(3 - \tfrac{3}{2}x\right)(-x + 2) - \tfrac{3}{4}(-x+2)^2\right] dx$$

$$= \int_0^2 \left(-3x + 6 + \tfrac{3}{2}x^2 - 3x - \tfrac{3}{4}x^2 + 3x - 3\right) dx$$

$$= \int_0^2 \left(3 + \tfrac{3}{4}x^2 - 3x\right) dx = \left[3x + \tfrac{1}{4}x^3 - \tfrac{3}{2}x^2\right]_0^2 = 6 + 2 - 6 = 2.$$

21.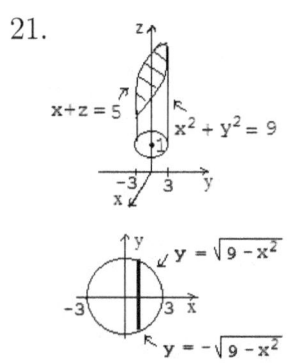

We note that z varies from the plane $z = 1$ up to the plane $z = 5 - x$, see upper figure. The projection of the solid onto the xy-plane yields the region of xy-integration, shown in the lower figure.

$$V = \int_{-3}^3 \int_{-\sqrt{9-x^2}}^{\sqrt{9-x^2}} \int_1^{5-x} dz\,dy\,dx = \int_{-3}^3 \int_{-\sqrt{9-x^2}}^{\sqrt{9-x^2}} \left[z\right]_{z=1}^{z=5-x} dy\,dx$$

$$= \int_{-3}^3 \int_{-\sqrt{9-x^2}}^{\sqrt{9-x^2}} (5 - x - 1)\, dy\,dx = \int_{-3}^3 \left[(4-x)y\right]_{y=-\sqrt{9-x^2}}^{y=\sqrt{9-x^2}} dx$$

$$= \int_{-3}^3 (4 - x)\left[\sqrt{9 - x^2} - (-\sqrt{9 - x^2})\right] dx$$

$$= 8\int_{-3}^3 \sqrt{9 - x^2}\, dx - 2\int_{-3}^3 x\sqrt{9 - x^2}\, dx$$

Now, $\displaystyle \int_{-3}^3 \sqrt{9 - x^2}\, dx = \int_{-\pi/2}^{\pi/2} 3\cos\theta \cdot 3\cos\theta\, d\theta = 9\int_{-\pi/2}^{\pi/2} \frac{1 + \cos 2\theta}{2}\, d\theta = \frac{9}{2}\left[\theta + \tfrac{1}{2}\sin 2\theta\right]_{-\pi/2}^{\pi/2}$

$x = 3\sin\theta,$	$x = -3 \Rightarrow \theta = -\frac{\pi}{2}$
$dx = 3\cos\theta\, d\theta,$	$x = 3 \Rightarrow \theta = \frac{\pi}{2}$

$$= \frac{9}{2}\left[\frac{\pi}{2} - \left(-\frac{\pi}{2}\right)\right] = \frac{9\pi}{2}$$

and, $\displaystyle \int_{-3}^3 x\sqrt{9 - x^2}\,dx = \int_0^0 = 0$

$u = 9 - x^2,$	$x = -3 \Rightarrow u = 0$
$du = -2x\,dx,$	$x = 3 \Rightarrow u = 0$

Notes: $\displaystyle \int_{-3}^3 \sqrt{9 - x^2}\, dx = \int_{-3}^3 y\, dx$ is the area of the top half of the circle: $\frac{1}{2}\pi(3^2) = \frac{9\pi}{2}$

$\displaystyle \int_{-a}^a$ of an odd function, is 0.

Therefore, $V = 8\left(\dfrac{9\pi}{2} + 0\right) = 36\pi.$

23.

We note that z varies from the plane $z = 0$ up to the plane $z = 2 - x - 2y$. The two planes $x = 2y$ and $x + 2y + z = 2$ intersect in the xy-plane ($z = 0$) where $2y = 2 - 2y \Rightarrow 4y = 2 \Rightarrow y = \frac{1}{2}$. There $x = 2y \Rightarrow x = 1$ (see upper figure). The projection of the solid onto the xy-plane yields the region of xy-integration, shown in the lower figure.

$$V = \int_0^1 \int_{\frac{1}{2}x}^{1-\frac{1}{2}x} \int_0^{2-x-2y} dz\,dy\,dx = \int_0^1 \int_{\frac{1}{2}x}^{1-\frac{1}{2}x} \left[z\right]_{z=0}^{z=2-x-2y} dy\,dx$$

$$= \int_0^1 \int_{\frac{1}{2}x}^{1-\frac{1}{2}x} (2 - x - 2y)\,dy\,dx = \int_0^1 \left[(2-x)y - y^2\right]_{y=\frac{1}{2}x}^{y=1-\frac{1}{2}x} dx$$

$$= \int_0^1 \left\{(2-x)\left(1-\tfrac{1}{2}x\right) - \left(1-\tfrac{1}{2}x\right)^2 - \left[(2-x)\cdot\tfrac{1}{2}x - \tfrac{1}{4}x^2\right]\right\} dx$$

$$= \int_0^1 \left(2 - x - x + \tfrac{1}{2}x^2 - 1 + x - \tfrac{1}{4}x^2 - x + \tfrac{1}{2}x^2 + \tfrac{1}{4}x^2\right) dx$$

$$= \int_0^1 (1 - 2x + x^2)dx = \left[x - x^2 + \tfrac{1}{3}x^3\right]_0^1 = 1 - 1 + \tfrac{1}{3} = \tfrac{1}{3}.$$

25.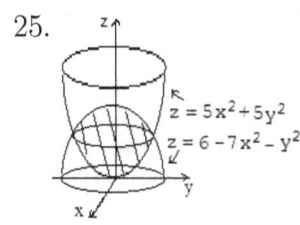

We note that z varies from the paraboloid $z = 5x^2 + 5y^2$ up to the paraboloid $z = 6 - 7x^2 - y^2$ (see upper figure). The two paraboloids intersect where $5x^2 + 5y^2 = 6 - 7x^2 - y^2 \Rightarrow 12x^2 + 6y^2 = 6 \Rightarrow 2x^2 + y^2 = 1$ which is an ellipse. The projection of the solid onto the xy-plane yields the elliptical region of xy-integration, shown in the lower figure.

$$V = \int_{-1/\sqrt{2}}^{1/\sqrt{2}} \int_{-\sqrt{1-2x^2}}^{\sqrt{1-2x^2}} \int_{5x^2+5y^2}^{6-7x^2-y^2} dz\,dy\,dx = \int_{-1/\sqrt{2}}^{1/\sqrt{2}} \int_{-\sqrt{1-2x^2}}^{\sqrt{1-2x^2}} \left[z\right]_{z=5x^2+5y^2}^{z=6-7x^2-y^2} dy\,dx$$

$$= \int_{-1/\sqrt{2}}^{1/\sqrt{2}} \int_{-\sqrt{1-2x^2}}^{\sqrt{1-2x^2}} [6 - 7x^2 - y^2 - (5x^2 + 5y^2)]\,dy\,dx$$

$$= \int_{-1/\sqrt{2}}^{1/\sqrt{2}} \int_{-\sqrt{1-2x^2}}^{\sqrt{1-2x^2}} (6 - 12x^2 - 6y^2)\,dy\,dx = \int_{-1/\sqrt{2}}^{1/\sqrt{2}} \left[(6-12x^2)y - 2y^3\right]_{y=-\sqrt{1-2x^2}}^{y=\sqrt{1-2x^2}} dx$$

$$= \int_{-1/\sqrt{2}}^{1/\sqrt{2}} \left\{6(1-2x^2)\sqrt{1-2x^2} - 2(1-2x^2)^{\frac{3}{2}} - \left[-6(1-2x^2)\sqrt{1-2x^2} + 2(1-2x^2)^{\frac{3}{2}}\right]\right\} dx$$

$$= \int_{-1/\sqrt{2}}^{1/\sqrt{2}} \left[12(1-2x^2)^{\frac{3}{2}} - 4(1-2x^2)^{\frac{3}{2}}\right] dx = 8\int_{-1/\sqrt{2}}^{1/\sqrt{2}} (1-2x^2)^{\frac{3}{2}}\,dx$$

Make a trigonometric substitution to evaluate the integral: we want, say, $1 - 2x^2 = \cos^2\theta \Rightarrow 2x^2 = 1 - \cos^2\theta = \sin^2\theta \Rightarrow \sin\theta = \sqrt{2}x$. Thus:

$$\boxed{\begin{array}{ll} x = \dfrac{1}{\sqrt{2}}\sin\theta, & x = -\dfrac{1}{\sqrt{2}} \Rightarrow \theta = -\dfrac{\pi}{2} \\[2mm] dx = \dfrac{1}{\sqrt{2}}\cos\theta\,d\theta, & x = \dfrac{1}{\sqrt{2}} \Rightarrow \theta = \dfrac{\pi}{2} \end{array}}$$

$$V = \frac{8}{\sqrt{2}} \int_{-\pi/2}^{\pi/2} \left(\cos^2 \theta\right)^{\frac{3}{2}} \cos \theta d\theta = \frac{8}{\sqrt{2}} \int_{-\pi/2}^{\pi/2} \cos^4 \theta \, d\theta = \frac{8}{\sqrt{2}} \int_{-\pi/2}^{\pi/2} \left(\frac{1 + \cos 2\theta}{2}\right)^2 d\theta$$

$$= \frac{2}{\sqrt{2}} \int_{-\pi/2}^{\pi/2} \left(1 + 2\cos 2\theta + \cos^2 2\theta\right) d\theta = \frac{2}{\sqrt{2}} \int_{-\pi/2}^{\pi/2} \left(1 + 2\cos 2\theta + \frac{1 + \cos 4\theta}{2}\right) d\theta$$

$$= \frac{2}{\sqrt{2}} \left[\theta + \sin 2\theta + \frac{1}{2}\theta + \frac{1}{8}\sin 4\theta\right]_{-\pi/2}^{\pi/2} = \frac{2}{\sqrt{2}} \cdot \frac{3}{2} \left(\frac{\pi}{2} + \frac{\pi}{2}\right) = \frac{3\pi}{\sqrt{2}}.$$

27.

$$M = \int_0^a \int_0^a \int_0^a (x^2 + y^2 + z^2) dz dy dx = \int_0^a \int_0^a \left[(x^2 + y^2)z + \frac{1}{3}z^3\right]_{z=0}^{z=a} dy dx$$

$$= \int_0^a \int_0^a \left(ax^2 + ay^2 + \frac{1}{3}a^3\right) dy dx = \int_0^a \left[ax^2y + \frac{1}{3}ay^3 + \frac{1}{3}a^3y\right]_{y=0}^{y=a} dx$$

$$= \int_0^a \left(a^2x^2 + \frac{1}{3}a^4 + \frac{1}{3}a^4\right) dx = \int_0^a \left(a^2x^2 + \frac{2}{3}a^4\right) dx = \left[\frac{1}{3}a^2x^3 + \frac{2}{3}a^4x\right]_0^a$$

$$= \frac{1}{3}a^5 + \frac{2}{3}a^5 = a^5.$$

29.

We note that z varies from the plane $z = 0$ to the plane $z = 1$, upper figure. The projection of the solid onto the xy-plane is the unit disk over which x and y vary (see the lower figure). Also, as $z \geq 0$, $\delta(x, y, z) = |xyz| = |xy|z \Rightarrow$

$$M = \int_{-1}^{1} \int_{-\sqrt{1-x^2}}^{\sqrt{1-x^2}} \int_0^1 |xy|z \, dz dy dx = \int_{-1}^{1} \int_{-\sqrt{1-x^2}}^{\sqrt{1-x^2}} \left[|xy| \cdot \frac{1}{2}z^2\right]_{z=0}^{z=1} dy dx$$

$$= \frac{1}{2} \int_{-1}^{1} \int_{-\sqrt{1-x^2}}^{\sqrt{1-x^2}} |x||y| \, dy dx = \frac{1}{2} \int_{-1}^{1} \left[|x| \int_{-\sqrt{1-x^2}}^{\sqrt{1-x^2}} |y| \, dy\right] dx$$

$$= \frac{1}{2} \int_{-1}^{1} \left[|x| \left\{\int_{-\sqrt{1-x^2}}^{0} (-y) dy + \int_0^{\sqrt{1-x^2}} y\right\} dy\right] dx$$

$$= \frac{1}{2} \int_{-1}^{1} |x| \left\{\left[-\frac{1}{2}y^2\right]_{y=-\sqrt{1-x^2}}^{y=0} + \left[\frac{1}{2}y^2\right]_{y=0}^{y=\sqrt{1-x^2}}\right\} dx = \frac{1}{4} \int_{-1}^{1} |x|(1 - x^2 + 1 - x^2) \, dx$$

$$= \frac{1}{2} \int_{-1}^{1} |x|(1 - x^2) \, dx = \frac{1}{2} \int_{-1}^{0} (-x)(1 - x^2) \, dx + \frac{1}{2} \int_0^1 x(1 - x^2) \, dx$$

$$= \frac{1}{2} \int_{-1}^{0} (-x + x^3) \, dx + \frac{1}{2} \int_0^1 (x - x^3) \, dx = \frac{1}{2} \left\{\left[-\frac{1}{2}x^2 + \frac{1}{4}x^4\right]_{-1}^{0} + \left[\frac{1}{2}x^2 - \frac{1}{4}x^4\right]_0^1\right\}$$

$$= \frac{1}{2} \left[-\left(-\frac{1}{2} + \frac{1}{4}\right) + \frac{1}{2} - \frac{1}{4}\right] = \frac{1}{2} \left(\frac{1}{2} - \frac{1}{4} + \frac{1}{2} - \frac{1}{4}\right) = \frac{1}{4}.$$

31.

We note that z varies from the plane $z = 0$ to the plane $z = x$, upper figure. The projection of the solid onto the xy-plane yields the region of xy-integration, shown in the lower figure.

$$M = \int_0^1 \int_{-\sqrt{x}}^{\sqrt{x}} \int_0^x k\,dz\,dy\,dx = k\int_0^1 \int_{-\sqrt{x}}^{\sqrt{x}} \Big[z\Big]_{z=0}^{z=x} dy\,dx = k\int_0^1 \int_{-\sqrt{x}}^{\sqrt{x}} x\,dy\,dx$$

$$= k\int_0^1 \Big[xy\Big]_{y=-\sqrt{x}}^{y=\sqrt{x}} dx = k\int_0^1 x\left[\sqrt{x} - (-\sqrt{x})\right] dx = 2k\int_0^1 x^{\frac{3}{2}}\,dx$$

$$= 2k\left[\frac{2}{5}x^{\frac{5}{2}}\right]_0^1 = \frac{4k}{5}.$$

33.

We note that z varies from the plane $z = 0$ to the paraboloid $z = 4 - x^2 - y^2$, upper figure. The projection of the solid onto the xy-plane yields the region of xy-integration, shown in the lower figure.

$$M = \int_{-2}^{2} \int_{-\sqrt{4-x^2}}^{\sqrt{4-x^2}} \int_0^{4-x^2-y^2} a\,dz\,dy\,dx = a\int_{-2}^{2} \int_{-\sqrt{4-x^2}}^{\sqrt{4-x^2}} \Big[z\Big]_{z=0}^{z=4-x^2-y^2} dy\,dx$$

$$= a\int_{-2}^{2} \int_{-\sqrt{4-x^2}}^{\sqrt{4-x^2}} (4 - x^2 - y^2)\,dy\,dx = a\int_{-2}^{2} \left[(4-x^2)y - \frac{1}{3}y^3\right]_{y=-\sqrt{4-x^2}}^{y=\sqrt{4-x^2}} dx$$

$$= a\int_{-2}^{2} \left\{ (4-x^2)\sqrt{4-x^2} - \frac{1}{3}(4-x^2)^{\frac{3}{2}} - \left[(4-x^2)(-\sqrt{4-x^2}) + \frac{1}{3}(4-x^2)^{\frac{3}{2}}\right]\right\} dx$$

$$= 2a\int_{-2}^{2} \frac{2}{3}(4-x^2)^{\frac{3}{2}}\,dx = \frac{4}{3}a\int_{-2}^{2}(4-x^2)^{\frac{3}{2}}\,dx = \frac{4}{3}a\int_{-\pi/2}^{\pi/2} 8\cos^3\theta \cdot 2\cos\theta\,d\theta$$

$$\boxed{\begin{array}{ll} x = 2\sin\theta, & x = -2 \Rightarrow \theta = -\dfrac{\pi}{2} \\[2mm] dx = 2\cos\theta\,d\theta, & x = 2 \Rightarrow \theta = \dfrac{\pi}{2} \end{array}}$$

$$= \frac{64}{3}a\int_{-\pi/2}^{\pi/2} \cos^4\theta\,d\theta = \frac{64}{3}a\int_{-\pi/2}^{\pi/2} \left(\frac{1+\cos 2\theta}{2}\right)^2 d\theta$$

$$= \frac{16}{3}a\int_{-\pi/2}^{\pi/2} \left(1 + 2\cos 2\theta + \frac{1+\cos 4\theta}{2}\right) d\theta = \frac{16}{3}a\int_{-\pi/2}^{\pi/2} \left(\frac{3}{2} + 2\cos 2\theta + \frac{1}{2}\cos 4\theta\right) d\theta$$

$$= \frac{16}{3}a\left[\frac{3}{2}\theta + \sin 2\theta + \frac{1}{8}\sin 4\theta\right]_{-\pi/2}^{\pi/2} = \frac{16}{3}a\left[\frac{3}{2}\left(\frac{\pi}{2} + \frac{\pi}{2}\right)\right] = 8\pi a$$

From the symmetry of the paraboloid, clearly $\bar{x} = \bar{y} = 0$.

$$\bar{z} = \frac{\iiint\limits_{W} z\delta(x,y,z)\,dV}{M} = \frac{1}{8\pi a}\int_{-2}^{2} \int_{-\sqrt{4-x^2}}^{\sqrt{4-x^2}} \int_0^{4-x^2-y^2} az\,dz\,dy\,dx = \frac{1}{8\pi}\int_{-2}^{2} \int_{-\sqrt{4-x^2}}^{\sqrt{4-x^2}} \left[\frac{1}{2}z^2\right]_{z=0}^{z=4-x^2-y^2} dy\,dx$$

$$= \frac{1}{16\pi}\int_{-2}^{2} \int_{-\sqrt{4-x^2}}^{\sqrt{4-x^2}} (4-x^2-y^2)^2\,dy\,dx = \frac{1}{16\pi}\int_{-2}^{2} \int_{-\sqrt{4-x^2}}^{\sqrt{4-x^2}} \left[(4-x^2)^2 - 2(4-x^2)y^2 + y^4\right] dy\,dx$$

$$= \frac{1}{16\pi}\int_{-2}^{2} \left[(4-x^2)^2 y - \frac{2}{3}(4-x^2)y^3 + \frac{1}{5}y^5\right]_{y=-\sqrt{4-x^2}}^{y=\sqrt{4-x^2}} dx$$

$$= \frac{1}{16\pi}\int_{-2}^{2} \left\{ (4-x^2)^2\sqrt{4-x^2} - \frac{2}{3}(4-x^2)(4-x^2)^{\frac{3}{2}} + \frac{1}{5}(4-x^2)^{\frac{5}{2}} - \right.$$
$$\left. \left[-(4-x^2)^2\sqrt{4-x^2} + \frac{2}{3}(4-x^2)(4-x^2)^{\frac{3}{2}} - \frac{1}{5}(4-x^2)^{\frac{5}{2}}\right]\right\} dx$$

$$= \frac{1}{8\pi} \int_{-2}^{2} \left(1 - \frac{2}{3} + \frac{1}{5}\right)(4 - x^2)^{\frac{5}{2}} \, dx = \frac{1}{15\pi} \int_{-2}^{2} (4 - x^2)^{\frac{5}{2}} \, dx = \frac{32}{15\pi} \int_{-\pi/2}^{\pi/2} \cos^5 \theta \cdot 2 \cos \theta \, d\theta$$

$$\boxed{\begin{array}{ll} x = 2\sin\theta, & x = -2 \Rightarrow \theta = -\frac{\pi}{2} \\ dx = 2\cos\theta \, d\theta, & x = 2 \Rightarrow \theta = \frac{\pi}{2} \end{array}}$$

$$= \frac{64}{15\pi} \int_{-\pi/2}^{\pi/2} \cos^6 \theta \, d\theta = \frac{64}{15\pi} \int_{-\pi/2}^{\pi/2} \left(\frac{1 + \cos 2\theta}{2}\right)^3 \, d\theta$$

$$= \frac{8}{15\pi} \int_{-\pi/2}^{\pi/2} \left(1 + 3\cos 2\theta + 3\cos^2 2\theta + \cos^3 2\theta\right) d\theta$$

$$= \frac{8}{15\pi} \int_{-\pi/2}^{\pi/2} \left[1 + 3\cos 2\theta + 3 \cdot \left(\frac{1 + \cos 4\theta}{2}\right) + (1 - \sin^2 2\theta)\cos 2\theta\right] d\theta$$

$$= \frac{8}{15\pi} \int_{-\pi/2}^{\pi/2} \left[\frac{5}{2} + 3\cos 2\theta + \frac{3}{2}\cos 4\theta + \cos 2\theta - \sin^2 2\theta \cos 2\theta\right] d\theta$$

$$= \frac{8}{15\pi} \left[\frac{5}{2}\theta + \frac{3}{2}\sin 2\theta + \frac{3}{8}\sin 4\theta + \frac{1}{2}\sin 2\theta - \frac{1}{3}\sin^3 2\theta\right]_{-\pi/2}^{\pi/2} = \frac{8}{15\pi} \left[\frac{5}{2}\left(\frac{\pi}{2} + \frac{\pi}{2}\right)\right] = \frac{4}{3}$$

Thus, the center of mass $(\bar{x}, \bar{y}, \bar{z}) = \left(0, 0, \frac{4}{3}\right)$.

35. From Exercise #31, $M = \frac{4k}{5}$, and from the symmetry, $\bar{y} = 0$.

$$\bar{x} = \frac{1}{M} \iiint_W xk \, dV = \frac{1}{M} \int_0^1 \int_{-\sqrt{x}}^{\sqrt{x}} \int_0^x kx \, dz \, dy \, dx = \frac{k}{M} \int_0^1 \int_{-\sqrt{x}}^{\sqrt{x}} \left[xz\right]_{z=0}^{z=x} dy \, dx$$

$$= \frac{k}{M} \int_0^1 \int_{-\sqrt{x}}^{\sqrt{x}} x^2 \, dy \, dx = \frac{k}{M} \int_0^1 \left[x^2 y\right]_{y=-\sqrt{x}}^{y=\sqrt{x}} dx = \frac{k}{M} \int_0^1 x^2 (\sqrt{x} + \sqrt{x}) \, dx$$

$$= \frac{2k}{M} \int_0^1 x^{\frac{5}{2}} \, dx = \frac{2k}{M} \left[\frac{2}{7} x^{\frac{7}{2}}\right]_0^1 = \frac{4k}{7M} = \frac{5}{7}$$

$$\bar{z} = \frac{1}{M} \iiint_W zk \, dV = \frac{1}{M} \int_0^1 \int_{-\sqrt{x}}^{\sqrt{x}} \int_0^x kz \, dz \, dy \, dx = \frac{k}{M} \int_0^1 \int_{-\sqrt{x}}^{\sqrt{x}} \left[\frac{1}{2} z^2\right]_{z=0}^{z=x} dy \, dx$$

$$= \frac{k}{2M} \int_0^1 \int_{-\sqrt{x}}^{\sqrt{x}} x^2 \, dy \, dx = \frac{1}{2} \cdot \frac{5}{7} = \frac{5}{14} \text{ [from 3 lines above]}$$

Thus, $(\bar{x}, \bar{y}, \bar{z}) = \left(\frac{5}{7}, 0, \frac{5}{14}\right)$.

37. From Exercise #27, $M = a^5$, and because of the symmetry, $\bar{x} = \bar{y} = \bar{z}$.

$$\bar{x} = \frac{1}{M} \iiint_W x(x^2 + y^2 + z^2) \, dV = \frac{1}{M} \int_0^a \int_0^a \int_0^a (x^3 + xy^2 + xz^2) \, dz \, dy \, dx$$

$$= \frac{1}{M} \int_0^a \int_0^a \left[(x^3 + xy^2)z + \frac{1}{3}xz^3\right]_{z=0}^{z=a} dy \, dx = \frac{1}{M} \int_0^a \int_0^a \left(ax^3 + axy^2 + \frac{1}{3}a^3 x\right) dy \, dx$$

$$= \frac{1}{M} \int_0^a \left[ax^3 y + \frac{1}{3}axy^3 + \frac{1}{3}a^3 xy\right]_{y=0}^{y=a} dx = \frac{1}{M} \int_0^a \left(a^2 x^3 + \frac{2}{3}a^4 x\right) dx = \frac{1}{M} \left[\frac{1}{4}a^2 x^4 + \frac{1}{3}a^4 x^2\right]_0^a$$

$$= \frac{1}{M} \left(\frac{1}{4}a^6 + \frac{1}{3}a^6\right) = \frac{7a^6}{12a^5} = \frac{7a}{12}$$

Thus, $(\bar{x}, \bar{y}, \bar{z}) = \left(\frac{7a}{12}, \frac{7a}{12}, \frac{7a}{12}\right)$.

39. $M = \int_1^2 \int_0^z \int_x^z (y-x)\, dy\, dx\, dz = \int_1^2 \int_0^z \left[\frac{1}{2}y^2 - xy \right]_{y=x}^{y=z} dx\, dz$

$\quad = \int_1^2 \int_0^z \left[\frac{1}{2}z^2 - xz - \left(\frac{1}{2}x^2 - x^2 \right) \right] dx\, dz = \int_1^2 \int_0^z \left(\frac{1}{2}z^2 - xz + \frac{1}{2}x^2 \right) dx\, dz$

$\quad = \int_1^2 \left[\frac{1}{2}z^2 x - \frac{1}{2}zx^2 + \frac{1}{6}x^3 \right]_{x=0}^{x=z} dz = \int_1^2 \left(\frac{1}{2}z^3 - \frac{1}{2}z^3 + \frac{1}{6}z^3 \right) dz = \left[\frac{1}{6} \cdot \frac{1}{4}z^4 \right]_1^2$

$\quad = \frac{1}{24}(2^4 - 1) = \frac{15}{24} = \frac{5}{8}$

$\bar{x} = \frac{1}{M} \iiint_W x(y-x)\, dV = \frac{1}{M} \int_1^2 \int_0^z \int_x^z (xy - x^2)\, dy\, dx\, dz = \frac{1}{M} \int_1^2 \int_0^z \left[\frac{1}{2}xy^2 - x^2 y \right]_{y=x}^{y=z} dx\, dz$

$\quad = \frac{1}{M} \int_1^2 \int_0^z \left[\frac{1}{2}xz^2 - x^2 z - \left(\frac{1}{2}x^3 - x^3 \right) \right] dx\, dz = \frac{1}{M} \int_1^2 \int_0^z \left(\frac{1}{2}xz^2 - x^2 z + \frac{1}{2}x^3 \right) dx\, dz$

$\quad = \frac{1}{M} \int_1^2 \left[\frac{1}{4}x^2 z^2 - \frac{1}{3}x^3 z + \frac{1}{8}x^4 \right]_{x=0}^{x=z} dz = \frac{1}{M} \int_1^2 \frac{1}{24}z^4\, dz = \frac{1}{24M} \left[\frac{1}{5}z^5 \right]_1^2 = \frac{31}{120} \cdot \frac{8}{5} = \frac{31}{75}$

$\bar{y} = \frac{1}{M} \iiint_W y(y-x)\, dV = \frac{1}{M} \int_1^2 \int_0^z \int_x^z (y^2 - xy)\, dy\, dx\, dz = \frac{1}{M} \int_1^2 \int_0^z \left[\frac{1}{3}y^3 - \frac{1}{2}xy^2 \right]_{y=x}^{y=z} dx\, dz$

$\quad = \frac{1}{M} \int_1^2 \int_0^z \left[\frac{1}{3}z^3 - \frac{1}{2}xz^2 - \left(\frac{1}{3}x^3 - \frac{1}{2}x^3 \right) \right] dx\, dz = \frac{1}{M} \int_1^2 \int_0^z \left(\frac{1}{3}z^3 - \frac{1}{2}xz^2 + \frac{1}{6}x^3 \right) dx\, dz$

$\quad = \frac{1}{M} \int_1^2 \left[\frac{1}{3}xz^3 - \frac{1}{4}x^2 z^2 + \frac{1}{24}x^4 \right]_{x=0}^{x=z} dz = \frac{1}{M} \int_1^2 \frac{1}{8}z^4\, dz = \frac{1}{8M} \left[\frac{1}{5}z^5 \right]_1^2 = \frac{31}{40} \cdot \frac{8}{5} = \frac{31}{25}$

$\bar{z} = \frac{1}{M} \iiint_W z(y-x)\, dV = \frac{1}{M} \int_1^2 \int_0^z \int_x^z (yz - xz)\, dy\, dx\, dz = \frac{1}{M} \int_1^2 \int_0^z \left[\frac{1}{2}y^2 z - xyz \right]_{y=x}^{y=z} dx\, dz$

$\quad = \frac{1}{M} \int_1^2 \int_0^z \left[\frac{1}{2}z^3 - xz^2 - \left(\frac{1}{2}x^2 z - x^2 z \right) \right] dx\, dz = \frac{1}{M} \int_1^2 \int_0^z \left(\frac{1}{2}z^3 - xz^2 + \frac{1}{2}x^2 z \right) dx\, dz$

$\quad = \frac{1}{M} \int_1^2 \left[\frac{1}{2}xz^3 - \frac{1}{2}x^2 z^2 + \frac{1}{6}x^3 z \right]_{x=0}^{x=z} dz = \frac{1}{M} \int_1^2 \frac{1}{6}z^4\, dz = \frac{1}{6M} \left[\frac{1}{5}z^5 \right]_1^2 = \frac{31}{30} \cdot \frac{8}{5} = \frac{124}{75}$

Thus, $(\bar{x}, \bar{y}, \bar{z}) = \left(\frac{31}{75}, \frac{31}{25}, \frac{124}{75} \right)$.

41.

Integrating first wrt z: As $y = -2z + 4 \Rightarrow 2z = 4 - y \Rightarrow z = 2 - \frac{1}{2}y$, we have $0 \le z \le 2 - \frac{1}{2}y$. Then, projecting the solid onto the xy-plane yields the region shown in figures (a) and (b). In figure (a), we're integrating first wrt x and then wrt y, and in (b), first wrt y then wrt x. Thus, respectively,

$$\int_0^4 \int_{-\sqrt{y}}^{\sqrt{y}} \int_0^{2-\frac{1}{2}y} k(x,y,z)\, dz\, dx\, dy, \quad \int_{-2}^2 \int_{x^2}^4 \int_0^{2-\frac{1}{2}y} k(x,y,z)\, dz\, dy\, dx$$

Integrating first wrt x: The figure of the solid, above, shows that x varies over the parabolic cylinder $y = x^2 \Rightarrow x = \pm\sqrt{y}$, so we have $-\sqrt{y} \le x \le \sqrt{y}$. Then, projecting the solid onto the yz-plane yields the region shown in figures (c) and (d). In figure (c), we're integrating first wrt y and then wrt z, and in (d), first wrt z then wrt y. Thus, respectively,

$$\int_0^2 \int_0^{-2z+4} \int_{-\sqrt{y}}^{\sqrt{y}} k(x,y,z)\,dx\,dy\,dz, \quad \int_0^4 \int_0^{2-\frac{1}{2}y} \int_{-\sqrt{y}}^{\sqrt{y}} k(x,y,z)\,dx\,dz\,dy$$

Integrating first wrt y: The figure of the solid, above, shows that y varies from $y = x^2$ to the plane $y = -2z + 4$, so we have $x^2 \le y \le -2z + 4$. Then, projecting the solid onto the xz-plane yields the region shown in figures (e) and (f). In figure (e), we're integrating first wrt x and then wrt z, and in (f), first wrt z then wrt x. Thus, respectively,

$$\int_0^2 \int_{-\sqrt{4-2z}}^{\sqrt{4-2z}} \int_{x^2}^{-2z+4} k(x,y,z)\,dy\,dx\,dz, \quad \text{and}$$

$$\int_{-2}^2 \int_0^{2-\frac{1}{2}x^2} \int_{x^2}^{-2z+4} k(x,y,z)\,dy\,dz\,dx.$$

43. 　Integrating first wrt z: As $y = 4 - x^2 - 4z^2 \Rightarrow 4z^2 = 4 - x^2 - y \Rightarrow z = \pm\frac{1}{2}\sqrt{4 - x^2 - y}$, we have $-\frac{1}{2}\sqrt{4 - x^2 - y} \le z \le \frac{1}{2}\sqrt{4 - x^2 - y}$. Then, projecting the solid onto the xy-plane yields the region shown in figures (a) and (b). In figure (a), we're integrating first wrt x and then wrt y, and in (b), first wrt y then wrt x. Thus, respectively,

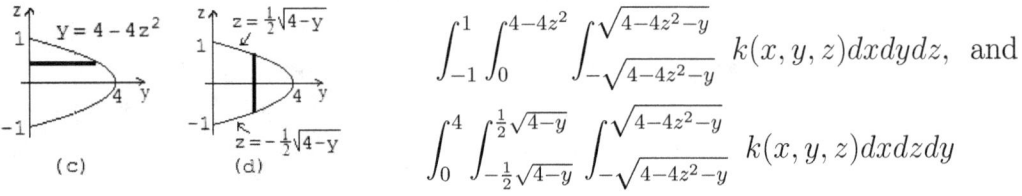

$$\int_0^4 \int_{-\sqrt{4-y}}^{\sqrt{4-y}} \int_{-\frac{1}{2}\sqrt{4-x^2-y}}^{\frac{1}{2}\sqrt{4-x^2-y}} k(x,y,z)\,dz\,dx\,dy, \quad \text{and}$$

$$\int_{-2}^2 \int_0^{4-x^2} \int_{-\frac{1}{2}\sqrt{4-x^2-y}}^{\frac{1}{2}\sqrt{4-x^2-y}} k(x,y,z)\,dz\,dy\,dx$$

Integrating first wrt x: The figure of the solid, above, shows that x varies over the paraboloid $y = 4 - x^2 - 4z^2 \Rightarrow x^2 = 4 - y - 4z^2 \Rightarrow x = \pm\sqrt{4 - y - 4z^2}$, so we have $-\sqrt{4 - y - 4z^2} \le x \le \sqrt{4 - y - 4z^2}$. Then, projecting the solid onto the yz-plane yields the region shown in figures (c) and (d). In figure (c), we're integrating first wrt y and then wrt z, and in (d), first wrt z then wrt y. Thus, respectively,

$$\int_{-1}^1 \int_0^{4-4z^2} \int_{-\sqrt{4-4z^2-y}}^{\sqrt{4-4z^2-y}} k(x,y,z)\,dx\,dy\,dz, \quad \text{and}$$

$$\int_0^4 \int_{-\frac{1}{2}\sqrt{4-y}}^{\frac{1}{2}\sqrt{4-y}} \int_{-\sqrt{4-4z^2-y}}^{\sqrt{4-4z^2-y}} k(x,y,z)\,dx\,dz\,dy$$

Integrating first wrt y: The figure of the solid, above, shows that y varies from $y = 0$ to the paraboloid $y = 4 - x^2 - 4z^2$, so we have $0 \le y \le 4 - x^2 - 4z^2$. Then, projecting the solid onto the xz-plane yields the region shown in figures (e) and (f). In figure (e), we're integrating first wrt x and then wrt z, and in (f), first wrt z then wrt x. Thus, respectively,

$$\int_{-1}^1 \int_{-\sqrt{4-4z^2}}^{\sqrt{4-4z^2}} \int_0^{4-x^2-4z^2} k(x,y,z)\,dy\,dx\,dz, \quad \text{and}$$

$$\int_{-2}^2 \int_{-\frac{1}{2}\sqrt{4-x^2}}^{\frac{1}{2}\sqrt{4-x^2}} \int_0^{4-x^2-4z^2} k(x,y,z)\,dy\,dz\,dx.$$

45.

$$\int_0^1 \int_y^1 \int_0^y k(x,y,z)\,dzdxdy \;\Rightarrow\; 0 \le z \le y,\; y \le x \le 1,\; \text{and } 0 \le y \le 1.$$

Integrating first wrt z again, we project the solid onto the xy-plane resulting in the region of figure (a), where as indicated we next integrate wrt y and finally wrt x.

$$\int_0^1 \int_0^x \int_0^y k(x,y,z)\,dzdydx$$

Integrating first wrt x: The figure of the solid, above, shows that x varies from $x = y$ to $x = 1 \Rightarrow y \le x \le 1$. Then, projecting the solid onto the yz-plane yields the region shown in figures (b) and (c). In figure (b), we're integrating first wrt y and then wrt z, and in (c), first wrt z then wrt y. Thus, respectively,

$$\int_0^1 \int_z^1 \int_y^1 k(x,y,z)\,dxdydz, \text{ and } \int_0^1 \int_0^y \int_y^1 k(x,y,z)\,dxdzdy$$

Integrating first wrt y: The figure of the solid, above, shows that y varies from $y = z$ to $y = x \Rightarrow z \le y \le x$. Then, projecting the solid onto the xz-plane yields the region shown in figures (d) and (e). In figure (d), we're integrating first wrt x and then wrt z, and in (c), first wrt z then wrt x. Thus, respectively,

$$\int_0^1 \int_z^1 \int_z^x k(x,y,z)\,dydxdz, \text{ and } \int_0^1 \int_0^x \int_z^x k(x,y,z)\,dydzdx.$$

47.

$$\int_0^1 \int_{-1}^0 \int_0^{y^2} k(x,y,z)\,dzdydx \;\Rightarrow\; 0 \le z \le y^2,\; -1 \le y \le 0,\; \text{and } 0 \le x \le 1.$$

Integrating first wrt z again, we project the solid onto the xy-plane resulting in the square region of figure (a), where as indicated we next integrate wrt x and finally wrt y.

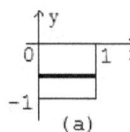

$$\int_{-1}^0 \int_0^1 \int_0^{y^2} k(x,y,z)\,dzdxdy$$

Integrating first wrt x: The figure of the solid, above, shows that x varies from $x = 0$ to $x = 1 \Rightarrow 0 \le x \le 1$. Then, projecting the solid onto the yz-plane yields the region shown in figures (b) and (c). In figure (b), we're integrating first wrt y and then wrt z, and in (c), first wrt z then wrt y. Thus, respectively,

$$\int_0^1 \int_{-1}^{-\sqrt{z}} \int_0^1 k(x,y,z)\,dxdydz, \text{ and } \int_{-1}^0 \int_0^{y^2} \int_0^1 k(x,y,z)\,dxdzdy$$

Integrating first wrt y: The figure of the solid, above, shows that y varies from $y = -1$ to $z = y^2, y \le 0, \Rightarrow y = -\sqrt{z} \Rightarrow -1 \le y \le -\sqrt{z}$. Then, projecting the solid onto the xz-plane yields the square region shown in figures (d) and (e). In figure (d), we're integrating first wrt x and then wrt z, and in (c), first wrt z then wrt x. Thus, respectively, (and this time identically equal)

(d)

(e)

$$\int_0^1 \int_0^1 \int_{-1}^{-\sqrt{z}} k(x,y,z)\,dy\,dx\,dz, \text{ and } \int_0^1 \int_0^1 \int_{-1}^{-\sqrt{z}} k(x,y,z)\,dy\,dz\,dx.$$

§11.6 Cylindrical and Spherical Coordinates

1. $(r, \theta, z) = \left(2, \frac{5\pi}{6}, 3\right) \Rightarrow r = 2, \theta = \frac{5\pi}{6}$ and $z = 3$. Then

$x = r\cos\theta = 2\cos\frac{5\pi}{6} = 2\left(-\frac{\sqrt{3}}{2}\right) = -\sqrt{3}$, and $y = r\sin\theta = 2\sin\frac{5\pi}{6} = 2 \cdot \frac{1}{2} = 1 \Rightarrow$ $(x,y,z) = (-\sqrt{3}, 1, 2)$.

3. $(r, \theta, z) = \left(6, \frac{\pi}{4}, 1\right) \Rightarrow r = 6, \theta = \frac{\pi}{4}$ and $z = 1$. Then

$x = r\cos\theta = 6\cos\frac{\pi}{4} = 6 \cdot \frac{1}{\sqrt{2}} = 3\sqrt{2}$, and $y = r\sin\theta = 6\sin\frac{\pi}{4} = 6 \cdot \frac{1}{\sqrt{2}} = 3\sqrt{2} \Rightarrow$ $(x,y,z) = (3\sqrt{2}, 3\sqrt{2}, 1)$.

5. $(x, y, z) = (-4, -4\sqrt{3}, -8) \Rightarrow x = -4, y = -4\sqrt{3}, z = -8$. Then $x^2 + y^2 = r^2 \Rightarrow$ $r^2 = (-4)^2 + (-4\sqrt{3})^2 = 16 + 48 = 64 \Rightarrow r = 8$. As $\tan\theta = \frac{y}{x} = \frac{-4\sqrt{3}}{-4} = \sqrt{3} \Rightarrow \theta = \frac{\pi}{3}$ or $\frac{4\pi}{3}$ and, $x < 0$ and $y < 0$, then $\theta = \frac{4\pi}{3} \Rightarrow (r, \theta, z) = \left(8, \frac{4\pi}{3}, -8\right)$.

7. $(x, y, z) = (0, 2, -2) \Rightarrow x = 0, y = 2, z = -2$. Then $x^2 + y^2 = r^2 \Rightarrow r^2 = 4 \Rightarrow r = 2$. As $\tan\theta = \frac{y}{x}$ is undefined, and in the xy-plane $(0, 2)$ lies on the y-axis, $\theta = \frac{\pi}{2}$, so that $(r, \theta, z) = \left(2, \frac{\pi}{2}, -2\right)$.

9. As $x^2 + y^2 = r^2$, $z = x^2 + y^2$ becomes $z = r^2$.

11. As $x^2 + y^2 = r^2$ and $y = r\sin\theta$, then $x^2 + y^2 = 2y \Rightarrow r^2 = 2r\sin\theta \Rightarrow r = 2\sin\theta$.

13.
$$\int_0^{2\pi} \int_0^1 \int_0^{\sqrt{1-r^2}} r\,dz\,dr\,d\theta = \int_0^{2\pi} \int_0^1 \left[rz\right]_{z=0}^{z=\sqrt{1-r^2}} dr\,d\theta = \int_0^{2\pi} \int_0^1 r\sqrt{1-r^2}\,dr\,d\theta$$

$$= \int_0^{2\pi} \int_1^0 -\frac{1}{2}u^{\frac{1}{2}}\,du\,d\theta \qquad \boxed{\begin{array}{ll} u = 1 - r^2, & r = 0 \Rightarrow u = 1 \\ du = -2r\,dr & r = 1 \Rightarrow u = 0 \end{array}}$$

$$= -\frac{1}{2}\int_0^{2\pi} \left[\frac{2}{3}u^{\frac{3}{2}}\right]_{u=1}^{u=0} d\theta$$

$$= -\frac{1}{3}\int_0^{2\pi} (0-1)\,d\theta = \frac{1}{3}\left[\theta\right]_0^{2\pi} = \frac{2\pi}{3}.$$

15. $\displaystyle\int_0^1\int_0^{\sqrt{z}}\int_0^{2\pi}(r^2\cos^2\theta+z^2)\,r\,d\theta drdz=\int_0^1\int_0^{\sqrt{z}}\int_0^{2\pi}\left[r^3\left(\frac{1+\cos2\theta}{2}\right)+rz^2\right]d\theta drdz$

$$=\int_0^1\int_0^{\sqrt{z}}\left[\frac{r^3}{2}\left(\theta+\tfrac{1}{2}\sin2\theta\right)+rz^2\theta\right]_{\theta=0}^{\theta=2\pi}drdz$$

$$=\int_0^1\int_0^{\sqrt{z}}\left(\frac{r^3}{2}\cdot2\pi+2\pi rz^2\right)drdz$$

$$=\int_0^1\left[\frac{\pi}{4}r^4+\pi r^2z^2\right]_{r=0}^{r=\sqrt{z}}dz=\int_0^1\left(\frac{\pi}{4}z^2+\pi z^3\right)dz$$

$$=\left[\frac{\pi}{12}z^3+\frac{\pi}{4}z^4\right]_0^1=\frac{\pi}{12}+\frac{\pi}{4}=\frac{\pi}{3}.$$

17.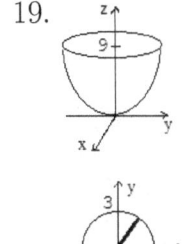
$\displaystyle\int_0^1\int_0^{\sqrt{1-x^2}}\int_0^{\sqrt{4-(x^2+y^2)}}dzdydx\ \Rightarrow\ 0\le y\le\sqrt{1-x^2}$ and $0\le x\le1$,
which is the first quadrant of the unit circle (see figure). As $x^2+y^2=r^2$,
$4-(x^2+y^2)=4-r^2$, and $dydx=rdrd\theta$.

$\displaystyle\int_0^1\int_0^{\sqrt{1-x^2}}\int_0^{\sqrt{4-(x^2+y^2)}}dzdydx=\int_0^{\pi/2}\int_0^1\int_0^{\sqrt{4-r^2}}rdzdrd\theta=\int_0^{\pi/2}\int_0^1\left[rz\right]_{z=0}^{z=\sqrt{4-r^2}}drd\theta$

$$=\int_0^{\pi/2}\int_0^1 r\sqrt{4-r^2}\,drd\theta=\int_0^{\pi/2}\int_4^3-\tfrac{1}{2}u^{\frac{1}{2}}dud\theta$$

$$\boxed{\begin{array}{ll}u=4-r^2,&r=0\ \Rightarrow\ u=4\\du=-2r\,dr&r=1\ \Rightarrow\ u=3\end{array}}$$

$$=-\frac{1}{2}\int_0^{\pi/2}\left[\frac{2}{3}u^{\frac{3}{2}}\right]_{u=4}^{u=3}d\theta=-\frac{1}{3}\int_0^{\pi/2}(3\sqrt{3}-8)d\theta$$

$$=\frac{1}{3}\left(8-3\sqrt{3}\right)\int_0^{\pi/2}d\theta=\frac{1}{3}\left(8-3\sqrt{3}\right)\cdot\frac{\pi}{2}=\frac{\pi}{6}\left(8-3\sqrt{3}\right).$$

19.
Since z varies from the paraboloid $z=x^2+y^2$, i.e. $z=r^2$, up to the
plane $z=9$ (see top figure), then $r^2\le z\le9$. The projection of the
solid paraboloid onto the xy-plane is the circle in the bottom figure. So
$$V=\int_0^{2\pi}\int_0^3\int_{r^2}^9 r\,dzdrd\theta=\int_0^{2\pi}\int_0^3\left[rz\right]_{z=r^2}^{z=9}drd\theta=\int_0^{2\pi}\int_0^3(9-r^2)rdrd\theta$$

$$=\int_0^{2\pi}\int_0^3(9r-r^3)drd\theta=\int_0^{2\pi}\left[\frac{9}{2}r^2-\frac{1}{4}r^4\right]_{r=0}^{r=3}d\theta=\int_0^{2\pi}\left(\frac{81}{2}-\frac{81}{4}\right)d\theta$$

$$=\frac{81}{4}\left[\theta\right]_0^{2\pi}=\frac{81\pi}{2}.$$

21.
Since z varies from $z=0$ up to the cone $z=\sqrt{x^2+y^2}$, i.e. $z=r$
(see top figure), then $0\le z\le r$. The projection of the solid onto the
xy-plane is the circle in the bottom figure. Thus,
$$V=\int_0^{2\pi}\int_0^2\int_0^r r\,dzdrd\theta=\int_0^{2\pi}\int_0^2\left[rz\right]_{z=0}^{z=r}drd\theta=\int_0^{2\pi}\int_0^2 r^2drd\theta$$

$$=\int_0^{2\pi}\left[\frac{1}{3}r^3\right]_{r=0}^{r=2}d\theta=\frac{8}{3}\int_0^{2\pi}d\theta=\frac{16\pi}{3}.$$

23.

Since z varies from $z = -1$ up to the plane $z = 4 - x$, i.e. $z = 4 - r\cos\theta$ (see top figure), then $-1 \le z \le 4 - r\cos\theta$. The projection of the solid onto the xy-plane is the circle in the bottom figure. Therefore,

$$V = \int_0^{2\pi}\int_0^2\int_{-1}^{4-r\cos\theta} r\,dz\,dr\,d\theta = \int_0^{2\pi}\int_0^2 \left[rz\right]_{-1}^{4-r\cos\theta} dr\,d\theta$$

$$= \int_0^{2\pi}\int_0^2 r(4 - r\cos\theta + 1)dr\,d\theta = \int_0^{2\pi}\int_0^2 (5r - r^2\cos\theta)dr\,d\theta$$

$$= \int_0^{2\pi}\left[\tfrac{5}{2}r^2 - \tfrac{1}{3}r^3\cos\theta\right]_{r=0}^{r=2} d\theta = \int_0^{2\pi}\left(10 - \tfrac{8}{3}\cos\theta\right)d\theta = \left[10\theta - \tfrac{8}{3}\sin\theta\right]_0^{2\pi} = 20\pi.$$

25.

Since z varies from the bottom of the sphere to the top (see figure), and $x^2 + y^2 + z^2 = 4 \Rightarrow r^2 + z^2 = 4 \Rightarrow z = \pm\sqrt{4 - r^2}$, then $-\sqrt{4 - r^2} \le z \le \sqrt{4 - r^2}$. The projection of the solid onto the xy-plane (in the bottom figure) shows that $0 \le r \le 2\sin\theta$, while $0 \le \theta \le \pi$. By symmetry, the volume is 4 times the volume in the first octant. Consequently,

$$V = 4\int_0^{\pi/2}\int_0^{2\sin\theta}\int_0^{\sqrt{4-r^2}} r\,dz\,dr\,d\theta = 4\int_0^{\pi/2}\int_0^{2\sin\theta}\left[rz\right]_{z=0}^{z=\sqrt{4-r^2}} dr\,d\theta$$

$$= 4\int_0^{\pi/2}\int_0^{2\sin\theta} r\sqrt{4 - r^2}\,dr\,d\theta = 2\int_0^{\pi/2}\int_4^{4\cos^2\theta} -u^{\frac{1}{2}}\,du\,d\theta$$

$$\boxed{\begin{array}{ll} u = 4 - r^2, & r = 0 \Rightarrow u = 4 \\ du = -2r\,dr & r = 2\sin\theta \Rightarrow u = 4\cos^2\theta \end{array}}$$

$$= -\frac{4}{3}\int_0^{\pi/2}\left[u^{\frac{3}{2}}\right]_{u=4}^{u=4\cos^2\theta} d\theta = -\frac{4}{3}\int_0^{\pi/2}(8\cos^3\theta - 8)d\theta =$$

$$= \frac{32}{3}\left[\int_0^{\pi/2} d\theta - \int_0^{\pi/2}\cos^3\theta\,d\theta\right] = \frac{32}{3}\left[\frac{\pi}{2} - \int_0^{\pi/2}(1 - \sin^2\theta)\cos\theta\,d\theta\right]$$

$$= \frac{16\pi}{3} - \frac{32}{3}\left[\sin\theta - \frac{1}{3}\sin^3\theta\right]_0^{\pi/2} = \frac{16\pi}{3} - \frac{32}{3}\left(1 - \frac{1}{3}\right) = \frac{48\pi - 64}{9}$$

$$= \frac{16}{9}(3\pi - 4)$$

[Note: Without using any symmetry, we would have had to evaluate $\int_0^\pi (\cos^2)^{\frac{3}{2}}\,d\theta = \int_0^\pi |\cos\theta|^3 d\theta = \int_0^{\pi/2}(\cos\theta)^3 d\theta + \int_{\pi/2}^\pi (-\cos\theta)^3 d\theta$, etc. But on $[0, \frac{\pi}{2}]$, $|\cos\theta| = \cos\theta$.]

27.

For the top half of the ellipsoid (top figure), z goes from $z = 0$ to the ellipsoid, $4x^2 + 4y^2 + z^2 = 16 \Rightarrow z = \sqrt{16 - 4x^2 - 4y^2} = \sqrt{16 - 4r^2}$, i.e. $0 \le z \le \sqrt{16 - 4r^2}$. The projection onto the xy-plane is the circle in the bottom figure. Thus,

$$M = \int_0^{2\pi}\int_0^2\int_0^{\sqrt{16-4r^2}} kz\,r\,dz\,dr\,d\theta = k\int_0^{2\pi}\int_0^2 \left[\tfrac{1}{2}z^2 r\right]_{z=0}^{z=\sqrt{16-4r^2}} dr\,d\theta$$

$$= \frac{k}{2}\int_0^{2\pi}\int_0^2 (16 - 4r^2)r\,dr\,d\theta = \frac{k}{2}\int_0^{2\pi}\int_0^2 (16r - 4r^3)\,dr\,d\theta$$

$$= \frac{k}{2}\int_0^{2\pi}\left[8r^2 - r^4\right]_{r=0}^{r=2} d\theta = \frac{k}{2}\int_0^{2\pi}(32 - 16)\,d\theta = 8k\left[\theta\right]_0^{2\pi} = 16\pi k.$$

29.

For the top half of the paraboloid (top figure), z goes from $z = 0$ to the paraboloid, $z = 1 - (x^2 + y^2) = 1 - r^2$, i.e. $0 \le z \le 1 - r^2$. The projection onto the xy-plane is the unit circle (bottom figure). Thus,

(a) $M = \displaystyle\int_0^{2\pi} \int_0^1 \int_0^{1-r^2} kz\, r\, dz\, dr\, d\theta = k \int_0^{2\pi} \int_0^1 \left[\tfrac{1}{2} z^2 r\right]_{z=0}^{z=1-r^2} dr\, d\theta$

$= \dfrac{k}{2} \displaystyle\int_0^{2\pi} \int_0^1 (1-r^2)^2 r\, dr\, d\theta = \dfrac{k}{2} \int_0^{2\pi} \int_0^1 (r - 2r^3 + r^5)\, dr\, d\theta$

$= \dfrac{k}{2} \displaystyle\int_0^{2\pi} \left[\tfrac{1}{2} r^2 - \tfrac{1}{2} r^4 + \tfrac{1}{6} r^6\right]_{r=0}^{r=1} d\theta = \dfrac{k}{2} \cdot \dfrac{1}{6} \int_0^{2\pi} d\theta = \dfrac{\pi k}{6}$

(b) $M = \displaystyle\int_0^{2\pi} \int_0^1 \int_0^{1-r^2} k(\sqrt{r^2 + z^2})^2\, r\, dz\, dr\, d\theta = k \int_0^{2\pi} \int_0^1 \int_0^{1-r^2} (r^3 + rz^2)\, dz\, dr\, d\theta$

$= k \displaystyle\int_0^{2\pi} \int_0^1 \left[r^3 z + \tfrac{1}{3} rz^3\right]_{z=0}^{z=1-r^2} = k \int_0^{2\pi} \int_0^1 \left[r^3(1-r^2) + \tfrac{1}{3} r(1-r^2)^3\right] dr\, d\theta$

$= k \displaystyle\int_0^{2\pi} \int_0^1 \left[r^3 - r^5 + \tfrac{1}{3} r(1 - 3r^2 + 3r^4 - r^6)\right] dr\, d\theta = \dfrac{k}{3} \int_0^{2\pi} \int_0^1 (r - r^7)\, dr\, d\theta$

$= \dfrac{k}{3} \displaystyle\int_0^{2\pi} \left[\tfrac{1}{2} r^2 - \tfrac{1}{8} r^8\right]_{r=0}^{r=1} d\theta = \dfrac{k}{3} \left(\dfrac{1}{2} - \dfrac{1}{8}\right) \int_0^{2\pi} d\theta = \dfrac{\pi k}{4}.$

31.

The values of z range from the paraboloid $z = 4x^2 + 4y^2 = 4r^2$ (top figure) to the plane $z = a$, so that $4r^2 \le z \le a$. The projection of the solid on the xy-plane is the circle: $4r^2 = a \Rightarrow r = \tfrac{1}{2}\sqrt{a}$ (bottom figure). Thus,

$M = \displaystyle\int_0^{2\pi} \int_0^{\sqrt{a}/2} \int_{4r^2}^{a} k\, r\, dz\, dr\, d\theta = k \int_0^{2\pi} \int_0^{\sqrt{a}/2} \left[rz\right]_{z=4r^2}^{z=a} dr\, d\theta$

$= k \displaystyle\int_0^{2\pi} \int_0^{\sqrt{a}/2} (ar - 4r^3)\, dr\, d\theta = k \int_0^{2\pi} \left[\tfrac{1}{2} ar^2 - r^4\right]_0^{\sqrt{a}/2} d\theta$

$= k \displaystyle\int_0^{2\pi} \left[\dfrac{a}{2}\left(\dfrac{a}{4}\right) - \left(\dfrac{a}{4}\right)^2\right] d\theta = \dfrac{ka^2}{16} \int_0^{2\pi} d\theta = \dfrac{k\pi a^2}{8}.$

By symmetry, $\bar{x} = \bar{y} = 0$.

$\bar{z} = \dfrac{1}{M} \displaystyle\int_0^{2\pi} \int_0^{\sqrt{a}/2} \int_{4r^2}^{a} z\, k\, r\, dz\, dr\, d\theta = \dfrac{k}{M} \int_0^{2\pi} \int_0^{\sqrt{a}/2} \left[\tfrac{1}{2} rz^2\right]_{z=4r^2}^{z=a} dr\, d\theta$

$= \dfrac{k}{2M} \displaystyle\int_0^{2\pi} \int_0^{\sqrt{a}/2} r[a^2 - (4r^2)^2]\, dr\, d\theta = \dfrac{k}{2M} \int_0^{2\pi} \int_0^{\sqrt{a}/2} (a^2 r - 16 r^5)\, dr\, d\theta$

$= \dfrac{k}{2M} \displaystyle\int_0^{2\pi} \left[\tfrac{1}{2} a^2 r^2 - \tfrac{16}{6} r^6\right]_0^{\sqrt{a}/2} d\theta = \dfrac{k}{2M} \int_0^{2\pi} \left[\dfrac{a^2}{2}\left(\dfrac{a}{4}\right) - \dfrac{8}{3}\left(\dfrac{a}{4}\right)^3\right] d\theta = \dfrac{k}{2M} \int_0^{2\pi} \dfrac{a^3}{12} d\theta$

$= \dfrac{ka^3}{24M} \displaystyle\int_0^{2\pi} d\theta = \dfrac{k\pi a^3}{12} \cdot \dfrac{8}{k\pi a^2} = \dfrac{2a}{3}$

Hence, $(\bar{x}, \bar{y}, \bar{z}) = \left(0, 0, \dfrac{2a}{3}\right).$

33.

The values of z range from $z = 0$ up to the hemisphere $x^2 + y^2 + z^2 = 1$, $z \geq 0$ $\Rightarrow$ $z = \sqrt{1 - (x^2 + y^2)} = \sqrt{1 - r^2}$ (top figure). The projection of the solid on the xy-plane is the unit circle (bottom figure). Thus,

$$M = \int_0^{2\pi} \int_0^1 \int_0^{\sqrt{1-r^2}} k\, r\, dz\, dr\, d\theta = k \int_0^{2\pi} \int_0^1 \left[rz \right]_{z=0}^{z=\sqrt{1-r^2}} dz\, dr\, d\theta$$

$$= k \int_0^{2\pi} \int_0^1 r\sqrt{1-r^2}\, dr\, d\theta = k \int_0^{2\pi} \int_1^0 -\tfrac{1}{2} u^{\frac{1}{2}}\, du\, d\theta$$

$$\boxed{\begin{array}{ll} u = 1 - r^2, & r = 0 \Rightarrow u = 1 \\ du = -2r\, dr & r = 1 \Rightarrow u = 0 \end{array}}$$

$$= -\frac{k}{2} \int_0^{2\pi} \left[\tfrac{2}{3} u^{\frac{3}{2}} \right]_{u=1}^{u=0} d\theta = -\frac{k}{2} \int_0^{2\pi} -\tfrac{2}{3}\, d\theta = \frac{2\pi k}{3}$$

By symmetry, $\bar{x} = \bar{y} = 0$.

$$\bar{z} = \frac{1}{M} \int_0^{2\pi} \int_0^1 \int_0^{\sqrt{1-r^2}} kz\, r\, dz\, dr\, d\theta = \frac{k}{M} \int_0^{2\pi} \int_0^1 \left[\tfrac{1}{2} rz^2 \right]_{z=0}^{z=\sqrt{1-r^2}} dz\, dr\, d\theta$$

$$= \frac{k}{2M} \int_0^{2\pi} \int_0^1 r(1 - r^2)\, dr\, d\theta = \frac{k}{2M} \int_0^{2\pi} \int_0^1 (r - r^3)\, dr\, d\theta = \frac{k}{2M} \int_0^{2\pi} \left[\tfrac{1}{2} r^2 - \tfrac{1}{4} r^4 \right]_{r=0}^{r=1} d\theta$$

$$= \frac{k}{2M} \int_0^{2\pi} \tfrac{1}{4}\, d\theta = \frac{k}{2M} \cdot \tfrac{1}{4} \cdot 2\pi = \frac{\pi k}{4M} = \frac{\pi k}{4} \cdot \frac{3}{2\pi k} = \frac{3}{8}$$

Hence, $(\bar{x}, \bar{y}, \bar{z}) = \left(0, 0, \tfrac{3}{8} \right)$.

35.

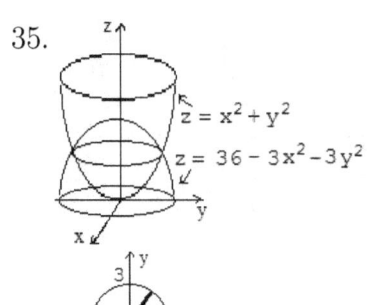

The values of z range from the paraboloid $z = x^2 + y^2 = r^2$ up to the paraboloid $z = 36 - 3x^2 - 3y^2 = 36 - 3r^2$ $\Rightarrow$ $r^2 \leq z \leq 36 - 3r^2$ (top figure). The projection of the solid on the xy-plane is the circle of intersection of the two paraboloids $x^2 + y^2 = 36 - 3x^2 - 3y^2$ $\Rightarrow$ $4x^2 + 4y^2 = 36$ $\Rightarrow$ $x^2 + y^2 = 9$ (bottom figure). Thus,

$$M = \int_0^{2\pi} \int_0^3 \int_{r^2}^{36-3r^2} k\, r\, dz\, dr\, d\theta = k \int_0^{2\pi} \int_0^3 \left[rz \right]_{z=r^2}^{z=36-3r^2} dr\, d\theta$$

$$= k \int_0^{2\pi} \int_0^3 r(36 - 3r^2 - r^2)\, dr\, d\theta = k \int_0^{2\pi} \int_0^3 (36r - 4r^3)\, dr\, d\theta$$

$$= k \int_0^{2\pi} \left[18r^2 - r^4 \right]_{r=0}^{r=3} d\theta = k \int_0^{2\pi} (162 - 81)\, d\theta = 162\pi k$$

By symmetry, $\bar{x} = \bar{y} = 0$.

$$\bar{z} = \frac{1}{M} \int_0^{2\pi} \int_0^3 \int_{r^2}^{36-3r^2} kz\, r\, dz\, dr\, d\theta = \frac{k}{M} \int_0^{2\pi} \int_0^3 \left[\tfrac{1}{2} rz^2 \right]_{z=r^2}^{z=36-3r^2} dr\, d\theta$$

$$= \frac{k}{2M} \int_0^{2\pi} \int_0^3 r[(36 - 3r^2)^2 - (r^2)^2]\, dr\, d\theta = \frac{k}{2M} \int_0^{2\pi} \int_0^3 \left(1296\, r - 216\, r^3 + 8r^5 \right) dr\, d\theta$$

$$= \frac{k}{2M} \int_0^{2\pi} \left[648r^2 - 54r^4 + \tfrac{4}{3} r^6 \right]_{r=0}^{r=3} d\theta = \frac{k}{2M} \left[648(9) - 54(81) + \tfrac{4}{3}(3^6) \right] \int_0^{2\pi} d\theta$$

$$= \frac{2430k}{2M} \cdot 2\pi = \frac{2430\pi k}{162\pi k} = 15. \quad \text{Hence, } (\bar{x}, \bar{y}, \bar{z}) = (0, 0, 15).$$

37. $(\rho, \phi, \theta) = \left(4, \frac{\pi}{3}, \frac{\pi}{4}\right) \implies x = \rho \sin \phi \cos \theta = 4 \sin \frac{\pi}{3} \cos \frac{\pi}{4} = 4 \cdot \frac{\sqrt{3}}{2} \cdot \frac{1}{\sqrt{2}} = \frac{2\sqrt{3}}{\sqrt{2}} \cdot \frac{\sqrt{2}}{\sqrt{2}} = \sqrt{6}$

$y = \rho \sin \phi \sin \theta = 4 \sin \frac{\pi}{3} \sin \frac{\pi}{4} = 4 \cdot \frac{\sqrt{3}}{2} \cdot \frac{1}{\sqrt{2}} = \sqrt{6}, \ z = \rho \cos \phi = 4 \cos \frac{\pi}{3} = 4 \cdot \frac{1}{2} = 2 \implies$

$(x, y, z) = \left(\sqrt{6}, \sqrt{6}, 2\right).$

39. $(\rho, \phi, \theta) = \left(7, \frac{\pi}{2}, \pi\right) \implies x = \rho \sin \phi \cos \theta = 7 \sin \frac{\pi}{2} \cos \pi = -7, \ y = \rho \sin \phi \sin \theta =$

$7 \sin \frac{\pi}{2} \sin \pi = 0, \ z = \rho \cos \phi = 7 \cos \frac{\pi}{2} = 0 \implies (x, y, z) = (-7, 0, 0).$

41. $(x, y, z) = \left(1, \sqrt{3}, 2\sqrt{3}\right) \implies \rho = \sqrt{x^2 + y^2 + z^2} = \sqrt{1 + 3 + 12} = \sqrt{16} = 4,$

$z = \rho \cos \phi \implies \cos \phi = \frac{z}{\rho} = \frac{2\sqrt{3}}{4} = \frac{\sqrt{3}}{2} \implies \phi = \frac{\pi}{6}, \ x = \rho \sin \phi \cos \theta \implies$

$\cos \theta = \frac{x}{\rho \sin \phi} = \frac{1}{4 \sin \frac{\pi}{6}} = \frac{1}{2} \implies \theta = \frac{\pi}{3} \implies (\rho, \phi, \theta) = \left(4, \frac{\pi}{6}, \frac{\pi}{3}\right).$

43. $(x, y, z) = (0, -1, -1) \implies \rho = \sqrt{x^2 + y^2 + z^2} = \sqrt{0 + 1 + 1} = \sqrt{2}, \ z = \rho \cos \phi \implies$

$\cos \phi = \frac{z}{\rho} = \frac{-1}{\sqrt{2}} \implies \phi = \frac{3\pi}{4}, \ x = \rho \sin \phi \cos \theta \implies \cos \theta = \frac{x}{\rho \sin \phi} = \frac{0}{\sqrt{2} \sin \frac{3\pi}{4}} = 0 \implies$

$\theta = \frac{\pi}{2}$ or $\theta = \frac{3\pi}{2}$. As $y = -1 < 0$, then $\theta = \frac{3\pi}{2} \implies (\rho, \phi, \theta) = \left(\sqrt{2}, \frac{3\pi}{4}, \frac{3\pi}{2}\right).$

For Exercises 45 and 47: From $x = \rho \sin \phi \cos \theta, \ y = \rho \sin \phi \sin \theta, \ z = \rho \cos \phi$:

45. $z^2 = x^2 + y^2 \implies \rho^2 \cos^2 \phi = \rho^2 \sin^2 \phi \cos^2 \phi + \rho^2 \sin^2 \phi \sin^2 \phi = \rho^2 \sin^2 \phi (\cos^2 \theta + \sin^2 \theta) =$

$\rho^2 \sin^2 \phi \implies \cos^2 \phi = \sin^2 \phi \implies \tan^2 \phi = 1 \implies \phi = \frac{\pi}{4}$ or $\frac{3\pi}{4}.$

47. $x^2 + z^2 = 9 \implies \rho^2 \sin^2 \phi \cos^2 \theta + \rho^2 \cos^2 \phi = 9 \implies \rho^2 (\sin^2 \phi \cos^2 \theta + \cos^2 \phi) = 9.$

49. $\displaystyle\int_0^{2\pi} \int_0^\pi \int_0^1 \rho^3 \sin \phi \, d\rho \, d\phi \, d\theta = \int_0^{2\pi} \int_0^\pi \left[\frac{1}{4}\rho^4 \sin \phi\right]_{\rho=0}^{\rho=1} d\phi \, d\theta = \frac{1}{4} \int_0^{2\pi} \int_0^\pi \sin \phi \, d\phi \, d\theta$

$\displaystyle = \frac{1}{4} \int_0^{2\pi} \left[-\cos \phi\right]_{\phi=0}^{\phi=\pi} d\theta = \frac{1}{4} \int_0^{2\pi} (-\cos \pi + \cos 0) \, d\theta$

$\displaystyle = \frac{1}{2} \int_0^{2\pi} d\theta = \frac{1}{2}\left[\theta\right]_0^{2\pi} = \pi.$

51. $\displaystyle\int_0^{\pi/2} \int_0^{\sin \phi} \int_0^{\pi/4} \rho^2 \sin \phi \, d\theta \, d\rho \, d\phi = \int_0^{\pi/2} \int_0^{\sin \phi} \left[(\rho^2 \sin \phi)\theta\right]_{\theta=0}^{\theta=\pi/4} d\rho \, d\phi$

$\displaystyle = \frac{\pi}{4} \int_0^{\pi/2} \int_0^{\sin \phi} \rho^2 \sin \phi \, d\rho \, d\phi = \frac{\pi}{4} \int_0^{\pi/2} \left[\frac{1}{3}\rho^3 \sin \phi\right]_{\rho=0}^{\rho=\sin \phi} d\phi$

$\displaystyle = \frac{\pi}{12} \int_0^{\pi/2} \sin^4 \phi \, d\phi = \frac{\pi}{12} \int_0^{\pi/2} \left(\frac{1 - \cos 2\phi}{2}\right)^2 d\phi$

$\displaystyle = \frac{\pi}{48} \int_0^{\pi/2} \left(1 - 2\cos 2\phi + \frac{1 + \cos 4\phi}{2}\right) d\phi$

$\displaystyle = \frac{\pi}{48} \left[\frac{3}{2}\phi - \sin 2\phi + \frac{1}{8}\sin 4\phi\right]_0^{\pi/2} = \frac{\pi}{48}\left(\frac{3\pi}{4}\right) = \frac{\pi^2}{64}.$

53.

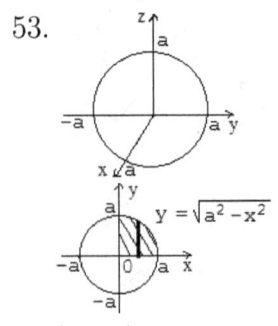

$$\int_0^a \int_0^{\sqrt{a^2-x^2}} \int_0^{\sqrt{a^2-x^2-y^2}} (x^2+y^2+z^2)\, dz\,dy\,dx \Rightarrow 0 \le z \le \sqrt{a^2-x^2-y^2},$$

$0 \le y \le \sqrt{a^2-x^2}$, and $0 \le x \le a$. Now, $z = \sqrt{a^2-x^2-y^2} \Rightarrow$ $x^2+y^2+z^2 = a^2$, $z \ge 0$ is the top hemisphere of the sphere in the top figure, and the range of values of x and y (see lower figure) restrict us to the 1st octant part of the sphere (i.e. $x \ge 0, y \ge 0, z \ge 0$). Therefore, $0 \le \rho \le a$, $0 \le \phi \le \frac{\pi}{2}$ and $0 \le \theta \le \frac{\pi}{2}$. Since $x^2+y^2+z^2 = \rho^2$, the integral in spherical coordinates becomes:

$$\int_0^{\pi/2} \int_0^{\pi/2} \int_0^a \rho^2\, \rho^2 \sin\phi\, d\rho\,d\phi\,d\theta = \int_0^{\pi/2}\int_0^{\pi/2} \left[\tfrac{1}{5}\rho^5 \sin\phi\right]_{\rho=0}^{\rho=a} d\phi\,d\theta = \frac{a^5}{5}\int_0^{\pi/2}\int_0^{\pi/2} \sin\phi\, d\phi\,d\theta$$

$$= \frac{a^5}{5}\int_0^{\pi/2}\left[-\cos\phi\right]_{\phi=0}^{\phi=\pi/2} d\theta = \frac{a^5}{5}\int_0^{\pi/2}\left(-\cos\tfrac{\pi}{2}+\cos 0\right) d\theta$$

$$= \frac{a^5}{5}\cdot\frac{\pi}{2} = \frac{\pi a^5}{10}.$$

55.

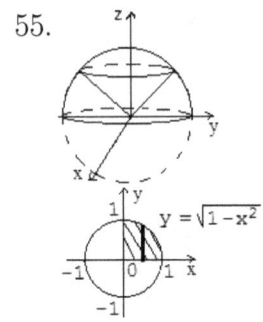

$$\int_0^1 \int_0^{\sqrt{1-x^2}} \int_{\sqrt{x^2+y^2}}^{\sqrt{2-x^2-y^2}} xy\, dz\,dy\,dx \Rightarrow \sqrt{x^2+y^2} \le z \le \sqrt{2-x^2-y^2},$$

$0 \le y \le \sqrt{1-x^2}$, and $0 \le x \le 1$. Now, $z = \sqrt{x^2+y^2} \Rightarrow$ $z^2 = x^2+y^2$, $z \ge 0$ which is the top nappe of the cone (see figure), and $z = \sqrt{2-x^2-y^2} \Rightarrow x^2+y^2+z^2 = 2$, $z \ge 0$ is the top hemisphere of radius $\sqrt{2}$ (see top figure). The cone and hemisphere intersect where $x^2+y^2 = 2-x^2-y^2 \Rightarrow x^2+y^2 = 1$, which is the unit circle (bottom figure), but as $x \ge 0$ and $y \ge 0$, (and $z \ge 0$) we are restricted to only the first octant.

Next, we find the spherical equations of the cone and sphere. The cone: $z^2 = x^2+y^2 \Rightarrow$ $\rho^2\cos^2\phi = \rho^2\sin^2\phi \Rightarrow \tan^2\phi = 1 \Rightarrow \phi = \frac{\pi}{4}$. The sphere: $x^2+y^2+z^2 = 2 \Rightarrow \rho = \sqrt{2}$. Therefore, $0 \le \rho \le \sqrt{2}$, $0 \le \theta \le \frac{\pi}{2}$, and $0 \le \phi \le \frac{\pi}{4}$. The integral in spherical coordinates follows.[1]

$$\int_0^{\pi/2}\int_0^{\pi/4}\int_0^{\sqrt{2}} (\rho\sin\phi\cos\theta)(\rho\sin\phi\sin\theta)\rho^2\sin\phi\, d\rho\,d\phi\,d\theta$$

$$= \int_0^{\pi/2}\int_0^{\pi/4}\int_0^{\sqrt{2}} \rho^4\sin^3\phi\cos\theta\sin\theta\, d\rho\,d\phi\,d\theta = \left[\int_0^{\pi/2}\sin\theta\cos\theta\, d\theta\right]\left[\int_0^{\pi/4}\sin^3\phi\, d\phi\right]\left[\int_0^{\sqrt{2}}\rho^4\, d\rho\right]$$

$$= \left[\tfrac{1}{2}\sin^2\theta\right]_0^{\pi/2} \cdot \int_0^{\pi/4}(1-\cos^2\phi)\sin\phi\, d\phi \cdot \left[\tfrac{1}{5}\rho^5\right]_0^{\sqrt{2}} = \tfrac{1}{2}\left[-\cos\phi+\tfrac{1}{3}\cos^3\phi\right]_0^{\pi/4} \cdot \frac{4\sqrt{2}}{5}$$

$$= \frac{2\sqrt{2}}{5}\left(-\cos\tfrac{\pi}{4}+\tfrac{1}{3}\cos^3\tfrac{\pi}{4}+\cos 0 - \tfrac{1}{3}\cos^3 0\right) = \frac{2\sqrt{2}}{5}\left(-\tfrac{1}{\sqrt{2}}+\tfrac{1}{3}\cdot\tfrac{1}{2\sqrt{2}}+1-\tfrac{1}{3}\right)$$

$$= \frac{2\sqrt{2}}{5}\left(-\frac{5}{6\sqrt{2}}+\frac{2}{3}\right) = \frac{2\sqrt{2}}{5}\left(\frac{-5+4\sqrt{2}}{6\sqrt{2}}\right) = \frac{1}{15}\left(4\sqrt{2}-5\right).$$

---

[1]Note: Since all of the limits of integration are constants and the variables in the integrand can be separated out, the value of each integration is a constant, and thus we can just separate the triple integral into the product of three integrals.

57. Let the sphere be centered at the origin. Then $0 \le \rho \le a$, $0 \le \phi \le \pi$, and $0 \le \theta \le 2\pi$, so that (see footnote on previous page)

$$V = \int_0^{2\pi} \int_0^{\pi} \int_0^a \rho^2 \sin\phi \, d\rho \, d\phi \, d\theta = \left(\int_0^{2\pi} d\theta\right) \left(\int_0^{\pi} \sin\phi \, d\phi\right) \left(\int_0^a \rho^2 \, d\rho\right) = 2\pi \left[-\cos\phi\right]_0^{\pi} \left[\frac{1}{3}\rho^3\right]_0^a$$

$$= \frac{2}{3}\pi a^3 \left(-\cos\pi + \cos 0\right) = \frac{4}{3}\pi a^3.$$

59. 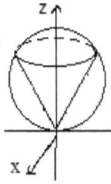 (To graph the sphere $\rho = 2a\cos\phi$, recall the polar graph of $r = 2a\cos\theta$.) We see from the figure that ρ varies from zero to the sphere: $0 \le \rho \le 2a\cos\phi$, ϕ varies from zero to the cone: $0 \le \phi \le \frac{\pi}{4}$, and θ varies from zero to 2π: $0 \le \theta \le 2\pi$:

$$V = \int_0^{2\pi} \int_0^{\pi/4} \int_0^{2a\cos\phi} \rho^2 \sin\phi \, d\rho \, d\phi \, d\theta = \int_0^{2\pi} \int_0^{\pi/4} \left[\frac{1}{3}\rho^3 \sin\phi\right]_{\rho=0}^{\rho=2a\cos\phi} d\phi \, d\theta$$

$$= \frac{1}{3} \int_0^{2\pi} \int_0^{\pi/4} (2a\cos\phi)^3 \sin\phi \, d\phi \, d\theta = \frac{8a^3}{3} \int_0^{2\pi} \left[-\frac{1}{4}\cos^4\phi\right]_0^{\pi/4} d\theta$$

$$= -\frac{2a^3}{3} \left[\cos^4\frac{\pi}{4} - \cos^4 0\right] \cdot 2\pi = -\frac{2a^3}{3}\left[\frac{1}{4} - 1\right] \cdot 2\pi = \pi a^3.$$

61. From the figure we see that ρ varies from zero to the sphere $x^2 + y^2 + z^2 = 9$ $\Rightarrow$ $\rho = 3$: $0 \le \rho \le 3$, while ϕ varies from zero to the cone: $z^2 = x^2 + y^2$ $\Rightarrow$ $\rho^2\cos^2\phi = \rho^2\sin^2\phi$ $\Rightarrow$ $\tan^2\phi = 1$ $\Rightarrow$ $\phi = \frac{\pi}{4}$: $0 \le \phi \le \frac{\pi}{4}$. θ clearly varies from zero to 2π: $0 \le \theta \le 2\pi$:

$$V = \int_0^{2\pi} \int_0^{\pi/4} \int_0^3 \rho^2 \sin\phi \, d\rho \, d\phi \, d\theta = \int_0^{2\pi} \int_0^{\pi/4} \left[\frac{1}{3}\rho^3 \sin\phi\right]_{\rho=0}^{\rho=3} d\phi \, d\theta$$

$$= 9 \int_0^{2\pi} \left[-\cos\phi\right]_0^{\pi/4} d\theta = 9 \int_0^{2\pi} \left(-\cos\frac{\pi}{4} + \cos 0\right) d\theta$$

$$= 9\left(-\frac{1}{\sqrt{2}} + 1\right) \cdot 2\pi = 9\pi\left(2 - \sqrt{2}\right).$$

63. Situate the sphere centered at the origin. Then the density function $\delta(\rho, \phi, \theta) = k\rho$. Clearly, $0 \le \rho \le a$, $0 \le \phi \le \pi$, and $0 \le \theta \le 2\pi$:

$$M = \int_0^{2\pi} \int_0^{\pi} \int_0^a k\rho \, \rho^2 \sin\phi \, d\rho \, d\phi \, d\theta = k \int_0^{2\pi} \int_0^{\pi} \left[\frac{1}{4}\rho^4 \sin\phi\right]_{\rho=0}^{\rho=a} d\phi \, d\theta = \frac{ka^4}{4} \int_0^{2\pi} \left[-\cos\phi\right]_{\phi=0}^{\phi=\pi} d\theta$$

$$= \frac{ka^4}{4} \int_0^{2\pi} (-\cos\pi + \cos 0) \, d\theta = \frac{ka^4}{4} \cdot 2 \cdot 2\pi = \pi ka^4.$$

65. From the figure we see that ρ varies from zero to the unit sphere $x^2 + y^2 + z^2 = 1$ $\Rightarrow$ $\rho = 1$: $0 \le \rho \le 1$, ϕ varies from zero to the cone : $z^2 = x^2 + y^2$ $\Rightarrow$ $\rho^2\cos^2\phi = \rho^2\sin^2\phi$ $\Rightarrow$ $\tan^2\phi = 1$ $\Rightarrow$ $\phi = \frac{\pi}{4}$: $0 \le \phi \le \frac{\pi}{4}$. θ clearly varies from zero to 2π: $0 \le \theta \le 2\pi$. The density function is $\delta(x, y, z) = \sqrt{x^2 + y^2 + z^2}$ $\Rightarrow$ $\delta(\rho, \phi, \theta) = \rho$:

$$M = \int_0^{2\pi} \int_0^{\pi/4} \int_0^1 \rho \rho^2 \sin \phi \, d\rho d\phi d\theta = \int_0^{2\pi} \int_0^{\pi/4} \left[\tfrac{1}{4} \rho^4 \sin \phi \right]_{\rho=0}^{\rho=1} d\phi d\theta$$

$$= \tfrac{1}{4} \int_0^{2\pi} \left[-\cos \phi \right]_0^{\pi/4} d\theta = \tfrac{1}{4} \int_0^{2\pi} \left(-\cos \tfrac{\pi}{4} + \cos 0 \right) d\theta$$

$$= \tfrac{1}{4} \left(-\tfrac{1}{\sqrt{2}} + 1 \right) \cdot 2\pi = \tfrac{\pi}{4} \left(2 - \sqrt{2} \right).$$

67. Position the unit ball with its center at the origin. Then $0 \le \rho \le 1$, $0 \le \phi \le \pi$, and $0 \le \theta \le 2\pi$. As distance from the center of the ball is ρ, $d = \rho \Rightarrow \delta(\rho, \phi, \theta) = \dfrac{1}{1+\rho^2}$ (See footnote page 248 regarding separating the integrals):

$$M = \int_0^{2\pi} \int_0^{\pi} \int_0^1 \frac{1}{1+\rho^2} \rho^2 \sin \phi \, d\rho d\phi d\theta = \left(\int_0^{2\pi} d\theta \right) \left(\int_0^{\pi} \sin \phi \, d\phi \right) \left(\int_0^1 \frac{\rho^2}{1+\rho^2} d\rho \right)$$

$$\int_0^{2\pi} d\theta = 2\pi, \quad \int_0^{\pi} \sin \phi \, d\phi = \left[-\cos \phi \right]_0^{\pi} = -\cos \pi + \cos 0 = 2, \quad \text{and}$$

$$\int_0^1 \frac{\rho^2}{1+\rho^2} d\rho = \int_0^1 \frac{\rho^2 + 1 - 1}{1+\rho^2} d\rho = \int_0^1 \left[1 - \frac{1}{1+\rho^2} \right] d\rho = \left[\rho - \tan^{-1} \rho \right]_0^1 = 1 - \tan^{-1} 1 = 1 - \frac{\pi}{4}$$

$$M = 2\pi(2) \left(1 - \tfrac{\pi}{4} \right) = 4\pi - \pi^2.$$

69.
From the figure we see that $\tan \phi = \frac{a}{h} \Rightarrow \phi = \tan^{-1} \frac{a}{h} \Rightarrow 0 \le \phi \le \tan^{-1} \frac{a}{h}$. The top of the cone is $z = h$, but $z = \rho \cos \phi$, so $h = \rho \cos \phi \Rightarrow \rho = \frac{h}{\cos \phi} \Rightarrow 0 \le \rho \le \frac{h}{\cos \phi}$. Clearly, $0 \le \theta \le 2\pi$. Call the density, k.

$$M = \int_0^{2\pi} \int_0^{\tan^{-1} \frac{a}{h}} \int_0^{\frac{h}{\cos \phi}} k \rho^2 \sin \phi \, d\rho d\phi d\theta = k \int_0^{2\pi} \int_0^{\tan^{-1} \frac{a}{h}} \left[\tfrac{1}{3} \rho^3 \sin \phi \right]_{\rho=0}^{\rho=\frac{h}{\cos \phi}} d\phi d\theta$$

$$= \frac{kh^3}{3} \int_0^{2\pi} \int_0^{\tan^{-1} \frac{a}{h}} \frac{\sin \phi}{\cos^3 \phi} d\phi d\theta. \quad \text{Call the double integral } I, \text{ so } M = \frac{kh^3}{3} I.$$

By symmetry, $\bar{x} = \bar{y} = 0$, and as $z = \rho \cos \phi$:

$$\bar{z} = \frac{1}{M} \int_0^{2\pi} \int_0^{\tan^{-1} \frac{a}{h}} \int_0^{\frac{h}{\cos \phi}} k \rho \cos \phi \rho^2 \sin \phi \, d\rho d\phi d\theta = \frac{k}{M} \int_0^{2\pi} \int_0^{\tan^{-1} \frac{a}{h}} \left[\tfrac{1}{4} \rho^4 \sin \phi \cos \phi \right]_{\rho=0}^{\rho=\frac{h}{\cos \phi}} d\phi d\theta$$

$$= \frac{kh^4}{4M} \int_0^{2\pi} \int_0^{\tan^{-1} \frac{a}{h}} \frac{\sin \phi}{\cos^3 \phi} d\phi d\theta = \frac{kh^4}{4M} I = \frac{kh^4 I}{4} \cdot \frac{3}{kh^3 I} = \frac{3}{4} h$$

Thus, $(\bar{x}, \bar{y}, \bar{z}) = \left(0, 0, \tfrac{3}{4} h \right).$

71. Call the density k. From Exercise #61 above, $M = kV = 9k\pi(2 - \sqrt{2})$. By symmetry, $\bar{x} = \bar{y} = 0$, and as $z = \rho \cos \phi$: (See footnote page 248, regarding separating the integrals.)

$$\bar{z} = \frac{1}{M} \int_0^{2\pi} \int_0^{\pi/4} \int_0^3 k \rho \cos \phi \rho^2 \sin \phi \, d\rho d\phi d\theta = \frac{k}{M} \left(\int_0^{2\pi} d\theta \right) \left(\int_0^{\pi/4} \sin \phi \cos \phi \, d\phi \right) \left(\int_0^3 \rho^3 d\rho \right)$$

$$= \frac{k}{M} (2\pi) \left[\tfrac{1}{2} \sin^2 \phi \right]_0^{\pi/4} \left[\tfrac{1}{4} \rho^4 \right]_0^3 = \frac{\pi k}{4M} \left[\sin^2 \tfrac{\pi}{4} \right] (3^4) = \frac{81\pi k}{8M} = \frac{81\pi k}{8 \cdot 9k\pi(2 - \sqrt{2})} \cdot \frac{2 + \sqrt{2}}{2 + \sqrt{2}} = \frac{18 + 9\sqrt{2}}{16}$$

Thus, $(\bar{x}, \bar{y}, \bar{z}) = \left(0, 0, \dfrac{18 + 9\sqrt{2}}{16} \right).$

CHAPTER 12
Vectors and Vector-Valued Functions

§12.1 Vectors in the Plane and Beyond

1.
The terminal point of the vector in standard position is
$[0-(-2), 1-(-2)] = (2,3)$.

3.
The terminal point of the vector in standard position is
$[-2] - 1, 3 - 1] = (-3, 2)$.

5.
The terminal point of the vector in standard position is
$[1 - (-2), -1 - (-1)] = (3, 0)$.

7. The vector from $A = (1, 2, 3)$ to $B = (3, 2, 1)$ is $\overrightarrow{AB} = (3 - 1, 2 - 2, 1 - 3) = (2, 0, -2)$.

9. The vector from $A = (0, 1, -9)$ to $B = (-9, 0, 2)$ is
$$\overrightarrow{AB} = [-9 - 0, 0 - 1, 2 - (-9)] = (-9, -1, 11).$$

11. $5(3, -2) + (0, 1) + (-2, -4) = (15, -10) + (0, 1) + (-2, -4) = (15 + 0 - 2, -10 + 1 - 4)$
$= (13, -13)$.

13. $-(2, 3, 1) + [-(1, -2, 0)] = (-2, -3, -1) - (1, -2, 0) = [-2 - 1, -3 - (-2), -1 - 0]$
$= (-3, -1, -1)$.

15. $4[(2\mathbf{i} - 4\mathbf{j}) - (\mathbf{i} + 3\mathbf{j})] = 4[2\mathbf{i} - 4\mathbf{j} - \mathbf{i} - 3\mathbf{j}] = 4(\mathbf{i} - 7\mathbf{j}) = 4\mathbf{i} - 28\mathbf{j}$.

17. $-2(3\mathbf{i} + 2\mathbf{k}) + (2\mathbf{j} - \mathbf{k}) - (\mathbf{i} + \mathbf{j} + \mathbf{k}) = -6\mathbf{i} - 4\mathbf{k} + 2\mathbf{j} - \mathbf{k} - \mathbf{i} - \mathbf{j} - \mathbf{k})$
$= (-6 - 1)\mathbf{i} + (2 - 1)\mathbf{j} + (-4 - 1 - 1)\mathbf{k} = -7\mathbf{i} + \mathbf{j} - 6\mathbf{k}$.

19. $\mathbf{v} = (5, 2) \Rightarrow \|\mathbf{v}\| = \sqrt{5^2 + 2^2} = \sqrt{29}$. A unit vector in the direction of $\mathbf{v}$ is $\left(\frac{5}{\sqrt{29}}, \frac{2}{\sqrt{29}}\right)$, and $\mathbf{v} - \sqrt{29}\left(\frac{5}{\sqrt{29}}, \frac{2}{\sqrt{29}}\right)$.

21. $\mathbf{v} = 2\mathbf{i} - 4\mathbf{j} + \mathbf{k} \Rightarrow \|\mathbf{v}\| = \sqrt{2^2 + 4^2 + 1^2} = \sqrt{21}$. A unit vector in the direction of $\mathbf{v}$ is $\frac{2}{\sqrt{21}}\mathbf{i} - \frac{4}{\sqrt{21}}\mathbf{j} + \frac{1}{\sqrt{21}}\mathbf{k}$, and $\mathbf{v} = \sqrt{21}\left(\frac{2}{\sqrt{21}}\mathbf{i} - \frac{4}{\sqrt{21}}\mathbf{j} + \frac{1}{\sqrt{21}}\mathbf{k}\right)$.

23. $\mathbf{v} = \sqrt{2}\mathbf{i} - \frac{1}{3}\mathbf{j} \Rightarrow \|\mathbf{v}\| = \sqrt{2 + \frac{1}{9}} = \frac{\sqrt{19}}{3}$. A unit vector in the direction of $\mathbf{v}$ is $\frac{3\sqrt{2}}{\sqrt{19}}\mathbf{i} - \frac{1}{\sqrt{19}}\mathbf{j}$, and $\mathbf{v} = \frac{\sqrt{19}}{3}\left(\frac{3\sqrt{2}}{\sqrt{19}}\mathbf{i} - \frac{1}{\sqrt{19}}\mathbf{j}\right)$.

25. Given $\mathbf{u} = (1, 3)$ $\mathbf{v} = (2, 4)$, $\mathbf{w} = (6, -2)$.

(a) To find r, s such that $r\mathbf{u} + s\mathbf{v} = \mathbf{w}$:
$$r(1, 3) + s(2, 4) = (6, -2) \Rightarrow (r + 2s, 3r + 4s) = (6, -2) \Rightarrow$$
$$(1) \quad r + 2s = 6 \text{ and } (2) \quad 3r + 4s = -2.$$
From (1) we get (3): $r = 6 - 2s$, which we substitute into (2):
$$3(6 - 2s) + 4s = -2 \Rightarrow 18 - 2s = -2 \Rightarrow 2s = 20 \Rightarrow s = 10$$
From (3): $r = 6 - 20 = -14$. Thus $r = -14$ and $s = 10$.

(b) To find r, s such that $-r\mathbf{u} + s\mathbf{w} = \mathbf{v}$:
$$-r(1, 3) + s(6, -2) = (2, 4) \Rightarrow (-r, -3r) + (6s, -2s) = (2, 4) \Rightarrow$$
$$(-r + 6s, -3r - 2s) = (2, 4) \Rightarrow$$
$$(1) \ -r + 6s = 2 \text{ and } (2) \ -3r - 2s = 4.$$
From (1) we get (3): $r = 6s - 2$, which we substitute into (2):
$$-3(6s - 2) - 2s = 4 \Rightarrow -20s = -2 \Rightarrow s = \frac{1}{10}$$
From (3): $r = 6\left(\frac{1}{10}\right) - 2 = \frac{3}{5} - 2 = -\frac{7}{5}$. Thus $r = -\frac{7}{5}$ and $s = \frac{1}{10}$.

(c) To find r, s such that $r\mathbf{v} + (-s\mathbf{w}) = \mathbf{u}$:
$$r(2, 4) + (-6s, 2s) = \mathbf{u} \Rightarrow (2r, 4r) + (-6s, 2s) = (1, 3) \Rightarrow$$
$$(2r - 6s, 4r + 2s) = (1, 3) \Rightarrow$$
$$(1) \ 2r - 6s = 1 \text{ and } (2) \ 4r + 2s = 3$$
From (1) we get (3): $r = \frac{1}{2}(1 + 6s)$, which we substitute into (2):
$$2(1 + 6s) + 2s = 3 \Rightarrow 14s = 1 \Rightarrow s = \frac{1}{14}$$
From (3): $r = \frac{1}{2}\left(1 + \frac{6}{14}\right) = \frac{5}{7}$. Thus $r = \frac{5}{7}$ and $s = \frac{1}{14}$.

27.   To find r, s, t such that $-r(1, 3, 0) + s(2, 1, 6) + [-t(1, 4, 6)] = (7, 5, 6)$:
$$(-r, -3r, 0) + (2s, s, 6s) + (-t, -4t, -6t) = (7, 5, 6) \Rightarrow$$
$$(-r + 2s - t, -3r + s - 4t, 6s - 6t) = (7, 5, 6) \Rightarrow$$
$$(1) \ -r + 2s - t = 7, \quad (2) \ -3r + s - 4t = 5, \text{ and } \ 6s - 6t = 6 \Rightarrow (3) \ s - t = 1$$
From (3), $s = 1 + t$, so substituting this in (1) and (2), resp.:
$$\text{In } (1), -r + 2(1 + t) - t = 7 \Rightarrow (4) - r + t = 5.$$
$$\text{In } (2), -3r + 1 + t - 4t = 5 \Rightarrow (5) - 3r - 3t = 4.$$
Adding 3 times equation (4) to equation (5): $-6r = 19 \Rightarrow r = -\frac{19}{6}$
Substituting this into equation (4), $\frac{19}{6} + t = 5 \Rightarrow t = 5 - \frac{19}{6} = \frac{11}{6}$
Finally, from equation (3), $s = 1 + t = 1 + \frac{11}{6} = \frac{17}{6}$.
Consequently, $r = -\frac{19}{6}$, $s = \frac{17}{6}$, and $t = \frac{11}{6}$.

29.   The vector $\mathbf{v}$ from the point $(1,3)$ to the point $(3,1)$ is $\mathbf{v} = (3 - 1, 1 - 3) = (2, -2)$. The magnitude of $\mathbf{v}$ is $\|\mathbf{v}\| = \sqrt{2^2 + (-2)^2} = \sqrt{8} = 2\sqrt{2}$. Thus, a unit vector with the same

direction as **v** is $\frac{1}{2\sqrt{2}}(2,-2) = \left(\frac{1}{\sqrt{2}}, -\frac{1}{\sqrt{2}}\right)$. Therefore the vector $(\mathbf{a}, \mathbf{b}) = 5\left(\frac{1}{\sqrt{2}}, -\frac{1}{\sqrt{2}}\right) = \left(\frac{5}{\sqrt{2}}, -\frac{5}{\sqrt{2}}\right)$.

31.

$\mathbf{F}_1 = \|\mathbf{F}_1\|(\cos 30°(-\mathbf{i}) + \sin 30°\mathbf{j})$
$\mathbf{F}_2 = \|\mathbf{F}_2\|(\cos 45°\mathbf{i} + \sin 45°\mathbf{j})$
$\mathbf{F}_1 + \mathbf{F}_2 - 10\mathbf{j} = 0$
Therefore,

$\|\mathbf{F}_1\|(-\cos 30°\mathbf{i} + \sin 30°\mathbf{j}) + \|\mathbf{F}_2\|(\cos 45°\mathbf{i} + \sin 45°\mathbf{j}) - 10\mathbf{j} = 0\mathbf{i} + 0\mathbf{j} \Rightarrow$

(*) $(-\|\mathbf{F}_1\|\cos 30° + \|\mathbf{F}_2\|\cos 45°)\mathbf{i} + (\|\mathbf{F}_1\|\sin 30° + \|\mathbf{F}_2\|\sin 45° - 10)\mathbf{j} = 0\mathbf{i} + 0\mathbf{j}$.

Equating coefficients of **i** on both sides of equation (*):

$-\|\mathbf{F}_1\|\cos 30° + \|\mathbf{F}_2\|\cos 45° = 0 \Rightarrow -\|\mathbf{F}_1\|\frac{\sqrt{3}}{2} + \|\mathbf{F}_2\|\frac{\sqrt{2}}{2} = 0 \Rightarrow$

(1) : $-\|\mathbf{F}_1\|\sqrt{3} + \|\mathbf{F}_2\|\sqrt{2} = 0$

Equating coefficients of **j** on both sides of equation (*):

$\|\mathbf{F}_1\|\sin 30° + \|\mathbf{F}_2\|\sin 45° - 10 = 0 \Rightarrow \|\mathbf{F}_1\|\frac{1}{2} + \|\mathbf{F}_2\|\frac{\sqrt{2}}{2} = 10 \Rightarrow$

(2) : $\|\mathbf{F}_1\| + \|\mathbf{F}_2\|\sqrt{2} = 20$

Subtracting equation (1) from equation (2) yields:

$\|\mathbf{F}_1\|(1 + \sqrt{3}) = 20 \Rightarrow \|\mathbf{F}_1\| = \frac{20}{1+\sqrt{3}} \cdot \frac{1-\sqrt{3}}{1-\sqrt{3}} = 10(\sqrt{3} - 1)$.

Then, from (1), $\|\mathbf{F}_2\| = \|\mathbf{F}_1\|\frac{\sqrt{3}}{\sqrt{2}} = 10(\sqrt{3} - 1) \cdot \frac{\sqrt{3}}{\sqrt{2}} = \frac{10(3 - \sqrt{3})}{\sqrt{2}}$

Finally: $\mathbf{F}_1 = \|\mathbf{F}_1\|(\cos 30°(-\mathbf{i}) + \sin 30°\mathbf{j}) = 10(\sqrt{3} - 1)\left(-\frac{\sqrt{3}}{2}\mathbf{i} + \frac{1}{2}\mathbf{j}\right) = 5(\sqrt{3} - 1)(-\sqrt{3}\mathbf{i} + \mathbf{j})$

and $\mathbf{F}_2 = \|\mathbf{F}_2\|(\cos 45°\mathbf{i} + \sin 45°\mathbf{j}) = \frac{10(3 - \sqrt{3})}{\sqrt{2}}\left(\frac{1}{\sqrt{2}}\mathbf{i} + \frac{1}{\sqrt{2}}\mathbf{j}\right) = 5(3 - \sqrt{3})(\mathbf{i} + \mathbf{j})$.

33.

$\mathbf{F}_1 = \|\mathbf{F}_1\|(\cos 30°(-\mathbf{i}) + \sin 30°\mathbf{j})$
$\mathbf{F}_2 = \|\mathbf{F}_2\|(\cos 45°\mathbf{i} + \sin 45°\mathbf{j})$
$\mathbf{F}_1 + \mathbf{F}_2 + 10[\cos 60°\mathbf{i} + \sin 60°(-\mathbf{j})] = 0$
Therefore,

$\|\mathbf{F}_1\|(-\cos 30°\mathbf{i} + \sin 30°\mathbf{j}) + \|\mathbf{F}_2\|(\cos 45°\mathbf{i} + \sin 45°\mathbf{j}) + 10(\cos 60°\mathbf{i} - \sin 60°\mathbf{j}) = 0\mathbf{i} + 0\mathbf{j} \Rightarrow$

(*) $(-\|\mathbf{F}_1\|\cos 30° + \|\mathbf{F}_2\|\cos 45° + 10\cos 60°)\mathbf{i} + (\|\mathbf{F}_1\|\sin 30° + \|\mathbf{F}_2\|\sin 45° - 10\sin 60°)\mathbf{j}$
     $= 0\mathbf{i} + 0\mathbf{j} \Rightarrow$

Equating coefficients of **i** on both sides of equation (*):

$-\|\mathbf{F}_1\|\cos 30° + \|\mathbf{F}_2\|\cos 45° + 10\cos 60° = 0 \Rightarrow -\|\mathbf{F}_1\|\frac{\sqrt{3}}{2} + \|\mathbf{F}_2\|\frac{\sqrt{2}}{2} + 10 \cdot \frac{1}{2} = 0 \Rightarrow$

$$(1): \ -\|\mathbf{F_1}\|\sqrt{3} + \|\mathbf{F_2}\|\sqrt{2} = -10$$

Equating coefficients of $\mathbf{j}$ on both sides of equation (*):

$$\|\mathbf{F_1}\|\sin 30° + \|\mathbf{F_2}\|\sin 45° - 10\sin 60° = 0 \ \Rightarrow \ \|\mathbf{F_1}\|\tfrac{1}{2} + \|\mathbf{F_2}\|\tfrac{\sqrt{2}}{2} = 10\tfrac{\sqrt{3}}{2} \ \Rightarrow$$

$$(2): \ \|\mathbf{F_1}\| + \|\mathbf{F_2}\|\sqrt{2} = 10\sqrt{3}$$

Subtracting equation (1) from equation (2) yields:

$$\|\mathbf{F_1}\|(1+\sqrt{3}) = 10(1+\sqrt{3}) \ \Rightarrow \ \|\mathbf{F_1}\| = 10$$

Then, from (2), $\|\mathbf{F_2}\| = \dfrac{10\sqrt{3}-10}{\sqrt{2}}$

Finally: $\mathbf{F_1} = \|\mathbf{F_1}\|(\cos 30°(-\mathbf{i}) + \sin 30°\mathbf{j}) = 10\left(-\tfrac{\sqrt{3}}{2}\mathbf{i} + \tfrac{1}{2}\mathbf{j}\right) = -5\sqrt{3}\mathbf{i} + 5\mathbf{j}$

and $\mathbf{F_2} = \|\mathbf{F_2}\|(\cos 45°\mathbf{i} + \sin 45°\mathbf{j}) = \dfrac{10\sqrt{3}-10)}{\sqrt{2}}\left(\tfrac{1}{\sqrt{2}}\mathbf{i} + \tfrac{1}{\sqrt{2}}\mathbf{j}\right) = (5\sqrt{3}-5)(\mathbf{i}+\mathbf{j}).$

35.

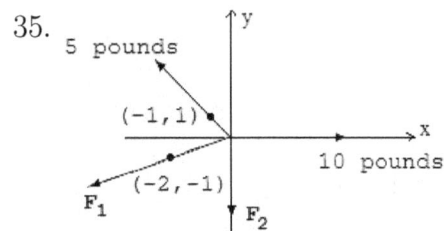

A unit vector in the direction of $\mathbf{F_1}$ is:

$$\frac{-2\mathbf{i} - \mathbf{j}}{\sqrt{2^2 + 1^2}} = \frac{1}{\sqrt{5}}(-2\mathbf{i} - \mathbf{j}), \text{ so } \mathbf{F_1} = \frac{\|\mathbf{F_1}\|}{\sqrt{5}}(-2\mathbf{i} - \mathbf{j})$$

$$\mathbf{F_2} = \|\mathbf{F_2}\|(-\mathbf{j}) = -\|\mathbf{F_2}\|\mathbf{j}$$

The vector along the positive x-axis is $10\mathbf{i}$, and a unit vector in the direction of the vector in QII is:

$\dfrac{-\mathbf{i}+\mathbf{j}}{\sqrt{1^2+1^2}} = \dfrac{1}{\sqrt{2}}(-\mathbf{i}+\mathbf{j})$, so that vector is $\dfrac{5}{\sqrt{2}}(-\mathbf{i}+\mathbf{j})$. As the sum of the 4 vectors is the zero vector:

$$(*) \quad \frac{\|\mathbf{F_1}\|}{\sqrt{5}}(-2\mathbf{i}-\mathbf{j}) - \|\mathbf{F_2}\|\mathbf{j} + 10\mathbf{i} + \frac{5}{\sqrt{2}}(-\mathbf{i}+\mathbf{j}) = 0\mathbf{i} + 0\mathbf{j}$$

Equating coefficients of $\mathbf{i}$ on both sides of equation (*):

$$\frac{\|\mathbf{F_1}\|}{\sqrt{5}}(-2) + 10 - \frac{5}{\sqrt{2}} = 0 \ \Rightarrow \ \|\mathbf{F_1}\| = -\frac{\sqrt{5}}{2}\left(\frac{5}{\sqrt{2}}\cdot\frac{\sqrt{2}}{\sqrt{2}} - 10\right) = \frac{5\sqrt{5}}{4}(-\sqrt{2}+4) = \frac{5\sqrt{5}}{4}(4-\sqrt{2})$$

Equating coefficients of $\mathbf{j}$ on both sides of equation (*):

$$\frac{\|\mathbf{F_1}\|}{\sqrt{5}}(-1) - \|\mathbf{F_2}\| + \frac{5}{\sqrt{2}} = 0 \ \Rightarrow \ -\frac{\frac{5\sqrt{5}}{4}(4-\sqrt{2})}{\sqrt{5}} - \|\mathbf{F_2}\| = -\frac{5}{\sqrt{2}} = -\frac{5\sqrt{2}}{2} \ \Rightarrow$$

$$\|\mathbf{F_2}\| = -5 + \frac{5}{4}\sqrt{2} + \frac{5\sqrt{2}}{2} = \frac{15\sqrt{2}-20}{4}.$$

Therefore,

$$\mathbf{F_1} = \left[\frac{5\sqrt{5}}{4}(4-\sqrt{2})\right]\cdot\frac{1}{\sqrt{5}}(-2\mathbf{i}-\mathbf{j}) = \left(\frac{20-5\sqrt{2}}{4}\right)(-2\mathbf{i}-\mathbf{j}), \text{ and}$$

$$\mathbf{F_2} = \left(\frac{15\sqrt{2}-20}{4}\right)(-\mathbf{j}).$$

37. 15 pounds

$\mathbf{F}_{\text{horiz.}} = 15\cos 30° \, \mathbf{i} = \frac{15\sqrt{3}}{2}\,\mathbf{i} \Rightarrow$ horiz. comp. is $\frac{15\sqrt{3}}{2}$ lbs.

$\mathbf{F}_{\text{vert.}} = 15\sin 30° \, \mathbf{i} = \frac{15}{2}\,\mathbf{j} \Rightarrow$ vert. comp. is $\frac{15}{2}$ lbs.

15 sin 30⁰

30⁰

15 cos 30⁰

39.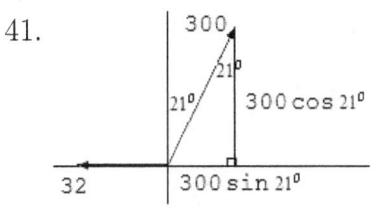

Let $\mathbf{v}$ denote the plane's velocity vector, $\mathbf{w}$ the wind's vector, and $\mathbf{r}$ the resulting sum vector giving the track and ground speed. Then, $\mathbf{w} = 80\cos 45°\,\mathbf{i} - 80\sin 45°\,\mathbf{j}$, and $\mathbf{v} = 400\,\mathbf{i}$. Since $\mathbf{r} = \mathbf{v} + \mathbf{w}$:

$\mathbf{r} = 400\,\mathbf{i} + 80\cos 45°\,\mathbf{i} - 80\sin 45°\,\mathbf{j} = (400 + 80\cos 45°)\,\mathbf{i} - 80\sin 45°\,\mathbf{j}$

The plane's ground speed is then

$$\|\mathbf{r}\| = \sqrt{(400 + 80\cos 45°)^2 + (80\sin 45°)^2} = \sqrt{\left(400 + \frac{80}{\sqrt{2}}\right)^2 + \left(\frac{80}{\sqrt{2}}\right)^2} \approx 460.06 \text{ km/hr}$$

41.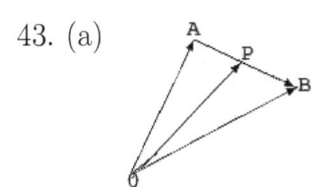

Without the wind, the plane's vector is
$300(\sin 21°\mathbf{i} + \cos 21°\mathbf{j})$
The wind's vector is $-32\,\mathbf{i}$, so the plane's track is
$$\mathbf{T} = (300\sin 21° - 32)\mathbf{i} + (300\cos 21°\mathbf{j}) \approx 75.5\mathbf{i} + 280.1\mathbf{j}$$

From the lower figure, we can determine the bearing θ of the track: $\tan\theta \approx \frac{75.5}{280.1} \Rightarrow \theta \approx \tan^{-1}\frac{75.5}{280.1} \Rightarrow \theta \approx 15.1°$. The ground speed is
$$\|\mathbf{T}\| \approx \sqrt{75.5^2 + 280.1^2} \approx 290.1 \text{ km/hr.}$$

43. (a)

If the length of $\overrightarrow{AP}$ equals the length of $\overrightarrow{PB}$, then the length of $\overrightarrow{AP}$ is half the length of $\overrightarrow{AB}$ so that
$$\overrightarrow{AP} = \tfrac{1}{2}\,\overrightarrow{AB} = \tfrac{1}{2}(\mathbf{B} - \mathbf{A}) \Rightarrow$$
$$\mathbf{P} = \mathbf{A} + \tfrac{1}{2}(\mathbf{B} - \mathbf{A})$$
$$= \mathbf{A} + \tfrac{1}{2}\mathbf{B} - \tfrac{1}{2}\mathbf{A}$$
$$= \tfrac{1}{2}\mathbf{A} + \tfrac{1}{2}\mathbf{B}.$$

(b)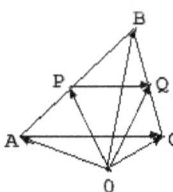

Let P and Q be the midpoints of AB and BC, respectively. From part (a),
$$\mathbf{P} = \tfrac{1}{2}\mathbf{A} + \tfrac{1}{2}\mathbf{B}, \text{ and } \mathbf{Q} = \tfrac{1}{2}\mathbf{B} + \tfrac{1}{2}\mathbf{C},$$

and then,

$$\vec{PQ} = Q - P = \left(\tfrac{1}{2}B + \tfrac{1}{2}C\right) - \left(\tfrac{1}{2}A + \tfrac{1}{2}B\right)$$
$$= \tfrac{1}{2}[(B + C) - (A + B)]$$
$$= \tfrac{1}{2}[B + C - A - B]$$
$$= \tfrac{1}{2}(C - A)$$
$$= \tfrac{1}{2}\,\vec{AC}$$

It follows that $\vec{PQ}$ has half the length of $\vec{AC}$ and is parallel to $\vec{AC}$.

45. 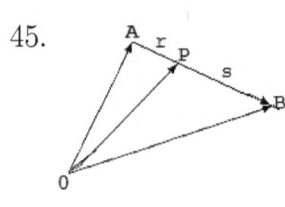   The length of $\vec{AP}$ is r, and the length of $\vec{PB}$ is s.

$$P = A + \vec{AP} = A + \left(\frac{r}{r+s}\right)\vec{AB} = A + \left(\frac{r}{r+s}\right)(B - A)$$
$$= \left(1 - \frac{r}{r+s}\right)A + \left(\frac{r}{r+s}\right)B = \left(\frac{r+s-r}{r+s}\right)A + \left(\frac{r}{r+s}\right)B$$
$$= \left(\frac{s}{r+s}\right)A + \left(\frac{r}{r+s}\right)B$$

47. Let $v = (v_1, v_2)$. Then $rv = r(v_1, v_2) = (rv_1, rv_2) = (0,0) \Rightarrow rv_1 = 0$ and $rv_2 = 0$. From $rv_1 = 0$, either $r = 0$ or $v_1 = 0$. Likewise, from $rv_2 = 0$, either $r = 0$ or $v_2 = 0$. Thus, either $r = 0$ or both $v_1 = 0$ and $v_2 = 0$, i.e. either $r = 0$ or $v = (v_1, v_2) = (0,0) = 0$. Conversely, if $r = 0$ or $v = 0$, then $r = 0$ or $v_1 = v_2 = 0$ so that $(rv_1, rv_2) = (0,0)$.

49. To prove: $(u + v) + w = u + (v + w)$ in $\Re^n$.

Let $u = (u_1, u_2, \ldots, u_n)$, $v = (v_1, v_2, \ldots, v_n)$, and $w = (w_1, w_2, \ldots, w_n)$. Then,

$$(u + v) + w = [(u_1, u_2, \ldots, u_n) + (v_1, v_2, \ldots, v_n)] + (w_1, w_2, \ldots, w_n)$$

(Definition 12.2) $= (u_1 + v_1, u_2 + v_2, \ldots, u_n + v_n) + (w_1, w_2, \ldots, w_n)$

(Definition 12.2) $= [(u_1 + v_1) + w_1, (u_2 + v_2) + w_2, \ldots, (u_n + v_n) + w_n]$

P of $\Re$ $= [u_1 + (v_1 + w_1), u_2 + (v_2 + w_2), \ldots, u_n + (v_n + w_n)]$

(Definition 12.2) $= (u_1, u_2, \ldots, u_n) + [(v_1 + w_1), (v_2 + w_2), \ldots, (v_n + w_n)]$

(Definition 12.2) $= (u_1, u_2, \ldots, u_n) + [(v_1, v_2, \ldots, v_n) + (w_1, w_2, \ldots, w_n)]$

$$= u + (v + w)$$

51. To prove: $(r + s)\mathbf{v} = r\mathbf{v} + s\mathbf{v}$ in $\Re^n$. Let $\mathbf{v} = (\mathbf{v_1}, \mathbf{v_2}, \ldots, \mathbf{v_n})$ Then,

$$\text{(Definition 12.1)} \quad (r+s)\mathbf{v} = [(r+s)\mathbf{v_1}, (r+s)\mathbf{v_2}, \ldots, (r+s)\mathbf{v_n}]$$
$$\text{P of } \Re \quad\quad\quad = (\mathbf{rv_1} + \mathbf{sv_1}, \mathbf{rv_2} + \mathbf{sv_2}, \ldots, \mathbf{rv_n} + \mathbf{sv_n})$$
$$\text{(Definition 12.2)} \quad\quad = (\mathbf{rv_1}, \mathbf{rv_2}, \ldots, \mathbf{rv_n}) + (\mathbf{sv_1}, \mathbf{sv_2}, \ldots, \mathbf{sv_n})$$
$$\text{(Definition 12.1)} \quad\quad = \mathbf{r}(\mathbf{v_1}, \mathbf{v_2}, \ldots, \mathbf{v_n}) + \mathbf{s}(\mathbf{v_1}, \mathbf{v_2}, \ldots, \mathbf{v_n})$$
$$\quad\quad\quad\quad = r\mathbf{v} + s\mathbf{v}$$

§12.2 Dot and Cross Products

1. $2(5\mathbf{i} + 2\mathbf{j}) \cdot (-3\mathbf{i} + \mathbf{j}) = (10\mathbf{i} + 4\mathbf{j}) \cdot (-3\mathbf{i} + \mathbf{j}) = 10(-3) + 4(1) = -26.$

3. $(3, 2, 1) \cdot (-4, 0, 5) = 3(-4) + 2(0) + 1(5) = -12 + 5 = -7.$

5. $4[(2\mathbf{i} - 4\mathbf{j}) - (\mathbf{i} + 3\mathbf{j})] \cdot (\mathbf{i} - 2\mathbf{j} + 3\mathbf{k}) = 4[2\mathbf{i} - 4\mathbf{j} - \mathbf{i} - 3\mathbf{j}] \cdot (\mathbf{i} - 2\mathbf{j} + 3\mathbf{k}) = 4(\mathbf{i} - 7\mathbf{j}) \cdot (\mathbf{i} - 2\mathbf{j} + 3\mathbf{k}) = $
$(4\mathbf{i} - 28\mathbf{j} + 0\mathbf{k}) \cdot (\mathbf{i} - 2\mathbf{j} + 3\mathbf{k}) = 4(1) + (-28)(-2) + (0)3 = 4 + 56 = 60.$

7. $\mathbf{u} = -\mathbf{i} + 2\mathbf{j}$ and $\mathbf{v} = 2\mathbf{i} + \mathbf{j} \Rightarrow \mathbf{u} \cdot \mathbf{v} = (-1)(2) + 2(1) = 0 \Rightarrow$ the vectors are orthogonal (Definition 12.6) $\Rightarrow \theta = 90°.$

9. $\mathbf{u} = (1, 4)$ and $\mathbf{v} = (3, -2) \Rightarrow \mathbf{u} \cdot \mathbf{v} = (1)(3) + 4(-2) = -5$, so that
$$\theta = \cos^{-1}\left(\frac{\mathbf{u} \cdot \mathbf{v}}{\|\mathbf{u}\|\|\mathbf{v}\|}\right) = \cos^{-1}\left(\frac{-5}{\sqrt{1+16}\sqrt{9+4}}\right) = \cos^{-1}\left(\frac{-5}{\sqrt{17}\sqrt{13}}\right) = \cos^{-1}\left(\frac{-5}{\sqrt{221}}\right) \approx 110°.$$

11. $\mathbf{u} = (2, -1, 1)$ and $\mathbf{v} = (1, 1, -1) \Rightarrow \mathbf{u} \cdot \mathbf{v} = (2)(1) + (-1)(1) + 1(-1) = 0 \Rightarrow$ the vectors are orthogonal (Definition 12.6) $\Rightarrow \theta = 90°.$

13. $\mathbf{u} = \mathbf{i} + \mathbf{j} = \mathbf{i} + \mathbf{j} + 0\mathbf{k}$ and $\mathbf{v} = \mathbf{j} + \mathbf{k} = 0\mathbf{i} + \mathbf{j} + \mathbf{k} \Rightarrow \mathbf{u} \cdot \mathbf{v} = (1)(0) + 1(1) + 0(1) = 1$, so that
$$\theta = \cos^{-1}\left(\frac{\mathbf{u} \cdot \mathbf{v}}{\|\mathbf{u}\|\|\mathbf{v}\|}\right) = \cos^{-1}\left(\frac{1}{\sqrt{1+1}\sqrt{1+1}}\right) = \cos^{-1}\left(\tfrac{1}{2}\right) = 60°.$$

15. $\mathbf{u} = (3, 0, 4)$ and $\mathbf{v} = (0, \sqrt{7}, -5) \Rightarrow \mathbf{u} \cdot \mathbf{v} = 3(0) + 0(\sqrt{7}) + 4(-5) = -20$, so that
$$\theta = \cos^{-1}\left(\frac{\mathbf{u} \cdot \mathbf{v}}{\|\mathbf{u}\|\|\mathbf{v}\|}\right) = \cos^{-1}\left(\frac{-20}{\sqrt{9+16}\sqrt{7+25}}\right) = \cos^{-1}\left(\frac{-20}{5 \cdot 4\sqrt{2}}\right) = \cos^{-1}\left(-\frac{1}{\sqrt{2}}\right) = 135°,$$
as the inverse cosine of a negative value lies in QII and the reference angle, having a cosine of $\frac{1}{\sqrt{2}}$, is $45°$.

17. $\mathbf{u} = 3\mathbf{i} - \mathbf{j} + 5\mathbf{k}$ and $\mathbf{v} = -2\mathbf{i} + 4\mathbf{j} + 3\mathbf{k} \Rightarrow \mathbf{u} \cdot \mathbf{v} = 3(-2) + (-1)(4) + 5(3) = -6 - 4 + 15 = 5$, so that
$$\theta = \cos^{-1}\left(\frac{\mathbf{u} \cdot \mathbf{v}}{\|\mathbf{u}\|\|\mathbf{v}\|}\right) = \cos^{-1}\left(\frac{5}{\sqrt{9+1+25}\sqrt{4+16+9}}\right) = \cos^{-1}\left(\frac{5}{\sqrt{35}\sqrt{29}}\right)$$
$$= \cos^{-1}\left(\frac{5}{\sqrt{1015}}\right) \approx 81°$$

19. $\mathbf{u} = (2, -1, 1)$ and $\mathbf{v} = (3, 1, -5)$. The two vectors are not parallel, as neither is a multiple of the other. As $\mathbf{u} \cdot \mathbf{v} = 2(3) + (-1)(1) + (1)(-5) = 6 - 1 - 5 = 0$, the vectors are orthogonal, by Definition 12.6.

21. $\mathbf{u} = 3\mathbf{i} - \mathbf{j} + 5\mathbf{k}$ and $\mathbf{v} = -2\mathbf{i} + 4\mathbf{j} + 3\mathbf{k}$. The two vectors are not parallel, as neither is a multiple of the other. As $\mathbf{u} \cdot \mathbf{v} = 3(-2) + (-1)(4) + 5(3) = -6 - 4 + 15 = 5 \neq 0$, the vectors are not orthogonal. Consequently, the two vectors are neither parallel nor orthogonal.

   [Note, in Exercise 17 we found the angle between the two vectors to be about $81°$, which confirms that they are neither parallel ($0°$ or $180°$ apart) nor orthogonal ($90°$ apart).]

23. $\mathbf{v} = (2, 1)$ and $\mathbf{u} = (-3, 2)$ $\Rightarrow$

$$\text{proj}_{\mathbf{u}}\mathbf{v} = \left(\frac{\mathbf{v} \cdot \mathbf{u}}{\mathbf{u} \cdot \mathbf{u}}\right)\mathbf{u} = \left[\frac{(2,1) \cdot (-3,2)}{(-3,2) \cdot (-3,2)}\right](-3,2) = \left[\frac{2(-3) + 1(2)}{9+4}\right](-3,2)$$

$$= -\frac{4}{13}(-3,2) = \left(\frac{12}{13}, -\frac{8}{13}\right)$$

$$\mathbf{v} - \text{proj}_{\mathbf{u}}\mathbf{v} = (2,1) - \left(\frac{12}{13}, -\frac{8}{13}\right) = \left(2 - \frac{12}{13}, 1 + \frac{8}{13}\right) = \left(\frac{14}{13}, \frac{21}{13}\right)$$

Thus, $\mathbf{v} = \text{proj}_{\mathbf{u}}\mathbf{v} + (\mathbf{v} - \text{proj}_{\mathbf{u}}\mathbf{v})$ becomes

$$(2,1) = \left(\frac{12}{13}, -\frac{8}{13}\right) + \left(\frac{14}{13}, \frac{21}{13}\right).$$

25. $\mathbf{v} = 3\mathbf{j} + 4\mathbf{k} = 0\mathbf{i} + 3\mathbf{j} + 4\mathbf{k}$ and $\mathbf{u} = \mathbf{i} + \mathbf{j} = \mathbf{i} + \mathbf{j} + 0\mathbf{k}$ $\Rightarrow$

$$\text{proj}_{\mathbf{u}}\mathbf{v} = \left(\frac{\mathbf{u} \cdot \mathbf{v}}{\mathbf{u} \cdot \mathbf{u}}\right)\mathbf{u} = \left[\frac{(\mathbf{i} + \mathbf{j} + 0\mathbf{k}) \cdot (0\mathbf{i} + 3\mathbf{j} + 4\mathbf{k})}{(\mathbf{i} + \mathbf{j}) \cdot (\mathbf{i} + \mathbf{j})}\right](\mathbf{i} + \mathbf{j}) = \left[\frac{1(0) + 1(3) + 0(4)}{1+1}\right](\mathbf{i} + \mathbf{j})$$

$$= \frac{3}{2}(\mathbf{i} + \mathbf{j}) = \left(\frac{3}{2}\mathbf{i} + \frac{3}{2}\mathbf{j}\right)$$

$$\mathbf{v} - \text{proj}_{\mathbf{u}}\mathbf{v} = 0\mathbf{i} + 3\mathbf{j} + 4\mathbf{k} - \left(\frac{3}{2}\mathbf{i} + \frac{3}{2}\mathbf{j} + 0\mathbf{k}\right) = -\frac{3}{2}\mathbf{i} + (3 - \frac{3}{2})\mathbf{j} + 4\mathbf{k} = -\frac{3}{2}\mathbf{i} + \frac{3}{2}\mathbf{j} + 4\mathbf{k}$$

Thus, $\mathbf{v} = \text{proj}_{\mathbf{u}}\mathbf{v} + (\mathbf{v} - \text{proj}_{\mathbf{u}}\mathbf{v})$ becomes

$$3\mathbf{j} + 4\mathbf{k} = \left(\frac{3}{2}\mathbf{i} + \frac{3}{2}\mathbf{j}\right) + \left(-\frac{3}{2}\mathbf{i} + \frac{3}{2}\mathbf{j} + 4\mathbf{k}\right).$$

27. $\mathbf{v} = (8, 4, -12)$ and $\mathbf{u} = (1, 2, -1)$ $\Rightarrow$

$$\text{proj}_{\mathbf{u}}\mathbf{v} = \left(\frac{\mathbf{u} \cdot \mathbf{v}}{\mathbf{u} \cdot \mathbf{u}}\right)\mathbf{u} = \left[\frac{(1,2,-1) \cdot (8,4,-12)}{(1,2,-1) \cdot (1,2,-1)}\right](1,2,-1) = \left[\frac{1(8) + 2(4) + (-1)(-12)}{1+4+1}\right](1,2,-1)$$

$$= \frac{28}{6}(1,2,-1) = \left(\frac{14}{3}, \frac{28}{3}, -\frac{14}{3}\right)$$

$$\mathbf{v} - \text{proj}_{\mathbf{u}}\mathbf{v} = (8, 4, -12) - \left(\frac{14}{3}, \frac{28}{3}, -\frac{14}{3}\right) = \left(8 - \frac{14}{3}, 4 - \frac{28}{3}, -12 + \frac{14}{3}\right) = \left(\frac{10}{3}, -\frac{16}{3}, -\frac{22}{3}\right)$$

Thus, $\mathbf{v} = \text{proj}_{\mathbf{u}}\mathbf{v} + (\mathbf{v} - \text{proj}_{\mathbf{u}}\mathbf{v})$ becomes

$$(8, 4, -12) = \left(\frac{14}{3}, \frac{28}{3}, -\frac{14}{3}\right) + \left(\frac{10}{3}, -\frac{16}{3}, -\frac{22}{3}\right).$$

29. $\mathbf{v} = \sqrt{12}\mathbf{i} + \sqrt{48}\mathbf{k} = \sqrt{12}\mathbf{i} + 0\mathbf{j} + \sqrt{48}\mathbf{k}$ and $\mathbf{u} = \frac{1}{\sqrt{3}}\mathbf{i} + \frac{1}{\sqrt{3}}\mathbf{j} - \frac{1}{\sqrt{3}}\mathbf{k} \Rightarrow$

$$\text{proj}_{\mathbf{u}}\mathbf{v} = \left(\frac{\mathbf{u}\cdot\mathbf{v}}{\mathbf{u}\cdot\mathbf{u}}\right)\mathbf{u} = \left[\frac{(\frac{1}{\sqrt{3}}\mathbf{i} + \frac{1}{\sqrt{3}}\mathbf{j} - \frac{1}{\sqrt{3}}\mathbf{k})\cdot(\sqrt{12}\mathbf{i} + 0\mathbf{j} + \sqrt{48}\mathbf{k})}{\frac{1}{3} + \frac{1}{3} + \frac{1}{3}}\right]\left(\frac{1}{\sqrt{3}}\mathbf{i} + \frac{1}{\sqrt{3}}\mathbf{j} - \frac{1}{\sqrt{3}}\mathbf{k}\right)$$

$$= \left[\frac{\sqrt{12}}{\sqrt{3}} + \frac{1}{\sqrt{3}}(0) - \frac{1}{\sqrt{3}}\sqrt{48}\right]\left(\frac{1}{\sqrt{3}}\mathbf{i} + \frac{1}{\sqrt{3}}\mathbf{j} - \frac{1}{\sqrt{3}}\mathbf{k}\right) = (2-4)\left(\frac{1}{\sqrt{3}}\mathbf{i} + \frac{1}{\sqrt{3}}\mathbf{j} - \frac{1}{\sqrt{3}}\mathbf{k}\right)$$

$$= \left(-\frac{2}{\sqrt{3}}\mathbf{i} - \frac{2}{\sqrt{3}}\mathbf{j} + \frac{2}{\sqrt{3}}\mathbf{k}\right)$$

$$\mathbf{v} - \text{proj}_{\mathbf{u}}\mathbf{v} = \sqrt{12}\mathbf{i} + 0\mathbf{j} + \sqrt{48}\mathbf{k} - \left(-\frac{2}{\sqrt{3}}\mathbf{i} - \frac{2}{\sqrt{3}}\mathbf{j} + \frac{2}{\sqrt{3}}\mathbf{k}\right)$$

$$= \left(\sqrt{12} + \frac{2}{\sqrt{3}}\right)\mathbf{i} + \frac{2}{\sqrt{3}}\mathbf{j} + \left(\sqrt{48} - \frac{2}{\sqrt{3}}\right)\mathbf{k} = \frac{8}{\sqrt{3}}\mathbf{i} + \frac{2}{\sqrt{3}}\mathbf{j} + \frac{10}{\sqrt{3}}\mathbf{k}$$

Thus, $\mathbf{v} = \text{proj}_{\mathbf{u}}\mathbf{v} + (\mathbf{v} - \text{proj}_{\mathbf{u}}\mathbf{v})$ becomes

$$\sqrt{12}\mathbf{i} + \sqrt{48}\mathbf{k} = \left(-\frac{2}{\sqrt{3}}\mathbf{i} - \frac{2}{\sqrt{3}}\mathbf{j} + \frac{2}{\sqrt{3}}\mathbf{k}\right) + \left(\frac{8}{\sqrt{3}}\mathbf{i} + \frac{2}{\sqrt{3}}\mathbf{j} + \frac{10}{\sqrt{3}}\mathbf{k}\right).$$

31. $(\mathbf{2}, -\mathbf{1}, \mathbf{6}) \times (-\mathbf{3}, \mathbf{4}, \mathbf{1}) = \det\begin{bmatrix} \mathbf{i} & \mathbf{j} & \mathbf{k} \\ 2 & -1 & 6 \\ -3 & 4 & 1 \end{bmatrix}$

$$= \det\begin{bmatrix} -1 & 6 \\ 4 & 1 \end{bmatrix}\mathbf{i} - \det\begin{bmatrix} 2 & 6 \\ -3 & 1 \end{bmatrix}\mathbf{j} + \det\begin{bmatrix} 2 & -1 \\ -3 & 4 \end{bmatrix}\mathbf{k}$$

$$= [-1(1) - 6(4)]\mathbf{i} - [2(1) - 6(-3)]\mathbf{j} + [2(4) - (-1)(-3)]\mathbf{k}$$

$$= -25\mathbf{i} - 20\mathbf{j} + 5\mathbf{k} = (-\mathbf{25}, -\mathbf{20}, \mathbf{5}).$$

33. Let $\mathbf{v} = [(-4\mathbf{i} + \mathbf{j}) \times (2\mathbf{i} - \mathbf{j} - 3\mathbf{k})] \times (3\mathbf{i} - 2\mathbf{j} - 3\mathbf{k})$.

$[(-4\mathbf{i} + \mathbf{j}) \times (2\mathbf{i} - \mathbf{j} - 3\mathbf{k})] = \det\begin{bmatrix} \mathbf{i} & \mathbf{j} & \mathbf{k} \\ -4 & 1 & 0 \\ 2 & -1 & -3 \end{bmatrix}$

$$= \det\begin{bmatrix} 1 & 0 \\ -1 & -3 \end{bmatrix}\mathbf{i} - \det\begin{bmatrix} -4 & 0 \\ 2 & -3 \end{bmatrix}\mathbf{j} + \det\begin{bmatrix} -4 & 1 \\ 2 & -1 \end{bmatrix}\mathbf{k}$$

$$= [1(-3) - 0(-1)]\mathbf{i} - [-4(-3) - 0(2)]\mathbf{j} + [-4(-1) - 1(2)]\mathbf{k}$$

$$= -3\mathbf{i} - 12\mathbf{j} + 2\mathbf{k}$$

Then, $\mathbf{v} = (-3\mathbf{i} - 12\mathbf{j} + 2\mathbf{k}) \times (3\mathbf{i} - 2\mathbf{j} - 3\mathbf{k}) = \det\begin{bmatrix} \mathbf{i} & \mathbf{j} & \mathbf{k} \\ -3 & -12 & 2 \\ 3 & -2 & -3 \end{bmatrix}$

$$= \det\begin{bmatrix} -12 & 2 \\ -2 & -3 \end{bmatrix}\mathbf{i} - \det\begin{bmatrix} -3 & 2 \\ 3 & -3 \end{bmatrix}\mathbf{j} + \det\begin{bmatrix} -3 & -12 \\ 3 & -2 \end{bmatrix}\mathbf{k}$$

$$= [(-12)(-3) - (2)(-2)]\mathbf{i} - [(-3(-3) - 2(3)]\mathbf{j} + [(-3)(-2) - (-12)(3)]\mathbf{k}$$

$$= 40\mathbf{i} - 3\mathbf{j} + 42\mathbf{k}$$

35. Let $\mathbf{v} = [(2, -1, 4) \times (7, 2, 3)] \cdot (-1, 1, 2)$.

$$(2, -1, 4) \times (7, 2, 3) = \det \begin{bmatrix} \mathbf{i} & \mathbf{j} & \mathbf{k} \\ 2 & -1 & 4 \\ 7 & 2 & 3 \end{bmatrix}$$

$$= \det \begin{bmatrix} -1 & 4 \\ 2 & 3 \end{bmatrix} \mathbf{i} - \det \begin{bmatrix} 2 & 4 \\ 7 & 3 \end{bmatrix} \mathbf{j} + \det \begin{bmatrix} 2 & -1 \\ 7 & 2 \end{bmatrix} \mathbf{k}$$

$$= [-1(3) - 4(2)]\,\mathbf{i} - [2(3) - 4(7)]\,\mathbf{j} + [2(2) - (-1)(7)]\,\mathbf{k}$$

$$= -11\mathbf{i} + 22\mathbf{j} + 11\mathbf{k}$$

Then, $\mathbf{v} = (-11, 22, 11) \cdot (-1, 1, 2) = (-11)(-1) + 22(1) + 11(2) = 11 + 22 + 22 = 55$.

37. (a) Let $\mathbf{u} = (u_1, u_2)$ and $\mathbf{v} = (v_1, v_2)$, then,

$$(\mathbf{u} - \mathbf{v}) \cdot (\mathbf{u} - \mathbf{v}) = (u_1 - v_1, u_2 - v_2) \cdot (u_1 - v_1, u_2 - v_2) = (u_1 - v_1)^2 + (u_2 - v_2)^2$$

$$= u_1^2 - 2u_1 v_1 + v_1^2 + u_2^2 - 2u_2 v_2 + v_2^2 \quad (*)$$

On the other hand,

$$\mathbf{u} \cdot (\mathbf{u} - \mathbf{v}) - \mathbf{v} \cdot (\mathbf{u} - \mathbf{v}) = (u_1, u_2) \cdot (u_1 - v_1, u_2 - v_2) - (v_1, v_2) \cdot (u_1 - v_1, u_2 - v_2)$$

$$= u_1(u_1 - v_1) + u_2(u_2 - v_2) - [v_1(u_1 - v_1) + v_2(u_2 - v_2)]$$

$$= u_1^2 - u_1 v_1 + u_2^2 - u_2 v_2 - [v_1 u_1 - v_1^2 + v_2 u_2 - v_2^2]$$

$$= u_1^2 - u_1 v_1 + u_2^2 - u_2 v_2 - u_1 v_1 + v_1^2 - u_2 v_2 + v_2^2$$

$$= u_1^2 - 2u_1 v_1 + v_1^2 + u_2^2 - 2u_2 v_2 + v_2^2 \ = \ (*)$$

which proves that $(\mathbf{u} - \mathbf{v}) \cdot (\mathbf{u} - \mathbf{v}) = \mathbf{u} \cdot (\mathbf{u} - \mathbf{v}) - \mathbf{v} \cdot (\mathbf{u} - \mathbf{v})$, in $\Re^2$.

(b) Let $\mathbf{u} = (u_1, u_2, u_3)$ and $\mathbf{v} = (v_1, v_2, v_3)$, then,

$$(\mathbf{u} - \mathbf{v}) \cdot (\mathbf{u} - \mathbf{v}) = (u_1 - v_1, u_2 - v_2, u_3 - v_3) \cdot (u_1 - v_1, u_2 - v_2, u_3 - v_3)$$

$$= (u_1 - v_1)^2 + (u_2 - v_2)^2 + (u_3 - v_3)^2$$

$$= u_1^2 - 2u_1 v_1 + v_1^2 + u_2^2 - 2u_2 v_2 + v_2^2 + u_3^2 - 2u_3 v_3 + v_3^2 \quad (*)$$

On the other hand,

$$\mathbf{u} \cdot (\mathbf{u} - \mathbf{v}) - \mathbf{v} \cdot (\mathbf{u} - \mathbf{v}) = (u_1, u_2, u_3) \cdot (u_1 - v_1, u_2 - v_2, u_3 - v_3)$$

$$- (v_1, v_2, v_3) \cdot (u_1 - v_1, u_2 - v_2, u_3 - v_3)$$

$$= u_1(u_1 - v_1) + u_2(u_2 - v_2) + u_3(u_3 - v_3)$$

$$- [v_1(u_1 - v_1) + v_2(u_2 - v_2) + v_3(u_3 - v_3)]$$

$$= u_1^2 - u_1 v_1 + u_2^2 - u_2 v_2 + u_3^2 - u_3 v_3$$

$$- [v_1 u_1 - v_1^2 + v_2 u_2 - v_2^2 + v_3 u_3 - v_3^2]$$

$$= u_1^2 - u_1 v_1 + u_2^2 - u_2 v_2 + u_3^2 - u_3 v_3$$

$$- u_1 v_1 + v_1^2 - u_2 v_2 + v_2^2 - u_3 v_3 + v_3^2$$

$$= u_1^2 - 2u_1 v_1 + v_1^2 + u_2^2 - 2u_2 v_2 + v_2^2 + u_3^2 - 2u_3 v_3 + v_3^2 \ = \ (*)$$

which proves that $(\mathbf{u} - \mathbf{v}) \cdot (\mathbf{u} - \mathbf{v}) = \mathbf{u} \cdot (\mathbf{u} - \mathbf{v}) - \mathbf{v} \cdot (\mathbf{u} - \mathbf{v})$, in $\Re^3$.

(c) Let $\mathbf{u} = (\mathbf{u_1}, \mathbf{u_2}, \ldots, \mathbf{u_n})$ and $\mathbf{v} = (\mathbf{v_1}, \mathbf{v_2}, \ldots, \mathbf{v_n})$, then,

$$(\mathbf{u} - \mathbf{v}) \cdot (\mathbf{u} - \mathbf{v}) = (\mathbf{u_1} - \mathbf{v_1}, \mathbf{u_2} - \mathbf{v_2}, \ldots, \mathbf{u_n} - \mathbf{v_n}) \cdot (\mathbf{u_1} - \mathbf{v_1}, \mathbf{u_2} - \mathbf{v_2}, \ldots, \mathbf{u_n} - \mathbf{v_n})$$
$$= (u_1 - v_1)^2 + (u_2 - v_2)^2 + \cdots + (u_n - v_n)^2$$
$$= u_1^2 - 2u_1v_1 + v_1^2 + u_2^2 - 2u_2v_2 + v_2^2 + \cdots + u_n^2 - 2u_nv_n + v_n^2 \quad (*)$$

On the other hand,

$$\mathbf{u} \cdot (\mathbf{u} - \mathbf{v}) - \mathbf{v} \cdot (\mathbf{u} - \mathbf{v}) = (\mathbf{u_1}, \mathbf{u_2}, \ldots, \mathbf{u_n}) \cdot (\mathbf{u_1} - \mathbf{v_1}, \mathbf{u_2} - \mathbf{v_2}, \ldots, \mathbf{u_n} - \mathbf{v_n})$$
$$- (\mathbf{v_1}, \mathbf{v_2}, \ldots, \mathbf{v_n}) \cdot (\mathbf{u_1} - \mathbf{v_1}, \mathbf{u_2} - \mathbf{v_2}, \ldots \mathbf{u_n} - \mathbf{v_n})$$
$$= u_1(u_1 - v_1) + u_2(u_2 - v_2) + \cdots + u_n(u_n - v_n)$$
$$- [v_1(u_1 - v_1) + v_2(u_2 - v_2) + \cdots + v_n(u_n - v_n)]$$
$$= u_1^2 - u_1v_1 + u_2^2 - u_2v_2 + \cdots + u_n^2 - u_nv_n$$
$$- [v_1u_1 - v_1^2 + v_2u_2 - v_2^2 + \cdots + v_nu_n - v_n^2]$$
$$= u_1^2 - u_1v_1 + u_2^2 - u_2v_2 + \cdots + u_n^2 - u_nv_n$$
$$- u_1v_1 + v_1^2 - u_2v_2 + v_2^2 - \cdots - u_nv_n + v_n^2$$
$$= u_1^2 - 2u_1v_1 + v_1^2 + u_2^2 - 2u_2v_2 + v_2^2 + \cdots + u_n^2 - 2u_nv_n + v_n^2 = (*)$$

which proves that $(\mathbf{u} - \mathbf{v}) \cdot (\mathbf{u} - \mathbf{v}) = \mathbf{u} \cdot (\mathbf{u} - \mathbf{v}) - \mathbf{v} \cdot (\mathbf{u} - \mathbf{v})$, in $\Re^n$.

39. (a) Let $\mathbf{u} = (\mathbf{u_1}, \mathbf{u_2})$ and $\mathbf{v} = (\mathbf{v_1}, \mathbf{v_2})$, then, $\dfrac{|\mathbf{u} \cdot \mathbf{v}|}{\|\mathbf{u}\|\|\mathbf{v}\|} = \dfrac{|u_1v_1 + u_2v_2|}{\sqrt{u_1^2 + u_2^2}\sqrt{v_1^2 + v_2^2}}$. To prove:

$|u_1v_1 + u_2v_2| \le \sqrt{u_1^2 + u_2^2}\sqrt{v_1^2 + v_2^2}$. As both sides of the inequality are positive, squaring them results in an equivalent inequality: $|u_1v_1 + u_2v_2|^2 \le \left(\sqrt{u_1^2 + u_2^2}\sqrt{v_1^2 + v_2^2}\right)^2$, i.e., to prove:

$$(u_1v_1 + u_2v_2)^2 \le (u_1^2 + u_2^2)(v_1^2 + v_2^2)$$

i.e. $\cancel{u_1^2}v_1^2 + 2u_1u_2v_1v_2 + \cancel{u_2^2}v_2^2 \le \cancel{u_1^2 v_1^2} + u_1^2v_2^2 + u_2^2v_1^2 + \cancel{u_2^2 v_2^2}$

i.e. $2u_1u_2v_1v_2 \le u_1^2v_2^2 + u_2^2v_1^2$

i.e. $u_1^2v_2^2 - 2u_1u_2v_1v_2 + u_2^2v_1^2 \ge 0$

i.e. $(u_1v_2 + u_2v_1)^2 \ge 0$ which is true, as squares are always nonnegative!

Since $|u_1v_1 + u_2v_2| \le \sqrt{u_1^2 + u_2^2}\sqrt{v_1^2 + v_2^2}$, then $\dfrac{|u_1v_1 + u_2v_2|}{\sqrt{u_1^2 + u_2^2}\sqrt{v_1^2 + v_2^2}} = \dfrac{|\mathbf{u} \cdot \mathbf{v}|}{\|\mathbf{u}\|\|\mathbf{v}\|} \le 1$, in $\Re^2$.

(b) Let $\mathbf{u} = (\mathbf{u_1}, \mathbf{u_2}, \mathbf{u_3})$ and $\mathbf{v} = (\mathbf{v_1}, \mathbf{v_2}, \mathbf{v_3})$, then,

$\dfrac{|\mathbf{u} \cdot \mathbf{v}|}{\|\mathbf{u}\|\|\mathbf{v}\|} = \dfrac{|u_1v_1 + u_2v_2 + u_3v_3|}{\sqrt{u_1^2 + u_2^2 + u_3^2}\sqrt{v_1^2 + v_2^2 + v_3^2}}$. To prove:

$$|u_1v_1 + u_2v_2 + u_3v_3| \le \sqrt{u_1^2 + u_2^2 + u_3^2}\sqrt{v_1^2 + v_2^2 + v_3^2}$$

As both sides of the inequality are positive, squaring them results in an equivalent inequality:

$|u_1v_1 + u_2v_2 + u_3v_3|^2 \le \left(\sqrt{u_1^2 + u_2^2 + u_3^2}\sqrt{v_1^2 + v_2^2 + v_3^2}\right)^2$, i.e., to prove:

$$(u_1v_1 + u_2v_2 + u_3v_3)^2 \le (u_1^2 + u_2^2 + u_3^2)(v_1^2 + v_2^2 + v_3^2)$$

i.e. $(u_1^2v_1^2 + u_1v_1u_2v_2 + u_1v_1u_3v_3) + (u_2v_2u_1v_1 + u_2^2v_2^2 + u_2v_2u_3v_3)$
$$+ (u_3v_3u_1v_1 + u_3v_3u_2v_2 + u_3^2v_3^2) \leq$$
$$(u_1^2v_1^2 + u_1^2v_2^2 + u_1^2v_3^2) + +(u_2^2v_1^2 + u_2^2v_2^2 + u_2^2v_3^2) + (u_3^2v_1^2 + u_3^2v_2^2 + u_3^2v_3^2)$$

i.e. $(2u_1v_1u_2v_2 + 2u_1v_1u_3v_3 + 2u_2v_2u_3v_3) \leq (u_1^2v_2^2 + u_1^2v_3^2) + (u_2^2v_1^2 + u_2^2v_3^2) + (u_3^2v_1^2 + u_3^2v_2^2)$

i.e. $(u_1^2v_2^2 - 2u_1v_1u_2v_2 + u_2^2v_1^2) + (u_1^2v_3^2 - 2u_1v_1u_3v_3 + u_3^2v_1^2) + (u_2^2v_3^2 - 2u_2v_2u_3v_3 + u_3^2v_2^2) \geq 0$

i.e. $(u_1v_2 - u_2v_1)^2 + (u_1v_3 - u_3v_1)^2 + (u_2v_3 - u_3v_2)^2 \geq 0$

which is true, as a sum of squares is always nonnegative!

Since $|u_1v_1 + u_2v_2 + u_3v_3| \leq \sqrt{u_1^2 + u_2^2 + u_3^2}\sqrt{v_1^2 + v_2^2 + v_3^2}$, then $\dfrac{|u_1v_1 + u_2v_2 + u_3v_3|}{\sqrt{u_1^2 + u_2^2 + u_3^2}\sqrt{v_1^2 + v_2^2 + v_3^2}} =$

$\dfrac{|\mathbf{u} \cdot \mathbf{v}|}{\|\mathbf{u}\|\|\mathbf{v}\|} \leq 1$, in $\Re^3$.

(c) Let $\mathbf{u} = (\mathbf{u_1}, \mathbf{u_2}, \ldots, \mathbf{u_n})$ and $\mathbf{v} = (\mathbf{v_1}, \mathbf{v_2}, \ldots, \mathbf{v_n})$, then,

$\dfrac{|\mathbf{u} \cdot \mathbf{v}|}{\|\mathbf{u}\|\|\mathbf{v}\|} = \dfrac{|u_1v_1 + u_2v_2 + \cdots + u_nv_n|}{\sqrt{u_1^2 + u_2^2 + \cdots + u_n^2}\sqrt{v_1^2 + v_2^2 + \cdots + v_n^2}}$. To prove:

$$|u_1v_1 + u_2v_2 + \cdots + u_nv_n| \leq \sqrt{u_1^2 + u_2^2 + \cdots + u_n^2}\sqrt{v_1^2 + v_2^2 + \cdots + v_n^2}.$$

As both sides of the inequality are positive, squaring them results in an equivalent inequality:

$$|u_1v_1 + u_2v_2 + \cdots + u_nv_n|^2 \leq \left(\sqrt{u_1^2 + u_2^2 + \cdots + u_n^2}\sqrt{v_1^2 + v_2^2 + \cdots + v_n^2}\right)^2,$$

i.e., to prove:

$$(u_1v_1 + u_2v_2 + \cdots + u_nv_n)^2 \leq (u_1^2 + u_2^2 + \cdots + u_n^2)(v_1^2 + v_2^2 + \cdots + v_n^2)$$

i.e. $(u_1^2v_1^2 + u_1v_1u_2v_2 + \cdots + u_1v_1u_nv_n) + (u_2v_2u_1v_1 + u_2^2v_2^2 + \ldots + u_2v_2u_nv_n) + \cdots$
$+ (u_nv_nu_1v_1 + u_nv_nu_2v_2 + \cdots + u_n^2v_n^2) \leq$
$$(u_1^2v_1^2 + u_1^2v_2^2 + \cdots + u_1^2v_n^2) + (u_2^2v_1^2 + u_2^2v_2^2 + \cdots + u_2^2v_n^2) + \cdots + (u_n^2v_1^2 + u_n^2v_2^2 + \cdots + u_n^2v_n^2)$$

i.e. $(2u_1v_1u_2v_2 + 2u_1v_1u_3v_3 + \cdots + 2u_1v_1u_nv_n) + (2u_2v_2u_3v_3 + \cdots 2u_2v_2u_nv_n) + \cdots +$
$(2u_{n-1}v_{n-1}u_nv_n) \leq$
$$(u_1^2v_2^2 + \cdots + u_1^2v_n^2) + (u_2^2v_1^2 + u_2^2v_3^2 + \cdots + u_2^2v_n^2) + \cdots + (u_n^2v_1^2 + u_n^2v_2^2 + \cdots + u_n^2v_{n-1}^2)$$

i.e. $[(u_1v_2 - u_2v_1)^2 + (u_1v_3 - u_3v_1)^2 + \cdots + (u_1v_n - u_nv_1)^2] + [(u_2v_3 - u_3v_2)^2 + (u_2v_4 - u_4v_2)^2 + \cdots + (u_2v_n - u_nv_2)^2] + \cdots + [(u_{n-1}v_n - v_{n-1}u_n)^2] \geq 0$

which is true, as a sum of squares is always nonnegative!

Since, $|u_1v_1 + u_2v_2 + \cdots + u_nv_n| \leq \sqrt{u_1^2 + u_2^2 + \cdots + u_n^2}\sqrt{v_1^2 + v_2^2 + \cdots + v_n^2}$, then

$\dfrac{|\mathbf{u} \cdot \mathbf{v}|}{\|\mathbf{u}\|\|\mathbf{v}\|} = \dfrac{|u_1v_1 + u_2v_2 + \cdots + u_nv_n|}{\sqrt{u_1^2 + u_2^2 + \cdots + u_n^2}\sqrt{v_1^2 + v_2^2 + \cdots + v_n^2}} \leq 1$, in $\Re^n$.

41. (a) To prove: $(\mathbf{u} + \mathbf{v}) \times \mathbf{w} = \mathbf{u} \times \mathbf{w} + \mathbf{v} \times \mathbf{w}$. Let $\mathbf{u} = (\mathbf{u_1}, \mathbf{u_2}, \mathbf{u_3})$, $\mathbf{v} = (\mathbf{v_1}, \mathbf{v_2}, \mathbf{v_3})$ and $\mathbf{w} = (\mathbf{w_1}, \mathbf{w_2}, \mathbf{w_3})$. Then

$$(\mathbf{u} + \mathbf{v}) \times \mathbf{w} = (\mathbf{u_1} + \mathbf{v_1}, \mathbf{u_2} + \mathbf{v_2}, \mathbf{u_3} + \mathbf{v_3}) \times (\mathbf{w_1}, \mathbf{w_2}, \mathbf{w_3}) = \det \begin{bmatrix} \mathbf{i} & \mathbf{j} & \mathbf{k} \\ u_1 + v_1 & u_2 + v_2 & u_3 + v_3 \\ w_1 & w_2 & w_3 \end{bmatrix}$$

$$= \det \begin{bmatrix} u_2 + v_2 & u_3 + v_3 \\ w_2 & w_3 \end{bmatrix} \mathbf{i} - \det \begin{bmatrix} u_1 + v_1 & u_3 + v_3 \\ w_1 & w_3 \end{bmatrix} \mathbf{j} + \det \begin{bmatrix} u_1 + v_1 & u_2 + v_2 \\ w_1 & w_2 \end{bmatrix} \mathbf{k}$$

$$= [(u_2 + v_2)w_3 - (u_3 + v_3)w_2]\,\mathbf{i} - [(u_1 + v_1)w_3 - (u_3 + v_3)w_1]\,\mathbf{j}$$
$$+ [(u_1 + v_1)w_2 - (u_2 + v_2)w_1]\,\mathbf{k}$$

$$= (u_2 w_3 + v_2 w_3 - u_3 w_2 - v_3 w_2)\,\mathbf{i} - (u_1 w_3 + v_1 w_3 - u_3 w_1 - v_3 w_1)\,\mathbf{j}$$
$$+ (u_1 w_2 + v_1 w_2 - u_2 w_1 - v_2 w_1)\,\mathbf{k}$$

$$= (u_2 w_3 - u_3 w_2)\,\mathbf{i} + (v_2 w_3 - v_3 w_2)\,\mathbf{i} - (u_1 w_3 - u_3 w_1)\,\mathbf{j} - (v_1 w_3 - v_3 w_1)\,\mathbf{j}$$
$$+ (u_1 w_2 - u_2 w_1)\,\mathbf{k} + (v_1 w_2 - v_2 w_1)\,\mathbf{k}$$

$$= [(u_2 w_3 - u_3 w_2)\,\mathbf{i} - (u_1 w_3 - u_3 w_1)\,\mathbf{j} + (u_1 w_2 - u_2 w_1)\,\mathbf{k}]$$
$$+ [(v_2 w_3 - v_3 w_2)\,\mathbf{i} - (v_1 w_3 - v_3 w_1)\,\mathbf{j} + (v_1 w_2 - v_2 w_1)\,\mathbf{k}]$$

$$= \det \begin{bmatrix} \mathbf{i} & \mathbf{j} & \mathbf{k} \\ u_1 & u_2 & u_3 \\ w_1 & w_2 & w_3 \end{bmatrix} + \det \begin{bmatrix} \mathbf{i} & \mathbf{j} & \mathbf{k} \\ v_1 & v_2 & v_3 \\ w_1 & w_2 & w_3 \end{bmatrix} = \mathbf{u} \times \mathbf{w} + \mathbf{v} \times \mathbf{w}.$$

(b) To prove: $c\,\mathbf{v} \times \mathbf{w} = \mathbf{v} \times c\,\mathbf{w} = c(\mathbf{v} \times \mathbf{w})$. Let $\mathbf{v} = (\mathbf{v_1}, \mathbf{v_2}, \mathbf{v_3})$ and $\mathbf{w} = (\mathbf{w_1}, \mathbf{w_2}, \mathbf{w_3})$. Then

$$c\,\mathbf{v} \times \mathbf{w} = \det \begin{bmatrix} \mathbf{i} & \mathbf{j} & \mathbf{k} \\ cv_1 & cv_2 & cv_3 \\ w_1 & w_2 & w_3 \end{bmatrix} = \det \begin{bmatrix} cv_2 & cv_3 \\ w_2 & w_3 \end{bmatrix} \mathbf{i} - \det \begin{bmatrix} cv_1 & cv_3 \\ w_1 & w_3 \end{bmatrix} \mathbf{j} + \det \begin{bmatrix} cv_1 & cv_2 \\ w_1 & w_2 \end{bmatrix} \mathbf{k}$$

$$= (cv_2 w_3 - cv_3 w_2)\,\mathbf{i} - (cv_1 w_3 - cv_3 w_1)\,\mathbf{j} + (cv_1 w_2 - cv_2 w_1)\,\mathbf{k}$$

$$= c[(v_2 w_3 - v_3 w_2)\,\mathbf{i} - (v_1 w_3 - v_3 w_1)\,\mathbf{j} + (v_1 w_2 - v_2 w_1)\,\mathbf{k}]$$

$$= c \det \begin{bmatrix} \mathbf{i} & \mathbf{j} & \mathbf{k} \\ v_1 & v_2 & v_3 \\ w_1 & w_2 & w_3 \end{bmatrix} = c(\mathbf{v} \times \mathbf{w})$$

$$\mathbf{v} \times c\,\mathbf{w} = \det \begin{bmatrix} \mathbf{i} & \mathbf{j} & \mathbf{k} \\ v_1 & v_2 & v_3 \\ cw_1 & cw_2 & cw_3 \end{bmatrix} = \det \begin{bmatrix} v_2 & v_3 \\ cw_2 & cw_3 \end{bmatrix} \mathbf{i} - \det \begin{bmatrix} v_1 & v_3 \\ cw_1 & cw_3 \end{bmatrix} \mathbf{j} + \det \begin{bmatrix} v_1 & v_2 \\ cw_1 & cw_2 \end{bmatrix} \mathbf{k}$$

$$= (v_2 cw_3 - v_3 cw_2)\,\mathbf{i} - (v_1 cw_3 - v_3 cw_1)\,\mathbf{j} + (v_1 cw_2 - v_2 cw_1)\,\mathbf{k}$$

$$= c[(v_2 w_3 - v_3 w_2)\,\mathbf{i} - (v_1 w_3 - v_3 w_1)\,\mathbf{j} + (v_1 w_2 - v_2 w_1)\,\mathbf{k}]$$

$$= c \det \begin{bmatrix} \mathbf{i} & \mathbf{j} & \mathbf{k} \\ v_1 & v_2 & v_3 \\ w_1 & w_2 & w_3 \end{bmatrix} = c(\mathbf{v} \times \mathbf{w})$$

Therefore,

as $c\,\mathbf{v} \times \mathbf{w} = c(\mathbf{v} \times \mathbf{w})$, and $\mathbf{v} \times c\,\mathbf{w} = c(\mathbf{v} \times \mathbf{w})$, then $c\,\mathbf{v} \times \mathbf{w} = \mathbf{v} \times c\,\mathbf{w} = c(\mathbf{v} \times \mathbf{w})$.

(c) To prove: $(\mathbf{v} \times \mathbf{u}) = -(\mathbf{u} \times \mathbf{v})$. Let $\mathbf{u} = (\mathbf{u_1}, \mathbf{u_2}, \mathbf{u_3})$ and $\mathbf{v} = (\mathbf{v_1}, \mathbf{v_2}, \mathbf{v_3})$, then

$$\mathbf{u} \times \mathbf{v} = \det \begin{bmatrix} \mathbf{i} & \mathbf{j} & \mathbf{k} \\ u_1 & u_2 & u_3 \\ v_1 & v_2 & v_3 \end{bmatrix} = \det \begin{bmatrix} u_2 & u_3 \\ v_2 & v_3 \end{bmatrix} \mathbf{i} - \det \begin{bmatrix} u_1 & u_3 \\ v_1 & v_3 \end{bmatrix} \mathbf{j} + \det \begin{bmatrix} u_1 & u_2 \\ v_1 & v_2 \end{bmatrix} \mathbf{k}$$

$$= (u_2 v_3 - u_3 v_2)\,\mathbf{i} - (u_1 v_3 - u_3 v_1)\,\mathbf{j} + (u_1 v_2 - u_2 v_1)\,\mathbf{k}$$

and,

$$\mathbf{v} \times \mathbf{u} = \det \begin{bmatrix} \mathbf{i} & \mathbf{j} & \mathbf{k} \\ v_1 & v_2 & v_3 \\ u_1 & u_2 & u_3 \end{bmatrix} = \det \begin{bmatrix} v_2 & v_3 \\ u_2 & u_3 \end{bmatrix} \mathbf{i} - \det \begin{bmatrix} v_1 & v_3 \\ u_1 & u_3 \end{bmatrix} \mathbf{j} + \det \begin{bmatrix} v_1 & v_2 \\ u_1 & u_2 \end{bmatrix} \mathbf{k}$$

$$= (v_2 u_3 - v_3 u_2)\,\mathbf{i} - (v_1 u_3 - v_3 u_1)\,\mathbf{j} + (v_1 u_2 - v_2 u_1)\,\mathbf{k}$$

$$= (u_3 v_2 - u_2 v_3)\,\mathbf{i} - (u_3 v_1 - u_1 v_3)\,\mathbf{j} + (u_2 v_1 - u_1 v_2)\,\mathbf{k}$$

$$= -[(-u_3 v_2 + u_2 v_3)\,\mathbf{i} - (-u_3 v_1 + u_1 v_3)\,\mathbf{j} + (-u_2 v_1 + u_1 v_2)\,\mathbf{k}]$$

$$= -[(u_2 v_3 - u_3 v_2)\,\mathbf{i} - (u_1 v_3 - u_3 v_1)\,\mathbf{j} + (u_1 v_2 - u_2 v_1)\,\mathbf{k}] = -(\mathbf{u} \times \mathbf{v})$$

43.

The diagonals are orthogonal if and only if their dot product is 0:

$(a\,\mathbf{i} + b\,\mathbf{j}) \cdot (a\,\mathbf{i} - b\,\mathbf{j}) = a^2 - b^2 = 0$ if and only if $a^2 = b^2 \Leftrightarrow a = \pm b$, but a and b are both positive, so, if and only if $a = b$, which means the rectangle is a square.

45. To prove: $(\mathbf{u} \times \mathbf{v}) \times \mathbf{w} + (\mathbf{v} \times \mathbf{w}) \times \mathbf{u} + (\mathbf{w} \times \mathbf{u}) \times \mathbf{v} = \mathbf{0}$, for $\mathbf{u}, \mathbf{v}, \mathbf{w}$ in $\Re^3$.

$$\mathbf{u} \times \mathbf{v} = \det \begin{bmatrix} \mathbf{i} & \mathbf{j} & \mathbf{k} \\ u_1 & u_2 & u_3 \\ v_1 & v_2 & v_3 \end{bmatrix} = \det \begin{bmatrix} u_2 & u_3 \\ v_2 & v_3 \end{bmatrix} \mathbf{i} - \det \begin{bmatrix} u_1 & u_3 \\ v_1 & v_3 \end{bmatrix} \mathbf{j} + \det \begin{bmatrix} u_1 & u_2 \\ v_1 & v_2 \end{bmatrix} \mathbf{k}$$

$$= (u_2 v_3 - u_3 v_2)\,\mathbf{i} - (u_1 v_3 - u_3 v_1)\,\mathbf{j} + (u_1 v_2 - u_2 v_1)\,\mathbf{k}$$

so that,

$$(\mathbf{u} \times \mathbf{v}) \times \mathbf{w} = \det \begin{bmatrix} \mathbf{i} & \mathbf{j} & \mathbf{k} \\ u_2 v_3 - u_3 v_2 & -u_1 v_3 + u_3 v_1 & u_1 v_2 - u_2 v_1 \\ w_1 & w_2 & w_3 \end{bmatrix}$$

$$= \det \begin{bmatrix} -u_1 v_3 + u_3 v_1 & u_1 v_2 - u_2 v_1 \\ w_2 & w_3 \end{bmatrix} \mathbf{i} - \det \begin{bmatrix} u_2 v_3 - u_3 v_2 & u_1 v_2 - u_2 v_1 \\ w_1 & w_3 \end{bmatrix} \mathbf{j}$$

$$+ \det \begin{bmatrix} u_2 v_3 - u_3 v_2 & -u_1 v_3 + u_3 v_1 \\ w_1 & w_2 \end{bmatrix} \mathbf{k}$$

$$= [(-u_1 v_3 + u_3 v_1)w_3 - (u_1 v_2 - u_2 v_1)w_2]\,\mathbf{i} - [(u_2 v_3 - u_3 v_2)w_3 - (u_1 v_2 - u_2 v_1)w_1]\,\mathbf{j}$$

$$+ [(u_2 v_3 - u_3 v_2)w_2 - (-u_1 v_3 + u_3 v_1)w_1]\,\mathbf{k}$$

$$= (-u_1 v_3 w_3 + u_3 v_1 w_3 - u_1 v_2 w_2 + u_2 v_1 w_2)\,\mathbf{i} - (u_2 v_3 w_3 - u_3 v_2 w_3 - u_1 v_2 w_1 + u_2 v_1 w_1)\,\mathbf{j}$$

$$+ (u_2 v_3 w_2 - u_3 v_2 w_2 + u_1 v_3 w_1 - u_3 v_1 w_1)\,\mathbf{k} \quad (*)$$

Then, just changing the names in each term in $(*)$ from u, v, w respectively to v, w, u gives us:

$(\mathbf{v} \times \mathbf{w}) \times \mathbf{u}$

$$= (-v_1 w_3 u_3 + v_3 w_1 u_3 - v_1 w_2 u_2 + v_2 w_1 u_2)\,\mathbf{i} - (v_2 w_3 u_3 - v_3 w_2 u_3 - v_1 w_2 u_1 + v_2 w_1 u_1)\,\mathbf{j}$$

$$+ (v_2 w_3 u_2 - v_3 w_2 u_2 + v_1 w_3 u_1 - v_3 w_1 u_1)\,\mathbf{k}, \quad \text{and then reordering the terms:}$$

$$= (-u_3 v_1 w_3 + u_3 v_3 w_1 - u_2 v_1 w_2 + u_2 v_2 w_1)\,\mathbf{i} - (u_3 v_2 w_3 - u_3 v_3 w_2 - u_1 v_1 w_2 + u_1 v_2 w_1)\,\mathbf{j}$$

$$+ (u_2 v_2 w_3 - u_2 v_3 w_2 + u_1 v_1 w_3 - u_1 v_3 w_1)\,\mathbf{k}$$

Lastly, changing the names in each term in $(*)$ from u, v, w respectively to w, u, v gives us:

$(\mathbf{w} \times \mathbf{u}) \times \mathbf{v}$

$$= (-w_1 u_3 v_3 + w_3 u_1 v_3 - w_1 u_2 v_2 + w_2 u_1 v_2)\,\mathbf{i} - (w_2 u_3 v_3 - w_3 u_2 v_3 - w_1 u_2 v_1 + w_2 u_1 v_1)\,\mathbf{j}$$

$$+ (w_2 u_3 v_2 - w_3 u_2 v_2 + w_1 u_3 v_1 - w_3 u_1 v_1)\,\mathbf{k}, \quad \text{and then reordering the terms:}$$

$$= (-u_3 v_3 w_1 + u_1 v_3 w_3 - u_2 v_2 w_1 + u_1 v_2 w_2)\,\mathbf{i} - (u_3 v_3 w_2 - u_2 v_3 w_3 - u_2 v_1 w_1 + u_1 v_1 w_2)\,\mathbf{j}$$

$$+ (u_3 v_2 w_2 - u_2 v_2 w_3 + u_3 v_1 w_1 - u_1 v_1 w_3)\,\mathbf{k}$$

Adding the three cross products together, we have:

$(\mathbf{u} \times \mathbf{v}) \times \mathbf{w} + (\mathbf{v} \times \mathbf{w}) \times \mathbf{u} + (\mathbf{w} \times \mathbf{u}) \times \mathbf{v}$

$$= (-u_1 v_3 w_3 + u_3 v_1 w_3 - u_1 v_2 w_2 + u_2 v_1 w_2)\,\mathbf{i} - (u_2 v_3 w_3 - u_3 v_2 w_3 - u_1 v_2 w_1 + u_2 v_1 w_1)\,\mathbf{j}$$

$$+ (u_2 v_3 w_2 - u_3 v_2 w_2 + u_1 v_3 w_1 - u_3 v_1 w_1)\,\mathbf{k}$$

$$+ (-u_3 v_1 w_3 + u_3 v_3 w_1 - u_2 v_1 w_2 + u_2 v_2 w_1)\,\mathbf{i} - (u_3 v_2 w_3 - u_3 v_3 w_2 - u_1 v_1 w_2 + u_1 v_2 w_1)\,\mathbf{j}$$

$$+ (u_2 v_2 w_3 - u_2 v_3 w_2 + u_1 v_1 w_3 - u_1 v_3 w_1)\,\mathbf{k}$$

$$+ (-u_3 v_3 w_1 + u_1 v_3 w_3 - u_2 v_2 w_1 + u_1 v_2 w_2)\,\mathbf{i} - (u_3 v_3 w_2 - u_2 v_3 w_3 - u_2 v_1 w_1 + u_1 v_1 w_2)\,\mathbf{j}$$

$$+ (u_3 v_2 w_2 - u_2 v_2 w_3 + u_3 v_1 w_1 - u_1 v_1 w_3)\,\mathbf{k}$$

Looking at the coefficient of $\mathbf{i}$:

$(-u_1 v_3 w_3 + u_3 v_1 w_3 - u_1 v_2 w_2 + u_2 v_1 w_2) + (-u_3 v_1 w_3 + u_3 v_3 w_1 - u_2 v_1 w_2 + u_2 v_2 w_1) + (-u_3 v_3 w_1 + u_1 v_3 w_3 - u_2 v_2 w_1 + u_1 v_2 w_2)$ (and reordering terms, first those with u_1, then with u_2 etc.)
$= (-u_1 v_3 w_3 - u_1 v_2 w_2 + u_1 v_3 w_3 + u_1 v_2 w_2) + (u_2 v_1 w_2 - u_2 v_1 w_2 + u_2 v_2 w_1 - u_2 v_2 w_1) + (u_3 v_1 w_3 - u_3 v_1 w_3 + u_3 v_3 w_1 - u_3 v_3 w_1) = 0.$

Looking at the coefficient of $-\mathbf{j}$:

$(u_2 v_3 w_3 - u_3 v_2 w_3 - u_1 v_2 w_1 + u_2 v_1 w_1) + (u_3 v_2 w_3 - u_3 v_3 w_2 - u_1 v_1 w_2 + u_1 v_2 w_1) + (u_3 v_3 w_2 - u_2 v_3 w_3 - u_2 v_1 w_1 + u_1 v_1 w_2)$ (and reordering terms, first those with u_1, then with u_2 etc.)
$(-u_1 v_2 w_1 - u_1 v_1 w_2 + u_1 v_2 w_1 + u_1 v_1 w_2) + (u_2 v_3 w_3 + u_2 v_1 w_1 - u_2 v_3 w_3 - u_2 v_1 w_1) + (-u_3 v_2 w_3 + u_3 v_2 w_3 - u_3 v_3 w_2 + u_3 v_3 w_2) = 0.$

Looking at the coefficient of $\mathbf{k}$:

$(u_2 v_3 w_2 - u_3 v_2 w_2 + u_1 v_3 w_1 - u_3 v_1 w_1) + (u_2 v_2 w_3 - u_2 v_3 w_2 + u_1 v_1 w_3 - u_1 v_3 w_1) + (u_3 v_2 w_2 - u_2 v_2 w_3 + u_3 v_1 w_1 - u_1 v_1 w_3)$ (and reordering terms, first those with u_1, then with u_2 etc.)

$(u_1v_3w_1 + u_1v_1w_3 - u_1v_3w_1 - u_1v_1w_3) + (u_2v_3w_2 + u_2v_2w_3 - u_2v_3w_2 - u_2v_2w_3) + (-u_3v_2w_2 - u_3v_1w_1 + u_3v_2w_2 + u_3v_1w_1) = 0$.

This proves that the sum of the three cross products is the zero vector.

47. Work $W = \mathbf{F} \cdot \mathbf{d}$. Here $\mathbf{F} = 20\mathbf{k}$, and $\mathbf{d} = (1,1,1) = \mathbf{i} + \mathbf{j} + \mathbf{k}$. Thus
$W = 20\mathbf{k} \cdot (\mathbf{i} + \mathbf{j} + \mathbf{k}) = 20$ ft-lbs.

49. Work $W = \mathbf{F} \cdot \mathbf{d} = \|\mathbf{F}\|\|\mathbf{d}\| \cos\theta = 200(100) \cos 30° = 20{,}000\frac{\sqrt{3}}{2} = 10{,}000\sqrt{3}$ ft-lbs.

51. As $\tau = \overrightarrow{OP} \times \mathbf{F} \Rightarrow \|\tau\| = \|\overrightarrow{OP} \times \mathbf{F}\| = \|\overrightarrow{OP}\| \|\mathbf{F}\| \sin\theta$, by Theorem 12.6.

53. By Exercise 51, $\|\tau\| = \|\overrightarrow{OP}\| \|\mathbf{F}\| \sin\theta$. Here, $\|\mathbf{F}\| = 30$ lbs, $\|\overrightarrow{OP}\| = \frac{9}{12} = \frac{3}{4}$ ft, and

(a) $\theta = 30° \Rightarrow \|\tau\| = \frac{3}{4}(30) \sin 30° = \frac{90}{4} \cdot \frac{1}{2} = \frac{45}{4} = 11.25$ ft-lbs.

(b) $\theta = 90° \Rightarrow \|\tau\| = \frac{3}{4}(30) \sin 90° = \frac{90}{4} \cdot 1 = \frac{90}{4} = 22.5$ ft-lbs.

(c) $\theta = 135° \Rightarrow \|\tau\| = \frac{3}{4}(30) \sin 135° = \frac{90}{4} \sin 45° = \frac{45}{2} \cdot \frac{1}{\sqrt{2}} = \frac{45}{2\sqrt{2}} \approx 15.9$ ft-lbs.

55. To prove: $\mathbf{u} \cdot (\mathbf{v} \times \mathbf{w}) = (\mathbf{u} \times \mathbf{v}) \cdot \mathbf{w}$. Let $\mathbf{u} = (u_1, u_2, u_3)$, $\mathbf{v} = (v_1, v_2, v_3)$, and $\mathbf{w} = (w_1, w_2, w_3)$. Then

$$\mathbf{v} \times \mathbf{w} = \det \begin{bmatrix} \mathbf{i} & \mathbf{j} & \mathbf{k} \\ v_1 & v_2 & v_3 \\ w_1 & w_2 & w_3 \end{bmatrix} = \det \begin{bmatrix} v_2 & v_3 \\ w_2 & w_3 \end{bmatrix} \mathbf{i} - \det \begin{bmatrix} v_1 & v_3 \\ w_1 & w_3 \end{bmatrix} \mathbf{j} + \det \begin{bmatrix} v_1 & v_2 \\ w_1 & w_2 \end{bmatrix} \mathbf{k}$$

$$= (v_2w_3 - v_3w_2)\,\mathbf{i} - (v_1w_3 - v_3w_1)\,\mathbf{j} + (v_1w_2 - v_2w_1)\,\mathbf{k}$$

Therefore, $\mathbf{u} \cdot (\mathbf{v} \times \mathbf{w}) = (u_1\,\mathbf{i} + u_2\,\mathbf{j} + u_3\,\mathbf{k}) \cdot [(v_2w_3 - v_3w_2)\,\mathbf{i} - (v_1w_3 - v_3w_1)\,\mathbf{j} + (v_1w_2 - v_2w_1)\,\mathbf{k}]$

$$= u_1(v_2w_3 - v_3w_2) - u_2(v_1w_3 - v_3w_1) + u_3(v_1w_2 - v_2w_1)$$

$$= u_1v_2w_3 - u_1v_3w_2 - u_2v_1w_3 + u_2v_3w_1 + u_3v_1w_2 - u_3v_2w_1$$

Now reorder with the terms containing w_1 first, then those with w_2, etc.

$$= (u_2v_3w_1 - u_3v_2w_1) + (-u_1v_3w_2 + u_3v_1w_2) + (u_1v_2w_3 - u_2v_1w_3)$$

$$= (u_2v_3 - u_3v_2)w_1 - (u_1v_3 - u_3v_1)w_2 + (u_1v_2 - u_2v_1)w_3$$

$$= \det \begin{bmatrix} u_2 & u_3 \\ v_2 & v_3 \end{bmatrix} w_1 - \det \begin{bmatrix} u_1 & u_3 \\ v_1 & v_3 \end{bmatrix} w_2 + \det \begin{bmatrix} u_1 & u_2 \\ v_1 & v_2 \end{bmatrix} w_3$$

$$= \det \begin{bmatrix} \mathbf{i} & \mathbf{j} & \mathbf{k} \\ u_1 & u_2 & u_3 \\ v_1 & v_2 & v_3 \end{bmatrix} \cdot (w_1\mathbf{i} + w_2\mathbf{j} + w_3\mathbf{k}) = (\mathbf{u} \times \mathbf{v}) \cdot \mathbf{w}.$$

§12.3 Lines and Planes

1. Slope of 3 (over 1 and up 3) gives us the direction vector $\mathbf{v} = (1, 3)$. The given point (1,3) gives us the translation vector $\mathbf{u} = (1, 3)$. Hence, a vector equation of the line is $\mathbf{w} = (1, 3) + t(1, 3)$, and parametric equations are $x = 1 + t,\ y = 3 + 3t, -\infty < t < \infty$.

3. A direction vector for the line is $(2, 3, 1) - (0, 1, -1) = (2, 2, 2)$, so a simpler one is $\mathbf{v} = (1, 1, 1)$. A translation vector is $\mathbf{u} = (0, 1, -1)$. Thus a vector equation of the line is $\mathbf{w} = (0, 1, -1) + t(1, 1, 1)$, and parametric equations are $x = t, y = 1 + t, z = -1 + t$, $-\infty < t < \infty$.

5. First, we will show that for any given t, there is a $\bar{t}$ such that

$$(1, 1) + \mathbf{t}[(1, 1) - (1, 0)] = (1, 0) + \bar{\mathbf{t}}[(1, 0) - (1, 1)]$$
$$\text{i.e.}\quad (1, 1) + (0, \mathbf{t}) = (1, 0) + (0, -\bar{\mathbf{t}})$$
$$\text{i.e.}\quad (1, 1 + \mathbf{t}) = (1, -\bar{\mathbf{t}})\ (*)$$

Conclusion: $\bar{t} = -(1 + t)$.

Next, we will show that for any given $\bar{t}$, there is a t such that

$$(1, 1) + \mathbf{t}[(1, 1) - (1, 0)] = (1, 0) + \bar{\mathbf{t}}[(1, 0) - (1, 1)]$$

From $(*)$ above, $1 + t = -\bar{t} \Rightarrow t = -1 - \bar{t}$
Conclusion: $t = -(1 + \bar{t})$

7. (a) A direction vector for the line is $\mathbf{u} = (2, 3) - (0, 4) = (2, -1)$. The vector from $(0, 4)$ to $P = (1, 5)$ is $\mathbf{v} = (1, 1)$. Thus

$$\text{proj}_\mathbf{u}\mathbf{v} = \left(\frac{\mathbf{u} \cdot \mathbf{v}}{\mathbf{u} \cdot \mathbf{u}}\right)\mathbf{u} = \left(\frac{(2, -1) \cdot (1, 1)}{(2, -1) \cdot (2, -1)}\right)(2, -1) = \tfrac{1}{5}(2, -1) = \left(\tfrac{2}{5}, -\tfrac{1}{5}\right).$$

Then, the distance from P to the line is

$$\|\mathbf{v} - \text{proj}_\mathbf{u}\mathbf{v}\| = \left\|(1, 1) - \left(\tfrac{2}{5}, -\tfrac{1}{5}\right)\right\| = \left\|\left(\tfrac{3}{5}, \tfrac{6}{5}\right)\right\| = \tfrac{1}{5}\|(3, 6)\| = \tfrac{1}{5}\sqrt{9 + 36} = \tfrac{\sqrt{45}}{5} = \tfrac{3\sqrt{5}}{5}.$$

(b) A direction vector for the line is $\mathbf{u} = (2, 0, 4) - (1, 2, 2) = (1, -2, 2)$. The vector from $(2, 0, 4)$ to $P = (2, 0, 3)$ is $\mathbf{v} = (0, 0, 1)$. Thus

$$\text{proj}_\mathbf{u}\mathbf{v} = \left(\frac{\mathbf{u} \cdot \mathbf{v}}{\mathbf{u} \cdot \mathbf{u}}\right)\mathbf{u} = \left(\frac{(1, -2, 2) \cdot (0, 0, 1)}{(1, -2, 2) \cdot (1, -2, 2)}\right)(1, -2, 2) = \tfrac{2}{9}(1, -2, 2) = \left(\tfrac{2}{9}, -\tfrac{4}{9}, \tfrac{4}{9}\right).$$

Then, the distance from P to the line is

$$\|\mathbf{v} - \text{proj}_\mathbf{u}\mathbf{v}\| = \left\|(0, 0, 1) - \left(\tfrac{2}{9}, -\tfrac{4}{9}, \tfrac{4}{9}\right)\right\| = \left\|\left(-\tfrac{2}{9}, \tfrac{4}{9}, \tfrac{5}{9}\right)\right\| = \tfrac{1}{9}\sqrt{4 + 16 + 25} = \tfrac{1}{9}\sqrt{45}$$
$$= \tfrac{3\sqrt{5}}{9} = \tfrac{\sqrt{5}}{3}.$$

(c)   A direction vector for the line is $\mathbf{u} = (2, 2, 1, 1) - (1, 0, 2, 1) = (1, 2, -1, 0)$. The vector from (1,0,2,1) to $P = (1, 0, 1, 0)$ is $\mathbf{v} = (0, 0, -1, -1)$. Thus

$$\text{proj}_\mathbf{u}\mathbf{v} = \left(\frac{\mathbf{u}\cdot\mathbf{v}}{\mathbf{u}\cdot\mathbf{u}}\right)\mathbf{u} = \left(\frac{(1, 2, -1, 0)\cdot(0, 0, -1, -1)}{(1, 2, -1, 0)\cdot(1, 2, -1, 0)}\right)(1, 2, -1, 0) = \tfrac{1}{6}(1, 2, -1, 0) = \left(\tfrac{1}{6}, \tfrac{2}{6}, -\tfrac{1}{6}, 0\right)$$

Then, the distance from P to the line is

$$\|\mathbf{v} - \text{proj}_\mathbf{u}\mathbf{v}\| = \left\|(0, 0, -1, -1) - \left(\tfrac{1}{6}, \tfrac{2}{6}, -\tfrac{1}{6}, 0\right)\right\| = \left\|\left(-\tfrac{1}{6}, -\tfrac{2}{6}, -\tfrac{5}{6}, -\tfrac{6}{6}\right)\right\|$$

$$= \tfrac{1}{6}\sqrt{1 + 4 + 25 + 36} = \frac{\sqrt{66}}{6}.$$

9.   Given $\mathbf{n} = (1, 4, 2)$ and Q=(2,0,3) is a point on the plane, then a vector equation of the plane is $(1, 4, 2)\cdot(\mathbf{x} - 2, \mathbf{y} - 0, \mathbf{z} - 3) = 0$. A scalar equation is $x - 2 + 4y + 2(z - 3) = 0$, and in general, $x + 4y + 2z = 2 + 6 = 8$.

11.   Given $\mathbf{n} = 3\mathbf{i} + 2\mathbf{j} - \mathbf{k}$ and Q=(1,1,2) is a point on the plane, then a vector equation of the plane is $(3, 2, -1)\cdot(\mathbf{x} - 1, \mathbf{y} - 1, \mathbf{z} - 2) = 0$. A scalar equation is $3(x - 1) + 2(y - 1) - (z - 2) = 0$, and in general, $3x + 2y - z = 3$.

13.   Let $A = (1, 2, 0)$, $B = (2, 1, -2)$ and $C = (0, 0, 3)$, then $\vec{AB} = (2 - 1, 1 - 2, -2 - 0) = (1, -1, -2)$ and $\vec{AC} = (0 - 1, 0 - 2, 3 - 0) = (-1, -2, 3)$.

Thus, $\mathbf{n} = \vec{AB} \times \vec{AC} = \det\begin{bmatrix} \mathbf{i} & \mathbf{j} & \mathbf{k} \\ 1 & -1 & -2 \\ -1 & -2 & 3 \end{bmatrix} = \det\begin{bmatrix} -1 & -2 \\ -2 & 3 \end{bmatrix}\mathbf{i} - \det\begin{bmatrix} 1 & -2 \\ -1 & 3 \end{bmatrix}\mathbf{j} + \det\begin{bmatrix} 1 & -1 \\ -1 & -2 \end{bmatrix}\mathbf{k}$

$$= [-1(3) - (-2)(-2)]\mathbf{i} - [1(3) - (-2)(-1)]\mathbf{j} + [1(-2) - (-1)(-1)]\mathbf{k}$$

$$= -7\mathbf{i} - \mathbf{j} - 3\mathbf{k}$$

Therefore,
a vector equation of the plane is $(-7, -1, -3)\cdot(\mathbf{x} - 1, \mathbf{y} - 2, \mathbf{z} - 0) = 0$. So,
$-7(x - 1) - (y - 2) - 3z = 0 \Rightarrow -7x + 7 - y + 2 - 3z = 0 \Rightarrow -7x - y - 3z = -9 \Rightarrow 7x + y + 3z = 9$ is the general equation of the plane containing the given points.

15.   Let $A = (1, 1, 0)$, $B = (0, -2, 1)$ and $C = (0, 0, 1)$, then $\vec{AB} = (0 - 1, -2 - 1, 1 - 0) = (-1, -3, 1)$ and $\vec{AC} = (0 - 1, 0 - 1, 1 - 0) = (-1, -1, 1)$.

Thus, $\vec{AB} \times \vec{AC} = \det\begin{bmatrix} \mathbf{i} & \mathbf{j} & \mathbf{k} \\ -1 & -3 & 1 \\ -1 & -1 & 1 \end{bmatrix} = \det\begin{bmatrix} -3 & 1 \\ -1 & 1 \end{bmatrix}\mathbf{i} - \det\begin{bmatrix} -1 & 1 \\ -1 & 1 \end{bmatrix}\mathbf{j} + \det\begin{bmatrix} -1 & -3 \\ -1 & -1 \end{bmatrix}\mathbf{k}$

$$= [-3(1) - (1)(-1)]\mathbf{i} - [-1(1) - (1)(-1)]\mathbf{j} + [-1(-1) - (-3)(-1)]\mathbf{k}$$

$$= -2\mathbf{i} - 2\mathbf{k}$$

and a normal vector is $\mathbf{n} = \mathbf{i} + \mathbf{k}$, so that a vector equation of the plane is
$(1, 0, 1)\cdot(\mathbf{x} - 0, \mathbf{y} - 0, \mathbf{z} - 1) = 0 \Rightarrow x + z - 1 = 0 \Rightarrow x + z = 1$ is the general equation of the plane containing the given points.

17. We want the distance between $P = (1, 2, 2)$ and the plane $x + y + 2z = 4$: A normal to the plane is $\mathbf{n} = (1, 1, 2)$, and a point on the plane is $Q = (4, 0, 0)$. Let $\mathbf{v} = \overrightarrow{QP}$ $= (1 - 4, 2 - 0, 2 - 0) = (-3, 2, 2)$. The distance between P and the plane is

$$d = \|\text{proj}_\mathbf{n}\mathbf{v}\| = \frac{|\mathbf{v} \cdot \mathbf{n}|}{\sqrt{\mathbf{n} \cdot \mathbf{n}}} = \frac{(-3, 2, 2) \cdot (1, 1, 2)}{\sqrt{1 + 1 + 4}} = \frac{|-3 + 2 + 4|}{\sqrt{6}} = \frac{3}{\sqrt{6}} = \frac{3\sqrt{6}}{6} = \frac{\sqrt{6}}{2}.$$

19. We want the distance between $P = (0, 0, 0)$ and the plane with normal $\mathbf{n} = (1, 4, 2)$ that contains the point $Q = (2, 1, 3)$: Let $\mathbf{v} = \overrightarrow{QP} = (-2, -1, -3)$. The distance between P and the plane is

$$d = \|\text{proj}_\mathbf{n}\mathbf{v}\| = \frac{|\mathbf{v} \cdot \mathbf{n}|}{\sqrt{\mathbf{n} \cdot \mathbf{n}}} = \frac{(-2, -1, -3) \cdot (1, 4, 2)}{\sqrt{1 + 16 + 4}} = \frac{|-2 - 4 - 6|}{\sqrt{21}} = \frac{12}{\sqrt{21}} = \frac{12\sqrt{21}}{21} = \frac{4\sqrt{21}}{7}.$$

21. Since the plane is parallel to the plane $3x - 2y + 3z = 1$, a normal is $\mathbf{n} = (3, -2, 3)$. With $P = (2, -3, 3)$ and $Q = (1, 2, 2)$, then $\mathbf{v} = \overrightarrow{QP} = (2 - 1, -3 - 2, 3 - 2) = (1, -5, 1)$. The distance between P and the plane is

$$d = \|\text{proj}_\mathbf{n}\mathbf{v}\| = \frac{|\mathbf{v} \cdot \mathbf{n}|}{\sqrt{\mathbf{n} \cdot \mathbf{n}}} = \frac{(1, -5, 1) \cdot (3, -2, 3)}{\sqrt{9 + 4 + 9}} = \frac{|3 + 10 + 3|}{\sqrt{22}} = \frac{16}{\sqrt{22}} = \frac{16\sqrt{22}}{22} = \frac{8\sqrt{22}}{11}.$$

23. Given $P = (3, 0, 1)$ and the distance d from P to the plane $ax + 2y + z = 3$ is 1 unit, then a normal to the plane is $\mathbf{n} = (a, 2, 1)$. Another point on the plane is $Q = (0, 0, 3)$. Let $\mathbf{v} = \overrightarrow{QP} = (3 - 0, 0 - 0, 1 - 3) = (3, 0, -2)$. Then,

$$d = \|\text{proj}_\mathbf{n}\mathbf{v}\| = \frac{|\mathbf{v} \cdot \mathbf{n}|}{\sqrt{\mathbf{n} \cdot \mathbf{n}}} = \frac{\|(3, 0, -2) \cdot (a, 2, 1)\|}{\sqrt{a^2 + 4 + 1}} = \frac{|3a - 2|}{\sqrt{a^2 + 5}} = 1 \Rightarrow |3a - 2| = \sqrt{a^2 + 5}.$$

As both sides of the equation are nonnegative, squaring them does not introduce anything extraneous:

$$(3a - 2)^2 = a^2 + 5 \Rightarrow 9a^2 - 12a + 4 = a^2 + 5 \Rightarrow 8a^2 - 12a - 1 = 0 \Rightarrow$$

$$a = \frac{12 \pm \sqrt{12^2 - 4(8)(-1)}}{2(8)} = \frac{12 \pm \sqrt{176}}{16} = \frac{12 \pm 4\sqrt{11}}{16} = \frac{3 \pm \sqrt{11}}{4}.$$

25. Given the line of intersection of two planes is the line $y = 3x + 2$, i.e. $3x - y = -2$, any two planes with equations $3x - y + az = -2$ and $3x - y + bz = -2$, $a \neq b$; a, b in R, will intersect in that line, in the x, y-plane ($z = 0$). Another choice is the x, y-plane itself, $z = 0$, and, another choice of plane comes from realizing that in $\Re^3$, $3x - y = -2$ is the equation of a plane parallel to the z-axis, and intersecting the x, y-plane ($z = 0$) in the given line.

27. The angle between the two planes $x + y + z = 1$ and $x - y + z = 1$ is given by $\theta = \cos^{-1}\left(\frac{\mathbf{n_1} \cdot \mathbf{n_2}}{\|\mathbf{n_1}\|\|\mathbf{n_2}\|}\right)$. Here, $\mathbf{n_1} = (1, 1, 1)$ and $\mathbf{n_2} = (1, -1, 1)$. Thus,

$$\theta = \cos^{-1}\left(\frac{(1, 1, 1) \cdot (1, -1, 1)}{\sqrt{1 + 1 + 1}\sqrt{1 + 1 + 1}}\right) = \cos^{-1}\left(\tfrac{1}{3}\right) \approx 70.5°.$$

29. The line we seek is perpendicular to the normal $(1, 1, 1)$ of the plane $x + y + z = 2$ and to the direction vector $(1, -1, 2)$ of the given line, so its direction vector is

$$(1, 1, 1) \times (1, -1, 2) = \det \begin{bmatrix} \mathbf{i} & \mathbf{j} & \mathbf{k} \\ 1 & 1 & 1 \\ 1 & -1 & 2 \end{bmatrix} = \det \begin{bmatrix} 1 & 1 \\ -1 & 2 \end{bmatrix} \mathbf{i} - \det \begin{bmatrix} 1 & 1 \\ 1 & 2 \end{bmatrix} \mathbf{j} + \det \begin{bmatrix} 1 & 1 \\ 1 & -1 \end{bmatrix} \mathbf{k}$$

$$= [1(2) - 1(-1)]\mathbf{i} - [1(2) - 1(1)]\mathbf{j} + [1(-1) - 1(1)]\mathbf{k} = 3\mathbf{i} - \mathbf{j} - 2\mathbf{k}$$

Therefore the vector equation of our line is $\mathbf{w} = (0, 1, 2) + t(3, -1, -2) = (3t, 1 - t, 2 - 2t)$, $-\infty < t < \infty$.

31. The line of intersection of the given planes $3x + y - z = 2$ and $2x + y + 4z = 1$ has a direction vector $\mathbf{v}$ orthogonal to the normals to the two planes: $\mathbf{v} = \mathbf{n_1} \times \mathbf{n_2}$, where $\mathbf{n_1} = (3, 1, -1)$ and $\mathbf{n_2} = (2, 1, 4)$, so

$$\mathbf{v} = \det \begin{bmatrix} \mathbf{i} & \mathbf{j} & \mathbf{k} \\ 3 & 1 & -1 \\ 2 & 1 & 4 \end{bmatrix} = \det \begin{bmatrix} 1 & -1 \\ 1 & 4 \end{bmatrix} \mathbf{i} - \det \begin{bmatrix} 3 & -1 \\ 2 & 4 \end{bmatrix} \mathbf{j} + \det \begin{bmatrix} 3 & 1 \\ 2 & 1 \end{bmatrix} \mathbf{k}$$

$$= [1(4) - (-1)(1)]\mathbf{i} - [3(4) - (-1)(2)]\mathbf{j} + [3(1) - 1(2)]\mathbf{k} = 5\mathbf{i} - 14\mathbf{j} + \mathbf{k}$$

We need another vector in the plane we seek. We are given one point in that plane, $Q = (2, 1, -3)$, another is found by setting $z = 0$ in the equations of the given planes and solving for x and y: $3x + y = 2 \Rightarrow y = 2 - 3x$ substituting this into the other equation, $2x + y = 1$ gives: $2x + 2 - 3x = 1 \Rightarrow x = 1 \Rightarrow y = 2 - 3(1) = -1 \Rightarrow$ another point is $P = (1, -1, 0)$, and the second vector is $\vec{PQ} = (2 - 1, 1 - (-1), -3 - 0) = (1, 2, -3)$. Thus, a normal to our plane is

$$\mathbf{n} = \mathbf{v} \times \vec{PQ} = \det \begin{bmatrix} \mathbf{i} & \mathbf{j} & \mathbf{k} \\ 5 & -14 & 1 \\ 1 & 2 & -3 \end{bmatrix} = \det \begin{bmatrix} -14 & 1 \\ 2 & -3 \end{bmatrix} \mathbf{i} - \det \begin{bmatrix} 5 & 1 \\ 1 & -3 \end{bmatrix} \mathbf{j} + \det \begin{bmatrix} 5 & -14 \\ 1 & 2 \end{bmatrix} \mathbf{k}$$

$$= [-14(-3) - 1(2)]\mathbf{i} - [5(-3) - 1(1)]\mathbf{j} + [5(2) - (-14)(1)]\mathbf{k} = 40\mathbf{i} + 16\mathbf{j} + 24\mathbf{k}$$

and the equation of our plane with this normal and containing the point $Q = (2, 1, -3)$ is $40(x - 2) + 16(y - 1) + 24(z + 3) = 0 \Rightarrow 40x + 16y + 24z = 80 + 16 - 72 \Rightarrow 40x + 16y + 24z = 24 \Rightarrow 5x + 2y + 3z = 3$.

33. The plane perpendicular to the two given planes $2x + y - z = -2$ and $x - y + 3z = 1$ has a normal perpendicular to both normals of those given planes, so $\mathbf{n} = \mathbf{n_1} \times \mathbf{n_2}$, where $\mathbf{n_1} = (2, 1, -1)$ and $\mathbf{n_2} = (1, -1, 3)$:

$$\mathbf{n} = \det \begin{bmatrix} \mathbf{i} & \mathbf{j} & \mathbf{k} \\ 2 & 1 & -1 \\ 1 & -1 & 3 \end{bmatrix} = \det \begin{bmatrix} 1 & -1 \\ -1 & 3 \end{bmatrix} \mathbf{i} - \det \begin{bmatrix} 2 & -1 \\ 1 & 3 \end{bmatrix} \mathbf{j} + \det \begin{bmatrix} 2 & 1 \\ 1 & -1 \end{bmatrix} \mathbf{k}$$

$$= [1(3) - (-1)(-1)]\mathbf{i} - [2(3) - (-1)(1)]\mathbf{j} + [2(-1) - 1(1)]\mathbf{k} = 2\mathbf{i} - 7\mathbf{j} - 3\mathbf{k}$$

Then, the equation of the plane through $(1, 3, -2)$ with normal $\mathbf{n}$ is:

$$2(x - 1) - 7(y - 3) - 3(z + 2) = 0 \Rightarrow 2x - 7y - 3z = 2 - 21 + 6 \Rightarrow 2x - 7y - 3z = -13.$$

35. First we show that the two lines are neither parallel nor intersecting.
The line $x = 1 + 4t$, $y = 5 - 4t$, $z = -1 + 5t$ has a direction vector of $\mathbf{v_1} = (4, -4, 5)$, and
the line $x = 2 + 8t$, $y = 4 - 3t$, $z = 5 + t$ has a direction vector of $\mathbf{v_2} = (8, -3, 1)$. The
vectors $\mathbf{v_1}$ and $\mathbf{v_2}$ are not multiples of one another, so the lines are not parallel. If the lines
intersect, then for some s and some t,

$$(1) \qquad 1 + 4t = 2 + 8s$$
$$(2) \qquad 5 - 4t = 4 - 3s$$
$$(3) \quad -1 + 5t = 5 + s$$

Adding (1) and (2): $6 = 6 + 5s \Rightarrow s = 0$. From (1) then, $1 + 4t = 2 \Rightarrow t = \frac{1}{4}$. From (3)
then, $-1 + \frac{5}{4} = 5$ impossible! Thus, the lines do not intersect.

A vector orthogonal to both lines is

$$\mathbf{n} = \mathbf{v_1} \times \mathbf{v_2} = \det \begin{bmatrix} \mathbf{i} & \mathbf{j} & \mathbf{k} \\ 4 & -4 & 5 \\ 8 & -3 & 1 \end{bmatrix} = \det \begin{bmatrix} -4 & 5 \\ -3 & 1 \end{bmatrix} \mathbf{i} - \det \begin{bmatrix} 4 & 5 \\ 8 & 1 \end{bmatrix} \mathbf{j} + \det \begin{bmatrix} 4 & -4 \\ 8 & -3 \end{bmatrix} \mathbf{k}$$

$$= [-4(1) - 5(-3)]\mathbf{i} - [4(1) - 5(8)]\mathbf{j} + [4(-3) - (-4)(8)]\mathbf{k} = 11\mathbf{i} + 36\mathbf{j} + 20\mathbf{k}$$

Choosing one point on each line, say, when $t = 0$, we get $P = (1, 5, -1)$ and $Q = (2, 4, 5)$.
We then form the vector $\vec{PQ} = (2 - 1, 4 - 5, 5 - (-1)) = (1, -1, 6)$, and take its projection
on $\mathbf{n}$ which is the minimum distance d between the two lines:

$$d = \|\text{proj}_{\mathbf{n}} \vec{PQ}\| = \frac{|\vec{PQ} \cdot \mathbf{n}|}{\sqrt{\mathbf{n} \cdot \mathbf{n}}} = \frac{|(1, -1, 6) \cdot (11, 36, 20)|}{\sqrt{11^2 + 36^2 + 20^2}} = \frac{|11 - 36 + 120|}{\sqrt{1817}} = \frac{95}{\sqrt{1817}}.$$

§12.4 Vector-Valued Functions

1. $\mathbf{r}(t) = t\mathbf{i}$, $0 \le t \le 1$, defines a line segment along the positive x-axis
from the origin to $x = 1$.

3. $\mathbf{r}(t) = t\mathbf{i} + t\mathbf{j} + 4\mathbf{k}$, $1 \le t \le 2$, defines a line segment along the line
$y - x$ in the plane $z = 4$ (the plane parallel to the x, y−plane and 4
units above it), as $x = y$ and $z = 4$. Moreover, $\mathbf{r}(1) = (1, 1, 4)$ and
$\mathbf{r}(2) = (2, 2, 4)$, see figure.

5. $\mathbf{r}(t) = 3\mathbf{i} + t\mathbf{j} + t^2\mathbf{k}$, $-1 \le t \le 2$, defines a parabolic path in the plane
$x = 3$, as $z = y^2$. Moreover, $\mathbf{r}(-1) = (3, -1, 1)$ and $\mathbf{r}(2) = (3, 2, 4)$,
see figure.

7. $\mathbf{r}(t) = 2t\,\mathbf{i} + 5t^2\,\mathbf{j} \;\Rightarrow\; \lim\limits_{t \to 1} \mathbf{r}(t) = [\lim\limits_{t \to 1}(2t)]\,\mathbf{i} + [\lim\limits_{t \to 1}(5t^2)]\,\mathbf{j} = (2 \cdot 1)\,\mathbf{i} + (5 \cdot 1^2)\,\mathbf{j} = 2\,\mathbf{i} + 5\,\mathbf{j}.$

   $\mathbf{r}'(t) = (2t)'\,\mathbf{i} + (5t^2)'\,\mathbf{j} = 2\,\mathbf{i} + (5 \cdot 2t)\,\mathbf{j} = 2\,\mathbf{i} + 10t\,\mathbf{j},$ and

   $\int \mathbf{r}(t)\,dt = \int (2t\,\mathbf{i} + 5t^2\,\mathbf{j})\,dt = (\int 2t\,dt)\mathbf{i} + (\int 5t^2\,dt)\mathbf{j} = t^2\,\mathbf{i} + \tfrac{5}{3}t^3\,\mathbf{j} + \mathbf{C}.$

9. $\mathbf{r}(t) = (\sin t)\,\mathbf{i} - t^2\,\mathbf{j} \;\Rightarrow\; \lim\limits_{t \to 1}\mathbf{r}(t) = [\lim\limits_{t \to 1}(\sin t)]\,\mathbf{i} - [\lim\limits_{t \to 1}(t^2)]\,\mathbf{j} = (\sin 1)\,\mathbf{i} - 1^2\,\mathbf{j} = (\sin 1)\,\mathbf{i} - \mathbf{j}.$

   $\mathbf{r}'(t) = (\sin t)'\,\mathbf{i} - (t^2)'\,\mathbf{j} = (\cos t)\,\mathbf{i} - 2t\,\mathbf{j},$ and

   $\int \mathbf{r}(t)\,dt = \int [(\sin t)\,\mathbf{i} - t^2\,\mathbf{j}]\,dt = [\int(\sin t)\,dt]\,\mathbf{i} - [\int t^2\,dt]\,\mathbf{j} = (-\cos t)\,\mathbf{i} - \tfrac{1}{3}t^3\,\mathbf{j} + \mathbf{C}.$

11. $\mathbf{r}(t) = t\,\mathbf{i} + \mathbf{j} + \mathbf{k} \;\Rightarrow\; \lim\limits_{t \to 1}\mathbf{r}(t) = (\lim\limits_{t \to 1} t)\,\mathbf{i} + (\lim\limits_{t \to 1} 1)\,\mathbf{j} + (\lim\limits_{t \to 1} 1)\,\mathbf{k} = \mathbf{i} + \mathbf{j} + \mathbf{k}.$

   $\mathbf{r}'(t) = (t)'\,\mathbf{i} + (1)'\,\mathbf{j} + (1)'\,\mathbf{k} = \mathbf{i} + (0)\,\mathbf{j} + (0)\,\mathbf{k} = \mathbf{i},$ and

   $\int \mathbf{r}(t)\,dt = \int (t\,\mathbf{i} + \mathbf{j} + \mathbf{k})\,dt = (\int t\,dt)\mathbf{i} + (\int 1\,dt)\mathbf{j} + (\int 1\,dt)\mathbf{k} = \tfrac{1}{2}t^2\,\mathbf{i} + t\,\mathbf{j} + t\,\mathbf{k} + \mathbf{C}.$

13. $\mathbf{r}(t) = t\,\mathbf{i} - 3\,\mathbf{j} + \dfrac{\ln t}{t}\,\mathbf{k} \;\Rightarrow$

   $\lim\limits_{t \to 1}\mathbf{r}(t) = (\lim\limits_{t \to 1} t)\,\mathbf{i} - (\lim\limits_{t \to 1} 3)\,\mathbf{j} + \left(\lim\limits_{t \to 1}\dfrac{\ln t}{t}\right)\mathbf{k} = \mathbf{i} - 3\,\mathbf{j} + (\ln 1)\,\mathbf{k} = \mathbf{i} - 3\,\mathbf{j}.$

   $\mathbf{r}'(t) = (t)'\,\mathbf{i} - (3)'\,\mathbf{j} + \left(\dfrac{\ln t}{t}\right)'\mathbf{k} = \mathbf{i} - (0)\,\mathbf{j} + \left(\dfrac{t \cdot \frac{1}{t} - \ln t}{t^2}\right)\mathbf{k} = \mathbf{i} + \left(\dfrac{1 - \ln t}{t^2}\right)\mathbf{k},$ and

   $\int \mathbf{r}(t)\,dt = \int (t\,\mathbf{i} - 3\,\mathbf{j} + \dfrac{\ln t}{t}\,\mathbf{k})\,dt = (\int t\,dt)\mathbf{i} - (\int 3\,dt)\mathbf{j} + (\int \dfrac{\ln t}{t}\,dt)\mathbf{k} = \tfrac{1}{2}t^2\,\mathbf{i} - 3t\,\mathbf{j} + \tfrac{1}{2}(\ln t)^2\,\mathbf{k} + \mathbf{C}.$

   $$\boxed{\begin{array}{c} u = \ln t \;\Rightarrow\; du = \tfrac{1}{t}\,dt \;\Rightarrow \\[4pt] \int \dfrac{\ln t}{t}\,dt = \int u\,du = \tfrac{1}{2}u^2 + C \end{array}}$$

15. $\mathbf{r}(t) = t\,\mathbf{i} - t^2\,\mathbf{j} + (\tan t)\,\mathbf{k} \;\Rightarrow$

   $\lim\limits_{t \to 1}\mathbf{r}(t) = (\lim\limits_{t \to 1} t)\,\mathbf{i} - (\lim\limits_{t \to 1} t^2)\,\mathbf{j} + (\lim\limits_{t \to 1}\tan t)\,\mathbf{k} = \mathbf{i} - \mathbf{j} + (\tan 1)\,\mathbf{k}.$

   $\mathbf{r}'(t) = (t)'\,\mathbf{i} - (t^2)'\,\mathbf{j} + (\tan t)'\,\mathbf{k} = \mathbf{i} - 2t\,\mathbf{j} + (\sec^2 t)\,\mathbf{k},$ and

   $\int \mathbf{r}(t)\,dt = \int [t\,\mathbf{i} - t^2\,\mathbf{j} + (\tan t)\,\mathbf{k}]\,dt = (\int t\,dt)\mathbf{i} - (\int t^2\,dt)\mathbf{j} + (\int \tan t\,dt)\mathbf{k}$

   $= \tfrac{1}{2}t^2\,\mathbf{i} - \tfrac{1}{3}t^3\,\mathbf{j} + (\ln|\sec t|)\,\mathbf{k} + \mathbf{C}.$

17. $\mathbf{r}(t) = 2t\,\mathbf{i} + 5t^2\,\mathbf{j} \;\Rightarrow\; \mathbf{r}'(t) = (2t)'\,\mathbf{i} + (5t^2)'\,\mathbf{j} = 2\,\mathbf{i} + 10t\,\mathbf{j}.$ At $t = 1$, the tangent vector is $\mathbf{r}'(1) = 2\,\mathbf{i} + 10\,\mathbf{j}$ which is the direction vector of the tangent line. A point on the tangent line at $t = 1$ is $\mathbf{r}(1) = (2,5)$. Thus parametric equations for the tangent line at $t = 1$ are $x = 2 + 2t, \; y = 5 + 10t, \; -\infty < t < \infty.$

19. $\mathbf{r}(t) = (\sin t)\,\mathbf{i} - (2\cos t)\,\mathbf{j} + \mathbf{k} \;\Rightarrow\; \mathbf{r}'(t) = (\sin t)'\,\mathbf{i} - (2\cos t)'\,\mathbf{j} + (1)'\,\mathbf{k} = (\cos t)\,\mathbf{i} + (2\sin t)\,\mathbf{j}.$ At $t = 0$, the tangent vector is $\mathbf{r}'(0) = (\cos 0)\,\mathbf{i} + (2\sin 0)\,\mathbf{j} = \mathbf{i}$ which is the direction vector of the tangent line. A point on the tangent line at $t = 0$ is $\mathbf{r}(0) = (0, -2, 1)$. Thus parametric equations for the tangent line at $t = 0$ are $x = 0 + t = t, \; y = -2, \; z = 1.$

21. $\mathbf{r}(t) = t\,\mathbf{i} + 2t\,\mathbf{j} \Rightarrow \mathbf{r}'(t) = \mathbf{i} + 2\,\mathbf{j} \Rightarrow \mathbf{r}''(t) = \mathbf{0}$. Hence, at $t = 1$:
The velocity is $\mathbf{v}(1) = \mathbf{r}'(1) = \mathbf{i} + 2\,\mathbf{j}$, the acceleration is $\mathbf{a}(1) = \mathbf{r}''(1) = \mathbf{0}$, and the speed is $\|\mathbf{v}(1)\| = \sqrt{1^2 + 2^2} = \sqrt{5}$.

23. $\mathbf{r}(t) = (\sin t)\,\mathbf{i} + t\,\mathbf{j} + (\cos t)\,\mathbf{k} \Rightarrow \mathbf{r}'(t) = (\cos t)\,\mathbf{i} + \mathbf{j} - (\sin t)\,\mathbf{k} \Rightarrow$
$\mathbf{r}''(t) = (-\sin t)\,\mathbf{i} - (\cos t)\,\mathbf{k}$. Hence, at $t = 0$:
The velocity is $\mathbf{v}(0) = \mathbf{r}'(0) = (\cos 0)\,\mathbf{i} + \mathbf{j} - (\sin 0)\,\mathbf{k} = \mathbf{i} + \mathbf{j}$, the acceleration is $\mathbf{a}(0) = \mathbf{r}''(0) = (-\sin 0)\,\mathbf{i} - (\cos 0)\,\mathbf{k} = -\mathbf{k}$, and the speed is $\|\mathbf{v}(0)\| = \sqrt{1^2 + 1^2} = \sqrt{2}$.

25. Given $\mathbf{v_0} = \mathbf{i} - 12\,\mathbf{j} + \mathbf{k}$, then
$\mathbf{v}(t) = \int \mathbf{a}(t)\,dt = \int (2\,\mathbf{i} + \mathbf{j} - \mathbf{k})\,dt = 2t\,\mathbf{i} + t\,\mathbf{j} - t\,\mathbf{k} + \mathbf{v_0} = (2t\,\mathbf{i} + t\,\mathbf{j} - t\,\mathbf{k}) + (\mathbf{i} - 12\,\mathbf{j} + \mathbf{k})$
$= (2t + 1)\,\mathbf{i} + (t - 12)\,\mathbf{j} + (1 - t)\,\mathbf{k}$
Speed $= \|\mathbf{v}(t)\| = \sqrt{(2t + 1)^2 + (t - 12)^2 + (1 - t)^2}$
$= \sqrt{4t^2 + 4t + 1 + t^2 - 24t + 144 + 1 - 2t + t^2} = \sqrt{6t^2 - 22t + 146}$
$\|\mathbf{v}(t)\|' = \frac{1}{2}(6t^2 - 22t + 146)^{-\frac{1}{2}}(12t - 22) = 0 \Rightarrow 12t = 22 \Rightarrow t = \frac{11}{6}$. Sign:

Since a continuous function achieves its maximum and minimum values on a closed interval at either critical points inside the interval or at endpoints, to determine the minimum and maximum speed, we need only evaluate the speed at $t = 0$, 6, and $\frac{11}{6}$:
$\|\mathbf{v}(0)\| = \sqrt{146} \approx 12.1$, $\quad \|\mathbf{v}(6)\| = \sqrt{6(6^2) - 22(6) + 146} = \sqrt{230} \approx 15.2$, and
$\left\|\mathbf{v}\left(\frac{11}{6}\right)\right\| = \sqrt{6\left(\frac{11}{6}\right)^2 - 22\left(\frac{11}{6}\right) + 146} = \sqrt{\frac{755}{6}} \approx 11.2$.

Therefore, we see that the minimum speed is about 11.2 ft/sec, occurring when $t = \frac{11}{6}$ seconds (agreeing with the sign chart), and the maximum speed is about 15.2 ft/sec, occurring when $t = 6$ seconds.

27. To prove Theorem 12.8(a): $[c\mathbf{u}(t)]' = c\mathbf{u}'(t)$:
$[c\mathbf{u}(t)]' = \{c[u_1(t)\,\mathbf{i} + u_2(t)\,\mathbf{j} + u_3(t)\,\mathbf{k}]\}' = [cu_1(t)\,\mathbf{i} + cu_2(t)\,\mathbf{j} + cu_3(t)\,\mathbf{k}]'$
$= [cu_1(t)]'\,\mathbf{i} + [cu_2(t)]'\,\mathbf{j} + [cu_3(t)]'\,\mathbf{k} = cu_1'(t)\,\mathbf{i} + cu_2'(t)\,\mathbf{j} + cu_3'(t)\,\mathbf{k}$
$= c[u_1'(t)\,\mathbf{i} + u_2'(t)\,\mathbf{j} + u_3'(t)\,\mathbf{k}] = c\mathbf{u}'(t)$
To prove Theorem 12.8(e): $[\mathbf{u}(t) - \mathbf{v}(t)]' = \mathbf{u}'(t) - \mathbf{v}'(t)$:
$\mathbf{u}(t) - \mathbf{v}(t) = [u_1(t)\,\mathbf{i} + u_2(t)\,\mathbf{j} + u_3(t)\,\mathbf{k}] - [v_1(t)\,\mathbf{i} + v_2(t)\,\mathbf{j} + v_3(t)\,\mathbf{k}]$
$= [u_1(t) - v_1(t)]\,\mathbf{i} + [u_2(t) - v_2(t)]\,\mathbf{j} + [u_3(t) - v_3(t)]\,\mathbf{k}$
$[\mathbf{u}(t) - \mathbf{v}(t)]' = [u_1(t) - v_1(t)]'\,\mathbf{i} + [u_2(t) - v_2(t)]'\,\mathbf{j} + [u_3(t) - v_3(t)]'\,\mathbf{k}$
$= [u_1'(t) - v_1'(t)]\,\mathbf{i} + [u_2'(t) - v_2'(t)]\,\mathbf{j} + [u_3'(t) - v_3'(t)]\,\mathbf{k}$
$= [u_1'(t)\,\mathbf{i} + u_2'(t)\,\mathbf{j} + u_3'(t)\,\mathbf{k}] - [v_1'(t)\,\mathbf{i} + v_2'(t)\,\mathbf{j} + v_3'(t)\,\mathbf{k}]$
$= \mathbf{u}'(t) - \mathbf{v}'(t)$

29. To prove Theorem 12.8(c): $\{\mathbf{u}[f(t)]\}' = f'(t)\mathbf{u}'[f(t)]$:

$$\{\mathbf{u}[f(t)]\}' = \{u_1[f(t)]\,\mathbf{i} + u_2[f(t)]\,\mathbf{j} + u_3[f(t)]\,\mathbf{k}\}'$$
$$= \{u_1[f(t)]\}'\,\mathbf{i} + \{u_2[f(t)]\}'\,\mathbf{j} + \{u_3[f(t)]\}'\,\mathbf{k}$$
$$= u_1'[f(t)]f'(t)\,\mathbf{i} + u_2'[f(t)]f'(t)\,\mathbf{j} + u_3'[f(t)]f'(t)\,\mathbf{k}$$
$$= f'(t)\{u_1'[f(t)]\,\mathbf{i} + u_2'[f(t)]\,\mathbf{j} + u_3'[f(t)]\,\mathbf{k}\} = f'(t)\mathbf{u}'[f(t)]$$

31. We first recall that for scalar functions $a(t)$, $\frac{d}{dt}\int a(t)dt = a(t)$: Indeed,

$\int a(t)dt = A(t) + C$, where $A'(t) = a(t)$ $\Rightarrow$ $\frac{d}{dt}\int a(t)dt = \frac{d}{dt}[A(t) + C] = A'(t) + 0 = a(t)$
We are to prove that if $\mathbf{r}(t) = f(t)\,\mathbf{i} + g(t)\,\mathbf{j} + h(t)\,\mathbf{k}$, then

$$\int \mathbf{r}(t)dt = [\int f(t)dt]\,\mathbf{i} + [\int g(t)dt]\,\mathbf{j} + [\int h(t)dt]\,\mathbf{k}$$

By Definition 12.8, $\int \mathbf{r}(t)dt = \mathbf{R}(t) + \mathbf{C}$, where $\mathbf{R}'(t) = \mathbf{r}(t)$, so we need only show that

$$\frac{d}{dt}\{[\int f(t)dt]\,\mathbf{i} + [\int g(t)dt]\,\mathbf{j} + [\int h(t)dt]\,\mathbf{k}\} = \mathbf{r}(t)$$

CYU 12.27 tells us that to differentiate a vector function, just differentiate its components, therefore:

$$\frac{d}{dt}\{[\int f(t)dt]\,\mathbf{i} + [\int g(t)dt]\,\mathbf{j} + [\int h(t)dt]\,\mathbf{k}\} = \left[\frac{d}{dt}\int f(t)dt\right]\,\mathbf{i} + \left[\frac{d}{dt}\int g(t)dt\right]\,\mathbf{j} + \left[\frac{d}{dt}\int h(t)dt\right]\,\mathbf{k}$$
$$= \quad f(t) \quad \mathbf{i} + \quad g(t) \quad \mathbf{j} + \quad h(t) \quad \mathbf{k}$$
$$= \mathbf{r}(t)$$

33.   Since the unit is meters rather than feet, the acceleration due to gravity is 9.8 meters/sec^2.

The initial velocity is $\mathbf{v}_0 = (500\cos 45°)\,\mathbf{j} + (500\sin 45°)\,\mathbf{k} = (250\sqrt{2})\,\mathbf{j} + (250\sqrt{2})\,\mathbf{k}$, and the initial position is $\mathbf{r}_0 = 30\,\mathbf{k}$.

Thus, $\mathbf{v}(t) = \int \mathbf{a}(t)\,dt = \int -9.8\,\mathbf{k}\,dt = -9.8t\,\mathbf{k} + \mathbf{v}_0 = -9.8t\,\mathbf{k} + (250\sqrt{2})\,\mathbf{j} + (250\sqrt{2})\,\mathbf{k}$
$$= (250\sqrt{2})\,\mathbf{j} + (-9.8t + 250\sqrt{2})\,\mathbf{k}$$

Integrating again:
$$\mathbf{r}(t) = \int \mathbf{v}(t)\,dt = \int[(250\sqrt{2})\,\mathbf{j} + (-9.8t + 250\sqrt{2})\,\mathbf{k}]\,dt$$
$$= (250\sqrt{2}\,t)\,\mathbf{j} + (-4.9t^2 + 250\sqrt{2}\,t)\,\mathbf{k} + \mathbf{r}_0$$
$$= (250\sqrt{2}\,t)\,\mathbf{j} + (-4.9t^2 + 250\sqrt{2}\,t + 30)\,\mathbf{k}$$

The projectile hits the ground when its $\mathbf{k}$-component (vertical component) is zero:
$$-4.9t^2 + 250\sqrt{2}\,t + 30 = 0$$
$$t = \frac{-250\sqrt{2} \pm \sqrt{(250\sqrt{2})^2 - 4(-4.9)(30)}}{-9.8} = \frac{-250\sqrt{2} \pm \sqrt{125588}}{-9.8}$$

As $t > 0$, $t = \frac{-250\sqrt{2} - \sqrt{125588}}{-9.8} \approx 72.24$ seconds. The speed on impact with the ground is

about $\|\mathbf{v}(72.24)\| \approx \sqrt{(250\sqrt{2})^2 + [-9.8(72.24) + 250\sqrt{2}]^2} \approx 500.6$ m/sec.

35. Given that the initial velocity of the projectile is $\mathbf{v}_0 = 100\mathbf{j} + 100\mathbf{k}$ and its initial position is ground level, i.e. $\mathbf{r}_0 = \mathbf{0}$, then

$$\mathbf{v}(t) = \int \mathbf{a}(t)\,dt = \int -32\mathbf{k}\,dt = -32t\,\mathbf{k} + \mathbf{v}_0 = -32t\,\mathbf{k} + 100\mathbf{j} + 100\mathbf{k} = 100\mathbf{j} + (-32t + 100)\mathbf{k}$$

Integrating again:

$$\mathbf{r}(t) = \int \mathbf{v}(t)\,dt = \int (100\mathbf{j} + (-32t + 100)\mathbf{k})\,dt = 100t\,\mathbf{j} + (-16t^2 + 100t)\mathbf{k} + \cancel{\mathbf{r}_0} \ (\text{as } \mathbf{r}_0 = \mathbf{0})$$

Maximum height is reached when the $\mathbf{k}$-component (vertical component) of the velocity is zero:

$$-32t + 100 = 0 \ \Rightarrow \ t = \frac{100}{32} = \frac{25}{8} \text{ seconds.}$$

The maximum height reached is the $\mathbf{k}$-component (vertical component) of $\mathbf{r}$ at that time:

$$-16\left(\frac{25}{8}\right)^2 + 100\left(\frac{25}{8}\right) = \frac{-625 + 2(625)}{4} = \frac{625}{4} = 156.25 \text{ ft.}$$

To find the range, we need the value of t when the $\mathbf{k}$-component (vertical component) of $\mathbf{r}$ is zero:

$$-16t^2 + 100t = 0 \ \Rightarrow \ 16t = 100 \ \Rightarrow \ t = \frac{100}{16} = \frac{25}{4} \ [\text{Of course! If it takes } \frac{25}{8} \text{ seconds}$$

to reach max height, it takes the same amount of time to come back down, for a total of $2\left(\frac{25}{8}\right) = \frac{25}{4}$ seconds.]

The range is the $\mathbf{j}$-component (horizontal component) of $\mathbf{r}$ at that time: $100\left(\frac{25}{4}\right) = 625$ ft.

The speed on impact with the ground is

$$\left\|\mathbf{v}\left(\frac{25}{4}\right)\right\| = \sqrt{100^2 + \left[100 - 32 \cdot \frac{25}{4}\right]^2} = \sqrt{2 \cdot 100^2} = 100\sqrt{2} \text{ ft/sec.}$$

37. Since the stone is thrown downward, the $\mathbf{k}$-component (vertical component) of the initial velocity is negative:

$$\mathbf{v}_0 = (80\cos 60)^\circ\,\mathbf{j} - (80\sin 60^\circ)\,\mathbf{k} = 40\mathbf{j} - 40\sqrt{3}\,\mathbf{k}$$

The initial position of the stone is $\mathbf{r}_0 = 168\mathbf{k}$. Therefore,

$$\mathbf{v}(t) = \int \mathbf{a}(t)\,dt = \int -32\mathbf{k}\,dt = -32t\,\mathbf{k} + \mathbf{v}_0 = -32t\,\mathbf{k} + 40\mathbf{j} - 40\sqrt{3}\,\mathbf{k} = 40\mathbf{j} + (-32t - 40\sqrt{3})\mathbf{k}$$

Integrating again:

$$\mathbf{r}(t) = \int \mathbf{v}(t)\,dt = \int (40\mathbf{j} + (-32t - 40\sqrt{3})\mathbf{k})\,dt = 40t\,\mathbf{j} + (-16t^2 - 40\sqrt{3}t)\mathbf{k} + \mathbf{r}_0$$
$$= 40t\,\mathbf{j} + (-16t^2 - 40\sqrt{3}t + 168)\mathbf{k}$$

The distance from the base of the building is the range, and that occurs when the $\mathbf{k}$-component (vertical component) of $\mathbf{r}$ is zero:

$$-16t^2 - 40\sqrt{3}t + 168 = 0 \ \Rightarrow \ 2t^2 + 5\sqrt{3}t - 21 = 0$$

$$t = \frac{-5\sqrt{3} + \sqrt{(5\sqrt{3})^2 - 4(2)(-21)}}{4}$$

$$= \frac{-5\sqrt{3} + \sqrt{243}}{4} = \frac{-5\sqrt{3} + 9\sqrt{3}}{4} = \sqrt{3}$$

The range is the $\mathbf{j}$-component (horizontal component) of $\mathbf{r}$ at that time: $40\sqrt{3}$, i.e. the stone lands $40\sqrt{3}$ feet from the base of the building.

39. Let s be the speed to be determined. Then:

The initial velocity is $\mathbf{v}_0 = (s\cos 30°)\mathbf{j} + (s\sin 30°)\mathbf{k} = \left(\frac{\sqrt{3}}{2}s\right)\mathbf{j} + \left(\frac{1}{2}s\right)\mathbf{k}$, and the initial position is $\mathbf{r}_0 = \mathbf{0}$. Therefore,

$$\mathbf{v}(t) = \int \mathbf{a}(t)\,dt = \int -32\,\mathbf{k}\,dt = -32t\,\mathbf{k} + \mathbf{v}_0$$
$$= -32t\,\mathbf{k} + \left(\frac{\sqrt{3}}{2}s\right)\mathbf{j} + \left(\frac{1}{2}s\right)\mathbf{k} = \left(\frac{\sqrt{3}}{2}s\right)\mathbf{j} + \left(-32t + \frac{1}{2}s\right)\mathbf{k}$$

Integrating again:

$$\mathbf{r}(t) = \int \mathbf{v}(t)\,dt = \int \left[\left(\frac{\sqrt{3}}{2}s\right)\mathbf{j} + \left(-32t + \frac{1}{2}s\right)\mathbf{k}\right]dt = \left(\frac{\sqrt{3}}{2}st\right)\mathbf{j} + \left(-16t^2 + \frac{1}{2}st\right)\mathbf{k} + \mathbf{r}_0$$

(as $\mathbf{r}_0 = \mathbf{0}$)

We want the vertical component of $\mathbf{r}$ to be 35 ft when the horizontal component is 135 ft, so:

$$\frac{\sqrt{3}}{2}st = 135 \Rightarrow t = \frac{270}{\sqrt{3}s} \cdot \frac{\sqrt{3}}{\sqrt{3}} = \frac{90\sqrt{3}}{s}$$

and then

$$-16t^2 + \frac{1}{2}st = 35 \Rightarrow -16\left(\frac{90\sqrt{3}}{s}\right)^2 + \frac{1}{2}s\left(\frac{90\sqrt{3}}{s}\right) = 35 \Rightarrow$$

$$-16 \cdot \frac{(90\sqrt{3})^2}{s^2} = 35 - 45\sqrt{3} \Rightarrow s^2 = \frac{-16(90\sqrt{3})^2}{35 - 45\sqrt{3}} \approx 9054 \Rightarrow s \approx 95 \text{ ft/sec}.$$

41. Let s be the speed to be determined. Then:

The initial velocity is $\mathbf{v}_0 = s\mathbf{j}$, since the stone is thrown horizontally. The initial position of the stone is $\mathbf{r}_0 = 25\,\mathbf{k}$. Therefore,

$$\mathbf{v}(t) = \int \mathbf{a}(t)\,dt = \int -32\,\mathbf{k}\,dt = -32t\,\mathbf{k} + \mathbf{v}_0 = s\mathbf{j} - 32t\,\mathbf{k}$$

Integrating again:

$$\mathbf{r}(t) = \int \mathbf{v}(t)\,dt = \int (s\mathbf{j} - 32t\,\mathbf{k})\,dt = st\,\mathbf{j} - 16t^2\,\mathbf{k} + \mathbf{r}_0 = st\,\mathbf{j} + (-16t^2 + 25)\,\mathbf{k}$$

We want the vertical component of $\mathbf{r}$ to be 4 (ft) when the horizontal component is 45 (ft), so:

$$st = 45 \Rightarrow t = \frac{45}{s}$$

and then

$$-16t^2 + 25 = 4 \Rightarrow -16\left(\frac{45}{s}\right)^2 + 25 = 4 \Rightarrow 16 \cdot \frac{45^2}{s^2} = 21 \Rightarrow s^2 = \frac{16(45)^2}{21} \Rightarrow s = \frac{4(45)}{\sqrt{21}} \approx 39$$

That is, the speed is about 39 ft/sec.

43. Let s be the speed to be determined. Then:

The initial velocity is $\mathbf{v}_0 = (s\cos 30°)\mathbf{j} + (s\sin 30°)\mathbf{k} = \left(\frac{\sqrt{3}}{2}s\right)\mathbf{j} + \left(\frac{1}{2}s\right)\mathbf{k}$, and the initial position is $\mathbf{r}_0 = \mathbf{0}$. Therefore,

$$\mathbf{v}(t) = \int \mathbf{a}(t)\,dt = \int -32\,\mathbf{k}\,dt = -32t\,\mathbf{k} + \mathbf{v}_0$$
$$= -32t\,\mathbf{k} + \left(\frac{\sqrt{3}}{2}s\right)\mathbf{j} + \left(\frac{1}{2}s\right)\mathbf{k} = \left(\frac{\sqrt{3}}{2}s\right)\mathbf{j} + \left(-32t + \frac{1}{2}s\right)\mathbf{k}$$

Integrating again:

$$\mathbf{r}(t) = \int \mathbf{v}(t)\,dt = \int \left[\left(\frac{\sqrt{3}}{2}s\right)\mathbf{j} + \left(-32t + \frac{1}{2}s\right)\mathbf{k}\right]dt = \left(\frac{\sqrt{3}}{2}st\right)\mathbf{j} + (-16t^2 + \frac{1}{2}st)\mathbf{k} + \mathbf{r}_0$$

(as $\mathbf{r}_0 = \mathbf{0}$)

When the vertical component of **r** is 0, we want the horizontal component to be 50 (ft), so:

$$-16t^2 + \tfrac{1}{2}st = 0 \implies -16t + \tfrac{1}{2}s = 0 \implies t = \tfrac{s}{32}$$

and then

$$\tfrac{\sqrt{3}}{2}st = 50 \implies \tfrac{\sqrt{3}}{2}s\left(\tfrac{s}{32}\right) = 50 \implies s^2 = \tfrac{50(64)}{\sqrt{3}} \implies s = \sqrt{\tfrac{3200}{\sqrt{3}}} \approx 43 \text{ ft/sec.}$$

45. Let α be the angle at which a projectile is fired at a speed of v_0 ft/sec from the origin, so $\mathbf{r}_0 = \mathbf{0}$. Then,

$$\mathbf{v}(t) = \int \mathbf{a}(t)\,dt = \int -32\,\mathbf{k}\,dt = -32t\,\mathbf{k} + \mathbf{v}_0 = -32t\,\mathbf{k} + (v_0\cos\alpha)\,\mathbf{j} + (v_0\sin\alpha)\,\mathbf{k}$$
$$= (v_0\cos\alpha)\,\mathbf{j} + (-32t + v_0\sin\alpha)\,\mathbf{k}$$

Integrating again:

$$\mathbf{r}(t) = \int \mathbf{v}(t)\,dt = \int[(v_0\cos\alpha)\,\mathbf{j} + (-32t + v_0\sin\alpha)\,\mathbf{k}]\,dt$$
$$= [(v_0\cos\alpha)t]\,\mathbf{j} + [-16t^2 + (v_0\sin\alpha)t]\,\mathbf{k} + \mathbf{r}_0 \quad (\text{as } \mathbf{r}_0 = \mathbf{0})$$

Maximum range occurs when the vertical component of $\mathbf{r}(t)$ is zero, so:

$$-16t^2 + (v_0\sin\alpha)t = 0 \implies 16t = v_0\sin\alpha \implies t = \frac{v_0\sin\alpha}{16}$$

The range at that time is the horizontal component of $\mathbf{r}(t)$:

$$(v_0\cos\alpha)\left(\frac{v_0\sin\alpha}{16}\right) = \frac{v_0^2\sin\alpha\cos\alpha}{16} = \frac{v_0^2\cdot 2\sin\alpha\cos\alpha}{32} = \frac{v_0^2\sin 2\alpha}{32}$$

The maximum value of the range occurs when $\sin 2\alpha = 1 \implies 2\alpha = 90° \implies \alpha = 45°$.

47. From Exercise 45, the range, d, when a projectile is fired at an angle α and initial speed v_0 ft/sec is $d = \dfrac{v_0^2\sin 2\alpha}{32}$. Doubling the initial speed to $2v_0$ results in the new range,

$$D = \frac{(2v_0)^2\sin 2\alpha}{32} = \frac{4v_0^2\sin 2\alpha}{32} = 4d$$

49. The acceleration $\mathbf{a}(t) = \mathbf{r}''(t) = \mathbf{0}$. Integrating, $\mathbf{r}'(t) = a\,\mathbf{j} + b\,\mathbf{k}$, for some a, b in $\Re$. Integrating again,

$$\mathbf{r}(t) = (at\,\mathbf{j} + bt\,\mathbf{k}) + (c\,\mathbf{j} + d\,\mathbf{k}), \text{ for some } c, d \text{ in } \Re.$$
$$= t(a\,\mathbf{j} + b\,\mathbf{k}) + (c\,\mathbf{j} + d\,\mathbf{k}) = (c\,\mathbf{j} + d\,\mathbf{k}) + t(a\,\mathbf{j} + b\,\mathbf{k})$$

which is the vector equation of a line.

51. Integrating $\mathbf{a}(t) = \mathbf{r}''(t) = -g\,\mathbf{k}$ gives $\mathbf{r}'(t) = (-gt)\,\mathbf{k} + (a\,\mathbf{j} + b\,\mathbf{k})$, for some a, b in $\Re$. Integrating again:

$$\mathbf{r}(t) = (-\tfrac{1}{2}gt^2)\,\mathbf{k} + (at\,\mathbf{j} + bt\,\mathbf{k}) + (c\,\mathbf{j} + d\,\mathbf{k}), \text{ for some } c, d \text{ in } \Re.$$
$$= (at + c)\,\mathbf{j} + (-\tfrac{1}{2}gt^2 + bt + d)\,\mathbf{k}$$

Letting $\mathbf{r}(t) = x\,\mathbf{j} + y\,\mathbf{k} \implies x = at + c$ and $y = -\tfrac{1}{2}gt^2 + bt + d$. Noting that $a \neq 0$, since the trajectory is not vertical ($x \neq c$), we can solve for t in terms of x: $t = \frac{x-c}{a}$. Substituting this for t in y: $y = -\frac{1}{2}g\left(\frac{x-c}{a}\right)^2 + b\left(\frac{x-c}{a}\right) + d$, shows that y is a quadratic function of x, so its graph is a parabola.

§12.5 Arc Length and Curvature

1. $\mathbf{r}(t) = (2\sin t)\,\mathbf{i} + 5t\,\mathbf{j} + (2\cos t)\,\mathbf{k}, \ -10 \leq t \leq 10 \ \Rightarrow$

$$L = \int_a^b \|\mathbf{r}'(t)\|\, dt = \int_{-10}^{10} \|(2\cos t)\,\mathbf{i} + 5\,\mathbf{j} - (2\sin t)\,\mathbf{k}\|\, dt$$

$$= \int_{-10}^{10} \sqrt{4\cos^2 t + 25 + 4\sin^2 t}\, dt = \int_{-10}^{10} \sqrt{29}\, dt = \sqrt{29}\, t\,\Big|_{-10}^{10} = 20\sqrt{29}.$$

3. $\mathbf{r}(t) = a(1 - \sin t)\,\mathbf{i} + a(1 - \cos t)\,\mathbf{j}, \ 0 \leq t \leq 2\pi \ \Rightarrow$

$$L = \int_a^b \|\mathbf{r}'(t)\|\, dt = \int_0^{2\pi} \|a(-\cos t)\,\mathbf{i} + a(\sin t)\,\mathbf{j}\|\, dt$$

$$= \int_0^{2\pi} |a|\sqrt{\cos^2 t + \sin^2 t}\, dt = \int_0^{2\pi} |a|\, dt = |a|\, t\,\Big|_0^{2\pi} = 2\pi |a|.$$

5. $\mathbf{r}(t) = (a\cos t)\,\mathbf{i} + (a\sin t)\,\mathbf{j} + (bt)\,\mathbf{k}, \ 0 \leq t \leq 2\pi \ \Rightarrow$

$$L = \int_a^b \|\mathbf{r}'(t)\|\, dt = \int_0^{2\pi} \|(-a\sin t)\,\mathbf{i} + (a\cos t)\,\mathbf{j} + b\,\mathbf{k}\|\, dt$$

$$= \int_0^{2\pi} \sqrt{a^2\sin^2 t + a^2\cos^2 t + b^2}\, dt = \int_0^{2\pi} \sqrt{a^2 + b^2}\, dt$$

$$= \left(\sqrt{a^2 + b^2}\right) t\,\Big|_0^{2\pi} = 2\pi\sqrt{a^2 + b^2}.$$

7. $\mathbf{r}(t) = (\mathbf{t}, \ln(\sec \mathbf{t}), \mathbf{3}), \ 0 \leq t \leq \frac{\pi}{4} \ \Rightarrow$

$$L = \int_a^b \|\mathbf{r}'(t)\|\, dt = \int_0^{\pi/4} \left\|\left(1, \frac{1}{\sec t}\cdot \sec t \tan t, 0\right)\right\|\, dt = \int_0^{\pi/4} \|(1, \tan t, 0)\|\, dt$$

$$= \int_0^{\pi/4} \sqrt{1 + \tan^2 t}\, dt = \int_0^{\pi/4} \sqrt{\sec^2 t}\, dt = \int_0^{\pi/4} \sec t\, dt$$

$$= \ln|\sec t + \tan t|\,\Big|_0^{\pi/4} = \ln\left|\sec\frac{\pi}{4} + \tan\frac{\pi}{4}\right| - \ln|\sec 0 + \tan 0|$$

$$= \ln\left|\sqrt{2} + 1\right| - \ln|1 + 0| = \ln\left(\sqrt{2} + 1\right)$$

9. $\mathbf{r}(t) = (3t - 2)\,\mathbf{i} + (4t + 3)\,\mathbf{j}, t = 0 \ \Rightarrow$

$$s(t) = \int_0^t \|\mathbf{r}'(u)\|\, du = \int_0^t \|3\,\mathbf{i} + 4\,\mathbf{j}\|\, du = \int_0^t \sqrt{9 + 16}\, du = 5u\,\Big|_0^t = 5t \ \Rightarrow \ t = \frac{1}{5}s.$$

Substituting into $\mathbf{r}$: $\mathbf{r} = \left(\frac{3}{5}s - 2\right)\mathbf{i} + \left(\frac{4}{5}s + 3\right)\mathbf{j}$, so parametric equations are: $x = \frac{3}{5}s - 2, \ y = \frac{4}{5}s + 3$.

11. $\mathbf{r}(t) = \left(\frac{1}{3}\mathbf{t^3}, \frac{1}{2}\mathbf{t^2}\right)$, $t = 0 \Rightarrow$

$s(t) = \int_0^t \|\mathbf{r}'(u)\| \, du = \int_0^t \|(\mathbf{u^2}, \mathbf{u})\| \, du = \int_0^t \sqrt{u^4 + u^2} \, du = \int_0^t u\sqrt{u^2 + 1} \, du$

$$\boxed{\begin{array}{ll} w = u^2 + 1 & u = 0 \;\Rightarrow\; w = 1 \\ dw = 2u \, du & u = t \;\Rightarrow\; w = t^2 + 1 \end{array}}$$

$= \frac{1}{2} \int_1^{1+t^2} w^{\frac{1}{2}} \, dw = \frac{1}{2} \cdot \frac{2}{3} w^{\frac{3}{2}} \Big|_1^{1+t^2} = \frac{1}{3}\left[(1 + t^2)^{\frac{3}{2}} - 1\right]$

Now solve for t in terms of s:

$s = \frac{1}{3}\left[(1 + t^2)^{\frac{3}{2}} - 1\right] \Rightarrow 3s + 1 = (1 + t^2)^{\frac{3}{2}} \Rightarrow 1 + t^2 = (3s + 1)^{\frac{2}{3}} \Rightarrow t^2 = (3s + 1)^{\frac{2}{3}} - 1$.

Therefore, $t^3 = [(3s + 1)^{\frac{2}{3}} - 1]^{\frac{3}{2}}$. Substituting into $\mathbf{r}$:

$$\mathbf{r} = \left(\frac{1}{3}\left[(3\mathbf{s} + 1)^{\frac{2}{3}} - 1\right]^{\frac{3}{2}}, \frac{1}{2}\left[(3\mathbf{s} + 1)^{\frac{2}{3}} - 1\right]\right),$$

so parametric equations are:

$$x = \frac{1}{3}\left[(3s + 1)^{\frac{2}{3}} - 1\right]^{\frac{3}{2}}, \quad y = \frac{1}{2}\left[(3s + 1)^{\frac{2}{3}} - 1\right]$$

13. $\mathbf{r}(t) = \sin t \, \mathbf{i} + \cos t \, \mathbf{j} + t \, \mathbf{k}, t = 0 \Rightarrow$

$s(t) = \int_0^t \|\mathbf{r}'(u)\| \, du = \int_0^t \|\cos t \, \mathbf{i} - \sin t \, \mathbf{j} + \mathbf{k}\| \, du$

$= \int_0^t \sqrt{\cos^2 t + \sin^2 t + 1^2} \, du = \sqrt{2} \, u \Big|_0^t = \sqrt{2} \, t \Rightarrow t = \frac{s}{\sqrt{2}}$.

Substituting into $\mathbf{r}$: $\mathbf{r} = \sin\left(\frac{s}{\sqrt{2}}\right) \mathbf{i} + \cos\left(\frac{s}{\sqrt{2}}\right) \mathbf{j} + \frac{s}{\sqrt{2}} \mathbf{k}$,   so parametric equations are:

$$x = \sin\left(\frac{s}{\sqrt{2}}\right), \quad y = \cos\left(\frac{s}{\sqrt{2}}\right), \quad z = \frac{s}{\sqrt{2}}.$$

15. $\mathbf{r}(t) = (\mathbf{3\sin t}, \mathbf{3\cos t}, \mathbf{4t}) \Rightarrow$

$$\mathbf{T}(t) = \frac{\mathbf{r}'(t)}{\|\mathbf{r}'(t)\|} = \frac{(\mathbf{3\cos t}, -\mathbf{3\sin t}, \mathbf{4})}{\sqrt{9\cos^2 t + 9\sin^2 t + 16}} = \left(\frac{3}{5}\cos t, -\frac{3}{5}\sin t, \frac{4}{5}\right)$$

$$\mathbf{N}(t) = \frac{\mathbf{T}'(t)}{\|\mathbf{T}'(t)\|} = \frac{\left(-\frac{3}{5}\sin t, -\frac{3}{5}\cos t, 0\right)}{\frac{3}{5}\sqrt{\sin^2 t + \cos^2 t}} = (-\sin t, -\cos t, 0)$$

$$\mathbf{B}(t) = \mathbf{T}(t) \times \mathbf{N}(t) = \det \begin{bmatrix} \mathbf{i} & \mathbf{j} & \mathbf{k} \\ \frac{3}{5}\cos t & -\frac{3}{5}\sin t & \frac{4}{5} \\ -\sin t & -\cos t & 0 \end{bmatrix}$$

$$= \left(\frac{4}{5}\cos t\right)\mathbf{i} - \left(\frac{4}{5}\sin t\right)\mathbf{j} + \left(-\frac{3}{5}\cos^2 t - \frac{3}{5}\sin^2 t\right)\mathbf{k} = \left(\frac{4}{5}\cos t\right)\mathbf{i} - \left(\frac{4}{5}\sin t\right)\mathbf{j} - \frac{3}{5}\mathbf{k}$$

To find the curvature, $\kappa = \dfrac{\|\mathbf{r}'(t) \times \mathbf{r}''(t)\|}{\|\mathbf{r}'(t)\|^3}$, we first need $\mathbf{r}'(t) \times \mathbf{r}''(t)$:

$$\mathbf{r}'(t) \times \mathbf{r}''(t) = \det \begin{bmatrix} \mathbf{i} & \mathbf{j} & \mathbf{k} \\ 3\cos t & -3\sin t & 4 \\ -3\sin t & -3\cos t & 0 \end{bmatrix} = (12\cos t)\,\mathbf{i} - (12\sin t)\,\mathbf{j} - 9\,\mathbf{k} \Rightarrow$$

$$\|\mathbf{r}'(t) \times \mathbf{r}''(t)\| = \sqrt{144\cos^2 t + 144\sin^2 t + 81} = \sqrt{225} = 15 \Rightarrow$$

$$\kappa = \frac{\|\mathbf{r}'(t) \times \mathbf{r}''(t)\|}{\|\mathbf{r}'(t)\|^3} = \frac{15}{5^3} = \frac{3}{25}.$$

17. $\mathbf{r}(t) = (e^t \cos t)\,\mathbf{i} + (e^t \sin t)\,\mathbf{j} + 2\,\mathbf{k} \Rightarrow$

$$\mathbf{T}(t) = \frac{\mathbf{r}'(t)}{\|\mathbf{r}'(t)\|} = \frac{\cancel{e^t}(-\sin t + \cos t)\,\mathbf{i} + \cancel{e^t}(\cos t + \sin t)\,\mathbf{j}}{\cancel{e^t}\sqrt{(\cos t - \sin t)^2 + (\cos t + \sin t)^2}}$$

$$= \frac{(-\sin t + \cos t)\,\mathbf{i} + (\cos t + \sin t)\,\mathbf{j}}{\sqrt{\sin^2 t - 2\cancel{\sin t \cos t} + \cos^2 t + \sin^2 t + 2\cancel{\sin t \cos t} + \cos^2 t}}$$

$$= \frac{1}{\sqrt{2}}[(\cos t - \sin t)\,\mathbf{i} + (\cos t + \sin t)\,\mathbf{j}]$$

$$\mathbf{N}(t) = \frac{\mathbf{T}'(t)}{\|\mathbf{T}'(t)\|} = \frac{\frac{1}{\sqrt{2}}[(-\sin t - \cos t)\,\mathbf{i} + (-\sin t + \cos t)\,\mathbf{j}]}{\frac{1}{\sqrt{2}}\sqrt{(-\sin t - \cos t)^2 + (-\sin t + \cos t)^2}}$$

$$= \frac{(-\sin t - \cos t)\,\mathbf{i} + (-\sin t + \cos t)\,\mathbf{j}}{\sqrt{\sin^2 t + 2\cancel{\sin t \cos t} + \cos^2 t + \sin^2 t - 2\cancel{\sin t \cos t} + \cos^2 t}}$$

$$= \frac{1}{\sqrt{2}}[(-\sin t - \cos t)\,\mathbf{i} + (-\sin t + \cos t)\,\mathbf{j}]$$

$$\mathbf{B}(t) = \mathbf{T}(t) \times \mathbf{N}(t) = \left(\frac{1}{\sqrt{2}} \cdot \frac{1}{\sqrt{2}}\right) \det \begin{bmatrix} \mathbf{i} & \mathbf{j} & \mathbf{k} \\ \cos t - \sin t & \cos t + \sin t & 0 \\ -\sin t - \cos t & -\sin t + \cos t & 0 \end{bmatrix}$$

$$= \tfrac{1}{2}[(\cos t - \sin t)(-\sin t + \cos t) - (\cos t + \sin t)(-\sin t - \cos t)]\,\mathbf{k}$$

$$= \tfrac{1}{2}(\cos^2 t - 2\cancel{\sin t \cos t} + \sin^2 t + \sin^2 t + 2\cancel{\sin t \cos t} + \cos^2 t)\,\mathbf{k}$$

$$= \tfrac{1}{2}(2)\,\mathbf{k} = \mathbf{k}$$

To find the curvature, $\kappa = \dfrac{\|\mathbf{r}'(t) \times \mathbf{r}''(t)\|}{\|\mathbf{r}'(t)\|^3}$, we first need $\mathbf{r}'(t) \times \mathbf{r}''(t)$:

$\mathbf{r}'(t) = e^t[(\cos t - \sin t)\,\mathbf{i} + (\cos t + \sin t)\,\mathbf{j}]$,   then, by Theorem 12.8(b),

$\mathbf{r}''(t) = e^t[(\cos t - \sin t)\,\mathbf{i} + (\cos t + \sin t)\,\mathbf{j}]' + (e^t)'[(\cos t - \sin t)\,\mathbf{i} + (\cos t + \sin t)\,\mathbf{j}]$

$\quad = e^t[(-\sin t - \cos t)\,\mathbf{i} + (-\sin t + \cos t)\,\mathbf{j}] + e^t[(\cos t - \sin t)\,\mathbf{i} + (\cos t + \sin t)\,\mathbf{j}]$

$\quad = e^t\{[(-\sin t - \cancel{\cos t}) + (\cancel{\cos t} - \sin t)]\,\mathbf{i} + [(-\cancel{\sin t} + \cos t) + (\cos t + \cancel{\sin t})]\,\mathbf{j}\}$

$\quad = e^t[(-2\sin t)\,\mathbf{i} + (2\cos t)\,\mathbf{j}] = 2e^t[(-\sin t)\,\mathbf{i} + (\cos t)\,\mathbf{j}]$

Then

$$\mathbf{r}'(t) \times \mathbf{r}''(t) = e^t \cdot 2e^t \cdot \det \begin{bmatrix} \mathbf{i} & \mathbf{j} & \mathbf{k} \\ \cos t - \sin t & \cos t + \sin t & 0 \\ -\sin t & \cos t & 0 \end{bmatrix}$$

$$= 2e^{2t}[(\cos t - \sin t)(\cos t) - (\cos t + \sin t)(-\sin t)]\,\mathbf{k}$$

$$= 2e^{2t}[\cos^2 t - \cancel{\cos t \sin t} + \cancel{\cos t \sin t} + \sin^2 t]\,\mathbf{k} = 2e^{2t}\,\mathbf{k} \implies$$

$$\|\mathbf{r}'(t) \times \mathbf{r}''(t)\| = 2e^{2t}.$$

From our calculation of the tangent vector, above, we found $\|\mathbf{r}'(t)\| = \sqrt{2}e^t$, so that

$$\kappa = \frac{\|\mathbf{r}'(t) \times \mathbf{r}''(t)\|}{\|\mathbf{r}'(t)\|^3} = \frac{2e^{2t}}{(\sqrt{2}e^t)^3} = \frac{\cancel{2}e^{2t}}{\cancel{2}\sqrt{2}e^{3t}} = \frac{1}{\sqrt{2}e^t}.$$

19. $\mathbf{r}(t) = \left(\dfrac{t^3}{3}, \dfrac{t^2}{2}, \mathbf{0}\right)$, $t > 0 \implies$

$$\mathbf{T}(t) = \frac{\mathbf{r}'(t)}{\|\mathbf{r}'(t)\|} = \frac{(t^2, t, 0)}{\sqrt{t^4 + t^2}} = \frac{\cancel{t}(t, 1, 0)}{\cancel{t}\sqrt{t^2 + 1}} = \frac{(t, 1, 0)}{\sqrt{t^2 + 1}} \implies$$

$$\mathbf{T}'(t) = [(t^2 + 1)^{-\frac{1}{2}}(t, 1, 0)]' = (t^2 + 1)^{-\frac{1}{2}}(1, 0, 0) - \tfrac{1}{2}(t^2 + 1)^{-\frac{3}{2}}(2t)(t, 1, 0)$$

$$= \frac{(t^2 + 1)(1, 0, 0) - t(t, 1, 0)}{(t^2 + 1)^{\frac{3}{2}}} = \frac{(1, -t, 0)}{(t^2 + 1)^{\frac{3}{2}}} = (t^2 + 1)^{-\frac{3}{2}}(1, -t, 0) \implies$$

$$\|\mathbf{T}'(t)\| = \frac{1}{(t^2 + 1)^{\frac{3}{2}}} \cdot \sqrt{1 + t^2} = \frac{1}{1 + t^2} = (1 + t^2)^{-1} \implies$$

$$\mathbf{N}(t) = \frac{\mathbf{T}'(t)}{\|\mathbf{T}'(t)\|} = \frac{(t^2 + 1)^{-\frac{3}{2}}(1, -t, 0)}{(1 + t^2)^{-1}} = \frac{(1, -t, 0)}{\sqrt{t^2 + 1}}$$

$$\mathbf{B}(t) = \mathbf{T}(t) \times \mathbf{N}(t) = \frac{1}{\sqrt{t^2 + 1}} \cdot \frac{1}{\sqrt{t^2 + 1}} \cdot \det \begin{bmatrix} \mathbf{i} & \mathbf{j} & \mathbf{k} \\ t & 1 & 0 \\ 1 & -t & 0 \end{bmatrix}$$

$$= \frac{1}{t^2 + 1} \cdot (-t^2 - 1)\,\mathbf{k} = -\mathbf{k} = (0, 0, -1).$$

To find the curvature, $\kappa = \dfrac{\|\mathbf{r}'(t) \times \mathbf{r}''(t)\|}{\|\mathbf{r}'(t)\|^3}$, we first need $\mathbf{r}'(t) \times \mathbf{r}''(t)$:

$$\mathbf{r}'(t) \times \mathbf{r}''(t) = \det \begin{bmatrix} \mathbf{i} & \mathbf{j} & \mathbf{k} \\ t^2 & t & 0 \\ 2t & 1 & 0 \end{bmatrix} = (t^2 - 2t^2)\,\mathbf{k} = (-t^2)\,\mathbf{k} = (0, 0, -t^2)$$

Therefore,

$$\kappa = \frac{\|\mathbf{r}'(t) \times \mathbf{r}''(t)\|}{\|\mathbf{r}'(t)\|^3} = \frac{t^2}{(\sqrt{t^4 + t^2})^3} = \frac{t^2}{(t\sqrt{t^2 + 1})^3} = \frac{t^2}{t^3(t^2 + 1)^{\frac{3}{2}}} = \frac{1}{t(t^2 + 1)^{\frac{3}{2}}}.$$

21. To find the normal plane for the curve $\mathbf{r}(t) = (2\sin 3t)\,\mathbf{i} + t\,\mathbf{j} + (2\cos 3t)\,\mathbf{k}$, at $t = \pi$, we find a normal to that plane, namely, a vector parallel to the tangent vector, $\mathbf{T}(\pi) = \dfrac{\mathbf{r}'(\pi)}{\|\mathbf{r}'(\pi)\|}$. Such a vector is $\mathbf{r}'(\pi)$:

$$\mathbf{r}'(t) = (6\cos 3t)\,\mathbf{i} + \mathbf{j} + (-6\sin 3t)\,\mathbf{k} \;\Rightarrow\; \mathbf{r}'(\pi) = (6\cos 3\pi)\,\mathbf{i} + \mathbf{j} + (-6\sin 3\pi)\,\mathbf{k} = -6\,\mathbf{i} + \mathbf{j}.$$

A point on the normal plane: $\mathbf{r}(\pi) = (2\sin 3\pi)\,\mathbf{i} + \pi\,\mathbf{j} + (2\cos 3\pi)\,\mathbf{k} = (0, \pi, -2)$, so $(0, \pi, -2)$ is such a point. Then, the equation of the normal plane is

$$(-6, 1, 0) \cdot (\mathbf{x} - 0, \mathbf{y} - \pi, \mathbf{z} - (-2)) = 0 \;\Rightarrow\; -6x + y = \pi.$$

To find the osculating plane, we find a normal to that plane, namely a vector parallel to the binormal, such as $\overline{\mathbf{B}(\pi)}$:

$$\overline{\mathbf{B}(\pi)} = \det \begin{bmatrix} \mathbf{i} & \mathbf{j} & \mathbf{k} \\ 6\cos 3\pi & 1 & -6\sin 3\pi \\ -18\sin 3\pi & 0 & -18\cos 3\pi \end{bmatrix} = \det \begin{bmatrix} \mathbf{i} & \mathbf{j} & \mathbf{k} \\ -6 & 1 & 0 \\ 0 & 0 & 18 \end{bmatrix} = 18\,\mathbf{i} + 18(6)\,\mathbf{j}$$

A simpler normal is $\mathbf{i} + 6\,\mathbf{j}$. Thus, the equation of the osculating plane is:

$$(1, 6, 0) \cdot (\mathbf{x} - 0, \mathbf{y} - \pi, \mathbf{z} - (-2)) = 0 \;\Rightarrow\; x + 6y = 6\pi.$$

23. To find the normal plane for the curve $\mathbf{r}(t) = (\cos t, \sin t, t)$, at $\left(\dfrac{1}{\sqrt{2}}, \dfrac{1}{\sqrt{2}}, \dfrac{\pi}{4}\right)$, i.e. when $t = \dfrac{\pi}{4}$, we find a normal to that plane, namely, a vector parallel to the tangent vector, $\mathbf{T}\left(\dfrac{\pi}{4}\right) = \dfrac{\mathbf{r}'\left(\frac{\pi}{4}\right)}{\|\mathbf{r}'\left(\frac{\pi}{4}\right)\|}$. Such a vector is $\mathbf{r}'\left(\dfrac{\pi}{4}\right)$:

$$\mathbf{r}'(t) = (-\sin t, \cos t, 1) \Rightarrow \mathbf{r}'\left(\tfrac{\pi}{4}\right) = \left(-\tfrac{1}{\sqrt{2}}, \tfrac{1}{\sqrt{2}}, 1\right).$$

Thus, the equation of the normal plane is:

$$\left(-\tfrac{1}{\sqrt{2}}, \tfrac{1}{\sqrt{2}}, 1\right) \cdot \left(\mathbf{x} - \tfrac{1}{\sqrt{2}}, \mathbf{y} - \tfrac{1}{\sqrt{2}}, \mathbf{z} - \tfrac{\pi}{4}\right) = 0 \Rightarrow$$

$$-\tfrac{1}{\sqrt{2}}\left(x - \tfrac{1}{\sqrt{2}}\right) + \tfrac{1}{\sqrt{2}}\left(y - \tfrac{1}{\sqrt{2}}\right) + z - \tfrac{\pi}{4} = 0 \Rightarrow$$

$$-\tfrac{1}{\sqrt{2}}x + \tfrac{1}{\sqrt{2}}y + z = \tfrac{\pi}{4} + \tfrac{1}{2} - \tfrac{1}{2} \Rightarrow -\tfrac{1}{\sqrt{2}}x + \tfrac{1}{\sqrt{2}}y + z = \tfrac{\pi}{4}$$

To find the osculating plane, we find a normal to that plane, namely a vector parallel to the binormal, such as $\overline{\mathbf{B}\left(\dfrac{\pi}{4}\right)}$:

$$\overline{\mathbf{B}\left(\tfrac{\pi}{4}\right)} = \det \begin{bmatrix} \mathbf{i} & \mathbf{j} & \mathbf{k} \\ -\tfrac{1}{\sqrt{2}} & \tfrac{1}{\sqrt{2}} & 1 \\ -\cos\tfrac{\pi}{4} & -\sin\tfrac{\pi}{4} & 0 \end{bmatrix} = \det \begin{bmatrix} \mathbf{i} & \mathbf{j} & \mathbf{k} \\ -\tfrac{1}{\sqrt{2}} & \tfrac{1}{\sqrt{2}} & 1 \\ -\tfrac{1}{\sqrt{2}} & -\tfrac{1}{\sqrt{2}} & 0 \end{bmatrix} = \tfrac{1}{\sqrt{2}}\,\mathbf{i} - \tfrac{1}{\sqrt{2}}\,\mathbf{j} + \mathbf{k}$$

Thus, the equation of the osculating plane is:

$$\left(\tfrac{1}{\sqrt{2}}, -\tfrac{1}{\sqrt{2}}, 1\right) \cdot \left(\mathbf{x} - \tfrac{1}{\sqrt{2}}, \mathbf{y} - \tfrac{1}{\sqrt{2}}, \mathbf{z} - \tfrac{\pi}{4}\right) = 0 \Rightarrow \tfrac{1}{\sqrt{2}}\left(x - \tfrac{1}{\sqrt{2}}\right) - \tfrac{1}{\sqrt{2}}\left(y - \tfrac{1}{\sqrt{2}}\right) + z - \tfrac{\pi}{4} = 0 \Rightarrow$$

$$\tfrac{1}{\sqrt{2}}x - \tfrac{1}{\sqrt{2}}y + z = \tfrac{\pi}{4}.$$

25. Since the center of the circle of curvature, lies along the normal to the curve, we first must find this normal vector $\mathbf{N}$ for the curve $\mathbf{r}(t) = t\,\mathbf{i} + (\sin 2t)\,\mathbf{j}$ at $\left(\frac{\pi}{4}, 1\right)$, i.e. at $t = \frac{\pi}{4}$. This requires finding the tangent vector, $\mathbf{T}$:

$$\mathbf{T}(t) = \frac{\mathbf{r}'(t)}{\|\mathbf{r}'(t)\|} = \frac{\mathbf{i} + (2\cos 2t)\,\mathbf{j}}{\sqrt{1 + 4\cos^2 2t}} \;\Rightarrow\; \mathbf{T}\left(\tfrac{\pi}{4}\right) = \mathbf{i}$$

Instead of differentiating the tangent vector (which in this case is very involved) to find the normal vector, we will follow the method of Example 12.26: We will find $\overline{\mathbf{B}}$, then $\mathbf{B}$, then apply Exercise 30, $\mathbf{N} = \mathbf{B} \times \mathbf{T}$, all at $t = \frac{\pi}{4}$.

Differentiating $\mathbf{r}'(t) = \mathbf{i} + (2\cos 2t)\,\mathbf{j}$, we get $\mathbf{r}''(t) = (-4\sin 2t)\,\mathbf{j}$. Evaluating at $t = \frac{\pi}{4}$, we find $\mathbf{r}'\left(\frac{\pi}{4}\right) = \mathbf{i} + \left(2\cos\frac{\pi}{2}\right)\mathbf{j} = \mathbf{i}$, and $\mathbf{r}''\left(\frac{\pi}{4}\right) = \left(-4\sin\frac{\pi}{2}\right)\mathbf{j} = -4\mathbf{j} \;\Rightarrow\;$

$$\overline{\mathbf{B}\left(\tfrac{\pi}{4}\right)} = \det\begin{bmatrix} \mathbf{i} & \mathbf{j} & \mathbf{k} \\ 1 & 0 & 0 \\ 0 & -4 & 0 \end{bmatrix} = -4\mathbf{k}, \;\; \text{so } \mathbf{B}\left(\tfrac{\pi}{4}\right) = -\mathbf{k} \;\Rightarrow$$

$$\mathbf{N}\left(\tfrac{\pi}{4}\right) = \mathbf{B}\left(\tfrac{\pi}{4}\right) \times \mathbf{T}\left(\tfrac{\pi}{4}\right) = \det\begin{bmatrix} \mathbf{i} & \mathbf{j} & \mathbf{k} \\ 0 & 0 & -1 \\ 1 & 0 & 0 \end{bmatrix} = -\mathbf{j}.$$

Next, expressing the curve as $y = \sin 2x = f(x)$, then $f'(x) = 2\cos 2x$ and $f''(x) = -4\sin 2x$, we find

$$\kappa = \frac{|f''(x)|}{[1 + (f'(x))^2]^{\frac{3}{2}}} = \frac{|-4\sin 2x|}{[1 + 4\cos^2 2x]^{\frac{3}{2}}}. \;\; \text{Evaluating at } t = \tfrac{\pi}{4}, \kappa = 4, \text{ so the radius}$$

of the circle of curvature at the given point is $r = \frac{1}{\kappa} = \frac{1}{4}$. We can now locate the center of the circle: start at the given point and move $\frac{1}{4}$ unit along the normal $(-\mathbf{j})$, arriving at $\left(\frac{\pi}{4}, 1 - \frac{1}{4}\right) = \left(\frac{\pi}{4}, \frac{3}{4}\right)$. Therefore the equation of the circle is $\left(x - \frac{\pi}{4}\right)^2 + \left(y - \frac{3}{4}\right)^2 = \left(\frac{1}{4}\right)^2$.

27. Applying CYU 12.36, $\kappa = \dfrac{|f''(x)|}{[1 + (f'(x))^2]^{\frac{3}{2}}}$, to $f(x) = ax^2 \Rightarrow f'(x) = 2ax \Rightarrow$
$f''(x) = 2a$:

$$\kappa = \frac{|2a|}{(1 + 4a^2x^2)^{\frac{3}{2}}}, \text{ and } \kappa \text{ is greatest when the denominator is smallest, and this occurs}$$

at $x = 0$, which is where the vertex of the parabola is located.

29. To prove: If $\mathbf{r}(t) = f(t)\,\mathbf{i} + g(t)\,\mathbf{j} + h(t)\,\mathbf{k}$, then $\overline{\mathbf{B}(t)} = \det\begin{bmatrix} \mathbf{i} & \mathbf{j} & \mathbf{k} \\ f'(t) & g'(t) & h'(t) \\ f''(t) & g''(t) & h''(t) \end{bmatrix}$

has the same direction as the unit binormal, $\mathbf{B}(t) = \mathbf{T}(t) \times \mathbf{N}(t)$. For ease of writing, henceforth we will omit the argument "(t)". Expanding the determinant, we have:

$$\overline{\mathbf{B}} = (g'h'' - g''h')\,\mathbf{i} - (f'h'' - f''h')\,\mathbf{j} + (f'g'' - f''g')\,\mathbf{k}$$

We need to determine the direction of $\mathbf{T} \times \mathbf{N}$. Now, $\mathbf{T} = \dfrac{(\mathbf{f}', \mathbf{g}', \mathbf{h}')}{\sqrt{f'^2 + g'^2 + h'^2}} = D(\mathbf{f}', \mathbf{g}', \mathbf{h}')$,

where $D = \dfrac{1}{\sqrt{f'^2 + g'^2 + h'^2}}$ is a positive scalar function of t. As $\mathbf{N} = \dfrac{\mathbf{T}'}{\|\mathbf{T}'\|}$, we need to

determine $\mathbf{T}'$:

$$\mathbf{T}' = \left[\frac{(\mathbf{f}', \mathbf{g}', \mathbf{h}')}{\sqrt{f'^2 + g'^2 + h'^2}} \right]' = \left(\frac{1}{\sqrt{f'^2 + g'^2 + h'^2}} \right)' (\mathbf{f}', \mathbf{g}', \mathbf{h}') + \left(\frac{1}{\sqrt{f'^2 + g'^2 + h'^2}} \right) (\mathbf{f}'', \mathbf{g}'', \mathbf{h}'')$$

Let $C = \left(\dfrac{1}{\sqrt{f'^2 + g'^2 + h'^2}} \right)'$, the scalar function D', then

$$\mathbf{T}' = C(\mathbf{f}', \mathbf{g}', \mathbf{h}') + D(\mathbf{f}'', \mathbf{g}'', \mathbf{h}''), \quad \text{so that} \quad \mathbf{N} = \frac{C}{\|\mathbf{T}'\|}(\mathbf{f}', \mathbf{g}', \mathbf{h}') + \frac{D}{\|\mathbf{T}'\|}(\mathbf{f}'', \mathbf{g}'', \mathbf{h}'')$$

Let $E = \dfrac{C}{\|\mathbf{T}'\|}$ and $P = \dfrac{D}{\|\mathbf{T}'\|}$, two scalar functions of t, with $P > 0$, so that

$$\mathbf{N} = E(\mathbf{f}', \mathbf{g}', \mathbf{h}') + P(\mathbf{f}'', \mathbf{g}'', \mathbf{h}'') = (Ef' + Pf'', Eg' + Pg'', Eh' + Ph'')$$

Then

$$\mathbf{B} = \mathbf{T} \times \mathbf{N} = \det \begin{bmatrix} \mathbf{i} & \mathbf{j} & \mathbf{k} \\ Df' & Dg' & Dh' \\ Ef' + Pf'' & Eg' + Pg'' & Eh' + Ph'' \end{bmatrix}$$

$$= D\{[g'(\not\!Eh' + Ph'') - h'(\not\!Eg' + Pg'')]\mathbf{i} - [f'(\not\!Eh' + Ph'') - h'(\not\!Ef' + Pf'')]\mathbf{j}$$

$$+ [f'(\not\!Eg' + Pg'') - g'(\not\!Ef' + Pf'')]\mathbf{k}\}$$

$$= DP\{g'h'' - g''h')\mathbf{i} - (f'h'' - f''h')\mathbf{j} + (f'g'' - f''g')\mathbf{k}\}$$

$$= (DP)\overline{\mathbf{B}}$$

Since both D and P are positive scalar functions of t, and $\mathbf{B} = (DP)\overline{\mathbf{B}}$, both $\mathbf{B}$ and $\overline{\mathbf{B}}$ have the same direction. It follows that they can only differ in magnitude, so that the unit vector $\mathbf{B} = \dfrac{\overline{\mathbf{B}}}{\|\overline{\mathbf{B}}\|}$.

CHAPTER 13
Differentiating Functions of Several Variables

§13.1 Partial Derivatives and Differentiability

1. $z = x^2 + y^2 \implies \frac{\partial z}{\partial x} = 2x, \ \frac{\partial z}{\partial y} = 2y.$

3. $z = (xy - 1)^2 \implies \frac{\partial z}{\partial x} = 2(xy-1)y = 2xy^2 - 2y, \ \frac{\partial z}{\partial y} = 2(xy-1)x = 2x^2y - 2x.$

5. $z = 4e^{x^2}y^3 \implies \frac{\partial z}{\partial x} = 4e^{x^2}(2x)y^3 = 8xy^3e^{x^2}, \ \frac{\partial z}{\partial y} = 4e^{x^2}(3y^2) = 12y^2e^{x^2}.$

7. $z = \dfrac{x+y}{xy-1} \implies \dfrac{\partial z}{\partial x} = \dfrac{(xy-1)\cdot 1 - (x+y)y}{(xy-1)^2} = \dfrac{\cancel{xy} - 1 - \cancel{xy} - y^2}{(xy-1)^2} = -\dfrac{1+y^2}{(xy-1)^2}.$

    $\dfrac{\partial z}{\partial y} = \dfrac{(xy-1)\cdot 1 - (x+y)x}{(xy-1)^2} = \dfrac{\cancel{xy} - 1 - x^2 - \cancel{xy}}{(xy-1)^2} = -\dfrac{1+x^2}{(xy-1)^2}.$

9. $z = e^{x\sin y} \implies \dfrac{\partial z}{\partial x} = e^{x\sin y}\sin y = (\sin y)e^{x\sin y}, \ $ and

    $\dfrac{\partial z}{\partial y} = e^{x\sin y}x\cos y = (x\cos y)e^{x\sin y}.$

11. $z = e^{x\cos y}\ln x \implies$

    $\dfrac{\partial z}{\partial x} = \left(\dfrac{\partial}{\partial x}e^{x\cos y}\right)\ln x + e^{x\cos y}\left(\dfrac{\partial}{\partial x}\ln x\right)$

    $= e^{x\cos y}\cos y\ln x + e^{x\cos y}\cdot\dfrac{1}{x} = e^{x\cos y}\left[\cos y\ln x + \dfrac{1}{x}\right]$

    $\dfrac{\partial z}{\partial y} = \left(\dfrac{\partial}{\partial y}e^{x\cos y}\right)\ln x = e^{x\cos y}(-x\sin y)\ln x = -(x\sin y\ln x)e^{x\cos y}.$

13. $f(x,y) = 6x^2 - 8xy + 9y^2 \implies$

    $f_{xx} = \dfrac{\partial}{\partial x}\left[\dfrac{\partial}{\partial x}(6x^2 - 8xy + 9y^2)\right] = \dfrac{\partial}{\partial x}(12x - 8y) = 12$

    $f_{yy} = \dfrac{\partial}{\partial y}\left[\dfrac{\partial}{\partial y}(6x^2 - 8xy + 9y^2)\right] = \dfrac{\partial}{\partial y}(-8x + 18y) = 18$

    $f_{xy} = \dfrac{\partial}{\partial y}\left[\dfrac{\partial}{\partial x}(6x^2 - 8xy + 9y^2)\right] = \dfrac{\partial}{\partial y}(12x - 8y) = -8$

    $f_{yx} = \dfrac{\partial}{\partial x}\left[\dfrac{\partial}{\partial y}(6x^2 - 8xy + 9y^2)\right] = \dfrac{\partial}{\partial x}(-8x + 18y) = -8 = f_{xy} \ \checkmark$

15. $f(x, y) = e^{2x+3y} \Rightarrow$

$$f_{xx} = \frac{\partial}{\partial x}\left[\frac{\partial}{\partial x}(e^{2x+3y})\right] = \frac{\partial}{\partial x}(e^{2x+3y} \cdot 2) = 2e^{2x+3y} \cdot 2 = 4e^{2x+3y}$$

$$f_{yy} = \frac{\partial}{\partial y}\left[\frac{\partial}{\partial y}(e^{2x+3y})\right] = \frac{\partial}{\partial y}(e^{2x+3y} \cdot 3) = 3e^{2x+3y} \cdot 3 = 9e^{2x+3y}$$

$$f_{xy} = \frac{\partial}{\partial y}\left[\frac{\partial}{\partial x}(e^{2x+3y})\right] = \frac{\partial}{\partial y}(e^{2x+3y} \cdot 2) = 2e^{2x+3y} \cdot 3 = 6e^{2x+3y}$$

$$f_{yx} = \frac{\partial}{\partial x}\left[\frac{\partial}{\partial y}(e^{2x+3y})\right] = \frac{\partial}{\partial x}(e^{2x+3y} \cdot 3) = 3e^{2x+3y} \cdot 2 = 6e^{2x+3y} = f_{xy} \ \checkmark$$

17. $f(x, y) = \sqrt{x^2 + y^2} \Rightarrow$

$$f_{xx} = \frac{\partial}{\partial x}\left[\frac{\partial}{\partial x}(x^2 + y^2)^{\frac{1}{2}}\right] = \frac{\partial}{\partial x}\left[\frac{1}{2}(x^2 + y^2)^{-\frac{1}{2}} \cdot 2x\right] = \frac{\partial}{\partial x}\left[\frac{x}{\sqrt{x^2 + y^2}}\right]$$

$$= \frac{\sqrt{x^2 + y^2} \cdot 1 - x \cdot \frac{1}{2}(x^2 + y^2)^{-\frac{1}{2}} \cdot 2x}{x^2 + y^2} = \frac{\sqrt{x^2 + y^2} - \frac{x^2}{\sqrt{x^2 + y^2}}}{x^2 + y^2} = \frac{\cancel{x^2} + y^2 - \cancel{x^2}}{(x^2 + y^2)^{\frac{3}{2}}}$$

$$= \frac{y^2}{(x^2 + y^2)^{\frac{3}{2}}}$$

$$f_{yy} = \frac{\partial}{\partial y}\left[\frac{\partial}{\partial y}(x^2 + y^2)^{\frac{1}{2}}\right] = \frac{\partial}{\partial y}\left[\frac{1}{2}(x^2 + y^2)^{-\frac{1}{2}} \cdot 2y\right] = \frac{\partial}{\partial y}\left[\frac{y}{(x^2 + y^2)^{\frac{1}{2}}}\right]$$

$$= \frac{\sqrt{x^2 + y^2} \cdot 1 - y \cdot \frac{1}{2}(x^2 + y^2)^{-\frac{1}{2}} \cdot 2y}{x^2 + y^2} = \frac{\sqrt{x^2 + y^2} - \frac{y^2}{\sqrt{x^2 + y^2}}}{x^2 + y^2} = \frac{x^2 + \cancel{y^2} - \cancel{y^2}}{(x^2 + y^2)^{\frac{3}{2}}}$$

$$= \frac{x^2}{(x^2 + y^2)^{\frac{3}{2}}}$$

$$f_{xy} = \frac{\partial}{\partial y}\left[\frac{\partial}{\partial x}(x^2 + y^2)^{\frac{1}{2}}\right] = \frac{\partial}{\partial y}\left[\frac{1}{2}(x^2 + y^2)^{-\frac{1}{2}} \cdot 2x\right] = x\frac{\partial}{\partial y}(x^2 + y^2)^{-\frac{1}{2}}$$

$$= x\left[-\frac{1}{2}(x^2 + y^2)^{-\frac{3}{2}} \cdot 2y\right] = -\frac{xy}{(x^2 + y^2)^{\frac{3}{2}}}$$

$$f_{yx} = \frac{\partial}{\partial x}\left[\frac{\partial}{\partial y}(x^2 + y^2)^{\frac{1}{2}}\right] = \frac{\partial}{\partial x}\left[\frac{1}{2}(x^2 + y^2)^{-\frac{1}{2}} \cdot 2y\right] = y\frac{\partial}{\partial x}(x^2 + y^2)^{-\frac{1}{2}}$$

$$= y\left[-\frac{1}{2}(x^2 + y^2)^{-\frac{3}{2}} \cdot 2x\right] = -\frac{xy}{(x^2 + y^2)^{\frac{3}{2}}} = f_{xy} \ \checkmark$$

19. $f(x, y) = \dfrac{\sin x}{\cos y} \Rightarrow$

$$f_{xx} = \frac{\partial}{\partial x}\left[\frac{\partial}{\partial x}\left(\frac{\sin x}{\cos y}\right)\right] = \left[\frac{1}{\cos y}\right]\frac{\partial}{\partial x}\left(\frac{\partial}{\partial x}\sin x\right) = \left[\frac{1}{\cos y}\right]\frac{\partial}{\partial x}\cos x = \frac{1}{\cos y}(-\sin x) = -\frac{\sin x}{\cos y}$$

$$f_{yy} = \frac{\partial}{\partial y}\left[\frac{\partial}{\partial y}\left(\frac{\sin x}{\cos y}\right)\right] = \frac{\partial}{\partial y}\left[\frac{(\cos y)(0) - (\sin x)(-\sin y)}{\cos^2 y}\right] = (\sin x)\frac{\partial}{\partial y}\left[\frac{\sin y}{\cos^2 y}\right]$$

$$= (\sin x)\left[\frac{(\cos^2 y)(\cos y) - (\sin y)(2\cos y)(-\sin y)}{\cos^{\cancel{4}\,3} y}\right]$$

$$= (\sin x)\left[\frac{\cos^2 y + 2\sin^2 y}{\cos^3 y}\right] = (\sin x)\left[\frac{(\cos^2 y + \sin^2 y) + \sin^2 y}{\cos^3 y}\right] = \frac{(\sin x)(1 + \sin^2 y)}{\cos^3 y}$$

$$f_{xy} = \frac{\partial}{\partial y}\left[\frac{\partial}{\partial x}\left(\frac{\sin x}{\cos y}\right)\right] = \frac{\partial}{\partial y}\left[\left(\frac{1}{\cos y}\right)\frac{\partial}{\partial x}\sin x\right] = \frac{\partial}{\partial y}\left[\frac{\cos x}{\cos y}\right] = \frac{(\cos y)(0) - (\cos x)(-\sin y)}{\cos^2 y}$$

$$= \frac{\cos x \sin y}{\cos^2 y}$$

$$f_{yx} = \frac{\partial}{\partial x}\left[\frac{\partial}{\partial y}\left(\frac{\sin x}{\cos y}\right)\right] = \frac{\partial}{\partial x}\left[\frac{(\cos y)(0) - (\sin x)(-\sin y)}{\cos^2 y}\right] = \frac{\partial}{\partial x}\left[(\sin x)\left(\frac{\sin y}{\cos^2 y}\right)\right]$$

$$= \frac{\cos x \sin y}{\cos^2 y} = f_{xy} \quad \checkmark$$

21. $f(x, y) = \ln(xy - y^2) \Rightarrow$

$$f_{xx} = \frac{\partial}{\partial x}\left\{\frac{\partial}{\partial x}[\ln(xy - y^2)]\right\} = \frac{\partial}{\partial x}\left[\frac{1}{xy - y^2}\cdot y\right] = \frac{\partial}{\partial x}\left(\frac{1}{x - y}\right) = \frac{\partial}{\partial x}(x - y)^{-1}$$

$$= -(x - y)^{-2} = -\frac{1}{(x - y)^2}$$

$$f_{yy} = \frac{\partial}{\partial y}\left\{\frac{\partial}{\partial y}[\ln(xy - y^2)]\right\} = \frac{\partial}{\partial y}\left[\frac{1}{xy - y^2}\cdot(x - 2y)\right] = \frac{(xy - y^2)(-2) - (x - 2y)(x - 2y)}{(xy - y^2)^2}$$

$$= \frac{-2xy + 2y^2 - x^2 + 4xy - 4y^2}{(xy - y^2)^2} = \frac{2xy - 2y^2 - x^2}{y^2(x - y)^2}$$

$$f_{xy} = \frac{\partial}{\partial y}\left\{\frac{\partial}{\partial x}[\ln(xy - y^2)]\right\} = \frac{\partial}{\partial y}\left[\frac{y}{xy - y^2}\right] = \frac{\partial}{\partial y}\left(\frac{1}{x - y}\right) = \frac{\partial}{\partial y}(x - y)^{-1}$$

$$= -(x - y)^{-2}(-1) = \frac{1}{(x - y)^2}$$

$$f_{yx} = \frac{\partial}{\partial x}\left\{\frac{\partial}{\partial y}[\ln(xy - y^2)]\right\} = \frac{\partial}{\partial x}\left[\frac{x - 2y}{xy - y^2}\right] = \frac{(xy - y^2)\cdot 1 - (x - 2y)y}{(xy - y^2)^2}$$

$$= \frac{xy - y^2 - xy + 2y^2}{y^2(x - y)^2} = \frac{y^2}{y^2(x - y)^2} = \frac{1}{(x - y)^2} = f_{xy} \quad \checkmark$$

23. As $f(x, y, z) = x^3 y^2 z$ is a polynomial function, it is continuous, as are all of its partial derivatives. By Theorem 13.1, corresponding mixed partial derivatives are equal: $f_{xz} = f_{zx}$, $f_{zzy} = f_{zyz}$, and $f_{xyz} = f_{zyx}$.

$$f_{xz} = \frac{\partial}{\partial z}\left[\frac{\partial}{\partial x}(x^3 y^2 z)\right] = \frac{\partial}{\partial z}(3x^2 y^2 z) = 3x^2 y^2 = f_{zx}$$

$$f_{yyy} = \frac{\partial}{\partial y}\left\{\frac{\partial}{\partial y}\left[\frac{\partial}{\partial y}(x^3 y^2 z)\right]\right\} = \frac{\partial^2}{\partial y^2}(x^3 \cdot 2y \cdot z) = \frac{\partial^2}{\partial y^2}(2x^3 yz) = \frac{\partial}{\partial y}(2x^3 z) = 0$$

$$f_{zzy} = \frac{\partial}{\partial y}\left\{\frac{\partial}{\partial z}\left[\frac{\partial}{\partial z}(x^3 y^2 z)\right]\right\} = \frac{\partial}{\partial y}\left[\frac{\partial}{\partial z}(x^3 y^2)\right] = \frac{\partial}{\partial y}(0) = 0 = f_{zyz}$$

$$f_{xyz} = \frac{\partial}{\partial z}\left\{\frac{\partial}{\partial y}\left[\frac{\partial}{\partial x}(x^3 y^2 z)\right]\right\} = \frac{\partial}{\partial z}\left[\frac{\partial}{\partial y}(3x^2 y^2 z)\right] = \frac{\partial}{\partial z}(3x^2 \cdot 2y \cdot z) = 6x^2 y = f_{zyx}$$

25. As $f(x, y, z) = e^{xyz}$ is a continuous function, as are all of its partial derivatives. By Theorem 13.1, corresponding mixed partial derivatives are equal: $f_{xz} = f_{zx}$, $f_{zzy} = f_{zyz}$, and $f_{xyz} = f_{zyx}$.

$$f_{xz} = \frac{\partial}{\partial z}\left[\frac{\partial}{\partial x}(e^{xyz})\right] = \frac{\partial}{\partial z}(e^{xyz} \cdot yz) = e^{xyz}\frac{\partial}{\partial z}(yz) + yz\frac{\partial}{\partial z}e^{xyz} = ye^{xyz} + yze^{xyz} \cdot xy$$
$$= e^{xyz}(y + xy^2 z) = f_{zx}$$

$$f_{yyy} = \frac{\partial}{\partial y}\left\{\frac{\partial}{\partial y}\left[\frac{\partial}{\partial y}(e^{xyz})\right]\right\} = \frac{\partial^2}{\partial y^2}(e^{xyz} \cdot xz) = \frac{\partial}{\partial y}(xz \cdot e^{xyz} \cdot xz) = x^2 z^2 \cdot e^{xyz} \cdot xz = x^3 z^3 e^{xyz}$$

$$f_{zzy} = \frac{\partial}{\partial y}\left\{\frac{\partial}{\partial z}\left[\frac{\partial}{\partial z}(e^{xyz})\right]\right\} = \frac{\partial}{\partial y}\left[\frac{\partial}{\partial z}(e^{xyz} \cdot xy)\right] = \frac{\partial}{\partial y}(xy \cdot e^{xyz} \cdot xy) = \frac{\partial}{\partial y}(x^2 y^2 e^{xyz})$$
$$= x^2\left[y^2(\frac{\partial}{\partial y}e^{xyz}) + e^{xyz}(\frac{\partial}{\partial y}y^2)\right] = x^2\left[y^2 \cdot e^{xyz} \cdot xz + e^{xyz} \cdot 2y\right] = x^2 y(xyz + 2)e^{xyz}$$
$$= f_{zyz}$$

$$f_{xyz} = \frac{\partial}{\partial z}\left\{\frac{\partial}{\partial y}\left[\frac{\partial}{\partial x}(e^{xyz})\right]\right\} = \frac{\partial}{\partial z}\left[\frac{\partial}{\partial y}(e^{xyz} \cdot yz)\right] = \frac{\partial}{\partial z}\left\{z\left[y\left(\frac{\partial}{\partial y}e^{xyz}\right) + e^{xyz}\left(\frac{\partial}{\partial y}y\right)\right]\right\}$$
$$= \frac{\partial}{\partial z}\left[z\left(ye^{xyz} \cdot xz + e^{xyz}\right)\right] = \frac{\partial}{\partial z}\left[(xyz^2 + z)e^{xyz}\right]$$
$$= (xyz^2 + z)\frac{\partial}{\partial z}(e^{xyz}) + e^{xyz}\frac{\partial}{\partial z}(xyz^2 + z) = (xyz^2 + z)(e^{xyz} \cdot xy) + e^{xyz}(2xyz + 1)$$
$$= e^{xyz}(x^2 y^2 z^2 + xyz + 2xyz + 1) = e^{xyz}(x^2 y^2 z^2 + 3xyz + 1) = f_{zyx}$$

27. As $f(x, y, z) = z^2 + y\cos x$ is a continuous function, as are all of its partial derivatives. By Theorem 13.1, corresponding mixed partial derivatives are equal: $f_{xz} = f_{zx}$, $f_{zzy} = f_{zyz}$, and $f_{xyz} = f_{zyx}$.

$$f_{xz} = \frac{\partial}{\partial z}\left[\frac{\partial}{\partial x}(z^2 + y\cos x)\right] = \frac{\partial}{\partial z}[y(-\sin x)] = 0 = f_{zx}$$

$$f_{yyy} = \frac{\partial}{\partial y}\left\{\frac{\partial}{\partial y}\left[\frac{\partial}{\partial y}(z^2 + y\cos x)\right]\right\} = \frac{\partial}{\partial y}\left[\frac{\partial}{\partial y}(\cos x)\right] = \frac{\partial}{\partial y}(0) = 0$$

$$f_{zzy} = \frac{\partial}{\partial y}\left\{\frac{\partial}{\partial z}\left[\frac{\partial}{\partial z}(z^2 + y\cos x)\right]\right\} = \frac{\partial}{\partial y}\left[\frac{\partial}{\partial z}(2z)\right] = \frac{\partial}{\partial y}(2) = 0 = f_{zyz}$$

$$f_{xyz} = \frac{\partial}{\partial z}\left\{\frac{\partial}{\partial y}\left[\frac{\partial}{\partial x}(z^2 + y\cos x)\right]\right\} = \frac{\partial}{\partial z}\left\{\frac{\partial}{\partial y}[y(-\sin x)]\right\} = \frac{\partial}{\partial z}(-\sin x) = 0 = f_{zyx}$$

29. $f(x, y) = x^2 y - 9 \Rightarrow f_x = 2xy$ and $f_y = x^2$. Since both partial derivatives, being polynomial functions, are defined and continuous everywhere, $f(x, y)$ is differentiable, by Theorem 13.2.

31. $f(x, y) = e^{2x+3y} \Rightarrow$

$$f_x = e^{2x+3y} \left[\frac{\partial}{\partial x}(2x + 3y) \right] = 2e^{2x+3y}, \text{ and } f_y = e^{2x+3y} \left[\frac{\partial}{\partial y}(2x + 3y) \right] = 3e^{2x+3y}$$

Since both partial derivatives are defined and continuous everywhere, $f(x, y)$ is differentiable, by Theorem 13.2.

33. $f(x, y) = \ln(xy - y^2) \Rightarrow f_x = \frac{1}{xy - y^2} \cdot \left[\frac{\partial}{\partial x}(xy - y^2) \right] = \frac{y}{xy - y^2} = \frac{1}{x - y}$ and

$$f_y = \frac{1}{xy - y^2} \cdot \left[\frac{\partial}{\partial y}(xy - y^2) \right] = \frac{x - 2y}{xy - y^2}$$

Since both partial derivatives are defined and continuous for $xy - y^2 > 0$ (the domain of f), $f(x, y)$ is differentiable on its domain, by Theorem 13.2.

35. $f(x, y, z) = x^3 y^2 z^4 \Rightarrow$

$$f_x = 3x^2 y^2 z^4, \ f_y = x^3 (2y)z^4 = 2x^3 yz^4, \text{ and } f_z = x^3 y^2 \cdot 4z^3 = 4x^3 y^2 z^3$$

Since both partial derivatives, being polynomial functions, are defined and continuous everywhere, $f(x, y)$ is differentiable, by Theorem 13.2.

37. $\frac{\partial}{\partial x}(x^2 z + 2y^2 z^2 + y) = 0$

$2xz + x^2 \frac{\partial z}{\partial x} + 2y^2 \cdot 2z \frac{\partial z}{\partial x} = 0$

$(x^2 + 4y^2 z) \frac{\partial z}{\partial x} = -2xz$

$\frac{\partial z}{\partial x} = -\frac{2xz}{x^2 + 4y^2 z}$

$\frac{\partial}{\partial y}(x^2 z + 2y^2 z^2 + y) = 0$

$x^2 \frac{\partial z}{\partial y} + 2\left(z^2 \cdot 2y + y^2 \cdot 2z \frac{\partial z}{\partial y} \right) + 1 = 0$

$(x^2 + 4y^2 z) \frac{\partial z}{\partial y} = -4yz^2 - 1$

$\frac{\partial z}{\partial y} = -\frac{4yz^2 + 1}{x^2 + 4y^2 z}$

39. $\frac{\partial}{\partial x}(x^3 + y^3 + z^3 + 6xyz) = 0$

$3x^2 + 3z^2 \frac{\partial z}{\partial x} + 6yz + 6xy \frac{\partial z}{\partial x} = 0$

$x^2 + z^2 \frac{\partial z}{\partial x} + 2yz + 2xy \frac{\partial z}{\partial x} = 0$

$(z^2 + 2xy) \frac{\partial z}{\partial x} = -(x^2 + 2yz)$

$\frac{\partial z}{\partial x} = -\frac{x^2 + 2yz}{z^2 + 2xy}$

$\frac{\partial}{\partial y}(x^3 + y^3 + z^3 + 6xyz) = 0$

$3y^2 + 3z^2 \frac{\partial z}{\partial y} + 6xz + 6xy \frac{\partial z}{\partial y} = 0$

$y^2 + z^2 \frac{\partial z}{\partial y} + 2xz + 2xy \frac{\partial z}{\partial y} = 0$

$(z^2 + 2xy) \frac{\partial z}{\partial y} = -(y^2 + 2xz)$

$\frac{\partial z}{\partial y} = -\frac{y^2 + 2xz}{z^2 + 2xy}$

41. We seek $f(x, y)$ for which $f_x - y^2$ and $f_y = 2xy$:

$f_x = y^2 \Rightarrow f(x, y) = xy^2 + g(y) \Rightarrow f_y = x(2y) + g'(y) = 2xy + g'(y)$. Since we are given $f_y = 2xy$, then $g'(y) = 0 \Rightarrow g(y) = $ constant, take it to be 0, so $f(x, y) = xy^2$.

43. $f(x, y) = x^2 - y^2 \Rightarrow f_x = 2x$ and $f_y = -2y$. $g(x, y) = 2xy \Rightarrow g_x = 2y$ and $g_y = 2x$.
Thus, $f_x = g_y = 2x$ and $f_y = -g_x = -2y$.

45. $z = \tan^{-1}\frac{x}{y} \Rightarrow z_x = \dfrac{1}{1 + \left(\frac{x}{y}\right)^2} \cdot \dfrac{1}{y} = \dfrac{1}{y + \frac{x^2}{y}} = \dfrac{y}{x^2 + y^2}$

$$z_{xx} = \frac{\partial}{\partial x}[y(x^2 + y^2)^{-1}] = y(-1)(x^2 + y^2)^{-2}(2x) = -\frac{2xy}{(x^2 + y^2)^2}$$

$$z_y = \frac{1}{1 + \left(\frac{x}{y}\right)^2} \cdot x\left(-\frac{1}{y^2}\right) = -\frac{x}{x^2 + y^2}$$

$$z_{yy} = \frac{\partial}{\partial y}[-x(x^2 + y^2)^{-1}] = -x(-1)(x^2 + y^2)^{-2}(2y) = \frac{2xy}{(x^2 + y^2)^2}$$

Thus, $z_{xx} + z_{yy} = -\dfrac{2xy}{(x^2 + y^2)^2} + \dfrac{2xy}{(x^2 + y^2)^2} = 0.$ ✓

47. As $f(x, y) = g(x) + h(y)$, then $f_{xy} = (f_x)_y = [g'(x)]_y = 0.$

49. $z = f(x, y) = x^4 + 2x^2y^2 + y^4 \Rightarrow \dfrac{\partial z}{\partial x} = 4x^3 + 2(2x)y^2 = 4x^3 + 4xy^2$, and
$\dfrac{\partial z}{\partial y} = 2x^2(2y) + 4y^3 = 4x^2y + 4y^3$. Therefore,

$$x\frac{\partial z}{\partial x} + y\frac{\partial z}{\partial y} = x(4x^3 + 4xy^2) + y(4x^2y + 4y^3)$$
$$= 4x^4 + 4x^2y^2 + 4x^2y^2 + 4y^4$$
$$= 4x^4 + 8x^2y^2 + 4y^4 = 4(x^4 + 2x^2y^2 + y^4) = 4z$$

51. $w = f(x, y, z) = x^2y + y^2z + z^2x \Rightarrow \dfrac{\partial w}{\partial x} = 2xy + z^2$, $\dfrac{\partial w}{\partial y} = x^2 + 2yz$, and $\dfrac{\partial w}{\partial z} = y^2 + 2zx$.
Therefore,

$$\frac{\partial w}{\partial x} + \frac{\partial w}{\partial y} + \frac{\partial w}{\partial z} = 2xy + z^2 + x^2 + 2yz + y^2 + 2zx$$
$$= x^2 + y^2 + z^2 + 2xy + 2yz + 2xz \quad \text{while,}$$
$$(x + y + z)^2 = (x + y + z)(x + y + z) = x(x + y + z) + y(x + y + z) + z(x + y + z)$$
$$= x^2 + xy + xz + xy + y^2 + yz + xz + yz + z^2$$
$$= x^2 + y^2 + z^2 + 2xy + 2yz + 2xz$$

which shows that $\dfrac{\partial w}{\partial x} + \dfrac{\partial w}{\partial y} + \dfrac{\partial w}{\partial z} = (x + y + z)^2.$

53. $z = \cos(x + y) + \cos(x - y) \Rightarrow$

$$z_x = -\sin(x + y) \cdot 1 - \sin(x - y) \cdot 1 = -\sin(x + y) - \sin(x - y)$$

$$z_{xx} = (z_x)_x = -\cos(x + y) \cdot 1 - \cos(x - y) \cdot 1 = -\cos(x + y) - \cos(x - y)$$

$$z_y = -\sin(x + y) \cdot 1 - \sin(x - y) \cdot (-1) = -\sin(x + y) + \sin(x - y)$$

$$z_{yy} = (z_y)_y = -\cos(x + y) \cdot 1 + \cos(x - y) \cdot (-1) = -\cos(x + y) - \cos(x - y)$$

which shows that $z_{xx} - z_{yy} = -\cos(x + y) - \cos(x - y) - [-\cos(x + y) - \cos(x - y)] = 0.$

55. Since $PV = kT$, considering $P(V,T) = \frac{kT}{V}$ and differentiating,

$$\frac{\partial P}{\partial V} = \frac{\partial}{\partial V}(kTV^{-1}) = kT(-V^{-2}) = -\frac{kT}{V^2} \Rightarrow V\frac{\partial P}{\partial V} = -\frac{kT}{V} = -P \quad \checkmark$$

$$\text{As } \frac{\partial P}{\partial T} = \frac{k}{V}, \quad V\frac{\partial P}{\partial V} + T\frac{\partial P}{\partial T} = -\frac{kT}{V} + T\left(\frac{k}{V}\right) = 0 \quad \checkmark$$

Considering $V(T,P) = \frac{kT}{P}$, $\frac{\partial V}{\partial T} = \frac{k}{P}$. Considering $T(V,P) = \frac{PV}{k}$, $\frac{\partial T}{\partial P} = \frac{V}{k}$. Therefore,

$$\left(\frac{\partial V}{\partial T}\right)\left(\frac{\partial T}{\partial P}\right)\left(\frac{\partial P}{\partial V}\right) = \left(\frac{k}{P}\right)\left(\frac{V}{k}\right)\left(-\frac{kT}{V^2}\right) = -\frac{kT}{PV} = -\frac{kT}{kT} = -1 \quad \checkmark$$

57. $R = \frac{R_1 R_2}{R_1 + R_2} \Rightarrow \frac{\partial R}{\partial R_1} = \frac{(R_1 + R_2)R_2 - (R_1 R_2)\cdot 1}{(R_1 + R_2)^2} = \frac{R_2{}^2}{(R_1 + R_2)^2}$

$$\Rightarrow \frac{\partial^2 R}{\partial R_1{}^2} = \frac{\partial R}{\partial R_1}[R_2{}^2(R_1 + R_2)^{-2}] = R_2{}^2 \cdot [-2(R_1 + R_2)^{-3}\cdot 1] = -\frac{2R_2{}^2}{(R_1 + R_2)^3}$$

$$\Rightarrow \frac{\partial R}{\partial R_2} = \frac{(R_1 + R_2)R_1 - (R_1 R_2)\cdot 1}{(R_1 + R_2)^2} = \frac{R_1{}^2}{(R_1 + R_2)^2}$$

$$\Rightarrow \frac{\partial^2 R}{\partial R_2{}^2} = \frac{\partial R}{\partial R_2}[R_1{}^2(R_1 + R_2)^{-2}] = R_1{}^2 \cdot [-2(R_1 + R_2)^{-3}\cdot 1] = -\frac{2R_1{}^2}{(R_1 + R_2)^3}$$

Thus,

$$\frac{\partial^2 R}{\partial R_1{}^2} \cdot \frac{\partial^2 R}{\partial R_2{}^2} = \left[-\frac{2R_2{}^2}{(R_1 + R_2)^3}\right]\cdot\left[-\frac{2R_1{}^2}{(R_1 + R_2)^3}\right]$$

$$= \frac{4R_1{}^2 R_2{}^2}{(R_1 + R_2)^6} = 4\left[\frac{R_1 R_2}{R_1 + R_2}\right]^2 \cdot \frac{1}{(R_1 + R_2)^4} = \frac{4R^2}{(R_1 + R_2)^4}$$

59. At $(2,1,2)$, $z = \sqrt{9 - x^2 - y^2}$. The slope of the tangent line is $\frac{\partial z}{\partial y}\Big|_{(2,1)}$:

$$\frac{\partial z}{\partial y} = \frac{1}{2}(9 - x^2 - y^2)^{-\frac{1}{2}}(-2y) = -\frac{y}{\sqrt{9 - x^2 - y^2}} \Rightarrow \frac{\partial z}{\partial y}\Big|_{(2,1)} = -\frac{1}{\sqrt{9 - 4 - 1}} = -\frac{1}{2}.$$

61. $z = xy$, $x = \cos t$, $y = \sin t$

By the Chain Rule: $\frac{dz}{dt} = \frac{\partial z}{\partial x}\frac{dx}{dt} + \frac{\partial z}{\partial y}\frac{dy}{dt} = y(-\sin t) + x\cos t = -\sin^2 t + \cos^2 t = \cos 2t$

Without CR: $z = \cos t \sin t \Rightarrow \frac{dz}{dt} = (-\sin t)\sin t + \cos t(\cos t) = -\sin^2 t + \cos^2 t = \cos 2t$

63. $w = xy + z$, $x = \cos t$, $y = \sin t$, $z = t$

By the Chain Rule: $\frac{dw}{dt} = \frac{\partial w}{\partial x}\frac{dx}{dt} + \frac{\partial w}{\partial y}\frac{dy}{dt} + \frac{\partial w}{\partial z}\frac{dz}{dt}$

$$= y(-\sin t) + x(\cos t) + 1\cdot 1 = -y\sin t + x\cos t + 1$$

$$= -\sin^2 t + \cos^2 + 1 = 1 + \cos 2t$$

Without CR: $w = \cos t \sin t + t \Rightarrow \frac{dz}{dt} = (-\sin t)\sin t + \cos t(\cos t) + 1$

$$= -\sin^2 t + \cos^2 t + 1 = 1 + \cos 2t$$

65. $z = x \sin y, \; x = s + 3t, \; y = t^2$

By the Chain Rule: $\dfrac{\partial z}{\partial t} = \dfrac{\partial z}{\partial x}\dfrac{\partial x}{\partial t} + \dfrac{\partial z}{\partial y}\dfrac{dy}{dt}$

$$= (\sin y) \cdot 3 + (x \cos y) \cdot 2t = 3\sin t^2 + 2t(s + 3t)\cos t^2$$

$$\dfrac{\partial z}{\partial s} = \dfrac{\partial z}{\partial x}\dfrac{\partial x}{\partial s} + \dfrac{\partial z}{\partial y}\dfrac{\partial y}{\partial s} = \dfrac{\partial z}{\partial x}\dfrac{\partial x}{\partial s}, \;\; \text{as } \dfrac{\partial y}{\partial s} = 0$$

$$= (\sin y) \cdot 1 = \sin y = \sin t^2$$

Without CR: $z = (s + 3t)\sin t^2 \;\Rightarrow\; \dfrac{\partial z}{\partial t} = (s + 3t)[(\cos t^2) \cdot 2t] + (\sin t^2) \cdot 3$

$$= 3\sin t^2 + 2t(s + 3t)\cos t^2$$

$$\dfrac{\partial z}{\partial s} = \sin t^2$$

67. $w = x + 2y - 3z, \; x = st, y = e^{st}, z = \sin t$

By the Chain Rule: $\dfrac{\partial w}{\partial t} = \dfrac{\partial w}{\partial x}\dfrac{\partial x}{\partial t} + \dfrac{\partial w}{\partial y}\dfrac{\partial y}{\partial t} + \dfrac{\partial w}{\partial z}\dfrac{dz}{dt}$

$$= 1(s) + 2(e^{st} \cdot s) - 3\cos t = s + 2se^{st} - 3\cos t$$

$$\dfrac{\partial w}{\partial s} = \dfrac{\partial w}{\partial x}\dfrac{\partial x}{\partial s} + \dfrac{\partial w}{\partial y}\dfrac{\partial y}{\partial s} + \dfrac{\partial w}{\partial z}\dfrac{\partial z}{\partial s} = \dfrac{\partial w}{\partial x}\dfrac{\partial x}{\partial s} + \dfrac{\partial w}{\partial y}\dfrac{\partial y}{\partial s}, \;\; \text{as } \dfrac{\partial z}{\partial s} = 0$$

$$= 1 \cdot t + 2(e^{st} \cdot t) = t + 2te^{st}$$

Without CR: $w = st + 2e^{st} - 3\sin t \;\Rightarrow\; \dfrac{\partial w}{\partial t} = s + 2(e^{st} \cdot s) - 3\cos t = s + 2se^{st} - 3\cos t$

$$\dfrac{\partial w}{\partial s} = t + 2(e^{st} \cdot t) = t + 2te^{st}$$

§13.2 Directional Derivatives, Gradient Vectors, and Tangent Planes

1. $\mathbf{u} = \cos 45°\, \mathbf{i} + \sin 45°\, \mathbf{j} = \dfrac{1}{\sqrt{2}}\mathbf{i} + \dfrac{1}{\sqrt{2}}\mathbf{j} \;\Rightarrow\; a = b = \dfrac{1}{\sqrt{2}}, \text{ and } f(x, y) = x^2 + xy \;\Rightarrow$

$$D_{\mathbf{u}}f(x, y) = \lim_{h \to 0} \frac{f(x + ha, y + hb) - f(x, y)}{h}$$

$$= \lim_{h \to 0} \frac{\left(x + \frac{h}{\sqrt{2}}\right)^2 + \left(x + \frac{h}{\sqrt{2}}\right)\left(y + \frac{h}{\sqrt{2}}\right) - (x^2 + xy)}{h}$$

$$= \lim_{h \to 0} \frac{\cancel{x^2} + \frac{2hx}{\sqrt{2}} + \frac{h^2}{2} + \cancel{xy} + \frac{xh}{\sqrt{2}} + \frac{hy}{\sqrt{2}} + \frac{h^2}{2} - \cancel{x^2} - \cancel{xy}}{h}$$

$$= \lim_{h \to 0} \frac{\cancel{h}\left[\frac{3}{\sqrt{2}}x + h + \frac{1}{\sqrt{2}}y\right]}{\cancel{h}} = \frac{3x + y}{\sqrt{2}}$$

3. As $f(x,y) = xy^2 + x^2$ is differentiable, $D_u f(x,y) = af_x(x,y) + bf_y(x,y)$ for $\mathbf{u} = a\mathbf{i} + b\mathbf{j}$, by Theorem 13.5. Here $\mathbf{u} = \cos\frac{\pi}{3}\mathbf{i} + \sin\frac{\pi}{3}\mathbf{j} = \frac{1}{2}\mathbf{i} + \frac{\sqrt{3}}{2}\mathbf{j} \Rightarrow a = \frac{1}{2}$ and $b = \frac{\sqrt{3}}{2}$. Since $f_x(x,y) = y^2 + 2x$, and $f_y(x,y) = 2xy$,

$$D_u f(x,y) = \frac{1}{2}(y^2 + 2x) + \frac{\sqrt{3}}{2}(2xy) = \frac{1}{2}y^2 + x + \sqrt{3}xy, \text{ so that}$$

$$D_u f(-1,2) = \frac{1}{2}(2^2) - 1 + \sqrt{3}(-2) = 1 - 2\sqrt{3}.$$

5. As $f(x,y) = xe^y + ye^x$ is differentiable, $D_u f(x,y) = af_x(x,y) + bf_y(x,y)$ for $\mathbf{u} = a\mathbf{i} + b\mathbf{j}$, by Theorem 13.5. Here $\mathbf{u} = \cos\frac{\pi}{6}\mathbf{i} + \sin\frac{\pi}{6}\mathbf{j} = \frac{\sqrt{3}}{2}\mathbf{i} + \frac{1}{2}\mathbf{j} \Rightarrow a = \frac{\sqrt{3}}{2}$ and $b = \frac{1}{2}$. Since $f_x(x,y) = e^y + ye^x$ and $f_y(x,y) = xe^y + e^x$,

$$D_u f(x,y) = \frac{\sqrt{3}}{2}(e^y + ye^x) + \frac{1}{2}(xe^y + e^x), \text{ so that}$$

$$D_u f(0,0) = \frac{\sqrt{3}}{2}(1+0) + \frac{1}{2}(0+1) = \frac{\sqrt{3}}{2} + \frac{1}{2} = \frac{\sqrt{3}+1}{2}.$$

7. As $f(x,y) = xe^y - ye^x$ is differentiable, $D_u f(x,y) = af_x(x,y) + bf_y(x,y)$ for $\mathbf{u} = a\mathbf{i} + b\mathbf{j}$, by Theorem 13.5. Here $\mathbf{u} = \frac{3\mathbf{i} + 4\mathbf{j}}{\sqrt{3^2 + 4^2}} = \frac{3}{5}\mathbf{i} + \frac{4}{5}\mathbf{j} \Rightarrow a = \frac{3}{5}$ and $b = \frac{4}{5}$. Since $f_x(x,y) = e^y - ye^x$ and $f_y(x,y) = xe^y - e^x$,

$$D_u f(x,y) = \frac{3}{5}(e^y - ye^x) + \frac{4}{5}(xe^y - e^x), \text{ so that}$$

$$D_u f(0,0) = \frac{3}{5}(1-0) + \frac{4}{5}(0-1) = \frac{3}{5} - \frac{4}{5} = -\frac{1}{5}.$$

9. As $f(x,y) = y^2 \ln x$ is differentiable, $D_u f(x,y) = af_x(x,y) + bf_y(x,y)$ for $\mathbf{u} = a\mathbf{i} + b\mathbf{j}$, by Theorem 13.5. Here $\mathbf{u} = \frac{-3\mathbf{i} + 3\mathbf{j}}{\sqrt{3^2 + 3^2}} = \frac{-3\mathbf{i} + 3\mathbf{j}}{3\sqrt{2}} = -\frac{1}{\sqrt{2}}\mathbf{i} + \frac{1}{\sqrt{2}}\mathbf{j} \Rightarrow a = -\frac{1}{\sqrt{2}}$ and $b = \frac{1}{\sqrt{2}}$. Since $f_x(x,y) = \frac{y^2}{x}$ and $f_y(x,y) = 2y\ln x$,

$$D_u f(x,y) = -\frac{1}{\sqrt{2}}\left(\frac{y^2}{x}\right) + \frac{2}{\sqrt{2}}y\ln x, \text{ so that}$$

$$D_u f(1,4) = -\frac{1}{\sqrt{2}}\left(\frac{4^2}{1}\right) + \frac{2}{\sqrt{2}}(4)\ln 1 = -\frac{16}{\sqrt{2}} \cdot \frac{\sqrt{2}}{\sqrt{2}} = -8\sqrt{2}.$$

11. As $f(x,y) = (2x + 3y)^2$ is differentiable, $D_u f(x,y) = af_x(x,y) + bf_y(x,y)$ for $\mathbf{u} = a\mathbf{i} + b\mathbf{j}$, by Theorem 13.5. Here $\mathbf{u} = \mathbf{j} \Rightarrow a = 0$ and $b = 1$, so

$$D_u f(x,y) = f_y(x,y) = 2(2x + 3y) \cdot 3 = 12x + 18y \text{ and } D_u f(2,2) = 12(2) + 18(2) = 60.$$

13. As $f(x,y) = \frac{1}{x^2 + y^2}$ is differentiable near $(2,2)$, $D_u f(2,2) = af_x(2,2) + bf_y(2,2)$ for $\mathbf{u} = a\mathbf{i} + b\mathbf{j}$, by Theorem 13.5. Here $\mathbf{u} = \mathbf{j} \Rightarrow a = 0$ and $b = 1$, so

$$D_u f(x,y) = f_y(x,y) = \frac{\partial}{\partial y}(x^2 + y^2)^{-1} = -(x^2 + y^2)^{-2}(2y) \text{ and}$$

$$D_u f(2,2) = -\frac{4}{(2^2 + 2^2)^2} = -\frac{4}{64} = -\frac{1}{16}.$$

15.   As $f(x,y) = x^3 - x^2y + y^2$ is differentiable, $D_u f(x,y) = af_x(x,y) + bf_y(x,y)$ for

$\mathbf{u} = a\mathbf{i} + b\mathbf{j}$, by Theorem 13.5. Here $\mathbf{u} = \dfrac{(4-1)\mathbf{i} + [3-(-1)]\mathbf{j}}{\|(4-1)\mathbf{i} + [3-(-1)]\mathbf{j}\|} = \dfrac{3\mathbf{i} + 4\mathbf{j}}{\sqrt{3^2 + 4^2}} = \frac{3}{5}\mathbf{i} + \frac{4}{5}\mathbf{j} \Rightarrow$

$a = \frac{3}{5}$ and $b = \frac{4}{5}$. Since $f_x(x,y) = 3x^2 - 2xy$ and $f_y(x,y) = -x^2 + 2y$,

$D_u f(x,y) = \frac{3}{5}(3x^2 - 2xy) + \frac{4}{5}(-x^2 + 2y) = \frac{1}{5}(9x^2 - 6xy - 4x^2 + 8y) = x^2 - \frac{6}{5}xy + \frac{8}{5}y$ and

$D_u f(1,-1) = 1 + \frac{6}{5} - \frac{8}{5} = \frac{3}{5}$.

17.   As $f(x,y) = \sin^3(x+y)$ is differentiable, $D_u f(x,y) = af_x(x,y) + bf_y(x,y)$ for

$\mathbf{u} = a\mathbf{i} + b\mathbf{j}$, by Theorem 13.5. Here $\mathbf{u} = -\mathbf{i} \Rightarrow a = -1$ and $b = 0$, so

$D_u f(x,y) = -f_x(x,y) = -3\sin^2(x+y)\cos(x+y)$ and

$D_u f\left(\frac{\pi}{3}, -\frac{\pi}{6}\right) = -3\sin^2\left(\frac{\pi}{3} - \frac{\pi}{6}\right)\cos(\frac{\pi}{3} - \frac{\pi}{6}) = -3\sin^2\left(\frac{\pi}{6}\right)\cos\left(\frac{\pi}{6}\right) = -3\left(\frac{1}{2}\right)^2\left(\frac{\sqrt{3}}{2}\right) = -\frac{3\sqrt{3}}{8}$

19.   As $f(x,y) = \tan^{-1}\left(\dfrac{y}{x}\right)$ is differentiable near $(-2,2)$,

$D_u f(-2,2) = af_x(-2,2) + bf_y(-2,2)$ for $\mathbf{u} = a\mathbf{i} + b\mathbf{j}$, by Theorem 13.5.

Here $\mathbf{u} = \dfrac{-\mathbf{i} - \mathbf{j}}{\sqrt{1+1}} = -\frac{1}{\sqrt{2}}\mathbf{i} - \frac{1}{\sqrt{2}}\mathbf{j} \Rightarrow a = b = -\frac{1}{\sqrt{2}}$. Since

$$f_x(x,y) = \dfrac{1}{1 + \left(\dfrac{y}{x}\right)^2} \cdot \dfrac{\partial}{\partial x}(yx^{-1}) = \dfrac{x^2}{x^2 + y^2} \cdot -yx^{-2} = -\dfrac{y}{x^2 + y^2} \quad \text{and}$$

$$f_y(x,y) = \dfrac{1}{1 + \left(\dfrac{y}{x}\right)^2} \cdot \dfrac{\partial}{\partial y}\left(\dfrac{y}{x}\right) = \dfrac{x^2}{x^2 + y^2} \cdot \dfrac{1}{x} = \dfrac{x}{x^2 + y^2}, \quad \text{then}$$

$$D_u f(x,y) = \dfrac{1}{\sqrt{2}}\left[\dfrac{y}{x^2 + y^2} - \dfrac{x}{x^2 + y^2}\right] = \dfrac{1}{\sqrt{2}}\left[\dfrac{y-x}{x^2 + y^2}\right] \Rightarrow D_u f(-2,2) = \dfrac{1}{\sqrt{2}}\left[\dfrac{2 - (-2)}{4+4}\right] = \dfrac{1}{2\sqrt{2}}.$$

21.   As $f(x,y,z) = xy + yz + zx$ is differentiable,

$D_u f(x,y,z) = af_x(x,y,z) + bf_y(x,y,z) + cf_z(x,y,z)$ for $\mathbf{u} = a\mathbf{i} + b\mathbf{j} + c\mathbf{k}$, by

Theorem 13.5. Here $\mathbf{u} = \dfrac{3\mathbf{i} + 6\mathbf{j} - 2\mathbf{k}}{\sqrt{3^2 + 6^2 + 2^2}} = \frac{3}{7}\mathbf{i} + \frac{6}{7}\mathbf{j} - \frac{2}{7}\mathbf{k} \Rightarrow a = \frac{3}{7}, b = \frac{6}{7}$ and $c = -\frac{2}{7}$. Since

$f_x(x,y,z) = y + z, \quad f_y(x,y,z) = x + z, \quad \text{and} \quad f_z(x,y,z) = y + x,$

$D_u f(x,y,z) = \frac{3}{7}(y+z) + \frac{6}{7}(x+z) - \frac{2}{7}(y+x) = \frac{3}{7}y + \frac{3}{7}z + \frac{6}{7}x + \frac{6}{7}z - \frac{2}{7}y - \frac{2}{7}x = \frac{4}{7}x + \frac{1}{7}y + \frac{9}{7}z \Rightarrow$

$D_u f(1,-1,2) = \dfrac{4 - 1 + 18}{7} = \dfrac{21}{7} = 3.$

23.   As $f(x,y,z) = 3e^x \cos yz$ is differentiable,

$D_u f(x,y,z) = af_x(x,y,z) + bf_y(x,y,z) + cf_z(x,y,z)$ for $\mathbf{u} = a\mathbf{i} + b\mathbf{j} + c\mathbf{k}$, by

Theorem 13.5. Here $\mathbf{u} = \dfrac{\mathbf{i} + 2\mathbf{j} + 2\mathbf{k}}{\sqrt{1^2 + 2^2 + 2^2}} = \frac{1}{3}\mathbf{i} + \frac{2}{3}\mathbf{j} + \frac{2}{3}\mathbf{k} \Rightarrow a = \frac{1}{3}$, and $b = c = \frac{2}{3}$. Since

$$f_x(x, y, z) = 3e^x \cos yz, \quad f_y(x, y, z) = 3e^x(-\sin yz)z, \quad \text{and} \quad f_z(x, y, z) = 3e^x(-\sin yz)y,$$

$$D_u f(x, y, z) = e^x \cos yz - 2ze^x \sin yz - 2ye^x \sin yz \Rightarrow D_u f\left(1, 0, \tfrac{1}{2}\right) = e\left[1 - 2\left(\tfrac{1}{2}\right)\sin 0 - 0\right] = e$$

25. As $f(x, y, z) = (x + y^2 + z^3)^2$ is differentiable,

$$D_u f(x, y, z) = af_x(x, y, z) + bf_y(x, y, z) + cf_z(x, y, z) \quad \text{for} \quad \mathbf{u} = a\mathbf{i} + b\mathbf{j} + c\mathbf{k}, \text{ by}$$

Theorem 13.5. Here $\mathbf{u} = \dfrac{\mathbf{i} + \mathbf{j}}{\sqrt{1^2 + 1^2}} = \dfrac{1}{\sqrt{2}}\mathbf{i} + \dfrac{1}{\sqrt{2}}\mathbf{j} \Rightarrow a = b = \dfrac{1}{\sqrt{2}}$, and $c = 0$. Since

$$f_x(x, y, z) = 2(x + y^2 + z^3), \quad \text{and} \quad f_y(x, y, z) = 2(x + y^2 + z^3)(2y),$$

$$D_u f(x, y, z) = \frac{2}{\sqrt{2}}(x + y^2 + z^3) + \frac{4y}{\sqrt{2}}(x + y^2 + z^3) = \frac{2 + 4y}{\sqrt{2}}(x + y^2 + z^3) \quad \text{and}$$

$$D_u f(1, 2, 1) = \frac{2 + 8}{\sqrt{2}}(1 + 4 + 1) = \frac{60}{\sqrt{2}} \cdot \frac{\sqrt{2}}{\sqrt{2}} = 30\sqrt{2}.$$

27.   By Theorem 13.6, the maximum value is $\left\| \nabla f\left(2, \tfrac{1}{2}\right) \right\|$ and the minimum value is $-\left\| \nabla f\left(2, \tfrac{1}{2}\right) \right\|$, where $f(x, y) = x^3 y^2 - xy$ and $\nabla f = f_x \mathbf{i} + f_y \mathbf{j}$:

$$f_x(x, y) = 3x^2 y^2 - y, \quad \text{and} \quad f_y(x, y) = 2x^3 y - x \Rightarrow \nabla f(x, y) = (3x^2 y^2 - y)\mathbf{i} + (2x^3 y - x)\mathbf{j} \Rightarrow$$

$$\nabla f\left(2, \tfrac{1}{2}\right) = \left(3 \cdot 4 \cdot \tfrac{1}{4} - \tfrac{1}{2}\right)\mathbf{i} + \left(2 \cdot 2^3 \cdot \tfrac{1}{2} - 2\right)\mathbf{j} = \tfrac{5}{2}\mathbf{i} + 6\mathbf{j} \Rightarrow$$

$$\left\| \nabla f\left(2, \tfrac{1}{2}\right) \right\| = \sqrt{\left(\tfrac{5}{2}\right)^2 + 6^2} = \sqrt{\tfrac{25}{4} + 36} = \tfrac{1}{2}\sqrt{169} = \tfrac{13}{2}$$

The maximum value is $\frac{13}{2}$, and the minimum value is $-\frac{13}{2}$.

29.   By Theorem 13.6, the maximum value is $\|\nabla f(0, 0)\|$ and the minimum value is $-\|\nabla f(0, 0)\|$, where $f(x, y) = xe^y + ye^x$ and $\nabla f = f_x \mathbf{i} + f_y \mathbf{j}$:

$$f_x(x, y) = e^y + ye^x, \quad \text{and} \quad f_y(x, y) = xe^y + e^x \Rightarrow \nabla f(x, y) = (e^y + ye^x)\mathbf{i} + (xe^y + e^x)\mathbf{j} \Rightarrow$$

$$\nabla f(0, 0) = (1 + 0)\mathbf{i} + (0 + 1)\mathbf{j} = \mathbf{i} + \mathbf{j} \Rightarrow \|\nabla f(0, 0)\| = \sqrt{1^2 + 1^2} = \sqrt{2}$$

The maximum value is $\sqrt{2}$, and the minimum value is $-\sqrt{2}$.

31.   By Theorem 13.6, the maximum value is $\left\| \nabla f\left(\tfrac{\pi}{6}, \tfrac{\pi}{4}\right) \right\|$ and the minimum value is $-\left\| \nabla f\left(\tfrac{\pi}{6}, \tfrac{\pi}{4}\right) \right\|$, where $f(x, y) = \cos(3x - y)$ and $\nabla f = f_x \mathbf{i} + f_y \mathbf{j}$:

$$f_x(x, y) = -\sin(3x - y) \cdot 3 = -3\sin(3x - y), \quad \text{and}$$

$$f_y(x, y) = -\sin(3x - y)(-1) = \sin(3x - y) \Rightarrow \nabla f(x, y) = -3\sin(3x - y)\mathbf{i} + \sin(3x - y)\mathbf{j} \Rightarrow$$

$$\nabla f\left(\tfrac{\pi}{6}, \tfrac{\pi}{4}\right) = -3\sin\left(\tfrac{3\pi}{6} - \tfrac{\pi}{4}\right)\mathbf{i} + \sin\left(\tfrac{3\pi}{6} - \tfrac{\pi}{4}\right)\mathbf{j} = -\left(3\sin\tfrac{\pi}{4}\right)\mathbf{i} + \left(\sin\tfrac{\pi}{4}\right)\mathbf{j} = -3 \cdot \tfrac{1}{\sqrt{2}}\mathbf{i} + \tfrac{1}{\sqrt{2}}\mathbf{j} \Rightarrow$$

$$\left\| \nabla f\left(\tfrac{\pi}{6}, \tfrac{\pi}{4}\right) \right\| = \sqrt{\left(-\tfrac{3}{\sqrt{2}}\right)^2 + \left(\tfrac{1}{\sqrt{2}}\right)^2} = \sqrt{5}.$$

The maximum value is $\sqrt{5}$, and the minimum value is $-\sqrt{5}$.

33.   By Theorem 13.6, the maximum value is $\left\|\nabla f\left(\frac{\pi}{2}, \frac{\pi}{4}\right)\right\|$ and the minimum value is $-\left\|\nabla f\left(\frac{\pi}{2}, \frac{\pi}{4}\right)\right\|$, where $f(x, y) = \dfrac{\sin(x+y)}{\cos(x-y)}$ and $\nabla f = f_x \mathbf{i} + f_y \mathbf{j}$:

$$f_x(x, y) = \frac{\cos(x-y)\cos(x+y) - \sin(x+y)[-\sin(x-y)]}{\cos^2(x-y)}$$

$$= \frac{\cos(x+y)\cos(x-y) + \sin(x+y)\sin(x-y)}{\cos^2(x-y)}$$

$$= \frac{\cos[(x+y) - (x-y)]}{\cos^2(x-y)} = \frac{\cos 2y}{\cos^2(x-y)}$$

$$f_y(x, y) = \frac{\cos(x-y)\cos(x+y) - \sin(x+y)[-\sin(x-y)(-1)]}{\cos^2(x-y)}$$

$$= \frac{\cos(x+y)\cos(x-y) - \sin(x+y)\sin(x-y)}{\cos^2(x-y)}$$

$$= \frac{\cos[(x+y) + (x-y)]}{\cos^2(x-y)} = \frac{\cos 2x}{\cos^2(x-y)} \Rightarrow$$

$$\nabla f(x, y) = \frac{\cos 2y}{\cos^2(x-y)}\mathbf{i} + \frac{\cos 2x}{\cos^2(x-y)}\mathbf{j} \Rightarrow$$

$$\nabla f\left(\frac{\pi}{2}, \frac{\pi}{4}\right) = \frac{\cos\frac{\pi}{2}}{\cos^2\left(\frac{\pi}{2} - \frac{\pi}{4}\right)}\mathbf{i} + \frac{\cos\pi}{\cos^2\left(\frac{\pi}{2} - \frac{\pi}{4}\right)}\mathbf{j} = -\frac{1}{\left(\frac{1}{\sqrt{2}}\right)^2}\mathbf{j} = -2\mathbf{j} \Rightarrow \left\|\nabla f\left(\frac{\pi}{2}, \frac{\pi}{4}\right)\right\| = 2.$$

The maximum value is 2, and the minimum value is -2.

35.   By Theorem 13.6, the maximum value is $\|\nabla f(4, 1, 1)\|$ and the minimum value is $-\|\nabla f(4, 1, 1)\|$, where $f(x, y, z) = \frac{x}{y} - yz$ and $\nabla f = f_x \mathbf{i} + f_y \mathbf{j} + f_z \mathbf{k}$:

$$f_x(x, y, z) = \frac{1}{y}, \ \ f_y(x, y, z) = \frac{\partial}{\partial y}(xy^{-1}) - z = -xy^{-2} - z = -\left(\frac{x}{y^2} + z\right), \ f_z(x, y, z) = -y \Rightarrow$$

$$\nabla f(x, y, z) = \frac{1}{y}\mathbf{i} - \left(\frac{x}{y^2} + z\right)\mathbf{j} - y\mathbf{k} \Rightarrow \nabla f(4, 1, 1) = \mathbf{i} - (4+1)\mathbf{j} - \mathbf{k} = \mathbf{i} - 5\mathbf{j} - \mathbf{k} \Rightarrow$$

$$\|\nabla f(4, 1, 1)\| = \sqrt{1 + 25 + 1} = \sqrt{27} = 3\sqrt{3}.$$

The maximum value is $3\sqrt{3}$, and the minimum value is $-3\sqrt{3}$.

37.   By Theorem 13.6, the maximum value is $\|\nabla f(3, 6, -2)\|$ and the minimum value is $-\|\nabla f(3, 6, -2)\|$, where $f(x, y, z) = \sqrt{x^2 + y^2 + z^2}$ and $\nabla f = f_x \mathbf{i} + f_y \mathbf{j} + f_z \mathbf{k}$:

$$f_x(x, y, z) = \frac{1}{2}(x^2 + y^2 + z^2)^{-\frac{1}{2}}(2x) = \frac{x}{\sqrt{x^2 + y^2 + z^2}}$$

$$f_y(x, y, z) = \frac{1}{2}(x^2 + y^2 + z^2)^{-\frac{1}{2}}(2y) = \frac{y}{\sqrt{x^2 + y^2 + z^2}}$$

$$f_z(x, y, z) = \tfrac{1}{2}(x^2 + y^2 + z^2)^{-\frac{1}{2}}(2z) = \frac{z}{\sqrt{x^2 + y^2 + z^2}} \;\Rightarrow$$

$$\nabla f(x, y, z) = \frac{x}{\sqrt{x^2 + y^2 + z^2}}\,\mathbf{i} + \frac{y}{\sqrt{x^2 + y^2 + z^2}}\,\mathbf{j} + \frac{z}{\sqrt{x^2 + y^2 + z^2}}\,\mathbf{k} \;\Rightarrow$$

$$\nabla f(3, 6, -2) = \frac{3}{\sqrt{3^2 + 6^2 + 2^2}}\,\mathbf{i} + \frac{6}{\sqrt{3^2 + 6^2 + 2^2}}\,\mathbf{j} - \frac{2}{\sqrt{3^2 + 6^2 + 2^2}}\,\mathbf{k} = \tfrac{3}{7}\mathbf{i} + \tfrac{6}{7}\mathbf{j} - \tfrac{2}{7}\mathbf{k} \;\Rightarrow$$

$$\|\nabla f(3, 6, -2)\| = \sqrt{\tfrac{9}{49} + \tfrac{36}{49} + \tfrac{4}{49}} = 1.$$

The maximum value is 1, and the minimum value is -1.

39.   By Theorem 13.6, the maximum value is $\|\nabla f(1, 1, 1)\|$ and the minimum value is $-\|\nabla f(1, 1, 1)\|$, where $f(x, y, z) = \ln xy + \ln yz + \ln xz$ and $\nabla f = f_x\mathbf{i} + f_y\mathbf{j} + f_z\mathbf{k}$:

$$f_x(x, y, z) = \tfrac{1}{xy} \cdot y + \tfrac{1}{xz} \cdot z = \tfrac{2}{x}, \quad f_y(x, y, z) = \tfrac{1}{xy} \cdot x + \tfrac{1}{yz} \cdot z = \tfrac{2}{y}, \text{ and}$$

$$f_z(x, y, z) = \tfrac{1}{yz} \cdot y + \tfrac{1}{xz} \cdot x = \tfrac{2}{z} \;\Rightarrow$$

$$\nabla f(x, y, z) = \tfrac{2}{x}\mathbf{i} + \tfrac{2}{y}\mathbf{j} + \tfrac{2}{z}\mathbf{k} \;\Rightarrow\; \nabla f(1, 1, 1) = 2\mathbf{i} + 2\mathbf{j} + 2\mathbf{k} \;\Rightarrow$$

$$\|\nabla f(1, 1, 1)\| = \sqrt{4 + 4 + 4} = \sqrt{12} = 2\sqrt{3}.$$

The maximum value is $2\sqrt{3}$, and the minimum value is $-2\sqrt{3}$.

41.   (a) By Definition 13.5, a normal to the tangent plane is $\mathbf{n} = f_x\mathbf{i} + f_y\mathbf{j} - \mathbf{k}$. Here $z = f(x, y) = xy - y^3 + x^2 \Rightarrow f_x(x, y) = y + 2x$, and $f_y(x, y) = x - 3y^2 \Rightarrow f_x(1, 4) = 4 + 2 = 6$, and $f_y(1, 4) = 1 - 48 = -47 \Rightarrow \mathbf{n} = 6\mathbf{i} - 47\mathbf{j} - \mathbf{k}$. The equation of the tangent plane at $(1, 4, -59)$ is therefore

$$6(x - 1) - 47(y - 4) - (z + 59) = 0 \;\Rightarrow\; 6x - 47y - z = 6 - 188 + 59 = -123.$$

   (b) Let $F(x, y, z) = xy - y^3 + x^2 - z$. Then, by Theorem 13.7, $\nabla F(1, 4, -59)$ is a normal to the tangent plane at $(1, 4, -59)$. $\nabla F = F_x\mathbf{i} + F_y\mathbf{j} + F_z\mathbf{k} = (y + 2x)\mathbf{i} + (x - 3y^2)\mathbf{j} - \mathbf{k} \Rightarrow \nabla F(1, 4, -59) = 6\mathbf{i} - 47\mathbf{j} - \mathbf{k}$.
As this is the same normal as in (a), the equation of the tangent plane at $(1, 4, -59)$ is $6x - 47y - z = -123$.

43.   (a) By Definition 13.5, a normal to the tangent plane is $\mathbf{n} = f_x\mathbf{i} + f_y\mathbf{j} - \mathbf{k}$. Here $z = f(x, y) = 4x^2 - y^2 + 2y \Rightarrow f_x(x, y) - 8x$, and $f_y(x, y) = -2y + 2 \Rightarrow f_x(-1, 2) = -8$, and $f_y(-1, 2) = -4 + 2 = -2 \Rightarrow \mathbf{n} = -8\mathbf{i} - 2\mathbf{j} - \mathbf{k}$.
The equation of the tangent plane at $(-1, 2, 4)$ is therefore

$$-8(x + 1) - 2(y - 2) - (z - 4) = 0 \;\Rightarrow\; -8x - 2y - z = 8 - 4 - 4 = 0 \;\Rightarrow\; 8x + 2y + z = 0.$$

   (b) Let $F(x, y, z) = 4x^2 - y^2 + 2y - z$. Then, by Theorem 13.7, $\nabla F(-1, 2, 4)$ is a normal to the tangent plane at $(-1, 2, 4)$. $\nabla F = F_x\mathbf{i} + F_y\mathbf{j} + F_z\mathbf{k} = 8x\mathbf{i} + (-2y + 2)\mathbf{j} - \mathbf{k} \Rightarrow \nabla F(-1, 2, 4) = -8\mathbf{i} - 2\mathbf{j} - \mathbf{k}$.
As this is the same normal as in (a), the equation of the tangent plane at $(-1, 2, 4)$ is $8x + 2y + z = 0$.

45.   By Definition 13.5, a normal to the tangent plane is $\mathbf{n} = f_x\mathbf{i} + f_y\mathbf{j} - \mathbf{k}$.   Here $z = f(x,y) = y\cos(x-y) \Rightarrow$
$$f_x(x,y) = y[-\sin(x-y)], \text{ and } f_y(x,y) = \cos(x-y) + y[-\sin(x-y)(-1)] \Rightarrow$$
$f_x(2,2) = -2\sin 0 = 0$, and $f_y(2,2) = \cos 0 + 2\sin 0 = 1 \Rightarrow \mathbf{n} = \mathbf{j} - \mathbf{k}$.
The equation of the tangent plane at $(2,2,2)$ is therefore $(y-2)-(z-2) = 0 \Rightarrow y - z = 0$.

47.   By Definition 13.5, a normal to the tangent plane is $\mathbf{n} = f_x\mathbf{i} + f_y\mathbf{j} - \mathbf{k}$.   Here $z = f(x,y) = e^{3y}\sin 3x \Rightarrow$
$$f_x(x,y) = e^{3y}(\cos 3x)(3), \text{ and } f_y(x,y) = 3e^{3y}\sin 3x \Rightarrow$$
$f_x\left(\frac{\pi}{6},0\right) = 3\cos\frac{3\pi}{6} = 0$, and $f_y\left(\frac{\pi}{6},0\right) = 3\sin\frac{3\pi}{6} = 3 \Rightarrow \mathbf{n} = 3\mathbf{j} - \mathbf{k}$.
The equation of the tangent plane at $\left(\frac{\pi}{6},0,1\right)$ is $3(y-0)-(z-1) = 0 \Rightarrow 3y - z = -1$.

49.   Let $F(x,y,z) = 2x^2 + 3y^2 - z^2$. By Theorem 13.7, a normal to the tangent plane is
$\nabla F = F_x\mathbf{i} + F_y\mathbf{j} + F_z\mathbf{k} = 4x\mathbf{i} + 6y\mathbf{j} - 2z\mathbf{k} \Rightarrow \nabla F(2,1,-1) = 8\mathbf{i} + 6\mathbf{j} + 2\mathbf{k}$.
The equation of the tangent plane at $(2,1,-1)$ is therefore
$8(x-2) + 6(y-1) + 2(z+1) = 0 \Rightarrow 8x + 6y + 2z = 16 + 6 - 2 = 20 \Rightarrow 4x + 3y + z = 10$.

51.   Let $F(x,y,z) = \dfrac{y}{x+z}$.   By Theorem 13.7, a normal to the tangent plane is
$\nabla F = F_x\mathbf{i} + F_y\mathbf{j} + F_z\mathbf{k}$.

$$F_x(x,y,z) = \frac{\partial}{\partial x}[y(x+z)^{-1}] = y[-(x+z)^{-2}] = -\frac{y}{(x+z)^2}, \quad F_y(x,y,z) = \frac{1}{x+z}, \text{ and}$$

$$F_z(x,y,z) = \frac{\partial}{\partial z}[y(x+z)^{-1}] = y[-(x+z)^{-2}] = -\frac{y}{(x+z)^2} \Rightarrow$$

$\nabla F(x,y,z) = \left[-\dfrac{y}{(x+z)^2}\right]\mathbf{i} + \left[\dfrac{1}{x+z}\right]\mathbf{j} - \left[\dfrac{y}{(x+z)^2}\right]\mathbf{k} \Rightarrow \nabla F(1,8,3) = -\dfrac{8}{4^2}\mathbf{i} + \dfrac{1}{4}\mathbf{j} - \dfrac{8}{4^2}\mathbf{k}$.
The equation of the tangent plane at $(1,8,3)$ is then $-\frac{1}{2}(x-1) + \frac{1}{4}(y-8) - \frac{1}{2}(z-3) = 0 \Rightarrow$
$-2(x-1) + (y-8) - 2(z-3) = 0 \Rightarrow -2x + y - 2z = -2 + 8 - 6 = 0 \Rightarrow 2x - y + 2z = 0$.

53.   Let $F(x,y,z) = \ln(1+x+y+z)$. By Theorem 13.7, a normal to the tangent plane is
$\nabla F = F_x\mathbf{i} + F_y\mathbf{j} + F_z\mathbf{k}$. Here, $F_x(x,y,z) = F_y(x,y,z) = F_z(x,y,z) = \dfrac{1}{1+x+y+z} \Rightarrow$
$$\nabla F(x,y,z) = \frac{1}{1+x+y+z}[\mathbf{i} + \mathbf{j} + \mathbf{k}] \Rightarrow$$
$$\nabla F(e^2 + 2, e^3 - e^2, -3) = \frac{1}{1+e^2+2+e^3-e^2-3}[\mathbf{i}+\mathbf{j}+\mathbf{k}] = e^{-3}[\mathbf{i}+\mathbf{j}+\mathbf{k}].$$
The equation of the tangent plane at $(e^2 + 2, e^3 - e^2, -3)$ is then
$$e^{-3}[(x - e^2 - 2) + (y - e^3 + e^2) + (z + 3)] = 0 \Rightarrow x + y + z = e^3 - 1.$$

55.   Let $f(x, y, z) = (x^2 + y^2 + z^2)^{\frac{1}{2}}$, and $\Delta f = f(100, 199, 201) - f(100, 200, 200)$. Then

$f(100, 199, 201) = f(100, 200, 200) + \Delta f \approx f(100, 200, 200) + df$, where

$df = f_x(100, 200, 200)(100-100)+f_y(100, 200, 200)(199-200)+f_z(100, 200, 200)(201-200)$

$\quad = f_x(100, 200, 200)(0) + f_y(100, 200, 200)(-1) + f_z(100, 200, 200)(1)$.

$f_x(x, y, z) = \frac{1}{2}(x^2+y^2+z^2)^{-\frac{1}{2}} \cdot 2x = \dfrac{x}{\sqrt{x^2 + y^2 + z^2}} \Rightarrow f_x(100, 200, 200) = \dfrac{100}{\sqrt{100^2 + 200^2 + 200^2}}$

$f_y(x, y, z) = \frac{1}{2}(x^2+y^2+z^2)^{-\frac{1}{2}} \cdot 2y = \dfrac{y}{\sqrt{x^2 + y^2 + z^2}} \Rightarrow f_x(100, 200, 200) = \dfrac{200}{\sqrt{100^2 + 200^2 + 200^2}}$

$f_z(x, y, z) = \frac{1}{2}(x^2+y^2+z^2)^{-\frac{1}{2}} \cdot 2z = \dfrac{z}{\sqrt{x^2 + y^2 + z^2}} \Rightarrow f_x(100, 200, 200) = \dfrac{200}{\sqrt{100^2 + 200^2 + 200^2}}$

so that

$$df = -\dfrac{200}{\sqrt{100^2 + 200^2 + 200^2}} + \dfrac{200}{\sqrt{100^2 + 200^2 + 200^2}} = 0 \Rightarrow$$

$$\sqrt{100^2 + 199^2 + 201^2} \approx \sqrt{100^2 + 200^2 + 200^2} = \sqrt{100^2(1 + 4 + 4)} = 300.00.$$

57. $z = e^x \ln xy$, $(x, y) = (1, 2)$, $\Delta x = -0.1$, $\Delta y = 0.1 \Rightarrow$

$\Delta z \approx dz = z_x(1, 2)(-0.1) + z_y(1, 2)(0.1)$

As $z_x = e^x \ln xy + e^x \cdot \frac{1}{xy} \cdot y = e^x \left(\ln xy + \frac{1}{x} \right)$ and $z_y = e^x \left(\frac{1}{xy} \cdot x \right) = \frac{e^x}{y}$,

$\qquad z_x(1, 2) = e(\ln 2 + 1)$ and $z_y(1, 2) = \frac{e}{2} \Rightarrow$

$\Delta z \approx e(\ln 2 + 1)(-0.1) + \frac{e}{2}(0.1) = e\left[-\frac{1}{10}\ln 2 - \frac{1}{10} + \frac{1}{20} \right] = \frac{e}{20}(-2\ln 2 - 1) \approx -0.32$

59. $A = \frac{1}{2}bh$, $(b, h) = (2, 5)$, $\Delta b = 0.05$, $\Delta h = 0.1 \Rightarrow$

$$\Delta A \approx dA = A_b(2, 5)(0.05) + A_h(2, 5)(0.1)$$

As $A_b = \frac{1}{2}h$ and $A_h = \frac{1}{2}b$, $A_b(2, 5) = \frac{5}{2}$ and $A_h(2, 5) = 1 \Rightarrow$

$$\Delta A \approx \frac{5}{2}(0.05) + 0.1 = 0.125 + 0.1 = 0.225 \approx 0.23 \text{ cm}^2.$$

61. $f(x, y, z) = 2xy^2z^3$, $(x, y, z) = (1, -1, 2)$, $\Delta x = -0.01, \Delta y = -0.02, \Delta z = 0.02 \Rightarrow$

$\qquad \Delta f \approx df = f_x(1, -1, 2)(-0.01) + f_y(1, -1, 2)(-0.02) + f_z(1, -1, 2)(0.02)$

As $f_x = 2y^2z^3$, $f_y = 4xyz^3$, and $f_z = 6xy^2z^2 \Rightarrow f_x(1, -1, 2) = 2(-1)^2(2^3) = 16$,

$f_y(1, -1, 2) = 4(1)(-1)(2^3) = -32$, and $f_z(1, -1, 2) = 6(1)(-1)^2(2^2) = 24$, then

$\qquad \Delta f \approx 16(-0.01) - 32(-0.02) + 24(0.02) = -0.16 - 0.64 + 0.48 = -0.32.$

63. $V = lwh$, $(l, w, h) = (3, 4, 5)$, $\Delta l = \Delta w = \Delta h = 0.05 \Rightarrow$

$\qquad \Delta V \approx dV = V_l(3, 4, 5)(0.05) + V_w(3, 4, 5)(0.05) + V_h(3, 4, 5)(0.05)$

As $V_l = wh$, $V_w = lh$, and $V_h = lw \Rightarrow V_l(3, 4, 5) = 4 \cdot 5$, $V_w = 3 \cdot 5$, and $V_h = 3 \cdot 4$, then

$\qquad \Delta V \approx 0.05(20 + 15 + 12) = 0.05(47) = 2.35 \text{ cm}^3.$

65. $\nabla f(x_0, y_0) = f_x(x_0, y_0)\,\mathbf{i} + f_y(x_0, y_0)\,\mathbf{j} = \mathbf{0} \;\Rightarrow\; f_x(x_0, y_0) = f_y(x_0, y_0) = 0$. It follows that for any $\mathbf{u} = a\,\mathbf{i} + b\,\mathbf{j}$: $D_{\mathbf{u}}f(x_0, y_0) = af_x(x_0, y_0) + bf_y(x_0, y_0) = a\cdot 0 + b\cdot 0 = 0$.

67. For $f(x, y, z) = \dfrac{x^2}{a^2} + \dfrac{y^2}{b^2} - \dfrac{z^2}{c^2}$, $\nabla f(x, y, z) = \left(\dfrac{2x}{a^2}\right)\mathbf{i} + \left(\dfrac{2y}{b^2}\right)\mathbf{j} - \left(\dfrac{2z}{c^2}\right)\mathbf{k}$. I follows that

$\nabla f(x_0, y_0, z_0) = \left(\dfrac{2x_0}{a^2}\right)\mathbf{i} + \left(\dfrac{2y_0}{b^2}\right)\mathbf{j} - \left(\dfrac{2z_0}{c^2}\right)\mathbf{k}$ is a normal to the surface $\dfrac{x^2}{a^2} + \dfrac{y^2}{b^2} - \dfrac{z^2}{c^2} = 1$ at

(x_0, y_0, z_0), and that, therefore $\dfrac{2x_0}{a^2}(x - x_0) + \dfrac{2y_0}{b^2}(y - y_0) - \dfrac{2z_0}{c^2}(z - z_0) = 0$ is the tangent

plane at that point – a plane which can be rewritten in the form:

$$\frac{xx_0}{a^2} + \frac{yy_0}{b^2} - \frac{zz_0}{c^2} = \frac{x_0^2}{a^2} + \frac{y_0^2}{b^2} - \frac{z_0^2}{c^2} \;\Rightarrow$$

$$\frac{xx_0}{a^2} + \frac{yy_0}{b^2} - \frac{zz_0}{c^2} = 1,$$

as (x_0, y_0, z_0) lies on the surface $\dfrac{x^2}{a^2} + \dfrac{y^2}{b^2} - \dfrac{z^2}{c^2} = 1$.

69. (a) $\nabla(rf) = (rf)_x\,\mathbf{i} + (rf)_y\,\mathbf{j} + (rf)_z\,\mathbf{k} = rf_x\,\mathbf{i} + rf_y\,\mathbf{j} + rf_z\,\mathbf{k} = r(f_x\,\mathbf{i} + f_y\,\mathbf{j} + f_z\,\mathbf{k}) = r\nabla f$.

(b) $\nabla(f \pm g) = (f \pm g)_x\,\mathbf{i} + (f \pm g)_y\,\mathbf{j} + (f \pm g)_z\,\mathbf{k} = (f_x \pm g_x)\,\mathbf{i} + (f_y \pm g_y)\,\mathbf{j} + (f_z \pm g_z)\,\mathbf{k}$
$= (f_x\,\mathbf{i} + f_y\,\mathbf{j} + f_z\,\mathbf{k}) \pm (g_x\,\mathbf{i} + g_y\,\mathbf{j} + g_z\,\mathbf{k}) = \nabla f \pm \nabla g$

(c) $\nabla(fg) = (fg)_x\,\mathbf{i} + (fg)_y\,\mathbf{j} + (fg)_z\,\mathbf{k} = (fg_x + f_xg)\,\mathbf{i} + (fg_y + f_yg)\,\mathbf{j} + (fg_z + f_zg)\,\mathbf{k}$
$= [(fg_x)\,\mathbf{i} + (fg_y)\,\mathbf{j} + (fg_z)\,\mathbf{k}] + [(f_xg)\,\mathbf{i} + (f_yg)\,\mathbf{j} + (f_zg)\,\mathbf{k}]$
$= f(g_x\,\mathbf{i} + g_y\,\mathbf{j} + g_z\,\mathbf{k}) + g(f_x\,\mathbf{i} + f_y\,\mathbf{j} + f_z\,\mathbf{k}) = f\nabla g + g\nabla f$

(d) $\nabla\left(\dfrac{f}{g}\right) = \left(\dfrac{f}{g}\right)_x\mathbf{i} + \left(\dfrac{f}{g}\right)_y\mathbf{j} + \left(\dfrac{f}{g}\right)_z\mathbf{k} = \left[\dfrac{gf_x - fg_x}{g^2}\right]\mathbf{i} + \left[\dfrac{gf_y - fg_y}{g^2}\right]\mathbf{j} + \left[\dfrac{gf_z - fg_z}{g^2}\right]\mathbf{k}$

$= \dfrac{g(f_x\,\mathbf{i} + f_y\,\mathbf{j} + f_z\,\mathbf{k}) - f(g_x\,\mathbf{i} + g_y\,\mathbf{j} + g_z\,\mathbf{k})}{g^2} = \dfrac{g\nabla f - f\nabla g}{g^2}$

§13.3 Extreme Values

1. $f(x, y) = y^2 - xy + 2x + y + 1$. Then $f_x = -y + 2 = 0 \;\Rightarrow\; y = 2$, and $f_y = 2y - x + 1 = 0 \;\Rightarrow\; 2(2) - x + 1 = 0 \;\Rightarrow\; x = 5$. Thus $(5, 2)$ is a critical point. As $f_{xx} = 0$, $f_{yy} = 2$, and $f_{xy} = -1$, $D(5, 2) = f_{xx}(5, 2)f_{yy}(5, 2) - [f_{xy}(5, 2)]^2 = 0\cdot 2 - (-1)^2 = -1 < 0$. By Theorem 13.9, f has a saddle point at (5,2).

3. $f(x, y) = xy + 3$. Then $f_x = y = 0 \;\Rightarrow\; y = 0$, and $f_y = x = 0 \;\Rightarrow\; x = 0$. Thus $(0, 0)$ is a critical point. As $f_{xx} = 0$, $f_{yy} = 0$, and $f_{xy} = 1$, $D(0, 0) = f_{xx}(0, 0)f_{yy}(0, 0) - [f_{xy}(0, 0)]^2 = 0\cdot 0 - 1^2 = -1 < 0$. By Theorem 13.9, f has a saddle point at (0,0).

5. $f(x, y) = xy - x^2 - y^2 - 2x - 2y$. Then $f_x = y - 2x - 2 = 0 \Rightarrow y = 2x + 2$, and $f_y = x - 2y - 2 = x - 2(2x + 2) - 2 = 0 \Rightarrow -3x - 6 = 0 \Rightarrow x = -2 \Rightarrow y = 2(-2) + 2 = -2$. Thus $(-2, -2)$ is a critical point. As $f_{xx} = -2$, $f_{yy} = -2$, and $f_{xy} = 1$, $D(-2, -2) = f_{xx}(-2, -2)f_{yy}(-2, -2) - [f_{xy}(-2, -2)]^2 = (-2)(-2) - 1^2 = 3 > 0$. Since $f_{xx}(-2, -2) = -2 < 0$, by Theorem 13.9, f has a local maximum at $(-2, -2)$.

7. $f(x, y) = x^4 + y^4 - 4xy + 5$. Then $f_x = 4x^3 - 4y = 0 \Rightarrow y = x^3$, and $f_y = 4y^3 - 4x = 4(x^3)^3 - 4x = 0 \Rightarrow x^9 - x = 0 \Rightarrow x(x^8 - 1) = 0 \Rightarrow x = 0$ or $x = \pm 1$. When $x = 0, y = 0^3 = 0 \Rightarrow (0, 0)$ is a critical point. When $x = 1, y = x^3 = 1$, and when $x = -1, y = x^3 = -1 \Rightarrow (1, 1)$ and $(-1, -1)$ are also critical points.

Testing: $f_{xx} = 12x^2$, $f_{yy} = 12y^2$, $f_{xy} = -4 \Rightarrow D(x, y) = 12x^2(12y^2) - (-4)^2 = 144x^2 y^2 - 16$:

$D(0, 0) = -16 < 0$. By Theorem 13.9, f has a saddle point at $(0,0)$.

$D(-1, -1) = 144 - 16 > 0$ and $f_{xx}(-1, -1) = 12 > 0$. By Theorem 13.9, f has a local minimum at $(-1, -1)$.

$D(1, 1) = 144 - 16 > 0$ and $f_{xx}(1, 1) = 12 > 0$. By Theorem 13.9, f has a local minimum at $(1, 1)$.

9. $f(x, y) = x^3 - 12xy + 8y^3$. Then $f_x = 3x^2 - 12y = 0 \Rightarrow y = \frac{1}{4}x^2$, and $f_y = -12x + 24y^2 = -12x + 24\left(\frac{1}{4}x^2\right)^2 = 0 \Rightarrow -x + \frac{1}{8}x^4 = 0 \Rightarrow x^4 - 8x = 0 \Rightarrow x(x^3 - 8) = 0 \Rightarrow x = 0$ or x = 2. When $x = 0, y = \frac{1}{4}x^2 = 0 \Rightarrow (0, 0)$ is a critical point. When $x = 2, y = \frac{1}{4}x^2 = 1 \Rightarrow (2, 1)$ is also a critical point.

Testing: $f_{xx} = 6x$, $f_{yy} = 48y$, and $f_{xy} = -12 \Rightarrow$
$$D(x, y) = 6x(48y) - (-12)^2 = 288xy - 144 :$$

$D(0, 0) = -144 < 0$. By Theorem 13.9, f has a saddle point at $(0,0)$.

$D(2, 1) = 576 - 144 > 0$, and $f_{xx}(2, 1) = 6x = 12 > 0$. By Theorem 13.9, f has a local minimum at $(2,1)$.

11. $f(x, y) = 2x^2 + 3xy + 4y^2 - 5x + 2y$. Then $f_x = 4x + 3y - 5 = 0 \Rightarrow y = \frac{1}{3}(5 - 4x)$, and $f_y = 3x + 8y + 2 = 3x + 8\left[\frac{1}{3}(5 - 4x)\right] + 2 = 0 \Rightarrow 9x + 40 - 32x + 6 = 0 \Rightarrow 23x = 46 \Rightarrow x = 2 \Rightarrow y = \frac{1}{3}[5 - 4(2)] = -1 \Rightarrow (2, -1)$ is a critical point. As $f_{xx} = 4$, $f_{yy} = 8$, and $f_{xy} = 3$, $D(2, -1) = f_{xx}(2, -1)f_{yy}(2, -1) - [f_{xy}(2, -1)]^2 = 4 \cdot 8 - 3^2 = 23 > 0$. Since $f_{xx}(2, -1) = 4 > 0$. By Theorem 13.9, f has a local minimum at $(2, -1)$.

13. $f(x, y) = e^x \sin y$. Then $f_x = e^x \sin y = 0 \Rightarrow \sin y = 0$, and $f_y = e^x \cos y = 0 \Rightarrow \cos y = 0$. As sine and cosine are never both zero, there are no critical points, and hence no local extrema or saddle points.

15. $f(x, y) = e^{-(x^2 + y^2 + 2x)}$. Then $f_x = e^{-(x^2 + y^2 + 2x)}(-2x - 2) - 0 \Rightarrow x = -1$, and $f_y = e^{-(x^2 + y^2 + 2x)}(-2y) = 0 \Rightarrow y = 0$. There is one critical point: $(-1, 0)$.

$f_{xx} = e^{-(x^2 + y^2 + 2x)}[(-2x - 2)^2 - 2] \Rightarrow f_{xx}(-1, 0) = e^{-(1 + 0 - 2)}\{[-2(-1) - 2]^2 - 2\} = -2e$

$$f_{yy} = e^{-(x^2+y^2+2x)}(4y^2 - 2) \Rightarrow f_{yy}(-1,0) = -2e$$

$$f_{xy} = e^{-(x^2+y^2+2x)}(-2x-2)(-2y) \Rightarrow f_{xy}(-1,0) = 0 \Rightarrow$$

$D(-1,0) = f_{xx}(-1,0)f_{yy}(-1,0) - [f_{xy}(-1,0)]^2 = -2e(-2e) - 0 > 0$ and $f_{xx}(-1,0) < 0$.
By Theorem 13.9, f has a local maximum at $(-1,0)$.

17. $f(x,y) = \sin x + \sin y$ with $0 < x, y < \pi$. Then $f_x = \cos x = 0 \Rightarrow x = \frac{\pi}{2}$, and
$f_y = \cos y = 0 \Rightarrow y = \frac{\pi}{2}$. There is one critical point: $\left(\frac{\pi}{2}, \frac{\pi}{2}\right)$. As $f_{xx} = -\sin x$,
$f_{yy} = -\sin y$ and $f_{xy} = 0$,

$$D\left(\frac{\pi}{2},\frac{\pi}{2}\right) = f_{xx}\left(\frac{\pi}{2},\frac{\pi}{2}\right)f_{yy}\left(\frac{\pi}{2},\frac{\pi}{2}\right) - \left[f_{xy}\left(\frac{\pi}{2},\frac{\pi}{2}\right)\right]^2 = \sin\frac{\pi}{2}\sin\frac{\pi}{2} - 0 = 1 > 0.$$

Since $f_{xx}\left(\frac{\pi}{2},\frac{\pi}{2}\right) = -\sin\frac{\pi}{2} = -1 < 0$, by Theorem 13.9, f has a local maximum at $\left(\frac{\pi}{2},\frac{\pi}{2}\right)$.

19. $f(x,y) = 3xy - 6x + 7$:

Inside D: $f_x = 3y - 6 = 0 \Rightarrow y = 2$, and $f_y = 3x = 0 \Rightarrow x = 0$, but the point $(0,2)$ is not inside D: There are no interior critical points.

On a: $x = 0$, so $g(y) = f(0,y) = 7$ is never 0 which means that only the endpoints are critical points: $(0,0)$, $(0,5)$.

On b: $y = 5$, so $g(x) = f(x,5) = 15x - 6x + 7 = 9x + 7 \Rightarrow g'(x) = 9 \neq 0 \Rightarrow$ no interior critical points. There is one additional critical point, the endpoint $(3,5)$.

On c: $x = 3$, so $g(y) = f(3,y) = 9y - 11 \Rightarrow g'(y) = 9 \neq 0 \Rightarrow$ no interior critical points. There is one additional critical point, the endpoint $(3,0)$.

On d: $y = 0$, so $g(x) = f(x,0) = -6x + 7 \Rightarrow g'(x) = -6 \neq 0 \Rightarrow$ no interior critical points, and there are no new boundary points.

Testing: $f(0,0) = 7$, $f(0,5) = 7$, $f(3,5) = 45 - 18 + 7 = 34$, $f(3,0) = -18 + 7 = -11 \Rightarrow$ the absolute maximum is $f(3,5) = 34$ and the absolute minimum is $f(3,0) = -11$.

21. $f(x,y) = x^3 - 3xy + 3y^2$:

Inside D: $f_x = 3x^2 - 3y = 0 \Rightarrow y = x^2$, and $f_y = -3x + 6y = 0 \Rightarrow$ $-3x + 6(x^2) = 0 \Rightarrow x(2x - 1) = 0 \Rightarrow x = 0$ or $x = \frac{1}{2}$, but a point with $x = 0$ is not in D. When $x = \frac{1}{2}$, $y = x^2 = \frac{1}{4}$. f has one interior critical point: $\left(\frac{1}{2},\frac{1}{4}\right)$.

On a: $x = \frac{1}{4}$, so $g(y) = f\left(\frac{1}{4},y\right) = \frac{1}{64} - \frac{3}{4}y + 3y^2 \Rightarrow g'(y) = -\frac{3}{4} + 6y = 0 \Rightarrow y = \frac{1}{8} \Rightarrow \left(\frac{1}{4},\frac{1}{8}\right)$ is a critical point. There are also 2 endpoint critical points: $\left(\frac{1}{4},0\right)$ and $\left(\frac{1}{4},2\right)$.

On b: $y = 2$, so $g(x) = f(x,2) = x^3 - 6x + 12 \Rightarrow g'(x) = 3x^2 - 6 = 0 \Rightarrow x^2 = 2 \Rightarrow x = \pm\sqrt{2}$, which yields points not in D. There is one new critical point, the endpoint $(1,2)$.

On c: $x = 1$, so $g(y) = f(1,y) = 1 - 3y + 3y^2 \Rightarrow g'(y) = -3 + 6y = 0 \Rightarrow y = \frac{1}{2}$, giving the critical point $\left(1, \frac{1}{2}\right)$. There is also a new endpoint critical point: $(1, 0)$.

On d: $y = 0$, so $g(x) = f(x, 0) = x^3 \Rightarrow g'(x) = 3x^2 = 0 \Rightarrow x = 0$ yielding a point not in D.

Testing: $f\left(\frac{1}{2}, \frac{1}{4}\right) = \frac{1}{8} - \frac{3}{8} + \frac{3}{16} = -\frac{1}{16}$, $f\left(\frac{1}{4}, \frac{1}{8}\right) = \frac{1}{64} - \frac{3}{32} + \frac{3}{64} = -\frac{1}{32}$, $f\left(\frac{1}{4}, 0\right) = \frac{1}{64}$,

$f\left(\frac{1}{4}, 2\right) = \frac{1}{64} - \frac{3}{2} + 12 = \frac{673}{64} \approx 10.5$, $f(1, 2) = 1 - 6 + 12 = 7$, $f\left(1, \frac{1}{2}\right) = 1 - \frac{3}{2} + \frac{3}{4} = \frac{1}{4}$

and $f(1, 0) = 1$.

The absolute maximum is $f\left(\frac{1}{4}, 2\right) = \frac{673}{64} \approx 10.5$, and the absolute minimum is $f\left(\frac{1}{2}, \frac{1}{4}\right) = -\frac{1}{16}$.

23. $y = -x+2$ $f(x, y) = x^3 - xy + y^2 - x$:

Inside D: $f_x = 3x^2 - y - 1 = 0 \Rightarrow y = 3x^2 - 1$, and $f_y = -x + 2y = 0 \Rightarrow$
$-x + 2(3x^2 - 1) = 0 \Rightarrow 6x^2 - x - 2 = 0 \Rightarrow (3x - 2)(2x + 1) = 0 \Rightarrow x = \frac{2}{3}$,

as $x = -\frac{1}{2}$ yields no point in D. Then $y = 3\left(\frac{2}{3}\right)^2 - 1 = \frac{1}{3} \Rightarrow f$ has an interior critical point: $\left(\frac{2}{3}, \frac{1}{3}\right)$.

On a: $x = 0$, so $g(y) = f(0, y) = y^2 \Rightarrow g'(y) = 2y = 0 \Rightarrow y = 0$. So, the two endpoints, $(0,0)$ and $(0,2)$, are critical points.

On b: $y = -x + 2 \Rightarrow$
$g(x) = f(x, -x + 2) = x^3 - x(-x + 2) + (-x + 2)^2 - x = x^3 + x^2 - 2x + x^2 - 4x + 4 - x$
$= x^3 + 2x^2 - 7x + 4$

Then, $g'(x) = 3x^2 + 4x - 7 = (3x + 7)(x - 1) = 0 \Rightarrow x = 1$ (as $x \geq 0$ in D) $\Rightarrow y = -(1) + 2 = 1$. Thus $(1,1)$ and the new endpoint $(2,0)$ are critical points.

On c: $y = 0 \Rightarrow g(x) = f(x, 0) = x^3 - x \Rightarrow g'(x) = 3x^2 - 1 = 0 \Rightarrow x = \frac{1}{\sqrt{3}} \Rightarrow \left(\frac{1}{\sqrt{3}}, 0\right)$ is the last critical point.

Testing: $f(0, 0) = 0$, $f(0, 2) = 4$, $f\left(\frac{2}{3}, \frac{1}{3}\right) = \frac{8}{27} - \frac{2}{9} + \frac{1}{9} - \frac{2}{3} = \frac{8 - 3 - 18}{27} = -\frac{13}{27}$, $f(1, 1) = 0$,

$f(2, 0) = 8 - 2 = 6$, $f\left(\frac{1}{\sqrt{3}}, 0\right) = \frac{1}{3\sqrt{3}} - \frac{1}{\sqrt{3}} = -\frac{2}{3\sqrt{3}} = -\frac{2\sqrt{3}}{9}$, $\left[\text{note: } -\frac{2\sqrt{3}}{9} = -\frac{6\sqrt{3}}{27} > -\frac{13}{27}\right]$.

The absolute maximum is $f(2, 0) = 6$, and the absolute minimum is $f\left(\frac{2}{3}, \frac{1}{3}\right) = -\frac{13}{27}$.

25. $f(x, y) = xe^y - x^2 - e^y$:

Inside D: $f_x = e^y - 2x = 0 \Rightarrow x = \frac{1}{2}e^y$, and $f_y = xe^y - e^y = 0 \Rightarrow$
$e^y(x - 1) = 0 \Rightarrow x = 1 \Rightarrow 1 = \frac{1}{2}e^y \Rightarrow y = \ln 2$, yielding the interior critical point: $(1, \ln 2)$.

On a: $x = 0$, so $g(y) = f(0, y) = -e^y = g'(y) \neq 0$. There are only endpoint critical points here: $(0,0)$ and $(0,1)$.

On b: $y = 1 \Rightarrow g(x) = f(x, 1) = ex - x^2 - e \Rightarrow g'(x) = e - 2x = 0 \Rightarrow x = \frac{e}{2}$, yielding the critical point $\left(\frac{e}{2}, 1\right)$, and the endpoint point $(2,1)$.

On c: $x = 2 \Rightarrow g(y) = f(2, y) = 2e^y - 4 - e^y = e^y - 4$, so $g'(y) = e^y \neq 0$, and the only new critical point is the endpoint (2,0).

On d: $y = 0 \Rightarrow g(x) = x - x^2 - 1 \Rightarrow g'(x) = 1 - 2x = 0 \Rightarrow x = \frac{1}{2}$, providing the critical point $\left(\frac{1}{2}, 0\right)$.

Testing: $f(1, \ln 2) = e^{\ln 2} - 1 - e^{\ln 2} = 2 - 1 - 2 = -1$, $f(0,0) = -e^0 = -1$, $f(0, 1) = -e \approx -2.7$, $f\left(\frac{e}{2}, 1\right) = \frac{e^2}{2} - \frac{e^2}{4} - e = \frac{e^2}{4} - e \approx -0.87$, $f(2, 0) = 2 - 4 - 1 = -3$, $f(2, 1) = 2e - 4 - e = e - 4 \approx -1.28$, $f\left(\frac{1}{2}, 0\right) = \frac{1}{2} - \frac{1}{4} - 1 = -\frac{3}{4} = -0.75$.

The absolute maximum is $f\left(\frac{1}{2}, 0\right) = -\frac{3}{4}$, and the absolute minimum is $f(2, 0) = -3$.

27.

$f(x, y) = -2x^2 - y^2 + 4x + 3y + 5$:

Inside D: $f_x = -4x + 4 = 0 \Rightarrow x = 1$, and $f_y = -2y + 3 = 0 \Rightarrow y = \frac{3}{2}$, so $\left(1, \frac{3}{2}\right)$ is an interior critical point.

On a: $y = -x \Rightarrow g(x) = f(x, -x) = -2x^2 - x^2 + 4x - 3x + 5 = -3x^2 + x + 5 \Rightarrow g'(x) = -6x + 1 = 0 \Rightarrow x = \frac{1}{6} \Rightarrow y = -\frac{1}{6} < 0$: the point is not in D. There are only endpoint critical points: (0,0) and (−2, 2).

On b: $y = 2 \Rightarrow g(x) = f(x, 2) = -2x^2 - 4 + 4x + 6 + 5 = -2x^2 + 4x + 7 \Rightarrow g'(x) = -4x + 4 = 0 \Rightarrow x = 1$, yielding the critical point (1,2) and the endpoint (2,2).

On c: $y = x \Rightarrow g(x) = f(x, x) = -2x^2 - x^2 + 4x + 3x + 5 = -3x^2 + 7x + 5 \Rightarrow g'(x) = -6x + 7 = 0 \Rightarrow x = \frac{7}{6} = y$, yielding the critical point $\left(\frac{7}{6}, \frac{7}{6}\right)$.

Testing: $f\left(1, \frac{3}{2}\right) = -2 - \frac{9}{4} + 4 + \frac{9}{2} + 5 = 7 + \frac{9}{4} = \frac{37}{4} = 9.25$, $f(0,0) = 5$, $f(-2, 2) = -8 - 4 - 8 + 6 + 5 = -9$, $f(1, 2) = -2 - 4 + 4 + 6 + 5 = 9$, $f(2, 2) = -8 - 4 + 8 + 6 + 5 = 7$, $f\left(\frac{7}{6}, \frac{7}{6}\right) = -\frac{98}{36} - \frac{49}{36} + \frac{14}{3} + \frac{7}{2} + 5 = \frac{109}{12} \approx 9.08$.

The absolute maximum is $f\left(1, \frac{3}{2}\right) = \frac{37}{4}$, and the absolute minimum is $f(-2, 2) = -9$.

29.

$f(x, y) = 2x^2 - y^2 + 6y$:

Inside D: $f_x = 4x = 0 \Rightarrow x = 0$, and $f_y = -2y + 6 = 0 \Rightarrow y = 3 \Rightarrow$ (0, 3) is an interior critical point.

On the boundary of D, $x^2 + y^2 = 16 \Rightarrow x^2 = 16 - y^2 \Rightarrow g(y) = 2(16 - y^2) - y^2 + 6y = -3y^2 + 6y + 32$, with $-4 \leq y \leq 4$. Then $g'(y) = -6y + 6 = 0 \Rightarrow y = 1 \Rightarrow x = \pm\sqrt{15}$, yielding the critical points $\left(\pm\sqrt{15}, 1\right)$ and the endpoints (0, ±4).

Testing: $f(0, 3) = -9 + 18 = 9$, $f\left(\pm\sqrt{15}, 1\right) = 30 - 1 + 6 = 35$, $f(0, -4) = -16 - 24 = -40$, $f(0, 4) = -16 + 24 = 8$.

The absolute maximum is $f\left(\pm\sqrt{15}, 1\right) = 35$, and the absolute minimum is $f(0, -4) = -40$.

31. With $f(x, y) = x + y$ and $g(x, y) = x^2 + y^2 = 1$, then $\nabla f(x, y) = \lambda \nabla g(x, y) \Rightarrow$
$\mathbf{i} + \mathbf{j} = \lambda[2x\,\mathbf{i} + 2y\,\mathbf{j}] \Rightarrow 2\lambda x = 1$ and $2\lambda y = 1 \Rightarrow x = \frac{1}{2\lambda}$ and $y = \frac{1}{2\lambda}$, so that $x = y$.
As $x^2 + y^2 = 1$, then $2x^2 = 1 \Rightarrow x = \pm\frac{1}{\sqrt{2}}$, and we have two critical points $\left(\frac{1}{\sqrt{2}}, \frac{1}{\sqrt{2}}\right)$
and $\left(-\frac{1}{\sqrt{2}}, -\frac{1}{\sqrt{2}}\right)$. We must also consider solutions of $\nabla g(x, y) = 0 \Rightarrow 2x = 2y = 0 \Rightarrow$
$x = y = 0$, but (0,0) is not on the circle. Since the circle is closed and bounded, the con-
tinuous function f will assume its maximum and minimum values there, and these occur at
the critical points. Testing:

$f\left(\frac{1}{\sqrt{2}}, \frac{1}{\sqrt{2}}\right) = \frac{1}{\sqrt{2}} + \frac{1}{\sqrt{2}} = \frac{2}{\sqrt{2}} = \sqrt{2}$ is the maximum value of f on the unit circle, and

$f\left(-\frac{1}{\sqrt{2}}, -\frac{1}{\sqrt{2}}\right) = -\frac{1}{\sqrt{2}} - \frac{1}{\sqrt{2}} = -\frac{2}{\sqrt{2}} = -\sqrt{2}$ is the minimum value of f on the unit circle.

33. With $f(x, y) = xy + 14$ and $g(x, y) = x^2 + y^2 = 18$, then $\nabla f(x, y) = \lambda \nabla g(x, y) \Rightarrow$
$y\,\mathbf{i} + x\,\mathbf{j} = \lambda[2x\,\mathbf{i} + 2y\,\mathbf{j}] \Rightarrow y = 2x\lambda$ and $x = 2y\lambda = 4\lambda^2 x \Rightarrow x = 0$ or $4\lambda^2 = 1 \Rightarrow \lambda = \pm\frac{1}{2}$.
If $x = 0$, then $y = 0$, but (0,0) is not on the circle. So $\lambda = \pm\frac{1}{2} \Rightarrow y = 2x\lambda = \pm x \Rightarrow$
$x^2 + x^2 = 18 \Rightarrow x^2 = 9 \Rightarrow x = \pm 3$. Then $y = \pm 3$, giving 4 critical points:
$(3, -3), (3, 3), (-3, -3)$ and $(-3, 3)$. We must also consider solutions of $\nabla g(x, y) = 0 \Rightarrow$
$2x = 2y = 0 \Rightarrow x = y = 0$, but (0,0) is not on the circle. Since the circle is closed and
bounded, the continuous function f will assume its maximum and minimum values there,
and these occur at the critical points.

Testing, we find that $f(3, 3) = f(-3, -3) = 23$ is the maximum value of f on the circle,
and $f(3, -3) = f(-3, 3) = 5$ is the minimum value of f on the circle.

35. With $f(x, y) = x^2 + 2y^2$ and $g(x, y) = x^2 + y^2 = 1$, then $\nabla f(x, y) = \lambda \nabla g(x, y) \Rightarrow$
$2x\,\mathbf{i} + 4y\,\mathbf{j} = \lambda[2x\,\mathbf{i} + 2y\,\mathbf{j}] \Rightarrow 2x = 2\lambda x$ and $4y = 2\lambda y$. Then from the first equation,
either $x = 0$ or $\lambda = 1$. If $x = 0$, from $x^2 + y^2 = 1, y = \pm 1$. If $\lambda = 1$, the second equa-
tion requires that $y = 0$, and then from $x^2 + y^2 = 1, x = \pm 1$. Consequently, there are 4
critical points: $(0, \pm 1)$ and $(\pm 1, 0)$. We must also consider solutions of $\nabla g(x, y) = 0 \Rightarrow$
$2x = 2y = 0 \Rightarrow x = y = 0$, but (0,0) is not on the circle. Since the circle is closed and
bounded, the continuous function f will assume its maximum and minimum values there,
and these occur at the critical points.

Testing: $f(0, 1) = f(0, -1) = 2$ and $f(-1, 0) = f(1, 0) = 1$. The maximum value of f on the
circle is $f(0, \pm 1) = 2$, and the minimum value of f on the circle is $f(\pm 1, 0) = 1$.

37. With $f(x, y) = 2x^2 + 3y^2 - 4x - 5$ and $g(x, y) = x^2 + y^2 = 16$, then $\nabla f(x, y) =$
$\lambda \nabla g(x, y) \Rightarrow (4x - 4)\,\mathbf{i} + 6y\,\mathbf{j} = \lambda[2x\,\mathbf{i} + 2y\,\mathbf{j}] \Rightarrow 4x - 4 = 2x\lambda$ and $6y = 2\lambda y$. From the last
equation, $y = 0$ or $\lambda = 3$. If $y = 0$, from $x^2 + y^2 = 16, x = \pm 4$, so $(\pm 4, 0)$ are critical points. If
$\lambda = 3$, from the first equation, $4x - 4 - 6x \Rightarrow x = -2 \Rightarrow (-2)^2 + y^2 = 16 \Rightarrow y = \pm 2\sqrt{3}$,
so $(-2, \pm 2\sqrt{3})$ are critical points. We must also consider solutions of $\nabla g(x, y) = 0 \Rightarrow$
$2x = 2y = 0 \Rightarrow x = y = 0$, but (0,0) is not on the circle. Since the circle is closed and

bounded, the continuous function f will assume its maximum and minimum values there, and these occur at the critical points.

Testing: $f(-4, 0) = 32 + 16 - 5 = 43$, $f(4, 0) = 32 - 16 - 5 = 11$,
$$f(-2, \pm 2\sqrt{3}) = 8 + 3(12) + 8 - 5 = 47$$
The maximum value of f on the circle is $f(-2, \pm 2\sqrt{3}) = 47$, and the minimum value of f on the circle is $f(4, 0) = 11$.

39. With $f(x, y, z) = x + y + 2z$ and $g(x, y, z) = x^2 + y^2 + z^2 = 3$, then $\nabla f(x, y, z) = \lambda \nabla g(x, y, z) \Rightarrow \mathbf{i} + \mathbf{j} + 2\mathbf{k} = \lambda[2x\,\mathbf{i} + 2y\,\mathbf{j} + 2z\,\mathbf{k}] \Rightarrow 1 = 2\lambda x$, $1 = 2\lambda y$, $2 = 2\lambda z$. From these equations, $\lambda \neq 0$, so $x = \frac{1}{2\lambda}$, $y = \frac{1}{2\lambda}$, $x = \frac{1}{\lambda}$. As $x^2 + y^2 + z^2 = 3$,
$$\left(\frac{1}{2\lambda}\right)^2 + \left(\frac{1}{2\lambda}\right)^2 + \left(\frac{1}{\lambda}\right)^2 = 3 \Rightarrow \lambda = \pm\frac{1}{\sqrt{2}},$$
and there are two critical points: $\left(\frac{\sqrt{2}}{2}, \frac{\sqrt{2}}{2}, \sqrt{2}\right)$ and $\left(-\frac{\sqrt{2}}{2}, -\frac{\sqrt{2}}{2}, -\sqrt{2}\right)$. We must also consider solutions of $\nabla g(x, y, z) = 0 \Rightarrow 2x = 2y = 2z = 0 \Rightarrow x = y = z = 0$, but $(0,0,0)$ is not on the sphere. Since the sphere is closed and bounded, the continuous function f will assume its maximum and minimum values there, and these occur at the critical points. After testing the 2 critical points: The maximum of f on the sphere is $f\left(\frac{\sqrt{2}}{2}, \frac{\sqrt{2}}{2}, \sqrt{2}\right) = 3\sqrt{2}$, and the minimum of f on the sphere is $f\left(-\frac{\sqrt{2}}{2}, -\frac{\sqrt{2}}{2}, -\sqrt{2}\right) = -3\sqrt{2}$.

41. With $f(x, y, z) = xyz$, $g(x, y, z) = x + y + z = 1$, and $x, y, z \geq 0$, then $\nabla f(x, y, z) = \lambda \nabla g(x, y, z) \Rightarrow yz\,\mathbf{i} + xz\,\mathbf{j} + xy\,\mathbf{k} = \lambda[\mathbf{i} + \mathbf{j} + \mathbf{k}] \Rightarrow yz = \lambda$, $xz = \lambda$, $xy = \lambda \Rightarrow yz = xz = xy$.

If $x = 0$, then $yz = 0 \Rightarrow y = 0$ or $z = 0$. If $y = 0$, then from $x + y + z = 1$, $z = 1$. If $z = 0$, then from $x + y + z = 1$, $y = 1$, and we have two critical points $(0,0,1)$ and $(0,1,0)$.

If $y = 0$, then $xz = 0 \Rightarrow x = 0$ or $z = 0$. If $x = 0$, then from $x + y + z = 1$, $z = 1$. If $z = 0$, then from $x + y + z = 1$, $x = 1$, and we have one old critical point $(0,0,1)$ and one new one: $(1,0,0)$.

If $z = 0$, then $xy = 0 \Rightarrow x = 0$ or $y = 0$. If $x = 0$, then from $x + y + z = 1$, $y = 1$. If $y = 0$, then from $x + y + z = 1$, $x = 1$, and we have two old critical points $(0,1,0)$ and $(1,0,0)$.

The remaining case is $x \neq 0, y \neq 0$ and $z \neq 0$. Dividing $yz = xz$ by z gives $y = x$, and dividing $xz = xy$ by x gives $z = y$, so that $x = y = z$. From $x + y + z = 1$, then $x = y = z = \frac{1}{3}$, and the final critical point is $\left(\frac{1}{3}, \frac{1}{3}, \frac{1}{3}\right)$.

Noting that $\nabla g(x, y)$ cannot be zero, there are no additional critical points. Since the constraint surface, being the first octant portion of the the plane, is closed and bounded, the continuous function f will assume its maximum and minimum values on that region, and these occur at the critical points.

After testing the 4 critical points: The maximum of f on C is $f\left(\frac{1}{3}, \frac{1}{3}, \frac{1}{3}\right) = \left(\frac{1}{3}\right)^3 = \frac{1}{27}$, and the minimum of f on C is $f(1, 0, 0) = f(0, 1, 0) = f(0, 0, 1) = 0$.

43. $f(x, y) = x^2 + 2y^2$ and $D : x^2 + y^2 \leq 1$. In Exercise 35, we found four critical points on the boundary of D, and $f(0, 1) = f(0, -1) = 2$ and $f(-1, 0) = f(1, 0) = 1$. Now, inside $D : f_x = 2x = 0$ and $f_y = 4y = 0 \Rightarrow (0,0)$ is an interior critical point, and $f(0,0) = 0$. Therefore the maximum of f on D remains $f(0, \pm 1) = 2$, but the minimum of f on D is $f(0,0) = 0$.

45. $f(x,y) = 2x^2 + 3y^2 - 4x - 5$ and $D : x^2 + y^2 \leq 16$. In Exercise 37, we found four critical points on the boundary of D, and $f(-4,0) = 43$, $f(4,0) = 11$, and $f\left(-2, \pm 2\sqrt{3}\right) = 47$. Now, inside $D : f_x = 4x - 4 = 0 \Rightarrow x = 1$, and $f_y = 6y = 0 \Rightarrow y = 0 \Rightarrow (1,0)$ is an interior critical point. As $f(1,0) = 2 - 4 - 5 = -7$, the maximum of f on D remains $f\left(-2, \pm 2\sqrt{3}\right) = 47$, but the minimum of f on D is $f(1,0) = -7$.

47. For $f(x,y,z) = 3x - y - 3z$, $g(x,y,z) = x + y - z = 0$ and $h(x,y,z) = x^2 + 2z^2 = 1$, $\nabla f(x,y,z) = \lambda \nabla g(x,y,z) + \mu \nabla h(x,y,z) \Rightarrow 3\mathbf{i} - \mathbf{j} - 3\mathbf{k} = \lambda[\mathbf{i} + \mathbf{j} - \mathbf{k}] + \mu[2x\mathbf{i} + 4z\mathbf{k}] \Rightarrow$
$$(1) : 3 = \lambda + 2\mu x, \qquad (2) : -1 = \lambda, \qquad (3) : -3 = -\lambda + 4\mu z$$
Substituting (2) into (1) and (3): $3 = -1 + 2\mu x \Rightarrow x = \frac{2}{\mu}$, and $-3 = 1 + 4\mu z \Rightarrow z = -\frac{1}{\mu}$. Thus $x = -2z$. From $x^2 + 2z^2 = 1$, then $4z^2 + 2z^2 = 1 \Rightarrow z = \pm\frac{1}{\sqrt{6}} \Rightarrow x = \mp\frac{2}{\sqrt{6}}$. Then $x + y - z = 0 \Rightarrow y = z - x = \pm\frac{3}{\sqrt{6}} \Rightarrow$ two critical points: $\left(-\frac{2}{\sqrt{6}}, \frac{3}{\sqrt{6}}, \frac{1}{\sqrt{6}}\right)$ and $\left(\frac{2}{\sqrt{6}}, -\frac{3}{\sqrt{6}}, -\frac{1}{\sqrt{6}}\right)$. Testing, we find $f\left(\frac{2}{\sqrt{6}}, -\frac{3}{\sqrt{6}}, -\frac{1}{\sqrt{6}}\right) = \frac{6}{\sqrt{6}} + \frac{3}{\sqrt{6}} + \frac{3}{\sqrt{6}} = 2\sqrt{6}$ to be the maximum, and $f\left(-\frac{2}{\sqrt{6}}, \frac{3}{\sqrt{6}}, \frac{1}{\sqrt{6}}\right) = -\frac{6}{\sqrt{6}} - \frac{3}{\sqrt{6}} - \frac{3}{\sqrt{6}} = -2\sqrt{6}$ to be the minimum.

49. For $f(x,y,z) = xy + yz$, $g(x,y,z) = xy = 1$ and $h(x,y,z) = y^2 + z^2 = 1$, $\nabla f(x,y,z) = \lambda \nabla g(x,y,z) + \mu \nabla h(x,y,z) \Rightarrow y\mathbf{i} + (x + z)\mathbf{j} + y\mathbf{k} = \lambda[y\mathbf{i} + x\mathbf{j}] + \mu[2y\mathbf{j} + 2z\mathbf{k}] \Rightarrow$
$$(1) : y = \lambda y, \qquad (2) : x + z = \lambda x + 2\mu y, \qquad (3) : y = 2\mu z$$
From (1), $\lambda = 1$, as $y \neq 0$ because $xy = 1$. Then (2) implies that $z = 2\mu y$, and substituting into (3), we have $y = 4\mu^2 y \Rightarrow 4\mu^2 = 1 \Rightarrow \mu = \pm\frac{1}{2} \Rightarrow z = \pm y$. From $y^2 + z^2 = 1$, $2y^2 = 1 \Rightarrow y = \pm\frac{1}{\sqrt{2}} \Rightarrow z = \pm\frac{1}{\sqrt{2}}$, and from $xy = 1$, then $x = \pm\sqrt{2}$. The critical points are: $\left(\sqrt{2}, \frac{1}{\sqrt{2}}, \frac{1}{\sqrt{2}}\right)$, $\left(-\sqrt{2}, -\frac{1}{\sqrt{2}}, \frac{1}{\sqrt{2}}\right)$, $\left(\sqrt{2}, \frac{1}{\sqrt{2}}, -\frac{1}{\sqrt{2}}\right)$, $\left(-\sqrt{2}, -\frac{1}{\sqrt{2}}, -\frac{1}{\sqrt{2}}\right)$. Testing, we find $f\left(\sqrt{2}, \frac{1}{\sqrt{2}}, \frac{1}{\sqrt{2}}\right) = f\left(-\sqrt{2}, -\frac{1}{\sqrt{2}}, -\frac{1}{\sqrt{2}}\right) = \frac{3}{2}$ to be the maximum value of f, and $f\left(-\sqrt{2}, -\frac{1}{\sqrt{2}}, \frac{1}{\sqrt{2}}\right) = f\left(\sqrt{2}, \frac{1}{\sqrt{2}}, -\frac{1}{\sqrt{2}}\right) = \frac{1}{2}$ to be the minimum value of f.

51. For $f(x,y,z) = 4y - 2z$, $g(x,y,z) = 2x - y - z = 2$ and $h(x,y,z) = x^2 + y^2 = 1$, $\nabla f(x,y,z) = \lambda \nabla g(x,y,z) + \mu \nabla h(x,y,z) \Rightarrow 4\mathbf{j} - 2\mathbf{k} = \lambda[2\mathbf{i} - \mathbf{j} - \mathbf{k}] + \mu[2x\mathbf{i} + 2y\mathbf{j}] \Rightarrow$
$$(1) : 0 = 2\lambda + 2\mu x, \qquad (2) : 4 = -\lambda + 2\mu y, \qquad (3) : -2 = -\lambda$$
From (3), $\lambda = 2 \Rightarrow$ [from (1) and (2)] that $x = -\frac{2}{\mu}$ and $y = \frac{3}{\mu}$. From $x^2 + y^2 = 1$, then $\frac{4}{\mu^2} + \frac{9}{\mu^2} = 1 \Rightarrow \mu^2 = 13 \Rightarrow \mu = \pm\sqrt{13}$.

When $\mu = \sqrt{13}$, $x = -\frac{2}{\sqrt{13}}$ and $y = \frac{3}{\sqrt{13}}$. As $2x - y - z = 2 \Rightarrow z = 2x - y - 2$, then $z = -2 - \frac{7}{\sqrt{13}}$, and we have the critical point $\left(-\frac{2}{\sqrt{13}}, \frac{3}{\sqrt{13}}, -2 - \frac{7}{\sqrt{13}}\right)$.

When $\mu = -\sqrt{13}$, $x = \frac{2}{\sqrt{13}}$ and $y = -\frac{3}{\sqrt{13}}$. As $2x - y - z = 2 \Rightarrow z = 2x - y - 2$, then $z = -2 + \frac{7}{\sqrt{13}}$, and we have the critical point $\left(\frac{2}{\sqrt{13}}, -\frac{3}{\sqrt{13}}, -2 + \frac{7}{\sqrt{13}}\right)$.

Testing, we find $f\left(-\frac{2}{\sqrt{13}}, \frac{3}{\sqrt{13}}, -2 - \frac{7}{\sqrt{13}}\right) = 4 + \frac{26}{\sqrt{13}} = 4 + 2\sqrt{13}$ is the maximum value of f, and $f\left(\frac{2}{\sqrt{13}}, -\frac{3}{\sqrt{13}}, -2 + \frac{7}{\sqrt{13}}\right) = 4 - \frac{26}{\sqrt{13}} = 4 - 2\sqrt{13}$ is the minimum value of f.

53. It suffices to minimize the square of the distance from the origin to a point (x, y, z) on the surface $z^2 = xy + 1$: So consider $D(x, y, z) = x^2 + y^2 + z^2$.

(a) Substituting $xy + 1$ for z^2 in D, we have $f(x, y) = x^2 + y^2 + xy + 1$. Then $\nabla f = 0 \Rightarrow$ (1): $f_x = 2x + y = 0$ and (2): $f_y = 2y + x = 0$. From (1), $y = -2x$, so that from (2), $2(-2x) + x = 0 \Rightarrow x = 0 \Rightarrow y = 0 \Rightarrow z^2 = 0 + 1 \Rightarrow z = \pm 1$, and there are two critical points, $(0, 0, \pm 1)$. Testing: $f_{xx}f_{yy} - f_{xy}^2 = 2 \cdot 2 - 1^2 = 3 > 0$ and $f_{xx} = 2 > 0$ shows that we have a minimum.

The points on the surface $z^2 = xy + 1$ closest to the origin are the points $(0, 0, \pm 1)$.

(b) Let $f(x, y, z)$ be the function D, with the constraint $g(x, y, z) = z^2 - xy = 1$. Then $\nabla f(x, y, z) = \lambda \nabla g(x, y, z) \Rightarrow 2x\,\mathbf{i} + 2y\,\mathbf{j} + 2z\,\mathbf{k} = \lambda[-y\,\mathbf{i} - x\,\mathbf{j} + 2z\,\mathbf{k}] \Rightarrow$

$$(1): \ 2x = -\lambda y, \quad (2): \ 2y = -\lambda x, \quad (3): \ 2z = 2\lambda z$$

From (3), either $z = 0$, or $\lambda = 1$:

If $z = 0$, then $z^2 - xy = 1 \Rightarrow xy = -1 \Rightarrow y = -\frac{1}{x}$. Putting this into (1), we get: $2x = -\lambda\left(-\frac{1}{x}\right) \Rightarrow \lambda = 2x^2$. Then from (2), $-\frac{2}{x} = -2x^2 \cdot x \Rightarrow x^4 = 1 \Rightarrow x = \pm 1 \Rightarrow y = -\frac{1}{x} = \mp 1$. We have two critical points: $(1, -1, 0)$ and $(-1, 1, 0)$.

If $\lambda = 1$, then from (1), $2x = -y \Rightarrow y = -2x$. Putting this into (2), we get: $2(-2x) = -x \Rightarrow x = 0 \Rightarrow y = 0$. Then $z^2 - xy = 1 \Rightarrow z = \pm 1$. We have two more critical points: $(0, 0, \pm 1)$.

Testing: $f(\pm 1, \mp 1, 0) = 1 + 1 + 0 = 2$ and $f(0, 0, \pm 1) = 0 + 0 + 1 = 1$.

Therefore, the points on the surface closest to the origin are $(0, 0, \pm 1)$.

55. Let the base of the box be x ft by y ft and the height be z ft. We want to minimize the cost, which in cents, is $C = 2(30xy) + 10(2xz + 2yz) = 60xy + 20xz + 20yz$, given the volume $V = xyz = 96$.

(a) Substituting from the volume equation, $z = \frac{96}{xy}$, into the cost function, call it now

$$f(x, y) = 60xy + 20x \cdot \frac{96}{xy} + 20y \cdot \frac{96}{xy} = 60xy + 1920y^{-1} + 1920x^{-1}$$

Then, $\nabla f = 0 \Rightarrow f_x = 60y - \frac{1920}{x^2} = 0 \Rightarrow y = \frac{32}{x^2}$, and

$$f_y = 60x - \frac{1920}{y^2} = 0 \implies 60x = \frac{1920}{\left(\frac{32}{x^2}\right)^2} \implies x = \frac{x^4}{32} \implies x^3 = 32 \implies x = 32^{\frac{1}{3}} = 2 \cdot 4^{\frac{1}{3}}.$$

Then, $y = \frac{32}{x^2} = \frac{32}{32^{\frac{2}{3}}} = 32^{\frac{1}{3}} = 2 \cdot 4^{\frac{1}{3}}$, and $z = \frac{96}{xy} = \frac{96}{32^{\frac{2}{3}}} = 3 \cdot 32^{\frac{1}{3}} = 6 \cdot 4^{\frac{1}{3}}$, and there is one critical point $(2 \cdot 4^{\frac{1}{3}}, 2 \cdot 4^{\frac{1}{3}}, 6 \cdot 4^{\frac{1}{3}})$.

Testing: $f_{xx} = 2(1920)x^{-3} = \frac{3840}{32} = 120$, $f_{yy} = 2(1920)y^{-3} = \frac{3840}{32} = 120$, and $f_{xy} = 60 \implies$ $f_{xx}f_{yy} - f_{xy}^2 = (120)(120) - (60^2) = 14400 - 3600 > 0$ and $f_{xx} = 120 > 0$ shows that we have a minimum.

Therefore, the dimensions of the most economical crate are a square base of side $2 \cdot 4^{\frac{1}{3}}$ ft and a height of $6 \cdot 4^{\frac{1}{3}}$ ft.

(b) $\nabla C = \lambda \nabla V \implies (60y + 20z)\mathbf{i} + (60x + 20z)\mathbf{j} + (20x + 20y)\mathbf{k} = \lambda(yz\,\mathbf{i} + xz\,\mathbf{j} + xy\,\mathbf{k}) \implies$

$\quad$ (1) $60y + 20z = \lambda yz$, $\quad$ (2) $60x + 20z = \lambda xz$, $\quad$ (3) $20x + 20y = \lambda xy$

Multiplying equation (1) by x, (2) by y, and (3) by z, we have

$$60xy + 20xz = 60xy + 20yz = 20xz + 20yz$$

Now, $60xy + 20xz = 60xy + 20yz \implies 20xz = 20yz \implies x = y$ and $60xy + 20yz = 20xz + 20yz \implies 60xy = 20xz \implies 3y = z$.

Substituting for x and z in terms of y in $V = xyz = 96$, then $y \cdot y \cdot 3y = 96 \implies y^3 = 32 \implies$ $y = 32^{\frac{1}{3}} = 2 \cdot 4^{\frac{1}{3}} = x$, and $z = 3y = 6 \cdot 4^{\frac{1}{3}}$, as in (a) above.

These dimensions result in minimum, not maximum cost, as, for example, a base of 1 ft by 1 ft and height of 96 ft would result in a cost of $60 + 20(96) + 20(96) = 3900$ cents, versus our cost of $60(32^{\frac{2}{3}}) + 20(3 \cdot 32^{\frac{2}{3}}) + 20(3 \cdot 32^{\frac{2}{3}}) = 180 \cdot 32^{\frac{2}{3}} \approx 1814$ cents.

---

57. It suffices to minimize the square of the distance from $(1,3,4)$ to the plane $2x - y + z = 1$. Let $D(x, y, z) = (x - 1)^2 + (y - 3)^2 + (z - 4)^2$.

(a) Substituting $1 - 2x + y$ for z in D, we have $f(x, y) = (x - 1)^2 + (y - 3)^2 + (-3 - 2x + y)^2$. Then $\nabla f = 0 \implies f_x = 2(x - 1) + 2(-3 - 2x + y)(-2) = 10x - 4y + 10 = 0 \implies$ (1) $5x - 2y = -5$ and $\qquad f_y = 2(y - 3) + 2(-3 - 2x + y) = -4x + 4y - 12 = 0 \implies$ (2) $-x + y = 3$. Adding equation (1) to 5 times equation (2), we have: $3y = 10 \implies y = \frac{10}{3}$. From (2), $x = y - 3 = \frac{1}{3}$, and $z = 1 - 2x + y = 1 - \frac{2}{3} + \frac{10}{3} = \frac{11}{3}$.

Testing: $f_{xx}f_{yy} - f_{xy}^2 = 10 \cdot 4 - (-4)^2 = 24 > 0$ and $f_{xx} = 10 > 0$ shows that we have a minimum. The required distance is

$$\sqrt{D\left(\frac{1}{3}, \frac{10}{3}, \frac{11}{3}\right)} = \sqrt{\left(\frac{1}{3} - 1\right)^2 + \left(\frac{10}{3} - 3\right)^2 + \left(\frac{11}{3} - 4\right)^2} = \sqrt{\frac{4}{9} + \frac{1}{9} + \frac{1}{9}} = \frac{\sqrt{6}}{3}$$

The shortest distance from $(1, 3, 4)$ to the plane $2x - y + z = 1$ is $\frac{\sqrt{6}}{3}$.

(b) Let $f(x, y, z) = D(x, y, z)$ in (a), and $g(x, y, z) = 2x - y + z = 1$. Now, $\nabla f(x, y, z) = \lambda \nabla g(x, y, z) \implies 2(x - 1)\mathbf{i} + 2(y - 3)\mathbf{j} + 2(z - 4)\mathbf{k} = \lambda[2\mathbf{i} - \mathbf{j} + \mathbf{k}] \implies$

$\qquad$ (1): $2(x - 1) = 2\lambda$, $\qquad$ (2): $2(y - 3) = -\lambda$, $\qquad$ (3): $2(z - 4) = \lambda$

Then $x = \lambda + 1$, $y = -\frac{\lambda}{2} + 3$ and $z = \frac{\lambda}{2} + 4$. Substituting into $2x - y + z = 1$,

$2(\lambda + 1) - (-\frac{\lambda}{2} + 3) + \frac{\lambda}{2} + 4 = 1 \Rightarrow 3\lambda + 3 = 1 \Rightarrow \lambda = -\frac{2}{3}$. Thus $x = -\frac{2}{3} + 1 = \frac{1}{3}$,

$y = -\frac{1}{2}(-\frac{2}{3}) + 3 = \frac{10}{3}$ and $z = \frac{1}{2}(-\frac{2}{3}) + 4 = \frac{11}{3}$, and the shortest distance is, as in (a), $\frac{\sqrt{6}}{3}$.
(This must be the minimum, as there is no maximum.)

59. We will maximize $f(x, y) = xy$ in the first quadrant, and then multiply that value by 4.

Thus we only consider $x > 0$ and $y > 0$. The constraint is that $g(x, y) = \dfrac{x^2}{a^2} + \dfrac{y^2}{b^2} = 1$. Take

$a > 0$ and $b > 0$. $\nabla f(x, y) = \lambda \nabla g(x, y) \Rightarrow y\mathbf{i} + x\mathbf{j} = \lambda\left[\left(\dfrac{2x}{a^2}\right)\mathbf{i} + \left(\dfrac{2y}{b^2}\right)\mathbf{j}\right] \Rightarrow y = \dfrac{2x\lambda}{a^2}$

and $x = \dfrac{2y\lambda}{b^2} \Rightarrow 2\lambda = \dfrac{b^2x}{y} = \dfrac{a^2y}{x} \Rightarrow x^2 = \dfrac{a^2y^2}{b^2}$. Substituting into $\dfrac{x^2}{a^2} + \dfrac{y^2}{b^2} = 1$ gives

$\dfrac{2y^2}{b^2} = 1 \Rightarrow y = \dfrac{b}{\sqrt{2}} \Rightarrow x = \dfrac{a}{\sqrt{2}} \Rightarrow$ the maximum area is $4\left(\dfrac{a}{\sqrt{2}} \cdot \dfrac{b}{\sqrt{2}}\right) = 2ab$.

61.   We want to minimize $C(x, y) = 2x^2 + xy + y^2 + 500$ with the constraint that $g(x, y) = x + y = 200$. Then $\nabla C(x, y) = \lambda \nabla g(x, y) \Rightarrow (4x + y)\mathbf{i} + (x + 2y)\mathbf{j} = \lambda[\mathbf{i} + \mathbf{j}] \Rightarrow 4x + y = \lambda$ and $x + 2y = \lambda$ which implies that $4x + y = x + 2y \Rightarrow y = 3x$. From $x + y = 200$, $x + (3x) = 200 \Rightarrow x = 50 \Rightarrow y = 150$.

63.   Let the lengths of the sides of the box be x, y, and z . We want to maximize the volume $V(x, y, z) = xyz$ subject to the constraints $x + y + z = 45$ and $y = 2x$. Let $g(x, y, z) = x + y + z = 45$ and $h(x, y, z) = y - 2x = 0$. Then

$$\nabla V(x, y, z) = \lambda \nabla g(x, y, z) + \mu \nabla h(x, y, z) \Rightarrow yz\mathbf{i} + xz\mathbf{j} + xy\mathbf{k} = \lambda[\mathbf{i} + \mathbf{j} + \mathbf{k}] + \mu[-2\mathbf{i} + \mathbf{j}] \Rightarrow$$

$$(1): yz = \lambda - 2\mu, \quad (2): xz = \lambda + \mu, \quad (3): xy = \lambda$$

Substituting from (3) into (2): $xz = xy + \mu \Rightarrow \mu = xz - xy$. Substituting into (1): $yz = xy - 2(xz - xy)$. As $y = 2x$, $2xz = x(2x) - 2xz + 2x(2x) \Rightarrow 4xz = 6x^2 \Rightarrow 2z = 3x \Rightarrow z = \frac{3}{2}x$. Substituting for both y and z in $x + y + z = 45$, then $x + 2x + \frac{3}{2}x = 45 \Rightarrow x = 10 \Rightarrow y = 2(10) = 20$ and $z = \frac{3}{2}x = 15$.
The dimensions of the box are 10 in. by 20 in. by 15 in.

65. Let x, y, z denote the width, length, and height, resp., of the box. We want to maximize the volume $V(x, y, z) = xyz$ subject to the constraint $g(x, y, z) = xy + 2xz + 2yz = S$. Then $\nabla V(x, y, z) = \lambda \nabla g(x, y, z) \Rightarrow yz\mathbf{i} + xz\mathbf{j} + xy\mathbf{k} = \lambda[(y + 2z)\mathbf{i} + (x + 2z)\mathbf{j} + (2x + 2y)\mathbf{k}] \Rightarrow$

$$yz = (y + 2z)\lambda, \quad xz = (x + 2z)\lambda, \quad xy = (2x + 2y)\lambda$$

Solve for λ in each equation, and then equate the results:

$$\dfrac{yz}{y + 2z} = \dfrac{xz}{x + 2z} = \dfrac{xy}{2x + 2y}$$

$\dfrac{yz}{y + 2z} = \dfrac{xz}{x + 2z} \Rightarrow yz(x + 2z) = xz(y + 2z) \Rightarrow xyz + 2yz^2 = xyz + 2xz^2 \Rightarrow x = y$.

$$\frac{xz}{x+2z} = \frac{xy}{2x+2y} \;\Rightarrow\; xz(2x+2y) = xy(x+2z) \;\Rightarrow\; 2x^2z+2xyz = x^2y+2xyz \;\Rightarrow\; y = 2z.$$

As $xy + 2xz + 2yz = S$, $(2z)(2z) + 2(2z)z + 2(2z)z = S \;\Rightarrow\; 12z^2 = S \Rightarrow z = \sqrt{\frac{S}{12}} = \frac{1}{2}\sqrt{\frac{S}{3}}$.

Then, $x = y = 2z = \sqrt{\frac{S}{3}}$. The dimensions are a base of $\sqrt{\frac{S}{3}}$ by $\sqrt{\frac{S}{3}}$, and a height of $\frac{1}{2}\sqrt{\frac{S}{3}}$.

67.   We want to maximize the area A of the trapezoid, where

$$A(x,\alpha) = 2 \cdot \frac{1}{2}(x\cos\alpha)(x\sin\alpha) + (x\sin\alpha)(l - 2x)$$
$$= x^2 \sin\alpha\cos\alpha + (xl - 2x^2)\sin\alpha, \quad x > 0,\; 0 < \alpha < \frac{\pi}{2}$$

The maximum occurs when $\nabla A = 0 \;\Rightarrow\; A_x = A_\alpha = 0$:

$\quad A_x = 2x\sin\alpha\cos\alpha + (l - 4x)\sin\alpha$, and

$\quad A_\alpha = x^2[(\sin\alpha)(-\sin\alpha) + \cos\alpha\cos\alpha] + (xl - 2x^2)\cos\alpha$

$\qquad = x\{x[(\sin\alpha)(-\sin\alpha) + \cos\alpha\cos\alpha] + (l - 2x)\cos\alpha\}$

$A_x = 0 \;\Rightarrow\; 2x\sin\alpha\cos\alpha + (l - 4x)\sin\alpha = 0 \;\Rightarrow\; x = \dfrac{l\sin\alpha}{4\sin\alpha - 2\sin\alpha\cos\alpha} = \dfrac{l}{4 - 2\cos\alpha}$

so that $A_\alpha = 0 \;\Rightarrow\; A_\alpha/x = 0$, as $x \neq 0$. Thus

$$0 = \left(\frac{l}{4 - 2\cos\alpha}\right)[(\sin\alpha)(-\sin\alpha) + \cos\alpha\cos\alpha] + \left[l - 2\left(\frac{l}{4 - 2\cos\alpha}\right)\right]\cos\alpha$$

$$0 = [(\sin\alpha)(-\sin\alpha) + \cos\alpha\cos\alpha] + [(4 - 2\cos\alpha) - 2]\cos\alpha$$

$$= -\sin^2\alpha + \cos^2\alpha + 2\cos\alpha - 2\cos^2\alpha = -(\sin^2\alpha + \cos^2\alpha) + 2\cos\alpha$$

$$= 2\cos\alpha - 1 \;\Rightarrow\; \cos\alpha = \frac{1}{2} \;\Rightarrow\; \alpha = 60° \;\Rightarrow\; x = \frac{l}{4 - 2\cdot\frac{1}{2}} = \frac{l}{3}.$$

CHAPTER 14
Vector Calculus

§14.1 Line Integrals

1. With $\mathbf{r}_0 = (0,0)$, and $\mathbf{r}_1 = (1,1)$, a parametrization of C is $\mathbf{r}(t) = (1-t)\mathbf{r}_0(t) + t\mathbf{r}_1(t) = (1-t)(0,0) + t(1,1) = (t,t)$, for $0 \le t \le 1$. Then $\frac{dx}{dt} = \frac{dy}{dt} = 1$, and $f(x,y) = x^2 + y \Rightarrow$

$$\int_C f(x,y)ds = \int_0^1 (t^2 + t)\sqrt{\left(\frac{dx}{dt}\right)^2 + \left(\frac{dy}{dt}\right)^2}\, dt = \int_0^1 (t^2 + t)\sqrt{1^2 + 1^2}\, dt = \sqrt{2}\left[\frac{1}{3}t^3 + \frac{1}{2}t^2\right]_0^1$$

$$= \sqrt{2}\left(\frac{1}{3} + \frac{1}{2}\right) = \frac{5\sqrt{2}}{6}.$$

3. Parametrizing C: $x = \cos t$, $y = \sin t$, $0 \le t \le \pi \Rightarrow \frac{dx}{dt} = -\sin t$ and $\frac{dy}{dt} = \cos t$. Then $f(x,y) = 2 + x^2 y \Rightarrow$

$$\int_C f(x,y)ds = \int_0^\pi (2 + \cos^2 t \sin t)\sqrt{\left(\frac{dx}{dt}\right)^2 + \left(\frac{dy}{dt}\right)^2}\, dt$$

$$= \int_0^\pi (2 + \cos^2 t \sin t)\sqrt{(-\sin t)^2 + (\cos t)^2}\, dt = \int_0^\pi (2 + \cos^2 t \sin t)\, dt$$

$$= \left[2t - \frac{1}{3}\cos^3 t\right]_0^\pi = 2\pi - \frac{1}{3}(\cos^3 \pi - \cos^3 0) = 2\pi + \frac{2}{3}.$$

5. Parametrizing C: $x = 2\cos t$, $y = 2\sin t$, $-\frac{\pi}{2} \le t \le \frac{\pi}{2} \Rightarrow \frac{dx}{dt} = -2\sin t$ and $\frac{dy}{dt} = 2\cos t$. Then $f(x,y) = xy^4 \Rightarrow$

$$\int_C f(x,y)ds = \int_{-\pi/2}^{\pi/2} (2\cos t)(2\sin t)^4 \sqrt{\left(\frac{dx}{dt}\right)^2 + \left(\frac{dy}{dt}\right)^2}\, dt$$

$$= \int_{-\pi/2}^{\pi/2} (2\cos t)(2\sin t)^4 \sqrt{(-2\sin t)^2 + (2\cos t)^2}\, dt = 2^6 \int_{-\pi/2}^{\pi/2} \sin^4 t \cos t\, dt$$

$$= 64\left[\frac{1}{5}\sin^5 t\right]_{-\pi/2}^{\pi/2} = \frac{64}{5}\left[\sin^5 \frac{\pi}{2} - \sin^5\left(-\frac{\pi}{2}\right)\right] = \frac{64}{5}[1 - (-1)] = \frac{128}{5}.$$

7. Since the curve C is $y = x^2$, from $(0,0)$ to $(3,9)$, a parametrization is: $x = t$, $y = t^2$, $0 \le t \le 3 \Rightarrow \frac{dx}{dt} = 1$ and $\frac{dy}{dt} = 2t$. Then $f(x,y) = 3x \Rightarrow$

$$\int_C f(x,y)ds = \int_0^3 3t\sqrt{\left(\frac{dx}{dt}\right)^2 + \left(\frac{dy}{dt}\right)^2}\, dt = \int_0^3 3t\sqrt{1 + 4t^2}\, dt = \frac{3}{8}\int_1^{37} u^{\frac{1}{2}}\, du = \frac{3}{8}\left[\frac{2}{3}u^{\frac{3}{2}}\right]_1^{37}$$

$$\boxed{\begin{array}{ll} u = 1 + 4t^2, & t = 0 \Rightarrow u = 1 \\ du = 8t\, dt & t = 3 \Rightarrow u = 37 \end{array}}$$

$$= \frac{1}{4}\left[37^{\frac{3}{2}} - 1\right] = \frac{1}{4}\left(37\sqrt{37} - 1\right).$$

9.

$C = C_1 \cup C_2$ (see figure), and parametrizations are: $C_1 : x = t$, $y = t$, for $0 \leq t \leq 1$, with $\frac{dx}{dt} = \frac{dy}{dt} = 1$ and $C_2 : x = 1$, $y = t$, for $1 \leq t \leq 2$, with $\frac{dx}{dt} = 0$ and $\frac{dy}{dt} = 1$. Then $f(x, y) = xy \Rightarrow$

$$\int_C f(x,y)ds = \int_{C_1} f(x,y)ds + \int_{C_2} f(x,y)ds = \int_{C_1} t^2\sqrt{\left(\frac{dx}{dt}\right)^2 + \left(\frac{dy}{dt}\right)^2}\, dt + \int_{C_2} t\sqrt{\left(\frac{dx}{dt}\right)^2 + \left(\frac{dy}{dt}\right)^2}\, dt$$

$$= \int_0^1 t^2\sqrt{1+1}\, dt + \int_1^2 t\sqrt{0+1}\, dt = \sqrt{2}\left[\tfrac{1}{3}t^3\right]_0^1 + \left[\tfrac{1}{2}t^2\right]_1^2 = \tfrac{\sqrt{2}}{3} + 2 - \tfrac{1}{2} = \tfrac{2\sqrt{2}+9}{6}.$$

11.

$C = C_1 \cup C_2 \cup C_3$ (see figure), and parametrizations are:

$$C_1 : x = t,\ y = -1,\quad -2 \leq t \leq 0,\quad \tfrac{dx}{dt} = 1,\ \tfrac{dy}{dt} = 0$$

$$C_2 : x = t,\ y = t^3 - 1,\quad 0 \leq t \leq 1,\quad \tfrac{dx}{dt} = 1,\ \tfrac{dy}{dt} = 3t^2$$

$$C_3 : x = 1,\ y = t,\quad 0 \leq t \leq 2,\quad \tfrac{dx}{dt} = 0,\ \tfrac{dy}{dt} = 1$$

Then $f(x, y) = 4x^3 \Rightarrow$

$$\int_C f(x,y)ds = \int_{C_1} f(x,y)ds + \int_{C_2} f(x,y)ds + \int_{C_3} f(x,y)ds$$

$$\int_{C_1} f(x,y)ds = \int_{-2}^0 4t^3\sqrt{\left(\frac{dx}{dt}\right)^2 + \left(\frac{dy}{dt}\right)^2}\, dt = \int_{-2}^0 4t^3\sqrt{1+0}\, dt = \left[t^4\right]_{-2}^0 = -16$$

$$\int_{C_2} f(x,y)ds = \int_0^1 4t^3\sqrt{\left(\frac{dx}{dt}\right)^2 + \left(\frac{dy}{dt}\right)^2}\, dt = \int_0^1 4t^3\sqrt{1+9t^4}\, dt = \tfrac{1}{9}\int_1^{10} u^{\frac{1}{2}}\, du$$

$$\boxed{\begin{array}{l} u = 1 + 9t^4,\ t = 0 \Rightarrow u = 1 \\ du = 36t^3\, dt \quad t = 1 \Rightarrow u = 10 \end{array}}$$

$$= \tfrac{1}{9}\left[\tfrac{2}{3}u^{\frac{3}{2}}\right]_1^{10} = \tfrac{2}{27}(10\sqrt{10} - 1)$$

$$\int_{C_3} f(x,y)ds = \int_0^2 4\sqrt{\left(\frac{dx}{dt}\right)^2 + \left(\frac{dy}{dt}\right)^2}\, dt = \int_0^2 4\sqrt{0+1}\, dt = 4 \cdot (2 - 0) = 8$$

Therefore,

$$\int_C f(x,y)ds = -16 + \tfrac{2}{27}\left(10\sqrt{10} - 1\right) + 8 = -8 + \tfrac{2}{27}\left(10\sqrt{10} - 1\right) = \tfrac{20\sqrt{10} - 218}{27}.$$

13. With $\mathbf{r}_0 = (0, 0, 0)$, and $\mathbf{r}_1 = (1, 1, 1)$, a parametrization of C is:

$$\mathbf{r}(t) = (1 - t)\mathbf{r}_0(t) + t\mathbf{r}_1(t) = (1 - t)(0, 0, 0) + t(1, 1, 1) = (t, t, t),\ \text{for } 0 \leq t \leq 1$$

Then $\frac{dx}{dt} = \frac{dy}{dt} = \frac{dz}{dt} = 1$, and $f(x, y, z) = x - 3y^2 + z \Rightarrow$

$$\int_C f(x,y,z)ds = \int_0^1 (t - 3t^2 + t)\sqrt{\left(\frac{dx}{dt}\right)^2 + \left(\frac{dy}{dt}\right)^2 + \left(\frac{dz}{dt}\right)^2}\, dt = \int_0^1 (2t - 3t^2)\sqrt{1+1+1}\, dt$$

$$= \sqrt{3}\int_0^1 (2t - 3t^2)\, dt = \sqrt{3}\left[t^2 - t^3\right]_0^1 = \sqrt{3}(1 - 1) = 0.$$

15.   With $f(x, y, z) = xyz$, and $x(t) = \cos t$, $y(t) = \sin t$, $z(t) = 3t$ for $0 \leq t \leq 4\pi$, then $\frac{dx}{dt} = -\sin t$, $\frac{dy}{dt} = \cos t$ and $\frac{dz}{dt} = 3$, so that

$$\int_C xyz \, ds = \int_0^{4\pi} 3t \sin t \cos t \sqrt{\left(\frac{dx}{dt}\right)^2 + \left(\frac{dy}{dt}\right)^2 + \left(\frac{dz}{dt}\right)^2} \, dt$$

$$= \int_0^{4\pi} 3t \sin t \cos t \sqrt{\sin^2 t + \cos^2 t + 9} \, dt = 3\sqrt{10} \int_0^{4\pi} t \sin t \cos t \, dt$$

$$\boxed{\begin{array}{ll} u = t, & dv = \sin t \cos t \, dt \\ du = dt & v = \frac{1}{2} \sin^2 t \end{array}}$$

$$= 3\sqrt{10} \left(\left[\tfrac{1}{2} t \sin^2 t \right]_0^{4\pi} - \int_0^{4\pi} \tfrac{1}{2} \sin^2 t \, dt \right) = -\frac{3\sqrt{10}}{2} \int_0^{4\pi} \left(\frac{1 - \cos 2t}{2} \right) dt$$

$$= -\frac{3\sqrt{10}}{4} \left[t - \tfrac{1}{2} \sin 2t \right]_0^{4\pi} = -3\sqrt{10}\,\pi.$$

17.   With $\delta(x, y) = y + 2$, and $x(t) = 2\cos t$, $y(t) = 2\sin t$ for $0 \leq t \leq \pi$, then $\frac{dx}{dt} = -2\sin t$, $\frac{dy}{dt} = 2\cos t$, so that

$$M = \int_C (y + 2) \, ds = \int_0^\pi (2\sin t + 2) \sqrt{\left(\frac{dx}{dt}\right)^2 + \left(\frac{dy}{dt}\right)^2} \, dt = \int_0^\pi (2\sin t + 2)\sqrt{4\sin^2 t + 4\cos^2 t} \, dt$$

$$= 4\int_0^\pi (\sin t + 1) \, dt = 4\left[-\cos t + t \right]_0^\pi = 4(-\cos\pi + \pi + \cos 0) = 4(\pi + 2) = 4\pi + 8.$$

19.   With $\delta(x, y, z) = x^2 + y^2 + z^2$, and $x(t) = 3\cos t$, $y(t) = 3\sin t$, $z(t) = 4t$ for $0 \leq t \leq \pi$, then $\frac{dx}{dt} = -3\sin t$, $\frac{dy}{dt} = 3\cos t$, and $\frac{dz}{dt} = 4$, so that

$$M = \int_C (x^2 + y^2 + z^2) \, ds = \int_0^\pi (9\cos^2 t + 9\sin^2 t + 16t^2) \sqrt{\left(\frac{dx}{dt}\right)^2 + \left(\frac{dy}{dt}\right)^2 + \left(\frac{dz}{dt}\right)^2} \, dt$$

$$= \int_0^\pi (9\cos^2 t + 9\sin^2 t + 16t^2)\sqrt{9\sin^2 t + 9\cos^2 t + 16} \, dt = \int_0^\pi (9 + 16t^2)\sqrt{9 + 16} \, dt$$

$$= 5\int_0^\pi (9 + 16t^2) \, dt = 5\left[9t + \tfrac{16}{3}t^3 \right]_0^\pi = 5\left(9\pi + \tfrac{16}{3}\pi^3 \right) = 45\pi + \tfrac{80}{3}\pi^3.$$

21.   For $\mathbf{F}(x, y) = xy\,\mathbf{i} + y^2\,\mathbf{j}$ and $\mathbf{r}(t) = \cos t\,\mathbf{i} + \sin t\,\mathbf{j}$, $0 \leq t \leq \frac{\pi}{3}$:

$$\int_C \mathbf{F} \cdot d\mathbf{r} = \int_C \mathbf{F}(\mathbf{r}(t)) \cdot \mathbf{r}'(t) \, dt = \int_0^{\pi/3} (\sin t \cos t\,\mathbf{i} + \sin^2 t\,\mathbf{j}) \cdot (-\sin t\,\mathbf{i} + \cos t\,\mathbf{j}) \, dt$$

$$= \int_0^{\pi/3} (-\sin^2 t \cos t + \sin^2 t \cos t) \, dt = \int_0^{\pi/3} 0 \, dt = 0.$$

23.   For $\mathbf{F}(x, y) = (e^x + y)\mathbf{i} + x\,\mathbf{j}$ and $\mathbf{r}(t) = t\,\mathbf{i} + t^2\,\mathbf{j}$, $0 \leq t \leq 1$:

$$\int_C \mathbf{F} \cdot d\mathbf{r} = \int_C \mathbf{F}(\mathbf{r}(t)) \cdot \mathbf{r}'(t) \, dt = \int_0^1 [(e^t + t^2)\mathbf{i} + t\,\mathbf{j}] \cdot (\mathbf{i} + 2t\,\mathbf{j}) \, dt$$

$$= \int_0^1 (e^t + t^2 + 2t^2) \, dt = \int_0^1 (e^t + 3t^2) \, dt = \left[e^t + t^3 \right]_0^1 = e + 1 - e^0 = e.$$

25. For $\mathbf{F}(x, y) = x^2\,\mathbf{i} - y\,\mathbf{j}$ and $\mathbf{r}(t) = t\,\mathbf{i} + e^t\,\mathbf{j},\ 0 \le t \le 2$:

$$\int_C \mathbf{F} \cdot d\mathbf{r} = \int_C \mathbf{F}(\mathbf{r}(t)) \cdot \mathbf{r}'(t)\,dt = \int_0^2 (t^2\,\mathbf{i} - e^t\,\mathbf{j}] \cdot (\mathbf{i} + e^t\,\mathbf{j})\,dt$$

$$= \int_0^2 (t^2 - e^{2t})\,dt = \left[\tfrac{1}{3}t^3 - \tfrac{1}{2}e^{2t}\right]_0^2 = \tfrac{8}{3} - \tfrac{1}{2}e^4 + \tfrac{1}{2}e^0 = \tfrac{19}{6} - \tfrac{1}{2}e^4.$$

27. For $\mathbf{F}(x, y, z) = 8x^2 yz\,\mathbf{i} + 5z\,\mathbf{j} - 4xy\,\mathbf{k}$ and $\mathbf{r}(t) = t\,\mathbf{i} + t^2\,\mathbf{j} + t^3\,\mathbf{k},\ 0 \le t \le 1$,
$\mathbf{F}(\mathbf{r}(t)) = 8t^2(t^2)(t^3)\mathbf{i} + 5t^3\,\mathbf{j} - 4t(t^2)\mathbf{k} = 8t^7\,\mathbf{i} + 5t^3\,\mathbf{j} - 4t^3\,\mathbf{k}$ and $\mathbf{r}'(t) = \mathbf{i} + 2t\,\mathbf{j} + 3t^2\,\mathbf{k} \Rightarrow$

$$\int_C \mathbf{F} \cdot d\mathbf{r} = \int_C \mathbf{F}(\mathbf{r}(t)) \cdot \mathbf{r}'(t)\,dt = \int_0^1 (8t^7\,\mathbf{i} + 5t^3\,\mathbf{j} - 4t^3\,\mathbf{k}) \cdot (\mathbf{i} + 2t\,\mathbf{j} + 3t^2\,\mathbf{k})\,dt$$

$$= \int_0^1 (8t^7 + 10t^4 - 12t^5)\,dt = \left[t^8 + 2t^5 - 2t^6\right]_0^1 = 1 + 2 - 2 = 1.$$

29. For $\mathbf{F}(x, y, z) = (2xz + y^2)\mathbf{i} + 2xy\,\mathbf{j} + (x^2 + 3z^2)\mathbf{k}$ and $\mathbf{r}(t) = t^2\,\mathbf{i} + (t + 1)\mathbf{j} + (2t - 1)\mathbf{k}$,
for $0 \le t \le 1$, then $\mathbf{r}'(t) = 2t\,\mathbf{i} + \mathbf{j} + 2\,\mathbf{k}$ and
$\mathbf{F}(\mathbf{r}(t)) = [2t^2(2t - 1) + (t + 1)^2]\mathbf{i} + 2t^2(t + 1)\mathbf{j} + [t^4 + 3(2t - 1)^2]\mathbf{k}$
$$= (4t^3 - 2t^2 + t^2 + 2t + 1)\mathbf{i} + (2t^3 + 2t^2)\mathbf{j} + (t^4 + 12t^2 - 12t + 3)\mathbf{k} \Rightarrow$$
$\mathbf{F}(\mathbf{r}(t)) \cdot \mathbf{r}'(t) = (8t^4 - 2t^3 + 4t^2 + 2t) + (2t^3 + 2t^2) + (2t^4 + 24t^2 - 24t + 6)$
$$= 10t^4 + 30t^2 - 22t + 6 \Rightarrow$$
$$\int_C \mathbf{F} \cdot d\mathbf{r} = \int_C \mathbf{F}(\mathbf{r}(t)) \cdot \mathbf{r}'(t)\,dt = \int_0^1 (10t^4 + 30t^2 - 22t + 6)\,dt = \left[2t^5 + 10t^3 - 11t^2 + 6t\right]_0^1$$
$$= 2 + 10 - 11 + 6 = 7.$$

31. For $\mathbf{F}(x, y, z) = \sin x\,\mathbf{i} + \cos y\,\mathbf{j} + xz\,\mathbf{k}$ and $\mathbf{r}(t) = t^3\,\mathbf{i} + t^2\,\mathbf{j} + t\,\mathbf{k},\ 0 \le t \le 1$,
$\mathbf{F}(\mathbf{r}(t)) = \sin t^3\,\mathbf{i} + \cos t^2\,\mathbf{j} + t^4\,\mathbf{k}$ and $\mathbf{r}'(t) = 3t^2\,\mathbf{i} + 2t\,\mathbf{j} + \mathbf{k} \Rightarrow$

$$\int_C \mathbf{F} \cdot d\mathbf{r} = \int_C \mathbf{F}(\mathbf{r}(t)) \cdot \mathbf{r}'(t)\,dt = \int_0^1 (\sin t^3\,\mathbf{i} + \cos t^2\,\mathbf{j} + t^4\,\mathbf{k}) \cdot (3t^2\,\mathbf{i} + 2t\,\mathbf{j} + \mathbf{k})\,dt$$

$$= \int_0^1 (3t^2 \sin t^3 + 2t \cos t^2 + t^4)\,dt = \int_0^1 \sin u\,du + \int_0^1 \cos v\,dv + \left[\tfrac{1}{5}t^5\right]_0^1$$

$u = t^3,\quad t = 0 \Rightarrow u = 0$	$v = t^2,\quad t = 0 \Rightarrow v = 0$
$du = 3t^2\,dt,\ t = 1 \Rightarrow u = 1$	$dv = 2t\,dt,\ t = 1 \Rightarrow v = 1$

$$= \left[-\cos u\right]_0^1 + \left[\sin v\right]_0^1 + \tfrac{1}{5} = -\cos 1 + \cos 0 + \sin 1 - \sin 0 + \tfrac{1}{5} = \tfrac{6}{5} - \cos 1 + \sin 1.$$

33. For $\mathbf{F}(x, y) = xy\,\mathbf{i} + x^3\,\mathbf{j}$ and $\mathbf{r}(t) = t^{\frac{1}{2}}\,\mathbf{i} + t^{\frac{1}{4}}\,\mathbf{j},\ 1 \le t \le 16 \Rightarrow \mathbf{r}'(t) = \tfrac{1}{2}t^{-\frac{1}{2}}\,\mathbf{i} + \tfrac{1}{4}t^{-\frac{3}{4}}\,\mathbf{j}$:

$$W = \int_a^b \mathbf{F}(\mathbf{r}(t)) \cdot \mathbf{r}'(t)\,dt = \int_1^{16} \left(t^{\frac{1}{2}}t^{\frac{1}{4}}\,\mathbf{i} + t^{\frac{3}{2}}\,\mathbf{j}\right) \cdot \left(\tfrac{1}{2}t^{-\frac{1}{2}}\,\mathbf{i} + \tfrac{1}{4}t^{-\frac{3}{4}}\,\mathbf{j}\right)\,dt = \int_1^{16} \left(\tfrac{1}{2}t^{\frac{1}{4}} + \tfrac{1}{4}t^{\frac{3}{4}}\right)\,dt$$

$$= \left[\tfrac{1}{2} \cdot \tfrac{4}{5}t^{\frac{5}{4}} + \tfrac{1}{4} \cdot \tfrac{4}{7}t^{\frac{7}{4}}\right]_1^{16} = \tfrac{2}{5}(16)^{\frac{5}{4}} + \tfrac{1}{7}(16)^{\frac{7}{4}} - \left(\tfrac{2}{5} + \tfrac{1}{7}\right) = \tfrac{2}{5}(32) + \tfrac{1}{7}(128) - \tfrac{19}{35} = \tfrac{1069}{35}.$$

35. For $\mathbf{F}(x,y) = x\mathbf{i} + (y+2)\mathbf{j}$ and $\mathbf{r}(t) = (t - \sin t)\mathbf{i} + (1 - \cos t)\mathbf{j}$, $0 \le t \le 2\pi \Rightarrow$
$\mathbf{r}'(t) = (1 - \cos t)\mathbf{i} + \sin t\,\mathbf{j}$:

$$W = \int_a^b \mathbf{F}(\mathbf{r}(t)) \cdot \mathbf{r}'(t)\, dt = \int_0^{2\pi} [(t - \sin t)\mathbf{i} + (3 - \cos t)\mathbf{j}] \cdot [(1 - \cos t)\mathbf{i} + \sin t\,\mathbf{j}]\, dt$$

$$= \int_0^{2\pi} [(t - \sin t)(1 - \cos t) + (3 - \cos t)\sin t]\, dt = \left[\tfrac{1}{2}(t - \sin t)^2 + \tfrac{1}{2}(3 - \cos t)^2\right]_0^{2\pi}$$

$$= \tfrac{1}{2}(2\pi - 0)^2 + \tfrac{1}{2}(3 - \cos 2\pi)^2 - \tfrac{1}{2}(3 - \cos 0)^2 = 2\pi^2 + 2 - 2 = 2\pi^2.$$

37. For $\mathbf{F}(x,y,z) = (y - x^2)\mathbf{i} + (z - y^2)\mathbf{j} + (x - z^2)\mathbf{k}$ and $\mathbf{r}(t) = t\mathbf{i} + t^2\mathbf{j} + t^3\mathbf{k}$, $0 \le t \le 1$,
$\mathbf{F}(\mathbf{r}(t)) = (t^2 - t^2)\mathbf{i} + (t^3 - t^4)\mathbf{j} + (t - t^6)\mathbf{k} = (t^3 - t^4)\mathbf{j} + (t - t^6)\mathbf{k}$ and $\mathbf{r}'(t) = \mathbf{i} + 2t\mathbf{j} + 3t^2\mathbf{k} \Rightarrow$

$$W = \int_a^b \mathbf{F}(\mathbf{r}(t)) \cdot \mathbf{r}'(t)\, dt = \int_0^1 [(t^3 - t^4)\mathbf{j} + (t - t^6)\mathbf{k}] \cdot [\mathbf{i} + 2t\mathbf{j} + 3t^2\mathbf{k}]\, dt$$

$$= \int_0^1 (2t^4 - 2t^5 + 3t^3 - 3t^8)\, dt = \left[\tfrac{2}{5}t^5 - \tfrac{1}{3}t^6 + \tfrac{3}{4}t^4 - \tfrac{1}{3}t^9\right]_0^1 = \tfrac{2}{5} - \tfrac{1}{3} + \tfrac{3}{4} - \tfrac{1}{3} = \tfrac{29}{60}.$$

39. For $\mathbf{F}(x,y,z) = x^2\mathbf{i} + y^2\mathbf{j} + z^2\mathbf{k}$ and $\mathbf{r}(t) = (1 + t^2)\mathbf{i} + t^2\mathbf{j} + (2 + \sin \pi t)\mathbf{k}$, $0 \le t \le 1$,
$\mathbf{F}(\mathbf{r}(t)) = (1 + t^2)^2\mathbf{i} + t^4\mathbf{j} + (2 + \sin \pi t)^2\mathbf{k}$ and $\mathbf{r}'(t) = 2t\mathbf{i} + 2t\mathbf{j} + (\pi \cos \pi t)\mathbf{k} \Rightarrow$
$\mathbf{F}(\mathbf{r}(t)) \cdot \mathbf{r}'(t) = 2t(1 + t^2)^2 + 2t^5 + (\pi \cos \pi t)(2 + \sin \pi t)^2 \Rightarrow$

$$W = \int_a^b \mathbf{F}(\mathbf{r}(t)) \cdot \mathbf{r}'(t)\, dt = \int_0^1 [2t(1 + t^2)^2 + 2t^5 + (\pi \cos \pi t)(2 + \sin \pi t)^2]\, dt$$

$$= \left[\tfrac{1}{3}(1 + t^2)^3 + \tfrac{1}{3}t^6 + \tfrac{1}{3}(2 + \sin \pi t)^3\right]_0^1 = \tfrac{8}{3} + \tfrac{1}{3} + \tfrac{8}{3} - \left(\tfrac{1}{3} + \tfrac{8}{3}\right) = \tfrac{8}{3}.$$

§14.2 Conservative Fields and Path Independence

1. The vector field $\mathbf{F}(x,y) = (x - y)\mathbf{i} + (x - 2)\mathbf{j} = P(x,y)\mathbf{i} + Q(x,y)\mathbf{j}$, with $P(x,y) = x - y$ and $Q(x,y) = x - 2$. Then $P_y = -1$ while $Q_x = 1 \Rightarrow \mathbf{F}$ is not conservative.

3. The vector field $\mathbf{F}(x,y) = 2xy^3\mathbf{i} + (1 + 3x^2y^2)\mathbf{j} = P(x,y)\mathbf{i} + Q(x,y)\mathbf{j}$, with $P(x,y) = 2xy^3$ and $Q(x,y) = 1 + 3x^2y^2$. Then $P_y = 6xy^2 = Q_x \Rightarrow \mathbf{F}$ is conservative, and there exists $f(x,y)$ for which $\mathbf{F} = \nabla f$: $2xy^3\mathbf{i} + (1 + 3x^2y^2)\mathbf{j} = f_x(x,y)\mathbf{i} + f_y(x,y)\mathbf{j} \Rightarrow f_x(x,y) = 2xy^3$ and $f_y(x,y) = 1 + 3x^2y^2$.
Integrating f_x wrt x: $f(x,y) = \int 2xy^3 dx = x^2y^3 + g(y) \Rightarrow f_y(x,y) = 3x^2y^2 + g'(y)$, but $f_y(x,y) = 1 + 3x^2y^2$, so that $g'(y) = 1 \Rightarrow g(y) = y + C$. Taking $C = 0 \Rightarrow f(x,y) = x^2y^3 + y$.

5. The vector field $\mathbf{F}(x,y) = (y\cos x + 2xe^y)\mathbf{i} + (\sin x + xe^y + 4)\mathbf{j} = P(x,y)\mathbf{i} + Q(x,y)\mathbf{j}$, with $P(x,y) = y\cos x + 2xe^y$ and $Q(x,y) = \sin x + xe^y + 4$. Then $P_y = \cos x + 2xe^y$ while $Q_x = \cos x + e^y \Rightarrow \mathbf{F}$ is not conservative.

7. The vector field $\mathbf{F}(x,y,z) = yz^2\mathbf{i} + (2xyz - y)\mathbf{j} + (xz^2 - z)\mathbf{k} = P(x,y,z)\mathbf{i} + Q(x,y,z)\mathbf{j} + R(x,y,z)\mathbf{k}$, with $P(x,y,z) = yz^2$, $Q(x,y,z) = 2xyz - y$, and $R(x,y,z) = xz^2 - z$. Then $P_y = z^2$ while $Q_x = 2yz \Rightarrow \mathbf{F}$ is not conservative.

9.   The vector field $\mathbf{F}(x,y,z) = (e^x \cos y + yz)\mathbf{i} + (xz - e^x \sin y)\mathbf{j} + (xy + z)\mathbf{k} = P(x,y,z)\mathbf{i} + Q(x,y,z)\mathbf{j} + R(x,y,z)\mathbf{k}$, with $P(x,y,z) = e^x \cos y + yz$, $Q(x,y,z) = xz - e^x \sin y$, and $R(x,y,z) = xy + z$. Then
$$P_y = (e^x \cos y + yz)_y = e^x(-\sin y) + z = -e^x \sin y + z = Q_x$$
$$P_z = (e^x \cos y + yz)_z = y = R_x, \quad \text{and} \quad Q_z = x = R_y \Rightarrow \mathbf{F} \text{ is conservative and there}$$
exists $f(x,y,z)$ for which $\mathbf{F} = \nabla f \Rightarrow$
$$f_x = P = e^x \cos y + yz, \ f_y = Q = xz - e^x \sin y, \text{ and } f_z = R = xy + z$$
Integrating f_x wrt x: $f(x,y,z) = \int(e^x \cos y + yz)dx = e^x \cos y + xyz + g(y,z) \Rightarrow$
$f_y = -e^x \sin y + xz + g_y(y,z)$, but $f_y = Q = xz - e^x \sin y \Rightarrow g_y(y,z) = 0 \Rightarrow g(y,z)$ is a function of only z, call it $h(z)$. Thus, $f(x,y,z) = e^x \cos y + xyz + h(z)$.
   Differentiating this wrt z: $f_z = xy + h'(z) = R = xy + z \Rightarrow h'(z) = z \Rightarrow h(z) = \frac{1}{2}z^2 + C$.
Taking $C = 0$, we have $f(x,y,z) = e^x \cos y + xyz + \frac{1}{2}z^2$.

11.   The vector field $\mathbf{F}(x,y,z) = y^2\mathbf{i} + (2xy + e^{3z})\mathbf{j} + 3ye^{3z}\mathbf{k} = P(x,y,z)\mathbf{i} + Q(x,y,z)\mathbf{j} + R(x,y,z)\mathbf{k}$, with $P(x,y,z) = y^2$, $Q(x,y,z) = 2xy + e^{3z}$, and $R(x,y,z) = 3ye^{3z}$. Then
$$P_y = 2y = Q_x, \ P_z = 0 = R_x, \quad \text{and} \quad Q_z = 3e^{3z} = R_y \Rightarrow \mathbf{F} \text{ is conservative and there}$$
exists $f(x,y,z)$ for which $\mathbf{F} = \nabla f \Rightarrow$
$$f_x = P = y^2, \ f_y = Q = 2xy + e^{3z}, \text{ and } f_z = R = 3ye^{3z}$$
Integrating f_x wrt x: $f(x,y,z) = \int y^2 dx = xy^2 + g(y,z) \Rightarrow f_y = 2xy + g_y(y,z)$, but $f_y = Q = 2xy + e^{3z} \Rightarrow g_y(y,z) = e^{3z}$.
   Integrating $g_y(y,z)$ wrt y : $g(y,z) = \int e^{3z}dy = ye^{3z} + h(z) \Rightarrow f(x,y,z) = xy^2 + ye^{3z} + h(z)$.
   Differentiating wrt z: $f_z = 3ye^{3z} + h'(z)$, but $f_z = R = 3ye^{3z} \Rightarrow h'(z) = 0 \Rightarrow h(z) = C$.
Taking $C = 0$, we have $f(x,y,z) = xy^2 + ye^{3z}$.

13.   The vector field $\mathbf{F}(x,y) = y\mathbf{i} + x\mathbf{j} = P(x,y)\mathbf{i} + Q(x,y)\mathbf{j}$, with $P(x,y) = y$ and $Q(x,y) = x$. Then $P_y = 1 = Q_x \Rightarrow \mathbf{F}$ is conservative. From $\mathbf{r}(t) = \sin t\,\mathbf{i} + 2t\,\mathbf{j}$, for $\frac{\pi}{2} \le t \le \pi$, $\mathbf{r}\left(\frac{\pi}{2}\right) = \sin\frac{\pi}{2}\mathbf{i} + \pi\mathbf{j} \Rightarrow$ the initial point of C is $(1, \pi)$, and $\mathbf{r}(\pi) = \sin\pi\,\mathbf{i} + 2\pi\,\mathbf{j} \Rightarrow$ the terminal point of C is $(0, 2\pi)$.

(a) As $\mathbf{F}$ is conservative, there exists $f(x,y)$ for which $\mathbf{F} = \nabla f$: $y\mathbf{i} + x\mathbf{j} = f_x(x,y)\mathbf{i} + f_y(x,y)\mathbf{j}$ $\Rightarrow f_x(x,y) = y$ and $f_y(x,y) = x$.
Integrating f_x wrt x: $f(x,y) = \int y dx = xy + g(y) \Rightarrow f_y(x,y) = x + g'(y)$, but $f_y(x,y) = x$, so that $g'(y) = 0 \Rightarrow g(y) = C$. Taking $C = 0 \Rightarrow f(x,y) = xy$. By Theorem 14.9(iii), $\int_C \mathbf{F} \cdot d\mathbf{r} = f(0, 2\pi) - f(1, \pi) = (0 \cdot 2\pi) - (1 \cdot \pi) = -\pi$.

(b) $C = C_1 \cup C_2$ (see figure), and parametrizations are:

$C_1 : \mathbf{r}(t) = (1-t)\mathbf{i} + \pi\mathbf{j}, \ 0 \le t \le 1 \Rightarrow \mathbf{r}'(t) = -\mathbf{i}$
$C_2 : \mathbf{r}(t) = t\mathbf{j}, \ \pi \le t \le 2\pi \Rightarrow \mathbf{r}'(t) = \mathbf{j}$

$$\int_C \mathbf{F} \cdot d\mathbf{r} = \int_{C_1} \mathbf{F} \cdot d\mathbf{r} + \int_{C_2} \mathbf{F} \cdot d\mathbf{r} :$$
$$\int_{C_1} \mathbf{F} \cdot d\mathbf{r} = \int_{C_1} (y\mathbf{i} + x\mathbf{j}) \cdot (-\mathbf{i})\,dt = \int_0^1 [\pi\mathbf{i} + (1-t)\mathbf{j}] \cdot (-\mathbf{i})\,dt = \int_0^1 -\pi\,dt = \left[-\pi t\right]_0^1 = -\pi$$
$$\int_{C_2} \mathbf{F} \cdot d\mathbf{r} = \int_{C_2} (y\mathbf{i} + x\mathbf{j}) \cdot \mathbf{j}\,dt = \int_\pi^{2\pi} t\mathbf{i} \cdot \mathbf{j}dt = \int_\pi^{2\pi} 0\,dt = 0 \Rightarrow \int_C \mathbf{F} \cdot d\mathbf{r} = -\pi + 0 = -\pi.$$

15. The vector field $\mathbf{F}(x,y,z) = (yz+1)\mathbf{i} + (xz+2)\mathbf{j} + (xy+3)\mathbf{k} = P(x,y,z)\mathbf{i} + Q(x,y,z)\mathbf{j} + R(x,y,z)\mathbf{k}$, with $P(x,y,z) = yz+1$, $Q(x,y,z) = xz+2$, and $R(x,y,z) = xy+3$. Then $P_y = z = Q_x$, $P_z = y = R_x$, and $Q_z = x = R_y$ $\Rightarrow$ $\mathbf{F}$ is conservative. C is any smooth curve from $(0,0,0)$ to $(1,2,3)$.

(a) As $\mathbf{F}$ is conservative, there exists $f(x,y,z)$ for which $\mathbf{F} = \nabla f$ $\Rightarrow$ $f_x = P = yz+1$, $f_y = Q = xz+2$, and $f_z = R = xy+3$.

Integrating f_x wrt x: $f(x,y,z) = \int(yz+1)dx = xyz+x+g(y,z)$ $\Rightarrow$ $f_y = xz+g_y(y,z)$, but $f_y = Q = xz+2$ $\Rightarrow$ $g_y(y,z) = 2$.

Integrating $g_y(y,z)$ wrt y: $g(y,z) = \int 2\,dy = 2y+h(z)$. Thus $f(x,y,z) = xyz+x+2y+h(z)$.

Differentiating this wrt z: $f_z = xy + h'(z)$, but $f_z = R = xy+3$ $\Rightarrow$ $h'(z) = 3$ $\Rightarrow$ $h(z) = 3z+C$. Taking $C = 0$, $f(x,y,z) = xyz+x+2y+3z$.

By Theorem 14.9(iii), $\int_C \mathbf{F} \cdot d\mathbf{r} = f(1,2,3) - f(0,0,0) = 1(2)(3) + 1 + 2(2) + 3(3) - 0 = 20$.

(b) $C = C_1 \cup C_2 \cup C_3$ (see figure), and parametrizations are:
$C_1 : \mathbf{r}(t) = t\mathbf{i}$, $0 \le t \le 1$ $\Rightarrow$ $\mathbf{r}'(t) = \mathbf{i}$
$C_2 : \mathbf{r}(t) = \mathbf{i} + t\mathbf{j}$, $0 \le t \le 2$ $\Rightarrow$ $\mathbf{r}'(t) = \mathbf{j}$
$C_3 : \mathbf{r}(t) = \mathbf{i} + 2\mathbf{j} + t\mathbf{k}$, $0 \le t \le 3$ $\Rightarrow$ $\mathbf{r}'(t) = \mathbf{k}$

$$\int_C \mathbf{F} \cdot d\mathbf{r} = \int_{C_1} \mathbf{F} \cdot d\mathbf{r} + \int_{C_2} \mathbf{F} \cdot d\mathbf{r} + + \int_{C_3} \mathbf{F} \cdot d\mathbf{r} :$$

$$\int_{C_1} \mathbf{F} \cdot d\mathbf{r} = \int_{C_1} (\mathbf{F} \cdot \mathbf{i})\,dt = \int_{C_1} (yz+1)dt = \int_0^1 1\,dt = \left[t\right]_0^1 = 1$$

$$\int_{C_2} \mathbf{F} \cdot d\mathbf{r} = \int_{C_2} (\mathbf{F} \cdot \mathbf{j})\,dt = \int_{C_2} (xz+2)dt = \int_0^2 2\,dt = \left[2t\right]_0^2 = 4$$

$$\int_{C_3} \mathbf{F} \cdot d\mathbf{r} = \int_{C_3} (\mathbf{F} \cdot \mathbf{k})\,dt = \int_{C_3} (xy+3)dt = \int_0^3 (1 \cdot 2 + 3)dt = \left[5t\right]_0^3 = 15 \Rightarrow$$

$$\int_C \mathbf{F} \cdot d\mathbf{r} = 1 + 4 + 15 = 20.$$

17. The vector field $\mathbf{F}(x,y) = (3+2xy)\mathbf{i} + (x^2-3y^2)\mathbf{j} = P(x,y)\mathbf{i} + Q(x,y)\mathbf{j}$, with $P(x,y) = 3+2xy$ and $Q(x,y) = x^2-3y^2$. Then $P_y = 2x = Q_x$ $\Rightarrow$ $\mathbf{F}$ is conservative. From $\mathbf{r}(t) = e^t\mathbf{i} + \frac{1}{t^2+2}\mathbf{j}$, for $0 \le t \le 1$, $\mathbf{r}(0) = e^0\mathbf{i} + \frac{1}{2}\mathbf{j}$ $\Rightarrow$ the initial point of C is $\left(1, \frac{1}{2}\right)$, and $\mathbf{r}(1) = e\mathbf{i} + \frac{1}{3}\mathbf{j}$ $\Rightarrow$ the terminal point of C is $\left(e, \frac{1}{3}\right)$.

As $\mathbf{F}$ is conservative, there exists $f(x,y)$ for which $\mathbf{F} = \nabla f$: $(3+2xy)\mathbf{i} + (x^2-3y^2)\mathbf{j} = f_x(x,y)\mathbf{i} + f_y(x,y)\mathbf{j}$ $\Rightarrow$ $f_x(x,y) = 3+2xy$ and $f_y(x,y) = x^2-3y^2$.

Integrating f_x wrt x: $f(x,y) = \int(3+2xy)dx = 3x+x^2y+g(y)$ $\Rightarrow$ $f_y(x,y) = x^2+g'(y)$, but $f_y(x,y) = x^2-3y^2$, so that $g'(y) = -3y^2$ $\Rightarrow$ $g(y) = \int(-3y^2)dy = -y^3+C$. Taking $C = 0$ $\Rightarrow$ $f(x,y) = 3x+x^2y-y^3$. By Theorem 14.9(iii),

$$\int_C \mathbf{F} \cdot d\mathbf{r} = f\left(e, \tfrac{1}{3}\right) - f\left(1, \tfrac{1}{2}\right) = 3e + \tfrac{1}{3}e^2 - \tfrac{1}{27} - \left(3 + \tfrac{1}{2} - \tfrac{1}{8}\right) = 3e + \tfrac{1}{3}e^2 - \tfrac{737}{216}.$$

19.   The vector field $\mathbf{F}(x, y) = xy^2\,\mathbf{i} + x^2y\,\mathbf{j} = P(x, y)\mathbf{i} + Q(x, y)\mathbf{j}$, with $P(x, y) = xy^2$ and $Q(x, y) = x^2y$. Then $P_y = 2xy = Q_x \Rightarrow \mathbf{F}$ is conservative.

From $\mathbf{r}(t) = \left(t + \sin\frac{\pi t}{2}\right)\mathbf{i} + \left(t + \cos\frac{\pi t}{2}\right)\mathbf{j}$, for $0 \le t \le 1$, $\mathbf{r}(0) = (\cos 0)\mathbf{j} \Rightarrow$ the initial point of C is $(0, 1)$, and $\mathbf{r}(1) = (1 + 1)\mathbf{i} + (1 + 0)\mathbf{j} \Rightarrow$ the terminal point of C is $(2, 1)$.

As $\mathbf{F}$ is conservative, there exists $f(x, y)$ for which $\mathbf{F} = \nabla f : \mathbf{F}(x, y) = xy^2\,\mathbf{i} + x^2y\,\mathbf{j} = f_x(x, y)\mathbf{i} + f_y(x, y)\mathbf{j} \Rightarrow f_x(x, y) = xy^2$ and $f_y(x, y) = x^2y$.

Integrating f_x wrt x: $f(x, y) = \int xy^2 dx = \frac{1}{2}x^2y^2 + g(y) \Rightarrow f_y(x, y) = x^2y + g'(y)$, but $f_y(x, y) = x^2y$, so that $g'(y) = 0 \Rightarrow g(y) = C$. Taking $C = 0 \Rightarrow f(x, y) = \frac{1}{2}x^2y^2$. By Theorem 14.9(iii),

$$\int_C \mathbf{F} \cdot d\mathbf{r} = f(2, 1) - f(0, 1) = \tfrac{1}{2}(4)(1) - 0 = 2.$$

21.   By Exercise #17 above, the vector field $\mathbf{F}(x, y) = (3 + 2xy)\mathbf{i} + (x^2 - 3y^2)\mathbf{j}$ is conservative, with potential function $f(x, y) = 3x + x^2y - y^3$. As $\mathbf{r}(t) = e^t \sin t\,\mathbf{i} + e^t \cos t\,\mathbf{j}$, $0 \le t \le \pi$, then $\mathbf{r}(0) = e^0 \sin 0\,\mathbf{i} + e^0 \cos 0\,\mathbf{j} = \mathbf{j} \Rightarrow$ the initial point of C is $(0, 1)$, and $\mathbf{r}(\pi) = e^\pi \sin \pi\,\mathbf{i} + e^\pi \cos \pi\,\mathbf{j} \Rightarrow$ the terminal point of C is $(0, -e^\pi)$. Therefore, by Theorem 14.9(iii),

$$\int_C \mathbf{F} \cdot d\mathbf{r} = f(0, -e^\pi) - f(0, 1) = -(-e^\pi)^3 - (-1) = e^{3\pi} + 1.$$

23.   The vector field $\mathbf{F}(x, y, z) = yz^2\,\mathbf{i} + (xz^2 - z)\mathbf{j} + (2xyz - y)\mathbf{k} = P(x, y, z)\mathbf{i} + Q(x, y, z)\mathbf{j} + R(x, y, z)\mathbf{k}$, with $P(x, y, z) = yz^2$, $Q(x, y, z) = xz^2 - z$, and $R(x, y, z) = 2xyz - y$. Then
$P_y = z^2 = Q_x$, $P_z = 2yz = R_x$, and $Q_z = 2xz - 1 = R_y \Rightarrow \mathbf{F}$ is conservative, and there exists $f(x, y, z)$ for which $\mathbf{F} = \nabla f \Rightarrow f_x = P = yz^2$, $f_y = Q = xz^2 - z$, and $f_z = R = 2xyz - y$.

Integrating f_x wrt x: $f(x, y, z) = \int yz^2 dx = xyz^2 + g(y, z) \Rightarrow f_y = xz^2 + g_y(y, z)$, but $f_y = Q = xz^2 - z \Rightarrow g_y(y, z) = -z$.

Integrating $g_y(y, z)$ wrt y: $g(y, z) = \int(-z)dy = -yz + h(z)$. Thus $f(x, y, z) = xyz^2 - yz + h(z)$.

Differentiating this wrt z: $f_z = 2xyz - y + h'(z)$, but $f_z = R = 2xyz - y \Rightarrow h'(z) = 0 \Rightarrow h(z) = C$. Taking $C = 0$, $f(x, y, z) = xyz^2 - yz$.

Since $\mathbf{r}(t) = \frac{1}{t^2 + 2}\mathbf{i} + t^2\mathbf{j} + \mathbf{k}$, for $0 \le t \le 1$, $\mathbf{r}(0) = \frac{1}{2}\mathbf{i} + \mathbf{k} \Rightarrow$ the initial point of C is $\left(\frac{1}{2}, 0, 1\right)$, and $\mathbf{r}(1) = \frac{1}{3}\mathbf{i} + \mathbf{j} + \mathbf{k} \Rightarrow$ the terminal point of C is $\left(\frac{1}{3}, 1, 1\right)$. By Theorem 14.9(iii),

$$\int_C \mathbf{F} \cdot d\mathbf{r} = f\left(\tfrac{1}{3}, 1, 1\right) - f\left(\tfrac{1}{2}, 0, 1\right) = \tfrac{1}{3} - 1 - (0 - 0) = -\tfrac{2}{3}.$$

25.   By Exercise #15 above, the vector field $\mathbf{F}(x, y, z) = (yz + 1)\mathbf{i} + (xz + 2)\mathbf{j} + (xy + 3)\mathbf{k}$ is conservative, with potential function $f(x, y, z) = xyz + x + 2y + 3z$. As $\mathbf{r}(t) = e^t\,\mathbf{i} + t^3\,\mathbf{j} + e^t t^2\mathbf{k}$, $0 \le t \le 1$, then $\mathbf{r}(0) = \mathbf{i} \Rightarrow$ the initial point of C is $(1, 0, 0)$, and $\mathbf{r}(1) = e\,\mathbf{i} + \mathbf{j} + e\,\mathbf{k} \Rightarrow$ the terminal point of C is $(e, 1, e)$. Therefore, by Theorem 14.9(iii),

$$\int_C \mathbf{F} \cdot d\mathbf{r} = f(e, 1, e) - f(1, 0, 0) = e^2 + e + 2 + 3e - (0 + 1 + 0 + 0) = e^2 + 4e + 1.$$

27. The vector field $\mathbf{F}(x,y) = 2xy^2\,\mathbf{i} + K(x,y)\,\mathbf{j} = P(x,y)\mathbf{i} + Q(x,y)\mathbf{j}$, with $P(x,y) = 2xy^2$ and $Q(x,y) = K(x,y)$. The field is conservative when $P_y = Q_x$. Here $P_y = 4xy$ and $Q_x = K_x \Rightarrow K_x = 4xy$. Integrating wrt x: $K(x,y) = \int 4xy\,dx = 2x^2y + g(y)$. Therefore, one possible solution, the simplest solution, is $K(x,y) = 2x^2y$.

29. The vector field $\mathbf{F}(x,y) = K(x,y)\mathbf{i} + x^2e^y\,\mathbf{j} = P(x,y)\mathbf{i} + Q(x,y)\mathbf{j}$, with $P(x,y) = K(x,y)$ and $Q(x,y) = x^2e^y$. The field is conservative when $P_y = Q_x$. Here $P_y = K_y$ and $Q_x = 2xe^y \Rightarrow K_y = 2xe^y$. Integrating wrt y: $K(x,y) = \int 2xe^y\,dy = 2xe^y + g(x)$. Therefore, one possible solution, the simplest solution, is $K(x,y) = 2xe^y$.

31. The vector field $\mathbf{F}(x,y,z) = M(x,y,z)\,\mathbf{i} + 2xy^3\,\mathbf{j} + N(x,y,z)\,\mathbf{k} = P(x,y,z)\mathbf{i} + Q(x,y,z)\mathbf{j} + R(x,y,z)\mathbf{k}$, with $P(x,y,z) = M(x,y,z)$, $Q(x,y,z) = 2xy^3$, and $R(x,y,z) = N(x,y,z)$. The field is conservative when $P_y = Q_x$, $P_z = R_x$, and $Q_z = R_y$.

$P_y = Q_x \Rightarrow M_y = 2y^3$. Integrating wrt y: $M(x,y,z) = \int 2y^3\,dy = \frac{1}{2}y^4 + g(x,z)$
$P_z = R_x \Rightarrow M_z = N_x \Rightarrow (*): N_x = g_z(x,z)$
$Q_z = R_y \Rightarrow 0 = N_y$ means that N is a function only of x and z, so that
$\quad$ (**): $N(x,y,z) = N(x,z)$.

Therefore, one possible solution is taking $g(x,z) = 0$. Then $M = \frac{1}{2}y^4$, and from (*) and (**), N is a function only of z, for example, take $N = z$. Then $\mathbf{F}(x,y,z) = \frac{1}{2}y^4\,\mathbf{i} + 2xy^3\,\mathbf{j} + z\,\mathbf{k}$.

[Let's check that $\mathbf{F}$ is indeed conservative: $P_y = 2y^3 = Q_x$, $P_z = 0 = R_x$, $Q_z = 0 = R_y\ \checkmark$.]

33. The force field $\mathbf{F}(x,y) = e^y\,\mathbf{i} + xe^y\,\mathbf{j} = P(x,y)\mathbf{i} + Q(x,y)\mathbf{j}$, with $P(x,y) = e^y$ and $Q(x,y) = xe^y$. Then $P_y = e^y = Q_x \Rightarrow \mathbf{F}$ is conservative. The circle C: $\mathbf{r}(t) = \cos t\,\mathbf{i} + \sin t\,\mathbf{j}$, for $0 \le t \le 2\pi$ is a smooth, closed curve, so that by Theorem 14.9(i), the work done, $\int_C \mathbf{F} \cdot \mathbf{dr} = 0$.

35. The force field $\mathbf{F}(x,y,z) = (x^2 + y)\mathbf{i} + (y^2 + x)\mathbf{j} + ze^z\,\mathbf{k} = P(x,y,z)\mathbf{i} + Q(x,y,z)\mathbf{j} + R(x,y,z)\mathbf{k}$, with $P(x,y,z) = x^2 + y$, $Q(x,y,z) = y^2 + x$, and $R(x,y,z) = ze^z$. The field is conservative when $P_y = Q_x$, $P_z = R_x$, and $Q_z = R_y$.

$P_y = 1 = Q_x$, $P_z = 0 = R_x$, and $Q_z = 0 = R_y \Rightarrow \mathbf{F}$ is conservative, and there exists $f(x,y,z)$ for which $\mathbf{F} = \nabla f \Rightarrow f_x = P = x^2 + y$, $f_y = Q = y^2 + x$, and $f_z = R = ze^z$.

Integrating f_x wrt x: $f(x,y,z) = \int(x^2 + y)dx = \frac{1}{3}x^3 + xy + g(y,z) \Rightarrow f_y = x + g_y(y,z)$, but $f_y = Q = y^2 + x \Rightarrow g_y(y,z) = y^2$.

Integrating $g_y(y,z)$ wrt y: $g(y,z) = \int y^2\,dy = \frac{1}{3}y^3 + h(z)$. Thus $f(x,y,z) = \frac{1}{3}x^3 + xy + \frac{1}{3}y^3 + h(z)$. Differentiating this wrt z: $f_z = h'(z)$, but $f_z = R = ze^z \Rightarrow h'(z) = ze^z \Rightarrow h(z) = \int ze^z\,dz = ze^z - \int e^z dz = ze^z - e^z = (z-1)e^z \Rightarrow f(x,y,z) = \frac{1}{3}x^3 + xy + \frac{1}{3}y^3 + (z-1)e^z$

$$\boxed{\begin{array}{ll} u = z, & dv = e^z\,dz \\ du = dz & v = e^z \end{array}}$$

Since $\mathbf{r}(t) = \cos t\,\mathbf{i} + \sin t\,\mathbf{j} + \frac{t}{2\pi}\mathbf{k}$, for $0 \le t \le 2\pi$, $\mathbf{r}(0) = \cos 0\,\mathbf{i} + \sin 0\,\mathbf{j} + 0\,\mathbf{k} = \mathbf{i} \Rightarrow$ the initial point of the helix is $(1,0,0)$, and $\mathbf{r}(2\pi) = \cos 2\pi\,\mathbf{i} + \sin 2\pi\,\mathbf{j} + \mathbf{k} \Rightarrow$ the terminal point is $(1,0,1)$. By Theorem 14.9(iii), Work $= \int_C \mathbf{F} \cdot \mathbf{dr} = f(1,0,1) - f(1,0,0) = \frac{1}{3} - \left(\frac{1}{3} - e^0\right) = \frac{1}{3} - \frac{1}{3} + 1 = 1$.

37. Since $f(x, y)$ and $h(x, y)$ are both potential functions for the same conservative vector field, $\nabla f = \nabla h$, which means that (1): $f_x(x, y) = h_x(x, y)$ and (2): $f_y(x, y) = h_y(x, y)$. Let $k(x, y) = f(x, y) - h(x, y)$. From (1), $k_x(x, y) = 0$ which means that k is a function only of y, so $k(x, y) = k(y)$. From (2), $k_y(y) = k'(y) = 0$ so $k(y) = C$, a constant. As $k = f - h$, then $f(x, y) - h(x, y) = C$.

§14.3 Green's Theorem

1. The vector field is $\mathbf{F}(x, y) = (x - y)\mathbf{i} + x\mathbf{j}$. We parametrize the positively oriented unit circle: $\mathbf{r}(t) = \cos t\,\mathbf{i} + \sin t\,\mathbf{j}$, for $0 \le t \le 2\pi$, $\Rightarrow \frac{d\mathbf{r}}{dt} = -\sin t\,\mathbf{i} + \cos t\,\mathbf{j}$.

(a) $\oint_C \mathbf{F} \cdot \mathbf{T}\,ds = \oint_C \mathbf{F} \cdot d\mathbf{r} = \int_0^{2\pi} \{[(\cos t - \sin t)\mathbf{i} + \cos t\,\mathbf{j}] \cdot [-\sin t\,\mathbf{i} + \cos t\,\mathbf{j}]\}\,dt$

$\qquad = \int_0^{2\pi} (-\sin t\cos t + \sin^2 t + \cos^2 t)dt = \int_0^{2\pi} (1 - \sin t\cos t)dt$

$\qquad = \left[t - \tfrac{1}{2}\sin^2 t\right]_0^{2\pi} = 2\pi.$

(b) $P = x - y$, and $Q = x \Rightarrow \iint_D \left(\dfrac{\partial Q}{\partial x} - \dfrac{\partial P}{\partial y}\right) dA = \iint_D [1 - (-1)]\,dA = 2A = 2(\pi \cdot 1^2) = 2\pi.$

3. (a)

To parametrize the line segment from $\mathbf{r_0}$ to $\mathbf{r_1}$, let $\mathbf{r}(t) = (1-t)\mathbf{r_0} + t\mathbf{r_1}$, with $0 \le t \le 1$. $C = C_1 \cup C_2 \cup C_3 \cup C_4$ (see figure), and parametrizations are:

$C_1 : \mathbf{r}(t) = t\,\mathbf{i}, \qquad\qquad 0 \le t \le 1, \quad \frac{d\mathbf{r}}{dt} = \mathbf{i}$

$C_2 : \mathbf{r}(t) = \mathbf{i} + t\,\mathbf{j}, \qquad\quad 0 \le t \le 1, \quad \frac{d\mathbf{r}}{dt} = \mathbf{j}$

$C_3 : \mathbf{r}(t) = (1-t)\mathbf{i} + \mathbf{j}, \quad 0 \le t \le 1, \quad \frac{d\mathbf{r}}{dt} = -\mathbf{i}$

$C_4 : \mathbf{r}(t) = (1-t)\mathbf{j}, \qquad\quad 0 \le t \le 1, \quad \frac{d\mathbf{r}}{dt} = -\mathbf{j}$

Then $\mathbf{F}(x, y) = xy\,\mathbf{i} + (x^2 - y^2)\mathbf{j} \Rightarrow$

$\oint_C \mathbf{F} \cdot \mathbf{T}\,ds = \oint_C \mathbf{F} \cdot d\mathbf{r} = \int_{C_1} \mathbf{F} \cdot d\mathbf{r} + \int_{C_2} \mathbf{F} \cdot d\mathbf{r} + \int_{C_3} \mathbf{F} \cdot d\mathbf{r} + \int_{C_4} \mathbf{F} \cdot d\mathbf{r}$

$\qquad = \int_{C_1} \{[xy\,\mathbf{i} + (x^2 - y^2)\mathbf{j}] \cdot \mathbf{i}\}\,dt + \int_{C_2} \{[xy\,\mathbf{i} + (x^2 - y^2)\mathbf{j}] \cdot \mathbf{j}\}\,dt$

$\qquad\quad + \int_{C_3} \{[xy\,\mathbf{i} + (x^2 - y^2)\mathbf{j}] \cdot [-\mathbf{i}]\}\,dt + \int_{C_4} \{[xy\,\mathbf{i} + (x^2 - y^2)\mathbf{j}] \cdot [-\mathbf{j}]\}\,dt$

$\qquad = \int_{C_1} xy\,dt + \int_{C_2} (x^2 - y^2)\,dt + \int_{C_3} (-xy)\,dt + \int_{C_4} [-(x^2 - y^2)]dt$

$\qquad = \int_0^1 0\,dt + \int_0^1 (1 - t^2)dt + \int_0^1 [-(1-t)]dt + \int_0^1 (1-t)^2 dt$

$\qquad = \int_0^1 (1 - t^2 - 1 + t + 1 - 2t + t^2)dt = \int_0^1 (1 - t)dt = \left[t - \tfrac{1}{2}t^2\right]_0^1 = 1 - \tfrac{1}{2} = \tfrac{1}{2}.$

(b) $P = xy$, and $Q = x^2 - y^2 \Rightarrow$

$$\iint_D \left(\frac{\partial Q}{\partial x} - \frac{\partial P}{\partial y} \right) dA = \iint_D (2x - x) \, dA = \int_0^1 \int_0^1 x \, dy dx = \int_0^1 \left[xy \right]_{y=0}^{y=1} dx = \int_0^1 x \, dx$$

$$= \left[\tfrac{1}{2} x^2 \right]_0^1 = \tfrac{1}{2}.$$

5. (a)

To parametrize the line segment from $\mathbf{r_0}$ to $\mathbf{r_1}$, let $\mathbf{r}(t) = (1-t)\mathbf{r_0} + t\mathbf{r_1}$, with $0 \le t \le 1$.
$C = C_1 \cup C_2 \cup C_3$ (see figure), and parametrizations are:

$$C_1 : \mathbf{r}(t) = t\mathbf{i} + t\mathbf{j}, \qquad 0 \le t \le 1, \quad \tfrac{d\mathbf{r}}{dt} = \mathbf{i} + \mathbf{j}$$

$$C_2 : \mathbf{r}(t) = (1 - t)\mathbf{i} + \mathbf{j}, \quad 0 \le t \le 1, \quad \tfrac{d\mathbf{r}}{dt} = -\mathbf{i}$$

$$C_3 : \mathbf{r}(t) = (1 - t)\mathbf{j}, \qquad 0 \le t \le 1, \quad \tfrac{d\mathbf{r}}{dt} = -\mathbf{j}$$

Then $\mathbf{F}(x, y) = x^2 \mathbf{i} - xy^2 \mathbf{j} \Rightarrow$

$$\oint_C \mathbf{F} \cdot \mathbf{T} \, ds = \oint_C \mathbf{F} \cdot d\mathbf{r} = \int_{C_1} \mathbf{F} \cdot d\mathbf{r} + \int_{C_2} \mathbf{F} \cdot d\mathbf{r} + \int_{C_3} \mathbf{F} \cdot d\mathbf{r}$$

$$= \int_{C_1} \{ [x^2 \mathbf{i} - xy^2 \mathbf{j}] \cdot (\mathbf{i} + \mathbf{j}) \} \, dt + \int_{C_2} \{ [x^2 \mathbf{i} - xy^2 \mathbf{j}] \cdot (-\mathbf{i}) \} \, dt$$

$$+ \int_{C_3} \{ [x^2 \mathbf{i} - xy^2 \mathbf{j}] \cdot (-\mathbf{j}) \} \, dt$$

$$= \int_{C_1} (x^2 - xy^2) \, dt + \int_{C_2} (-x^2) \, dt + \int_{C_3} xy^2 \, dt$$

$$= \int_0^1 (t^2 - t^3) \, dt + \int_0^1 [-(1 - t)^2] \, dt + \int_0^1 0 \, dt$$

$$= \int_0^1 (t^2 - t^3 - 1 + 2t - t^2) \, dt = \int_0^1 (-t^3 - 1 + 2t) dt$$

$$= \left[-\tfrac{1}{4} t^4 - t + t^2 \right]_0^1 = -\tfrac{1}{4} - 1 + 1 = -\tfrac{1}{4}.$$

(b) $P = x^2$, and $Q = -xy^2$, and from the figure, the line segment C_1 is $y = x$, so that
$D : 0 \le x \le y, 0 \le y \le 1 \Rightarrow$

$$\iint_D \left(\frac{\partial Q}{\partial x} - \frac{\partial P}{\partial y} \right) dA = \iint_D (-y^2 - 0) \, dA = - \int_0^1 \int_0^y y^2 \, dx dy = - \int_0^1 \left[xy^2 \right]_{x=0}^{x=y} dy$$

$$= - \int_0^1 y^3 \, dy = - \left[\tfrac{1}{4} y^4 \right]_0^1 = -\tfrac{1}{4}.$$

7. (a)

$C = C_1 \cup C_2 \cup C_3$ (see figure), and parametrizations are:

$$C_1 : \mathbf{r}(t) = t\mathbf{i}, \qquad\qquad 0 \le t \le 1, \quad \tfrac{d\mathbf{r}}{dt} = \mathbf{i}$$

$$C_2 : \mathbf{r}(t) = \cos t\mathbf{i} + \sin t\mathbf{j}, \quad 0 \le t \le \tfrac{\pi}{2}, \quad \tfrac{d\mathbf{r}}{dt} = -\sin t\mathbf{i} + \cos t\mathbf{j}$$

$$C_3 : \mathbf{r}(t) = (1 - t)\mathbf{j}, \qquad\quad 0 \le t \le 1, \quad \tfrac{d\mathbf{r}}{dt} = -\mathbf{j}$$

Then $\mathbf{F}(x, y) = -y\mathbf{i} + x\mathbf{j} \Rightarrow$

$$\oint_C \mathbf{F} \cdot \mathbf{T} \, ds = \oint_C \mathbf{F} \cdot d\mathbf{r} = \int_{C_1} \mathbf{F} \cdot d\mathbf{r} + \int_{C_2} \mathbf{F} \cdot d\mathbf{r} + \int_{C_3} \mathbf{F} \cdot d\mathbf{r}$$

$$= \int_{C_1} \{[-y\,\mathbf{i} + x\,\mathbf{j}] \cdot \mathbf{i}\} \, dt + \int_{C_2} \{[-y\,\mathbf{i} + x\,\mathbf{j}] \cdot [-\sin t\,\mathbf{i} + \cos t\,\mathbf{j}]\} \, dt$$

$$+ \int_{C_3} \{[-y\,\mathbf{i} + x\,\mathbf{j}] \cdot [-\mathbf{j}]\} \, dt$$

$$= \int_{C_1} (-y) \, dt + \int_{C_2} (y \sin t + x \cos t) \, dt + \int_{C_3} (-x) \, dt$$

$$= \int_0^1 0 \, dt + \int_0^{\pi/2} (\sin^2 t + \cos^2 t) \, dt + \int_0^1 0 \, dt = \int_0^{\pi/2} dt = \frac{\pi}{2}.$$

(b) $P = -y$, and $Q = x$ $\Rightarrow$

$$\iint_D \left(\frac{\partial Q}{\partial x} - \frac{\partial P}{\partial y} \right) dA = \iint_D [1 - (-1)] \, dA = 2A = 2 \cdot \frac{1}{4}(\pi \cdot 1^2) = \frac{\pi}{2}.$$

9. (a)

$C = C_1 \cup C_2 \cup C_3$ (see figure), and parametrizations are:

$$C_1 : \mathbf{r}(t) = t\,\mathbf{i}, \qquad\qquad 0 \le t \le 1, \quad \frac{d\mathbf{r}}{dt} = \mathbf{i}$$

$$C_2 : \mathbf{r}(t) = \mathbf{i} + t\,\mathbf{j}, \qquad\qquad 0 \le t \le 2, \quad \frac{d\mathbf{r}}{dt} = \mathbf{j}$$

$$C_3 : \mathbf{r}(t) = (1-t)\mathbf{i} + 2(1-t)\mathbf{j}, \quad 0 \le t \le 1, \quad \frac{d\mathbf{r}}{dt} = -\mathbf{i} - 2\mathbf{j}$$

$$\oint_C xy \, dx + x^2 y^3 \, dy = \int_{C_1} xy \, dx + x^2 y^3 \, dy + \int_{C_2} xy \, dx + x^2 y^3 \, dy + \int_{C_3} xy \, dx + x^2 y^3 \, dy$$

$$= \int_0^1 0 \, dt + \int_0^2 [t \cdot 0 + t^3 \cdot 1] \, dt$$

$$+ \int_0^1 [(1-t) \cdot 2(1-t) \cdot (-1) + (1-t)^2 \cdot 2^3 (1-t)^3 \cdot (-2)] \, dt$$

$$= \left[\frac{1}{4}t^4 \right]_0^2 + \int_0^1 [-2(1-t)^2 - 16(1-t)^5] \, dt = 4 + \left[\frac{2}{3}(1-t)^3 + \frac{16}{6}(1-t)^6 \right]_0^1$$

$$= 4 - \left(\frac{2}{3} + \frac{8}{3} \right) = \frac{2}{3}.$$

(b) $P = xy$, and $Q = x^2 y^3$, and from the figure, the line segment C_3 is $y = 2x$, so that $D : 0 \le y \le 2x, 0 \le x \le 1$ $\Rightarrow$

$$\iint_D \left(\frac{\partial Q}{\partial x} - \frac{\partial P}{\partial y} \right) dA = \iint_D (2xy^3 - x) \, dA = \int_0^1 \int_0^{2x} (2xy^3 - x) \, dy dx$$

$$= \int_0^1 \left[\frac{1}{2}xy^4 - xy \right]_{y=0}^{y=2x} dx = \int_0^1 \left[\frac{1}{2}x(2x)^4 - x(2x) \right] dx$$

$$= \int_0^1 (8x^5 - 2x^2) \, dx = \left[\frac{4}{3}x^6 - \frac{2}{3}x^3 \right]_0^1 = \frac{4}{3} - \frac{2}{3} = \frac{2}{3}.$$

11. (a)

$$\mathbf{r}(t) = 2\cos t\,\mathbf{i} + 2\sin t\,\mathbf{j}, \quad 0 \le t \le 2\pi,$$

$$\frac{d\mathbf{r}}{dt} = -2\sin t\,\mathbf{i} + 2\cos t\,\mathbf{j}$$

$$\oint_C y^3\,dx - x^3\,dy = \int_0^{2\pi} [(8\sin^3 t)(-2\sin t) - (8\cos^3 t)(2\cos t)]\,dt = -16\int_0^{2\pi} (\sin^4 t + \cos^4 t)\,dt$$

$$= -16\int_0^{2\pi} [(\sin^2 t)^2 + (\cos^2 t)^2]\,dt = -16\int_0^{2\pi}\left[\left(\frac{1-\cos 2t}{2}\right)^2 + \left(\frac{1+\cos 2t}{2}\right)^2\right]dt$$

$$= -4\int_0^{2\pi} (1 - 2\cos 2t + \cos^2 2t + 1 + 2\cos 2t + \cos^2 2t)\,dt$$

$$= -4\int_0^{2\pi} (2 + 2\cos^2 2t)\,dt = -8\int_0^{2\pi}\left(1 + \frac{1+\cos 4t}{2}\right)dt$$

$$= -8\left[t + \tfrac{1}{2}t + \tfrac{1}{8}\sin 4t\right]_0^{2\pi} = -8\left[\tfrac{3}{2}t\right]_0^{2\pi} = -24\pi.$$

(b) $P = y^3$, and $Q = -x^3$

$$\iint_D \left(\frac{\partial Q}{\partial x} - \frac{\partial P}{\partial y}\right)dA = \iint_D (-3x^2 - 3y^2)\,dA = -3\iint_D (x^2 + y^2)\,dA = -3\int_0^{2\pi}\int_0^2 r^2 \cdot r\,dr\,d\theta$$

$$= -3\int_0^{2\pi}\left[\tfrac{1}{4}r^4\right]_{r=0}^{r=2}\,d\theta = -\tfrac{3}{4}(2^4)\int_0^{2\pi}\,d\theta = -12(2\pi) = -24\pi.$$

13.

$$\oint_C y^4\,dx + 2xy^3\,dy \;\Rightarrow\; P = y^4, \text{ and } Q = 2xy^3$$

$$\oint_C y^4\,dx + 2xy^3\,dy = \iint_D \left(\frac{\partial Q}{\partial x} - \frac{\partial P}{\partial y}\right)dA = \iint_D (2y^3 - 4y^3)\,dA$$

$$= -2\int_{-\sqrt{2}}^{\sqrt{2}}\int_{-\sqrt{1-x^2/2}}^{\sqrt{1-x^2/2}} y^3\,dy\,dx = -2\int_{-\sqrt{2}}^{\sqrt{2}}\left[\tfrac{1}{4}y^4\right]_{-\sqrt{1-x^2/2}}^{\sqrt{1-x^2/2}}\,dx$$

$$= -\frac{1}{2}\int_{-\sqrt{2}}^{\sqrt{2}}\left[\left(\sqrt{1 - \frac{x^2}{2}}\right)^4 - \left(-\sqrt{1 - \frac{x^2}{2}}\right)^4\right]dx = 0.$$

15.

$$\oint_C y\,dx - x\,dy \;\Rightarrow\; P = y, \text{ and } Q = -x$$

$$\oint_C y\,dx - x\,dy = \iint_D \left(\frac{\partial Q}{\partial x} - \frac{\partial P}{\partial y}\right)dA = \iint_D (-1 - 1)\,dA = \int_0^1\int_0^{1-x^2} (-1 - 1)\,dy\,dx$$

$$= -2\int_0^1 (1 - x^2)\,dx = -2\left[x - \tfrac{1}{3}x^3\right]_0^1 = -2\left(1 - \tfrac{1}{3}\right) = -\tfrac{4}{3}.$$

17.

$$\oint_C y \cos x \, dx + x^2 \, dy \;\Rightarrow\; P = y \cos x, \;\; \text{and} \;\; Q = x^2$$

$$\oint_C y \cos x \, dx + x^2 \, dy = \iint_D \left(\frac{\partial Q}{\partial x} - \frac{\partial P}{\partial y} \right) dA = \iint_D (2x - \cos x) \, dA$$

$$= \int_0^1 \int_0^1 (2x - \cos x) \, dy dx = \int_0^1 \Big[(2x - \cos x)y \Big]_{y=0}^{y=1} dx$$

$$= \int_0^1 (2x - \cos x) \, dx = \Big[x^2 - \sin x \Big]_0^1 = 1 - \sin 1.$$

19.

$$\mathbf{F}(x,y) = (x^2 + xy)\mathbf{i} + xy^2 \mathbf{j} \;\Rightarrow\; P = x^2 + xy \;\; \text{and} \;\; Q = xy^2$$

$$W = \oint_C \mathbf{F} \cdot \mathbf{T} \, ds = \iint_D \left(\frac{\partial Q}{\partial x} - \frac{\partial P}{\partial y} \right) dA = \iint_D (y^2 - x) dA = \int_0^1 \int_0^{-x+1} (y^2 - x) \, dy dx$$

$$= \int_0^1 \Big[\tfrac{1}{3}y^3 - xy \Big]_{y=0}^{y=-x+1} dx = \int_0^1 \Big[\tfrac{1}{3}(-x+1)^3 - x(-x+1) \Big] \, dx$$

$$= \left[\frac{1}{3} \cdot \frac{(-x+1)^4}{-4} + \frac{1}{3}x^3 - \frac{1}{2}x^2 \right]_0^1 = -\frac{1}{12}(0) + \frac{1}{3} - \frac{1}{2} - \left(-\frac{1}{12} \right) = -\frac{1}{12}.$$

21.

$$\mathbf{F}(x,y) = 7x\,\mathbf{i} + (x^3 + 3xy^2)\mathbf{j} \;\Rightarrow\; P = 7x \;\; \text{and} \;\; Q = x^3 + 3xy^2$$

$$W = \oint_C \mathbf{F} \cdot \mathbf{T} \, ds = \iint_D \left(\frac{\partial Q}{\partial x} - \frac{\partial P}{\partial y} \right) dA = \iint_D (3x^2 + 3y^2) dA = \int_0^\pi \int_0^2 3r^2 \cdot r \, dr d\theta$$

$$= \int_0^\pi \int_0^2 3r^3 \, dr d\theta = \frac{3}{4} \int_0^\pi \Big[r^4 \Big]_{r=0}^{r=2} d\theta = \frac{3}{4}(2^4)\Big[\theta \Big]_0^\pi = 12\pi.$$

23. Parametrize C by: $\mathbf{r}(t) = r \cos t\,\mathbf{i} + r \sin t\,\mathbf{j}, \; 0 \le t \le 2\pi, \;\Rightarrow$
$\frac{d\mathbf{r}}{dt} = -r \sin t\,\mathbf{i} + r \cos t\,\mathbf{j}$. Then the area A of the region D is:

$$A = \frac{1}{2} \oint_C -y \, dx + x \, dy = \frac{1}{2} \int_0^{2\pi} \big[(-r \sin t)(-r \sin t) + (r \cos t)(r \cos t) \big] \, dt$$

$$= \frac{1}{2} \int_0^{2\pi} r^2 (\sin^2 t + \cos^2 t) \, dt = \frac{1}{2}r^2 \int_0^{2\pi} dt = \pi r^2.$$

25.

$C = C_1 \cup C_2$ (see figure), and parametrizations are:

$C_1 : \mathbf{r}(t) = t\,\mathbf{i} + t^3\,\mathbf{j}$, $0 \le t \le 1$, $\frac{d\mathbf{r}}{dt} = \mathbf{i} + 3t^2\,\mathbf{j}$

$C_2 : \mathbf{r}(t) = (1-t)\mathbf{i} + (1-t)^2\,\mathbf{j}$, $0 \le t \le 1$, $\frac{d\mathbf{r}}{dt} = -\mathbf{i} - 2(1-t)\mathbf{j}$

Then the area A of the region D is:

$$A = \frac{1}{2}\oint_C -y\,dx + x\,dy = \frac{1}{2}\int_{C_1} -y\,dx + x\,dy + \frac{1}{2}\int_{C_2} -y\,dx + x\,dy$$

$$= \frac{1}{2}\left\{\int_0^1 [-t^3(1) + t(3t^2)]\,dt + \int_0^1 [-(1-t)^2(-1) + (1-t)(-2)(1-t)]\,dt\right\}$$

$$= \frac{1}{2}\left[\int_0^1 2t^3\,dt - \int_0^1 (1-t)^2\,dt\right] = \left[\tfrac{1}{4}t^4\right]_0^1 - \frac{1}{2}\int_0^1 (1 - 2t + t^2)\,dt = \frac{1}{4} - \frac{1}{2}\left[t - t^2 + \tfrac{1}{3}t^3\right]_0^1$$

$$= \frac{1}{4} - \frac{1}{2}\left(1 - 1 + \tfrac{1}{3}\right) = \frac{1}{12}.$$

27.

Find end of arch ($y = 0, x = ?$): $1 - \cos t = 0 \Rightarrow t = 2\pi \Rightarrow$ $x = a(2\pi - \sin 2\pi) = 2\pi a$. $C = C_1 \cup C_2$ (see figure), and parametrizations are:

$C_1 : \mathbf{r}(t) = t\,\mathbf{i}$, $0 \le t \le 2\pi a \Rightarrow \frac{d\mathbf{r}}{dt} = \mathbf{i}$

$C_2 : \mathbf{r}(t) = a(t - \sin t)\mathbf{i} + a(1 - \cos t)\mathbf{j}$, t from 2π to 0 (positive orientation), and $\frac{d\mathbf{r}}{dt} = a(1 - \cos t)\mathbf{i} + a\sin t\,\mathbf{j}$

Then the area A of the region D is:

$$A = \frac{1}{2}\oint_C -y\,dx + x\,dy = \frac{1}{2}\int_{C_1} -y\,dx + x\,dy + \frac{1}{2}\int_{C_2} -y\,dx + x\,dy$$

$$= \frac{1}{2}\int_0^{2\pi a} [-(0)(1) + t(0)]\,dt + \frac{1}{2}\int_{2\pi}^0 [-a(1 - \cos t)a(1 - \cos t) + a(t - \sin t)(a\sin t)]\,dt$$

$$= \frac{a^2}{2}\int_{2\pi}^0 \left(-1 + 2\cos t - \cos^2 t + t\sin t - \sin^2 t\right)\,dt = \frac{a^2}{2}\int_{2\pi}^0 (-2 + 2\cos t + t\sin t)\,dt$$

$$= \frac{a^2}{2}\left[-2t + 2\sin t\right]_{2\pi}^0 + \frac{a^2}{2}\int_{2\pi}^0 t\sin t\,dt = \frac{a^2}{2}(4\pi) + \frac{a^2}{2}\left\{\left[-t\cos t\right]_{2\pi}^0 + \int_{2\pi}^0 \cos t\,dt\right\}$$

$u = t$,	$dv = \sin t\,dt$
$du = dt$	$v = -\cos t$

$$= 2\pi a^2 + \frac{a^2}{2}\left(2\pi + \left[\sin t\right]_{2\pi}^0\right) = 2\pi a^2 + \frac{a^2}{2}(2\pi) = 3\pi a^2.$$

29.

$\mathbf{F}(x, y) = (x - y)\mathbf{i} + x\mathbf{j} \Rightarrow P = x - y$ and $Q = x$

(a) $\displaystyle\iint_D \left(\frac{\partial P}{\partial x} + \frac{\partial Q}{\partial y}\right) dA = \iint_D (1 + 0)\,dA = A = \pi(1^2) = \pi$

(b) Parametrize $C : \mathbf{r}(t) = \cos t\,\mathbf{i} + \sin t\,\mathbf{j}$, for $0 \le t \le 2\pi$, $\Rightarrow \frac{d\mathbf{r}}{dt} = -\sin t\,\mathbf{i} + \cos t\,\mathbf{j} \Rightarrow$

$$\oint_C P\,dy - Q\,dx = \int_0^{2\pi} [(\cos t - \sin t)\cos t - \cos t(-\sin t)]\,dt$$

$$= \int_0^{2\pi} (\cos^2 t - \sin t \cos t + \sin t \cos t)\,dt = \int_0^{2\pi} \cos^2 t\,dt = \int_0^{2\pi} \left(\frac{1 + \cos 2t}{2}\right)\,dt$$

$$= \frac{1}{2}\left[t + \frac{1}{2}\sin 2t\right]_0^{2\pi} = \frac{1}{2}(2\pi) = \pi.$$

31.

$$\mathbf{F}(x, y) = (4x - 2y)\mathbf{i} + (2x - 4y)\mathbf{j} \;\Rightarrow\; P = 4x - 2y \text{ and } Q = 2x - 4y$$

(a) $\displaystyle\iint_D \left(\frac{\partial P}{\partial x} + \frac{\partial Q}{\partial y}\right)dA = \iint_D (4 - 4)\,dA = 0.$

(b) Parametrize C : $x - 2 = 2\cos t \;\Rightarrow\; x = 2 + 2\cos t$, and $y - 2 = 2\sin t \;\Rightarrow\; y = 2 + 2\sin t$
for $0 \le t \le 2\pi$, $\;\Rightarrow\; \frac{dx}{dt} = -2\sin t$ and $\frac{dy}{dt} = 2\cos t$.

$$\oint_C P\,dy - Q\,dx = \int_0^{2\pi} \{[4(2 + 2\cos t) - 2(2 + 2\sin t)](2\cos t)$$

$$- [2(2 + 2\cos t) - 4(2 + 2\sin t)](-2\sin t)\}\,dt$$

$$= \int_0^{2\pi} \Big(16\cos t + 16\cos^2 t - 8\cos t - 8\sin t \cos t + 8\sin t + 8\sin t \cos t$$

$$- 16\sin t - 16\sin^2 t\Big)\,dt$$

$$= \int_0^{2\pi} (8\cos t - 8\sin t + 16\cos 2t)\,dt = \Big[8\sin t + 8\cos t + 8\sin 2t\Big]_0^{2\pi}$$

$$= 8(\cos 2\pi - \cos 0) = 0.$$

§14.4 Curl and Div

1. $\mathbf{F}(x, y) = (x^2 - y)\mathbf{i} + (xy - y^2)\mathbf{j} = P\mathbf{i} + Q\mathbf{j}$, with $P = x^2 - y$ and $Q = xy - y^2 \;\Rightarrow$
curl $\mathbf{F} = (Q_x - P_y)\mathbf{k} = \left[\frac{\partial}{\partial x}(xy - y^2) - \frac{\partial}{\partial y}(x^2 - y)\right]\mathbf{k} = [y - (-1)]\mathbf{k} = (y + 1)\mathbf{k}.$

3. $\mathbf{F}(x, y) = \sin xy\,\mathbf{i} + \cos xy\,\mathbf{j} = P\mathbf{i} + Q\mathbf{j}$, with $P = \sin xy$ and $Q = \cos xy \;\Rightarrow$
curl $\mathbf{F} = (Q_x - P_y)\mathbf{k} = \left[\frac{\partial}{\partial x}(\cos xy) - \frac{\partial}{\partial y}(\sin xy)\right]\mathbf{k} = (-y\sin xy - x\cos xy)\mathbf{k}.$

5. For $\mathbf{F}(x, y, z) = y^2 z\,\mathbf{i} - x^3\,\mathbf{j} + xy\,\mathbf{k}$:

$$\text{curl } \mathbf{F} = \det\begin{bmatrix} \mathbf{i} & \mathbf{j} & \mathbf{k} \\ \frac{\partial}{\partial x} & \frac{\partial}{\partial y} & \frac{\partial}{\partial z} \\ y^2 z & -x^3 & xy \end{bmatrix} = \left[\frac{\partial}{\partial y}(xy) - \frac{\partial}{\partial z}(-x^3)\right]\mathbf{i} - \left[\frac{\partial}{\partial x}(xy) - \frac{\partial}{\partial z}(y^2 z)\right]\mathbf{j}$$

$$+ \left[\frac{\partial}{\partial x}(-x^3) - \frac{\partial}{\partial y}(y^2 z)\right]\mathbf{k}$$

$$= x\,\mathbf{i} - (y - y^2)\mathbf{j} + (-3x^2 - 2yz)\mathbf{k}.$$

7. For $\mathbf{F}(x, y, z) = xz\,\mathbf{i} + xyz\,\mathbf{j} - y^2\,\mathbf{k}$:

$$\text{curl } \mathbf{F} = \det \begin{bmatrix} \mathbf{i} & \mathbf{j} & \mathbf{k} \\ \frac{\partial}{\partial x} & \frac{\partial}{\partial y} & \frac{\partial}{\partial z} \\ xz & xyz & -y^2 \end{bmatrix} = \left[\frac{\partial}{\partial y}(-y^2) - \frac{\partial}{\partial z}(xyz)\right]\mathbf{i} - \left[\frac{\partial}{\partial x}(-y^2) - \frac{\partial}{\partial z}(xz)\right]\mathbf{j}$$
$$+ \left[\frac{\partial}{\partial x}(xyz) - \frac{\partial}{\partial y}(xz)\right]\mathbf{k}$$

$$= (-2y - xy)\mathbf{i} + x\,\mathbf{j} + yz\,\mathbf{k}.$$

9. For $\mathbf{F}(x, y, z) = \ln(x + z)\,\mathbf{i} - e^{yz}\,\mathbf{j} + xy\,\mathbf{k}$:

$$\text{curl } \mathbf{F} = \det \begin{bmatrix} \mathbf{i} & \mathbf{j} & \mathbf{k} \\ \frac{\partial}{\partial x} & \frac{\partial}{\partial y} & \frac{\partial}{\partial z} \\ \ln(x + z) & -e^{yz} & xy \end{bmatrix} = \left[\frac{\partial}{\partial y}(xy) - \frac{\partial}{\partial z}(-e^{yz})\right]\mathbf{i} - \left[\frac{\partial}{\partial x}(xy) - \frac{\partial}{\partial z}\ln(x + z)\right]\mathbf{j}$$
$$+ \left[\frac{\partial}{\partial x}(-e^{yz}) - \frac{\partial}{\partial y}\ln(x + z)\right]\mathbf{k}$$

$$= (x + ye^{yz})\mathbf{i} - \left(y - \frac{1}{x + z}\right)\mathbf{j}.$$

11. For $\mathbf{F}(x, y, z) = \sin yz\,\mathbf{i} + \sin zx\,\mathbf{j} - \sin xy\,\mathbf{k}$:

$$\text{curl } \mathbf{F} = \det \begin{bmatrix} \mathbf{i} & \mathbf{j} & \mathbf{k} \\ \frac{\partial}{\partial x} & \frac{\partial}{\partial y} & \frac{\partial}{\partial z} \\ \sin yz & \sin zx & -\sin xy \end{bmatrix}$$
$$= \left[\frac{\partial}{\partial y}(-\sin xy) - \frac{\partial}{\partial z}(\sin zx)\right]\mathbf{i} - \left[\frac{\partial}{\partial x}(-\sin xy) - \frac{\partial}{\partial z}(\sin yz)\right]\mathbf{j}$$
$$+ \left[\frac{\partial}{\partial x}(\sin zx) - \frac{\partial}{\partial y}(\sin yz)\right]\mathbf{k}$$

$$= (-x\cos xy - x\cos zx)\mathbf{i} - (-y\cos xy - y\cos yz)\mathbf{j} + (z\cos zx - z\cos yz)\mathbf{k}$$
$$= -x(\cos xy + \cos zx)\mathbf{i} + y(\cos xy + \cos yz)\mathbf{j} + z(\cos zx - \cos yz)\mathbf{k}.$$

13. For $\mathbf{F}(x, y, z) = \tan yz\,\mathbf{i} + \cos xy\,\mathbf{j} - \sin xy\,\mathbf{k}$:

$$\text{curl } \mathbf{F} = \det \begin{bmatrix} \mathbf{i} & \mathbf{j} & \mathbf{k} \\ \frac{\partial}{\partial x} & \frac{\partial}{\partial y} & \frac{\partial}{\partial z} \\ \tan yz & \cos xy & -\sin xy \end{bmatrix}$$
$$= \left[\frac{\partial}{\partial y}(-\sin xy) - \frac{\partial}{\partial z}(\cos xy)\right]\mathbf{i} - \left[\frac{\partial}{\partial x}(-\sin xy) - \frac{\partial}{\partial z}(\tan yz)\right]\mathbf{j}$$
$$+ \left[\frac{\partial}{\partial x}(\cos xy) - \frac{\partial}{\partial y}(\tan yz)\right]\mathbf{k}$$

$$= -x\cos xy\,\mathbf{i} - (-y\cos xy - y\sec^2 yz)\mathbf{j} + (-y\sin xy - z\sec^2 yz)\mathbf{k}$$
$$= -x\cos xy\,\mathbf{i} + y(\cos xy + \sec^2 yz)\mathbf{j} - (y\sin xy + z\sec^2 yz)\mathbf{k}.$$

15. $\text{div}F = \nabla \cdot \mathbf{F} = \left(\frac{\partial}{\partial x}\mathbf{i} + \frac{\partial}{\partial y}\mathbf{j} + \frac{\partial}{\partial z}\mathbf{k}\right) \cdot (x^2y\,\mathbf{i} + \sin x\,\mathbf{j}) = \frac{\partial}{\partial x}(x^2y) + \frac{\partial}{\partial y}(\sin x) + \frac{\partial}{\partial z}(0)$

$= 2xy.$

17. $\text{div}F = \nabla \cdot \mathbf{F} = \left(\frac{\partial}{\partial x}\mathbf{i} + \frac{\partial}{\partial y}\mathbf{j} + \frac{\partial}{\partial z}\mathbf{k}\right) \cdot (xy\,\mathbf{i} + yz\,\mathbf{j} - 2xz\,\mathbf{k})$

$= \frac{\partial}{\partial x}(xy) + \frac{\partial}{\partial y}(yz) + \frac{\partial}{\partial z}(-2xz) = y + z - 2x.$

19. $\text{div}F = \nabla \cdot \mathbf{F} = \left(\frac{\partial}{\partial x}\mathbf{i} + \frac{\partial}{\partial y}\mathbf{j} + \frac{\partial}{\partial z}\mathbf{k}\right) \cdot (e^{x-z}\,\mathbf{i} + e^{z-y}\,\mathbf{j} - e^{y-x}\,\mathbf{k})$

$= \frac{\partial}{\partial x}(e^{x-z}) + \frac{\partial}{\partial y}(e^{z-y}) + \frac{\partial}{\partial z}(-e^{y-x}) = e^{x-z} - e^{z-y}.$

21. $\text{div}F = \nabla \cdot \mathbf{F} = \left(\frac{\partial}{\partial x}\mathbf{i} + \frac{\partial}{\partial y}\mathbf{j} + \frac{\partial}{\partial z}\mathbf{k}\right) \cdot (\cos xy\,\mathbf{i} - \sin yz\,\mathbf{j} + \cos y \sin x\,\mathbf{k})$

$= \frac{\partial}{\partial x}(\cos xy) + \frac{\partial}{\partial y}(-\sin yz) + \frac{\partial}{\partial z}(\cos y \sin x) = -y\sin xy - z\cos yz.$

23.   It suffices to show that $\text{curl } \mathbf{F} = \mathbf{0}$, where $\mathbf{F} = x^2 y\,\mathbf{i} + \frac{x^3}{3}\,\mathbf{j} = P\mathbf{i} + Q\mathbf{j}$, with $P = x^2 y$
and $Q = \frac{x^3}{3}$: $\text{curl } \mathbf{F} = (Q_x - P_y)\mathbf{k} = \left[\frac{\partial}{\partial x}\left(\frac{x^3}{3}\right) - \frac{\partial}{\partial y}(x^2 y)\right]\mathbf{k} = (x^2 - x^2)\mathbf{k} = \mathbf{0}.$

25.   It suffices to show that $\text{curl } \mathbf{F} = \mathbf{0}$, where $\mathbf{F} = y^2 z^3\,\mathbf{i} + 2xyz^3\,\mathbf{j} + 3xy^2 z^2\,\mathbf{k}$:

$\text{curl } \mathbf{F} = \det \begin{bmatrix} \mathbf{i} & \mathbf{j} & \mathbf{k} \\ \frac{\partial}{\partial x} & \frac{\partial}{\partial y} & \frac{\partial}{\partial z} \\ y^2 z^3 & 2xyz^3 & 3xy^2 z^2 \end{bmatrix}$

$= \left[\frac{\partial}{\partial y}(3xy^2 z^2) - \frac{\partial}{\partial z}(2xyz^3)\right]\mathbf{i} - \left[\frac{\partial}{\partial x}(3xy^2 z^2) - \frac{\partial}{\partial z}(y^2 z^3)\right]\mathbf{j}$

$+ \left[\frac{\partial}{\partial x}(2xyz^3) - \frac{\partial}{\partial y}(y^2 z^3)\right]\mathbf{k}$

$= (6xyz^2 - 6xyz^2)\mathbf{i} - (3y^2 z^2 - 3y^2 z^2)\mathbf{j} + (2yz^3 - 2yz^3)\mathbf{k} = 0\mathbf{i} + 0\mathbf{j} + 0\mathbf{k} = \mathbf{0}.$

27.   To prove $\text{div}\,(f\mathbf{F}) = f(\text{div } \mathbf{F}) + (\mathbf{F} \cdot \nabla f)$. Let $\mathbf{F} = P\mathbf{i} + Q\mathbf{j} + R\mathbf{k}$:
$\text{div}(f\mathbf{F}) = \nabla \cdot (f\mathbf{F}) = \left(\frac{\partial}{\partial x}\mathbf{i} + \frac{\partial}{\partial y}\mathbf{j} + \frac{\partial}{\partial z}\mathbf{k}\right) \cdot (fP\mathbf{i} + fQ\mathbf{j} + fR\mathbf{k})$

$= \frac{\partial}{\partial x}(fP) + \frac{\partial}{\partial y}(fQ) + \frac{\partial}{\partial z}(fR) = (f_x P + f P_x) + (f_y Q + f Q_y) + (f_z R + f R_z)$

$= f(P_x + Q_y + R_z) + (P f_x + Q f_y + R f_z)$

$= f\left[\left(\frac{\partial}{\partial x}\mathbf{i} + \frac{\partial}{\partial y}\mathbf{j} + \frac{\partial}{\partial z}\mathbf{k}\right) \cdot (P\mathbf{i} + Q\mathbf{j} + R\mathbf{k})\right] + [(P\mathbf{i} + Q\mathbf{j} + R\mathbf{k}) \cdot (f_x\mathbf{i} + f_y\mathbf{j} + f_z\mathbf{k})]$

$= f(\text{div}\mathbf{F}) + (\mathbf{F} \cdot \nabla f).$

29.   To prove $\text{div}\,(\nabla f \times \nabla g) = 0$:

$\nabla f \times \nabla g = \det \begin{bmatrix} \mathbf{i} & \mathbf{j} & \mathbf{k} \\ f_x & f_y & f_z \\ g_x & g_y & g_z \end{bmatrix} = (f_y g_z - f_z g_y)\mathbf{i} - (f_x g_z - f_z g_x)\mathbf{j} + (f_x g_y - f_y g_x)\mathbf{k} \Rightarrow$

$$\text{div}(\nabla f \times \nabla g) = \nabla \cdot (\nabla f \times \nabla g)$$

$$= \left[\frac{\partial}{\partial x}\mathbf{i} + \frac{\partial}{\partial y}\mathbf{j} + \frac{\partial}{\partial z}\mathbf{k} \right] \cdot [(f_y g_z - f_z g_y)\mathbf{i} - (f_x g_z - f_z g_x)\mathbf{j} + (f_x g_y - f_y g_x)\mathbf{k}]$$

$$= \frac{\partial}{\partial x}(f_y g_z - f_z g_y) + \frac{\partial}{\partial y}(-f_x g_z + f_z g_x) + \frac{\partial}{\partial z}(f_x g_y - f_y g_x)$$

$$= [(f_{yx} g_z + f_y g_{zx}) - (f_{zx} g_y + f_z g_{yx})] + [(-f_{xy} g_z - f_x g_{zy}) + (f_{zy} g_x + f_z g_{xy})]$$

$$+[(f_{xz} g_y + f_x g_{yz}) - (f_{yz} g_x + f_y g_{xz})]$$

$$= 0 \text{ (as the second-order mixed partial derivatives are equal, e.g. } f_{yx} = f_{xy}).$$

31.   To prove $\text{div}(\mathbf{F} \times \mathbf{G}) = \mathbf{G} \cdot \text{curl } \mathbf{F} - \mathbf{F} \cdot \text{curl } \mathbf{G}$. Let $\mathbf{F} = a\mathbf{i} + b\mathbf{j} + c\mathbf{k}$ and $\mathbf{G} = A\mathbf{i} + B\mathbf{j} + C\mathbf{k}$, then determine the LHS:

$$\mathbf{F} \times \mathbf{G} = \det \begin{bmatrix} \mathbf{i} & \mathbf{j} & \mathbf{k} \\ a & b & c \\ A & B & C \end{bmatrix} = (bC - Bc)\mathbf{i} - (aC - Ac)\mathbf{j} + (aB - Ab)\mathbf{k} \Rightarrow$$

$$\text{div}(\mathbf{F} \times \mathbf{G}) = (bC - Bc)_x - (aC - Ac)_y + (aB - Ab)_z$$

Now, determine the RHS:

$$\text{curl } \mathbf{F} = \det \begin{bmatrix} \mathbf{i} & \mathbf{j} & \mathbf{k} \\ \frac{\partial}{\partial x} & \frac{\partial}{\partial y} & \frac{\partial}{\partial z} \\ a & b & c \end{bmatrix} = (c_y - b_z)\mathbf{i} - (c_x - a_z)\mathbf{j} + (b_x - a_y)\mathbf{k} \Rightarrow$$

$$\mathbf{G} \cdot \text{curl } \mathbf{F} = Ac_y - Ab_z - Bc_x + Ba_z + Cb_x - Ca_y = (Cb_x - Bc_x) + (Ac_y - Ca_y) + (Ba_z - Ab_z)$$

and

$$\text{curl } \mathbf{G} = \det \begin{bmatrix} \mathbf{i} & \mathbf{j} & \mathbf{k} \\ \frac{\partial}{\partial x} & \frac{\partial}{\partial y} & \frac{\partial}{\partial z} \\ A & B & C \end{bmatrix} = (C_y - B_z)\mathbf{i} - (C_x - A_z)\mathbf{j} + (B_x - A_y)\mathbf{k} \Rightarrow$$

$$\mathbf{F} \cdot \text{curl } \mathbf{G} = aC_y - aB_z - bC_x + bA_z + cB_x - cA_y = (cB_x - bC_x) + (aC_y - cA_y) + (bA_z - aB_z)$$

so that

$$\mathbf{G} \cdot \text{curl } \mathbf{F} - \mathbf{F} \cdot \text{curl } \mathbf{G} = (Cb_x - Bc_x - cB_x + bC_x) + (Ac_y - Ca_y - aC_y + cA_y)$$

$$+ (Ba_z - Ab_z - bA_z + aB_z)$$

$$= (bC - Bc)_x - (aC - Ac)_y + (aB - Ab)_z = \text{div } (\mathbf{F} \times \mathbf{G}).$$

33.   To prove $\nabla \cdot [\nabla(fg)] = f[\nabla \cdot (\nabla g)] + g[\nabla \cdot (\nabla f)] + 2(\nabla f \cdot \nabla g)$:

$$\nabla(fg) = \frac{\partial}{\partial x}(fg)\mathbf{i} + \frac{\partial}{\partial y}(fg)\mathbf{j} + \frac{\partial}{\partial z}(fg)\mathbf{k} = (f_x g + fg_x)\mathbf{i} + (f_y g + fg_y)\mathbf{j} + (f_z g + fg_z)\mathbf{k}$$

$$\nabla \cdot [\nabla(fg)] = [\frac{\partial}{\partial x}\mathbf{i} + \frac{\partial}{\partial y}\mathbf{j} + \frac{\partial}{\partial z}\mathbf{k}] \cdot [(f_x g + fg_x)\mathbf{i} + (f_y g + fg_y)\mathbf{j} + (f_z g + fg_z)\mathbf{k}]$$

$$= \frac{\partial}{\partial x}(f_x g + fg_x) + \frac{\partial}{\partial y}(f_y g + fg_y) + \frac{\partial}{\partial z}(f_z g + fg_z)$$

$$= (f_{xx} g + f_x g_x + f_x g_x + fg_{xx}) + (f_{yy} g + f_y g_y + f_y g_y + fg_{yy})$$

$$+ (f_{zz} g + f_z g_z + f_z g_z + fg_{zz})$$

$$= f(g_{xx} + g_{yy} + g_{zz}) + (f_{xx} + f_{yy} + f_{zz})g + 2(f_x g_x + f_y g_y + f_z g_z)$$

$$= f\left[\frac{\partial}{\partial x}g_x + \frac{\partial}{\partial y}g_y + \frac{\partial}{\partial z}g_z \right] + \left[\frac{\partial}{\partial x}f_x + \frac{\partial}{\partial y}f_y + \frac{\partial}{\partial z}f_z \right]g$$

$$+ 2[(f_x\mathbf{i} + f_y\mathbf{j} + f_z\mathbf{k}) \cdot (g_x\mathbf{i} + g_y\mathbf{j} + g_z\mathbf{k})]$$

$$= f[\nabla \cdot (\nabla g)] + g[\nabla \cdot (\nabla f)] + 2(\nabla f \cdot \nabla g).$$

35. $\nabla\mathbf{F}$ is not meaningful, as you can only take the gradient of a scalar function.

37. curl(curl $\mathbf{F}$) is meaningful, as the curl applies to a vector field, and the result of the curl is again a vector field. The final result is a vector field.

39. div(∇f) is meaningful, as the divergence is applied to a vector field, and the gradient of a scalar function is a vector field.

41. div[curl(∇f)] is meaningful, as ∇f is a vector field and the curl applies to such. Moreover, the result of the curl is a vector field, and the divergence is always taken on a vector field. The final result is the result of a divergence, namely a scalar field. [In fact, the result is 0, because (see Theorem 14.14) curl(∇f) = $\mathbf{0}$ ⟹ div $\mathbf{0}$ = 0.]

§14.5 Surface Integrals

1. The surface is $z = \sqrt{x^2 + y^2} = f(x,y)$, and $D = \{(x,y)|x^2 + y^2 \le 4\}$. Then $f_x = \frac{1}{2}(x^2 + y^2)^{-\frac{1}{2}}(2x) = \dfrac{x}{\sqrt{x^2 + y^2}}$ and $f_y = \frac{1}{2}(x^2 + y^2)^{-\frac{1}{2}}(2y) = \dfrac{y}{\sqrt{x^2 + y^2}}$ ⟹

$$A(S) = \iint_D \sqrt{1 + [f_x(x,y)]^2 + [f_y(x,y)]^2}\, dA = \iint_D \sqrt{1 + \left[\frac{x}{\sqrt{x^2+y^2}}\right]^2 + \left[\frac{y}{\sqrt{x^2+y^2}}\right]^2}\, dA$$

$$= \iint_D \sqrt{1 + \frac{x^2+y^2}{x^2+y^2}}\, dA = \sqrt{2}\iint_D dA = \sqrt{2}(\pi \cdot 2^2) = 4\sqrt{2}\pi.$$

3. The surface is $z = x + \frac{2}{3}y^{\frac{3}{2}} = f(x,y)$, and $D = \{(x,y)|1 \le x \le 4,\ 2 \le y \le 7\}$. Then $f_x = 1$ and $f_y = y^{\frac{1}{2}}$ ⟹

$$A(S) = \iint_D \sqrt{1 + [f_x(x,y)]^2 + [f_y(x,y)]^2}\, dA = \iint_D \sqrt{1 + 1 + y}\, dA = \int_1^4 \int_2^7 \sqrt{2+y}\, dy\, dx$$

$$= \int_1^4 dx \int_2^7 (2+y)^{\frac{1}{2}}\, dy = \left[x\right]_1^4 \cdot \left[\frac{2}{3}(2+y)^{\frac{3}{2}}\right]_2^7 = (4-1)\cdot\frac{2}{3}(9^{\frac{3}{2}} - 4^{\frac{3}{2}}) = 2(27-8) = 38.$$

5.

The surface is $z = y^2 - x^2 = f(x,y)$, and D is the shaded region between the two circles, in the figure. Then $f_x = -2x$ and $f_y = 2y$ ⟹

$$A(S) = \iint_D \sqrt{1 + [f_x(x,y)]^2 + [f_y(x,y)]^2}\, dA = \iint_D \sqrt{1 + 4x^2 + 4y^2}\, dA$$

$$= \int_0^{2\pi} \int_1^2 \sqrt{1 + 4r^2}\, r\, dr\, d\theta = \int_0^{2\pi} d\theta \int_1^2 \sqrt{1 + 4r^2}\, r\, dr = 2\pi \left[\frac{2}{3\cdot 8}(1 + 4r^2)^{\frac{3}{2}}\right]_1^2$$

$$= \frac{\pi}{6}\left(17^{\frac{3}{2}} - 5^{\frac{3}{2}}\right) = \frac{\pi}{6}\left(17\sqrt{17} - 5\sqrt{5}\right).$$

7.   The surface is $z = \sqrt{25 - x^2 - y^2} = f(x, y)$, and $D = \{(x, y) | x^2 + y^2 \le 9\}$. Then

$$f_x = \tfrac{1}{2}(25 - x^2 - y^2)^{-\frac{1}{2}}(-2x) = -\frac{x}{\sqrt{25 - x^2 - y^2}}, \text{ and}$$

$$f_y = \tfrac{1}{2}(25 - x^2 - y^2)^{-\frac{1}{2}}(-2y) = -\frac{y}{\sqrt{25 - x^2 - y^2}} \Rightarrow$$

$$A(S) = \iint_D \sqrt{1 + [f_x(x, y)]^2 + [f_y(x, y)]^2}\, dA = \iint_D \sqrt{1 + \frac{x^2}{25 - x^2 - y^2} + \frac{y^2}{25 - x^2 - y^2}}\, dA$$

$$= \iint_D \sqrt{\frac{25}{25 - x^2 - y^2}}\, dA = 5 \iint_D (25 - x^2 - y^2)^{-\frac{1}{2}}\, dA = 5 \int_0^{2\pi} \int_0^3 (25 - r^2)^{-\frac{1}{2}}\, r\, dr\, d\theta$$

$$= 5 \int_0^{2\pi} d\theta \int_0^3 (25 - r^2)^{-\frac{1}{2}}\, r\, dr\, d\theta = 10\pi \left[-(25 - r^2)^{\frac{1}{2}}\right]_0^3 = -10\pi \left[\sqrt{16} - \sqrt{25}\right] = 10\pi.$$

9. The surface is $z = \tfrac{2}{3}\left(x^{\frac{3}{2}} + y^{\frac{3}{2}}\right) = f(x, y)$, and $D = \{(x, y) | 2 \le x \le 4,\ 1 \le y \le 4\}$. Then $f_x = x^{\frac{1}{2}}$ and $f_y = y^{\frac{1}{2}} \Rightarrow$

$$A(S) = \iint_D \sqrt{1 + [f_x(x, y)]^2 + [f_y(x, y)]^2}\, dA = \iint_D \sqrt{1 + x + y}\, dA = \int_2^4 \int_1^4 (1 + x + y)^{\frac{1}{2}}\, dy\, dx$$

$$= \int_2^4 \left[\tfrac{2}{3}(1 + x + y)^{\frac{3}{2}}\right]_{y=1}^{y=4}\, dx = \tfrac{2}{3} \int_2^4 \left[(5 + x)^{\frac{3}{2}} - (2 + x)^{\frac{3}{2}}\right]\, dx = \tfrac{4}{15} \left[(5 + x)^{\frac{5}{2}} - (2 + x)^{\frac{5}{2}}\right]_2^4$$

$$= \tfrac{4}{15} \left[9^{\frac{5}{2}} - 6^{\frac{5}{2}} - \left(7^{\frac{5}{2}} - 4^{\frac{5}{2}}\right)\right] = \tfrac{4}{15} \left[3^5 - 36\sqrt{6} - 49\sqrt{7} + 2^5\right] = \tfrac{4}{15} \left[275 - 36\sqrt{6} - 49\sqrt{7}\right].$$

11.

$S: \ z = 2 - 3y + x^2 = f(x, y) \Rightarrow f_x = 2x$ and $f_y = -3$, and D is shown in the figure.

$$\iint_S (z + 3y - x^2)\, dS = \iint_D [(2 - 3y + x^2) + 3y - x^2]\sqrt{1 + (2x)^2 + 3^2}\, dA = \iint_D 2\sqrt{4x^2 + 10}\, dA$$

$$= \int_0^2 \int_{-2x}^0 2\sqrt{4x^2 + 10}\, dy\, dx = \int_0^2 \left[2\sqrt{4x^2 + 10}\, y\right]_{y=-2x}^{y=0}\, dx$$

$$= \int_0^2 4x\sqrt{4x^2 + 10}\, dx = \left[4 \cdot \tfrac{2}{3} \cdot \tfrac{1}{8}(4x^2 + 10)^{\frac{3}{2}}\right]_0^2$$

$$= \tfrac{1}{3}\left(26^{\frac{3}{2}} - 10^{\frac{3}{2}}\right) = \tfrac{1}{3}\left(26\sqrt{26} - 10\sqrt{10}\right).$$

13. $S: \ z = x + y^2 = f(x, y) \Rightarrow f_x = 1$ and $f_y = 2y$, and $D = \{(x, y) | 0 \le x \le 1,\ 0 \le y \le 2\}$.

$$\iint_S y\, dS = \iint_D y\sqrt{1 + 1 + (2y)^2}\, dA = \int_0^1 \int_0^2 y\sqrt{2 + 4y^2}\, dy\, dx = \int_0^1 \left[\tfrac{2}{3} \cdot \tfrac{1}{8}(2 + 4y^2)^{\frac{3}{2}}\right]_{y=0}^{y=2}\, dx$$

$$= \tfrac{1}{12} \int_0^1 \left(18^{\frac{3}{2}} - 2^{\frac{3}{2}}\right)\, dx = \tfrac{1}{12}\left(18\sqrt{18} - 2\sqrt{2}\right) \int_0^1 dx = \tfrac{1}{6}\left(9 \cdot 3\sqrt{2} - \sqrt{2}\right) = \tfrac{13\sqrt{2}}{3}.$$

15.

$S: \quad z = \sqrt{9-y^2} = f(x,y) \quad \Rightarrow \quad f_x = 0 \text{ and } f_y = -\dfrac{y}{\sqrt{9-y^2}}, \text{ and}$

$D = \{(x,y)|0 \le x \le 4, \ 0 \le y \le 3\}.$

$$\iint_S (x+z)\,dS = \iint_D \left[x + \sqrt{9-y^2}\right]\sqrt{1 + \frac{y^2}{9-y^2}}\,dA = \int_0^4 \int_0^3 \left[x + \sqrt{9-y^2}\right] \cdot \frac{3}{\sqrt{9-y^2}}\,dy\,dx$$

$$= 3\int_0^4 \int_0^3 \frac{x + \sqrt{9-y^2}}{\sqrt{9-y^2}}\,dy\,dx = 3\int_0^4 \int_0^3 \frac{x}{\sqrt{9-y^2}}\,dy\,dx + 3\int_0^4 \int_0^3 dy\,dx$$

$$= 3\int_0^4 x\,dx \int_0^3 \frac{dy}{\sqrt{9-y^2}} + 3\cdot 3\cdot 4 = 3\left[\tfrac{1}{2}x^2\right]_0^4 \cdot \left[\int_0^{\pi/2} \frac{3\cos\theta\,d\theta}{\sqrt{9-9\sin^2\theta}}\right] + 36$$

$$\boxed{\begin{array}{ll} y = 3\sin\theta, & y = 0 \Rightarrow \theta = 0 \\ dy = 3\cos\theta\,d\theta & y = 3 \Rightarrow \theta = \frac{\pi}{2} \end{array}}$$

$$= 24\int_0^{\pi/2} d\theta + 36 = 24 \cdot \frac{\pi}{2} + 36 = 12\pi + 36.$$

17. $S: \ z = \sqrt{x^2+y^2} = f(x,y) \Rightarrow f_x = \dfrac{x}{\sqrt{x^2+y^2}} \text{ and } f_y = \dfrac{y}{\sqrt{x^2+y^2}}, \text{ and } D: \ x^2+y^2 \le 1.$

$$\iint_S x^2 y^2 z^2\,dS = \iint_D x^2 y^2 (x^2+y^2)\sqrt{1 + \frac{x^2}{x^2+y^2} + \frac{y^2}{x^2+y^2}}\,dA = \sqrt{2}\iint_D x^2 y^2 (x^2+y^2)\,dA$$

$$= \sqrt{2}\int_0^{2\pi} \int_0^1 r^2\cos^2\theta \cdot r^2\sin^2\theta \cdot r^2 \cdot r\,dr\,d\theta = \sqrt{2}\int_0^{2\pi} \int_0^1 r^7 \left(\tfrac{1}{2}\sin 2\theta\right)^2 dr\,d\theta$$

$$= \sqrt{2}\left[\int_0^{2\pi} \tfrac{1}{4}\sin^2 2\theta\,d\theta \int_0^1 r^7\,dr\right] = \frac{\sqrt{2}}{32}\int_0^{2\pi} \frac{1-\cos 4\theta}{2}\,d\theta = \frac{\sqrt{2}}{64}\left[\theta - \tfrac{1}{4}\sin 4\theta\right]_0^{2\pi}$$

$$= \frac{\pi\sqrt{2}}{32}.$$

19. $S: \ z = 9 - y^2 = f(x,y) \Rightarrow f_x = 0, \ f_y = -2y, \text{ and } \delta(x,y,z) = y:$

$$M = \int_0^3 \int_0^3 y\sqrt{1+4y^2}\,dy\,dx = \frac{1}{8}\int_0^3 \left[\tfrac{2}{3}(1+4y^2)^{\frac{3}{2}}\right]_0^3 dx = \frac{1}{12}\int_0^3 \left(37^{\frac{3}{2}} - 1\right)dx$$

$$= \frac{1}{12}\left(37\sqrt{37} - 1\right)\int_0^3 dx = \frac{1}{4}\left(37\sqrt{37} - 1\right).$$

21.

$S: \ x + y + z = a \Rightarrow z = a - x - y = f(x,y) \Rightarrow f_x = f_y = -1,$
$\delta(x,y,z) = kx^2, \text{ and } D \text{ is shown in the figure.}$

$$M = \int_0^a \int_0^{-x+a} kx^2\sqrt{1+1+1}\,dy\,dx = \sqrt{3}k\int_0^a \left[x^2 y\right]_{y=0}^{y=-x+a} dx$$

$$= \sqrt{3}k\int_0^a x^2(-x+a)\,dx = \sqrt{3}k\int_0^a (-x^3 + ax^2)\,dx$$

$$= \sqrt{3}k\left[-\tfrac{1}{4}x^4 + \tfrac{a}{3}x^3\right]_0^a = \sqrt{3}k\left(-\tfrac{1}{4}a^4 + \tfrac{a^4}{3}\right) = \sqrt{3}ka^4 \cdot \frac{1}{12} = \frac{\sqrt{3}}{12}ka^4.$$

23. $S: \mathbf{r}(u,v) = (u+v)\mathbf{i} + v\mathbf{j} + (u-v)\mathbf{k},\ 0 \le u \le 1,\ 0 \le v \le 2,$ and $g(x,y,z) = xyz$:

$$\|\mathbf{r_u} \times \mathbf{r_v}\| = \left\| \det \begin{bmatrix} \mathbf{i} & \mathbf{j} & \mathbf{k} \\ 1 & 0 & 1 \\ 1 & 1 & -1 \end{bmatrix} \right\| = \|-\mathbf{i} - (-2)\mathbf{j} + (1)\mathbf{k}\| = \sqrt{1+4+1} = \sqrt{6}$$

$$\iint_S xyz\,dS = \int_0^2 \int_0^1 (u+v)v(u-v)\sqrt{6}\,du\,dv = \sqrt{6}\int_0^2 \int_0^1 (u^2 v - v^3)\,du\,dv$$

$$= \sqrt{6}\int_0^2 \left[\tfrac{1}{3}u^3 v - uv^3\right]_{u=0}^{u=1} = \sqrt{6}\int_0^2 \left(\tfrac{1}{3}v - v^3\right)\,dv = \sqrt{6}\left[\tfrac{1}{6}v^2 - \tfrac{1}{4}v^4\right]_0^2$$

$$= \sqrt{6}\left(\tfrac{2}{3} - 4\right) = -\tfrac{10\sqrt{6}}{3}.$$

25. $S: \mathbf{r}(u,v) = (u\cos v)\mathbf{i} + (u\sin v)\mathbf{j} + v\mathbf{k},\ 0 \le u \le 1,\ 0 \le v \le \pi,$ and $g(x,y,z) = \sqrt{1+x^2+y^2}$:

$$\|\mathbf{r_u} \times \mathbf{r_v}\| = \left\| \det \begin{bmatrix} \mathbf{i} & \mathbf{j} & \mathbf{k} \\ \cos v & \sin v & 0 \\ -u\sin v & u\cos v & 1 \end{bmatrix} \right\| = \|(\sin v\mathbf{i} - (\cos v)\mathbf{j} + u\mathbf{k}\|$$

$$= \sqrt{\sin^2 v + \cos^2 v + u^2} = \sqrt{1+u^2}, \text{ so that the surface integral is:}$$

$$\iint_S \sqrt{1+x^2+y^2}\,dS = \int_0^\pi \int_0^1 \sqrt{1+u^2}\sqrt{1+u^2}\,du\,dv = \int_0^\pi \int_0^1 (1+u^2)\,du\,dv$$

$$= \int_0^\pi \left[u + \tfrac{1}{3}u^3\right]_{u=0}^{u=1}\,dv = \int_0^\pi \left(1 + \tfrac{1}{3}\right)\,dv = \tfrac{4}{3}\int_0^\pi dv = \tfrac{4\pi}{3}.$$

27.

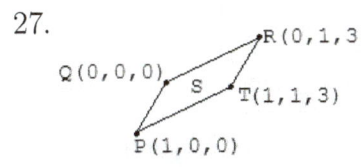

Determining the equation of the plane PQRT: $\overrightarrow{QP} = (1,0,0)$, and $\overrightarrow{QR} = (0,1,3)$, then

$$\mathbf{N} = \overrightarrow{QP} \times \overrightarrow{QR} = \det \begin{bmatrix} \mathbf{i} & \mathbf{j} & \mathbf{k} \\ 1 & 0 & 0 \\ 0 & 1 & 3 \end{bmatrix} = -3\mathbf{j} + \mathbf{k} \text{ is a}$$

normal to the plane, so the equation of the plane is $0(x-0) - 3(y-0) + (z-0) = 0 \Rightarrow z = 3y = f(x,y) \Rightarrow f_x = 0,\ f_y = 3,$ and D is shown in the lower figure. Thus,

$$\text{Flux } (\mathbf{F}_S) = \iint_S \mathbf{F} \cdot \mathbf{n}\,dS = \iint_D \mathbf{F} \cdot (-f_x\mathbf{i} - f_y\mathbf{j} + \mathbf{k})\,dA = \int_0^1 \int_0^1 z\mathbf{k} \cdot (-3\mathbf{j} + \mathbf{k})\,dy\,dx$$

$$= \int_0^1 \int_0^1 z\,dy\,dx = \int_0^1 \int_0^1 3y\,dy\,dx = \int_0^1 \left[\tfrac{3}{2}y^2\right]_{y=0}^{y=1}\,dx = \tfrac{3}{2}\int_0^1 dx = \tfrac{3}{2}.$$

29. $S: \ z = 16 - x^2 - y^2 = f(x, y) \ \Rightarrow \ f_x = -2x$ and $f_y = -2y, D: \ x^2 + y^2 \le 16$, and $\mathbf{F}(x, y, z) = y\,\mathbf{i} + x\,\mathbf{j} + z\,\mathbf{k}$.

Flux $(\mathbf{F}_S) = \displaystyle\iint_S \mathbf{F} \cdot \mathbf{n} \, dS = \iint_D \mathbf{F} \cdot (-f_x\,\mathbf{i} - f_y\,\mathbf{j} + \mathbf{k}) \, dA$

$\qquad = \displaystyle\iint_D (y\,\mathbf{i} + x\,\mathbf{j} + z\,\mathbf{k}) \cdot (-2x\,\mathbf{i} - 2y\,\mathbf{j} + \mathbf{k}) \, dA = \iint_D (-2xy - 2xy + z) \, dA$

$\qquad = \displaystyle\iint_D (-4xy + 16 - x^2 - y^2) \, dA = \int_0^{2\pi} \int_0^4 [-4(r\cos\theta)(r\sin\theta) + 16 - r^2)] \, r \, dr \, d\theta$

$\qquad = \displaystyle\int_0^{2\pi} \int_0^4 [r^3(-4\sin\theta\cos\theta - 1) + 16r] \, dr \, d\theta$

$\qquad = \displaystyle\int_0^{2\pi} \left[\tfrac{1}{4}r^4(-4\sin\theta\cos\theta - 1) + 8r^2\right]_{r=0}^{r=4} d\theta = \int_0^{2\pi} [64(-4\sin\theta\cos\theta - 1) + 128] \, d\theta$

$\qquad = \displaystyle\int_0^{2\pi} (-256\sin\theta\cos\theta + 64) \, d\theta = \left[-128\sin^2\theta + 64\theta\right]_0^{2\pi} = 128\pi.$

§14.6 Stoke's Theorem

1. For $\mathbf{F}(x, y, z) = -3z\,\mathbf{i} + (x + y)\,\mathbf{j} + y\,\mathbf{k}$,

$$\text{curl } \mathbf{F} = \det\begin{bmatrix} \mathbf{i} & \mathbf{j} & \mathbf{k} \\ \frac{\partial}{\partial x} & \frac{\partial}{\partial y} & \frac{\partial}{\partial z} \\ -3z & x+y & y \end{bmatrix} = \mathbf{i} - 3\mathbf{j} + \mathbf{k}.$$

S is the surface $z = f(x, y) = 4x - 8y + 5 \ \Rightarrow \ f_x = 4$ and $f_y = -8$. By Stoke's Theorem,

$\displaystyle\oint_C \mathbf{F} \cdot d\mathbf{r} = \iint_S \text{curl } \mathbf{F} \cdot \mathbf{n} \, dS = \iint_D \text{curl } \mathbf{F} \cdot (-f_x\,\mathbf{i} - f_y\,\mathbf{j} + \mathbf{k}) \, dA$

$\qquad = \displaystyle\iint_D [\mathbf{i} - 3\mathbf{j} + \mathbf{k}] \cdot [-4\mathbf{i} + 8\mathbf{j} + \mathbf{k}] \, dA = \iint_D (-4 - 24 + 1) \, dA = -27 \iint_D dA$

$\qquad = -27 \, (\text{area of the ellipse } D) = -27(\pi \cdot 6 \cdot 2) = -324\pi.$

[since the area of an ellipse whose equation is $\dfrac{(x - x_0)^2}{a^2} + \dfrac{(y - y_0)^2}{b^2} = 1$ is πab, and

$D: \ (x - 1)^2 + 9(y - 3)^2 = 36 \ \Rightarrow \ \dfrac{(x - 1)^2}{6^2} + \dfrac{(y - 3)^2}{2^2} = 1]$

3.   As a surface S whose boundary is C, take $x^2 + y^2 \le 4$ in the plane $z = -3$, so that $\mathbf{n} \, dS = \mathbf{k} \, dA$, and D is the disk: $x^2 + y^2 \le 4$. For $\mathbf{F}(x, y, z) = y\,\mathbf{i} + xz^3\,\mathbf{j} - zy^3\,\mathbf{k}$,

$$\text{curl } \mathbf{F} = \det\begin{bmatrix} \mathbf{i} & \mathbf{j} & \mathbf{k} \\ \frac{\partial}{\partial x} & \frac{\partial}{\partial y} & \frac{\partial}{\partial z} \\ y & xz^3 & -zy^3 \end{bmatrix} = \mathbf{i}(-3y^2z - 3xz^2)\mathbf{i} + (z^3 - 1)\mathbf{k}.$$

By Stoke's Theorem,

$$\oint \mathbf{F} \cdot d\mathbf{r} = \iint_S \text{curl } \mathbf{F} \cdot \mathbf{n} \, dS = \iint_D [(-3y^2z - 3xz^2)\mathbf{i} + (z^3 - 1)\mathbf{k}] \cdot \mathbf{k} \, dA = \iint_D (z^3 - 1) \, dA$$

$$= \iint_D [(-3)^3 - 1] \, dA = -28 \, (\text{area of the disk } D) = -28(\pi \cdot 2^2) = -112\pi.$$

5. As a surface S whose boundary is C, take the part of the plane $x + y + z = 1$ enclosed by C. Then $\Rightarrow z = 1 - x - y = f(x, y)$, with $D : x^2 + y^2 \leq 1$. For $\mathbf{F}(x, y, z) = -y^3 \mathbf{i} + x^3 \mathbf{j} - z^3 \mathbf{k}$,

$$\text{curl } \mathbf{F} = \det \begin{bmatrix} \mathbf{i} & \mathbf{j} & \mathbf{k} \\ \frac{\partial}{\partial x} & \frac{\partial}{\partial y} & \frac{\partial}{\partial z} \\ -y^3 & x^3 & -z^3 \end{bmatrix} = (3x^2 + 3y^2)\mathbf{k}. \quad \text{By Stoke's Theorem,}$$

$$\oint \mathbf{F} \cdot d\mathbf{r} = \iint_S \text{curl } \mathbf{F} \cdot \mathbf{n} \, dS = \iint_D \text{curl } \mathbf{F} \cdot (-f_x \mathbf{i} - f_y \mathbf{j} + \mathbf{k}) \, dA$$

$$= \iint_D [(3x^2 + 3y^2)\mathbf{k}] \cdot [\mathbf{i} + \mathbf{j} + \mathbf{k}] \, dA = \iint_D (3x^2 + 3y^2) \, dA = \int_0^{2\pi} \int_0^1 3r^2 \cdot r \, dr d\theta$$

$$= \int_0^{2\pi} \left[\frac{3}{4}r^4 \right]_{r=0}^{r=1} d\theta = \frac{3}{4} \int_0^{2\pi} d\theta = \frac{3\pi}{2}.$$

7. As a surface S whose boundary is C, take the circle $y - 2 = 2\cos t$, $z - 2 = 2\sin t$, $0 \leq t \leq 2\pi$, and its interior D, in the plane $x = 0$. Since S is flat and lying in the plane $x = 0$, $\mathbf{n} \, dS = \mathbf{i} \, dA$. For $\mathbf{F}(x, y, z) = x^2 e^{5z} \mathbf{i} + x \cos y \, \mathbf{j} + 3y \, \mathbf{k}$,

$$\text{curl } \mathbf{F} = \det \begin{bmatrix} \mathbf{i} & \mathbf{j} & \mathbf{k} \\ \frac{\partial}{\partial x} & \frac{\partial}{\partial y} & \frac{\partial}{\partial z} \\ x^2 e^{5z} & x \cos y & 3y \end{bmatrix} = 3\mathbf{i} + 5x^2 e^{5z} \mathbf{j} + \cos y \, \mathbf{k}. \quad \text{By Stoke's Theorem,}$$

$$\oint \mathbf{F} \cdot d\mathbf{r} = \iint_S \text{curl } \mathbf{F} \cdot \mathbf{n} \, dS = \iint_D [3\mathbf{i} + 5x^2 e^{5z} \mathbf{j} + \cos y \, \mathbf{k}] \cdot \mathbf{i} \, dA$$

$$= = \iint_D 3 \, dA = 3 \, (\text{area of the disk} D) = 3(\pi \cdot 2^2) = 12\pi.$$

9.

As a surface S whose boundary is C, take the plane $2x + y + z = 2$ $\Rightarrow z = 2 - 2x - y = f(x, y)$, for $x \geq 0, y \geq 0, z \geq 0$. For $\mathbf{F}(x, y, z) = xz \, \mathbf{i} + xy \, \mathbf{j} + 3xz \, \mathbf{k}$,

$$\text{curl } \mathbf{F} = \det \begin{bmatrix} \mathbf{i} & \mathbf{j} & \mathbf{k} \\ \frac{\partial}{\partial x} & \frac{\partial}{\partial y} & \frac{\partial}{\partial z} \\ xz & xy & 3xz \end{bmatrix} = (x - 3z)\mathbf{j} + y \, \mathbf{k}. \quad \text{By Stoke's Theorem,}$$

$$\oint \mathbf{F} \cdot d\mathbf{r} = \int\!\!\int_S \operatorname{curl} \mathbf{F} \cdot \mathbf{n}\, dS = \int\!\!\int_D \operatorname{curl} \mathbf{F} \cdot (-f_x \mathbf{i} - f_y \mathbf{j} + \mathbf{k})\, dA$$

$$= \int\!\!\int_D [(x - 3z)\mathbf{j} + y\,\mathbf{k}] \cdot [2\mathbf{i} + \mathbf{j} + \mathbf{k}]\, dA = \int\!\!\int_D (x - 3z + y)\, dA$$

$$= \int\!\!\int_D [x - 3(2 - 2x - y) + y]\, dA = \int_0^1 \int_0^{-2x+2} (7x + 4y - 6)\, dy\, dx$$

$$= \int_0^1 \left[(7x - 6)y + 2y^2 \right]_{y=0}^{y=-2x+2} dx = \int_0^1 [(7x - 6)(-2x + 2) + 2(-2x + 2)^2]\, dx$$

$$= \int_0^1 (-14x^2 + 26x - 12 + 8x^2 - 16x + 8)\, dx = \int_0^1 (-6x^2 + 10x - 4)\, dx$$

$$= \left[-2x^3 + 5x^2 - 4x \right]_0^1 = -2 + 5 - 4 = -1.$$

11. The bounding curve C is: $1 = 5 - x^2 - y^2 \Rightarrow x^2 + y^2 = 4$ which we can parametrize as $x = 2\cos t$, $y = 2\sin t$ in the plane $z = 1$. Thus, $\mathbf{r}(t) = 2\cos t\,\mathbf{i} + 2\sin t\,\mathbf{j} + \mathbf{k}$, with $0 \le t \le 2\pi$. As $\mathbf{F}(x, y, z) = z^2\,\mathbf{i} - 3xy\,\mathbf{j} + x^3 y^3\,\mathbf{k}$, $\mathbf{F}(\mathbf{r}(t)) = \mathbf{i} - 3(2\cos t)(2\sin t)\mathbf{j} + (2\cos t)^3(2\sin t)^3\mathbf{k}$, and $\mathbf{r}'(t) = -2\sin t\,\mathbf{i} + 2\cos t\,\mathbf{j}$. By Stoke's Theorem,

$$\int\!\!\int_S \operatorname{curl} \mathbf{F} \cdot \mathbf{n}\, dS = \oint_C \mathbf{F} \cdot d\mathbf{r} = \int_0^{2\pi} \mathbf{F}(\mathbf{r}(t)) \cdot \mathbf{r}'(t)\, dt$$

$$= \int_0^{2\pi} [\mathbf{i} - 3(2\cos t)(2\sin t)\mathbf{j} + (2\cos t)^3(2\sin t)^3\mathbf{k}] \cdot [-2\sin t\,\mathbf{i} + 2\cos t\,\mathbf{j}]\, dt$$

$$= \int_0^{2\pi} (-2\sin t - 24\sin t\cos^2 t)\, dt = \left[2\cos t + 8\cos^3 t \right]_0^{2\pi} = 0.$$

13.

The bounding curve C is made up of 4 line segments in the plane $z = 1$, and we can parametrize them as follows.

$$C_1 : \mathbf{r}(t) = t\mathbf{i} + \mathbf{k}, \qquad\qquad 0 \le t \le 1, \quad \mathbf{r}'(t) = \mathbf{i}$$

$$C_2 : \mathbf{r}(t) = \mathbf{i} + t\mathbf{j} + \mathbf{k}, \qquad\quad 0 \le t \le 1, \quad \mathbf{r}'(t) = \mathbf{j}$$

$$C_3 : \mathbf{r}(t) = (1 - t)\mathbf{i} + \mathbf{j} + \mathbf{k}, \quad 0 \le t \le 1, \quad \mathbf{r}'(t) = -\mathbf{i}$$

$$C_4 : \mathbf{r}(t) = (1 - t)\mathbf{j} + \mathbf{k}, \qquad\quad 0 \le t \le 1, \quad \mathbf{r}'(t) = -\mathbf{j}$$

As $\mathbf{F}(x, y, z) = z^2\,\mathbf{i} + 5x\,\mathbf{j}$,

$$\text{On } C_1 : \mathbf{F}(\mathbf{r}(t)) = \mathbf{i} + 5t\mathbf{j} \qquad\qquad \text{On } C_2 : \mathbf{F}(\mathbf{r}(t)) = \mathbf{i} + 5\mathbf{j}$$

$$\text{On } C_3 : \mathbf{F}(\mathbf{r}(t)) = \mathbf{i} + (5 - 5t)\mathbf{j} \qquad \text{On } C_4 : \mathbf{F}(\mathbf{r}(t)) = \mathbf{i}$$

By Stoke's Theorem,

$$\int\!\!\int_S \operatorname{curl} \mathbf{F} \cdot \mathbf{n}\, dS = \oint_C \mathbf{F} \cdot d\mathbf{r} = \oint_C \mathbf{F}(\mathbf{r}(t)) \cdot \mathbf{r}'(t)\, dt$$

$$= \int_{C_1} \mathbf{F}(\mathbf{r}(t)) \cdot \mathbf{r}'(t)\, dt + \int_{C_2} \mathbf{F}(\mathbf{r}(t)) \cdot \mathbf{r}'(t)\, dt + \int_{C_3} \mathbf{F}(\mathbf{r}(t)) \cdot \mathbf{r}'(t)\, dt$$

$$+ \int_{C_4} \mathbf{F}(\mathbf{r}(t)) \cdot \mathbf{r}'(t)\, dt$$

$$= \int_0^1 [\mathbf{i} + 5t\,\mathbf{j}] \cdot \mathbf{i}\,dt + \int_0^1 [\mathbf{i} + 5\,\mathbf{j}] \cdot \mathbf{j}\,dt + \int_0^1 [\mathbf{i} + (5-5t)\mathbf{j}] \cdot [-\mathbf{i}]\,dt + \int_0^1 \mathbf{i} \cdot [-\mathbf{j}]\,dt$$

$$= \int_0^1 dt + \int_0^1 5\,dt + \int_0^1 (-1)\,dt + \int_0^1 0\,dt = \Big[t\Big]_0^1 + \Big[5t\Big]_0^1 + \Big[-t\Big]_0^1$$

$$= 1 + 5 - 1 = 5.$$

15. The bounding curve C is the circle $x^2 + y^2 = 9$ in the xy-plane. which we can parametrize as $x = 3\cos t$, $y = 3\sin t$ in the plane $z = 0$. Thus, $\mathbf{r}(t) = 3\cos t\,\mathbf{i} + 3\sin t\,\mathbf{j} + 0\mathbf{k}$, with $0 \le t \le 2\pi \Rightarrow \mathbf{r}'(t) = -3\sin t\,\mathbf{i} + 3\cos t\,\mathbf{j}$. As $\mathbf{F}(x, y, z) = 2y\cos z\,\mathbf{i} + e^x \sin z\,\mathbf{j} + xe^y\,\mathbf{k}$,

$$\mathbf{F}(\mathbf{r}(t)) = 2(3\sin t)(\cos 0)\mathbf{i} + e^{3\cos t}(\sin 0)\mathbf{j} + (3\cos t)e^{3\sin t}\mathbf{k} = 6\sin t\,\mathbf{i} + (3\cos t)e^{3\sin t}\mathbf{k}.$$

By Stoke's Theorem,

$$\iint_{S'} \operatorname{curl} \mathbf{F} \cdot \mathbf{n}\,dS = \oint_C \mathbf{F} \cdot d\mathbf{r} = \int_0^{2\pi} \mathbf{F}(\mathbf{r}(t)) \cdot \mathbf{r}'(t)\,dt$$

$$= \int_0^{2\pi} [6\sin t\,\mathbf{i} + (3\cos t)e^{3\sin t}\mathbf{k}] \cdot [-3\sin t\,\mathbf{i} + 3\cos t\,\mathbf{j}]\,dt = \int_0^{2\pi} (-18\sin^2 t)dt$$

$$= -18 \int_0^{2\pi} \frac{1 - \cos 2t}{2}\,dt = -9\Big[t - \frac{1}{2}\sin 2t\Big]_0^{2\pi} = -18\pi.$$

17. The bounding curve C is the elliptical region which is the intersection of the plane $y + z = 2$ and the cylinder $x^2 + y^2 = 1$. Since $x^2 + y^2 = 1$, a parametrization is $x = \cos t$, $y = \sin t$, and as $z = 2 - y$, then $z = 2 - \sin t$, so that $\mathbf{r}(t) = \cos t\,\mathbf{i} + \sin t\,\mathbf{j} + (2 - \sin t)\mathbf{k}$, for $0 \le t \le 2\pi$, and $\mathbf{r}'(t) = -\sin t\,\mathbf{i} + \cos t\,\mathbf{j} - \cos t\,\mathbf{k}$. As $\mathbf{F}(x, y, z) = -y^2\,\mathbf{i} + x\,\mathbf{j} + z^2\,\mathbf{k}$, $\mathbf{F}(\mathbf{r}(t)) = -\sin^2 t\,\mathbf{i} + \cos t\,\mathbf{j} + (2 - \sin t)^2\,\mathbf{k}$.

By Stoke's Theorem,

$$\iint_{S'} \operatorname{curl} \mathbf{F} \cdot \mathbf{n}\,dS = \oint_C \mathbf{F} \cdot d\mathbf{r} = \int_0^{2\pi} \mathbf{F}(\mathbf{r}(t)) \cdot \mathbf{r}'(t)\,dt$$

$$= \int_0^{2\pi} [-\sin^2 t\,\mathbf{i} + \cos t\,\mathbf{j} + (2 - \sin t)^2\,\mathbf{k}] \cdot [-\sin t\,\mathbf{i} + \cos t\,\mathbf{j} - \cos t\,\mathbf{k}]\,dt$$

$$= \int_0^{2\pi} [\sin^3 t + \cos^2 t - (2 - \sin t)^2 \cos t]\,dt$$

Now, $\displaystyle \int_0^{2\pi} \sin^3 t\,dt - \int_0^{2\pi} (1 - \cos^2 t)\sin t\,dt = \int_0^{2\pi} \sin t\,dt - \int_0^{2\pi} \cos^2 t \sin t\,dt$

$$= \Big[-\cos t\Big]_0^{2\pi} + \Big[\frac{1}{3}\cos^3 t\Big]_0^{2\pi} = 0$$

and, $\displaystyle \int_0^{2\pi} \cos^2 t\,dt = \int_0^{2\pi} \frac{1 + \cos 2t}{2}\,dt = \frac{1}{2}\Big[t + \frac{1}{2}\sin 2t\Big]_0^{2\pi} = \pi$

and, $\displaystyle \int_0^{2\pi} (2 - \sin t)^2 \cos t\,dt = \Big[-\frac{1}{3}(2 - \sin t)^3\Big]_0^{2\pi} = 0$

Thus, $\displaystyle \iint_{S'} \operatorname{curl} \mathbf{F} \cdot \mathbf{n}\,dS = 0 + \pi - 0 = \pi.$

19. S is the surface $z = 1 - x^2 - y^2 = f(x, y) \Rightarrow f_x = -2x$ and $f_y = -2y$, with $z \geq 0$, so that C is the bounding curve: $0 = 1 - x^2 - y^2 \Rightarrow x^2 + y^2 = 1$ in the plane $z = 0$, and $D : x^2 + y^2 \leq 1$. For $\mathbf{F}(x, y, z) = y\,\mathbf{i} + z\,\mathbf{j} + x\,\mathbf{k}$,

$$\text{curl } \mathbf{F} = \det \begin{bmatrix} \mathbf{i} & \mathbf{j} & \mathbf{k} \\ \frac{\partial}{\partial x} & \frac{\partial}{\partial y} & \frac{\partial}{\partial z} \\ y & z & x \end{bmatrix} = -\mathbf{i} - \mathbf{j} - \mathbf{k}.$$

$$\iint_S \text{curl } \mathbf{F} \cdot \mathbf{n}\, dS = \iint_D \text{curl } \mathbf{F} \cdot (-f_x\,\mathbf{i} - f_y\,\mathbf{j} + \mathbf{k})\, dA$$

$$= \iint_D [-\mathbf{i} - \mathbf{j} - \mathbf{k}] \cdot [2x\,\mathbf{i} + 2y\,\mathbf{j} + \mathbf{k}]\, dA = \iint_D (-2x - 2y - 1)\, dA$$

$$= \int_0^{2\pi} \int_0^1 (-2r \cos\theta - 2r \sin\theta - 1)\, r\, dr\, d\theta = \int_0^{2\pi} \int_0^1 [-2r^2(\cos\theta + \sin\theta) - r]\, dr\, d\theta$$

$$= \int_0^{2\pi} \left[-\tfrac{2}{3} r^3 (\cos\theta + \sin\theta) - \tfrac{1}{2} r^2 \right]_{r=0}^{r=1} d\theta = \int_0^{2\pi} \left[-\tfrac{2}{3}(\cos\theta + \sin\theta) - \tfrac{1}{2} \right] d\theta$$

$$= \left[-\tfrac{2}{3}(\sin\theta - \cos\theta) - \tfrac{1}{2}\theta \right]_0^{2\pi} = -\pi.$$

To evaluate $\oint_C \mathbf{F} \cdot d\mathbf{r} = \oint_C \mathbf{F}(\mathbf{r}(t)) \cdot \mathbf{r}'(t)\, dt$, parametrize C: $\mathbf{r}(t) = \cos t\,\mathbf{i} + \sin t\,\mathbf{j} + 0\,\mathbf{k}$, $0 \leq t \leq 2\pi$. Then, $\mathbf{r}'(t) = -\sin t\,\mathbf{i} + \cos t\,\mathbf{j}$, and $\mathbf{F}(\mathbf{r}(t)) = \sin t\,\mathbf{i} + 0\,\mathbf{j} + \cos t\,\mathbf{k} \Rightarrow$

$$\oint_C \mathbf{F} \cdot d\mathbf{r} = \oint_C \mathbf{F}(\mathbf{r}(t)) \cdot \mathbf{r}'(t)\, dt = \int_0^{2\pi} [\sin t\,\mathbf{i} + \cos t\,\mathbf{k}] \cdot [-\sin t\,\mathbf{i} + \cos t\,\mathbf{j}]\, dt = \int_0^{2\pi} -\sin^2 t\, dt$$

$$= -\int_0^{2\pi} \frac{1 - \cos 2t}{2}\, dt = -\tfrac{1}{2} \left[t - \tfrac{1}{2} \sin 2t \right]_0^{2\pi} = -\pi.$$

21. S is the surface $z = \sqrt{1 - x^2 - y^2} = f(x, y) \Rightarrow f_x = -\dfrac{x}{\sqrt{1 - x^2 - y^2}}$ and $f_y = -\dfrac{y}{\sqrt{1 - x^2 - y^2}}$, so that C is the bounding curve $x^2 + y^2 = 1$ in the plane $z = 0$, and $D : x^2 + y^2 \leq 1$. For $\mathbf{F}(x, y, z) = z^2\,\mathbf{i} - 2x\,\mathbf{j} + y^3\,\mathbf{k}$,

$$\text{curl } \mathbf{F} = \det \begin{bmatrix} \mathbf{i} & \mathbf{j} & \mathbf{k} \\ \frac{\partial}{\partial x} & \frac{\partial}{\partial y} & \frac{\partial}{\partial z} \\ z^2 & -2x & y^3 \end{bmatrix} = 3y^2\,\mathbf{i} + 2z\,\mathbf{j} - 2\,\mathbf{k}.$$

$$\iint_S \text{curl } \mathbf{F} \cdot \mathbf{n}\, dS = \iint_D \text{curl } \mathbf{F} \cdot (-f_x\,\mathbf{i} - f_y\,\mathbf{j} + \mathbf{k})\, dA$$

$$= \iint_D [3y^2\,\mathbf{i} + 2z\,\mathbf{j} - 2\,\mathbf{k}] \cdot \left[\frac{x}{\sqrt{1 - x^2 - y^2}}\,\mathbf{i} + \frac{y}{\sqrt{1 - x^2 - y^2}}\,\mathbf{j} + \mathbf{k} \right] dA$$

$$= \iint_D \left(\frac{3xy^2}{\sqrt{1 - x^2 - y^2}} + \frac{2yz}{\sqrt{1 - x^2 - y^2}} - 2 \right) dA$$

$$= \iint_D \left(\frac{3xy^2}{\sqrt{1 - x^2 - y^2}} + 2y - 2 \right) dA$$

Now, $\displaystyle\iint_D \frac{3xy^2}{\sqrt{1-x^2-y^2}}\,dA = \int_{-1}^{1}\int_{-\sqrt{1-y^2}}^{\sqrt{1-y^2}} \frac{3xy^2}{\sqrt{1-x^2-y^2}}\,dx\,dy$

$$= \int_{-1}^{1}\left[3y^2\cdot\frac{\sqrt{1-x^2-y^2}}{-1}\right]_{x=-\sqrt{1-y^2}}^{x=\sqrt{1-y^2}}dy = \int_{-1}^{1}3y^2\cdot 0\,dy = 0$$

So,

$$\iint_S \text{curl }\mathbf{F}\cdot\mathbf{n}\,dS = \iint_D (2y-2)\,dA = \int_0^{2\pi}\int_0^1 (2r\sin\theta - 2)r\,dr\,d\theta = \int_0^{2\pi}\left[\tfrac{2}{3}r^3\sin\theta - r^2\right]_{r=0}^{r=1}d\theta$$

$$= \int_0^{2\pi}\left(\tfrac{2}{3}\sin\theta - 1\right)d\theta = \left[-\tfrac{2}{3}\cos\theta - \theta\right]_0^{2\pi} = -2\pi$$

To evaluate $\displaystyle\oint_C \mathbf{F}\cdot d\mathbf{r} = \oint_C \mathbf{F}(\mathbf{r}(t))\cdot\mathbf{r}'(t)\,dt$, parametrize C: $\mathbf{r}(t) = \cos t\,\mathbf{i} + \sin t\,\mathbf{j} + 0\,\mathbf{k}$, $0\le t\le 2\pi$. Then, $\mathbf{r}'(t) = -\sin t\,\mathbf{i} + \cos t\,\mathbf{j}$, and $\mathbf{F}(\mathbf{r}(t)) = 0\,\mathbf{i} - 2\cos t\,\mathbf{j} + \sin^3 t\,\mathbf{k} \Rightarrow$

$$\oint_C \mathbf{F}\cdot d\mathbf{r} = \oint_C \mathbf{F}(\mathbf{r}(t))\cdot\mathbf{r}'(t)\,dt = \int_0^{2\pi}[-2\cos t\,\mathbf{j} + \sin^3 t\,\mathbf{k}]\cdot[-\sin t\,\mathbf{i} + \cos t\,\mathbf{j}]\,dt$$

$$= \int_0^{2\pi}-2\cos^2 t\,dt = -2\int_0^{2\pi}\frac{1+\cos 2t}{2}\,dt = -\left[t + \tfrac{1}{2}\sin 2t\right]_0^{2\pi} = -2\pi.$$

23. S is the surface $z = \sqrt{1-x^2-y^2} = f(x,y) \Rightarrow f_x = -\dfrac{x}{\sqrt{1-x^2-y^2}}$ and $f_y = -\dfrac{y}{\sqrt{1-x^2-y^2}}$, so that C is the bounding curve $x^2+y^2 = 1$ in the plane $z = 0$, and $D: x^2+y^2\le 1$. For $\mathbf{F}(x,y,z) = 6xz\,\mathbf{i} - x^2\,\mathbf{j} - 3y^2\,\mathbf{k}$,

$$\text{curl }\mathbf{F} = \det\begin{bmatrix}\mathbf{i} & \mathbf{j} & \mathbf{k}\\ \frac{\partial}{\partial x} & \frac{\partial}{\partial y} & \frac{\partial}{\partial z}\\ 6xz & -x^2 & -3y^2\end{bmatrix} = -6y\,\mathbf{i} + 6x\,\mathbf{j} - 2x\,\mathbf{k}.$$

$$\iint_S \text{curl }\mathbf{F}\cdot\mathbf{n}\,dS = \iint_D \text{curl }\mathbf{F}\cdot(-f_x\,\mathbf{i} - f_y\,\mathbf{j} + \mathbf{k})\,dA$$

$$= \iint_D [-6y\,\mathbf{i} + 6x\,\mathbf{j} - 2x\,\mathbf{k}]\cdot\left[\frac{x}{\sqrt{1-x^2-y^2}}\,\mathbf{i} + \frac{y}{\sqrt{1-x^2-y^2}}\,\mathbf{j} + \mathbf{k}\right]dA$$

$$= \iint_D \left(-\frac{6xy}{\sqrt{1-x^2-y^2}} + \frac{6xy}{\sqrt{1-x^2-y^2}} - 2x\right)dA = \iint_D (-2x)\,dA$$

$$= \int_0^{2\pi}\int_0^1 (-2r\cos\theta)r\,dr\,d\theta = \int_0^{2\pi}\left[-\tfrac{2}{3}r^3\cos\theta\right]_{r=0}^{r=1}d\theta = -\tfrac{2}{3}\int_0^{2\pi}\cos\theta\,d\theta$$

$$= -\tfrac{2}{3}\left[\sin\theta\right]_0^{2\pi} = 0$$

To evaluate $\displaystyle\oint_C \mathbf{F}\cdot d\mathbf{r} = \oint_C \mathbf{F}(\mathbf{r}(t))\cdot\mathbf{r}'(t)\,dt$, parametrize C: $\mathbf{r}(t) = \cos t\,\mathbf{i} + \sin t\,\mathbf{j} + 0\,\mathbf{k}$, $0\le t\le 2\pi$. Then, $\mathbf{r}'(t) = -\sin t\,\mathbf{i} + \cos t\,\mathbf{j}$, and $\mathbf{F}(\mathbf{r}(t)) = 6(\cos t)(0)\,\mathbf{i} - \cos^2 t\,\mathbf{j} - 3\sin^2 t\,\mathbf{k} \Rightarrow$

$$\oint_C \mathbf{F} \cdot d\mathbf{r} = \oint_C \mathbf{F}(\mathbf{r}(t)) \cdot \mathbf{r}'(t)\,dt = \int_0^{2\pi} \left[-\cos^2 t\,\mathbf{j} - 3\sin^2 t\,\mathbf{k} \right] \cdot \left[-\sin t\,\mathbf{i} + \cos t\,\mathbf{j} \right] dt$$

$$= \int_0^{2\pi} (-\cos^3 t)\,dt = -\int_0^{2\pi} (1 - \sin^2 t)\cos t\,dt = \int_0^{2\pi} (-\cos t + \sin^2 t \cos t)\,dt$$

$$= \left[-\sin t + \tfrac{1}{3}\sin^3 t \right]_0^{2\pi} = 0.$$

25.

The surface S is the plane of the triangle: $x + y + z = 2 \;\Rightarrow\; z = 2 - x - y = f(x, y)$, for $0 \le x, y, z \le 2 \;\Rightarrow\; f_x = f_y = -1$.

For $\mathbf{F}(x, y, z) = (x^2 + y^2)\mathbf{i} + y^2\,\mathbf{j} + (x^2 + z^2)\mathbf{k}$,

$$\text{curl } \mathbf{F} = \det \begin{bmatrix} \mathbf{i} & \mathbf{j} & \mathbf{k} \\ \frac{\partial}{\partial x} & \frac{\partial}{\partial y} & \frac{\partial}{\partial z} \\ x^2 + y^2 & y^2 & x^2 + z^2 \end{bmatrix} = 0\mathbf{i} - 2x\,\mathbf{j} - 2y\,\mathbf{k}.$$

$$\iint_S \text{curl } \mathbf{F} \cdot \mathbf{n}\,dS = \iint_D \text{curl } \mathbf{F} \cdot (-f_x\,\mathbf{i} - f_y\,\mathbf{j} + \mathbf{k})\,dA = \iint_D [-2x\,\mathbf{j} - 2y\,\mathbf{k}] \cdot [\mathbf{i} + \mathbf{j} + \mathbf{k}]\,dA$$

$$= \int_0^2 \int_0^{-x+2} (-2x - 2y)\,dy\,dx = \int_0^2 \left[-2xy - y^2 \right]_{y=0}^{y=-x+2} dx$$

$$= \int_0^2 \left[-2x(-x + 2) - (-x + 2)^2 \right] dx = \int_0^2 (2x^2 - 4x - x^2 + 4x - 4)\,dx$$

$$= \int_0^2 (x^2 - 4)\,dx = \left[\tfrac{1}{3}x^3 - 4x \right]_0^2 = \tfrac{1}{3}(8) - 8 = -\tfrac{16}{3}$$

To evaluate $\oint_C \mathbf{F} \cdot d\mathbf{r} = \oint_C \mathbf{F}(\mathbf{r}(t)) \cdot \mathbf{r}'(t)\,dt$, parametrize C:

The bounding curve C is made up of 3 line segments:

$C_1 : \mathbf{r}(t) = (1 - t)(2, 0, 0) + t(0, 2, 0) = (2 - 2t)\mathbf{i} + 2t\,\mathbf{j}, \quad 0 \le t \le 1,$
$\quad \mathbf{r}'(t) = -2\mathbf{i} + 2\mathbf{j}$

$C_2 : \mathbf{r}(t) = (1 - t)(0, 2, 0) + t(0, 0, 2) = (2 - 2t)\mathbf{j} + 2t\,\mathbf{k}, \quad 0 \le t \le 1,$
$\quad \mathbf{r}'(t) = -2\mathbf{j} + 2\mathbf{k}$

$C_3 : \mathbf{r}(t) = (1 - t)(0, 0, 2) + t(2, 0, 0) = 2t\,\mathbf{i} + (2 - 2t)\mathbf{k}, \quad 0 \le t \le 1,$
$\quad \mathbf{r}'(t) = 2\mathbf{i} - 2\mathbf{k}$

As $\mathbf{F}(x, y, z) = (x^2 + y^2)\mathbf{i} + y^2\,\mathbf{j} + (x^2 + z^2)\mathbf{k}$,

On $C_1 : \mathbf{F}(\mathbf{r}(t)) = \left[(2 - 2t)^2 + 4t^2 \right]\mathbf{i} + 4t^2\,\mathbf{j} + (2 - 2t)^2\mathbf{k}$

On $C_2 : \mathbf{F}(\mathbf{r}(t)) = (2 - 2t)^2\mathbf{i} + (2 - 2t)^2\,\mathbf{j} + 4t^2\,\mathbf{k}$

On $C_3 : \mathbf{F}(\mathbf{r}(t)) = 4t^2\,\mathbf{i} + \left[4t^2 + (2 - 2t)^2 \right]\mathbf{k}$

So

$$\oint_C \mathbf{F} \cdot d\mathbf{r} = \oint_C \mathbf{F}(\mathbf{r}(t)) \cdot \mathbf{r}'(t)\,dt$$

$$= \oint_{C_1} \mathbf{F}(\mathbf{r}(t)) \cdot \mathbf{r}'(t)\,dt + \oint_{C_2} \mathbf{F}(\mathbf{r}(t)) \cdot \mathbf{r}'(t)\,dt + \oint_{C_3} \mathbf{F}(\mathbf{r}(t)) \cdot \mathbf{r}'(t)\,dt$$

$$\oint_{C_1} \mathbf{F}(\mathbf{r}(t)) \cdot \mathbf{r}'(t)\, dt = \int_0^1 \left(\left[(2 - 2t)^2 + 4t^2 \right] \mathbf{i} + 4t^2\, \mathbf{j} + (2 - 2t)^2 \mathbf{k} \right) \cdot (-2\mathbf{i} + 2\mathbf{j})\, dt$$

$$= \int_0^1 [(4 - 8t + 8t^2)(-2) + 8t^2]\, dt = \int_0^1 (-8t^2 + 16t - 8)\, dt$$

$$= \left[-\tfrac{8}{3} t^3 + 8t^2 - 8t \right]_0^1 = -\tfrac{8}{3} + 8 - 8 = -\tfrac{8}{3}$$

$$\oint_{C_2} \mathbf{F}(\mathbf{r}(t)) \cdot \mathbf{r}'(t)\, dt = \int_0^1 \left[(2 - 2t)^2 \mathbf{i} + (2 - 2t)^2\, \mathbf{j} + 4t^2\, \mathbf{k} \right] \cdot [-2\mathbf{j} + 2\mathbf{k}]\, dt$$

$$= \int_0^1 \left[-2(4 - 8t + 4t^2) + 8t^2 \right] dt = \int_0^1 (-8 + 16t)\, dt$$

$$= \left[-8t + 8t^2 \right]_0^1 = -8 + 8 = 0$$

$$\oint_{C_3} \mathbf{F}(\mathbf{r}(t)) \cdot \mathbf{r}'(t)\, dt = \int_0^1 \left(4t^2\, \mathbf{i} + \left[4t^2 + (2 - 2t)^2 \right] \mathbf{k} \right) \cdot (2\mathbf{i} - 2\mathbf{k})\, dt$$

$$= \int_0^1 \left[2(4t^2) - 2(4t^2 + 4 - 8t + 4t^2) \right] dt = \int_0^1 (-8 + 16t - 8t^2)\, dt$$

$$= \left[-8t + 8t^2 - \tfrac{8}{3} t^3 \right]_0^1 = -8 + 8 - \tfrac{8}{3} = -\tfrac{8}{3}$$

Thus, $\oint_C \mathbf{F} \cdot d\mathbf{r} = -\tfrac{8}{3} + 0 - \tfrac{8}{3} = -\tfrac{16}{3}$.

27. The surface S is $z = 1 - x^2 - y^2 = f(x, y) \Rightarrow f_x = -2x$, and $f_y = -2y$, and $z \geq 0$. C is the bounding curve $x^2 + y^2 = 1$ in the plane $z = 0$, and $D : x^2 + y^2 \leq 1$.

For $\mathbf{F}(x, y, z) = (z - y)\mathbf{i} + (z + x)\mathbf{j} - (x + y)\mathbf{k}$,

$$\text{curl } \mathbf{F} = \det \begin{bmatrix} \mathbf{i} & \mathbf{j} & \mathbf{k} \\ \frac{\partial}{\partial x} & \frac{\partial}{\partial y} & \frac{\partial}{\partial z} \\ z - y & z + x & -(x + y) \end{bmatrix} = -2\mathbf{i} + 2\mathbf{j} + 2\mathbf{k}.$$

$$\iint_S \text{curl } \mathbf{F} \cdot \mathbf{n}\, dS = \iint_D \text{curl } \mathbf{F} \cdot (-f_x\, \mathbf{i} - f_y\, \mathbf{j} + \mathbf{k})\, dA$$

$$= \iint_D [-2\mathbf{i} + 2\mathbf{j} + 2\mathbf{k}] \cdot [2x\, \mathbf{i} + 2y\, \mathbf{j} + \mathbf{k}]\, dA$$

$$= \iint_D (-4x + 4y + 2)\, dA = \int_0^{2\pi} \int_0^1 (-4r \cos\theta + 4r \sin\theta + 2)\, r\, dr\, d\theta$$

$$= \int_0^{2\pi} \int_0^1 [-4r^2 (\cos\theta - \sin\theta) + 2r]\, dr\, d\theta$$

$$= \int_0^{2\pi} \left[-\tfrac{4}{3} r^3 (\cos\theta - \sin\theta) + r^2 \right]_{r=0}^{r=1} d\theta = \int_0^{2\pi} \left[-\tfrac{4}{3} (\cos\theta - \sin\theta) + 1 \right] d\theta$$

$$= \left[-\tfrac{4}{3} (\sin\theta + \cos\theta) + \theta \right]_0^{2\pi} = 2\pi$$

To evaluate $\oint_C \mathbf{F} \cdot d\mathbf{r} = \oint_C \mathbf{F}(\mathbf{r}(t)) \cdot \mathbf{r}'(t)\, dt$, parametrize C: $\mathbf{r}(t) = \cos t\, \mathbf{i} + \sin t\, \mathbf{j} + 0\, \mathbf{k}$,

$0 \leq t \leq 2\pi$. Then, $\mathbf{r}'(t) = -\sin t\,\mathbf{i} + \cos t\,\mathbf{j}$, and $\mathbf{F}(\mathbf{r}(t)) = -\sin t\,\mathbf{i} + \cos t\,\mathbf{j} - (\cos t + \sin t)\mathbf{k} \Rightarrow$

$$\oint_C \mathbf{F} \cdot d\mathbf{r} = \oint_C \mathbf{F}(\mathbf{r}(t)) \cdot \mathbf{r}'(t)\,dt = \int_0^{2\pi} [-\sin t\,\mathbf{i} + \cos t\,\mathbf{j} - (\cos t + \sin t)\mathbf{k}] \cdot [-\sin t\,\mathbf{i} + \cos t\,\mathbf{j}]\,dt$$

$$= \int_0^{2\pi} (\sin^2 t + \cos^2 t)\,dt = \int_0^{2\pi} dt = \Big[t\Big]_0^{2\pi} = 2\pi.$$

29. The surface S is the part of the plane $x + y + z = 1$ enclosed by C, and $z = 1 - x - y = f(x,y) \Rightarrow f_x = f_y = -1$. Since

$$\oint_C z\,dx - 2x\,dy + 3y\,dz = \oint_C (z\,\mathbf{i} - 2x\,\mathbf{j} + 3y\,\mathbf{k}) \cdot d\mathbf{r},$$

let $\mathbf{F}(x,y,z) = z\,\mathbf{i} - 2x\,\mathbf{j} + 3y\,\mathbf{k}$. By Stoke's Theorem,

$$\oint_C (z\,\mathbf{i} - 2x\,\mathbf{j} + 3y\,\mathbf{k}) \cdot d\mathbf{r} = \oint_C \mathbf{F} \cdot d\mathbf{r} = \iint_S \text{curl } \mathbf{F} \cdot \mathbf{n}\,dS.$$

Now, $\mathbf{n} = \frac{1}{\sqrt{3}}(\mathbf{i} + \mathbf{j} + \mathbf{k})$, and

$$\text{curl }\mathbf{F} = \det \begin{bmatrix} \mathbf{i} & \mathbf{j} & \mathbf{k} \\ \frac{\partial}{\partial x} & \frac{\partial}{\partial y} & \frac{\partial}{\partial z} \\ z & -2x & 3y \end{bmatrix} = 3\,\mathbf{i} + \mathbf{j} - 2\,\mathbf{k}.$$

so that

$$\iint_S \text{curl }\mathbf{F} \cdot \mathbf{n}\,dS = \iint_S (3\,\mathbf{i} + \mathbf{j} - 2\,\mathbf{k}) \cdot \frac{1}{\sqrt{3}}(\mathbf{i} + \mathbf{j} + \mathbf{k})\,dS = \frac{2}{\sqrt{3}} \iint_S dS$$

$$= \frac{2}{\sqrt{3}}\ (\text{ surface area of } S) = \frac{2}{\sqrt{3}}\ (\text{area of the region enclosed by } C).$$

§14.7 The Divergence Theorem

1. For $\mathbf{F}(x,y,z) = x^3\,\mathbf{i} + y^3\,\mathbf{j} + z^3\,\mathbf{k}$, and $E: x^2 + y^2 + z^2 \leq a^2$,

$$\text{div}(\mathbf{F}) = (x^3)_x + (y^3)_y + (z^3)_z = 3x^2 + 3y^2 + 3z^2$$

By the Divergence Theorem, and changing to spherical coordinates:

$$\iint_S \mathbf{F} \cdot \mathbf{n}\,dS = \iiint_E \text{div}(\mathbf{F})dV = \iiint_E 3(x^2 + y^2 + z^2)dV = \int_0^{2\pi} \int_0^\pi \int_0^a 3\rho^2 \rho^2 \sin\phi\,d\rho d\phi d\theta$$

$$= 3\int_0^{2\pi} \int_0^\pi \left[\frac{\rho^5}{5}\sin\phi\right]_{\rho=0}^{\rho=a} d\phi d\theta = \frac{3a^5}{5} \int_0^{2\pi} \int_0^\pi \sin\phi\,d\phi d\theta$$

$$= \frac{3a^5}{5} \int_0^{2\pi} \Big[-\cos\phi\Big]_{\phi=0}^{\phi=\pi} d\theta = \frac{3a^5}{5} \int_0^{2\pi} (-\cos\pi + \cos 0)\,d\theta = \frac{6a^5}{5} \int_0^{2\pi} d\theta = \frac{12\pi a^5}{5}.$$

3. For $\mathbf{F}(x,y,z) = (x+y)\mathbf{i} + z^2\,\mathbf{j} + x^2\,\mathbf{k}$, and $E: x^2 + y^2 + z^2 \leq 1$ with $z \geq 0$,

$$\text{div}(\mathbf{F}) = (x+y)_x + (z^2)_y + (x^2)_z = 1 + 0 + 0 = 1.$$

By the Divergence Theorem:

$$\iint_S \mathbf{F} \cdot \mathbf{n}\,dS = \iiint_E \text{div}(\mathbf{F})dV = \iiint_E dV = \tfrac{1}{2} \text{ the volume of the sphere} = \tfrac{1}{2}\left[\tfrac{4}{3}\pi(1)^3\right] = \frac{2\pi}{3}.$$

5. For $\mathbf{F}(x, y, z) = ye^{z^2}\mathbf{i} + y^2\mathbf{j} + e^{xy}\mathbf{k}$, and $E : x^2 + y^2 \leq 9$, with $0 \leq z \leq y - 3$,
$$\text{div}(\mathbf{F}) = (ye^{z^2})_x + (y^2)_y + (e^{xy})_z = 0 + 2y + 0 = 2y.$$

By the Divergence Theorem, and changing to cylindrical coordinates:

$$\iint_S \mathbf{F} \cdot \mathbf{n}\, dS = \iiint_E \text{div}(\mathbf{F})dV = \int_{-3}^3 \int_{-\sqrt{9-x^2}}^{\sqrt{9-x^2}} \int_0^{y-3} 2y\, dzdydx$$

$$= \int_0^{2\pi} \int_0^3 \int_0^{r\sin\theta-3} (2r\sin\theta)r\, dzdrd\theta = \int_0^{2\pi} \int_0^3 \left[(2r^2\sin\theta)z\right]_{z=0}^{z=r\sin\theta-3} drd\theta$$

$$= 2\int_0^{2\pi} \int_0^3 (r^2\sin\theta)(r\sin\theta - 3)\, drd\theta = 2\int_0^{2\pi} \int_0^3 (r^3\sin^2\theta - 3r^2\sin\theta)\, drd\theta$$

$$= 2\int_0^{2\pi} \left[\tfrac{1}{4}r^4\sin^2\theta - r^3\sin\theta\right]_{r=0}^{r=3} d\theta = 2\int_0^{2\pi} \left(\tfrac{81}{4}\sin^2\theta - 27\sin\theta\right) d\theta$$

$$= \tfrac{81}{2}\int_0^{2\pi} \frac{1 - \cos 2\theta}{2}\, d\theta + \left[54\cos\theta\right]_0^{2\pi} = \tfrac{81}{4}\left[\theta - \frac{\sin 2\theta}{2}\right]_0^{2\pi} + 0 = \tfrac{81}{4}(2\pi) = \tfrac{81\pi}{2}.$$

7. For $\mathbf{F}(x, y, z) = y^2\mathbf{i} + xz^3\mathbf{j} + (z-1)^2\mathbf{k}$, and $E : x^2 + y^2 \leq 16$, with $1 \leq z \leq 5$,
$$\text{div}(\mathbf{F}) = (y^2)_x + (xz^3)_y + [(z-1)^2]_z = 0 + 0 + 2(z-1).$$

By the Divergence Theorem, and changing to cylindrical coordinates:

$$\iint_S \mathbf{F} \cdot \mathbf{n}\, dS = \iiint_E \text{div}(\mathbf{F})dV = 2\int_0^{2\pi} \int_0^4 \int_1^5 (z-1)r\, dzdrd\theta$$

$$= 2\int_0^{2\pi} \int_0^4 \left[\left(\tfrac{1}{2}z^2 - z\right)r\right]_{z=1}^{z=5} drd\theta = 2\int_0^{2\pi} \int_0^4 \left[\tfrac{25}{2} - 5 - \left(\tfrac{1}{2} - 1\right)\right] rdrd\theta$$

$$= 16\int_0^{2\pi} \int_0^4 rdrd\theta = 16\int_0^{2\pi} \left[\tfrac{1}{2}r^2\right]_{r=0}^{r=4} d\theta = 16\int_0^{2\pi} 8\, d\theta = 128\int_0^{2\pi} d\theta = 256\pi.$$

9. For $\mathbf{F}(x, y, z) = (xz\sin yz + x^3)\mathbf{i} + \cos yz\,\mathbf{j} + (3zy^2 - e^{x^2+y^2})\mathbf{k}$, and $E : 0 \leq z \leq 4 - x^2 - y^2$,
$$\text{div}(\mathbf{F}) = (xz\sin yz + x^3)_x + (\cos yz)_y + (3zy^2 - e^{x^2+y^2})_z$$
$$= (z\sin yz + 3x^2) + (-z\sin yz) + (3y^2) = 3x^2 + 3y^2$$

By the Divergence Theorem, and changing to cylindrical coordinates:

$$\iint_S \mathbf{F} \cdot \mathbf{n}\, dS = \iiint_E \text{div}(\mathbf{F})dV = \int_0^{2\pi} \int_0^2 \int_0^{4-r^2} 3r^2 \cdot r\, dzdrd\theta = 3\int_0^{2\pi} \int_0^2 \left[r^3 z\right]_{z=0}^{z=4-r^2} drd\theta$$

$$= 3\int_0^{2\pi} \int_0^2 r^3(4 - r^2)\, drd\theta = 3\int_0^{2\pi} \left[r^4 - \tfrac{1}{6}r^6\right]_{r=0}^{r=2} d\theta = 3\int_0^{2\pi} \left(16 - \tfrac{64}{6}\right) d\theta$$

$$= \int_0^{2\pi} (48 - 32)\, d\theta = 16(2\pi) = 32\pi.$$

11.

For $\mathbf{F}(x, y, z) = xy\,\mathbf{i} + (y^2 + e^{xz^2})\mathbf{j} + \sin xy\,\mathbf{k}$, and E (see figure)

$$\text{div}(\mathbf{F}) = (xy)_x + (y^2 + e^{xz^2})_y + (\sin xy)_z$$
$$= y + 2y + 0 = 3y$$

By the Divergence Theorem, :

$$\iint_S \mathbf{F} \cdot \mathbf{n}\,dS = \iiint_E \text{div}(\mathbf{F})dV = \int_{-1}^1 \int_0^{1-x^2} \int_0^{2-z} 3y\,dy dz dx = 3\int_{-1}^1 \int_0^{1-x^2} \left[\tfrac{1}{2}y^2\right]_{y=0}^{y=2-z} dz dx$$

$$= \frac{3}{2}\int_{-1}^1 \int_0^{1-x^2} (2-z)^2\,dz dx = \frac{3}{2}\int_{-1}^1 \left[-\tfrac{1}{3}(2-z)^3\right]_{z=0}^{z=1-x^2} dx$$

$$= -\frac{1}{2}\int_{-1}^1 \left[(2-1+x^2)^3 - 8\right] dx = -\frac{1}{2}\int_{-1}^1 (x^6 + 3x^4 + 3x^2 - 7)\,dx$$

$$= -\frac{1}{2}\left[\tfrac{1}{7}x^7 + \tfrac{3}{5}x^5 + x^3 - 7x\right]_{-1}^1 = -\frac{1}{2}\left[\left(\tfrac{1}{7} + \tfrac{3}{5} + 1 - 7\right) - \left(-\tfrac{1}{7} - \tfrac{3}{5} - 1 + 7\right)\right]$$

$$= -\frac{1}{7} - \frac{3}{5} - 1 + 7 = \frac{-5 - 21 - 35 + 245}{35} = \frac{184}{35}.$$

13.

For $\mathbf{F}(x, y, z) = x^2\,\mathbf{i} + y^2\,\mathbf{j} + z^2\,\mathbf{k}$, and E (see figure):

(a) Evaluating $\iint_S \mathbf{F} \cdot \mathbf{n}\,dS$ directly. As noted in the exercise, the value is the sum of 6 surface integrals, one for each face of the cube.

Face	$\mathbf{n}$	$\mathbf{F} \cdot \mathbf{n}$
S_1 (top): $0 \le x \le 1$, $0 \le y \le 1$, $z = 1$	$\mathbf{k}$	$z^2 = 1^2 = 1$
S_2 (bottom): $0 \le x \le 1$, $0 \le y \le 1$, $z = 0$	$-\mathbf{k}$	$-z^2 = 0^2 = 0$
S_3 (left side): $0 \le x \le 1$, $0 \le z \le 1$, $y = 0$	$-\mathbf{j}$	$-y^2 = 0^2 = 0$
S_4 (right side): $0 \le x \le 1$, $0 \le z \le 1$, $y = 1$	$\mathbf{j}$	$y^2 = 1^2 = 1$
S_5 (back): $0 \le y \le 1$, $0 \le z \le 1$, $x = 0$	$-\mathbf{i}$	$-x^2 = 0^2 = 0$
S_6 (front): $0 \le y \le 1$, $0 \le z \le 1$, $x = 1$	$\mathbf{i}$	$x^2 = 1^2 = 1$

Thus, $\iint_{S_2} \mathbf{F} \cdot \mathbf{n}\,dS = \iint_{S_3} \mathbf{F} \cdot \mathbf{n}\,dS = \iint_{S_5} \mathbf{F} \cdot \mathbf{n}\,dS = \iint 0\,dS = 0$, and

$$\iint_{S_1} \mathbf{F} \cdot \mathbf{n}\,dS = \iint_{S_4} \mathbf{F} \cdot \mathbf{n}\,dS = \iint_{S_6} \mathbf{F} \cdot \mathbf{n}\,dS = \iint 1\,dS = \text{the area of a face} = 1.$$

Therefore, $\iint_S \mathbf{F} \cdot \mathbf{n}\,dS = 1 + 0 + 0 + 1 + 0 + 1 = 3.$

(b) Evaluating $\iiint_E \text{div}(\mathbf{F})\,dV$: $\text{div}(\mathbf{F}) = (x^2)_x + (y^2)_y + (z^2)_z = 2x + 2y + 2z \Rightarrow$

$$\iiint\limits_{E} \operatorname{div}(\mathbf{F})\, dV = \int_0^1 \int_0^1 \int_0^1 (2x + 2y + 2z)\, dz\, dy\, dx = \int_0^1 \int_0^1 \left[(2x + 2y)z + z^2 \right]_{z=0}^{z=1} dy\, dx$$

$$= \int_0^1 \int_0^1 (2x + 2y + 1)\, dy\, dx = \int_0^1 \left[(2x + 1)y + y^2 \right]_{y=0}^{y=1} dx$$

$$= \int_0^1 (2x + 1 + 1)\, dx = \int_0^1 (2x + 2)\, dx = \left[x^2 + 2x \right]_0^1 = 1 + 2 = 3.$$

15. For $\mathbf{F}(x, y, z) = x^2\, \mathbf{i} + xz\, \mathbf{j} + 3z\, \mathbf{k}$, and $E : x^2 + y^2 + z^2 \le 4$.

(a) Evaluating $\iint\limits_{S} \mathbf{F} \cdot \mathbf{n}\, dS$ directly: $\mathbf{n} = \dfrac{x\,\mathbf{i} + y\,\mathbf{j} + z\,\mathbf{k}}{\sqrt{x^2 + y^2 + z^2}} \Rightarrow$

$\mathbf{F} \cdot \mathbf{n} = \dfrac{x^3 + xyz + 3z^2}{\sqrt{x^2 + y^2 + z^2}} = \frac{1}{2}(x^3 + xyz + 3z^2)$ on S, as there $x^2 + y^2 + z^2 = 4$.

Next, we parametrize S using spherical coordinates with $\rho = 2$:

$x = 2\sin\phi\cos\theta,\ y = 2\sin\phi\sin\theta,\ z = 2\cos\phi,$ with $0 \le \theta \le 2\pi$, and $0 \le \phi \le \pi \Rightarrow$

$\mathbf{r}(\phi, \theta) = 2\sin\phi\cos\theta\,\mathbf{i} + 2\sin\phi\sin\theta\,\mathbf{j} + 2\cos\phi\,\mathbf{k} \Rightarrow$ from Example 14.24 on page 639,

$\|\mathbf{r}_\phi \times \mathbf{r}_\theta\| = r^2\sin\phi = 4\sin\phi$, as $r^2 = 4 \Rightarrow$

$$\iint\limits_{S} \mathbf{F} \cdot \mathbf{n}\, dS = \frac{1}{2}\int_0^{2\pi}\int_0^\pi \left[(2\sin\phi\cos\theta)^3 + 8\sin^2\phi\cos\phi\sin\theta\cos\theta + 3\cdot4\cos^2\phi \right] 4\sin\phi\, d\phi\, d\theta$$

$$= \int_0^{2\pi}\int_0^\pi 16\sin^4\phi\cos^3\theta\, d\phi\, d\theta + \int_0^{2\pi}\int_0^\pi 16\sin^3\phi\cos\phi\sin\theta\cos\theta\, d\phi\, d\theta$$

$$+ \int_0^{2\pi}\int_0^\pi 24\cos^2\phi\sin\phi\, d\phi\, d\theta$$

Evaluating each integral separately, we find:

$$\int_0^{2\pi}\int_0^\pi 16\sin^4\phi\cos^3\theta\, d\phi\, d\theta = 16\left(\int_0^{2\pi}\cos^3\theta\, d\theta\right)\left(\int_0^\pi \sin^4\phi\, d\phi\right), \text{ and}$$

$$\int_0^{2\pi}\cos^3\theta\, d\theta = \int_0^{2\pi}(1 - \sin^2\theta)\cos\theta\, d\theta = \left[\sin\theta - \tfrac{1}{3}\sin^3\theta\right]_0^{2\pi} = 0$$

$$\int_0^{2\pi}\int_0^\pi 16\sin^3\phi\cos\phi\sin\theta\cos\theta\, d\phi\, d\theta = \left(\int_0^{2\pi}\sin\theta\cos\theta\, d\theta\right)\left(16\int_0^\pi \sin^3\phi\cos\phi\, d\phi\right), \text{ and}$$

$$\int_0^{2\pi}\sin\theta\cos\theta\, d\theta = \left[\tfrac{1}{2}\sin^2\theta\right]_0^{2\pi} = 0$$

$$\int_0^{2\pi}\int_0^\pi 24\cos^2\phi\sin\phi\, d\phi\, d\theta = \left(\int_0^{2\pi}d\theta\right)\left(24\int_0^\pi \cos^2\phi\sin\phi\, d\phi\right) = 2\pi\left[-8\cos^3\phi\right]_0^\pi$$

$$= -16\pi(\cos^3\pi - \cos^3 0) = 32\pi \Rightarrow$$

$$\iint\limits_{S} \mathbf{F} \cdot \mathbf{n}\, dS = 0 + 0 + 32\pi = 32\pi.$$

(b) Evaluating $\iiint\limits_{E} \operatorname{div}(\mathbf{F})\, dV$: $\operatorname{div}(\mathbf{F}) = (x^2)_x + (xz)_y + (3z)_z = 2x + 0 + 3 = 2x + 3 \Rightarrow$

using spherical coordinates:

$$\iiint_E \operatorname{div}(\mathbf{F})\, dV = \iiint_E (2x + 3)\, dV = \int_0^{2\pi} \int_0^{\pi} \int_0^2 (2\rho \sin \phi \cos \theta + 3)\rho^2 \sin \phi \, d\rho d\phi d\theta$$

$$= \int_0^{2\pi} \int_0^{\pi} \int_0^2 (2\rho^3 \sin^2 \phi \cos \theta + 3\rho^2 \sin \phi)\, d\rho d\phi d\theta$$

$$= \int_0^{2\pi} \int_0^{\pi} \left[\frac{1}{2}\rho^4 \sin^2 \phi \cos \theta + \rho^3 \sin \phi \right]_{\rho=0}^{\rho=2} d\phi d\theta$$

$$= \int_0^{2\pi} \int_0^{\pi} (8 \sin^2 \phi \cos \theta + 8 \sin \phi)\, d\phi d\theta$$

$$= \int_0^{2\pi} \int_0^{\pi} \left[8 \cdot \frac{1 - \cos 2\phi}{2} \cos \theta + 8 \sin \phi \right] d\phi d\theta$$

$$= \int_0^{2\pi} \int_0^{\pi} [4(1 - \cos 2\phi) \cos \theta + 8 \sin \phi]\, d\phi d\theta$$

$$= \int_0^{2\pi} \left[4\left(\phi - \frac{\sin 2\phi}{2} \right) \cos \theta - 8 \cos \phi \right]_{\phi=0}^{\phi=\pi} d\theta$$

$$= \int_0^{2\pi} [4\pi \cos \theta - 8 \cos \pi - (-8 \cos 0)]\, d\theta = \int_0^{2\pi} (4\pi \cos \theta + 8 + 8)\, d\theta$$

$$= \left[4\pi \sin \theta + 16\theta \right]_0^{2\pi} = 16(2\pi) = 32\pi.$$

17. $\mathbf{F} = \mathbf{c} \Rightarrow \operatorname{div}(\mathbf{F}) = 0 \Rightarrow \iint_S \mathbf{F} \cdot \mathbf{n}\, dS = \iiint_E \operatorname{div}(\mathbf{F})\, dV = \iiint_E 0\, dV = 0.$